Mathematical Analysis and Optimization for Economists

Mathematical Analysis and Optimization for Economists

Michael J. Panik

CRC Press
Taylor & Francis Group
Boca Raton London New York

CRC Press is an imprint of the
Taylor & Francis Group, an **informa** business
A CHAPMAN & HALL BOOK

First edition published 2022
by CRC Press
6000 Broken Sound Parkway NW, Suite 300, Boca Raton, FL 33487-2742

and by CRC Press
2 Park Square, Milton Park, Abingdon, Oxon, OX14 4RN

CRC Press is an imprint of Taylor & Francis Group, LLC

Access the Support Material: www.routledge.com/9780367759018

Library of Congress Cataloging-in-Publication Data
A catalog record has been requested for this book

ISBN: 978-0-367-75901-8 (hbk)
ISBN: 978-0-367-75902-5 (pbk)
ISBN: 978-1-003-16449-4 (ebk)

Typeset in Minion
by MPS Limited, Dehradun

To my wife, Paula,
who makes all things possible.

Contents

Preface

My objective in writing this textbook is two-fold: first, to introduce students of economics to the power and versatility of traditional as well as contemporary methodologies in mathematics and optimization theory; and, second, to illustrate how these techniques can be applied in solving microeconomic problems. Presuming only a course in the calculus, this text is suitable for advanced undergraduates as well as first-year graduate students. It is comprehensive and contains numerous and detailed example problems in both mathematics and economic analysis. Each major section contains some basic, as well as thought-provoking exercises and the solutions to the same, are provided in full detail.

The specific strengths of the book are:

i. its readability and pace – Chapters 1-4 are foundational in that they develop a whole host of mathematical tools from the "ground up," i.e., essential notions in the calculus and in linear algebra are introduced and explored fully before items such as implicit function theorems, functional dependence, multivariate optimality conditions, and so on, are introduced.

ii. the inclusion of numerous as well as diverse economics examples and exercises. The modeling of utility maximization, profit maximization, and cost-minimizing performance is accompanied by the development of *duality concepts and applications.*

iii. particular emphasis is placed upon linear transformations; convex (concave) functions and generalizations thereof; Fritz-John as well as Karush-Kuhn-Tucker optimality conditions; Lagrangian saddle-points and duality; conjugate functions; generalized (Everett) Lagrange multipliers; homogeneous, homothetic, and almost homogeneous functions; generalized directional derivatives and generalized gradients; simplicial mappings and fixed points; comparative statics analysis (using traditional as well as duality approaches); and optimal control modeling with a survey of differential equations.

iv. applications in economics include: household equilibrium, demand and expenditure functions, and the income and substitution effects; production and cost functions; profit maximization under various market conditions; conjugate production functions and normalized long-run profit functions; constrained output maximization via the Lagrangian dual; envelope results for expenditure, profit, and cost functions; the elasticity of substitution for homogeneous and almost homogeneous production functions; the existence of a competitive equilibrium for pure exchange and production economies; neoclassical optimal growth modeling; and comparative statics analysis applied to constrained utility maximization and cost minimization as well as profit maximization (including the Le Châtelier effect).

v. a wealth of example problems dealing with optimization techniques and their applications in economics are included. Additionally, detailed solutions to the odd-numbered exercise are provided, with "hints" offered for the solution of the more challenging problems.

This text can be used for a one-or two-semester course in quantitative methods for economists, depending upon the backgrounds of the students as well as upon the amount of microeconomic applications that are to be covered. If the economic applications are omitted, this book can serve as a beginning text in optimization for applied mathematics students.

An accompaniment to this textbook is a set of electronic chapters (termed e-Chapters) that are available online and can be accessed for free on the book's official webpage.[§] These chapters contain selected topics in linear algebra and matrix theory, nonsmooth optimization, economic efficiency, and distance functions

The scope and variety of intellectual inputs into this work are numerous and diverse, not the least of which are my collaborations over the years with Bharat Kolluri, Rao Singamsetti, Farhad Rassekh, Demetrios Giannaros, Ed Gullason, and Mahmoud Wahab; along with my many stimulating conversations with Jim Peta, John Sullivan, Dominick Armentano, Ken Friedman, John Speir, David Rosenberg, Frank DelloIacono, Greg Wassell, and Jeff Cohen.

I also wish to thank Alice Schoenrock for her steadfast, accurate, and prompt typing of the entire manuscript. Her efforts are greatly appreciated. Additional notes of thanks are extended to Callum Fraser, Mathematic Editor, CRC Press/Taylor & Francis Group, for his editorial expertise and professionalism and to his editorial assistant, Mansi Kabra, for her manifest considerations regarding this project.

[§] The link to the webpage is www.routledge.com/9780367759018.

Author

Michael J. Panik is Professor Emeritus in the Department of Economics and Finance at the University of Hartford, CT. He has taught courses in economic and business statistics, introductory and advanced quantitative methods, and econometrics. Dr. Panik is the author of several textbooks, monographs, and numerous articles in professional journals.

Symbols and Abbreviations

$\boldsymbol{R}$	the set of real numbers
$\boldsymbol{R}_+$	the nonnegative real numbers
$\boldsymbol{R}_{++}$	the positive real numbers
$\boldsymbol{R}^n$	the set of ordered n-tuples $(x_1, \ldots, x_n)$ or n-dimensional Euclidean space
$\boldsymbol{R}^*$	extended real number system
$\varnothing$	null or empty set
$\boldsymbol{x} = (x_i)$	the vector whose ith component is x_i
$\boldsymbol{1}$	the sum vector containing all 1's
$\boldsymbol{O}$	the null vector/matrix containing all 0's
$\boldsymbol{I}_n$	identity matrix of order n
$\boldsymbol{e}_i$	ith unit column vector
$\boldsymbol{x}^T$	transpose of a vector/matrix
$\boldsymbol{x} \cdot \boldsymbol{y}$	scalar product of vectors $\boldsymbol{x}$, $\boldsymbol{y}$
$\lVert\cdot\rVert$	Euclidean norm
$\boldsymbol{x}/\lVert\boldsymbol{x}\rVert$	direction of vector $\boldsymbol{x}$
$\lVert\boldsymbol{x} - \boldsymbol{y}\rVert$	distance between points $\boldsymbol{x}$, $\boldsymbol{y}$
$\boldsymbol{A} = [a_{ij}]$	an $(m \times n)$ matrix $\boldsymbol{A}$, where $i = 1, \ldots, m$; $j = 1, \ldots, n$
$\rho(A)$	rank of a matrix $\boldsymbol{A}$
$\lvert A\rvert$	determinant of an nth order matrix $\boldsymbol{A}$
$\boldsymbol{A}^{-1}$	inverse of an nth order matrix $\boldsymbol{A}$
δ_{ij}	the Kronecker delta
$\mathcal{M}$	set of all square matrices
M_{ij}	minor of the element a_{ij}
C_{ij}	cofactor of the element a_{ij}
M_K	kth naturally ordered principal minor of an $(n \times n)$ matrix $\boldsymbol{A}$
$\rightarrow$	transformation
$\mathcal{F}_{\mathrm{R}}$	field of real scalars
V_n	n-dimensional vector (linear) space
$\mathcal{A}\times\mathcal{B}$	Cartesian product of sets $\mathcal{A}$, $\mathcal{B} \in R^n$
$\boldsymbol{A}^+$	adjoint of an nth order matrix $\boldsymbol{A}$
dim (V_n)	dimension of a vector space V_n
$\boldsymbol{x} \in \mathcal{S}$	x is an element of set $\mathcal{S}$
$\boldsymbol{x} \notin \mathcal{S}$	x is not an element of set $\mathcal{S}$
$\mathcal{A}\subseteq\mathcal{B}$	set $\mathcal{A}$ is a subset of set $\mathcal{B}$ (possibly $\mathcal{A} = \mathcal{B}$)
$\mathcal{A}\subset\mathcal{B}$	set $\mathcal{A}$ is a proper subset of set $\mathcal{B}$
$\mathcal{H}$	hyperplane
$[\mathcal{H}^+],[\mathcal{H}^-]$	closed half-planes
$(\mathcal{H}^+),(\mathcal{H}^-)$	open half-planes
sup	supremum or least upper bound
inf	infinium or greatest lower bound
$\mathcal{C}$	cone
$-\mathcal{C}$	negative cone of $\mathcal{C}$
$\mathcal{C}^{\perp}$	orthogonal cone of $\mathcal{C}$
dim $(\mathcal{C})$	dimension of cone $\mathcal{C}$

$\mathcal{L}$	line
$\mathcal{C}^*$	dual cone of $\mathcal{C}$
$\mathcal{C}^+$	polar cone of $\mathcal{C}$
$(\boldsymbol{a})$	half-line
$(\boldsymbol{a})^{\perp}$	orthogonal cone to $(\boldsymbol{a})$
$(\boldsymbol{a})^*$	dual of $(\boldsymbol{a})$
$(\boldsymbol{a})^+$	polar of $(\boldsymbol{a})$
T	linear transformation
T^{-1}	inverse linear transformation
D_T	domain of T
Range (T)	range space of T
$\mathcal{L}$ $(\mathcal{K},)$	set of linear transformations from $\mathcal{K}$ to $\mathcal{Y}$, $\mathcal{K}$ and $\mathcal{Y}$ real vector spaces
Null (T)	null space of T
Nullity(T)	nullity of T
f: $X \rightarrow Y$	point-to-point mapping (or single-valued function) from set X to set Y
D_f	domain of f
R_f	range of f
G_f	graph of f
f^{-1}: $Y \rightarrow X$	inverse mapping of f
$g \circ f$	composition of f and g
$\{x_n\}$	sequence of points
$\{x_n\}_{n\in J}$	subsequence of points, where J is a subset of positive integers
$\hat{x}$ **or** $\bar{x}$	limit (cluster) point of a sequence
$\lim\limits_{n\to\infty} \sup x_n$	limit superior
$\lim\limits_{n\to\infty} \inf x_n$	limit inferior
$\boldsymbol{\delta}(\boldsymbol{x}_o)$	δ – neighborhood of a point x_o
$\mathcal{X}^c$	complimentary set of $\mathcal{X}$
$\boldsymbol{d}$ $(\mathcal{X})$	diameter of a set $\mathcal{X}$
$\bar{\mathcal{X}}$	closure of a set $\mathcal{X}$
$\mathcal{X}^{\mathbf{d}}$	derived set
$\overline{\lim} \mathcal{A}_i$	limit superior of a sequence $\{\mathcal{A}_i\}$
$\underline{\lim}_i \mathcal{A}_i$	limit inferior of a sequence $\{\mathcal{A}_i\}$
$\nabla f(\boldsymbol{x}_o)$	gradient vector of f at $\boldsymbol{x}_o$
$D_u f(\boldsymbol{x}_o)$	directional derivative with respect to $\boldsymbol{u}$ of f at $\boldsymbol{x}_o$
$C^{(q)}$	f is a function of class $C^{(q)}$
$H_f(\boldsymbol{x}_o)$	Hessian matrix of f at $\boldsymbol{x}_o$
$J_f(\boldsymbol{x}_o)$	Jacobian matrix of $\boldsymbol{f}^T = (f^1, \ldots, f^m)$ at $\boldsymbol{x}_o$
$\nabla_{\mathrm{u}} f$	partial gradient of f with respect to $\boldsymbol{u}$
∇^2_{uu}	second-order partial gradient of f with respect to $\boldsymbol{u}$
$d^j f(\boldsymbol{x}_o, \boldsymbol{h})$	jth order differential of f at $\boldsymbol{x}_o$
$U_f(\alpha)$	upper level set
$L_f(\alpha)$	lower level set

***hyp* (*f*)**	hypograph of f
***epi* (*f*)**	epigraph of f
x_c	convex combination of points
***co* (*A*)**	convex hull of set A
$\partial f(x_o)$	subdifferential or superdifferential of f at x_o
***dom* (*f*)**	effective domain of f
***l* (*f*)**	lower support set for f
***u* (*f*)**	upper support set for f
***cl* (*f*)**	closure of f
***lsc* (*f*)**	lower semicontinuous hull of f
***usc* (*f*)**	upper semicontinuous hull of f
f^*	conjugate function of f
f^{**}	conjugate function of f^*
H_f^g	bordered Hessian matrix of f (g is the bordering function)
$\mathcal{D}(x_o)$	set of feasible directions at x_o
$\mathcal{F}$	cone of ascent directions
FJ	Fritz-John
CS	complementary slackness
KKT	Karush-Kuhn-Tucker
CQ	constraint qualification
FM	Fiacco-McCormick
$(x_o)^+$	polar support cone
$(x_o)^*$	dual support cone
$\Phi(y)$	perturbation function
AE	Arrow-Enthoven
LLC	locally Lipschitz continuous
HB	Hahn-Banach
X^*	dual space of an arbitrary space X
$f^o(x_o; d)$	Clarke's generalized directional derivative of f
$\partial^o f(x)$	Clarke's generalized subdifferental
uhc	upper hemicontinuous
lhc	lower hemicontinuous
A^-	Moore-Penrose generalized inverse of $\mathbf{A}$
$A \otimes B$	Kronecker product of matrices $\mathbf{A}$ and $\mathbf{B}$
Δ^k	k-dimensional simplex
KKM	Knaster-Kuratowski-Mazurkiewicz
$H(\cdot)$	Hamiltonian
$H_{cv}(\cdot)$	current value Hamiltonian
$\dot{x}(t)$	law of motion for x
int $(\mathcal{A})$	interior of set $\mathcal{A}$
$\dot{u}(t)$	law of motion for u
$\dot{\lambda}(t)$	law of motion for λ

Mathematical Foundations 1

1.1 MATRICES AND DETERMINANTS

We start with

Definition 1.1.1: A **matrix** is an ordered set of elements arranged in a rectangular array of rows and columns.

That is, the matrix $\boldsymbol{A}$ may appear as

$$\boldsymbol{A} = \begin{bmatrix} a_{11} & a_{12} & \cdots & a_{1n} \\ a_{21} & a_{22} & \cdots & a_{2n} \\ \vdots & \vdots & & \vdots \\ a_{m1} & a_{m2} & \cdots & a_{mn} \end{bmatrix},$$

where a_{ij} represents the element in the ith row and jth column of $\boldsymbol{A}$, $i = 1, \ldots, m$; $j = 1, \ldots, n$. Since there are m rows and n columns in $\boldsymbol{A}$, this matrix is said to be of **order** $(\boldsymbol{m} \times \boldsymbol{n})$ ("read m by n"). When $i, j=1,\ldots,n$, the matrix is square and will simply be referred to as an $\boldsymbol{n}$**th order matrix**. The matrix $\boldsymbol{A}$ may be written in a more compact fashion as $\boldsymbol{A} = [a_{ij}]$, $i = 1, \ldots, m$; $j = 1, \ldots, n$.

We next have

Definition 1.1.2: The **sum** of two $(m \times n)$ matrices $\boldsymbol{A} = [a_{ij}]$, $\boldsymbol{B} = [b_{ij}]$ is the $(m \times n)$ matrix $\boldsymbol{C} = [c_{ij}]$, where $c_{ij} = a_{ij} + b_{ij}$ (we add corresponding elements), i.e.,

$$\boldsymbol{A} + \boldsymbol{B} = \boldsymbol{C} \text{ or } [a_{ij}] + [b_{ij}] = [a_{ij} + b_{ij}],$$
$$i = 1, \ldots, m; j = 1, \ldots, n.$$

Definition 1.1.3: The **product** of a (real) scalar λ and an $(m \times n)$ matrix $\boldsymbol{A} = [a_{ij}]$ is the $(m \times n)$ matrix $\lambda\boldsymbol{A} = [\lambda a_{ij}]$, $i = 1, \ldots, m$; $j = 1, \ldots, n$, obtained by multiplying each element of $\boldsymbol{A}$ by λ. The properties of matrix addition and scalar multiplication may be summarized as:

a. $\boldsymbol{A}+\boldsymbol{B}=\boldsymbol{B}+\boldsymbol{A}$ (**commutative law**);

b. $(\boldsymbol{A}+\boldsymbol{B})+\boldsymbol{C}=\boldsymbol{A}+(\boldsymbol{B}+\boldsymbol{C})$ (**associative law**);

c. $\lambda\boldsymbol{A}=\boldsymbol{A}\lambda$;

d. $\lambda(\boldsymbol{A}+\boldsymbol{B})=\lambda\boldsymbol{A}+\lambda\boldsymbol{B}$

e. $(\lambda_1+\lambda_2)\boldsymbol{A}=\lambda_1\boldsymbol{A}+\lambda_2\boldsymbol{A}$

(d–e: **distributive laws**);

f. $\boldsymbol{A}+\boldsymbol{O}=\boldsymbol{A}$ (here $\boldsymbol{O}$ represents the **zero (null) matrix** and serves as the additive identity),

where the λs are real scalars and the indicated matrices are **conformable for addition**, i.e., they all have the same order.

Definition 1.1.4: The **transpose** of an $(m \times n)$ matrix $\boldsymbol{A}$, $\boldsymbol{A}^T$ is the $(n \times m)$ matrix formed from $\boldsymbol{A}$ by interchanging its rows and columns, i.e., row i of $\boldsymbol{A}$ becomes column i of the transposed matrix.

The essential properties of matrix transposition are:

a. $(\boldsymbol{A}^T)^T = \boldsymbol{A}$;

b. $(\lambda_1\boldsymbol{A} + \lambda_2\boldsymbol{B})^T = \lambda_1\boldsymbol{A}^T + \lambda_2\boldsymbol{B}^T$;

c. $(\boldsymbol{AB})^T = \boldsymbol{B}^T\boldsymbol{A}^T$, where the product of $\boldsymbol{A}$ and $\boldsymbol{B}$, $\boldsymbol{AB}$, is assumed to exist.

Additionally,

Definition 1.1.5: A matrix of order n is **symmetric** if it equals its transpose, i.e., $\boldsymbol{A} = \boldsymbol{A}^T$ or $a_{ij} = a_{ij}$, $i \neq j$.

We next look to specific sets of elements within square or $(n \times n)$ matrices.

Definition 1.1.6: The **principal diagonal** of an nth-order matrix

$$\boldsymbol{A} = \begin{bmatrix} a_{11} & a_{12} & \cdots & a_{1n} \\ a_{21} & a_{22} & \cdots & a_{2n} \\ \vdots & \vdots & & \vdots \\ a_{n1} & a_{n2} & \cdots & a_{nn} \end{bmatrix},$$

is the set of elements extending from the upper left to lower right corner of $\boldsymbol{A}$, namely $a_{11}, a_{22}, \ldots, a_{nn}$. Hence, it consists of the elements a_{ii}, $i = 1, \ldots, n$. The **trace** of an nth-order matrix $\boldsymbol{A}$, denoted $tr(\boldsymbol{A})$, is the sum of the diagonal elements or $tr(\boldsymbol{A}) = \sum_{i=1}^{n} a_{ii}$.

We now turn to a few special types of square matrices.

Definition 1.1.7: A **diagonal matrix** of order n is one for which all elements off the principal diagonal are zero, i.e., $\boldsymbol{A} = [a_{ij}]$ is a diagonal matrix if $a_{ij} = 0$, $i \neq j$.

Definition 1.1.8: The **identity matrix** of order n, $\boldsymbol{I}_n$, is a diagonal matrix having ones along its principal diagonal and zeros elsewhere:

$$\boldsymbol{I}_n = \begin{bmatrix} 1 & 0 & \cdots & 0 \\ 0 & 1 & \cdots & 0 \\ \vdots & \vdots & & \vdots \\ 0 & 0 & \cdots & 1 \end{bmatrix} = [\delta_{ij}], \quad \text{where } \delta_{ij} = \begin{cases} 1, & i = j; \\ 0, & i \neq j. \end{cases}$$

Here, δ_{ij} denotes the **Kronecker delta.**

Definition 1.1.9: A **triangular matrix** of order n is one for which all elements on one side of the principal diagonal are zero, i.e., $\boldsymbol{A} = [a_{ij}]$ is **upper triangular** if $a_{ij} = 0$ for $i > j$; it is **lower triangular** if $a_{ij} = 0$ when $i < j$.

An important operation involving matrices is that of matrix multiplication. Definition 1.1.10 describes the process of forming the product of two matrices.

Definition 1.1.10: Given an $(m \times n)$ matrix $\boldsymbol{A}$ and an $(n \times p)$ matrix $\boldsymbol{B}$, the product $\boldsymbol{AB}$ is the $(m \times p)$ matrix $\boldsymbol{C}$ where elements are computed from the elements of $\boldsymbol{A}$, $\boldsymbol{B}$ according to the rule

$$c_{ij} = \textstyle\sum_{k=1}^{n} a_{ik} b_{kj}, \; i = 1, \ldots, m; \; j = 1, \ldots, p.$$

The product $\boldsymbol{AB}$ exists if and only if the matrices $\boldsymbol{A}$, $\boldsymbol{B}$ are conformable for multiplication, i.e., if and only if the number of columns in $\boldsymbol{A}$ is the same as the number of rows in $\boldsymbol{B}$. Here, $\boldsymbol{A}$ is termed the **pre-multiplier**, while $\boldsymbol{B}$ is referred to as the **post-multiplier**. In general, matrix multiplication is not commutative or $\boldsymbol{AB} \neq \boldsymbol{BA}$. In fact, $\boldsymbol{BA}$ may not exist. Matrix multiplication possesses the following properties:

a. $(\boldsymbol{AB})\boldsymbol{C} = \boldsymbol{A}(\boldsymbol{BC})$ **(associative law)**;

b. $\lambda(\boldsymbol{AB}) = \boldsymbol{A}(\lambda\boldsymbol{B}) = (\lambda\boldsymbol{A})\boldsymbol{B}$;

c. $\boldsymbol{A}(\boldsymbol{B}+\boldsymbol{C}) = \boldsymbol{AB}+\boldsymbol{AC}$ } **distributive laws;**

d. $(\boldsymbol{A}+\boldsymbol{B})\boldsymbol{C} = \boldsymbol{AC}+\boldsymbol{BC}$

e. $\boldsymbol{I}_m\boldsymbol{A} = \boldsymbol{AI}_m = \boldsymbol{A}$ ($\boldsymbol{I}_m$ **is the multiplicative identity)**;

f. $\boldsymbol{OA} = \boldsymbol{AO} = \boldsymbol{O}$,

where λ is a real scalar and the indicated products exist.

Example 1.1.1: The reader can easily check that:

a. For

$$\boldsymbol{A} = \begin{bmatrix} 3 & 2 \\ 1 & 4 \end{bmatrix}, \; \boldsymbol{B} = \begin{bmatrix} -1 & 0 \\ 1 & 2 \end{bmatrix}, \text{ and } \boldsymbol{C} = \begin{bmatrix} 3 & 3 \\ 4 & 1 \end{bmatrix},$$

$$\boldsymbol{A} + \boldsymbol{B} - 3\boldsymbol{C} = \begin{bmatrix} -7 & -7 \\ -10 & 3 \end{bmatrix}.$$

b. For

$$A = \begin{bmatrix} 3 & 1 & 0 \\ 1 & -1 & 4 \\ 5 & 1 & 1 \end{bmatrix}, \text{ we have } A^T = \begin{bmatrix} 3 & 1 & 5 \\ 1 & -1 & 1 \\ 0 & 4 & 1 \end{bmatrix}.$$

Clearly, A is not a symmetric matrix.

c. For

$$A = \begin{bmatrix} 3 & 2 \\ 1 & 4 \end{bmatrix}, B = \begin{bmatrix} -1 & 0 \\ 1 & 2 \end{bmatrix}, \text{ we have}$$

$$AB = \begin{bmatrix} 3(-1) + 2(1) & 3(0) + 2(2) \\ 1(-1) + 4(1) & 1(0) + 4(2) \end{bmatrix} = \begin{bmatrix} -1 & 4 \\ 3 & 8 \end{bmatrix} = C.$$

And, for

$$A = \begin{bmatrix} 1 & 0 & 1 \\ 6 & 3 & 2 \\ -1 & -4 & 2 \end{bmatrix}, B = \begin{bmatrix} -2 & 0 & 5 \\ 1 & 0 & 3 \\ 4 & 6 & 1 \end{bmatrix}, \text{ we have}$$

$$AB = \begin{bmatrix} 1(-2) + 0(1) + 1(4) & 1(0) + 0(0) + 1(6) & 1(5) + 0(4) + 1(1) \\ 6(-2) + 3(1) + 2(4) & 6(0) + 3(0) + 2(6) & 6(5) + 3(3) + 2(1) \\ (-1)(-2) + (-4)(1) + 2(4) & (-1)(0) + (-4)(0) + 2(6) & (-1)(5) + (-4)3 + 2(1) \end{bmatrix}$$

$$= \begin{bmatrix} 2 & 6 & 6 \\ -1 & 12 & 41 \\ 6 & 12 & -15 \end{bmatrix} = C.$$

d. For

$$A = \begin{bmatrix} 1 & 3 & 9 \\ 6 & 0 & 0 \\ 1 & 5 & 1 \end{bmatrix}, B = \begin{bmatrix} 2 & 1 & 6 \\ 1 & 4 & -1 \end{bmatrix}, \text{ we have}$$

AB = tilt! A and B are not conformable for multiplication.

Quite often, we will need to compute the determinant of a square matrix.

Definition 1.1.11: A **determinant** is a scalar-valued function D defined on the set of all square matrices $\mathcal{U}$.

Thus, D associates with elements in $\mathcal{U}$ unique scalar quantities called determinants, i.e., corresponding to each $(n \times n)$ matrix $A \in \mathcal{U}$ is the scalar $D(A) = |A|$. How is $|A|$ found? For the (2×2) case, let us define

$$|A| = \begin{vmatrix} a_{11} & a_{12} \\ a_{21} & a_{22} \end{vmatrix} \equiv a_{11}a_{22} - a_{12}a_{21}.$$

Then, for

$$A = \begin{bmatrix} 3 & 2 \\ 5 & 4 \end{bmatrix}, |A| = \begin{vmatrix} 3 & 2 \\ 5 & 4 \end{vmatrix} = 3(4) - 2(5) = 2.$$

(What is the distinction between A and $|A|$?) In order to calculate or "expand" a higher-order determinant, two additional concepts are needed. Specifically:

Definition 1.1.12: A **submatrix** is the $(k \times s)$ matrix B obtained by deleting all but k rows and s columns of an $(m \times n)$ matrix A.

Definition 1.1.13: A **minor** is the determinant of a square submatrix; and a **cofactor** is a signed minor.

To elaborate, the minor of the element a_{ij}, denoted M_{ij}, is the determinant of the submatrix formed by deleting row i and column j of an $(n \times n)$ matrix A, while the cofactor of a_{ij}, C_{ij} is calculated as $(-1)^{i+j}M_{ij}$. So, for an nth-order matrix A,

$$|A| = \sum_{j=1}^{n} a_{ij}C_{ij} \text{ (expansion by row } i) \\ = \sum_{j=1}^{n} a_{ij}(-1)^{i+j}M_{ij} \text{ for any } i; \tag{1.1}$$

or

$$|A| = \sum_{i=1}^{n} a_{ij}C_{ij} \textbf{ (expansion by column } j) \\ = \sum_{i=1}^{n} a_{ij}(-1)^{i+j}M_{ij} \textbf{ for any } j. \tag{1.2}$$

Here, equations (1.1) and (1.2) depict, respectively, the **cofactor expansion** of a determinant by a row (i is fixed and we sum over columns or over the j index) or by a column (j is fixed and we sum over rows or over the i index).

An assortment of important properties of determinants now follows:

a. interchanging any two rows or columns of a matrix $\boldsymbol{A}$ changes the sign of $|\boldsymbol{A}|$;

b. if a matrix $\boldsymbol{A}$ has two rows or two columns which are identical, then $|\boldsymbol{A}| = 0$;

c. if a row or column of a matrix $\boldsymbol{A}$ has all zero elements, then $|\boldsymbol{A}| = 0$;

d. if every element of a row or column of a matrix $\boldsymbol{A}$ is multiplied by a non-zero scalar λ to give a new matrix $\boldsymbol{B}$, then $|\boldsymbol{B}| = \lambda|\boldsymbol{A}|$;

e. if every element of an nth-order matrix $\boldsymbol{A}$ is multiplied by a non-zero scalar λ to give a new matrix $\boldsymbol{B}$, then $|\boldsymbol{B}| = |\lambda \boldsymbol{A}| = \lambda^n|\boldsymbol{A}|$ (property (d) is applied n times in succession);

f. if a row or column of a matrix $\boldsymbol{A}$ is a multiple of any other row or column, then $|\boldsymbol{A}| = 0$;

g. for an nth-order matrix $\boldsymbol{A}$, $|\boldsymbol{A}| = |\boldsymbol{A}^T|$;

h. if $\boldsymbol{A}$ is a diagonal or triangular matrix, $|\boldsymbol{A}| = a_{11}a_{22}\ldots a_{nn}$;

i. $|\boldsymbol{I}_n| = 1$ and $|\boldsymbol{O}| = 0$;

j. if $\boldsymbol{A}$, $\boldsymbol{B}$ are of order n so that the product $\boldsymbol{AB}$ exists, then $|\boldsymbol{AB}| = |\boldsymbol{A}|\cdot|\boldsymbol{B}|$;

k. if $\boldsymbol{A}$ is $(m \times n)$ and $\boldsymbol{B}$ is $(n \times m)$, $|\boldsymbol{AB}| = 0$ if $m > n$;

l. if $\boldsymbol{A}$, $\boldsymbol{B}$ are conformable for addition (their sum exists), then generally $|\boldsymbol{A} + \boldsymbol{B}| \neq |\boldsymbol{A}| + |\boldsymbol{B}|$;

m. if $\boldsymbol{A}$, $\boldsymbol{B}$ are of order n, $|\boldsymbol{AB}| = |\boldsymbol{A}^T\boldsymbol{B}| = |\boldsymbol{AB}^T| = |\boldsymbol{A}^T\boldsymbol{B}^T|$.

While the cofactor expansion technique can always be used to evaluate a determinant, it might not be computationally efficient, especially if we have a determinant of order greater than three. An alternative and more expedient method for evaluating high-order determinants is the "sweep-out process," to which we now turn. To execute this process, we need

Definition 1.1.14: An **elementary row (column) operation** may be performed on a matrix $\boldsymbol{A} = [a_{ij}]$ by:

a. interchanging any two rows (columns) of $\boldsymbol{A}$ (Type I);

b. multiplying a row (column) of $\boldsymbol{A}$ by a real scalar $\lambda \neq 0$ (Type II);

c. adding to the ith row (jth column) of $\boldsymbol{A}$ λ times any other row (column) (Type III).

Example 1.1.2: For the matrix

$$\boldsymbol{A} = \begin{bmatrix} 1 & 2 & -1 \\ 3 & 6 & 1 \\ 4 & -1 & 3 \end{bmatrix},$$

we may:

a. transform $\boldsymbol{A}$ into a matrix $\boldsymbol{B}$ via a Type I elementary row operation by interchanging, say, rows 1 and 3 of $\boldsymbol{A}$ or

$$\boldsymbol{A} = \begin{bmatrix} 1 & 2 & -1 \\ 3 & 6 & 1 \\ 4 & -1 & 3 \end{bmatrix} \underset{I}{\rightarrow} \begin{bmatrix} 4 & -1 & 3 \\ 3 & 6 & 1 \\ 1 & 2 & -1 \end{bmatrix} = \boldsymbol{B},$$

where the arrow "→" indicates "transformation";

b. transform $\boldsymbol{A}$ into a matrix $\boldsymbol{C}$ by a Type II elementary row operation by multiplying, say, row 2 of $\boldsymbol{A}$ by 5 or

$$\boldsymbol{A} = \begin{bmatrix} 1 & 2 & -1 \\ 3 & 6 & 1 \\ 4 & -1 & 3 \end{bmatrix} \underset{II}{\rightarrow} \begin{bmatrix} 1 & 2 & -1 \\ 15 & 30 & 5 \\ 4 & -1 & 3 \end{bmatrix} = C; \text{ and}$$

c. obtain from $\boldsymbol{A}$ a new matrix $\boldsymbol{D}$ via a Type III elementary row operation by adding to, say, row 1 of $\boldsymbol{A}$ three times row 3 of $\boldsymbol{A}$ or

$$\boldsymbol{A} = \begin{bmatrix} 1 & 2 & -1 \\ 3 & 6 & 1 \\ 4 & -1 & 3 \end{bmatrix} \underset{III}{\rightarrow} \begin{bmatrix} 13 & -1 & 8 \\ 3 & 6 & 1 \\ 4 & -1 & 3 \end{bmatrix} = D,$$

$$12 \quad -3 \quad 9$$

where the entries 12, −3, and 9 beneath $\boldsymbol{A}$ constitute a "work row." ■

Given an $(m \times n)$ matrix $\boldsymbol{A}$, how do these elementary row (column) operations affect $|\boldsymbol{A}|$? Using the preceding properties of determinants, we see that:

a. Type I operation – see property (a) above;

b. Type II operation – see property (d) above;

c. Type III operation – if in $\boldsymbol{A}$ we add any non-zero multiple of one row (column) to a different row (column) to get a new matrix $\boldsymbol{B}$, then $|\boldsymbol{B}| = |\boldsymbol{A}|$.

Thus, only Type III elementary row or column operations leave $|\boldsymbol{A}|$ unchanged. Equipped with these elementary row and column operations, we can now state the

Sweep-out Process: If |A| can be transformed, by elementary row operations, into the product of a constant k and the determinant of an upper-triangular matrix $\boldsymbol{B}$,

$$|\boldsymbol{A}| = k|\boldsymbol{B}| = k\begin{vmatrix} b_{11} & b_{12} & b_{13} & \cdots & b_{1n} \\ 0 & b_{22} & b_{23} & \cdots & b_{2n} \\ 0 & 0 & b_{33} & & b_{3n} \\ \vdots & \vdots & \vdots & & \vdots \\ 0 & 0 & 0 & \cdots & b_{nn} \end{vmatrix},$$

then $|\boldsymbol{A}| = k|\boldsymbol{B}| = k\, b_{11}b_{22}b_{33}\ldots b_{nn}$ (see property (h) of determinants), where k is chosen to compensate for the cumulative effects on $|\boldsymbol{A}|$ of successive Type I, II elementary row operations.

Example 1.1.3: (a) Find $|\boldsymbol{A}|$ via cofactor expansion by first using row one and then by using column two when

$$\boldsymbol{A} = \begin{bmatrix} 4 & 3 & 5 \\ 6 & 1 & 2 \\ 0 & 7 & 1 \end{bmatrix}.$$

Set i=1 in equation (1.1). Then,

$$|\boldsymbol{A}| = \Sigma_{j=1}^{3} a_{1j}(-1)^{1+j}M_{1j} = 4(-1)^{1+1}M_{11} + 3(-1)^{1+2}M_{12} + 5(-1)^{1+3}M_{13} = 4M_{11} - 3M_{12} + 5M_{13} = 4(-13) - 3(6) + 5(42) = 140.$$

Set $j = 2$ in equation (1.2). Then,

$$|\boldsymbol{A}| = \Sigma_{i=1}^{3} a_{i2}(-1)^{i+2}M_{i2} = -3M_{12} + M_{22} - 7M_{32} = -3(6) + 4 - 7(-22) = 140.$$

(b) Find $|\boldsymbol{A}|$ using the sweep-out process when

$$\boldsymbol{A} = \begin{bmatrix} 3 & 1 & 4 & 2 \\ 2 & 1 & 0 & 1 \\ 0 & 0 & 1 & 2 \\ 1 & 4 & 1 & 0 \end{bmatrix}.$$

■

$$|\boldsymbol{A}| = \begin{vmatrix} 3 & 1 & 4 & 2 \\ 2 & 1 & 0 & 1 \\ 0 & 0 & 1 & 2 \\ 1 & 4 & 1 & 0 \end{vmatrix} = -\begin{vmatrix} 1 & 4 & 1 & 0 \\ 2 & 1 & 0 & 1 \\ 0 & 0 & 1 & 2 \\ 3 & 1 & 4 & 2 \end{vmatrix} = -\begin{vmatrix} 1 & 4 & 1 & 0 \\ 0 & -7 & -2 & 1 \\ 0 & 0 & 1 & 2 \\ 0 & -11 & 1 & 2 \end{vmatrix}$$

$$= -(-7)\begin{vmatrix} 1 & 4 & 1 & 0 \\ 0 & 1 & \frac{2}{7} & -\frac{1}{7} \\ 0 & 0 & 1 & 2 \\ 0 & -11 & 1 & 2 \end{vmatrix} = 7\begin{vmatrix} 1 & 4 & 1 & 0 \\ 0 & 1 & \frac{2}{7} & -\frac{1}{7} \\ 0 & 0 & 1 & 2 \\ 0 & 0 & \frac{29}{7} & \frac{3}{7} \end{vmatrix} = 7\begin{vmatrix} 1 & 4 & 1 & 0 \\ 0 & 1 & \frac{2}{7} & -\frac{1}{7} \\ 0 & 0 & 1 & 2 \\ 0 & 0 & 0 & -\frac{55}{7} \end{vmatrix}$$

$$= -55.$$

We close this section by introducing a special type of minor associated with the principal diagonal of an nth-order matrix.

Definition 1.1.15: The kth naturally ordered **principal minor** of an nth-order matrix $\boldsymbol{A}$,

$$M_k = \begin{vmatrix} a_{11} & \cdots & a_{1k} \\ \vdots & & \vdots \\ a_{k1} & \cdots & a_{kk} \end{vmatrix}, \; k = 1, \ldots, n,$$

is the determinant of the $(k \times k)$ submatrix obtained from $\boldsymbol{A}$ by retaining only its first k rows and columns, i.e., for

$$\boldsymbol{A} = \begin{vmatrix} a_{11} & \cdots & a_{1n} \\ \vdots & & \vdots \\ a_{n1} & \cdots & a_{nn} \end{vmatrix},$$

it follows that

$$M_1 = a_{11}, \; M_2 = \begin{vmatrix} a_{11} & a_{12} \\ a_{21} & a_{22} \end{vmatrix},$$

$$M_3 = \begin{vmatrix} a_{11} & a_{12} & a_{13} \\ a_{21} & a_{22} & a_{23} \\ a_{31} & a_{32} & a_{33} \end{vmatrix}, \; \ldots, M_n = |\boldsymbol{A}|.$$

Example 1.1.4: Given

$$\boldsymbol{A} = \begin{bmatrix} 2 & 1 & -1 \\ 5 & 2 & 3 \\ 4 & 0 & 6 \end{bmatrix},$$

the naturally ordered principal minors of $\boldsymbol{A}$ are:

$$M_1 = 2, \; M_2 = \begin{vmatrix} 2 & 1 \\ 5 & 2 \end{vmatrix} = -1, \; M_3 = |\boldsymbol{A}| = 24.$$

Section 1.1 Exercises:

1.1.1 For

$$\boldsymbol{A} = \begin{bmatrix} 3 & 3 & -1 \\ 0 & 1 & 6 \\ 1 & 2 & -1 \end{bmatrix}, \; \boldsymbol{B} = \begin{bmatrix} 1 & 2 & 3 \\ 3 & -1 & 2 \\ 0 & 1 & 3 \end{bmatrix},$$

find:

a. $\boldsymbol{A} - 2\boldsymbol{B}$.

b. $\boldsymbol{AB}$, $\boldsymbol{BA}$.

c. $|\boldsymbol{A}|$, $|\boldsymbol{B}|$ using cofactor expansion via row 2 and by the sweep-out process.

1.1.2 For $\boldsymbol{A}$ and $\boldsymbol{B}$ given in Exercise 1.1.1, verify that:

$$|\boldsymbol{AB}| = |\boldsymbol{A}| \cdot |\boldsymbol{B}| = |\boldsymbol{A}^T \boldsymbol{B}^T|.$$

1.1.3 For $\boldsymbol{A}$ given in Exercise 1.1.1, select $\lambda = 2$ and verify that $|\lambda \boldsymbol{A}| = \lambda^n |\boldsymbol{A}|$.

1.1.4 For $\boldsymbol{B}$ given in Exercise 1.1.1, determine the naturally ordered principal minors of $\boldsymbol{B}$.

1.1.5. For

$$\boldsymbol{C} = \begin{bmatrix} 1 & 2 & -4 \\ 0 & 3 & -7 \\ 0 & 0 & 6 \end{bmatrix},$$

find $|\boldsymbol{C}|$.

1.1.6 For

$$\boldsymbol{A} = \begin{bmatrix} 1 & 5 & 1 \\ 2 & 2 & 4 \\ 3 & 0 & 6 \end{bmatrix},$$

find $|\boldsymbol{A}|$ using cofactor expansion via column 3 and by the sweep-out process.

1.2 VECTOR SPACES AND SUBSPACES

When specifying points in n-dimensional coordinate space, reference is typically made to the concept of a vector, i.e.,

Definition 1.2.1: A **vector** is an ordered n-tuple of (real) numbers expressed as a column $(n \times 1)$ or row $(1 \times n)$ matrix

$$\boldsymbol{x} = \begin{bmatrix} x_1 \\ x_2 \\ \vdots \\ x_n \end{bmatrix}, \; \boldsymbol{x}^T = (x_1, x_2, \ldots, x_n),$$

respectively. The n elements of a vector are called its **components**.

Hence, the notion of a point in n-dimensional space and the concept of an n-component vector (emanating from the origin or **null vector** $\boldsymbol{O}$) are one and the same.

Having defined a vector, we can now introduce the concept of an n-dimensional vector space. To this end, we state

Definition 1.2.2: A **vector (linear) space** $\mathcal{V}_n$ over a field[1] of (real) scalars $\mathcal{F}_R$ is a collection of n-component vectors, which is closed under the operations of addition and scalar multiplication. That is, if $\boldsymbol{x}, \boldsymbol{y} \in \mathcal{V}_n$, then $\boldsymbol{x} + \boldsymbol{y} \in \mathcal{V}_n$ and

a. $\boldsymbol{x} + \boldsymbol{y} = \boldsymbol{y} + \boldsymbol{x}$ (**commutative law**);

b. $\boldsymbol{x} + (\boldsymbol{y} + \boldsymbol{z}) = (\boldsymbol{x} + \boldsymbol{y}) + \boldsymbol{z}$ (**associative law**);

c. there is a unique element $\boldsymbol{O} \in \mathcal{V}_n$ such that $\boldsymbol{x} + \boldsymbol{O} = \boldsymbol{x}$ for each $\boldsymbol{x} \in \mathcal{V}_n$ (**additive identity**);

d. for each $\boldsymbol{x} \in \mathcal{V}_n$ there is a unique element $-\boldsymbol{x} \in \mathcal{V}_n$ such that $\boldsymbol{x} + (-\boldsymbol{x}) = \boldsymbol{O}$ (**additive inverse**);
while for $c \in \mathcal{F}_R$, $\boldsymbol{x} \in \mathcal{V}_n$, it follows that $c\boldsymbol{x} \in \mathcal{V}_n$ and

a′. $c(\boldsymbol{x}+\boldsymbol{y}) = c\boldsymbol{x} + c\boldsymbol{y}$

b′. $(c_1 + c_2)\boldsymbol{x} = c_1\boldsymbol{x} + c_2\boldsymbol{x}$ } (**distributive laws**);

c′. $(c_1c_2)\boldsymbol{x} = c_1(c_2\boldsymbol{x})$ (**associative law**);

d′. $1\boldsymbol{x} = \boldsymbol{x}$ for every $\boldsymbol{x} \in \mathcal{V}_n$ (**multiplicative identity**).

How does one actually add two vectors or multiply a vector by a scalar? We have

Definition 1.2.3: Let $\boldsymbol{x}, \boldsymbol{y} \in \mathcal{V}_n$. The **sum** of $\boldsymbol{x}$ and $\boldsymbol{y}$ is the n-component vector

$$\boldsymbol{x} + \boldsymbol{y} = \begin{bmatrix} x_1 + y_1 \\ x_2 + y_2 \\ \vdots \\ x_n + y_n \end{bmatrix} \in \mathcal{V}_n,$$

where x_i, y_i, $i = 1, \ldots, n$, denote, respectively, the ith components of $\boldsymbol{x}, \boldsymbol{y}$.

Definition 1.2.4: Let $\boldsymbol{x} \in \mathcal{V}_n$. The **product** of a scalar $c \in R_1$ and $\boldsymbol{x}$ is the n-component vector

$$c\boldsymbol{x} = \begin{bmatrix} cx_1 \\ cx_2 \\ \vdots \\ cx_n \end{bmatrix} \in \mathcal{V}_n.$$

Suppose $\mathcal{F}_R$ is taken to be the set of real scalars R. Then, $\mathcal{V}_n$ (R) is an **n-dimensional real vector space** and is simply denoted as R^n. In this regard, we have

Definition 1.2.5: The **Carterian product** (or **product set**) $\mathcal{A} \times \mathcal{B}$ of two sets $\mathcal{A}, \mathcal{B}$ in R^n is the set of ordered pairs (a, b), where $a \in \mathcal{A}$ and $b \in \mathcal{B}$. The product set $R \times R \times \ldots \times R = R^n = \{\boldsymbol{x}^T = (x_1, x_2, \ldots x_n) | x_1, x_2, \ldots x_n \in R\}$.

To specify a **metric** or concept of distance between two points $\boldsymbol{x}, \boldsymbol{y} \in R^n$, let us first define the **scalar product** of $\boldsymbol{x}, \boldsymbol{y}$ as the scalar

$$\boldsymbol{x} \cdot \boldsymbol{y} = \boldsymbol{x}^T \boldsymbol{y} = \textstyle\sum_{i=1}^{n} x_i y_i, \tag{1.3}$$

where

a. $\boldsymbol{x} \cdot \boldsymbol{x} \geq 0$ and $\boldsymbol{x} \cdot \boldsymbol{x} = 0$ if and only if $\boldsymbol{x} = \boldsymbol{O}$;

b. $\boldsymbol{x} \cdot \boldsymbol{y} = \boldsymbol{y} \cdot \boldsymbol{x}$ (**commutative law**);

c. $(\boldsymbol{x} + \boldsymbol{y}) \cdot \boldsymbol{z} = \boldsymbol{x} \cdot \boldsymbol{z} + \boldsymbol{y} \cdot \boldsymbol{z}$ (**distributive law**); and

d. $(c\boldsymbol{x}) \cdot \boldsymbol{y} = c(\boldsymbol{x} \cdot \boldsymbol{y})$, $c \in \mathcal{R}$.

Next, a **norm** on R^n is a function which assigns to each $\boldsymbol{x} \in R^n$ a number $\|x\|$ such that:

a. $\|\boldsymbol{x}\| \geq 0$ and $\|\boldsymbol{x}\| = 0$ if and only if $\boldsymbol{x} = \boldsymbol{O}$;

b. $\|\boldsymbol{x} + \boldsymbol{y}\| \leq \|\boldsymbol{x}\| + \|\boldsymbol{y}\|$ (**triangle inequality**);

c. $\|c\boldsymbol{x}\| = |c| \|\boldsymbol{x}\|$ (**homogeneity**); and

d. $|\boldsymbol{x} \cdot \boldsymbol{y}| \leq \|\boldsymbol{x}\| \|\boldsymbol{y}\|$ (**Cauchy-Schwarz inequality**).

In this regard, the **distance** between points $\boldsymbol{x}, \boldsymbol{y} \in R^n$ induced by the norm "$\|\cdot\|$" on R^n is

$$\|\boldsymbol{x} - \boldsymbol{y}\| = [(\boldsymbol{x} - \boldsymbol{y})^T (\boldsymbol{x} - \boldsymbol{y})]^{\frac{1}{2}} = [\textstyle\sum_{i=1}^{n} |x_i - y_i|^2]^{\frac{1}{2}}. \tag{1.4}$$

An important special case of (1.4) occurs when we consider the distance between any $\boldsymbol{x} \in R^n$ and the origin $\boldsymbol{O}$, i.e., the **length (magnitude)** of $\boldsymbol{x}$ or norm of $\boldsymbol{x}$ is

$$\|\boldsymbol{x}\| = (\boldsymbol{x}^T\boldsymbol{x})^{\frac{1}{2}} = [\textstyle\sum_{i=1}^{n} |x_i|^2]^{\frac{1}{2}}. \qquad (1.5)$$

The **direction** of $\boldsymbol{x}\,(\neq \boldsymbol{O})$ is a **unit vector** $\boldsymbol{x}/\|\boldsymbol{x}\|$ which "points in the direction of" and whose length is unity.

We next look to the specification of the angle between two non-zero vectors.

Definition 1.2.6: Let $\boldsymbol{x}, \boldsymbol{y} \in R^n$ with $\boldsymbol{x}, \boldsymbol{y} \neq \boldsymbol{O}$.The cosine of the angle θ between $\boldsymbol{x}, \boldsymbol{y}$ is

$$\cos\theta = (\boldsymbol{x}\cdot\boldsymbol{y})/\|\boldsymbol{x}\|\|\boldsymbol{y}\|,\ 0 \leq \theta \leq \pi.$$

Example 1.2.1: Find: (a) the distance and (b) the cosine of the angle θ between the vectors

$$\boldsymbol{x} = \begin{bmatrix} 2 \\ 1 \\ 2 \end{bmatrix}, \boldsymbol{y} = \begin{bmatrix} 2 \\ 0 \\ 5 \end{bmatrix};$$

and (c) the direction of $\boldsymbol{x}$.

a. $\|\boldsymbol{x} - \boldsymbol{y}\| = (0^2 + 1^2 + 3^2)^{\frac{1}{2}} = \sqrt{10}$;

b. Since $\boldsymbol{x}\cdot\boldsymbol{y} = 4 + 0 + 10 = 14$, $\|\boldsymbol{x}\| = (2^2 + 1^2 + 2^2)^{\frac{1}{2}} = 3$, and $\|\boldsymbol{y}\| = (2^2 + 0^2 + 5^2)^{\frac{1}{2}} = \sqrt{29}$, it follows that $\cos\theta = 14/3\sqrt{29}$.

c. The direction of $\boldsymbol{x}$ is the unit vector $\boldsymbol{x}/\|\boldsymbol{x}\|$ or

$$\frac{1}{3}\boldsymbol{x} = \begin{bmatrix} 2/3 \\ 1/3 \\ 2/3 \end{bmatrix}.$$

Does $\boldsymbol{x}/\|\boldsymbol{x}\|$ have length unity? It is readily shown that $\|\boldsymbol{x}/\|\boldsymbol{x}\|\| = (1/\|\boldsymbol{x}\|)\|\boldsymbol{x}\| = 1$ for any $\boldsymbol{x} \neq \boldsymbol{O}$, or we can directly find $\|\frac{1}{3}\boldsymbol{x}\| = \frac{1}{3}\|\boldsymbol{x}\| = 1$. ∎

It is important to note that the angle between $\boldsymbol{x}, \boldsymbol{y}$ is acute ($<\pi/2$), right ($=\pi/2$), or obtuse ($>\pi/2$) according to whether the scalar product $\boldsymbol{x}\cdot\boldsymbol{y}$ is positive, zero, or negative. If we employ the converse of this observation, we have

Definition 1.2.7: The vectors $\boldsymbol{x}, \boldsymbol{y} \in R^n$, $\boldsymbol{x}, \boldsymbol{y} \neq \boldsymbol{O}$, are **orthogonal** (i.e., mutually perpendicular) if $\boldsymbol{x}\cdot\boldsymbol{y} = 0$.

(Since $\boldsymbol{x}\cdot\boldsymbol{y} = \|\boldsymbol{x}\|\|\boldsymbol{y}\|\cos\theta$, $\boldsymbol{x}\cdot\boldsymbol{y} = 0$ if and only if $\theta = \pi/2$, and thus $\cos\theta = 0$.)

Finally,

Definition 1.2.8: A non-empty subset $\mathcal{U}$ of R^n is termed a **subspace** if it is closed under the operations of vector addition and scalar multiplication, i.e., given vectors $\boldsymbol{x}_1, \boldsymbol{x}_2 \in \mathcal{U}$, the vector $\lambda_1\boldsymbol{x}_1 + \lambda_2\boldsymbol{x}_2$ is also a member of $\mathcal{U}$ for every pair of scalars $\lambda_1, \lambda_2 \in R$.

Both the origin $\{\boldsymbol{O}\}$ and R^n are themselves (trivially) subspaces. In fact, since $\boldsymbol{O} = 0\boldsymbol{x}$ for any $\boldsymbol{x} \in R^n$, it follows that $\boldsymbol{O}$ is (trivially) a member of every subspace.

Section 1.2 Exercises:

1.2.1 Let $\boldsymbol{x}^T = (2, 3, 1)$, $\boldsymbol{y}^T = (1, 4, 5)$. Find $\boldsymbol{x}\cdot\boldsymbol{y}$, $\|\boldsymbol{x}\|$, $\|\boldsymbol{y}\|$, $\|\boldsymbol{x} - \boldsymbol{y}\|$, and the cosine of the angle θ between $\boldsymbol{x}$ and $\boldsymbol{y}$. What is the direction of $\boldsymbol{y}$?

1.2.2 Are the vectors $\boldsymbol{x}^T = (6, 1, 4)$, $\boldsymbol{y}^T = (3, -1, 5)$ orthogonal?

1.2.3 Let $\boldsymbol{e}_i$, $i = 1, \ldots, n$, denote the **ith unit column vector**, which has a "1" as its ith component and "0s" elsewhere. Verify that:

a. $\|\boldsymbol{e}_i\| = \boldsymbol{e}_i^T\boldsymbol{e}_i = 1$ and $\boldsymbol{e}_i^T\boldsymbol{e}_j = 0$, $i \neq j$;

b. for $\boldsymbol{x}^T = (x_1, x_2, \ldots, x_n)$, $\boldsymbol{e}_i^T\boldsymbol{x} = x_i$, $i = 1, \ldots, n$;

c. for $\boldsymbol{I}_n = [\boldsymbol{e}_1, \boldsymbol{e}_2, \ldots \boldsymbol{e}_n]$, $\boldsymbol{I}_n\boldsymbol{x} = \boldsymbol{x}$; and

d. $\boldsymbol{x} = \sum_{i=1}^{n} x_i\boldsymbol{e}_i$.

1.2.4. Prove the Cauchy-Schwarz inequality $|\boldsymbol{x}\cdot\boldsymbol{y}| \leq \|\boldsymbol{x}\|\|\boldsymbol{y}\|$. (Hint: (a) assume initially that $\boldsymbol{x}, \boldsymbol{y}$ are unit vectors or $\|\boldsymbol{x}\| = \|\boldsymbol{y}\| = 1$; (b) next calculate $0 \leq \|\boldsymbol{x} - \boldsymbol{y}\|^2$ and replace $\boldsymbol{x}, \boldsymbol{y}$ by $\boldsymbol{x}/\|\boldsymbol{x}\|$, $\boldsymbol{y}/\|\boldsymbol{y}\|$, respectively; and, finally, (c) replace $\boldsymbol{x}$ by $-\boldsymbol{x}$.)

1.2.5. Use the Cauchy-Schwarz inequality to derive the triangle inequality $\|\boldsymbol{x} + \boldsymbol{y}\| \leq \|\boldsymbol{x}\| + \|\boldsymbol{y}\|$. (Hint: start with $\|\boldsymbol{x} + \boldsymbol{y}\|^2$ and demonstrate that this expression is less than or equal to $(\|\boldsymbol{x}\| + \|\boldsymbol{y}\|)^2$).

1.3 MATRIX INVERSION

We begin with

Definition 1.3.1: Given an nth-order matrix $\boldsymbol{A}$, if there exists an nth-order matrix $\boldsymbol{A}^{-1}$ which satisfies the relation $\boldsymbol{AA}^{-1} = \boldsymbol{A}^{-1}\boldsymbol{A} = \boldsymbol{I}_n$, then $\boldsymbol{A}^{-1}$ is termed the **inverse** of $\boldsymbol{A}$.

Before we consider the calculation of the inverse of a matrix, two additional definitions are warranted. First,

Definition 1.3.2: The **adjoint** of an nth-order matrix $\boldsymbol{A}$, $\boldsymbol{A}^+$, is the nth-order matrix obtained from $\boldsymbol{A} = [a_{ij}]$ by replacing each element a_{ij} by its cofactor C_{ij} and transposing.

In addition,

Definition 1.3.3: An nth-order matrix $\boldsymbol{A}$ is **singular** if $|\boldsymbol{A}| = 0$ and **non-singular** if $|\boldsymbol{A}| \neq 0$.

In this regard, the inverse of the nth-order matrix $\boldsymbol{A}$ is calculated as

$$\boldsymbol{A}^{-1} = \boldsymbol{A}^+/|\boldsymbol{A}|. \tag{1.6}$$

Under what circumstance will $\boldsymbol{A}^{-1}$ exist? Since $|\boldsymbol{A}|$ appears in the denominator of (1.6), it follows that $\boldsymbol{A}^{-1}$ exists if and only if $\boldsymbol{A}$ is non-singular, i.e., if and only if $|\boldsymbol{A}| \neq 0$. Thus, only non-singular matrices have inverses; moreover, every non-singular matrix has an inverse. Additionally, if $\boldsymbol{A}^{-1}$ exists, it is unique.

The essential properties of matrix inversion are:

a. $(\boldsymbol{A}^{-1})^{-1} = \boldsymbol{A}$;

b. $|\boldsymbol{A}^{-1}| = 1/|\boldsymbol{A}|$;

c. $(\boldsymbol{A}^T)^{-1} = (\boldsymbol{A}^{-1})^T$; and

d. given that $\boldsymbol{A}$, $\boldsymbol{B}$ are non-singular and conformable for multiplication, $(\boldsymbol{AB})^{-1} = \boldsymbol{B}^{-1}\boldsymbol{A}^{-1}$.

Example 1.3.1: Find $\boldsymbol{A}^{-1}$ for

$$\boldsymbol{A} = \begin{bmatrix} 2 & 1 & -1 \\ 3 & 0 & 1 \\ 1 & 4 & 2 \end{bmatrix}.$$

The first step in finding $\boldsymbol{A}^{-1}$ is to determine if it exists. Since $|\boldsymbol{A}| = -25$, $\boldsymbol{A}$ is non-singular and thus possesses an inverse. We next find $\boldsymbol{A}^+$. With the **cofactor matrix** appearing as

$$\begin{bmatrix} C_{11} & C_{12} & C_{13} \\ C_{21} & C_{22} & C_{23} \\ C_{31} & C_{32} & C_{33} \end{bmatrix} = \begin{bmatrix} -4 & -5 & 12 \\ -6 & 5 & -7 \\ 1 & -5 & -3 \end{bmatrix},$$

it follows that

$$\boldsymbol{A}^+ = \begin{bmatrix} -4 & -6 & 1 \\ -5 & 5 & -5 \\ 12 & -7 & -3 \end{bmatrix}$$

and thus, from (1.6),

$$\boldsymbol{A}^{-1} = \frac{\boldsymbol{A}^+}{|\boldsymbol{A}|} = \begin{bmatrix} \frac{4}{25} & \frac{6}{25} & -\frac{1}{25} \\ \frac{1}{5} & -\frac{1}{5} & \frac{1}{5} \\ -\frac{12}{25} & \frac{7}{25} & \frac{3}{25} \end{bmatrix}.$$

One can readily check that $\boldsymbol{AA}^{-1} = \boldsymbol{I}_3$. In fact, this latter equality provides us with a check that $\boldsymbol{A}^{-1}$ has been calculated correctly. ■

Looking at computational efficiency, it should be obvious that the adjoint method of inverting a non-singular matrix may prove tedious when the order of $\boldsymbol{A}$ is large. An alternative and more effective procedure for finding $\boldsymbol{A}^{-1}$ utilizes elementary row operations. As a first step, let us form the $(n \times n + n)$ matrix

$$[\boldsymbol{A}|\boldsymbol{I}_n] = \left[\begin{array}{cccc|cccc} a_{11} & a_{12} & \cdots & a_{1n} & 1 & 0 & \cdots & 0 \\ a_{21} & a_{22} & \cdots & a_{2n} & 0 & 1 & \cdots & 0 \\ \vdots & \vdots & & \vdots & \vdots & \vdots & & \vdots \\ a_{n1} & a_{n2} & \cdots & a_{nn} & 0 & 0 & \cdots & 1 \end{array}\right].$$

Now, if a sequence of elementary row operations applied to $[\boldsymbol{A}|\boldsymbol{I}_n]$ reduces $\boldsymbol{A}$ to $\boldsymbol{I}_n$, that same sequence of operations transforms $\boldsymbol{I}_n$ to $\boldsymbol{A}^{-1}$, i.e.,

$$[\boldsymbol{A}|\boldsymbol{I}_n] \rightarrow [\boldsymbol{I}_n|\boldsymbol{A}^{-1}].$$

Example 1.3.2: Use elementary row operations to find the inverse of the matrix $\boldsymbol{A}$ given in Example 1.3.1. We start with

$$[A|I_3] = \left[\begin{array}{ccc|ccc} 2 & 1 & -1 & 1 & 0 & 0 \\ 3 & 0 & 1 & 0 & 1 & 0 \\ 1 & 4 & 2 & 0 & 0 & 1 \end{array}\right].$$

Then,

$$\left[\begin{array}{ccc|ccc} 2 & 1 & -1 & 1 & 0 & 0 \\ 3 & 0 & 1 & 0 & 1 & 0 \\ 1 & 4 & 2 & 0 & 0 & 1 \end{array}\right] \to \left[\begin{array}{ccc|ccc} 1 & 4 & 2 & 0 & 0 & 1 \\ 3 & 0 & 1 & 0 & 1 & 0 \\ 2 & 1 & -1 & 1 & 0 & 0 \end{array}\right] \to \left[\begin{array}{ccc|ccc} 1 & 4 & 2 & 0 & 0 & 1 \\ 0 & -12 & -5 & 0 & 1 & -3 \\ 0 & -7 & -5 & 1 & 0 & -2 \end{array}\right] \to$$

$$\left[\begin{array}{ccc|ccc} 1 & 4 & 2 & 0 & 0 & 1 \\ 0 & 1 & \frac{5}{12} & 0 & -\frac{1}{12} & \frac{1}{4} \\ 0 & -7 & -5 & 1 & 0 & -2 \end{array}\right] \to \left[\begin{array}{ccc|ccc} 1 & 0 & \frac{1}{3} & 0 & \frac{1}{3} & 0 \\ 0 & 1 & \frac{5}{12} & 0 & -\frac{1}{12} & \frac{1}{4} \\ 0 & 0 & -\frac{25}{12} & 0 & -\frac{7}{12} & -\frac{1}{4} \end{array}\right] \to \left[\begin{array}{ccc|ccc} 1 & 0 & \frac{1}{3} & 0 & \frac{1}{3} & 0 \\ 0 & 1 & \frac{5}{12} & 0 & -\frac{1}{12} & \frac{1}{4} \\ 0 & 0 & 1 & -\frac{12}{25} & \frac{7}{25} & \frac{3}{25} \end{array}\right] \to$$

$$\left[\begin{array}{ccc|ccc} 1 & 0 & 0 & \frac{4}{25} & \frac{6}{25} & -\frac{1}{25} \\ 0 & 1 & 0 & \frac{1}{5} & -\frac{1}{5} & \frac{1}{5} \\ 0 & 0 & 1 & -\frac{12}{25} & \frac{7}{25} & \frac{3}{25} \end{array}\right] = [I_3|A^{-1}].$$

Section 1.3 Exercises:

1.3.1 Find the inverse of each of the following matrices:

a. $A = \begin{bmatrix} 3 & 4 \\ 1 & 6 \end{bmatrix}$, b. $B = \begin{bmatrix} 2 & 1 & 1 \\ 4 & 1 & 3 \\ 0 & 1 & 3 \end{bmatrix}$.

First use the adjoint method (1.6) and then use elementary row operations. Perform a check on your calculations.

1.3.2 Using B in the preceding exercise, verify that:

a. $|B^{-1}| = \frac{1}{|B|}$;

b. $(B^T)^{-1} = (B^{-1})^T$.

1.3.3 For

$$A = \begin{bmatrix} 1 & 0 & 1 \\ 1 & 1 & 4 \\ -1 & 0 & 1 \end{bmatrix}, \quad B = \begin{bmatrix} 3 & 3 & 1 \\ 4 & 0 & 0 \\ 1 & -1 & 0 \end{bmatrix},$$

verify that $(AB^{-1}) = B^{-1}A^{-1}$.

1.4 SOLUTION SET OF A SYSTEM OF SIMULTANEOUS LINEAR EQUATIONS

The system of m linear equations in n unknowns

$$\begin{aligned} a_{11}x_1 + a_{12}x_2 + \cdots + a_{1n}x_n &= c_1 \\ a_{21}x_1 + a_{22}x_2 + \cdots + a_{2n}x_n &= c_2 \\ \cdots\ \cdots\ \cdots\ \cdots\ \cdots\ &\cdots \\ a_{m1}x_1 + a_{m2}x_2 + \cdots + a_{mn}x_n &= c_m \end{aligned}$$

may be written in matrix for as

$$\begin{bmatrix} a_{11} & a_{12} & \cdots & a_{1n} \\ a_{21} & a_{22} & \cdots & a_{2n} \\ \vdots & \vdots & & \vdots \\ a_{m1} & a_{m2} & \cdots & a_{mn} \end{bmatrix} \begin{bmatrix} x_1 \\ x_2 \\ \vdots \\ x_n \end{bmatrix} = \begin{bmatrix} c_1 \\ c_2 \\ \vdots \\ c_m \end{bmatrix} \text{ or } Ax = C, \quad (1.7)$$

where A is an $(m \times n)$ matrix of (constant) coefficients a_{ij}, x is an $(n \times 1)$ matrix (vector) of unknown variables x_i, and C is an $(m \times 1)$ matrix of constants c_i, $i = 1, \ldots, m$; $j = 1, \ldots, n$.

Does a solution to system (1.7) exist? If a solution does exist, is it unique? And exactly how do we obtain any such solution! To set the stage for developing a solution technique, let us start with

Definition 1.4.1: A system of equations is **consistent** if it has at least one solution; it is **inconsistent** if it does not possess a solution.

Definition 1.4.2: The **rank** of an $(m \times n)$ matrix A, denoted $\rho(A)$, is the order of the largest non-singular submatrix of A, i.e., it is the order of the largest non-vanishing determinant in A; moreover, if $\rho(A) = k$, all submatrices of order $k + 1$ are singular.

Additionally,

Definition 1.4.3: An nth order matrix $\boldsymbol{A}$ is of **full rank** if $\rho(\boldsymbol{A}) = n$, i.e., if $|\boldsymbol{A}| \neq 0$ or $\boldsymbol{A}$ is non-singular. So for an nth order matrix, $\rho(\boldsymbol{A}) < n$, if and only if $\boldsymbol{A}$ is singular.

As we shall now see, the **solution set** for $\boldsymbol{Ax} = \boldsymbol{C}$ involves three mutually exclusive and collectively exhaustive cases:

1. no solution exists ($\boldsymbol{Ax} = \boldsymbol{C}$ is inconsistent);
2. exactly one solution exists
3. infinitely many solutions exist

(cases 2 and 3: $\boldsymbol{Ax} = \boldsymbol{C}$ is consistent.)

To relate the concept of rank to the consistency concept, we have

Theorem 1.4.1: Given the system $\boldsymbol{Ax} = \boldsymbol{C}$, where $\boldsymbol{A}$ is of order $(m \times n)$ and $[\boldsymbol{A}, \boldsymbol{C}]$ denotes the **augmented matrix** of $\boldsymbol{Ax} = \boldsymbol{C}$, if:

1. $\rho[\boldsymbol{A}, \boldsymbol{C}] > \rho(\boldsymbol{A})$, the system is inconsistent;
2. $\rho[\boldsymbol{A}, \boldsymbol{C}] = \rho(\boldsymbol{A})=$ number of unknowns n, the system is consistent and possesses a unique solution;
3. $\rho[\boldsymbol{A}, \boldsymbol{C}] = \rho(\boldsymbol{A}) = k<$ number of unknowns n, the system is consistent and possesses an infinity of solutions, where arbitrary values may be assigned to $n- k$ of the variables.

If case (2) obtains and $\boldsymbol{A}$ is of order $(n \times n)$ and non-singular, then $\boldsymbol{x} = \boldsymbol{A}^{-1}\boldsymbol{C}$.

To solve system (1.7), we shall employ the Gauss elimination technique. By way of an introduction to this method, we have

Definition 1.4.4: An $(m \times n)$ matrix $\boldsymbol{E}$ (obtained from a matrix of the same order by a series of elementary row operations) is in **row-reduced echelon normal form** (or simply called an **echelon matrix**) if:

a. the first k rows ($k \geq 0$) are non-zero while the last $m - k$ rows contain only zero elements;

b. the first non-zero element in the ith row ($i = 1,\ldots,k$, $k \geq 1$) equals unity;

c. if c_i denotes the column in which the unity element appears, we require that $c_1 < c_2 < \cdots < c_k$; and lastly, (a), (b), and (c) imply that

d. the lower triangle of elements e_{ij}, where $j < i$, are all zero.

For instance, a typical echelon matrix is

$$\boldsymbol{E} = \begin{bmatrix} 0 & 1 & e_{13} & e_{14} & e_{15} \\ 0 & 0 & 1 & e_{24} & e_{25} \\ 0 & 0 & 0 & 1 & e_{35} \\ 0 & 0 & 0 & 0 & 0 \end{bmatrix}.$$

Armed with the concept of an echelon matrix, we can now state an alternative and equivalent definition of the rank of a matrix.

Definition 1.4.5: The **rank** of an $(m \times n)$ matrix $\boldsymbol{A}$, $\rho(\boldsymbol{A})$, is the number of non-zero rows in the row-reduced echelon normal form of $\boldsymbol{A}$.

Example 1.4.1: Find the rank of

$$\boldsymbol{A} = \begin{bmatrix} 1 & 1 & 2 & 3 \\ 3 & 0 & 1 & 5 \\ 2 & 1 & 4 & 1 \\ 4 & 1 & 0 & 1 \end{bmatrix}.$$

Our goal is to first transform $\boldsymbol{A}$ into an echelon matrix. To this end,

$$\boldsymbol{A} = \begin{bmatrix} 1 & 1 & 2 & 3 \\ 3 & 0 & 1 & 5 \\ 2 & 1 & 4 & 1 \\ 4 & 1 & 0 & 1 \end{bmatrix} \rightarrow \begin{bmatrix} 1 & 1 & 2 & 3 \\ 0 & -3 & -5 & -4 \\ 0 & -1 & 0 & -5 \\ 0 & -3 & -8 & -11 \end{bmatrix} \rightarrow \begin{bmatrix} 1 & 1 & 2 & 3 \\ 0 & 1 & \frac{5}{3} & \frac{4}{3} \\ 0 & -1 & 0 & -5 \\ 0 & -3 & -8 & -11 \end{bmatrix} \rightarrow$$

$$\begin{bmatrix} 1 & 1 & 2 & 3 \\ 0 & 1 & \frac{5}{3} & \frac{4}{3} \\ 0 & 0 & \frac{5}{3} & -\frac{11}{3} \\ 0 & 0 & -3 & -7 \end{bmatrix} \rightarrow \begin{bmatrix} 1 & 1 & 2 & 3 \\ 0 & 1 & \frac{5}{3} & \frac{4}{3} \\ 0 & 0 & 1 & -\frac{11}{5} \\ 0 & 0 & -3 & -7 \end{bmatrix} \rightarrow \begin{bmatrix} 1 & 1 & 2 & 3 \\ 0 & 1 & \frac{5}{3} & \frac{4}{3} \\ 0 & 0 & 1 & -\frac{11}{5} \\ 0 & 0 & 0 & \frac{68}{5} \end{bmatrix} \rightarrow$$

$$\begin{bmatrix} 1 & 1 & 2 & 3 \\ 0 & 1 & \frac{5}{3} & \frac{4}{3} \\ 0 & 0 & 1 & -\frac{11}{5} \\ 0 & 0 & 0 & 1 \end{bmatrix} = \boldsymbol{E}.$$

Clearly $\rho(\boldsymbol{A}) = 4$, the number of non-zero rows in the echelon form of $\boldsymbol{A}$. ■

Example 1.4.2: Is the equation system

$$4x_1 + 2x_2 = 5$$
$$2x_1 + x_2 = 7$$

consistent? Here,

$$|A| = \begin{vmatrix} 4 & 2 \\ 2 & 1 \end{vmatrix} = 0$$

so that $\rho(A) = 1$. To find $\rho[A, C]$, let us obtain the echelon form of $[A, C]$ or

$$[A, C] = \begin{bmatrix} 4 & 2 & 5 \\ 2 & 1 & 7 \end{bmatrix} \rightarrow \begin{bmatrix} 1 & \frac{1}{2} & \frac{5}{4} \\ 2 & 1 & 7 \end{bmatrix}$$
$$\rightarrow \begin{bmatrix} 1 & \frac{1}{2} & \frac{5}{4} \\ 0 & 0 & \frac{9}{2} \end{bmatrix} \rightarrow \begin{bmatrix} 1 & \frac{1}{2} & \frac{5}{4} \\ 0 & 0 & 1 \end{bmatrix} = E.$$

Hence, $\rho[A, C] = 2 > \rho(A) = 1$ so that the aforementioned equation system is inconsistent. ∎

an equivalent system of a prescribed form – the row-reduced echelon normal form. Armed with these considerations, we can now offer the

Gauss elimination technique. If the augmented matrix $[A, C]$ can be transformed, by a series of elementary row operations, into an echelon matrix E, the system of equations corresponding to $E$is (row) equivalent to that represented by $[A, C]$. Hence any solution of the system associated with E is a solution of the system associated with $[A, C]$, and conversely.

Example 1.4.3: Find the values x_1, x_2, and x_3 which simultaneously satisfy the system

$$x_1 + 4x_2 + x_3 = 3$$
$$3x_1 + x_2 \quad = 6$$
$$x_1 + 2x_2 + x_3 = 4.$$

Using the Gauss elimination routine, we obtain

$$[A, C] = \begin{bmatrix} 1 & 4 & 1 & 3 \\ 3 & 1 & 0 & 6 \\ 1 & 2 & 1 & 4 \end{bmatrix} \rightarrow \begin{bmatrix} 1 & 4 & 1 & 2 \\ 0 & -11 & -3 & -3 \\ 0 & -2 & 0 & 1 \end{bmatrix} \rightarrow \begin{bmatrix} 1 & 4 & 1 & 3 \\ 0 & 1 & \frac{3}{11} & \frac{3}{11} \\ 0 & -2 & 0 & 1 \end{bmatrix} \rightarrow$$

(leading non−zero element in row 1 is unity) (x₁ has been eliminated from rows 2,3) (leading non−zero element in row 2 is unity)

$$\begin{bmatrix} 1 & 4 & 1 & 3 \\ 0 & 1 & \frac{3}{11} & \frac{3}{11} \\ 0 & 0 & \frac{6}{11} & \frac{17}{11} \end{bmatrix} \rightarrow \begin{bmatrix} 1 & 4 & 1 & 3 \\ 0 & 1 & \frac{3}{11} & \frac{3}{11} \\ 0 & 0 & 1 & \frac{17}{6} \end{bmatrix} = E.$$

(x_1,x_2 have been eliminated from rows 3) (leading non−zero element in row 3 is unity)

Determining the rank of a matrix is not the only use of an echelon matrix. It is also useful in generating a solution to a simultaneous system of linear equations $Ax = C$, where A is of order $(m \times n)$. The solution method which we shall consider is that of successive (Gaussian) elimination. But, first, we state

Definition 1.4.6: Two systems of linear equations are **equivalent** if every particular solution of either one is also a solution of the other.

Hence, the process of solving a system such as $Ax = C$, with A of order $(m \times n)$, amounts to deducing from it

The values of the unknowns x_1, x_2, and x_3 may now be obtained from E by the process of **back substitution**: from row 3 of E, $x_3 = 17/6$; from the second row of E,

$$x_2 + \frac{3}{11}x_3 = \frac{3}{11}, \; x_2 = \frac{3}{11} - \frac{3}{11}\left(\frac{17}{6}\right) = -\frac{1}{2};$$

and from the first row of E,

$$x_1 + 4x_2 + x_3 = 3, \; x_1 = 3 - 4\left(-\frac{1}{2}\right) - \frac{17}{6} = 13/6.$$

An important feature of the Gauss elimination technique is that it can handle situations where case three of Theorem 1.4.1 holds.

Example 1.4.4: Is the system

$$\begin{aligned} 8x_1 + 10x_2 &= 15 \\ 16x_1 + 20x_2 &= 30 \end{aligned}$$

consistent? Under Gauss elimination, we obtain

$$[\boldsymbol{A}, \boldsymbol{C}] = \begin{bmatrix} 8 & 10 & 15 \\ 16 & 20 & 30 \end{bmatrix} \rightarrow \begin{bmatrix} 1 & \frac{5}{4} & \frac{15}{8} \\ 16 & 20 & 30 \end{bmatrix}$$

$$\rightarrow \begin{bmatrix} 1 & \frac{5}{4} & \frac{15}{8} \\ 0 & 0 & 0 \end{bmatrix} = \boldsymbol{E}.$$

Since $\rho(\boldsymbol{A}) = \rho[\boldsymbol{A}, \boldsymbol{C}] = 1$ (given that $|\boldsymbol{A}| = 0$), the answer is yes. Moreover, from the first row of $\boldsymbol{E}$, we have $x_1 = \frac{15}{8} - \frac{5}{4}x_2$. Hence, an infinity of particular solutions exists, one for every value of x_2. ∎

If $\boldsymbol{C} = \boldsymbol{O}$, then $\boldsymbol{Ax} = \boldsymbol{O}$ is termed a **homogenous** linear equation system. Obviously, this type of system can never be inconsistent, i.e., $\boldsymbol{Ax} = \boldsymbol{O}$ always has a solution since $\rho(\boldsymbol{A}) = \rho[\boldsymbol{A}, \boldsymbol{O}]$. Moreover, $\boldsymbol{x} = \boldsymbol{O}$ is always a (trivial) solution. What about the existence of non-trivial solutions? The answer is provided by a special case of Theorem 1.4.1, namely

Theorem 1.4.2: Given the system $\boldsymbol{Ax} = \boldsymbol{O}$, where $\boldsymbol{A}$ is of order $(m \times n)$, if:

1. $\rho(\boldsymbol{A})$= number of unknowns n, the system has a unique (trivial) solution $\boldsymbol{x} = \boldsymbol{O}$;
2. $\rho(\boldsymbol{A}) = k<$ number of unknowns n, the system has an infinity of non-trivial solutions, where arbitrary values may be assigned to $n - k$ of the variables.

If $\boldsymbol{A}$ is $(m \times n)$ and $m < n$, then $\boldsymbol{Ax} = \boldsymbol{O}$ always has an infinity of non-trivial solutions. In general, for $\boldsymbol{A}$ of order $(m \times n)$ and $\rho(\boldsymbol{A}) = k$, the solution set $\mathcal{S}$ of the homogeneous linear system $\boldsymbol{Ax} = \boldsymbol{O}$, $\boldsymbol{x} \in R^n$, is an $(n - k)$-dimensional subspace of R^n.

Section 1.4 Exercises:

1.4.1 Find the rank (two ways) of each of the following matrices

a. $\boldsymbol{A} = \begin{bmatrix} 3 & 1 \\ 4 & 4 \end{bmatrix}$, b. $\boldsymbol{B} = \begin{bmatrix} 2 & 3 & 3 \\ 1 & 4 & 1 \\ 0 & 6 & 2 \end{bmatrix}$,

c. $\boldsymbol{C} = \begin{bmatrix} \frac{1}{2} & 2 & 1 \\ 4 & 3 & 4 \\ 1 & 4 & 2 \end{bmatrix}$.

1.4.2 Solve the following simultaneous linear equation systems using Gauss elimination.

a. $\begin{aligned} 4x_1 + x_2 &= 1 \\ 3x_1 + 2x_2 &= 7 \end{aligned}$ b. $\begin{aligned} 3x_1 + 2x_2 + x_3 &= 10 \\ x_1 + 3x_2 + 3x_3 &= 16 \\ 2x_1 + x_2 + 4x_3 &= 16 \end{aligned}$ c. $\begin{aligned} 2x_1 + x_2 + x_3 &= 4 \\ x_1 + 3x_2 + 5x_3 &= 2 \\ 3x_1 + 2x_2 + 4x_3 &= 7 \end{aligned}$

1.4.3 A "look it up" exercise:

a. What is the Gauss-Jordan-elimination method for solving simultaneous linear equations? How does it differ from the Gauss-elimination technique?

b. What is Cramer's rule for solving simultaneous linear equations?

Use these two techniques to resolve problem 1.4.2.

1.4.4. Do the following homogeneous linear equation systems possess non-trivial solutions? If so, solve the system.

a. $\begin{aligned} 3x_1 + 4x_2 &= 0 \\ \frac{9}{4}x_1 + 3x_2 &= 0 \end{aligned}$ b. $\begin{aligned} x_1 + x_2 &= 0 \\ x_1 - 2x_2 &= 0 \\ x_1 + 2x_3 &= 0 \end{aligned}$ c. $\begin{aligned} x_1 + 2x_2 + x_3 &= 0 \\ 3x_1 + x_2 + 4x_3 &= 0 \end{aligned}$

1.5 LINEAR DEPENDENCE, DIMENSION, AND RANK

The discussion which follows involves the class of n-component vectors within R^n. In this regard, we have

Definition 1.5.1: The vector $\boldsymbol{x} \in R^n$ is a **linear combination** of the vectors $\boldsymbol{x}_j \in R^n$, $j = 1, \ldots, m$, if there exists scalars λ_j, $j = 1, \ldots, m$, such that

$$x = \sum_{j=1}^{m} \lambda_j x_j. \tag{1.8}$$

Additionally,

Definition 1.5.2: A set of vectors $\{x_j \in R^n, j = 1,\dots,m\}$ is **linearly dependent** if there exist scalars λ_j, $j = 1, \dots, m$, not all zero such that

$$\sum_{j=1}^{m} \lambda_j x_j = O, \tag{1.9}$$

i.e., the null vector is a linear combination of the vectors x_j. If the only set of scalars λ_j for which (1.9) holds is $\lambda_j = 0, j = 1, \dots, m$, the vectors x_j are **linearly independent**, i.e., the trivial combination $0x_1 + \cdots + 0x_m$ is the only linear combination of the x_j which equals the null vector.

(If, say, two vectors x_1, x_2 are linearly dependent, they are **collinear** – they point in the same direction – and one is simply a scalar multiple of the other, e.g., if $\lambda_1 x_1 + \lambda_2 x_2 = O$, with $\lambda_1, \lambda_2 \neq 0$, then

$$x_1 = \frac{\lambda_2}{\lambda_1} x_2 \text{ or } x_2 = -\frac{\lambda_1}{\lambda_2} x_1.$$

Clearly x_1, x_2 are essentially the same except for a scale factor.)

Some important features of a set of linearly dependent (independent) vectors $\{x_j \in R^n, j = 1,\dots,m\}$ are:

a. if the set $\{x_j, j = 1,\dots,m\}$ is linearly dependent, then one of the vectors x_j is a linear combination of the others; if no vector in the set can be written as a linear combination of the others, the set is linearly independent;

b. the set $\{x_j \in R^n, j = 1,\dots,m\}$ is linearly dependent if at least one of the x_j is the null vector. Conversely, if $\{x_j \in R^n, j = 1,\dots,m\}$ is a linearly independent set, it cannot contain the null vector;

c. the set $\{x \in R^n\}$ containing a single vector x is linearly independent if and only if $x \neq O$;

d. if the set $\{x_j \in R^n, j = 1,\dots,m\}$ is linearly independent, then any non-empty subset of this set is linearly independent.

e. if the set of m vectors $\{x_j \in R^n, j = 1,\dots,m\}$ is linearly independent, then the set containing $m + 1$ vectors $\{x, x_j \in R^n, j = 1,\dots,m\}$ is linearly dependent and conversely;

f. if $k < m$ is the maximum number of linearly independent vectors in the set $\{x_j \in R^n, j = 1,\dots,m\}$, then every other vector in the set is linearly dependent upon (i.e., can be expressed as a linear combination of) the k linearly independent vectors.

Moreover,

Theorem 1.5.1: [Steinitz, 1913]. Let $\mathcal{S}_1, \mathcal{S}_2$ be linearly independent sets of vectors in R^n with $\mathcal{S}_1$ containing fewer elements than $\mathcal{S}_2$. Then there exists at least one element $x \in \mathcal{S}_2$ such that the augmented set $\mathcal{S}_1 \cup \{x\}$ is also linearly independent.

Example 1.5.1: Find scalars λ_1, λ_2 such that the vector $x^T = (4, 6)$ is a linear combination of $x_1^T = (3, 5)$, $x_2^T = (2, 4)$. From (1.8), we have

$$\lambda_1 x_1 + \lambda_2 x_2 = (x_1, x_2)\begin{bmatrix} \lambda_1 \\ \lambda_2 \end{bmatrix} = A\lambda = x,$$

or,

$$\begin{aligned} 3\lambda_1 + 2\lambda_2 &= 4 \\ 5\lambda_1 + 4\lambda_2 &= 6. \end{aligned}$$

Since $\rho(A) = \rho(A, x) = 2$, this system possesses the unique solution $\lambda_1 = 2, \lambda_2 = -1$. With $\lambda_1, \lambda_2 \neq 0$, the set $\{x_1, x_2, x\}$ is linearly dependent. ■

Example 1.5.2: Are the vectors $x_1^T = (1, 1), x_2^T = (4, 3)$ linearly independent? From (1.9) we obtain the homogeneous system

$$\lambda_1 x_1 + \lambda_2 x_2 = (x_1, x_2)\begin{bmatrix} \lambda_1 \\ \lambda_2 \end{bmatrix} = A\lambda = O$$

or

$$\begin{aligned} \lambda_1 + 4\lambda_2 &= 0 \\ \lambda_1 + 3\lambda_2 &= 0. \end{aligned}$$

Since $\rho(\boldsymbol{A}) = 2$, the only solution is the trivial solution $\lambda = \boldsymbol{O}$. Hence, $\lambda_1 = \lambda_2 = 0$ and the set $\{\boldsymbol{x}_1, \boldsymbol{x}_2\}$ is linearly independent. ■

Example 1.5.3: Are the vectors $x_1^T = (2, 0, 8)$, $x_2^T = (2, 4, 8)$, and $x_3^T = (1, 1, 4)$ linearly independent? From the homogeneous system

$$\lambda_1\boldsymbol{x}_1 + \lambda_2\boldsymbol{x}_2 + \lambda_3\boldsymbol{x}_3 = (\boldsymbol{x}_1, \boldsymbol{x}_2, \boldsymbol{x}_3)\begin{bmatrix}\lambda_1\\ \lambda_2\\ \lambda_3\end{bmatrix} = \boldsymbol{A}\lambda = \boldsymbol{O}$$

or

$$\begin{aligned} 2\lambda_1 + 2\lambda_2 + \lambda_3 &= 0\\ 4\lambda_2 + \lambda_3 &= 0\\ 8\lambda_1 + 8\lambda_2 + 4\lambda_3 &= 0 \end{aligned}$$

we can express the echelon matrix of the same as

$$\boldsymbol{E} = \begin{bmatrix} 1 & 1 & \frac{1}{2} & 0\\ 0 & 1 & \frac{1}{4} & 0\\ 0 & 0 & 0 & 0 \end{bmatrix}.$$

Since $\rho(\boldsymbol{A}) = 2 <$ number of unknowns $= 3$, Theorem 1.4.2 informs us that the system has an infinity of nontrivial solutions, with $\lambda_1 = \lambda_2 = -\frac{1}{4}\lambda_3$, e.g., for $\lambda_3 = 1$,

$$-\frac{1}{4}\boldsymbol{x}_1 - \frac{1}{4}\boldsymbol{x}_2 + \boldsymbol{x}_3 = \boldsymbol{O}.$$

Hence, the set $\{\boldsymbol{x}_1, \boldsymbol{x}_2, \boldsymbol{x}_3\}$ is linearly dependent. ■

A generalization of our test for the linear independence of a set of vectors is the following. Suppose the set in question has the form $\{\boldsymbol{x}_j \in R^m, j = 1,\ldots,n\}$. If $\boldsymbol{x}_j$ represents the jth column of an $(m \times n)$ matrix $\boldsymbol{A}$ (here $\boldsymbol{A} = [\boldsymbol{x}_1, \boldsymbol{x}_2,\ldots,\boldsymbol{x}_n]$), then:

1. if $m \geq n$ and $\rho(\boldsymbol{A}) = n$, the set of vectors $\{x_j \in R^m, j = 1,\ldots,n\}$ is linearly independent (see part 1 of Theorem 1.4.2); but
2. if $m \geq n$ and $\rho(\boldsymbol{A}) < n$ or if $m < n$, the set of vectors $\{x_j \in R^m, j = 1,\ldots,n\}$ is linearly dependent (see part 2 of Theorem 1.4.2).

We next state

Definition 1.5.3: A set of vectors $\{\boldsymbol{x}_j \in R^n, j = 1,\ldots,m\}$ is a **spanning set** for R^n if every vector $\boldsymbol{x} \in R^n$ can be written as a linear combination of the vectors $\boldsymbol{x}_j$, i.e., for $\boldsymbol{x} \in R^n$,

$$\boldsymbol{x} = \sum_{j=1}^{m} \lambda_j\boldsymbol{x}_j.$$

Here, the set $\{\boldsymbol{x}_j \in R^n, j = 1,\ldots,m\}$ is said to **span** or **generate**R^n since every vector in R^n is (uniquely) linearly dependent on the spanning set. We note further that

a. the vectors which span R^n need not be linearly independent; however,

b. any set of vectors spanning R^n which contains the smallest possible number of vectors is linearly independent.

Definition 1.5.4: A **basis** for R^n is a linearly independent subset of vectors from R^n which spans R^n.

A basis for R^n has the following properties:

a. a basis for R^n is not unique; but the vectors in the basis are unique;

b. every basis for R^n has the same number of basis vectors; and there are precisely n vectors in every basis for R^n.

c. any set of n linearly independent vectors from R^n forms a basis for R^n; and

d. any set of $n + 1$ vectors from R^n is linearly dependent.

We can now state an alternative definition of the rank of a matrix. Specifically,

Definition 1.5.5: The **rank** of an $(m \times n)$ matrix $\boldsymbol{A}$, $\rho(\boldsymbol{A})$, is the maximum number of linearly independent vectors $\boldsymbol{x}_j \in R^m, j = 1, \ldots, n$, which span the columns of $\boldsymbol{A}$, i.e., it is at most the number of vectors in a basis for R^m.

Example 1.5.4: The unit column vectors $\boldsymbol{e}_j = [\delta_{ij}] \in R^n$, $j = 1,\ldots,n$, form a **natural basis** for R^n since they are linearly independent and span R^n. That is, any vector $\boldsymbol{x} \in R^n$ can be uniquely expressed as a linear combination of the $\boldsymbol{e}_j$. For instance, suppose $\boldsymbol{x}^T = (1, 4, -2, 2) \in R^4$. Then

$$\boldsymbol{x} = \boldsymbol{e}_1 + 4\boldsymbol{e}_2 - 2\boldsymbol{e}_3 + 2\boldsymbol{e}_4 = \begin{bmatrix} 1 \\ 0 \\ 0 \\ 0 \end{bmatrix} + 4\begin{bmatrix} 0 \\ 1 \\ 0 \\ 0 \end{bmatrix} - 2\begin{bmatrix} 0 \\ 0 \\ 1 \\ 0 \end{bmatrix} + 2\begin{bmatrix} 0 \\ 0 \\ 0 \\ 1 \end{bmatrix}.$$

■

We previously noted that R^n is an n-dimensional space. In what context should we consider the notion of the dimension of R^n? Generally,

Definition 1.5.6: The **dimension** of a vector space $\mathcal{V}_n$, dim $(\mathcal{V}_n)$, is the maximum number of linearly independent vectors that span the space, i.e., it is the number of vectors in a basis for $\mathcal{V}_n$.

Given the notion of the linear independence of a set of vectors in R^n, we can relate this concept to the existence of a solution to a system of simultaneous linear equations via

Theorem 1.5.2: Let $A^T = [\boldsymbol{x}_1, \boldsymbol{x}_2, \ldots, \boldsymbol{x}_m]$ be of order $(n \times m)$. If the vectors $\boldsymbol{x}_1, \ldots, \boldsymbol{x}_m$ are linearly independent in R^n, then there exists a vector $\lambda \in R^n$ such that $A\lambda = \boldsymbol{C}$ for any $\boldsymbol{C} \in R^m$

Section 1.5 Exercises:

1.5.1 Express the vector $\boldsymbol{x}^T = (1, 0, 1)$ as a linear combination of the vectors

$$\boldsymbol{x}_1 = \begin{bmatrix} 3 \\ 1 \\ 1 \end{bmatrix}, \boldsymbol{x}_2 = \begin{bmatrix} -1 \\ 2 \\ 0 \end{bmatrix}, \boldsymbol{x}_3 = \begin{bmatrix} 5 \\ 1 \\ 2 \end{bmatrix}.$$

1.5.2 Are the vectors $\boldsymbol{x}_i$, $i = 1, 2, 3$ in the preceding exercise linearly independent? Verify your answer.

1.5.3 Do the following vectors form a basis for R^3?

$$\boldsymbol{x}_1 = \begin{bmatrix} 1 \\ 2 \\ 4 \end{bmatrix}, \boldsymbol{x}_2 = \begin{bmatrix} 2 \\ 2 \\ 8 \end{bmatrix}, \boldsymbol{x}_3 = \begin{bmatrix} 1 \\ 0 \\ 4 \end{bmatrix}.$$

1.6 HYPERPLANES AND HALF-PLANES (-SPACES)

The generalization in R^n of the concept of a line in R^2 or of a plane in R^3 is provided by

Definition 1.6.1: A **linear form** or **hyperplane** in R^n is the set $\mathcal{H}$ of all points $\boldsymbol{x}$ such that $c_1x_1 + \cdots + c_nx_n = \boldsymbol{C}^T\boldsymbol{x} = \alpha$, where $\boldsymbol{C} \in R^n$ is a non-null vector and α is a scalar, i.e., $\mathcal{H} = \{\boldsymbol{x} | \boldsymbol{C}^T\boldsymbol{x} = \alpha, \boldsymbol{C}\,(\neq \boldsymbol{O}), \boldsymbol{x} \in R^n\}$.

Any hyperplane $\mathcal{H} \in R^n$ generates the two **closed half-planes** (- **spaces**)

$$[\mathcal{H}^+] = \{\boldsymbol{x} | \boldsymbol{C}^T\boldsymbol{x} \geq \alpha, \boldsymbol{C}\,(\neq \boldsymbol{O}), \boldsymbol{x} \in R^n\},$$

$$[\mathcal{H}^-] = \{\boldsymbol{x} | \boldsymbol{C}^T\boldsymbol{x} \leq \alpha, \boldsymbol{C}\,(\neq \boldsymbol{O}), \boldsymbol{x} \in R^n\}.$$

A point in R^n lies in $[\mathcal{H}^+]$, $[\mathcal{H}^-]$, or in both of these sets. With $\boldsymbol{C}^T\boldsymbol{x} > \alpha\,(<\alpha)$, we have the two **open half-planes** (- **spaces**)

$$(\mathcal{H}^+) = \{\boldsymbol{x} | \boldsymbol{C}^T\boldsymbol{x} > \alpha, \boldsymbol{C}\,(\neq \boldsymbol{O}), \boldsymbol{x} \in R^n\},$$

$$(\mathcal{H}^-) = \{\boldsymbol{x} | \boldsymbol{C}^T\boldsymbol{x} < \alpha, \boldsymbol{C}\,(\neq \boldsymbol{O}), \boldsymbol{x} \in R^n\}.$$

Note that $[\mathcal{H}^+]$, $[\mathcal{H}^-]$ are closed sets while $(\mathcal{H}^+)$, $(\mathcal{H}^-)$ are open sets.[2]

If $\boldsymbol{x}_o$ is taken to be a fixed reference point in $\mathcal{H}$, then $\boldsymbol{C}^T\boldsymbol{x}_o = \alpha$ and, for any other $\boldsymbol{x} \in \mathcal{H}$, $\boldsymbol{C}^T\boldsymbol{x} = \alpha$ so that $\boldsymbol{C}^T\boldsymbol{x} - \boldsymbol{C}^T\boldsymbol{x}_o = \boldsymbol{C}^T(\boldsymbol{x} - \boldsymbol{x}_o) = \alpha - \alpha = 0$. Thus a hyperplane has the alternative representation

$$\mathcal{H} = \{\boldsymbol{x} | \boldsymbol{C}^T(\boldsymbol{x} - \boldsymbol{x}_o) = 0, \boldsymbol{x} = \boldsymbol{x}_o \text{ fixed}, \boldsymbol{x} \in R^n\}.$$

Clearly, $\boldsymbol{C}$ is orthogonal to $\boldsymbol{x} - \boldsymbol{x}_o$ for any $\boldsymbol{x} \in \mathcal{H}$ and is termed the **normal** to the hyperplane. Additionally, for fixed $\boldsymbol{x}_o \in \mathcal{H}$, we now have an alternative representation of the preceding closed half-planes:

$$[\mathcal{H}^+] = \{x | C^T(x - x_o) \geq 0, x = x_o \text{ fixed}, x \in R^n\},$$

$$[\mathcal{H}^-] = \{x | C^T(x - x_o) \leq 0, x = x_o \text{ fixed}, x \in R^n\}.$$

We next have

Definition 1.6.2: A region $\mathcal{K} \subseteq R^n$ is termed **convex** if for points $x_j \in \mathcal{K} j = 1, \ldots, m$, the **convex combination** of the x_j,

$$x_c = \sum_{j=1}^{m} \theta_j x_j, \theta_j \geq 0, \sum_{j=1}^{m} \theta_j = 1,$$

is also a member of $\mathcal{K}$.

It can be readily determined that $\mathcal{H}$, $(\mathcal{H}^+)$, $(\mathcal{H}^-)$, $[\mathcal{H}^+]$, and $[\mathcal{H}^-]$ are all convex. For instance, if $x_1, x_1 \in \mathcal{H}$ (i.e., $C^T x_1 = \alpha = C^T x_2$), then, $\theta_1 = \theta, \theta_2 = 1 - \theta$, and $0 \leq \theta \leq 1$,

$$\begin{aligned} C^T x_c &= C^T(\theta x_2 + (1 - \theta)x_1) = \theta C^T x_2 + (1 - \theta) C^T x_1 \\ &= \theta\alpha + (1 - \theta)\alpha = \alpha \end{aligned}$$

and thus $x_c \in \mathcal{H}$ so that $\mathcal{H}$ is a convex set.

Let $\mathcal{S}_1, \mathcal{S}_2$ be non-empty sets in R^n Then the hyperplane $\mathcal{H} = \{x | C^T x = \alpha, x \in R^n\}$ **weakly separates** $\mathcal{S}_1, \mathcal{S}_2$ if $C^T x \geq \alpha$ for $x \in \mathcal{S}_1$, and $C^T x \leq \alpha$ for $x \in \mathcal{S}_2$. The hyperplane $\mathcal{H}$ is said to **strictly separate** $\mathcal{S}_1, \mathcal{S}_2$ if $C^T x > \alpha$ for $x \in \mathcal{S}_1$, and $C^T x < \alpha$ for $x \in \mathcal{S}_2$. A moment's reflection reveals that: strict separation implies weak separation.

Generally speaking, separability does not imply that $\mathcal{S}_1 \cap \mathcal{S}_2 = \emptyset$ since $\mathcal{S}_1, \mathcal{S}_2$ may have boundary points in common. Separability simply conveys the notion that all points in $\mathcal{S}_1$ lie in one of the closed half-planes determined by $\mathcal{H}$ and all points in $\mathcal{S}_2$ lie in the other.

In the light of the preceding discussion, we can now state

Definition 1.6.3: The hyperplane $\mathcal{H} = \{x | C^T x = \alpha, C(\neq O), x \in R^n\}$ is a **separating hyperplane** for the non-empty sets $\mathcal{S}_1$, $\mathcal{S}_2 \subseteq R^n$ if, for $x \in \mathcal{S}_1$, $C^T x \geq \alpha$, while, for $x \in \mathcal{S}_2$, $C^T x \leq \alpha$.

If $\mathcal{H}$ exists, then $\mathcal{S}_1, \mathcal{S}_2$ are said to be **separable** with

$$\sup_{x \in \mathcal{S}_2} C^T x \leq \alpha \leq \inf_{x \in \mathcal{S}_1} C^T x.$$

Additionally[3], if $\mathcal{H}$ exists and $C^T x > \alpha$ for $x \in \mathcal{S}_1$ and $C^T x < \alpha$ for $x \in \mathcal{S}_2$, then $\mathcal{S}_1$ and $\mathcal{S}_2$ are said to be **strictly separable**, i.e.,

$$\sup_{x \in \mathcal{S}_2} C^T x < \alpha < \inf_{x \in \mathcal{S}_1} C^T x.$$

Definition 1.6.4: The hyperplane $\mathcal{H} = \{x | C^T x = \alpha, C(\neq O), x \in R^n\}$ is a **supporting hyperplane** for a non-empty set $\mathcal{S} \subset R^n$ if $\mathcal{H}$ contains at least one boundary point of $\mathcal{S}$ and $C^T x \leq \alpha (\geq \alpha)$ for all $x \in \mathcal{S}$.

If $\mathcal{H}$ exists, all points in $\mathcal{S}$ lie in one of the closed half-planes determined by $\mathcal{H}$,

What are the requisite conditions for the existence of a separating or supporting hyperplane? The answer depends upon the following set of existence theorems.

Theorem 1.6.1: (**Weak Separation Theorem**). If $\mathcal{S}_1$, $\mathcal{S}_2 \subseteq R^n$ are two non-empty disjoint convex sets, then there exists a hyperplane $\mathcal{H} = \{x | C^T x = \alpha, C(\neq O), x \in R^n\}$ which weakly separates them, i.e., there exists a nonzero vector C and a scalar α such that for $x \in \mathcal{S}_1$, $C^T x \geq \alpha$, and for $x \in \mathcal{S}_2$, $C^T x \leq \alpha$.

We next have

Theorem 1.6.2: (**Strong Separation Theorem**). Let $\mathcal{S}$ be a non-empty closed convex set in R^n with $y \in R^n$ a vector not in $\mathcal{S}$. Then there exists a hyperplane

$$\mathcal{H} = \{x | C^T x = \alpha, C(\neq O), x \in R^n\}$$

which strictly separates $\mathcal{S}$ and y, i.e., there exists a vector $C(\neq O) \in R^n$ and a scalar $\alpha \in R$ such that $C^T x \leq \alpha < C^T y$ for all $x \in \mathcal{S}$.

This theorem states that if $\mathcal{S}$ is a closed convex set in R^n and if $y \in R^n$ is not a member of $\mathcal{S}$, then there exists a strictly separating hyperplane $\mathcal{H}$ such that y is in the open half-space $(\mathcal{H}^+)$ and $\mathcal{S}$ is in the closed half-space $[\mathcal{H}^-]$ determined by $\mathcal{H}$.

Theorem 1.6.2 was given under the assumption that $\mathcal{S}$ is a non-empty closed convex set. However, if $\mathcal{S}$ is not closed and $y \notin \overline{\mathcal{S}}$, then the basic result of this theorem still holds with $\boldsymbol{C}^T\boldsymbol{y} > \alpha = \sup\{\boldsymbol{C}^T\boldsymbol{x} | \boldsymbol{x} \in \mathcal{S}\}$. Hence a less restrictive form of Theorem 1.6.2 is

Theorem 1.6.3: (**Strong Separation Theorem**). Let $\mathcal{S}$ be a non-empty convex set in R^n with $\boldsymbol{y} \in R^n$ an exterior point of $\overline{\mathcal{S}}$. Then there exists a hyperplane

$$\mathcal{H} = \{\boldsymbol{x} | \boldsymbol{C}^T\boldsymbol{x} = \alpha, \boldsymbol{C}(\neq \boldsymbol{O}), \boldsymbol{x} \in R^n\}$$

which strictly separates $\mathcal{S}$ and $\boldsymbol{y}$, i.e., there exists a vector $\boldsymbol{C}(\neq \boldsymbol{O}) \in R^n$ and a scalar $\alpha \in R$ such that $\boldsymbol{C}^T\boldsymbol{y} > \alpha = \sup\{\boldsymbol{C}^T\boldsymbol{x} | \boldsymbol{x} \in \mathcal{S}\}$.

This result reveals that if $\mathcal{S}$ is a convex set in R^n and $\boldsymbol{y} \in R^n$ is not a member of the closure of $\mathcal{S}$, $\overline{\mathcal{S}}$, then there exists a strictly separating hyperplane $\mathcal{H}$ passing through a point in the boundary of $\mathcal{S}$ with $\mathcal{S}$ located on or beneath it with $\boldsymbol{y}$ an element of the open half-space ($\mathcal{H}^+$). (Note that if $\mathcal{S}$ is on or above $\mathcal{H}$, then the conclusion of the theorem is that $\inf\{\boldsymbol{C}^T\boldsymbol{x} | \boldsymbol{x} \in \mathcal{S}\} \geq \boldsymbol{C}^T\boldsymbol{y}$).

A slightly different take on strict separation is provided by

Theorem 1.6.4: (**Strong Separation Theorem**). Let $\mathcal{S}_1$, $\mathcal{S}_2$ be two non-empty disjoint convex sets in R^n with $\mathcal{S}_1$ compact and $\mathcal{S}_2$ closed. Then there exists a hyperplane

$$\mathcal{H} = \{\boldsymbol{x} | \boldsymbol{C}^T\boldsymbol{x} = \alpha, \boldsymbol{C}(\neq \boldsymbol{O}), \boldsymbol{x} \in R^n\}$$

which strictly separates $\mathcal{S}_1$ and $\mathcal{S}_2$, i.e., there exists a vector $\boldsymbol{C}(\neq \boldsymbol{O}) \in R^n$ and a scalar $\alpha \in R$ such that $\boldsymbol{C}^T\boldsymbol{x} < \alpha$, $\boldsymbol{x} \in \mathcal{S}_1$ and $\boldsymbol{C}^T\boldsymbol{x} > 0$, $\boldsymbol{x} \in \mathcal{S}_2$

We next have

Theorem 1.6.5: (**Plane of Support Theorem**). Let $\mathcal{S}$ be a non-empty closed convex set in R^n and let $\boldsymbol{y}$ be a boundary point of $\mathcal{S}$. Then there exists a supporting hyperplane $\boldsymbol{C}^T\boldsymbol{y} = \alpha$, $\boldsymbol{C}(\neq \boldsymbol{O}) \in R^n$, at $\boldsymbol{y}$, i.e., there exists a vector $\boldsymbol{C}(\neq \boldsymbol{O}) \in R^n$ and a scalar $\alpha \in R$ such that $\boldsymbol{C}^T\boldsymbol{x} \leq \alpha = \boldsymbol{C}^T\boldsymbol{y}(=\sup\{\boldsymbol{C}^T\boldsymbol{x} | \boldsymbol{x} \in \mathcal{S}\})$ for all $\boldsymbol{x} \in \mathcal{S}$.

If $\mathcal{S}$ is not a closed subset of R^n, then the conclusion of the preceding theorem may be stated as

Theorem 1.6.6: (**Plane of Support Theorem**). Let $\mathcal{S}$ be a non-empty convex set with interior points in R^n and let $\boldsymbol{y}$ be a boundary point of $\mathcal{S}$. Then, there exists a supporting hyperplane $\boldsymbol{C}^T\boldsymbol{y} = \alpha$, $\boldsymbol{C}(\neq \boldsymbol{O}) \in R^n$, at $\boldsymbol{y}$, i.e., there exists a vector $\boldsymbol{C}(\neq \boldsymbol{O}) \in R^n$ such that $\boldsymbol{C}^T\boldsymbol{x} \leq \alpha = \boldsymbol{C}^T\boldsymbol{y}(=\sup\{\boldsymbol{C}^T\boldsymbol{x} | \boldsymbol{x} \in \overline{\mathcal{S}}\})$ for each $\boldsymbol{x}$ contained within the closure of $\mathcal{S}$, $\overline{\mathcal{S}}$.

Theorems 1.6.5 and 1.6.6 state that a convex subset of R^n (which may or may not be closed) has a supporting hyperplane at each of its boundary points, i.e., given a boundary point of a convex set $\mathcal{S}$ in R^n, there exists a hyperplane that: (1) contains the boundary point ($\mathcal{H} = \{\boldsymbol{x} | \boldsymbol{C}^T\boldsymbol{x} = \alpha = \boldsymbol{C}^T\boldsymbol{y}, \boldsymbol{x} \in R^n\}$); and (2) contains $\mathcal{S}$ in one of its supporting closed half-spaces.

Section 1.6 Exercises:

1.6.1 Suppose a hyperplane is given by $\boldsymbol{C}^T\boldsymbol{x} = \alpha$. Then, $(\lambda\boldsymbol{C})^T\boldsymbol{x} = \lambda\alpha$, $\lambda \neq 0$, defines the same hyperplane. Both $\boldsymbol{C}$ and $\lambda\boldsymbol{C}$ are normal to the hyperplane. What does the unit normal to the hyperplane look like? What is the equation of the hyperplane in terms of the unit normal? Let $2x_1 + 3x_2 = 4$. Rewrite this hyperplane in terms of the unit normal.

1.6.2 Given the hyperplanes $\boldsymbol{C}_1^T\boldsymbol{x} = \alpha_1$, $\boldsymbol{C}_2^T\boldsymbol{x} = \alpha_2$, what is the connection between $\boldsymbol{C}_1$, $\boldsymbol{C}_2$ if these hyperplanes are parallel?

1.6.3 Given the closed half-planes

$$\begin{aligned} 2x_1 + 4x_2 &\leq 4 \\ x_1 - x_2 &\leq 2 \\ x_2 &\geq -2 \end{aligned}$$

determine their **solution set**, i.e., the set of points satisfying all three inequalities simultaneously. A graphical approach will do.

1.6.4 (Thought-provoking exercise.) Under what conditions will the vectors $\boldsymbol{x}_1$, $\boldsymbol{x}_2$, and $\boldsymbol{x}_3$ uniquely define a hyperplane in R^3? Under what conditions will the point $\boldsymbol{x}$ lie on the hyperplane?

1.7 CONVEX AND FINITE CONES

We begin with

Definition 1.7.1: A **cone** $\mathcal{C} \subset R^n$ is a set of points such that if $\boldsymbol{x} \in \mathcal{C}$, then so is every non-negative scalar

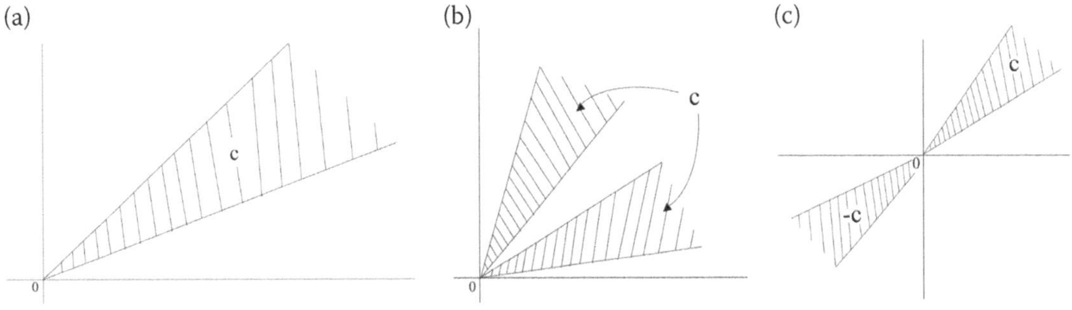

FIGURE 1.1 a. A cone in R^2; b. $\mathcal{C}$ is not convex; c. A cone $\mathcal{C}$ and its negative $-\mathcal{C}$.

multiple of $\boldsymbol{x}$, i.e., if $\boldsymbol{x} \in \mathcal{C}$, then $\lambda \boldsymbol{x} \in \mathcal{C}$, $\lambda \geq 0$, $\boldsymbol{x} \in R^n$ (see Figure 1.1a).

If we consider the set of points $\mathcal{X} = \{x\}$, then the **cone generated by** $\mathcal{X}$ is $\mathcal{C} = \{\boldsymbol{y} | \boldsymbol{y} = \lambda \boldsymbol{x},\ 0 \leq \lambda \in R,\ \boldsymbol{x} \in R^n\}$. And if $\boldsymbol{O} \notin \mathcal{X}$ and for each $\boldsymbol{y}\,(\neq \boldsymbol{O}) \in \mathcal{C}$ there is a unique $\boldsymbol{x} \in \mathcal{X}$ with $\lambda > 0$ such that $\boldsymbol{y} = \lambda \boldsymbol{x}$, then $\mathcal{X}$ is termed a **base** of $\mathcal{C}$. Also, the point $\boldsymbol{O} \in R^n$ is termed the **vertex** of a cone and is an element of every cone $\boldsymbol{y} = 0\boldsymbol{x} = \boldsymbol{O} \in \mathcal{C}$. In general, $\mathcal{C}$ is not a subspace of R^n since $\mathcal{C}$ is defined only for $\lambda \geq 0$ and not all λ.

In what follows we shall consider cones which are convex. Generally speaking, a convex cone is a cone which is a convex set. Specifically,

Definition 1.7.2: A cone $\mathcal{C} \subset R^n$ is **a convex cone** if and only if it is closed under the operations of addition and multiplication by a non-negative scalar, i.e., $\mathcal{C}$ is a convex cone if and only if: (a) for $\boldsymbol{x}, \boldsymbol{y} \in \mathcal{C}$, $\boldsymbol{x} + \boldsymbol{y} \in \mathcal{C}$; and (b) for $\boldsymbol{x} \in \mathcal{C}$, $\lambda \boldsymbol{x} \in \mathcal{C}$, $\lambda \geq 0$, where $\boldsymbol{x}, \boldsymbol{y} \in R^n$

Equivalently, $\mathcal{C}$ in R^n is a convex cone if any non-negative linear combination of points $\boldsymbol{x}_1, \boldsymbol{x}_2$ of $\mathcal{C}$ also belongs to $\mathcal{C}$, i.e., if $(\lambda_1 \boldsymbol{x}_1 + \lambda_2 \boldsymbol{x}_2) \in \mathcal{C}$ for all $0 \leq \lambda_i \in R$ and all $\boldsymbol{x}_i \in \mathcal{C}$, $i = 1, 2$.

It is important to note that not every cone is convex (Figure 1.1b). However, a hyperplane passing through the origin is a convex cone as are the closed half-spaces $[\mathcal{H}^+]$, $[\mathcal{H}^-]$ determined by $\mathcal{H}$. Moreover, the cone $\mathcal{C} = \{\boldsymbol{y} | \boldsymbol{y} = \lambda \boldsymbol{x},\ \lambda \geq 0,\ \boldsymbol{x} \in \mathcal{X}\}$ generated by the set of points $\mathcal{X} = \{x\}$ is convex if $\mathcal{X}$ is itself a convex set.

For a convex cone $\mathcal{C} = \{\boldsymbol{y} | \boldsymbol{y} = \lambda \boldsymbol{x},\ \lambda \geq 0,\ \boldsymbol{x} \in R^n\}$, the **negative cone** of $\mathcal{C}$ is the convex cone – (Figure 1.1c) while the **orthogonal cone** of $\mathcal{C}$ is the convex cone $\mathcal{C}^{\perp} = \{\boldsymbol{v} | \boldsymbol{y}^T \boldsymbol{v} = 0$ for all $\boldsymbol{y} \in \mathcal{C}\}$. Moreover, $\mathcal{C}^{\perp}$ is a subspace of R^n since if $\boldsymbol{v} \in \mathcal{C}^{\perp}$, then $\boldsymbol{y}^T \boldsymbol{v} = 0$ and $\lambda \boldsymbol{y}^T \boldsymbol{v} = 0$ for "all" λ (not just $\lambda \geq 0$) so that $\lambda \boldsymbol{y}^T \boldsymbol{v} \in \mathcal{C}^{\perp}$ for any λ.

Definition 1.7.3: The **dimension** (or **rank**) of a convex cone $\mathcal{C}$, dim ($\mathcal{C}$), is the dimension of the *smallest* subspace of R^n which contains $\mathcal{C}$ and thus amounts to the dimension of $\mathcal{C}$ + (-$\mathcal{C}$).

For a convex cone $\mathcal{C}$ in R^n:

a. The closure of $\mathcal{C}$, $\overline{\mathcal{C}}$, is convex;

b. $\mathcal{C}$ is never a strictly bounded set (although it may be bounded from above or from below) except if, trivially, $\boldsymbol{O}$ is its only element.

A convex cone $\mathcal{C}$ in R^n is **pointed** if it contains no subspace other than $\{\boldsymbol{O}\}$, i.e., given a vector $\alpha \in R^n$, $\mathcal{C}$ contains no **line** $\mathcal{L} = \{\boldsymbol{x} | \boldsymbol{x} = \lambda \alpha$ for all $\lambda \in R$, $\boldsymbol{x} \in R^n\}$. Given this definition, it follows that if a convex cone has a vertex, it is unique and must be located at the origin. Hence, a convex cone with a vertex $\boldsymbol{O}$ is pointed. It is also true that:

a. a convex cone $\mathcal{C}$ in R^n is pointed if and only if $\mathcal{C} \bigcap$ (-$\mathcal{C}$) = $\{\boldsymbol{O}\}$;

b. if $\mathcal{C}$ is a pointed convex cone in R^n, there exists a hyperplane $\mathcal{H}$ passing through $\boldsymbol{O}$ which separates $\mathcal{C}$, -$\mathcal{C}$ and supports both simultaneously;

c. a convex cone $\mathcal{C}$ in R^n which is also a linear subspace has no vertex and consequently is not pointed.

Fundamental operations on convex cones in R^n are addition, intersection, and duality:

a. if C_1, C_2 are convex cones, their **sum** $C_1 + C_2 = \{x|x = x_1 + x_2, x_1 \in C_1, x_2 \in C_2\}$ is a convex cone (Figure 1.2a).

b. if C_1, C_2 are convex cones, their **intersection** $C_1 \cap C_2 = \{x|x \in C_1$ and $x \in C_2\}$ is a convex cone (Figure 1.2b).

c. if C is a convex cone, the **dual cone** C^* $=\{y|x^T y \leq 0$ for all $x \in C\}$ is a convex cone (Figure 1.2c). (Note that C^* consists of all vectors making a non-acute angle ($\geq \pi/2$) with all vectors in C.

Looking at the properties of these operations on convex cones in R^n, we have:

1. if $C_1 \subset C_2$, then $C_2^* \subset C_1^*$;
2. $(C_1 + C_2)^* = C_1^* \cap C_2^*$;
3. $C_1^* + C_2^* \subset (C_1 \cap C_2)^*$; and
4. $C \subset (C^*)^* = C^{**}$.

Relative to fundamental operation (c), we may take the negative of the dual cone C^* of C in R^n to form the **polar cone** $C^+ = \{y|x^T y \geq 0$ for all $x \in C\}$ (Figure 1.2d) consisting of all vectors making a non-obtuse angle ($\leq \pi/2$) with all vectors of C. C^+ is a convex cone if C is. For convex cones in R^n, the polarity operator satisfies:

1. if $C_1 \subset C_2$, then $C_2^+ \subset C_1^+$;
2. $(C_1 + C_2)^+ = C_1^+ \cap C_2^+$;
3. $C_1^+ + C_2^+ \subset (C_1 \cap C_2)^+$; and
4. $C \subset (C^+)^+ = C^{++}$.

Given Definition 1.7.2, we have

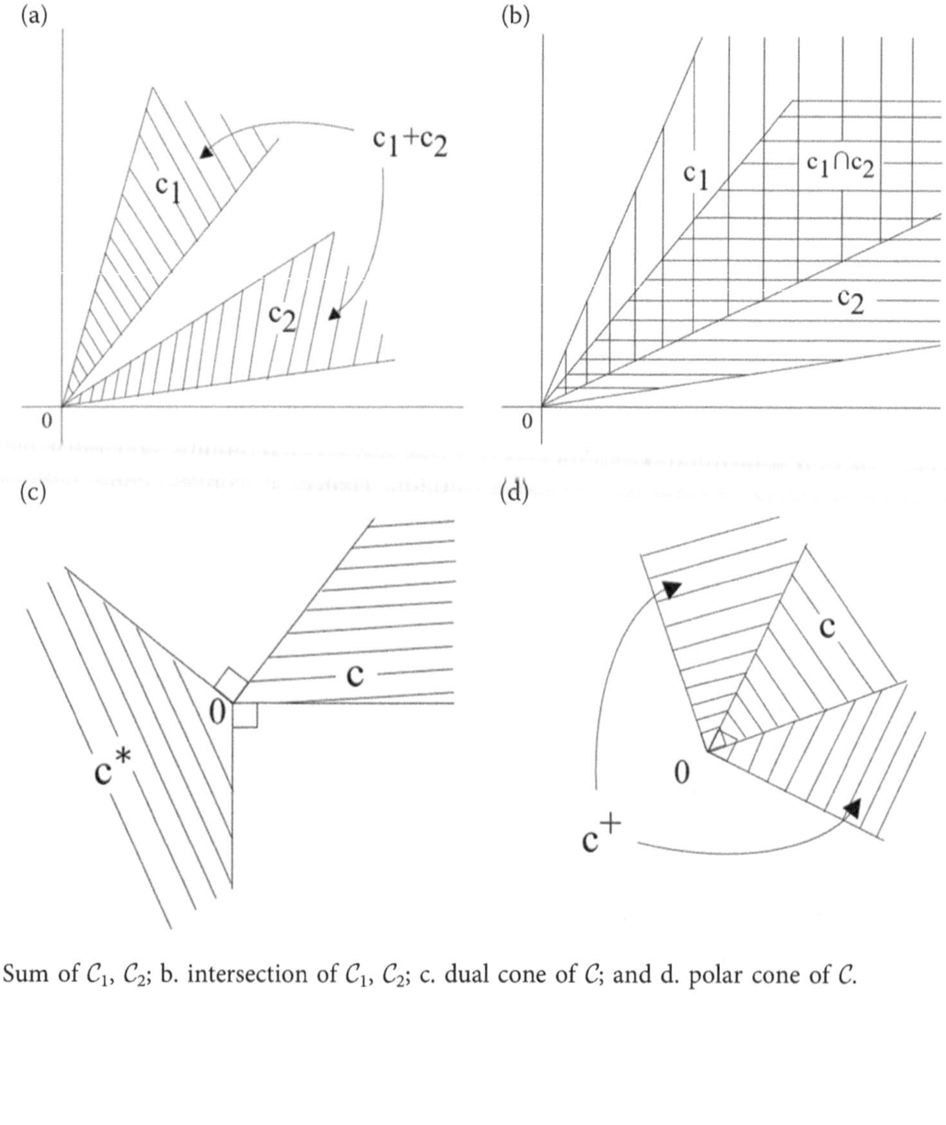

FIGURE 1.2 a. Sum of C_1, C_2; b. intersection of C_1, C_2; c. dual cone of C; and d. polar cone of C.

Definition 1.7.4: A convex cone $\mathcal{C} \subset R^n$ is termed a **finite cone** if it consists of the set of all non-negative linear combinations of a finite set of vectors, i.e., for points $\boldsymbol{x}_j \in R^n$, $j = 1, \ldots, m$,

$$\mathcal{C} = \{\boldsymbol{x} | \boldsymbol{x} = \sum_{j=1}^{m} \lambda_j \boldsymbol{x}_j, \lambda_j \geq 0\}. \quad (1.10)$$

Here $\mathcal{C}$ is said to be **generated** or **spanned** by the points $\boldsymbol{x}_j$. Hence, any vector which can be expressed as a non-negative linear combination of a finite set of vectors $\boldsymbol{x}_j$, $j = 1, \ldots, m$, lies in the finite convex cone spanned by those vectors. Another way of structuring (1.10) is provided by

Definition 1.7.5: A convex cone $\mathcal{C}$ in R^n is a **finite cone** if for some $(n \times m)$ matrix $\boldsymbol{A}^T = [\boldsymbol{a}_1, \ldots, \boldsymbol{a}_m]$, every $\boldsymbol{x} \in \mathcal{C}$ is a non-negative linear combination of the $(n \times 1)$ column vectors $\boldsymbol{a}_j$, $j = 1, \ldots, m$, of $\boldsymbol{A}^T$, i.e.,

$$\mathcal{C} = \{\boldsymbol{x} | \boldsymbol{x} = \boldsymbol{A}^T \lambda = \sum_{j=1}^{m} \lambda_j \boldsymbol{a}_j, 0 \leq \lambda \in R^m, \boldsymbol{x} \in R^n\}. \quad (1.11)$$

A cone such as (1.11) is often referred to as a **closed convex cone**. Moreover, a finite cone (one generated by a finite set of vectors) is synonymous with what will be termed a **polyhedral convex cone**, the latter being formed as the intersection of a finite number of half-planes, each of whose associated hyperplane passes through the origin. Polyhedral convex cones are usually written as the set $\{\boldsymbol{x} | \boldsymbol{A}\boldsymbol{x} \leq \boldsymbol{O}\}$ in R^n.

Given (1.11), the **dual** and **polar cones** may be written, respectively, as

$$\mathcal{C}^* = \{\boldsymbol{y} | \boldsymbol{A}\boldsymbol{y} \geq \boldsymbol{O}, \boldsymbol{y} \in R^n\}; \quad (1.12)$$

$$\mathcal{C}^+ = \{\boldsymbol{y} | \boldsymbol{A}\boldsymbol{y} \leq \boldsymbol{O}, \boldsymbol{y} \in R^n\}; \quad (1.13)$$

A special case of definitions 1.7.3, 1.7.4 occurs when there is only a single generator. That is, a convex cone $\mathcal{C}$ in R^n generated by a single vector $\boldsymbol{a} (\neq \boldsymbol{O})$ is termed a **half-line** (denoted $(\boldsymbol{a})$). So for $\boldsymbol{a} \in R^n$,

$$(\boldsymbol{a}) = \{\boldsymbol{x} | \boldsymbol{x} = \lambda \boldsymbol{a}, 0 \leq \lambda \in R, \boldsymbol{x} \in R^n\} \quad (1.14)$$

(Figure 1.3a). Given $(\boldsymbol{a})$, the **orthogonal cone** to $(\boldsymbol{a})$ is the hyperplane $(\boldsymbol{a})^{\perp} = \mathcal{H} = \{\boldsymbol{y} | \boldsymbol{a}^T \boldsymbol{y} = 0\}$ passing through

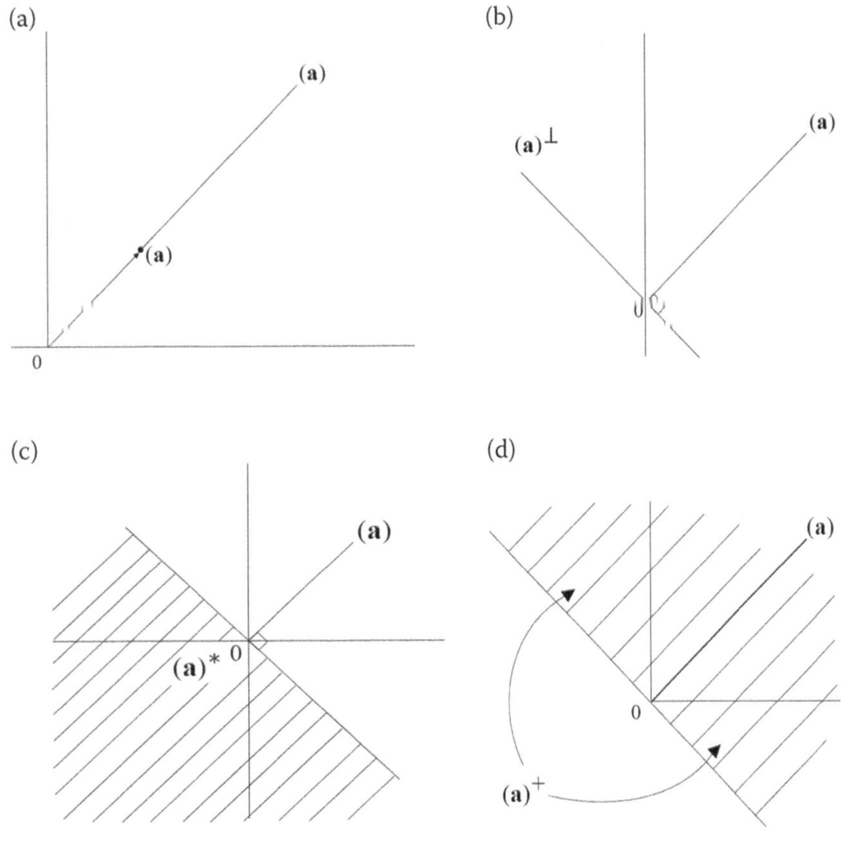

FIGURE 1.3 a. Half-line $(\boldsymbol{a})$; b. orthogonal cone to $(\boldsymbol{a})$; c. dual of $(\boldsymbol{a})$; and d. polar of $(\boldsymbol{a})$.

the origin (Figure 1.3b) while the **dual of** $(\boldsymbol{a})$ is the half-space $(a)^* = [\mathcal{H}^-] = \{y|\boldsymbol{a}^T\boldsymbol{y} \leq 0\}$; it consists of all vectors making a non-acute angle with $\boldsymbol{a}$ (Figure 1.3c). Additionally, the **polar of** $(\boldsymbol{a})$ is the half-space $(a)^+ = [\mathcal{H}^+] = \{\boldsymbol{y}|\boldsymbol{a}^T\boldsymbol{y} \geq 0\}$; it consists of all vectors making a non-obtuse angle with $\boldsymbol{a}$ (Figure 1.3d).

If we admit to our discussion of cones a finite number of half-lines, then we may arrive at an alternative specification of a finite cone provided by

Definition 1.7.6: A cone C in R^n is finite if there exists a finite number of half-lines $(\boldsymbol{a}_j)$in R^n, $j = 1, \ldots, m$, such that $C = \sum_{j=1}^{m} (\boldsymbol{a}_j)$, i.e., since C is the sum of the m cones $(\boldsymbol{a}_j)$, it may be written as

$$C = \{\boldsymbol{x}|\boldsymbol{x} = \sum_{j=1}^{m} \boldsymbol{x}_j,\ \boldsymbol{x}_j \in (\boldsymbol{a}_j),\ \boldsymbol{x} \in R^n\}. \quad (1.15)$$

Since $\boldsymbol{x} = \sum_{j=1}^{m} \boldsymbol{x}_j = \sum_{j=1}^{m} \lambda_j \boldsymbol{a}_j = \boldsymbol{A}^T\lambda,\ \lambda \geq \boldsymbol{O}$(1.15) is just $C = \{\boldsymbol{x}|\boldsymbol{A}^T\lambda,\ \lambda \geq \boldsymbol{O},\ \boldsymbol{x} \in R^n\}$, thus demonstrating the equivalence of (1.10), (1.11), and (1.15).

To summarize, a cone C in R^n is **finitely generated** if:

a. each element of C is expressible as a non-negative linear combination of a finite set of vectors; or, equivalently,

b. C can be written as the sum of a finite number of half-lines.

In specifying a finite cone no mention was made of the linear independence of the spanning set $\{\boldsymbol{x}_j, j = 1, \ldots, m\}$. After all, the vectors $\boldsymbol{x}_j$ need not be linearly independent. However, if some subset of the vectors generating a finite cone C is linearly independent, then we have

Definition 1.7.7: The **dimension** of a finite cone C, dim (C), is the maximum number of linearly independent vectors in C.The essential operations on finite cones are:

a. if C_1, C_2 are finite cones in R^n, their **sum** $C_1 + C_2 = \{\boldsymbol{x}|\boldsymbol{x} = \boldsymbol{x}_1 + \boldsymbol{x}_2,\ \boldsymbol{x}_1 \in C_1,\ \boldsymbol{x}_2 \in C_2\}$ is a finite cone (Figure 1.4a); and

b. if C_1, C_2 are finite cones in R^n, their **intersection** $C_1 \cap C_2$ is a finite cone (Figure 1.4b).

Example 1.7.1: Is the vector $\boldsymbol{x}^T = (1, 2, 1)$ an element of the cone spanned by the vectors $\boldsymbol{x}_1^T = \left(1, 0, \frac{1}{2}\right)$, $\boldsymbol{x}_2^T = (1, 4, 1)$ and $\boldsymbol{x}_3^T = \left(\frac{1}{2}, 1, 1\right)$? Solving the system

$$\boldsymbol{x} = \boldsymbol{A}^T\lambda = [\boldsymbol{x}_1, \boldsymbol{x}_2, \boldsymbol{x}_3]\lambda \text{ or } \begin{bmatrix} 1 & 1 & \frac{1}{2} \\ 0 & 4 & 1 \\ \frac{1}{2} & 1 & 1 \end{bmatrix} \begin{bmatrix} \lambda_1 \\ \lambda_2 \\ \lambda_3 \end{bmatrix} = \begin{bmatrix} 1 \\ 2 \\ 1 \end{bmatrix}$$

simultaneously yields $\lambda^T = (2/5, 2/5, 2/5)$. Hence $\boldsymbol{x} = \frac{2}{5}\boldsymbol{x}_1 + \frac{2}{5}\boldsymbol{x}_2 + \frac{2}{5}\boldsymbol{x}_3$, a non-negative linear combination of $\boldsymbol{x}_1$, $\boldsymbol{x}_2$, and $\boldsymbol{x}_3$, lies in the finite cone spanned

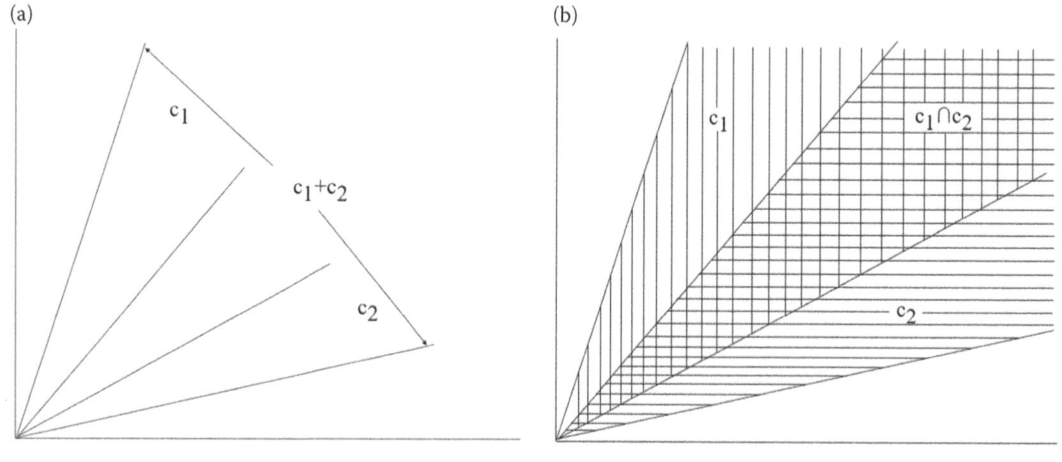

FIGURE 1.4 a. $C_1 + C_2$; b. $C_1 \cap C_2$.

by these vectors. Moreover, since $|A^T| \neq 0$, the vectors x_1, x_2, and x_3 generating $\mathcal{C}$ are linearly independent so that dim $(\mathcal{C}) = 3$. ■

Example 1.7.2: Using the above concepts, verify that: (1) if $A^T = [a_1,\ldots,a_m]$ is $(n \times m)$, the solution set $\mathcal{C}$ of the system of homogeneous linear inequalities $A^T\lambda \leq O$ is a finite cone; and (2) if $A^T = [a_1,\ldots,a_m]$ is $(n \times m)$, the set of non-negative solutions of the homogeneous linear system $A^T\lambda = O$ forms a finite cone.

To answer (1), we first note that $A^T\lambda \leq O$ consists of the n inequalities $\alpha_i\lambda \leq 0$, $i = 1, \ldots, n$, each specifying a closed half-plane of the form $[\mathcal{H}_i^-] = (\alpha_i)^*$ $= \{\lambda|\alpha_i\lambda \leq 0, \lambda \in R^m\}$, where α_i is the ith row of A^T. Since each $(\alpha_i)^*$ is a finite cone, $\mathcal{C} = \cap_{i=1}^n(\alpha_i)^*$ is a finite cone. (It is also true that the solution set $\mathcal{C}$ of $A^T\lambda \geq O$ is a finite cone since $(\alpha_i)^+ = \{\lambda|\alpha_i\lambda \geq 0, \lambda \in R^m\}$, $i = 1, \ldots, n$, is a finite cone and $\mathcal{C} = \cap_{i=1}^n(a_i)^+$.)

Looking to (2), since each of the hyperplanes $\mathcal{H}_i = (\alpha_i)^{\perp} = \{\lambda|\alpha_i\lambda = 0, \lambda \in R^m\}$ constituting $A^T\lambda = O$ is a finite cone, as is the non-negative orthant $\mathcal{N} = \{x|x \geq O, x \in R^m\}$ it follows that $\mathcal{N} \cap (\cap_{i=1}^n(\alpha_i)^{\perp})$ is a finite cone. ■

Section 1.7 Exercises:

1.7.1 Illustrate: (a) the half-lines generated by the points

$$a_1 = \begin{bmatrix} 3 \\ 1 \end{bmatrix}, a_2 = \begin{bmatrix} 1 \\ 3 \end{bmatrix}, a_3 = \begin{bmatrix} -2 \\ 2 \end{bmatrix};$$

(b) the cone $\mathcal{C} = (a_1) + (a_2) + (a_3)$ and $-\mathcal{C}$; (c) $\mathcal{C}^*$, $\mathcal{C}^+$.

1.7.2 Is the vector $x^T = (2, 0, 1)$ an element of the finite cone spanned by the vectors $x_1^T = (1, 4, 1)$, $x_2^T = (0, 1, 3)$, and $x_3^T = \left(1, \frac{1}{2}, 2\right)$?

1.7.3 Verify that $(\mathcal{C}_1+\mathcal{C}_2)^* = \mathcal{C}_1^* \cap \mathcal{C}_2^*$.

1.7.4 Let

$$A = \begin{bmatrix} 1 & 3 & 2 \\ 0 & -1 & 0 \\ 4 & 1 & 6 \end{bmatrix}.$$

What is the dimension of the cone $\mathcal{C}$ generated by A?

1.8 THEOREMS OF THE ALTERNATIVE FOR LINEAR SYSTEMS

We state first

Theorem 1.8.1: (**Strong Separation**). Let $\mathcal{S}$ be a non-empty closed convex set in R^n with $y \in R^n$ a vector not in $\mathcal{S}$. Then there exists a vector $C(\neq 0) \in R^n$ and a scalar $\alpha \in R$ such that $C^Tx \leq \alpha < C^Ty$ for all $x \in \mathcal{S}$.

This theorem states that if $\mathcal{S}$ is a closed convex set in R^n and if $y \in R^n$ is not a member of $\mathcal{S}$, then there exists a strongly separating hyperplane $\mathcal{H} = \{x|C^Tx = \alpha, x \in R^n\}$ such that y is in the open half-space $(\mathcal{H}^+)$ and $\mathcal{S}$ is in the closed half-space $[\mathcal{H}^-]$ determined by $\mathcal{H}$. (Clearly $\mathcal{H}$ also strictly separates $\mathcal{S}$ and y.)

The preceding theorem assumed that $\mathcal{S}$ is a non-empty closed convex set. If $\mathcal{S}$ is not closed and $y \notin \bar{\mathcal{S}}$, then the result of Theorem 1.8.1 still holds with $C^Ty > \alpha = \sup\{C^Tx|x \in \mathcal{S}\}$. Under this circumstance Theorem 1.8.1 can be restated in a less restrictive form as

Theorem 1.8.1a: (**Strong Separation**). Let $\mathcal{S}$ be a non-empty convex set in R^n with $y \in R^n$ an exterior point of $\bar{\mathcal{S}}$. Then there exists a vector $C(\neq 0) \in R^n$ and a scalar $\alpha \in R$ such that $C^Ty > \alpha = \sup\{C^Tx|x \in \mathcal{S}\}$.

This result indicates that if $\mathcal{S}$ is a convex set in R^n and $y \in R^n$ is not a member of $\mathcal{S}$, $\bar{\mathcal{S}}$, then there exists a strongly separating hyperplane $\mathcal{H} = \{x|C^Tx = \alpha, x \in R^n\}$ passing through a point $\bar{x}$ within the boundary of $\mathcal{S}$ and having $\mathcal{S}$ on or beneath it with $y$an element of the open half-space $(\mathcal{H}^+)$.

We next look to

Theorem 1.8.2: (**Plane of Support**). Let $\mathcal{S}$ be a non-empty closed convex set in R^n and let $\bar{x}$ be a boundary point of $\mathcal{S}$. Then there exists a vector $C(\neq 0) \in R^n$ and a scalar $\alpha \in R$ such that $C^Tx \leq \alpha = C^T\bar{x}$ ($=\sup\{C^Tx|x \in \bar{\mathcal{S}}\}$) for all $x \in \mathcal{S}$.

If $\mathcal{S}$ is not a closed subset of R^n, then the basic result of the preceding theorem may be framed as

Theorem 1.8.2a: (Plane of Support). Let $\mathcal{S}$ be a non-empty convex set with interior points in R^n and let $\bar{x}$ be a boundary point of $\mathcal{S}$. Then there exists a vector $C\,(\neq 0) \in R^n$ such that $C^T x \leq \alpha = C^T \bar{x}$ $(= sup\{C^T x | x \in \bar{\mathcal{S}}\})$ for each x contained within $\bar{\mathcal{S}}$.

Theorems 1.8.2 and 1.8.2a inform us that a convex subset of R^n (which may or may not be closed) has a supporting hyperplane at each of its boundary points. So given a boundary point of a convex set $\mathcal{S}$ in R^n, there exists a hyperplane that: (1) contains the boundary point (i.e., $\mathcal{H} = \{x | C^T x = \alpha = C^T \bar{x},\ x \in R^n\}$); and (2) contains $\mathcal{S}$ in one of its supporting closed half-spaces.

The preceding separation theorems involved the existence of a hyperplane that separates a convex set $\mathcal{S}$ (it may be closed) in R^n and a point exterior to $\mathcal{S}$. We now turn to a set of theorems that address the separation of two convex sets $\mathcal{S}_1$, $\mathcal{S}_2$ in R^n. We have first

Theorem 1.8.3: (Strong Separation). Let $\mathcal{S}_1$, $\mathcal{S}_2$ be two non-empty disjoint convex sets in R^n with $\mathcal{S}_1$ compact and $\mathcal{S}_2$ closed. Then there exists a vector $C\,(\neq 0) \in R^n$ and a scalar $\alpha \in R$ such that the following equivalent conditions hold:

a. $C^T x_1 < C^T x_2$ for $x_1 \in \mathcal{S}_1$, $x_2 \in \mathcal{S}_2$;

b. $C^T x < \alpha$, $x \in \mathcal{S}_1$; $C^T x > \alpha$, $x \in \mathcal{S}_2$.

Next comes

Theorem 1.8.4: (Weak Separation). Let $\mathcal{S}_1$, $\mathcal{S}_2$ be two non-empty disjoint convex sets in R^n. Then there exists a vector $C\,(\neq 0) \in R^n$ and a scalar $\alpha \in R$ such that the following equivalent conditions hold:

a. $C^T x_1 \leq C^T x_2$ for $x_1 \in \mathcal{S}_1$, $x_2 \in \mathcal{S}_2$;

b. $C^T x \leq \alpha$, $x \in \mathcal{S}_1$; $C^T x \geq \alpha$, $x \in \mathcal{S}_2$.

An important application of *strong separation* is that it implies *disjoint alternatives*, i.e., it enables us to develop what will be called a *theorem of the alternative*.[4] A fundamental theorem of the alternative is Farkas' theorem. As we shall now see, this theorem can be construed as being based upon the existence of a hyperplane that strongly separates a closed convex set and a vector exterior to the set. While Farkas' theorem appears in a variety of different forms, it is essentially concerned with the conditions underlying the existence of a non-negative solution to a linear system of non-homogeneous equations. Specifically,

Theorem 1.8.5: (Farkas). For A an $(m \times n)$ matrix with $b \in R^n$, $x \in R^n$, and $y \in R^m$, a necessary and sufficient condition for the system $A^T y = b$, $y \geq O$ to have a solution is that $b^T x > 0$ for all x satisfying $Ax \leq O$.

This theorem can be restated in the usual "theorem of the alternative format" as

Theorem 1.8.5a: (Farkas' Theorem of the Alternative). For any $(m \times n)$ matrix A and a vector $b \in R^n$, either

I. $Ax \leq O$, $b^T x > 0$ has a solution $x \in R^n$; or

II. $A^T y = b$, $y \geq O$ has a solution $y \in R^m$, but never both.

(Note that the sense of the inequalities in (I) may be reversed if we set $A = -C$, $b = -d$. Hence, an equivalent version of Theorem 1.8.5a is

Theorem 1.8.5b: (Farkas' Theorem of the Alternative). For any $(m \times n)$ matrix C and a vector $d \in R^n$, either

(I′) $Cx \geq O$, $d^T x < 0$ has a solution $x \in R^n$; or

(II′) $C^T y = d$, $y \geq O$ has a solution $y \in R^m$, but never both.)

According to Farkas' theorem 1.8.5a, if we are interested in finding a non-negative solution to $A^T y = b$, then we can do so by examining the conditions under which (II) does not hold. This will be the case if there exists an x such that $Ax \leq O$, $b^T x > 0$ since any solution of (II) must also satisfy $x^T A^T y = x^T b$. But since the left-hand side of this equality is negative and the right-hand side is positive, a contradiction emerges. If no such contradiction occurs, i.e., if no x satisfying (I) exists, then (II) has a solution.

Looking to the geometry of Theorem 1.8.5a, either: (a) there exists a vector $\boldsymbol{x}$ that makes an obtuse angle ($\geq \pi/2$) with the rows α_i, $i = 1, \ldots, m$, of $\boldsymbol{A}$ and a strictly acute angle ($< \pi/2$) with $\boldsymbol{b}$ (Figure 1.5a); or (b) $\boldsymbol{b}$ is expressible as a non-negative linear combination of the columns α_i^T, $i = 1, \ldots, m$, of $\boldsymbol{A}^T$, i.e., $\boldsymbol{b}$ is an element of the polyhedral convex cone $\mathcal{C}$ spanned by the columns of $\boldsymbol{A}^T$ (Figure 1.5b). Stated alternatively, if (II) has no solution, i.e., $\boldsymbol{b} \notin \mathcal{C}$, there exists a vector $\boldsymbol{x}$ which makes an obtuse angle with $\boldsymbol{b}$ and a non-obtuse angle with the rows of $\boldsymbol{A}$. Thus the hyperplane $\mathcal{H}$ has the polyhedral cone $\mathcal{C}$ on one side and $\boldsymbol{b}$ on the other. Given this observation, Farkas' theorem 1.8.5a is sometimes called the **theorem of the separating hyperplane** (Figure 1.6). (Note also that in Figure 1.5 b the hyperplane $\boldsymbol{b}^T\boldsymbol{x} = 0$ separates the solution set of (II) and $\{\boldsymbol{x} | \boldsymbol{A}\boldsymbol{x} \leq \boldsymbol{O}\}$.

(a)

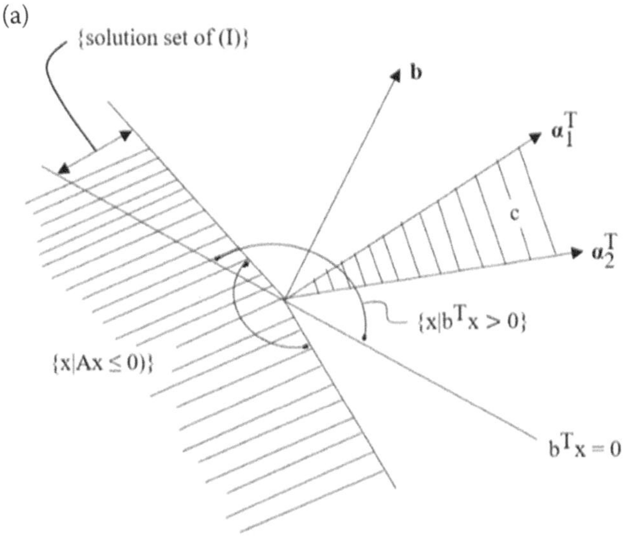

$\{x | Ax \leq 0\} \cap \{x | b^T x > 0\} \neq \emptyset$, i.e,

(I) has a solution; (II) does not.

(b)

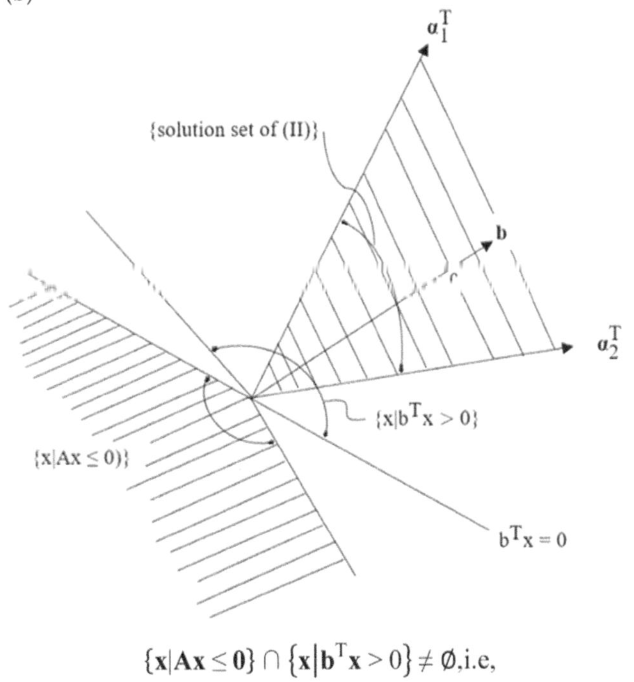

$\{\mathbf{x} | \mathbf{A}\mathbf{x} \leq \mathbf{0}\} \cap \{\mathbf{x} | \mathbf{b}^T \mathbf{x} > 0\} \neq \emptyset$, i.e,

(II) has a solution; (I) does not.

FIGURE 1.5 a. Farkas' (I) has a solution; b. Farkas' (II) does not.

Section 1.8 Exercises:

1.8.1 Given the systems

I. $\boldsymbol{A}\boldsymbol{x} = \boldsymbol{b}$;

II. $\boldsymbol{A}^T\boldsymbol{y} = \boldsymbol{O}$, $\boldsymbol{b}^T\boldsymbol{y} = 1$,
where $\boldsymbol{A}$ is of order $(m \times n)$ and $\boldsymbol{b} \in R^m$, demonstrate that these systems are inconsistent.

1.8.2. Are the following systems consistent?

I. $\boldsymbol{A}\boldsymbol{x} = \boldsymbol{b}$;

II. $\boldsymbol{A}^T\boldsymbol{y} = \boldsymbol{O}$, $\boldsymbol{b}^T\boldsymbol{y} = 0$.

1.8.3. Separating and supporting hyperplanes are two different types of bounding hyperplanes. Explain this statement.

1.8.4. Suppose that $\mathcal{S}$ in Exercise 1.8.3 is also convex and that a point $\boldsymbol{x} \in R^n$ is *not* an interior point of $\mathcal{S}$. Suppose also that a bounding hyperplane passes through $\boldsymbol{x}$. Distinguish between a separating and supporting hyperplane for $\mathcal{S}$ relative to the position of $\boldsymbol{x}$. Why is the convexity assumption needed?

1.8.5. Theorem 1.8.2 essentially stated that if $\boldsymbol{x}$ is a boundary point of a closed convex set $\mathcal{S} \subset R^n$, then there exists a supporting hyperplane of $\mathcal{S}$ at $\boldsymbol{x}$. An important implication of this theorem is that it allows us to represent a closed convex set in terms of its associated set of supporting hyperplanes at boundary points of $\mathcal{S}$. Specifically, we have a

Representation Theorem: A closed convex set $\mathcal{S} \subset R^n$ is the intersection of all supporting closed half-spaces containing it, e.g., $\mathcal{S} = \cap_i [\mathcal{H}_s^+]_i$.

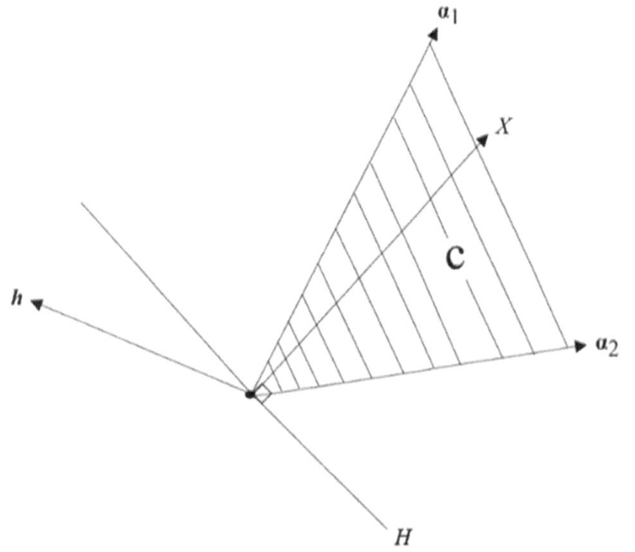

FIGURE 1.6 $\mathcal{H}$ separates $\boldsymbol{b}$ and C.

Prove this representation theorem.

1.8.6 (A Challenging Problem) Prove Theorem 1.8.5a.

1.9 QUADRATIC FORMS

1.9.1 Basic Structure

A quadratic form is a quadratic function. Specifically, we have

Definition 1.9.1: Let Q be a real-valued function of the n variables $x_1, \ldots, x_n$. Then Q is called a **quadradic form** in $x_1, \ldots, x_n$ if

$$Q(x_1, \ldots, x_n) = \sum_{i=1}^{n} \sum_{j=1}^{n} a_{ij} x_i x_j, \tag{1.16}$$

where at least one of the coefficients $a_{ij} \neq 0$, with $i, j = 1, \ldots, n$.

Here, the finite double sum (1.16) may be written explicitly as

$$\begin{aligned} \textstyle\sum_{i=1}^{n} \sum_{j=1}^{n} a_{ij} x_i x_j &= \textstyle\sum_{j=1}^{n} a_{1j} x_1 x_j + \sum_{j=1}^{n} a_{2j} x_2 x_j \\ &+ \cdots + \textstyle\sum_{j=1}^{n} a_{nj} x_n x_j \\ &= a_{11} x_1^2 + a_{12} x_1 x_2 + \cdots + a_{1n} x_1 x_n \\ &+ a_{21} x_2 x_1 + a_{22} x_2^2 + \cdots + a_{2n} x_2 x_n \\ &+ a_{n1} x_n x_1 + a_{n2} x_n x_2 + \cdots + a_{nn} x_{nn}^2, \end{aligned} \tag{1.17}$$

Looking at the properties of Q, we see that:

a. Q is a homogeneous polynomial of the second degree. (A form **is homogeneous of degree *t*** if, when each variable in the form is multiplied by a scalar λ, the whole form is multiplied by λ^t, i.e., $Q(\lambda x_1, \ldots, \lambda x_2) = \lambda^t Q(x_1, \ldots, x_2)$.)

b. Q contains n^2 distinct terms.

c. Q is continuous for all values of the variables x_i, $i = 1, \ldots, n$.

d. Q equals zero when all of the $x_i = 0$, $i = 1, \ldots, n$.

To economize on notation, (1.16) and (1.17) can be written in matrix form, for all vectors $\boldsymbol{x} \in R^n$, as the scalar quantity

$$Q(x_1, \ldots, x_n) = \boldsymbol{x}^T \boldsymbol{A} \boldsymbol{x}, \tag{1.18}$$

where

$$\underset{(n\times 1)}{\boldsymbol{x}} = \begin{bmatrix} x_1 \\ \vdots \\ x_n \end{bmatrix}, \quad \underset{(n\times n)}{\boldsymbol{A}} = \begin{bmatrix} a_{11} & \cdots & a_{1n} \\ \vdots & & \vdots \\ a_{n1} & \cdots & a_{nn} \end{bmatrix}.$$

(The reader is encouraged to perform the indicated multiplication in (1.18) so as to demonstrate that this expression is equivalent to (1.16) or (1.17).) From (1.17), it can be seen that $a_{ij} + a_{ji}$ is the coefficient on $x_i x_j = x_j x_i$, $i \neq j$.

Example 1.9.1: Find the quadratic form $Q(x_1, x_2, x_3)$ associated with the matrix

$$A = \begin{bmatrix} 2 & 1 & 1 \\ 5 & 3 & 0 \\ 1 & 4 & 1 \end{bmatrix}$$

From (1.18),

$$\begin{aligned} \boldsymbol{x}^T \boldsymbol{A} \boldsymbol{x} = \boldsymbol{x}^T \begin{bmatrix} 2 & 1 & 1 \\ 5 & 3 & 0 \\ 1 & 4 & 1 \end{bmatrix} \begin{bmatrix} x_1 \\ x_2 \\ x_3 \end{bmatrix} &= (x_1, x_2, x_3) \begin{bmatrix} 2x_1 + x_2 + x_3 \\ 5x_1 + 3x_2 \\ x_1 + 4x_2 + x_3 \end{bmatrix} \\ &= x_1(2x_1 + x_2 + x_3) + x_2(5x_1 + 3x_2) + x_3(x_1 + 4x_2 + x_3) \\ &= 2x_1^2 + 6x_1x_2 + 2x_1x_3 + 3x_2^2 + 4x_2x_3 + x_3^2. \end{aligned}$$

Clearly $a_{11} = 2$ is the coefficient on x_1^2, $a_{22} = 3$ is the coefficient on x_2^2, and $a_{33} = 1$ is the coefficient on x_3^2. In addition:

$a_{12} + a_{21} = 6$ is the cofficient on $x_1x_2 = x_2x_1$;
$a_{13} + a_{31} = 2$ is the cofficient on $x_1x_3 = x_3x_1$;
$a_{23} + a_{32} = 4$ is the cofficient on $x_2x_3 = x_3x_2$.

■

1.9.2 Symmetric Quadratic Forms

A matrix $\mathbf{A}$ is **symmetric** if $\mathbf{A} = \mathbf{A}^T$ or $a_{ij} = a_{ji}, i \neq j$. In this regard we have

Definition 1.9.2: A quadratic form $\mathbf{x}^T\mathbf{A}\mathbf{x}$ is **symmetric** if matrix $\mathbf{A}$ is symmetric.

Thus, $a_{ij} + a_{ji} = 2a_{ij}$ is the coefficient on x_ix_j since $a_{ij} = a_{ji}, i \neq j$. If $\mathbf{A}$ is not symmetric ($a_{ij} \neq a_{ji}, i \neq j$), it can always be transformed into a symmetric matrix $\mathbf{B}$ by defining new coefficients

$$b_{ij} = b_{ji} = \frac{a_{ij} + a_{ji}}{2} \quad \text{for all } i, j. \tag{1.19}$$

Then $b_{ij} + b_{ji} = 2b_{ij}$ is the coefficient on x_ix_j, $i \neq j$, in

$$\mathbf{x}^T\mathbf{B}\mathbf{x} = \sum_{i=1}^{n}\sum_{j=1}^{n}\left(\frac{a_{ij} + a_{ji}}{2}\right)x_ix_j. \tag{1.20}$$

Since by definition $b_{ij} + b_{ji} = a_{ij} + a_{ji}$, this redefinition of coefficients leaves the value of Q invariant so that $\mathbf{x}^T\mathbf{B}\mathbf{x} = \mathbf{x}^T\mathbf{A}\mathbf{x}$ for any $\mathbf{x} \in R^n$. In sum, given a quadratic form $\mathbf{x}^T\mathbf{A}\mathbf{x}$, the matrix $\mathbf{A}$ may be assumed to be symmetric; if it is not, use (1.19) to transform it into a symmetric matrix.

Example 1.9.2: Given the matrix

$$\mathbf{A} = \begin{bmatrix} 1 & 1 & 3 \\ 3 & -1 & 5 \\ 2 & 6 & 4 \end{bmatrix},$$

transform $\mathbf{A}$ into a symmetric matrix $\mathbf{B}$. From (1.19),

$$\begin{aligned}
b_{11} &= a_{11} = 1 \\
b_{12} &= b_{21} = \tfrac{a_{12} + a_{21}}{2} = 2 \\
b_{13} &= b_{31} = \tfrac{a_{13} + a_{31}}{2} = \tfrac{5}{2} \\
b_{22} &= a_{22} = -1 \\
b_{23} &= b_{32} = \tfrac{a_{23} + a_{32}}{2} = \tfrac{11}{2} \\
b_{33} &= a_{33} = 4
\end{aligned}$$

so that

$$B = \begin{bmatrix} 1 & 2 & 5/2 \\ 2 & -1 & 11/2 \\ 5/2 & 11/2 & 4 \end{bmatrix}.$$

■

Example 1.9.3: Find the matrix $\mathbf{A}$ associated with the quadratic form

$$\mathbf{x}^T\mathbf{A}\mathbf{x} = 2x_1^2 + 5x_1x_2 - \tfrac{9}{2}x_1x_3 + 4x_1x_4 + 3x_2x_3 + 4x_2x_4 + x_3^2 + 8x_3x_4 + 3x_4^2.$$

Since $\mathbf{A}$ may be taken to be symmetric,

$$\mathbf{A} = \begin{bmatrix} 2 & 5/2 & -9/4 & 2 \\ 5/2 & 0 & 3/2 & 2 \\ -9/4 & 3/2 & 1 & 4 \\ 2 & 2 & 4 & 3 \end{bmatrix}.$$

■

1.9.3 Classification of Quadratic Forms

As we shall now see, there are five mutually exclusive and collectively exhaustive categories of quadradic forms. First,

Definition 1.9.3: A quadratic form is said to be **positive definite** (**negative definite**) if it is positive (negative) at every point $\mathbf{x} \in R^n$ except $\mathbf{x} = \mathbf{O}$, i.e.,

a. $\mathbf{x}^T\mathbf{A}\mathbf{x}$ is positive definite if $\mathbf{x}^T\mathbf{A}\mathbf{x} > 0$ for every $\mathbf{x} \neq \mathbf{O}$;

b. $\mathbf{x}^T\mathbf{A}\mathbf{x}$ is negative definite if $\mathbf{x}^T\mathbf{A}\mathbf{x} < 0$ for every $\mathbf{x} \neq \mathbf{O}$.

It should be obvious that a quadratic form which is either positive or negative definite cannot assume both positive and negative values. (For example, the quadratic form $\mathbf{x}^T\mathbf{A}\mathbf{x} = x_1^2 + x_2^2$ is positive definite while $\mathbf{x}^T(-\mathbf{A})\mathbf{x} = -x_1^2 - x_2^2$ is negative definite since each vanishes only at the point $\mathbf{x} = \mathbf{O}$.)

We next have

Definition 1.9.4: A quadratic form is termed **positive semi-definite** (**negative semi-definite**) if it is non-negative (non-positive) at every point $x \in R^n$, and there exist points except $x \neq O$ for which it equals zero, i.e.,

a. $x^T Ax$ is positive semi-definite if $x^T Ax \geq 0$ for every$x$ and $x^T Ax = 0$ for some points $x \neq O$;

b. $x^T Ax$ is negative semi-definite if $x^T Ax \leq 0$ for every$x$ and $x^T Ax = 0$ for some points $x \neq O$.

So if the quadratic form $x^T Ax$ is positive definite (positive semi-definite), then $x^T(-A)x$ is negative definite (negative semi-definite) and conversely. (For instance, the quadratic form $x^T Ax = x_1^2 - 2x_1x_2 =(x_1 - x_2)^2$ is positive semi-definite and $x^T(-A)x = -x_1^2 + 2x_1x_2 - x_2^2 = -(x_1 - x_2)^2$ is negative semi-definite since the former is never negative while the latter is never positive, yet both equal zero for $x_1 = x_2 \neq 0$.)

Additionally, we have

Definition 1.9.5: A quadratic form $x^T Ax$ is **indefinite** if it is positive for some points $x \in R^n$ and negative for others.

(The quadratic form $x^T Ax = x_1x_2 + x_2^2$ is indefinite since $x^T Ax < 0$ for $x_1 = -2, x_2 = 1$ and $x^T Ax > 0$ for $x_1 = 2, x_2 = 1$).

It is important to note that we can classify matrices in terms of the *sign definiteness* or *sign semi-definiteness* of their associated quadratic forms. Specifically,

Definition 1.9.6: An nth-order symmetric matrix A is positive definite (negative definite) if and only if $x^T Ax > 0$ (<0) for all $x (\neq 0) \in R^n$.

Definition 1.9.7: An nth-order symmetric matrix A is positive semi-definite (negative semi-definite) if and only if $x^T Ax \geq 0$ (≤ 0) for all $x \in R^n$.

Some of the properties of definite matrices are:

1. if A is an nth-order positive (negative) definite symmetric matrix, then $|A| > 0(<0)$ and thus $\rho(A) = n$;
2. if A is an nth-order positive (negative) definite symmetric matrix, then so is A^{-1} (clearly A is non-singular).

(Note that if the symmetric matrix A is positive (negative) semi-definite, then $|A| = 0$ so that A is singular with $\rho(A) < n$.)

1.9.4 Necessary Conditions for the Definiteness and Semi-Definiteness of Quadratic Forms

In this section (and the next), our objective is to offer a set of theorems with which to identify the various types of quadratic forms. We consider first

Theorem 1.9.1: If a quadratic form $x^T Ax$, $x \in R^n$, is positive (negative) definite, all the terms involving the second powers of the variables must have positive (negative) coefficients.

(Note, however, that the converse of this theorem does not hold – a quadratic form may have positive (negative) coefficients on all its terms involving second powers, yet not be definite.) Similarly,

Theorem 1.9.2: If a quadratic form $x^T Ax$, $x \in R^n$, is positive (negative) semi-definite, all of its terms involving second powers of the variables must have non-negative (non-positive) coefficients.

(Here, too, the converse of this theorem does not hold -- $Q = x_1^2 - x_1x_2$ has non-negative coefficients associated with its second-degree terms, yet clearly is indefinite.)

To set the stage for discussing conditions that are necessary and sufficient for the definiteness or semi-definiteness of a quadratic form, we need a specialized type of minor of A. In this regard, we have

Definition 1.9.8: The kth **naturally ordered principal minor** of an nth order matrix A,

$$M_k = \begin{vmatrix} a_{11} & \cdots & a_{1k} \\ \vdots & & \vdots \\ a_{k1} & \cdots & a_{kk} \end{vmatrix}, \ k = 1, \ldots, n,$$

is the determinant of the $(k \times k)$ submatrix obtained from $\boldsymbol{A}$ by retaining only its first k rows and columns, i.e., given

$$A = \begin{bmatrix} a_{11} & \cdots & a_{1n} \\ \vdots & & \vdots \\ a_{n1} & \cdots & a_{nn} \end{bmatrix},$$

$$M_1 = a_{11},\ M_2 = \begin{vmatrix} a_{11} & a_{12} \\ a_{21} & a_{22} \end{vmatrix},\ M_3 = \begin{vmatrix} a_{11} & a_{12} & a_{13} \\ a_{21} & a_{22} & a_{23} \\ a_{31} & a_{32} & a_{33} \end{vmatrix}, \ldots, M_n = |A|.$$

Our first application of the naturally ordered principal minor concept is incorporated in

Theorem 1.9.3: If the quadratic form $\boldsymbol{x}^T\boldsymbol{A}\boldsymbol{x}$, $\boldsymbol{x} \in R^n$, is definite, the naturally ordered principal minors of $\boldsymbol{A}$ are all different from zero.

So if any $M_k = 0$, $k = 1, \ldots, n$, the form is not definite; it may be semi-definite or indefinite. (For example, given $\boldsymbol{x}^T\boldsymbol{A}\boldsymbol{x} = x_1^2 - 2x_1x_2 + x_2^2$ with

$$\boldsymbol{A} = \begin{bmatrix} 1 & -1 \\ -1 & 1 \end{bmatrix},$$

$M_1 = 1 \neq 0$, and $M_2 = |\boldsymbol{A}| = 0$, it follows that this form is not definite. In fact, it is semi-definite.)

1.9.5 Necessary and Sufficient Conditions for the Definiteness and Semi-Definiteness of Quadratic Forms

A modification of Theorem 1.9.3 renders

Theorem 1.9.4: The quadratic form $\boldsymbol{x}^T\boldsymbol{A}\boldsymbol{x}$, $\boldsymbol{x} \in R^n$, is positive definite if and only if the naturally ordered principal minors of $\boldsymbol{A}$ are all positive, i.e.,

$$M_k = \begin{vmatrix} a_{11} & \cdots & a_{1k} \\ \vdots & & \vdots \\ a_{k1} & \cdots & a_{kk} \end{vmatrix} > 0,\ k = 1, \ldots, n,$$

or

$$M_1 = a_{11} > 0,\ M_2 = \begin{vmatrix} a_{11} & a_{12} \\ a_{21} & a_{22} \end{vmatrix} > 0,$$

$$M_3 = \begin{vmatrix} a_{11} & a_{12} & a_{13} \\ a_{21} & a_{22} & a_{23} \\ a_{31} & a_{32} & a_{33} \end{vmatrix} > 0, \ldots, M_n = |A| > 0.$$

In addition,

Theorem 1.9.5: The quadratic form $\boldsymbol{x}^T\boldsymbol{A}\boldsymbol{x}$, $\boldsymbol{x} \in R^n$, is negative definite if and only if the naturally ordered principal minors of $\boldsymbol{A}$alternate in sign, the first being negative, i.e.,

$$(-1)^k M_k = (-1)^k \begin{vmatrix} a_{11} & \cdots & a_{1k} \\ \vdots & & \vdots \\ a_{k1} & \cdots & a_{kk} \end{vmatrix} > 0,\ k = 1, \ldots, n,$$

or

$$M_1 = a_{11} < 0,\ M_2 = \begin{vmatrix} a_{11} & a_{12} \\ a_{21} & a_{22} \end{vmatrix} > 0,$$

$$M_3 = \begin{vmatrix} a_{11} & a_{12} & a_{13} \\ a_{21} & a_{22} & a_{23} \\ a_{31} & a_{32} & a_{33} \end{vmatrix} < 0, \ldots, M_n = (-1)^n|\boldsymbol{A}| > 0.$$

We close this section with a similar set of theorems pertaining to semi-definite forms. But first,

Definition 1.9.9: Suppose $\boldsymbol{A}$ is an nth order matrix. A **principal submatrix** of $\boldsymbol{A}$ is the $(n - r) \times (n - r)$ submatrix obtained from $\boldsymbol{A}$ by deleting r of its rows along with the corresponding r columns. The determinant of this principal submatrix is called a **principal minor** of $\boldsymbol{A}$.

A principal minor of order k can be expressed as

$$M\begin{pmatrix} i_1, i_2, i_3, \ldots, i_k \\ i_1, i_2, i_3, \ldots, i_k \end{pmatrix} = \begin{vmatrix} a_{i_1 i_1} & a_{i_1 i_2} & \cdots & a_{i_1 i_k} \\ a_{i_2 i_1} & a_{i_2 i_2} & \cdots & a_{i_2 i_k} \\ \vdots & \vdots & & \vdots \\ a_{i_k i_1} & a_{i_k i_2} & \cdots & a_{i_k i_k} \end{vmatrix},$$

where $i_1, \ldots, i_k$ are the row and column indices for the rows and columns *retained* (the row index values appear above the column index values). For example, consider the following 4th-order matrix

$$\boldsymbol{A} = \begin{bmatrix} a_{11} & a_{12} & a_{13} & a_{14} \\ a_{21} & a_{22} & a_{23} & a_{24} \\ a_{31} & a_{32} & a_{33} & a_{34} \\ a_{41} & a_{42} & a_{43} & a_{44} \end{bmatrix}.$$

A few of its principal minors are:

$$M\begin{pmatrix}1,2,3,4\\1,2,3,4\end{pmatrix} = |\boldsymbol{A}|, \quad M\begin{pmatrix}1,3,4\\1,3,4\end{pmatrix} = \begin{vmatrix}a_{11} & a_{13} & a_{14}\\ a_{31} & a_{33} & a_{34}\\ a_{41} & a_{43} & a_{44}\end{vmatrix}$$

(we retain all rows and columns) ***(we retain rows** 1,3,4, **and columns** 1,3,4**)***

$$M\begin{pmatrix}2,4\\2,4\end{pmatrix} = \begin{vmatrix}a_{22} & a_{24}\\ a_{42} & a_{44}\end{vmatrix}, \qquad M\begin{pmatrix}2\\2\end{pmatrix} = a_{22}$$

(we retain rows** 2,4 **and columns** 2,4**) ***(we retain only row 2 and column 2)***

In all, an nth order matrix $\boldsymbol{A}$ has

$$\begin{pmatrix}n\\p\end{pmatrix} = \frac{n!}{p!(n-p)!}$$

principal minors of order p, e.g., a (3×3) matrix has 1 principal minor of order 3;

$$\begin{pmatrix}3\\2\end{pmatrix} = \frac{3!}{2!1!} = 3$$

principal minors of order 2; and

$$\begin{pmatrix}3\\1\end{pmatrix} = \frac{3!}{1!2!} = 3$$

principal minors of order 1. This yields a total of 7 principal minors. In general, for an $(n \times n)$ matrix, the total number of principal minors is $2^n - 1$.

Now for some additional theorems.

Theorem 1.9.6: An nth order symmetric matrix $\boldsymbol{A}$ is positive semi-definite if and only if all principal minors (not only the naturally ordered ones) are non-negative.

Theorem 1.9.7: An nth order symmetric matrix $\boldsymbol{A}$ is negative semi-definite if and only if

$$(-1)^k M\begin{pmatrix}i_1, i_2, \ldots, i_k\\ i_1, i_2, \ldots, i_k\end{pmatrix} \geq 0,$$

$$1 \leq i_1 \leq i_2 \leq \cdots \leq i_k \leq n, \; k = 1, 2, \ldots, n,$$

where

$$M\begin{pmatrix}i_1, i_2, \ldots, i_k\\ i_1, i_2, \ldots, i_k\end{pmatrix} = \begin{vmatrix}a_{i_1 i_1} & a_{i_1 i_2} & \cdots & a_{i_1 i_k}\\ a_{i_2 i_1} & a_{i_2 i_2} & \cdots & a_{i_2 i_k}\\ \vdots & \vdots & & \vdots\\ a_{i_k i_1} & a_{i_k i_2} & \cdots & a_{i_k i_k}\end{vmatrix}$$

is the determinant of the submatrix of $\boldsymbol{A}$ having k rows and columns from $\boldsymbol{A}$. That is, the principal minors alternate in sign, with those of odd order ≤ 0 and those of even order ≥ 0.

Example 1.9.4: Verify that the quadratic form $\boldsymbol{x}^T\boldsymbol{A}\boldsymbol{x} = 4x_1^2 + 2x_1x_2 + 6x_2^2 + 4x_2x_3 + x_3^2$ is positive definite. Given that

$$\boldsymbol{A} = \begin{bmatrix}4 & 1 & 0\\ 1 & 6 & 2\\ 0 & 2 & 1\end{bmatrix},$$

we have $M_1 = 4$, $M_2 = 23$, and $M_3 = |\boldsymbol{A}| = 7$ Hence $\boldsymbol{x}^T\boldsymbol{A}\boldsymbol{x}$ is positive definite by virtue of Theorem 1.9.4. ■

Example 1.9.5: Is the quadratic form $\boldsymbol{x}^T\boldsymbol{A}\boldsymbol{x} = -2x_1^2 - 2x_1x_2 + 2x_1x_3 - 2x_2^2 + 4x_2x_3 - 2x_3^2$ negative semi-definite? Here,

$$\boldsymbol{A} = \begin{bmatrix}-2 & -1 & 1\\ -1 & -2 & 2\\ 1 & 2 & -2\end{bmatrix}$$

with

$$M\begin{pmatrix}1\\1\end{pmatrix} = -2, \; M\begin{pmatrix}2\\2\end{pmatrix} = -2, \; M\begin{pmatrix}3\\33\end{pmatrix}$$

$$= -2, \; M\begin{pmatrix}1,2\\1,2\end{pmatrix} = \begin{vmatrix}-2 & -1\\ -1 & -2\end{vmatrix}$$

$$=3, \; M\begin{pmatrix}1,3\\1,3\end{pmatrix} = \begin{vmatrix}-2 & 1\\ 1 & -2\end{vmatrix} = 3,$$

$$M\begin{pmatrix}2,3\\2,3\end{pmatrix} = \begin{vmatrix}-2 & 2\\ 2 & -2\end{vmatrix} = 0, \text{ and } M\begin{pmatrix}1,2,3\\1,2,3\end{pmatrix} = |\boldsymbol{A}| = 0.$$

Hence, the conditions of Theorem 1.9.7 hold and thus $\boldsymbol{x}^T\boldsymbol{A}\boldsymbol{x}$ is indeed negative semi-definite. ■

To summarize, an nth order matrix $\boldsymbol{A}$ is:

i. *positive definite* if and only if the n naturally ordered principal are all positive (>0);

ii. negative definite if and only if the n naturally ordered principal minors alternate in sign, beginning negative (i.e., the signs of odd-ordered ones are negative (< 0); and signs of even-ordered ones are positive (> 0));

iii. *positive semi-definite* if and only if its 2^n -1 principal minors are non-negative (≥ 0); and

iv. *negative semi-definite* if and only if its 2^n -1 principal minors alternate in sign with those of odd order non-positive (≤ 0) and those of even order non-negative (≥ 0).

1.9.6 Constrained Quadratic Forms

Suppose $\boldsymbol{x}^T\boldsymbol{A}\boldsymbol{x}$ is a quadratic form, where $\boldsymbol{A}$ is an nth-order symmetric matrix. Additionally, let us assume that the variables $x_1, \ldots, x_n$ must satisfy the m ($<n$) linear **side relations** or **orthogonality constraints**

$$\begin{aligned} b_{11}x_1 + b_{21}x_2 + \cdots + b_{n1}x_n &= \boldsymbol{b}_1^T\boldsymbol{x} = 0 \\ b_{12}x_1 + b_{22}x_2 + \cdots + b_{n2}x_n &= \boldsymbol{b}_2^T\boldsymbol{x} = 0 \\ \cdots\cdots\cdots\cdots\cdots\cdots & \cdots\cdots\cdots\cdots \\ b_{1m}x_1 + b_{2m}x_2 + \cdots + b_{nm}x_n &= \boldsymbol{b}_m^T\boldsymbol{x} = 0 \end{aligned}$$

or, in matrix form,

$$\boldsymbol{x}^T\begin{bmatrix} b_{11} & b_{12} & \cdots & b_{1m} \\ b_{21} & b_{22} & \cdots & b_{2m} \\ \vdots & \vdots & & \vdots \\ b_{n1} & b_{n2} & \cdots & b_{nm} \end{bmatrix} = \boldsymbol{x}^T[\boldsymbol{b}_1, \boldsymbol{b}_2, \ldots, \boldsymbol{b}_m] = \boldsymbol{x}^T\boldsymbol{B} = \boldsymbol{O},$$

where $\boldsymbol{B}$ is of order ($n \times m$) and the $\boldsymbol{b}_j$, $j = 1, \ldots, m$, are taken to be linearly independent or $\rho(\boldsymbol{B}) = m$.

To address the issue of the sign definiteness of $\boldsymbol{x}^T\boldsymbol{A}\boldsymbol{x}$ subject to the aforementioned orthogonality constraints, let us form the bordered matrix of $\boldsymbol{A}$ as

$$\boldsymbol{H} = \left[\begin{array}{c|c} \boldsymbol{A} & \boldsymbol{B} \\ \hline \boldsymbol{B}^T & \boldsymbol{O} \end{array}\right].$$

In addition, let us define the $(r + m) \times (r + m)$ **bordered minor** of A as

$$\Delta_r = \left|\begin{array}{c|c} \boldsymbol{A}_{rr} & \boldsymbol{B}_{rm} \\ \hline \boldsymbol{B}_{rm}^T & \boldsymbol{O} \end{array}\right| = \left|\begin{array}{ccc|ccc} a_{11} & \cdots & a_{1r} & b_{11} & \cdots & b_{1m} \\ \vdots & & \vdots & \vdots & & \vdots \\ a_{r1} & \cdots & a_{rr} & b_{r1} & \cdots & b_{rm} \\ \hline b_{11} & \cdots & b_{r1} & & & \\ \vdots & & \vdots & & \boldsymbol{O} & \\ b_{1m} & \cdots & b_{rm} & & & \end{array}\right|, r = m+1, \ldots, n,$$

where r denotes the number of rows and columns selected from $\boldsymbol{A}$. To specify the sign definiteness of $\boldsymbol{x}^T\boldsymbol{A}\boldsymbol{x}$ (or simply of $\boldsymbol{A}$) subject to $\boldsymbol{x}^T\boldsymbol{B} = \boldsymbol{O}$, we must check the signs of the naturally ordered bordered principal minors of $\boldsymbol{H}$, beginning with the minor that has $m + 1$ rows and columns from $\boldsymbol{A}$ along with the requisite borders for these rows and columns, e.g., we check the signs of the sequence of determinants

$$\left|\begin{array}{cc|ccc} a_{11} & a_{12} & b_{11} & \cdots & b_{1m} \\ a_{21} & a_{22} & b_{21} & \cdots & b_{2m} \\ \hline b_{11} & b_{21} & & & \\ \vdots & \vdots & & \boldsymbol{O} & \\ b_{1m} & b_{2m} & & & \end{array}\right|, \left|\begin{array}{ccc|ccc} a_{11} & a_{12} & a_{13} & b_{11} & \cdots & b_{1m} \\ a_{21} & a_{22} & a_{23} & b_{21} & \cdots & b_{2m} \\ a_{31} & a_{32} & a_{33} & b_{31} & \cdots & b_{3m} \\ \hline b_{11} & b_{21} & b_{31} & & & \\ \vdots & \vdots & \vdots & & \boldsymbol{O} & \\ b_{1m} & b_{2m} & b_{3m} & & & \end{array}\right|, \ldots, |\boldsymbol{H}|.$$

Armed with these considerations we can now state

Theorem 1.9.8: Let $\boldsymbol{A}$ be an nth-order symmetric matrix with the vectors $\{\boldsymbol{b}_1, \ldots, \boldsymbol{b}_m\}$ linearly independent. Then:

1. $\boldsymbol{A}$ is positive definite subject to $\boldsymbol{x}^T\boldsymbol{B} = \boldsymbol{O}$ if and only if every rth-order naturally ordered bordered principal minor of $\boldsymbol{A}$ has sign $(-1)^m$ for $r > m$, i.e.,

$$\hat{\Delta}_r = (-1)^m \left|\begin{array}{c|c} \boldsymbol{A}_{rr} & \boldsymbol{B}_{rm} \\ \hline \boldsymbol{B}_{rm}^T & \boldsymbol{O} \end{array}\right| > 0, r = m+1, \ldots, n.$$

2. $\boldsymbol{A}$ is negative definite subject to $\boldsymbol{x}^T\boldsymbol{B} = \boldsymbol{O}$ if and only if every rth-order naturally ordered bordered principal minor of $\boldsymbol{A}$ has sign $(-1)^r$ for $r > m$, i.e.,

$$\hat{\Delta}_r = (-1)^r \left|\begin{array}{c|c} \boldsymbol{A}_{rr} & \boldsymbol{B}_{rm} \\ \hline \boldsymbol{B}_{rm}^T & \boldsymbol{O} \end{array}\right| > 0, r = m+1, \ldots, n.$$

Note that $\boldsymbol{A}$ subject to $\boldsymbol{x}^T\boldsymbol{B} = \boldsymbol{O}$ will be positive definite if the $\Delta_r s$ of order greater than m have the same sign, with the sign depending on m. For $\boldsymbol{A}$ subject to $\boldsymbol{x}^T\boldsymbol{B} = \boldsymbol{O}$ to be negative definite, the $\Delta_r s$ must alternate in sign. (When $m = 1$, $\boldsymbol{A}$ subject to $\boldsymbol{x}^T\boldsymbol{b} = 0$, $\boldsymbol{b}$ an $(n\times1)$ vector, is positive definite if the $\Delta_r s$are all negative; while $\boldsymbol{A}$ subject to $\boldsymbol{x}^T\boldsymbol{b} = 0$ is negative definite if all $\Delta_r s$ of even order are positive, and all $\Delta_r s$ of odd order are negative.)

Section 1.9 Exercises:

1.9.1. Find the quadratic forms associated with the following matrices:

a. $\begin{bmatrix} 1 & 2 \\ 2 & 0 \end{bmatrix}$; b. $\begin{bmatrix} 2 & 1 & 5 \\ 1 & 1 & 2 \\ 5 & 2 & 4 \end{bmatrix}$; c. $\begin{bmatrix} 2 & 1 & 4 & 1 \\ 1 & 2 & 6 & -2 \\ 4 & 6 & 4 & 0 \\ 1 & -2 & 0 & 5 \end{bmatrix}$.

1.9.2 Given the matrix A, transform A into a symmetric matrix.

$$A = \begin{bmatrix} 5 & 1 & 7 \\ 2 & 0 & 6 \\ 3 & 3 & 4 \end{bmatrix}.$$

1.9.3 Find the matrix A associated with the quadratic forms:

a. $x^TAX = x_1^2 + 4x_1x_2 + \frac{7}{3}x_1x_3 + 2x_2^2 + x_2x_3 + x_3^2$;

b. $x^TAX = 2x_1^2 + x_1x_2 - \frac{6}{5}x_1x_3 + 2x_1x_4 + x_2^2 + 2x_2x_3 + x_3^2 + 8x_3x_4 - x_4^2$.

1.9.4 Determine the sequence of naturally ordered principal minors of the following matrices.

a. $A = \begin{bmatrix} 1 & 4 \\ 1 & 6 \end{bmatrix}$; b. $B = \begin{bmatrix} 3 & 1 & 7 \\ 4 & 4 & 2 \\ 1 & -1 & 5 \end{bmatrix}$;

c. $C = \begin{bmatrix} 6 & 1 & 1 & 0 \\ 0 & 1 & 4 & 1 \\ 2 & 1 & -1 & 3 \\ 1 & 0 & 2 & 1 \end{bmatrix}$.

1.9.5 Use Theorems 1.9.4, 1.9.5 to determine if the following quadratic forms are definite:

a. $Q = 4x_1^2 + 6x_1x_2 + 6x_2^2$;

b. $Q = 3x_1^2 + x_2^2 + 8x_3^2 + 6x_1x_3 - 4x_2x_3$;

c. $Q = x_1^2 - 2x_1x_2 + 6x_2^2 - 4x_2x_3 + 3x_3^2$;

d. $Q = 2x_1^2 + 6x_1x_2 - 8x_1x_3 + 3x_2^2 - 2x_2x_3 - x_3^2$;

e. $Q = -4x_1^2 - 2x_1x_2 + 4x_1x_3 - 5x_2^2 + 6x_2x_3 - 6x_3^2$.

1.9.6. Verify that the following quadratic forms are semi-definite or indefinite:

a. $Q = 4x_1^2 + x_2^2 - 4x_1x_2 + 3x_3^2$;

b. $Q = 4x_1^2 - 3x_2^2$;

c. $Q = x_1^2 + 2x_1x_2 + x_2^2$;

d. $Q = -4x_1^2 + 12x_1x_2 - 9x_2^2$;

e. $Q = -x_1^2 - 9x_2^2 - 2x_3^2 + 6x_1x_2$.

1.9.7. Determine the sign definiteness of the following quadratic forms subject to (s.t.) linear constraints:

a. $Q = x_1^2 - 8x_1x_2 + 10x_2^2$ **s.t.** $x_1 = 2x_2$;

b. $Q = 3x_1^2 + x_2^2 + 2x_3^2$ **s. t.** $x_1 + x_2 + 2x_3 = 0$;

c. $Q = x_1^2 + 2x_1x_2 + x_2^2 + 2x_3^2$ **s. t.** $4x_1 - x_2 - 2x_3 = 0$.

1.9.8. (A modest challenge) For the quadratic form $Q = a_{11}x_1^2 + 2a_{12}x_1x_2 + a_{22}x_2^2$ subject to the linear constraint $b_1x_1 + b_2x_2 = 0$, $b_1 \neq 0$, demonstrate that this quadratic form is positive (negative) definite for all $x\,(\neq O)$ satisfying the constraint if and only if the bordered determinant

$$\Delta = \begin{vmatrix} a_{11} & a_{12} & b_1 \\ a_{21} & a_{22} & b_2 \\ b_1 & b_2 & 0 \end{vmatrix} < 0(>0).$$

1.10 LINEAR TRANSFORMATIONS

1.10.1 Matrix Transformations

Simply stated, a linear transformation is a mapping of variables in which the new variables are linear combinations of the original variables. For instance, if x is a point in R^n and y in a point in R^m, then the linear transformation of x into y can be written as

$$\begin{aligned} y_1 &= a_{11}x_1 + a_{12}x_2 + \cdots + a_{1n}x_n \\ y_2 &= a_{21}x_1 + a_{22}x_2 + \cdots + a_{2n}x_n \\ &\cdots \\ y_m &= a_{m1}x_1 + a_{m2}x_2 + \cdots + a_{mn}x_n \end{aligned}$$

or, in matrix terms

$$y = Ax, \quad \text{[Matrix Transformation]} \tag{1.21}$$

where

$$\underset{(m\times 1)}{y} = \begin{bmatrix} y_1 \\ y_2 \\ \vdots \\ y_m \end{bmatrix}, \quad \underset{(m\times n)}{A} = \begin{bmatrix} a_{11} & a_{12} & \cdots & a_{1n} \\ a_{21} & a_{22} & \cdots & a_{2n} \\ \vdots & \vdots & & \vdots \\ a_{m1} & a_{m2} & \cdots & a_{mn} \end{bmatrix},$$

$$\text{and } x = \begin{bmatrix} x_1 \\ x_2 \\ \vdots \\ x_n \end{bmatrix}.$$

This change of variables is a mapping T (induced by A) of R^n into all or part of R^m; it will be written as $y = T(x)$. Here R^n is called the **domain** of T, denoted D_T, (the set of elements subject to transformation or the inputs of the linear transformation), y is the **image** of x under T (the output of the transformation), and the **range space** of T, Range (T), is the set of elements which results by T operating on the elements in the domain. A linear transformation T is said to be **single-valued** if only one y corresponds to a given x.

Example 1.10.1: Suppose the linear transformation $T: R^3 \to R^3$ is defined as the function

$$T(x_1, x_2, x_3) = (x_1 - 2x_2, x_2 + x_3, x_1 + x_2 + x_3).$$

Then the matrix representation of T is

$$T(x_1, x_2, x_3) = \begin{bmatrix} 1 & -2 & 0 \\ 0 & 1 & 1 \\ 1 & 1 & 1 \end{bmatrix} \begin{bmatrix} x_1 \\ x_2 \\ x_3 \end{bmatrix} = Ax.$$

Note that since T is a linear transformation, we can associate with it the matrix A. Can we move in the opposite direction? That is, given a matrix B, can we obtain a function (denote it S). The answer is yes. For instance, let

$$\underset{(4\times 2)}{B} = \begin{bmatrix} 6 & -1 \\ 0 & 1 \\ 3 & -2 \\ 0 & -1 \end{bmatrix}.$$

Then the function $S(x_1, x_2) = Bx$, $x^T = (x_1, x_2)$, where

$$S(x_1, x_1) = (6x_1 - x_2, x_2, 3x_1 - 2x_2, -x_2).$$

Here $S: R^2 \to R^4$ or

$$\begin{aligned} y_1 &= 6x_1 - x_2 \\ y_2 &= \quad\;\; x_2 \\ y_3 &= 3x_1 - 2x_2 \\ y_4 &= \quad -x_2. \end{aligned}$$

■

Suppose A in (1.21) is $(n \times n)$ and non-singular (A^{-1} exists) so that $x = A^{-1}y$ -- every point y has a corresponding unique x. In this instance T is termed a **one-to-one linear transformation** from E^n into E^n -- one and only one x corresponds to each y, and one and only one y corresponds to each x. For A non-singular, Range (T) is the entirety of R^n.

Generally speaking, transformations are linear if: (a) the sum of any two inputs is transformed into the sum of their individual images or outputs; and (b) a multiple λ of any input is transformed into that same multiple of its output. More formally (abstractly) we have

Definition 1.10.1: Let $\mathcal{X}$, $\mathcal{Y}$ be real vector spaces. A linear transformation T from $\mathcal{X}$ to $\mathcal{Y}$ is a mapping or correspondence that assigns to each vector $x \in \mathcal{X}$ a vector $T(x) \in \mathcal{Y}$ such that:

1. $T(x_1 + x_2) = T(x_1) + T(x_2)$ for all vectors x_1, $x_2 \in \mathcal{X}$;
2. $T(\lambda x) = \lambda T(x)$ for all vectors $x \in \mathcal{X}$ and all scalars λ (**homogeneity property** of T).

Equivalently, we can write $T(\lambda_1 x_1 + \lambda_2 x_2) = \lambda_1 T(x_1) + \lambda_2 T(x_2)$ -- the **linearity principle.**

In terms of (1.10.1), let $\mathcal{X} = R^n$ and $\mathcal{Y} = R^m$ with A of order $(m \times n)$. Then, T defined by $T(x) = Ax$ is a linear transformation from $\mathcal{X}$ to $\mathcal{Y}$ if:

1. $T(x_1 + x_2) = A(x_1 + x_2) = Ax_1 + Ax_2 = T(x_1) + T(x_2)$;
2. $T(\lambda x) = A(\lambda x) = \lambda Ax = \lambda T(x)$.

Clearly, any linear transformation is a matrix transformation and conversely.

Example 1.10.2: Verify via the linearity principle that the transformation: (a) $T(\boldsymbol{x}) = ax$ is linear; (b) $T(\boldsymbol{x}) = ax^2$ is not linear; and $T(\boldsymbol{x}) = ax + b$, $b \neq 0$, is not linear.

a. $T(\lambda_1 x_1 + \lambda_2 x_2) = a(\lambda_1 x_1 + \lambda_2 x_2) = a\lambda_1 x_1 + a\lambda_2 x_2$
$= \lambda_1 a x_1 + \lambda_2 a x_2 = \lambda_1 T(x_1) + \lambda_2 T(x_2).$

b. $T(\lambda_1 x_1 + \lambda_2 x_2) = a(\lambda_1 x_1 + \lambda_2 x_2)^2 = a(\lambda_1 x_1)^2 + 2a\lambda_1\lambda_2 x_1 x_2$
$+ a(\lambda_2 x_2)^2 \neq \lambda_1 T(x_1) + \lambda_2 T(x_2)=$
$\lambda_1 a x_1^2 + \lambda_2 a x_2^2.$

c. $T(\lambda_1 x_1 + \lambda_2 x_2) = a(\lambda_1 x_1 + \lambda_2 x_2) + b = a\lambda_1 x_1 + a\lambda_2 x_2 + b$
$\neq \lambda T(x_1) + \lambda_2 T(x_2) = \lambda_1(ax_1 + b)$
$+ \lambda_2(ax_2 + b).$

(Clearly, the constant b compromises the linearity of the transformation. For a transformation to be linear, any constant must be the coefficient on a variable.) ■

For $\mathcal{X}$, $\mathcal{Y}$ real vector spaces, let $\mathcal{L}(\mathcal{X}, \mathcal{Y}) = \mathcal{L}(\cdot)$ represent the set of all linear transformations from $\mathcal{X}$ to $\mathcal{Y}$. Suppose $T_1, T_2 \in \mathcal{L}(\cdot)$. Let the sum $T_1 + T_2$ be that linear transformation for which $(T_1 + T_2)(\boldsymbol{x}) = T_1(\boldsymbol{x}) + T_2(\boldsymbol{x})$ Then $T_1 + T_2 \in \mathcal{L}(\cdot)$. And for λ a scalar, let λT be defined as $(\lambda T)(\boldsymbol{x}) = \lambda T(\boldsymbol{x})$ so that $\lambda T \in \mathcal{L}(\cdot)$ also. Then $\mathcal{L}(\mathcal{X}, \mathcal{Y})$ is a vector space. (If T is a linear transformation from $\mathcal{X}$ to $\mathcal{X}$, i.e., the domain and the range space are one and the same, then we can write $T \in \mathcal{L}(\mathcal{X}, \mathcal{X})$.)

1.10.2 Properties of Linear Transformations

We now turn to an examination of an assortment of additional definitions and characteristics pertaining to linear transformations. Given that $\mathcal{L}(\mathcal{X}, \mathcal{Y}) = \mathcal{L}(\cdot)$ is the set of all linear transformations from $\mathcal{X}$ to $\mathcal{Y}$, we have:

1. The linear transformations $T_1, T_2 \in \mathcal{L}(\cdot)$ are **equal** if for every $\boldsymbol{x} \in \mathcal{X}$, $T_1(\boldsymbol{x}) = T_2(\boldsymbol{x})$.

2. A linear transformation $\mathcal{O} \in \mathcal{L}(\cdot)$ is the **zero transformation** if for any $\boldsymbol{x} \in \mathcal{X}$, $\mathcal{O}(\boldsymbol{x}) = \boldsymbol{O}$.

3. A linear transformation $\mathcal{J} \in \mathcal{L}(\cdot)$ is the **identity transformation** if for any $\boldsymbol{x} \in \mathcal{X}$, $\mathcal{J}(\boldsymbol{x}) = \boldsymbol{x}$.

4. Let $T \in \mathcal{L}(\cdot)$. Then a linear transformation $-T$ (the **negative** of T) exists such that for any $\boldsymbol{x} \in \mathcal{X}$, $-TT(\boldsymbol{x}) = T(-\boldsymbol{x})$

5. Let $\mathcal{X}$, $\mathcal{Y}$, and $\mathcal{Z}$ be real vector spaces. For T_1, T_2 linear transformations let $T_1 \in \mathcal{L}(\mathcal{X}, \mathcal{Y})$ and $T_2 \in \mathcal{L}(\mathcal{Y}, \mathcal{Z})$. Then the **composition** of T_1 and T_2, denoted $T_1 T_2 \in \mathcal{L}(\mathcal{Y}, \mathcal{Z})$ is the linear transformation $T_2 T_1(\boldsymbol{x}) = T_2(T_1(\boldsymbol{x}))$, $\boldsymbol{x} \in \mathcal{X}$. (Note that $T_2 T_1$ is defined only when T_1 maps elements in $\mathcal{X}$ into the domain of T_2. Note also that : (a) the composition operation is *not commutative* or $T_1 T_2 = T_2 T_1$ is not necessarily true; (b) however, for $T \in \mathcal{L}(\mathcal{X}, \mathcal{Y})$, $T\mathcal{J} = \mathcal{J}T = T$, where, for $T\mathcal{J}=T$, $\mathcal{J} \in \mathcal{L}(\mathcal{X}, \mathcal{X})$; and for $\mathcal{J}T = T$, $\mathcal{J} \in \mathcal{L}(\mathcal{Y}, \mathcal{Y})$.)

6. A linear transformation T^{-1} from a subspace of $\mathcal{Y}$ to $\mathcal{X}$ is the (unique) **inverse** of $T \in \mathcal{L}(\cdot)$ if and only if $TT^{-1} = \mathcal{J} \in \mathcal{L}(\mathcal{Y}, \mathcal{Y})$ and $T^{-1}T = \mathcal{J} \in \mathcal{L}(\mathcal{X}, \mathcal{X})$, where $\mathcal{X}$ and $\mathcal{Y}$ must be different vector spaces. The subspace of $\mathcal{Y}$ mentioned above must be Range (T) so that T^{-1}**:** Range$(T) \rightarrow \mathcal{X}$. A linear transformation which has an inverse is termed **non-singular**.

   (If T: $R^n \rightarrow R^n$ is the matrix transformation $\boldsymbol{Ax} = \boldsymbol{y}$, then T has an inverse if and only if $\boldsymbol{A}$ is non-singular with $T^{-1} = \boldsymbol{A}^{-1}\boldsymbol{y}$.)

   Given $T \in \mathcal{L}(\mathcal{X}, \mathcal{Y})$, for each $\boldsymbol{y}$ define the **pre-image** of $\boldsymbol{y}$ to be the subject of $\mathcal{X}$ given by $T^{-1}(\boldsymbol{y}) = \{\boldsymbol{x} \in \mathcal{X} | T(\boldsymbol{x}) = \boldsymbol{y}\}$; it is the set of all those vectors $x \in \mathcal{X}$ that get sent to $\boldsymbol{y}$. We know that for each $x \in \mathcal{X}$ there is exactly one $\boldsymbol{y}$ in Range (T). But T need not display this property in the other direction, i.e. $y \in \mathcal{Y}$ could have many different $\boldsymbol{x}$s from $\mathcal{X}$ which T maps into $\boldsymbol{y}$, or there could be no inputs at all which T sends to $\boldsymbol{y}$.

7. Suppose $T \in \mathcal{L}(\mathcal{X}, \mathcal{Y})$ with $\{\boldsymbol{x}_1, \ldots, \boldsymbol{x}_n\}$ a basis for $\mathcal{X}$ and $\mathcal{Y}$ contains the vectors $\boldsymbol{y}_1, \ldots, \boldsymbol{y}_n$. If $x \in \mathcal{X}$, then $\boldsymbol{x} = a_1\boldsymbol{x}_1 + \cdots + a_n\boldsymbol{x}_n$. And if $T(\boldsymbol{x}_j) = \boldsymbol{y}_j$, $1 \leq j \leq n$, then

$$T(\boldsymbol{x}) = T(\Sigma_{j=1}^n a_j \boldsymbol{x}_j) = \Sigma_{j=1}^n a_j T(\boldsymbol{x}_j)$$
$$= \Sigma_{j=1}^n a_j \boldsymbol{y}_j = \boldsymbol{y}$$

due to the linearity of T. Hence any $\boldsymbol{y} \in \mathcal{Y}$ can be written as a linear combination of the $\boldsymbol{y}_j$s, $j = 1, \ldots, n$.

Let us explore the concept of a non-singular transformation in greater detail. To set the stage for this discussion we have the following general concepts for $T \in \mathcal{L}(\mathcal{X}, \mathcal{Y})$. So given a linear transformation $T \in \mathcal{L}(\cdot)$:

1. The set of all images under T is a subspace of $\mathcal{Y}$ called the **range space** of T, denoted Range $(T) = \{y \in \mathcal{Y} | y = T(x), x \in \mathcal{X}\}$. The dimension of Range (T) is called the rank of T, denoted Rank (T). Thus Rank (T) = dim (Range (T)).
2. T is **injective** (or **one-to-one**) if whenever $\boldsymbol{x}_1, \boldsymbol{x}_2 \in \mathcal{X}$ and $T(\boldsymbol{x}_1) = T(\boldsymbol{x}_2)$, we have $\boldsymbol{x}_1 = \boldsymbol{x}_2$.
3. T is **surjective** (or **onto**) if Range $(T) = \mathcal{Y}$.
4. The **null space** (or **kernel**) of T, denoted Null (T), is the subspace of $\mathcal{X}$ consisting of those vectors that T maps to $\boldsymbol{O}$. Hence, we can write Null $(T) = \{\boldsymbol{x} \in \mathcal{X} | T(\boldsymbol{x}) = \boldsymbol{O}\}$.

The dimension of the null space of T is called the **nullity** of T and denoted Nullity (T) = dim (Null(T)). Note that a test to determine if T is injective is: $T \in \mathcal{L}(\cdot)$ is injective if and only if Null$(T) = \{\boldsymbol{O}\}$, i.e., $\boldsymbol{O}$ is the only vector in $\mathcal{X}$ that gets transformed into $\boldsymbol{O}$.

Now, suppose $\boldsymbol{A}$ is $(m \times n)$ with $\rho(\boldsymbol{A}) = r$ and let $T(\boldsymbol{x}) = \boldsymbol{A}\boldsymbol{x}$. Then $\dim(\text{Null}(T)) = n - r$. If $r \geq m$, the image of R^n is all of R^m so that T is surjective; if $r < m$, the image of R^n is a plane of dimension r embedded in R^m and passing through the origin $\boldsymbol{O}$. Hence T is injective only if $r = n$.

Example 1.10.3: Let the matrix

$$\underset{(3\times4)}{A} = \begin{bmatrix} 1 & 4 & 2 & 4 \\ 0 & 0 & 3 & 6 \\ 0 & 0 & 2 & 4 \end{bmatrix}.$$

Since $\boldsymbol{A}$ is (3×4), T: $R^4 \to R^3$ is the transformation on R^4 induced by A and can be written as

$$\begin{aligned} \boldsymbol{y} = \boldsymbol{A}\boldsymbol{x} &= x_1 \begin{bmatrix} 1 \\ 0 \\ 0 \end{bmatrix} + x_2 \begin{bmatrix} 4 \\ 0 \\ 0 \end{bmatrix} + x_3 \begin{bmatrix} 2 \\ 3 \\ 2 \end{bmatrix} + x_4 \begin{bmatrix} 4 \\ 6 \\ 4 \end{bmatrix} \\ &= (x_1 + 4x_2) \begin{bmatrix} 1 \\ 0 \\ 0 \end{bmatrix} + (x_3 + 2x_4) \begin{bmatrix} 2 \\ 3 \\ 2 \end{bmatrix}. \end{aligned}$$

Hence, the range space of the transformation is a two-dimensional subspace of R^3 spanned by the vectors

$$\begin{bmatrix} 1 \\ 0 \\ 0 \end{bmatrix}, \begin{bmatrix} 2 \\ 3 \\ 2 \end{bmatrix}.$$

The transformation T: $R^3 \to R^2$ represented by

$$\boldsymbol{y} = \boldsymbol{A}\boldsymbol{x} = \begin{bmatrix} 3 & 0 & 0 \\ 0 & 3 & 0 \end{bmatrix} \boldsymbol{x} = x_1 \begin{bmatrix} 3 \\ 0 \end{bmatrix} + x_2 \begin{bmatrix} 0 \\ 3 \end{bmatrix} + x_3 \begin{bmatrix} 0 \\ 0 \end{bmatrix}$$

maps the point (x_1, x_2, x_3) into a point (x_1, x_2) in the x_1, x_2–plane. T is not injective (one – to – one) since every point with the first two components x_1, x_2 transforms into (x_1, x_2) no matter what the value of x_3 is. Hence $\boldsymbol{A}$ maps R^3 into the entirety of R^2.

The transformation T: $R^2 \to R^3$ induced by the matrix

$$\underset{(3\times2)}{\boldsymbol{A}} = \begin{bmatrix} 1 & 0 \\ 1 & 0 \\ 0 & 1 \end{bmatrix}$$

can be represented as $T(\boldsymbol{x}) = (x_1, x_2, x_3)$. Thus, T maps any point $(x_1, x_2) \in R^2$ into $(x_1, x_2, x_3) \in R^3$ so that the range space is a two-dimensional subspace of R^3. ■

Example 1.10.4: Let T_1: $R^n \to R^r$, T_2: $R^r \to R^m$, and $T_3 = T_2 T_1$: $R^n \to R^m$ Verify that T_3 is a linear transformation

$$\begin{aligned} T_3(\lambda_1 \boldsymbol{x}_1 + \lambda_2 \boldsymbol{x}_2) &= T_2(\lambda_1 T_1(\boldsymbol{x}_1) + \lambda_2 T_1(\boldsymbol{x}_2)) \\ &= \lambda_1 T_2(T_1(\boldsymbol{x}_1)) + \lambda_2 T_2(T_1(\boldsymbol{x}_2)) \\ &= \lambda_1 T_3(\boldsymbol{x}_1) + \lambda_2 T_3(\boldsymbol{x}_2). \end{aligned}$$

Example 1.10.5: Let T, S each be linear transformations from R^2 to R^2, where

$$T: \begin{cases} 2x_1 + x_2 = y_1 \\ 3x_1 + x_2 = y_2 \end{cases}$$

$$S: \begin{cases} 2x_1 - 3x_2 = y_1 \\ x_1 + x_2 = y_2 \end{cases}$$

and

$$\boldsymbol{A}_T = \begin{bmatrix} 2 & 1 \\ 3 & 1 \end{bmatrix}, \boldsymbol{A}_S = \begin{bmatrix} 2 & -3 \\ 1 & 1 \end{bmatrix}.$$

Then, since

$$\boldsymbol{A}_S \boldsymbol{A}_T = \begin{bmatrix} -5 & -1 \\ 5 & 2 \end{bmatrix},$$

the composition of T and S is the linear transformation

$$ST: \begin{cases} -5x_1 - x_2 = y_1 \\ 5x_1 - 2x_2 = y_2 \end{cases}$$

and, with

$$A_T A_S = \begin{bmatrix} 5 & -5 \\ 7 & -8 \end{bmatrix},$$

the composition of S and T is the linear transformation

$$TS: \begin{cases} -5x_1 - 5x_2 = y_1 \\ 7x_1 - 8x_2 = y_2 \end{cases}$$

■

Armed with these considerations we can state the following important theorem.

Theorem 1.10.2: (Rank-Nullity Theorem). If $\mathcal{X}$ is finite dimensional and $T \in \mathcal{L}(\cdot)$, then Range (T) is a finite-dimensional subspace of $\mathcal{Y}$ and

$$\begin{aligned} dim(\mathcal{X}) &= \text{Nullity}(T) + \text{Rank}(T) \\ &= \dim(\text{Null}(T)) + \dim(\text{Range}(T)) \end{aligned} \tag{1.22}$$

Thus, the dimension of the domain of T equals the dimension of the null space of T plus the dimension of the range of T. From this theorem, it follows that if $\mathcal{X}$, $\mathcal{Y}$ are finite-dimensional vector spaces, then: (a) if dim $(\mathcal{X})$ > dim $(\mathcal{Y})$, then $T \in \mathcal{L}(\cdot)$ cannot be injective or one-to-one; and (b) if dim $(\mathcal{X})$ < dim $(\mathcal{Y})$, then $T \in \mathcal{L}(\cdot)$ cannot be surjective or onto.

Let us now return to the issue of non-singular or invertible transformations $T \in \mathcal{L}(\mathcal{X}, \mathcal{Y})$. Specifically, T is invertible if and only if it is injective and surjective, i.e., one-to-one and onto. Suppose T is non-singular. Given (1.10.2), if Nullity $(T) = 0$, then $\mathcal{X}$ and Range (T) have the same dimension. In this regard, $T \in \mathcal{L}(\cdot)$ is non-singular if for *any* $\boldsymbol{x} \neq \boldsymbol{O}$, $T(\boldsymbol{x}) \neq \boldsymbol{O}$; it is **singular** if there exists an $\boldsymbol{x} \neq \boldsymbol{O}$ with $T(\boldsymbol{x}) = \boldsymbol{O}$. (This result is true since, for T non-singular, Null $(T) = \{\boldsymbol{O}\}$.) We note further that, for non-singular transformations $T \in \mathcal{L}(\cdot)$:

1. If T is non-singular, then $(T^{-1})^{-1} = T$.
2. If T is non-singular, then, for a scalar $\lambda \neq 0$, $(\lambda T)^{-1} = \lambda^{-1}T^{-1}$.
3. Suppose the linear transformations T_1, T_2 are both non-singular. Then the composite linear transformation T_2, T_1 is also non-singular (provided $T_2 T_1$ is defined).

1.10.3 Solvability of Equations

Let $T \in \mathcal{L}(\mathcal{X}, \mathcal{Y}) = \mathcal{L}(\cdot)$ be written as $T(\boldsymbol{x}) = \boldsymbol{y}$ Obviously, $T(\boldsymbol{x}) = \boldsymbol{y}$ has a solution $\boldsymbol{x}$ if and only if $\boldsymbol{y}$ is the image of some $\boldsymbol{x} \in \mathcal{X}$. To address the issue of the solvability of $T(\boldsymbol{x}) = \boldsymbol{y}$, we have

Theorem 1.10.3: (Solvability of Equations). Let $T \in \mathcal{L}(\cdot)$:

a. given $\boldsymbol{y}$, $T(\boldsymbol{x}) = \boldsymbol{y}$ is solvable if and only if $\boldsymbol{y} \in$ Range (T); and

b. for arbitrary $\boldsymbol{y} \in \mathcal{Y}$, $\boldsymbol{x}$ solves $T(\boldsymbol{x}) = \boldsymbol{y}$ if and only if T is surjective or an onto transformation, i.e., Range $(T) = \mathcal{Y}$.

Under what circumstances does a *unique* solution to $T(\boldsymbol{x}) = \boldsymbol{y}$ exist? Suppose $\boldsymbol{x}_1, \boldsymbol{x}_2 \in \mathcal{X}$ both solve $T(\boldsymbol{x}) = \boldsymbol{y}$. Then $T(\boldsymbol{x}_1) = y$ and $T(\boldsymbol{x}_2) = \boldsymbol{y}$ or $\boldsymbol{O} = \boldsymbol{y} - \boldsymbol{y} = T(\boldsymbol{x}_1) - T(\boldsymbol{x}_2) = T(\boldsymbol{x}_1 - \boldsymbol{x}_2)$ or $\boldsymbol{x} = \boldsymbol{x}_1 - \boldsymbol{x}_2$ solves the homogeneous equation system $T(\boldsymbol{x}) = \boldsymbol{O}$. Obviously the issue of uniqueness depends upon Null (T), the set (subspace) of all solutions to $T(\boldsymbol{x}) = \boldsymbol{O}$. In this regard, $T(\boldsymbol{x}) = \boldsymbol{y}$ has at most one solution if and only if Null $(T) = \{\boldsymbol{O}\}$. To summarize this discussion, we have:

Theorem 1.10.4: Let $T \in \mathcal{L}(\cdot)$. $T(\boldsymbol{x}) = \boldsymbol{y}$ has a unique solution $\boldsymbol{x} \in \mathcal{X}$ for each $\boldsymbol{y} \in \mathcal{Y}$ if and only if : (a) Range $(T) = \mathcal{Y}$; and (b) Null $(T) = \{\boldsymbol{O}\}$.
When (a) and (b) hold, the mapping from $\mathcal{Y}$ to the solution $\boldsymbol{x}$ is linear and denoted $\boldsymbol{x} = T^{-1}(\boldsymbol{y})$ (the inverse of T) so that $T(T^{-1}(\boldsymbol{y})) = \boldsymbol{y}$, and $T^{-1}(T(\boldsymbol{x})) = \boldsymbol{x}$ for all $\boldsymbol{x} \in \mathcal{X}$ and $\boldsymbol{y} \in \mathcal{Y}$.

1.10.4 Matrix transformations Revisited

For a matrix transformation T (equation (1.21), the domain is R^n and the range space Range $(T) \subseteq R^m$, i.e. Range $(T) = \{\boldsymbol{y} | \boldsymbol{y} = \boldsymbol{Ax}, \boldsymbol{x} \in R^n\}$ is a subset of R^m. Let us write $\boldsymbol{A} = [\boldsymbol{a}_1, \ldots, \boldsymbol{a}_n]$, where $\boldsymbol{a}_j, j = 1, \ldots, n$, is an $(m \times 1)$ vector. Here $\boldsymbol{y} = \boldsymbol{Ax} = \sum_{j=1}^{n} \boldsymbol{a}_j \boldsymbol{x}_j$ is a linear combination of the $\boldsymbol{a}_j$ and thus Range (T) is a subspace of R^m spanned by the columns of $\boldsymbol{A}$. Also, dim (Range (T)) is the maximum number of linearly independent columns of $\boldsymbol{A}$. Since $\boldsymbol{A}$ has n columns, it follows that dim (Range (T)) cannot exceed n. We shall refer to the maximum number of linearly independent columns in $\boldsymbol{A}$ as the (column) rank of $\boldsymbol{A}$, $\rho(\boldsymbol{A})$. Thus, Rank (T) = dim (Range (T)) $(=\rho(\boldsymbol{A}))$.

Let us express the matrix transformation $T \in \mathcal{L}(R^n, R^m)$ as

$$T(\boldsymbol{x}) = \left(\sum_{j=1}^{n} a_{1j}x_j, \ldots, \sum_{j=1}^{n} a_{mj}x_j \right), \boldsymbol{x} \in R^n. \quad (1.23)$$

Suppose we set $T(\boldsymbol{x}) = \boldsymbol{O}$. Then $T(\boldsymbol{x})$ can be rewritten as a homogeneous system of m linear equations in n unknowns $\boldsymbol{Ax} = \boldsymbol{O}$. Clearly, $\boldsymbol{x} = \boldsymbol{O}$ is a (trivial) solution to this system. Is $\boldsymbol{x} = \boldsymbol{O}$ the only solution to $\boldsymbol{Ax} = \boldsymbol{O}$? To answer this question we need to determine if Null $(T) = \{\boldsymbol{x} \in R^n | \boldsymbol{Ax} = \boldsymbol{O}\} = \{\boldsymbol{O}\}$ or if Null (T) is larger than the singleton $\{\boldsymbol{O}\}$. This latter situation emerges if T *is not* injective (one-to-one) or if $n > m$. So if $\boldsymbol{Ax} = \boldsymbol{O}$ has more unknowns than equations, then it must have non-trivial or non-zero solutions.

For a vector $\boldsymbol{C} \in R^m$, let us write the matrix transformation as $T(\boldsymbol{x}) = \boldsymbol{C}$ or $\boldsymbol{Ax} = \boldsymbol{C}$ (a non-homogeneous system of m linear equations in n unknowns). Give an arbitrary $\boldsymbol{C}$, does there exist at least one solution $\boldsymbol{x}$ to $\boldsymbol{Ax} = \boldsymbol{C}$? To answer this, we need to check if Range $(T) = R^m$. If $n = \dim(\mathcal{X}) <$ Rank $(T) = m$ (T is not surjective), then we have more equations than unknowns and thus $\boldsymbol{Ax} = \boldsymbol{C}$ has no solution for some $\boldsymbol{C} \in R^m$.

Section 1.10 Exercises:

1.10.1 Verify that the transformation $y = 3x_1 + 4x_2$ is linear.

1.10.2 Let T be the transformation

$$A = \begin{bmatrix} 2 & 0 & -1 \\ -1 & 3 & 1 \end{bmatrix}$$

from R^3 to R^2 which sends points $\boldsymbol{x}^T = (x_1, x_2, x_3)$ into $\boldsymbol{y}^T = (y_1, y_2)$. Find the image of the point $\boldsymbol{x}^T = (1, 2, 1)$.

1.10.3 Let the linear transformations:

$$T_1: R^3 \to R^2,$$
$$T_2: R^3 \to R^2,$$
$$T_3: R^2 \to R^2,$$

be defined, respectively, as

$$T_1(x_1, x_2, x_3) = (x_1, x_2 + x_3),$$
$$T_2(x_1, x_2, x_3) = (x_1, x_2 + 2x_3),$$
$$T_3(x_1, x_2) = (x_2, x_1).$$

Find: $T_1 + 4T_2$ and T_3T_1.

1.10.4. Demonstrate that if $T \in \mathcal{L}(\mathcal{X}, \mathcal{Y})$, then Null (T) is a subspace of $\mathcal{X}$. (Hint: show that Null (T) is a non-empty subset of $\mathcal{X}$ that is closed under addition and scalar multiplication.)

1.10.5 Demonstrate that if $T \in \mathcal{L}(\mathcal{X}, \mathcal{Y})$, then Range (T) is a subspace of $\mathcal{Y}$. (Hint: show that Range (T) is a non-empty subset of $\mathcal{Y}$ that is closed under addition and scalar multiplication.)

1.10.6 Given the linear transformation $T: R^3 \to R^3$ defined by $T(x_1, x_2, x_3) = (x_1, x_2, 0)$, determine: Null (T), Range (T), and Rank (T).

1.10.7 Demonstrate that $T^{-1}T$ and TT^{-1} represent identity mappings in two different spaces, i.e., $T^{-1}T$ depicts the identity transformation on $\mathcal{X}$ and TT^{-1} depicts the identity transformation on $\mathcal{Y}$.

1.10.8 Suppose $\boldsymbol{A}$ is a given $(m \times n)$ matrix, $\boldsymbol{B}$ is a given $(n \times n)$ matrix, and $T(\boldsymbol{Q}) = \boldsymbol{AQB}$ is a function from the class of real $(m \times n)$ matrices into itself. Verify that T is a linear transformation.

1.10.9 Verify that if $T: \mathcal{X} \to \mathcal{Y}$ is non-singular, then: (a) Null $(T) = \{\boldsymbol{O}\}$; (b) Nullity (T) = dim (Null (T)) = 0; and (c) dim $(\mathcal{X})$ = Rank (T) + dim (Range (T)).

1.10.10 Suppose $\mathcal{X}, \mathcal{Y}$ are real vector spaces and T is a one-to-one transformation of $\mathcal{X}$ onto $\mathcal{Y}$ (i.e., $\boldsymbol{x}_1 \neq \boldsymbol{x}_2$ implies $T(\boldsymbol{x}_1) \neq T(\boldsymbol{x}_2)$). The T is termed an **isomorphism of $\mathcal{X}$ onto $\mathcal{Y}$**. Answer the following: (a) is $\mathcal{X}$ isomorphic to itself? (b) if $\mathcal{X}$ is isomorphic to $\mathcal{Y}$ for an isomorphism T, is $\mathcal{Y}$ isomorphic to $\mathcal{X}$?

NOTES

1 A set R_1 of elements together with the operations of addition and multiplication on R_1 constitutes a **field** if – both addition and multiplication are associative and commutative; additive and multiplicative inverses and identity elements exist; and multiplication distributes over addition.

2 Readers not familiar with point-set theory should examine Section 2.5.

3 See Section 2.1 for the concepts of *supremum* and *infimum*.

4 A **theorem of the alternative** involves two mutually exclusive systems of linear inequalities and/or equalities denoted simply as (I), (II). Moreover, it asserts that either system (I) has a solution or system (II) has a solution, but never both. Theorems of the alternative possess the following structure:(I) if and only if not (II), or, equivalently, not (I) if and only if (II).

Mathematical Foundations 2

2.1 REAL AND EXTENDED REAL NUMBERS

The **real number system** may be characterized as a complete, ordered field, where, as indicated in Section 1.2, a **field** is a set $\mathcal{F}_R$ of elements together with the operations of addition and multiplication. Moreover, both addition and multiplication are associative and commutative, additive and multiplicative inverse and identity elements exist, and multiplication distributes over addition. Set $\mathcal{F}_R$ is **ordered** if there is a binary order relation "<" in $\mathcal{F}_R$ which satisfies:

1. for any elements x, y in $\mathcal{F}_R$ either $x < y$, $y < x$, or $x = y$;
2. for any elements x, y, and z in $\mathcal{F}_R$, if $x < y$ and $y < z$, then $x < z$.

Now, if $\mathcal{F}_R$ is an **ordered field**, then the order relations must be connected to the field operations according to:

1. if $x < y$, then $x + z < y + z$;
2. if $x < y$ and z is positive, then $zx < zy$.

Looking to the completeness property of the real number system, let us note first that a set $\mathcal{A}$ ($\neq\emptyset$) of real numbers is **bounded above** if there is a real number b (the **upper bound** for $\mathcal{A}$) such that $a \le b$ for every $a \in \mathcal{A}$. The **least upper bound** or **supremum** of $\mathcal{A}$ (denoted sup $\mathcal{A}$) is a real number b such that: (1) $a \le b$ for every $a \in \mathcal{A}$; and (2) if $a \le c$ for every $a \in \mathcal{A}$, then $b \le c$. So, if b is an upper bound for $\mathcal{A}$ such that no smaller element of $\mathcal{A}$ is also an upper bound for $\mathcal{A}$, then b is the least upper bound for $\mathcal{A}$. In a similar vein, we can state that a set

$\mathcal{A}$ ($\neq\emptyset$) of real number is **bound below** if there is a real number b (the **lower bound** for $\mathcal{A}$) such that $b \le a$ for every $a \in \mathcal{A}$. The **greatest lower bound** or **infimum** of $\mathcal{A}$ (written inf $\mathcal{A}$) is a real number b such that (1) $b \le a$ for every $a \in \mathcal{A}$; and (2) if $c \le a$ for every $a \in \mathcal{A}$, then $c \le b$. Hence if b is a lower bound for $\mathcal{A}$ such that no larger element of $\mathcal{A}$ is also a lower bound for $\mathcal{A}$, then b is the greatest lower bound for $\mathcal{A}$. Clearly, the supremum and infimum for $\mathcal{A}$ must be unique.

Armed with these considerations, we can state the **completeness property** as: every non-empty subset $\mathcal{A}$ of the ordered field $\mathcal{F}_R$ of real numbers which has an upper bound in $\mathcal{F}_R$ has a least upper bound in F_R.

If we admit the elements $\{-\infty\}$ and $\{+\infty\}$ to our discussion of real numbers R, then the **extended real number system** (denoted R^*) consists of the set of real numbers R together with $\pm\infty$, i.e., $R^* = R \cup \{-\infty\} \cup \{+\infty\}$.

Section 2.1 Exercises:

2.1.1 Verify that a non-empty set $\mathcal{A}$ has at most one supremum.

2.1.2 Does the set $\mathcal{S} = \{5, 6, 7, 8,\ldots\}$ have upper and lower bounds? What is inf $\mathcal{S}$?

2.1.3 Does the set $\mathcal{S} = \{x|0<x<2\}$ exhibit a maximal or minimal element? What are inf $\mathcal{S}$, sup $\mathcal{S}$?

2.2 SINGLE-VALUED FUNCTIONS

Given two non-empty sets $\mathcal{X}$ and $\mathcal{Y}$ (which may or may not be equal), a **single-valued function** or **point-**

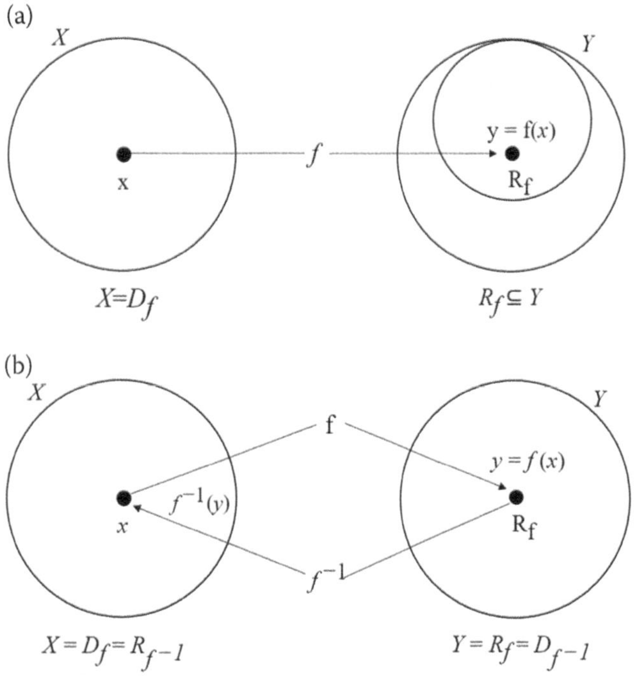

FIGURE 2.1 (a) f an into mapping; (b) f is one-to-one and onto.

to-point mapping $f: \mathcal{X} \rightarrow \mathcal{Y}$ is a rule or law of correspondence which associates with point $x \in X$ a unique point $y \in Y$. Here, $y = f(x)$ is the **image of x** under rule f. While set $\mathcal{X}$ is called the **domain of f** (denoted $\mathcal{D}_f$ or simply D), the collection of those ys which are the image of at least one $x \in X$ is called the **range of f** and denoted R_f. Clearly, the range of f is a subset of Y (Figure 2.1a). If $R_f \subset Y$, then f is an **into mapping**. And if $R_f = Y$ (i.e., *every* $y \in Y$ is the image of at least one $x \in X$ or all the ys are accounted for in the mapping process), then f is termed an **onto** or **surjective mapping**. Moreover, f is said to be **one-to-one** or **injective** if no $y \in Y$ is the image of more than one $x \in X$ (i.e., $x_1 \neq x_2$ implies $f(x_1) \neq f(x_2)$). Finally, f is called **bijective** if it is both one-to-one and onto or both surjective and injective. If the range of f consists of but a single element, then f is termed a **constant function**.

Given a non-empty set $\mathcal{X}$, if $\mathcal{Y}$ consists entirely of real numbers or $Y = R$, then $f: \mathcal{X} \rightarrow \mathcal{Y}$ is termed a **real-valued function** or **mapping** of a point $x \in X$ into a unique real number $y \in R$. Hence, the image of each point $x \in X$ is a real scalar $y = f(x) \in R$.

For sets $\mathcal{X}$ and $\mathcal{Y}$ with set $A \subset X$, let $f_1: A \rightarrow Y$ be a point-to-point mapping of A into Y and $f_2: \mathcal{X} \rightarrow \mathcal{Y}$ be a point-to-point mapping of $\mathcal{X}$ into $\mathcal{Y}$. Then, f_1 is said to be a **restriction** of f_2 and f_2 is termed an **extension** of f_1 if and only if for each $x \in A$, $f_1(x) = f_2(x)$.

Let $X_1, X_2, \ldots, X_n$ represent a class of non-empty sets. The **product set** of $X_1, X_2, \ldots, X_n$ (denoted $X_1 \times X_2 \times \cdots \times X_n$) is the set of all order n-tuples $(x_1, x_2, \ldots, x_n)$, where $x_i \in X_i$ for each $i = 1, \ldots, n$. Familiar particularizations of this definition are: $R^1 = R$ **(the real line)**; $R^2 = R \times R$ is the two-dimensional coordinate plane (made up of all ordered pairs (x_1, x_2), where both $x_1 \in R$ and $x_2 \in R$) and $R^n = R \times R \times \cdots \times R$ (the product is taken n times) depicts the collection of ordered n-tuples of real numbers. In this regard, for f a point-to-point mapping of $\mathcal{X}$ into $\mathcal{Y}$, the subset $G_f = \{(x, y) | x \in X, y = f(x) \in Y\}$ of $\mathcal{X} \times \mathcal{Y}$ is called the **graph of f**.

If the point-to-point mapping f is bijective (f is one-to-one and onto), then its single-valued **inverse mapping** $f^{-1}: \mathcal{Y} \rightarrow \mathcal{X}$ exists. Thus, to each point $y \in Y$, there corresponds a unique **inverse image** point $x \in X$ such that $x = f^{-1}(y) = f^{-1}(f(x))$ so that x is termed the **inverse function** of y. Here, the domain $\mathcal{D}_{f^{-1}}$ of f^{-1} is Y, and its range $R_{f^{-1}}$ is $\mathcal{X}$. Clearly, f^{-1} must also be bijective (Figure 2.1b).

Given two real-valued functions f and g, what might their *product* look like? That is, can we *multiply* functions? In general, suppose $f: \mathcal{X} \rightarrow \mathcal{Y}$ and $g: \mathcal{Y} \rightarrow \mathcal{Z}$ are two real-valued functions. The product of these mappings will be termed the **composition of f and g** and written $g \circ f: \mathcal{X} \rightarrow \mathcal{Z}$ or $g \circ f(x) = g(f(x))$ for all $x \in X$ and $f(x) \in Y$. Thus, an element $x \in X$ is taken by f to the element $f(x) \in Y$, and then g maps $f(x)$ to $g(f(x)) \in \mathcal{Z}$. Thus, $\mathcal{Y}$ contains the range of f, and the range of f is contained within the domain of g (see Figure 2.2 for f, g both one-to-one mappings).

At times, we shall find it necessary to explicitly determine the inverse of a real-valued function $y = f(x)$. Under what conditions does the inverse of f f^{-1}, exist? Also, does the derivative of f^{-1} exist? The answer is provided by the inverse function theorem.

Theorem 2.1: (Inverse Function Theorem). Let $y = f(x): \mathcal{D} = (a, b) \rightarrow R$ be differentiable on $\mathcal{D}$. The inverse of f, f^{-1}, exists if and only if f is strictly monotonic on $\mathcal{D}$.[1] In addition, f^{-1} is differentiable provided $f'(x) \neq 0$, $x \in \mathcal{D}$, and its derivative is

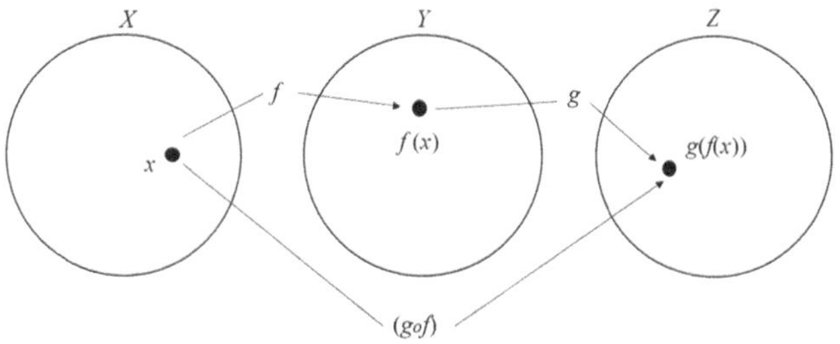

FIGURE 2.2 The composition of f and g.

$$\frac{df^{-1}(y)}{dy} = \frac{1}{f'(x)}.$$

For instance, let $y = f(x) = 1 + 3x$, $x \in \mathcal{D} = [5, 10]$. Then, $x = f^{-1}(y) = -\frac{1}{3} + \frac{1}{3}y$ and

$$\frac{df^{-1}(y)}{dy} = \frac{1}{3}\left(=\frac{1}{f'(x)}\right).$$

Section 2.2 Exercises:

2.2.1 If $f(x) = f(-x)$, the function f is said to be **even**; if $f(x) = -f(-x)$, the function f is said to be **odd**. Provide a graphical example of each of these functions. Demonstrate graphically that the function $f(x) = x^n$ is even or odd according to whether n is an even or odd integer.

2.2.2 Suppose f is a mapping with $X = \mathcal{D}_f = \{1, 2, 3, 4, 5, 6, 7\}$ and $Y = \{1, 3, 5, 7, 9\}$. The images of f are:

$$f(1) = f(4) = f(5) = 3,$$
$$f(2) = f(7) = 5,$$
$$f(3) = 7,$$
$$f(6) = 9.$$

Is f an into or onto mapping?

2.2.3 Let $X = \{1, 2, 3, 4\}$, $Y = \{1, 2, 3, 4, 5\}$, and $f: X \to Y$ be defined by $f(x) = x + 1$. What is the domain and range of f? Is f an into mapping?

2.2.4 Find $g \circ f$ for:

a. $g = 2x + 1$, $x \in R$, and $f = \sqrt{x}$, $x \geq 0$.

b. $g = \frac{x}{x+1}$, $x \neq -1$, and $f = \frac{1}{x}$, $x \neq 0$.

2.2.5 Given $y = f(x)$, find $f^{-1}(y)$ and $df^{-1}(y)/dy$ for:

a. $y = f(x) = \sqrt{x}$, $x > 0$.

b. $y = f(x) = x^2$, $0 < x < x_o$.

2.2.6 Suppose $y = f(x) = x^2 + 1: R \to R$. Then, the range of f is $\{x|x \geq 1\}$. Let $\mathcal{A} = [1, 4] \in R$. Clearly, $\mathcal{A}$ is a subset of the range of f. What is the set $f^{-1}(\mathcal{A})$?

2.2.7 We know that if the function F is suitably restricted, the equation $F(x_1, x_2) = 0$ defines a set of real-valued functions such that if $f(x_1)$ is one of them, then $F(x_1, f(x_1)) = 0$ for all x_1 in the domain of f. Given $f(x_1, x_2) = x_1^2 + x_2^2 - 10 = 0$, what does "suitably restricted" mean?

2.3 METRIC SPACES

Given a space $\mathcal{S}$, a metric defined on $\mathcal{S}$ is an everywhere finite real-valued function d of ordered pairs $(\boldsymbol{x}, \boldsymbol{y})$ of points of $\mathcal{S}$ or $d(\boldsymbol{x}, \boldsymbol{y})$: $\mathcal{S} \times \mathcal{S} \to [0, +\infty)$ satisfying:

1. for $\boldsymbol{x} \in S$, $d(\boldsymbol{x}, \boldsymbol{x}) = 0$ (**reflexitivity**);

2. for $\boldsymbol{x}, \boldsymbol{y} \in S$, $d(\boldsymbol{x}, \boldsymbol{y}) > 0$ and $d(\boldsymbol{x}, \boldsymbol{y}) = 0$ if and only if $\boldsymbol{x} = \boldsymbol{y}$;

3. for $\boldsymbol{x}, \boldsymbol{y} \in S$, $d(\boldsymbol{x}, \boldsymbol{y}) = d(\boldsymbol{y}, \boldsymbol{x})$ (**symmetry**); and

4. for $\boldsymbol{x}, \boldsymbol{y}, \boldsymbol{z} \in S$, $d(\boldsymbol{x}, \boldsymbol{y}) \leq d(\boldsymbol{x}, \boldsymbol{z}) + d(\boldsymbol{z}, \boldsymbol{y})$ (**triangle inequality for distances**).

Here, d serves to define the **distance** between $\boldsymbol{x}$ and $\boldsymbol{y}$. A **metric space** consists of the space $\mathcal{S}$ and a metric d defined on $\mathcal{S}$. Hence, a metric space will be denoted $(\mathcal{S}, d)$. For instance, if $\mathcal{S} = R$, then R is a metric space if $d(\boldsymbol{x}, \boldsymbol{y}) = |\boldsymbol{x} - \boldsymbol{y}|$ (the distance between points x and y on the real line). And if $\mathcal{S} = R^n$, then R^n can be considered a metric space if

$$d(\boldsymbol{x}, \boldsymbol{y}) = [\textstyle\sum_{i=1}^{n} |x_i - y_i|^2]^{1/2}, \tag{2.1}$$

where again $d(\boldsymbol{x}, \boldsymbol{y})$ is interpreted as the distance between $\boldsymbol{x}, \boldsymbol{y} \in R^n$.[2]

Suppose $\mathcal{S}$ is a metric space with metric d and $\mathcal{X}$ ($\neq \emptyset$) is an arbitrary subset of $\mathcal{S}$. If d is defined only for points in $\mathcal{X}$, then $(\mathcal{X}, d)$ is also a metric space. Then under this restriction on d, $\mathcal{X}$ is termed a **subspace** of $\mathcal{S}$.

The importance of a metric space is that it incorporates a concept of distance (d) which is applicable to the points within $\mathcal{S}$. And this distance function will enable us to tackle issues concerning the convergence of sequences in $\mathcal{S}$ and continuous functions defined on $\mathcal{S}$.

Section 2.3 Exercises:

2.3.1 For the vectors $\boldsymbol{x}^T = (1, 2, 1)$ *and* $\boldsymbol{y}^T = (0, 2, 5)$, find: $\|\boldsymbol{x}\|$, $\|\boldsymbol{y}\|$, $\boldsymbol{x}\cdot\boldsymbol{y} = \mathrm{x}^T\boldsymbol{y}$, and $\|\boldsymbol{x} - \boldsymbol{y}\|$.

2.3.2 The **direction** of a vector $\boldsymbol{x}$ was previously defined as a *unit vector that points in the direction of* $\boldsymbol{x}$ or $\boldsymbol{x}/\|\boldsymbol{x}\|$. For $\boldsymbol{x}^T = (1, 2, 1)$, find the direction of $\boldsymbol{x}$ and verify that its length is unity.

2.3.3 The **cosine of the angle** θ between the vectors $\boldsymbol{x}, \boldsymbol{y} \in R^n$ (with $\boldsymbol{x}, \boldsymbol{y} \neq O$) is

$$\cos\theta = (\boldsymbol{x}\cdot\boldsymbol{y}/\|\boldsymbol{x}\|\|\boldsymbol{y}\|),\ 0 \le \theta \le \pi.$$

Find $\cos\theta$ for $\boldsymbol{x}, \boldsymbol{y}$ given earlier in Exercise 2.3.1. What does the Cauchy-Schwarz inequality imply about $\cos\theta$?

2.3.4 Vectors $\boldsymbol{x}, \boldsymbol{y} \in R^n$ are said to be **orthogonal** (mutually perpendicular) if $\boldsymbol{x}\cdot\boldsymbol{y} = 0$. Use the equation from $\cos\theta$ to verify this notion.

2.3.5 (For the ambitious). Verify that the triangle inequality for distances holds.

2.4 LIMITS OF SEQUENCES

Let $\mathcal{X}$ be a subset of R^n. A **sequence of points** in $\mathcal{X}$ is a function whose domain is the set of all positive integers I and whose range appears in $\mathcal{X}$. If the value of the function $n \in I$ is $x_n \in \mathcal{X}$, then the range of the sequence will be denoted by $\{x_n\} = (x_1, x_2, \ldots)$ and interpreted as "the sequence of points $x_1, x_2, \ldots$ in $\mathcal{X}$." (Note that the sequence points $\{x_n\}$ mapped into $\mathcal{X}$ is not a subset of $\mathcal{X}$.) By deleting certain elements of the sequence $\{x_n\}$, we obtain the **subsequence** $\{x_n\}_{n\in J}$, where J is a subset of the positive integers.

A sequence $\{x_n\}$ in R^n **converges to a limit** $\bar{x}$ if and only if $\lim_{n\to\infty} d(x_n, \bar{x}) = \lim_{n\to\infty}\|x_n - \bar{x}\| = 0$. (This is alternatively expressed as $\lim_{n\to\infty} x_n = \bar{x}$ or $x_n \to \bar{x}$ as $n \to \infty$.) That is, $\bar{x}$ is the limit of $\{x_n\}$ if, for each $\varepsilon > 0$, there exists an index value $\bar{n}_\varepsilon$ such that $n > \bar{n}_\varepsilon$ implies $\|x_n - \bar{x}\| < \varepsilon$. If we think of the condition $\|x_n - \bar{x}\| < \varepsilon$ as defining an open sphere of radius ε about $\bar{x}$, then we can say that $\{x_n\}$ converges to $\bar{x}$ if, for each open sphere of radius $\varepsilon > 0$ centered on $\bar{x}$, there exists an $\bar{n}_\varepsilon$ such that $\{x_n\}$ is within this open sphere for all $n > \bar{n}_\varepsilon$. Hence, the said sphere contains all points of $\{x_n\}$ from $x_{\bar{n}_\varepsilon}$ on, i.e., $\bar{x}$ is the limit of the sequence $\{x_n\}$ in R^n if, given $\varepsilon > 0$, all but a finite number of terms of the sequence are within ε of $\bar{x}$.

A point $\hat{x} \in R^n$ is a **limit (cluster) point** of an infinite sequence $\{x_k\}$ if and only if there exists an infinite sequence $\{x_k\}_{k\in K}$ of $\{x_k\}$ which converges to $\hat{x}$, i.e., there exists an infinite sequence $\{x_k\}$ such that $\lim_{j\to\infty}\|x_{k_j} - \hat{x}\| = 0$ or $x_{k_j} \to \hat{x}$ as $j \to \infty$. Stated alternatively, $\hat{x}$ is a limit point of $\{x_k\}$ if, given a $\delta > 0$ and an index value $\bar{k}$, there exists some $k > \bar{k}$ such that $\|x_k - \hat{x}\| < \delta$ for infinitely many terms of $\{x_k\}$.

What is the distinction between the limit of a sequence and a limit point of a sequence? To answer this question, we state:

a. $\bar{x}$ is a limit of a sequence $\{x_k\}$ in R^n if, given a small positive $\varepsilon \in R$, all but a finite number of terms of the sequence are within ε *of* $\bar{x}$.

b. $\hat{x}$ is a limit point of $\{x_k\}$ in R^n if, given a real scalar $\varepsilon > 0$ and given $\bar{k}$, infinitely many terms of the sequence are within ε *of* $\hat{x}$.

Thus, a sequence $\{x_k\}$ in R^n may have a limit but no limit point. However, if a convergent sequence $\{x_k\}$ in R^n has infinitely many distinct points, then its limit is a limit point of $\{x_k\}$. Likewise, $\{x_k\}$ may possess a limit point but no limit. In fact, if the sequence $\{x_k\}$ in R^n has a limit point $\hat{x}$, then there is a subsequence $\{x_k\}_{k\in K}$ of $\{x_k\}$ which has x as a limit; but this does not necessarily mean that the entire $\{x_k\}$ converges to $\hat{x}$,[3] where k is a subset of the positive integers.

A sufficient condition that at least one limit point of an infinite sequence $\{x_k\}$ in R^n exists is that $\{x_k\}$ is **bounded**, i.e., there exists a scalar $M \in R$ such that $\|x_k\| \le M$ for all k. In this regard, if an infinite sequence of points $\{x_k\}$ in R^n is bounded and it has only one limit point, then the sequence converges and has as its limit that single limit point.

The preceding definition of the limit of the sequence $\{x_n\}$ explicitly incorporated the actual limit $\bar{x}$. If one does not know the actual value of $\bar{x}$, then the following theorem enables us to prove that a sequence converges even if its actual limit is unknown. To this end, we state first that a sequence is a **Cauchy sequence** if, for each $\varepsilon > 0$, there exists an index value $N_{\varepsilon/2}$ such that $m, n > N_{\varepsilon/2}$ implies $d(x_m, x_n) = \|x_m - x_n\| < \varepsilon$.[4] Second, R^n is said to be **complete** in that to every Cauchy sequence $\{x_n\}$ defined on R^n there corresponds a point $\bar{x}$ such that $\lim_{n\to 0} x_n = \bar{x}$. Given these concepts, we may now state.

Theorem 2.2: (Cauchy Convergence Criterion). Given that R^n is complete, a sequence $\{x_n\}$ in R^n converges to a limit $\bar{x}$ if and only if it is a Cauchy sequence, i.e., a necessary and sufficient condition for $\{x_n\}$ to be convergent in R^n is that $d(x_m, x_n) \to 0$ as $m, n \to \infty$.

Hence, every convergent sequence on R^n is a Cauchy sequence. The implication of this statement is that if the terms of a sequence approach a limit, then, beyond some point, the distance between pairs of terms diminishes.

It should be evident from the preceding discussion that a **complete metric space** is a metric space in which every Cauchy sequence converges, i.e., the space contains a point $\bar{x}$ to which the sequence converges or $\lim_{n\to\infty} x_n = \bar{x}$. In this regard, it should also be evident that the real line R is a complete metric space as in R^n.

We next define the **limit superior** and **limit inferior** of a sequence $\{x_n\}$ of real numbers as, respectively,

$$\begin{aligned} &\text{(a)} \quad \lim_{n\to\infty} \sup x_n = \lim_{n\to\infty}\left(\sup_{m\geq n} x_m\right); \\ &\text{(b)} \quad \lim_{n\to\infty} \mathit{inf}\ x_n = \lim_{n\to\infty}\left(\inf_{m\geq n} x_m\right); \end{aligned} \tag{2.2}$$

Hence, the limit superior of the sequence $\{x_n\}$ is the largest number $\bar{x}$ such that there is a subsequence of $\{x_n\}$ which converges to $\bar{x}$ - - and no subsequence converges to a higher value. Similarly, the limit inferior is the smallest limit attainable for some convergent subsequence of $\{x_n\}$ - - and no subsequence converges to a lower value. Looked at in another fashion, for, say, equation (2.2a), a number $\bar{x}$ is the limit superior of a sequence $\{x_n\}$ if: (1) for every $x < \bar{x}$, we have $x < x_n$ for infinitely many n's; and (2) for every $x > \bar{x}$, we have $x < x_n$ for only finitely many n's. Generally speaking, when there are multiple limit points around which the terms of a sequence tend to "pile up," the limit superior and limit inferior select the largest and smallest of these limit points, respectively.

We noted earlier that a sequence defined on a subset $\mathcal{X}$ of R^n is a function whose range is $\{x_n\}$. If this function is **bounded**, then its range $\{x_n\}$ is bounded from both above and below. In fact, if $\{x_n\}$ is a bounded sequence of real numbers, then the limit superior and limit inferior both exist. It is also important to note that $\lim_{n\to\infty} x_n$ exists if and only if the limit superior and limit inferior are equal. We end this discussion of limits by mentioning that since any set of extended real numbers has both a supremum and an infimum, it follows that every sequence of extended real numbers has both a limit superior and a limit inferior.

Section 2.4 Exercises:

2.4.1 Let $\{x_n\}$ be defined as

$$x_n = \begin{cases} \frac{1}{n} \text{ if } n \text{ is odd;} \\ 1 - \frac{1}{n} \text{ if } n \text{ is even.} \end{cases}$$

Does this sequence have any limit points? Is $\{x_n\}$ convergent?

2.4.2 Let the sequence $\{x_n\}$ be defined for each positive integer n as

$$x_n = \begin{cases} 1 \text{ if } n \text{ is odd;} \\ 0 \text{ if } n \text{ is even.} \end{cases}$$

Does this sequence have an infimum and supremum? Is the sequence bounded? Is the sequence convergent?

2.4.3 Suppose the sequence $\{x_n\}$ is defined for each positive integer n as

$$x_n = \begin{cases} 1 \text{ if } n \text{ is odd;} \\ 0 \text{ if } n \text{ is even} \end{cases}$$

and the sequence $\{y_n\}$ is defined for each positive integer n as

$$y_n = \begin{cases} 0 \text{ if } n \text{ is odd;} \\ 1 \text{ if } n \text{ is even.} \end{cases}$$

Does either sequence converge? Does their sum $\{z_n\} = \{x_n + y_n\}$ converge?

2.4.4 Which of the following statements are true?

i. Every convergent sequence in R is a Cauchy sequence.

ii. Every Cauchy sequence in R is bounded.

iii. Every Cauchy sequence in R has a convergent subsequence.

iv. Every Cauchy sequence in R is convergent.

2.4.5 Do the following sequences converge? If so, to what values?

a. $x_n = [2 + (-1)^n]/n,\ n = 1, 2, 3, \ldots$

b. $x_n = n,\ n = 1, 2, 3, \ldots$

c. $x_n = \frac{1}{2}[1 + (-1)^n],\ n = 1, 2, 3, \ldots$

d. $x_n = \left(n^{-1}, \frac{n}{n+1}\right),\ n = 1, 2, 3, \ldots$

2.4.6 Verify that a convergent sequence $\{x_n\}$ is bounded.

2.4.7 A well-known theorem about limits is that: *a bounded monotone sequence* $\{x_n\}$ *is convergent.* Is this statement applicable to the following sequences?

a. $x_n = (n + 1)/n,\ n = 1, 2, 3, \ldots$

b. $x_n = n^{1/n},\ n = 1, 2, 3, \ldots$

c. $x_n = \left(1 + \frac{1}{n}\right)^n,\ n = 1, 2, 3, \ldots$

2.4.8 Find $\limsup_{n\to\infty} x_n$ *and* $\liminf_{n\to\infty} x_n$ when:

a. $x_n = (-1)^n,\ n = 1, 2, 3, \ldots$

b. $x_n = (-1)^n\left(2 + \frac{2}{n}\right),\ n = 1, 2, 3, \ldots$

c. $x_n = \frac{n + (-1)^n(2n + 1)}{n},\ n = 1, 2, 3, \ldots$

2.5 POINT-SET THEORY

Let δ be any positive scalar. A **δ-neighborhood of a point $x_o \in R^n$** or **sphere of radius δ about x_o** is the set $\delta(x_o) = \{x \mid \|x - x_o\| < \delta, \delta > 0\}$. A point $\bar{x}$ is an **interior point** of a set $\mathcal{X}$ in R^n, denoted int $(\bar{x})$, if there exists a δ – neighborhood about $\bar{x}$ which contains only points of $\mathcal{X}$.

A set $\mathcal{X}$ in R^n is said to be open if, given any point $x_o \in \mathcal{X}$, there exists a positive scalar δ such that $\delta(x_o) \subseteq \mathcal{X}$. Hence, $\mathcal{X}$ is open if it contains only interior points. Moreover:

a. $\emptyset$, $\delta(x_o)$, and R^n are all open sets.

b. Any union of open sets in R^n is open; and any finite intersection of open sets in R^n is open.

Let $\mathcal{X}$ be a set in R^n. The **complementary set of** $\mathcal{X}$, denoted $\mathcal{X}^c$, is the collection of all points of R^n lying outside of $\mathcal{X}$. A point $\bar{x} \in \mathcal{X}^c$ is an **exterior point** of $\mathcal{X}$ in R^n if there exists a δ – neighborhood of $\bar{x}$ which contains only points of $\mathcal{X}^c$. A point $\bar{x}$ is a **boundary point** of a set $\mathcal{X}$ in R^n if every δ – neighborhood of $\bar{x}$ encompasses points in $\mathcal{X}$ and in $\mathcal{X}^c$.

A set $\mathcal{X}$ in R^n is **bounded** if there exists a scalar $M \in R$ such that $\|x\| \le M$ for all $\bar{x} \in \mathcal{X}$. Stated alternatively, $\mathcal{X}$ is bounded if it has a finite **diameter** $d(\mathcal{X}) = \sup\{\|x - y\| \mid x, y \in\}$.

A set $\mathcal{X}$ in R^n has an **open cover** if there exists a collection $\{G_i\}$ of open subsets from R^n such that $\mathcal{X} \subseteq \cup_i G_i$. The open cover $\{G_i\}$ of $\mathcal{X}$ in R^n is said to contain a **finite subcover** if there are finitely many indices $i_1, \ldots, i_m$ for which $\mathcal{X} \subseteq \cup_{j=1}^{m} G_{i_j}$.

A point $\bar{x}$ is termed a **point of closure** of a set $\mathcal{X}$ in R^n if every δ – neighborhood of $\bar{x}$ contains at least one point of $\mathcal{X}$, i.e., $\delta(\bar{x}) \cap \neq \emptyset$. It is important to note that a point of closure of $\mathcal{X}$ need not be a member of $\mathcal{X}$; however, every element within $\mathcal{X}$ is also a point of closure of $\mathcal{X}$. A subset of $\mathcal{X}$ in R^n is **closed** if every point of closure of $\mathcal{X}$ is contained in $\mathcal{X}$. The **closure** of a set $\mathcal{X}$ in R^n, denoted $\overline{\mathcal{X}}$, is the set of points of closure of $\mathcal{X}$. Clearly, a set $\mathcal{X}$ in R^n is closed if and only if $\mathcal{X} = \overline{\mathcal{X}}$. A set $\mathcal{X}$ in R^n has a **closed cover** if there exists a collection $\{G_i\}$ of closed subsets from R^n such that $\mathcal{X} \subseteq \cup_i G_i$.

Closely related to the concept of a point of closure of $\mathcal{X}$ is the notion of a **limit (cluster) point** of a set $\mathcal{X}$ in R^n. Specifically, $\bar{x}$ is a limit point of $\mathcal{X}$ if each δ – neighborhood about $\bar{x}$ contains at least one point of $\mathcal{X}$ different from $\bar{x}$, i.e., points of $\mathcal{X}$ different from $\bar{x}$ tend to "pile up" at $\bar{x}$. So, if $\bar{x}$ is a limit point of a set $\mathcal{X}$ in R^n, then $\mathcal{X} \cap \delta(\bar{x})$ is an infinite set - - every δ – neighborhood of $\bar{x}$ contains infinitely many points of $\mathcal{X}$. Moreover,

a. If $\mathcal{X}$ is a finite set in R^n, then it has no limit point.

b. The limit point of $\mathcal{X}$ need not be an element of $\mathcal{X}$.

c. The collection of all limit points of $\mathcal{X}$ in R^n is called the **derived set** and will be denoted as $\mathcal{X}^d$.

On the basis of the preceding discussion, we can alternatively characterize a set $\mathcal{X}$ in R^n as **closed** if it contains each of its limit point or if $\mathcal{X}^d \subseteq \mathcal{X}$. In addition, we can equivalently state that the **closure** of a set $\mathcal{X}$ in R^n is $\mathcal{X}$ together with its collection of limit points or $\overline{\mathcal{X}} = \mathcal{X} \cup \mathcal{X}^d$. Furthermore,

a. $\emptyset$, a single point, and R^n are all closed sets.

b. Any finite union of closed sets in R^n is closed; and any intersection of closed sets in R^n is closed.

c. The closure of any set $\mathcal{X}$ in R^n is the smallest closed set containing $\mathcal{X}$.

d. A subset $\mathcal{X}$ in R^n is closed if and only if its complementary set $\mathcal{X}^c$ is open.

e. A subset $\mathcal{X}$ in R^n is closed if and only if $\mathcal{X}$ contains its boundary.

Let's now briefly relate the concepts of a limit and a limit point of a sequence in R^n to some of the preceding point-set notions that we just developed. In particular, we shall take another look at the *point of closure* concept. To this end, a limit point (as well as a limit) of a sequence $\{x_k\}$ in R^n is a **point of closure** of a set $\mathcal{X}$ in R^n if $\mathcal{X}$ contains $\{x_k\}$. Conversely, if $\hat{x}$ is a point of closure of a set $\mathcal{X}$ in R^n, then there exists a sequence $\{x_k\}$ in $\mathcal{X}$ (and hence also a subsequence $\{x_k\}_{k\in K}$ in $\mathcal{X}$) such that $\hat{x}$ is a limit point of $\{x_k\}$ (and thus a limit of $\{x_k\}_{k\in K}$). Hence, the closure of $\mathcal{X}$, $\bar{\mathcal{X}}$, consists of all limit points of convergent sequences $\{x_k\}$ from $\mathcal{X}$.

Similarly, we note that a subset $\mathcal{X}$ in R^n is **closed** if and only if every convergent sequence of points $\{x_k\}$ from $\mathcal{X}$ has a limit in $\mathcal{X}$, i.e., $\mathcal{X}$ is closed if for $\{x_k\}$ in $\mathcal{X}$, $\lim_{k\to\infty} x_k = \hat{x} \in \mathcal{X}$. Also, a set $\mathcal{X}$ in R^n is **bounded** if every sequence of points $\{x_k\}$ formed from $\mathcal{X}$ is bounded. And if a set $\mathcal{X}$ in R^n is both closed and bounded, then it is termed **compact**. Equivalently, a set $\mathcal{X}$ in R^n is compact: (1) if it has the finite intersection property (every finite subclass has a nonempty intersection); (2) if it has the **Bolzano-Weierstrass property** (every infinite subset of $\mathcal{X}$ has at least one limit point in $\mathcal{X}$ or if every infinite sequence of points $\{x_n\}$ in $\mathcal{X}$ has at least one limit point in $\mathcal{X}$); and (3) if it possesses the **Heine-Borel property** (every open covering of $\mathcal{X}$ has a finite subcovering). We mention briefly that,

a. A closed subset of a compact set $\mathcal{X}$ in R^n is compact.

b. The union of a finite number of compact sets in R^n is compact; the intersection of any number of compact sets in R^n is compact.

c. A set $\mathcal{X}$ in R^n is compact if and only if it is closed and bounded.

d. Any finite set of points in R^n is compact.

e. If $\mathcal{X}$ in R^n is a set consisting of a convergent sequence $\{x_k\}$ and its limit $\bar{x} = \lim_{k\to\infty} x_k$, then $\mathcal{X}$ is compact. Conversely, if $\mathcal{X}$ in R^n is compact, every sequence $\{x_k\}$ has a convergent subsequence $\{x_k\}_{k\in K}$ whose limit belongs to $\mathcal{X}$.

A set $\mathcal{X}$ in R^n is **locally compact** if each of its points has a δ – neighborhood with compact closure, i.e., for each $x \in \mathcal{X}$, $\overline{\delta(x)}$ is compact. In this regard, any compact space is locally compact but not conversely, e.g., R^n is locally compact but not compact.

Section 2.5 Exercises:

2.5.1 Consider the sequence of open intervals

$$\left(-\frac{1}{n}, \frac{1}{n}\right), \; n = 1, 2, 3, \ldots.$$

Find the intersection of these open sets. Is the intersection open?

2.5.2 Verify that a set $\mathcal{A} \subset R$ is **bounded** if and only if there exists a non-negative number M such that $|x| \le M$ for every $x \in \mathcal{A}$.

2.5.3 Verify that the intersection of the open intervals (a, b), (c, d) is either the null set or an open interval.

2.5.4 Verify that: (i) an open set contains neither its supremum nor its infimum; and (ii) the set $\{x\}$ is not open.

2.5.5 Verify that the union of any collection of open sets is itself an open set.

2.5.6 Which of the following statements are true:

a. If $\mathcal{X}^d \neq \varnothing$, then $\mathcal{X}$ is an infinite set.

b. If a set is finite, then it has no limit point.

c. A subset $\mathcal{X}$ of R is not closed if and only if at least one limit point of $\mathcal{X}$ is not in $\mathcal{X}$.

d. A subset $\mathcal{X} \subset R$ is closed if and only if every limit point of $\mathcal{X}$ is in $\mathcal{X}$.

e. A subset $\mathcal{X} \subset R$ is closed if and only if $\mathcal{X}^d \subset \mathcal{X}$.

f. A subset $\mathcal{X} \subset R$ is closed if and only if $\mathcal{X} = \mathcal{X} \cup \mathcal{X}^d$, i.e., if and only if $\mathcal{X} = \overline{\mathcal{X}}$.

g. If $\mathcal{X} \subset R$, then both $\mathcal{X}^d$ and $\overline{\mathcal{X}}$ are closed.

2.5.7 A set $\mathcal{S} \subset R^2$ is termed **connected** if it cannot be covered by the union of two sets $\mathcal{A}$ and $\mathcal{B}$, with $\mathcal{A} \cap \mathcal{B} \neq \varnothing$, unless $\mathcal{S} \subset \mathcal{A}$ or $\mathcal{S} \subset \mathcal{B}$. Alternatively, an open set $\mathcal{S} \subset R^2$ is connected if any two points within $\mathcal{S}$ may be joined by a **polygonal path** or sequence of broken line segments lying entirely within $\mathcal{S}$.

Are the following sets connected?

a. $S = \{\mathbf{x} \in R^2 | \|\mathbf{x}\| < 1\}$.

b. $S = \{(x_1, x_2) \in R^2 | x_1 \geq 0\}$.

c. $S = \{(x_1, x_2) \in R^2 | |x_1| = |x_2|\}$.

d. $S = \{(x_1, x_2) \in R^2 | x_1^2 - x_2^2 \geq 1\}$.

e. $S = \{(x_1, x_2) \in R^2 | x_1, x_2 \text{ integers}\}$

f. $S = R^2, \varnothing$.

2.6 CONTINUOUS SINGLE-VALUED FUNCTIONS

For metric spaces $\mathcal{X}$ and $\mathcal{Y}$ with metrics d_1 and d_2, respectively, let f: $\mathcal{X} \to \mathcal{Y}$ be a point-to-point mapping of $\mathcal{X}$ into $\mathcal{Y}$. f is said to be **continuous at a point $x_o \in \mathcal{X}$** if either

a. for any $\varepsilon > 0$ there exists a $\delta_\varepsilon > 0$ such that $d_1(x, x_o) < \delta_\varepsilon$ implies $d_2(f(x), f(x_o)) < \varepsilon$. (Note that the subscript on δ means that "δ depends upon the ε chosen."); or

b. for each ε-neighborhood of $f(x_o)$, $\varepsilon(f(x_o))$, there exists a δ_ε – neighborhood about x_o, $\delta_\varepsilon(x_o)$, such that $f(\delta_\varepsilon(x_o)) \subseteq \varepsilon(f(x_o))$, i.e., points "near" x_o are mapped by f into points "near" $f(x_o)$ so that $\lim_{x \to x_o} f(x) = f(x_0)$.

In general, the point-to-point mapping f: $\mathcal{X} \to \mathcal{Y}$ is **continuous on** $\mathcal{X}$ if it is continuous at each point of $\mathcal{X}$.

Theorems 2.3 and 2.4 provide us with a set of necessary and sufficient conditions for the continuity of a point-to-point mapping at a specific point $x_o \in \mathcal{X}$ and at any arbitrary $x \in \mathcal{X}$, respectively. Specifically, we start with Theorem 2.3.

Theorem 2.3: (**Continuity in Terms of Convergent Sequences**). For metric spaces $\mathcal{X}$ and $\mathcal{Y}$, the point-to-point mapping f of $\mathcal{X}$ into $\mathcal{Y}$ is continuous at $x_o \in \mathcal{X}$ if and only if $x_k \to x_o$ implies $f(x_k) \to f(x_o)$ for every subsequence $\{x_k\}$ in $\mathcal{X}$.

Hence, f is a continuous mapping of $\mathcal{X}$ into $\mathcal{Y}$ if it "sends convergent sequences in $\mathcal{X}$ into convergent sequences in $\mathcal{Y}$". Next comes Theorem 2.4.

Theorem 2.4: (**Continuity in Terms of Open (resp., Closed) Sets**). For metric spaces $\mathcal{X}$ and $\mathcal{Y}$, let f be a point-to-point mapping of $\mathcal{X}$ into $\mathcal{Y}$. Then: (a) f is continuous if and only if $f^{-1}(\mathcal{A})$ is open in $\mathcal{X}$ whenever set $\mathcal{A}$ is open in $\mathcal{Y}$; (b) f is continuous if and only if $f^{-1}(\mathcal{A})$ is closed in $\mathcal{X}$ whenever $\mathcal{A}$ is closed in $\mathcal{Y}$.

Thus, f is continuous if it "pulls open (resp., closed) sets back to open (resp., closed) sets," i.e., the inverse images of open (resp., closed) sets are open (resp., closed).

We next consider a theorem (Theorem 2.5) which states that continuous mappings preserve compactness. That is,

Theorem 2.5: For metric spaces $\mathcal{X}$ and $\mathcal{Y}$, let f be a continuous point-to-point mapping of $\mathcal{X}$ into $\mathcal{Y}$. If $\mathcal{A}$ is a compact subset of $\mathcal{X}$, then so is its range $f(\mathcal{A})$.

Next, let $\mathcal{X}$ be a subset of R^n. A continuous point-to-point mapping g: $R^n \to \mathcal{X}$ is termed a **retraction mapping** on R^n if $g(x) = x$ for all $x \in \mathcal{X}$. Here, $\mathcal{X}$ is called a **retraction** of R^n. If $\mathcal{X}$ is contained within an arbitrary subset $\mathcal{A}$ of R^n, then g: $\mathcal{A} \to \mathcal{X}$ is a retraction of $\mathcal{A}$ onto $\mathcal{X}$ if $g(x) = x$ for all $x \in \mathcal{X}$.

Section 2.6 Exercises:

2.6.1 Verify that a constant function is continuous on R.

2.6.2 Verify via an example that if the domain of a real-valued function $y = f(x)$ is open, then it is not necessarily the case that the range of f is an open set.

2.6.3 Verify via an example that if the domain of a real-valued function $y = f(x)$ is closed, then it is not necessarily the case that the range of f is a closed set.

2.6.4 Verify that the real-valued function $f(x) = x^2 + 2, x \in R,$ is continuous using a δ, ε argument.

2.6.5 It was mentioned earlier that $f(x)$ is continuous at a point $x = x_o$ if $\lim_{x \to x_o} f(x) = f(x_0)$. What assumptions are we implicitly making when we use this definition of continuity? Determine if $f(x) = x^2$ is continuous at 1; and determine if $f(x) = (x^2 - 4)/(x - 2)$ is continuous at 2.

2.6.6 It is well known that a real-valued function $y = f(x)$ is **bounded** on a set $\mathcal{A}$ if there exists a number M such that $|f(x)| \leq M$ for all $x \in \mathcal{A}$. Which of the following functions are bounded over their given domains?

a. $f(x) = x^{-1}, 0 < x \leq 1.$

b. $f(x) = \frac{1-x}{1+x}, |x| < 1.$

c. $f(x) = \frac{|x|}{x}, 0 < x < 1.$

2.7 OPERATIONS ON SEQUENCES OF SETS

Let $\{\mathcal{A}_i\}$, $i = 1, 2, \ldots$, represent a sequence of sets in a metric space $\mathcal{X}$. If $\{\mathcal{A}_i\}$ is such that $\mathcal{A}_i \subseteq \mathcal{A}_{i+1}, i = 1, 2, \ldots$, then $\{\mathcal{A}_i\}$ is said to be a **non-decreasing sequence** (if $\mathcal{A} \subset \mathcal{A}_{i+1}$, then $\{\mathcal{A}_i\}$ is said to be an **expanding sequence**). And if $\{\mathcal{A}_i\}$ is such that $\mathcal{A}_i \supseteq \mathcal{A}_{i+1}, i = 1, 2, \ldots$, then $\{\mathcal{A}_i\}$ is called a **non-increasing sequence** (if $\mathcal{A}_i \supset \mathcal{A}_{i+1}, i = 1, 2, \ldots$, then $\{\mathcal{A}_i\}$ is termed a **contracting sequence**). A **monotone sequence** of sets is one which is either an expanding or contracting sequence.

If a sequence $\{\mathcal{A}_i\}$, $i = 1, 2, \ldots$, in $\mathcal{X}$ is non-decreasing, then its limit exists and we have,

$$\lim_{i \to \infty} \mathcal{A}_i = \cup_{i=1}^{\infty} \mathcal{A}_i \text{ (if } \{\mathcal{A}_i\} \text{ is non-decreasing);}$$

$$\lim_{i \to \infty} \mathcal{A}_i = \cap_{i=1}^{\infty} \mathcal{A}_i \text{ (if } \{\mathcal{A}_i\} \text{ is non-increasing).}$$

And for any sequence of sets $\{\mathcal{A}_i\}$, $i = 1, 2, \ldots$, in $\mathcal{X}$,

$$\sup\{\mathcal{A}_i\} = \cup_{i=1}^{\infty} \mathcal{A}_i, \ \inf\{\mathcal{A}_i\} = \cap_{i=1}^{\infty} \mathcal{A}_i;$$

with

$$\sup(\cup_{i=1}^{\infty} \mathcal{A}_i) = \sup\{\sup \mathcal{A}_i, i = 1, 2, \ldots\},$$

$$\inf(\cup_{i=1}^{\infty} \mathcal{A}_i) = \inf\{\inf \mathcal{A}_i, i = 1, 2, \ldots\}$$

$$\sup(\cap_{i=1}^{\infty} \mathcal{A}_i) \leq \inf\{\sup \mathcal{A}_i, i = 1, 2, \ldots\},$$

$$\inf(\cap_{i=1}^{\infty} \mathcal{A}_i) \geq \sup\{\inf \mathcal{A}_i, i = 1, 2, \ldots\}.$$

Let $\{\mathcal{A}_i\}$, $i = 1, 2, \ldots$, again depict a sequence of sets in $\mathcal{X}$. Then, there are subsets $E_i \subset \mathcal{A}_i$ of disjoint sets, with $E_j \cap E_k = \emptyset$ for $j \neq k$, such that

$$\cup_{i=1}^{\infty} E_i = \cup_{i=1}^{\infty} \mathcal{A}_i.$$

We next consider the concepts of the limit superior and limit inferior of a sequence of sets $\{\mathcal{A}_i\}$, $i = 1, 2, \ldots$, in a metric space $\mathcal{X}$. To this end, the **limit superior of a sequence** $\{\mathcal{A}_i\}$ is defined as

$$\begin{aligned} \overline{\lim_{i}} \ \mathcal{A}_i &= \lim_{i \to \infty} \sup \mathcal{A}_i = \cap_{i=1}^{\infty}(\cup_{k \geq i} \mathcal{A}_k) \\ &= (\mathcal{A}_1 \cup \mathcal{A}_2 \cup \cdots) \cap (\mathcal{A}_2 \cup \mathcal{A}_3 \cup \cdots) \cap \cdots \\ &= \{x \in \mathcal{X} | x \in \mathcal{A}_i \quad \text{for infinitely many } i\}. \end{aligned}$$

Hence, lim sup $\mathcal{A}_i$ is the set S of points such that, for every positive integer i, there exists a positive integer $k \geq i$ such that $S \in \mathcal{A}_i$; thus S consists of those points which belong to $\mathcal{A}_i$ for an infinite number of i values. Looked at in another fashion, if $x \in S$, then x is in all of $\cup_{k \geq i} \mathcal{A}_k$. Hence, no matter how large an i value is chosen, you can find a $k \geq i$ for which x is a member of $\mathcal{A}_i$.

Similarly, the **limit inferior of a sequence** $\{\mathcal{A}_i\}$ is

$$\begin{aligned} \underline{\lim_{i}} \ \mathcal{A}_i &= \lim_{i \to \infty} \inf \mathcal{A}_i = \cup_{i=1}^{\infty}(\cap_{k \geq i} \mathcal{A}_k) \\ &= (\mathcal{A}_1 \cap \mathcal{A}_2 \cap \cdots) \cup (\mathcal{A}_2 \cap \mathcal{A}_3 \cap \cdots) \cup \cdots \\ &= \{x \in \mathcal{X} | x \in \mathcal{A}_i \text{ for all but finitely many } i\}. \end{aligned}$$

Thus, lim inf $\mathcal{A}_i$ is the set I of points such that, for some positive integer i, $I \in \mathcal{A}_i$ for all positive integers $k \geq i$; hence, I consists of those points which belong to $\mathcal{A}_i$ for all except a finite number of i values. Stated alternatively, if $x \in I$, then x is an element of $\cap_{k \geq i} \mathcal{A}_k$ so that $x \in \mathcal{A}_i$ for $k \geq i$ - - x must be in I with only finitely many exceptions, i.e., for $x \in I$, there is an index value such that x is in every $\mathcal{A}_i$ in the remaining portion of the limit.[5]

We note briefly that if $\{\mathcal{A}_i\}$, $i = 1, 2, \ldots$, is any sequence of sets in a metric space $\mathcal{X}$, then lim inf $\mathcal{A}_i \subset$ lim sup $\mathcal{A}_i$. A sequence of sets $\{\mathcal{A}_i\}$ is **convergent** (or a subset $\mathcal{A}$ of $\mathcal{X}$ is the **limit** of $\{\mathcal{A}_i\}$) if

$$\lim_{i\to\infty} \sup \mathcal{A}_i = \lim_{i\to\infty} \inf \mathcal{A}_i = \lim_{i\to\infty} \mathcal{A}_i = \mathcal{A}.$$

Here $\mathcal{A}$ is termed the **limit set**. In this vein, any monotone sequence of sets $\{\mathcal{A}_i\}$, $i = 1, 2, \ldots.$, is convergent.

Section 2.7 Exercises:

2.7.1 Let $\{\mathcal{A}_i\}$, i = 1, 2, ..., be a sequence of sets in a metric space $\mathcal{X}$. Verify that there exists a sequence $\{E_i\}$, $i = 1, 2, \ldots$ of disjoint sets such that $\cup_{i=1}^{\infty} = \cup_{i=1}^{\infty} E_i$.

2.7.2 Verify that if $\{\mathcal{A}_i\}$, i = 1, 2, ..., is a sequence of sets in a metric space $\mathcal{X}$, then $\underset{i}{lim}\, \mathcal{A}_i \subset \underset{i}{\overline{lim}}\, \mathcal{A}_i$

2.7.3 (A challenge) Verify that every monotone sequence of sets $\{\mathcal{A}_i\}$, i = 1, 2,..., in a metric space $\mathcal{X}$ is convergent.

NOTES

1 If f is either an **increasing** ($f(x_1) < f(x_2)$, $x_1 < x_2$) or **decreasing** ($f(x_1) > f(x_2)$, $x_1 < x_2$) **function**, then f is termed **strictly monotonic** over D.

2 Equation (2.1) is actually a generalization of the absolute value function $|x-y|$. To see this, let us define on R^n a **norm** (denoted $\|\cdot\|$) – a function $\|\cdot\|: R^n \to [0, +\infty)$ which assigns to each $x \in R^n$ some number $\|x\|$ such that:

a. $\|x\| \geq 0$ and $\|x\| = 0$ if and only if $x = 0$;

b. $\|x + y\| \leq \|x\| + \|y\|$ (**triangle inequality**);

c. for a scalar c, $\|cx\| = |c|\|x\|$ (**homogeneity**); and

d. $|\sum_{i=1}^{n} x_i y_i| \leq \|x\|\|y\|$ (**Cauchy-Schwarz inequality**).

Then the distance between points $x, y \in R^n$ induced by the norm "$\|\cdot\|$" on R^n is

$$\|x - y\| = [\textstyle\sum_{i=1}^{n} |x_i - y_i|^2]^{1/2} \tag{2.1.1}$$

or equation (2.1). So if $S = R^n$ and d are given by (2.3.1), then R^n is metric space with metric (2.3.1).

3 If $x_k = n =$ ***constant*** for all k, then $\{x_k\}$ converges to the limit n. But since the range of this sequence contains only a single point, it is evident that the sequence has no limit point. If $x_k = \frac{1}{k}$, then the sequence $\{x_k\}$ converges to a limit of zero, which is also a limit point. And if $x_k = (-1)^k$, then the sequence $\{x_k\}$ has limit points at ± 1, but has no limit.

4 That is, for $\varepsilon > 0$ there exists a positive integer $N_{\varepsilon/2}$ such that: $m \geq N_{\varepsilon/2}$ implies $d(x_m, \bar{x}) < \frac{\varepsilon}{2}$; ***and*** $n \geq N_{\varepsilon/2}$ implies $d(x_n, \bar{x}) < \frac{\varepsilon}{2}$. Hence, both $m, n > N_{\varepsilon/2}$ imply, via the triangle inequality, that$d(x_m, x_n) \leq d(x_m, \bar{x}) + d(x_n, \bar{x}) < \frac{\varepsilon}{2} + \frac{\varepsilon}{2} = \varepsilon$.

5 Alternative definitions of the limit superior and limit inferior of a sequence of sets are the following. Again, let $\{\mathcal{A}_i\}$ be a sequence of sets in a metric space $\mathcal{X}$. Then,

$$\lim_{i\to\infty} \sup \mathcal{A}_i = \left\{ x \in \mathcal{X} \middle| \lim_{i\to\infty} \inf d(x, \mathcal{A}_i) = 0 \right\};$$

$$\lim_{i\to\infty} \inf \mathcal{A}_i = \{x \in \mathcal{X} | d(x, \mathcal{A}_i) = 0\}$$

where $d(x, \mathcal{A}_i)$ is the distance from x to $\mathcal{A}_i$.

Mathematical Foundations 3

3.1 BEYOND SINGLE-VALUED FUNCTIONS

In the preceding chapter (Sections 2.2 and 2.6), we encountered the concept of a single-valued function or point-to-point mapping. We discussed therein the varieties of point-to-point mappings and the continuity of the same. In this chapter, we expand the definition of a function beyond that of a single-valued mapping and introduce concepts such as a scalar-valued function of a vector, a vector-valued function of a scalar, and a vector-valued function of a vector or simply a transformation.

It was mentioned in Section 2.2 that for two non-empty sets $X = R$ and $Y = R$, a **scalar-valued function of a scalar** is a point-to-point mapping $f: X \to Y$ or $f: R \to R$. Thus, f is a rule which associates with each scalar $x \in X = R$ a scalar image $y = f(x) \in Y = R$.

If, now, $X = R^n$ and $Y = R$, then $f: X \to Y$ or $f: R^n \to R$ is a **scalar-valued function of a vector** $\mathbf{x}^T = (x_1, \ldots, x_n)$. Hence. the image of $\mathbf{x} \in X = R^n$ is the scalar $y = f(\mathbf{x}) \in Y = R$. (This rule is what is commonly called a *multivariate function*.)

Next, if $X = R$ and $Y = R^m$, then $f: X \to Y$ or $f: R \to R^m$ is a **vector-valued function of a scalar**, i.e., the image of the scalar $x \in X = R$ is the vector $f(x) = y^T = (y_1, \ldots, y_m) = (f^1(x), \ldots, f^m(x)) \in Y = R^m$. For instance, if $t \in R$ and the components of f are $f^1, \ldots, f^m$, then we can write $f(t) = (f^1(t), \ldots, f^m(t))$. Thus, f defines the ordered m-tuple of real-valued functions of t.

The most general case, and one which we shall discuss in considerable detail, is when $X = R^n$, $Y = R^m$, and $\mathbf{f}: X \to Y$ or $\mathbf{f}: R^n \to R^m$. Then, f is a **vector-valued function of a vector** or simply a **transformation** in that the image of each vector $\mathbf{x}^T = (x_1, \ldots, x_n) \in X = R^n$ is a vector $f(\mathbf{x}) = \mathbf{y}^T = (y_1, \ldots, y_m) = (f^1(\mathbf{x}), \ldots, f^m(\mathbf{x})) \in Y = R^m$.

Section 3.1 Exercises:

3.1.1 Determine the domain of definition of the following scalar-valued functions of a vector $\mathbf{x}^T = (x_1, x_2)$:

a. $f(x_1, x_2) = \ln(3 - x_1^2 - x_2^2)$.

b. $f(x_1, x_2) = (x_1^2 + x_2^2 - 6)^{1/2} + \ln(10 - x_1^2 - x_2^2)$.

c. $f(x_1, x_2) = x_1(x_2^2 - 4x_1)^{-1}$.

3.1.2 One way of describing a curve in two dimensions is to view it as a one-dimensional set of points embedded in two-dimensional space. That is, it can be represented by the two parametric equations $x = x(t)$, $y = y(t)$ with $a \le t \le b$, where t is the parameter. In this regard, let $f(t) = f(x(t), y(t))$, with initial point $(x(a), y(a))$ and terminal point $(x(b), y(b))$. Suppose $x = 3t^2 + 2$ and $y = 4t$, $1 \le t \le 3$. Find the parametric representation of $f(x, y) = (3x - 1, y^2)$ along with its initial and terminal points.

3.1.3 Determine the domain of the vector-valued function $f(t) = f(x(t), y(t), z(t)) = (t^2 + 1, \ln(4 - t), \sqrt{t + 2})$.

3.1.4 Express the vector-valued function $f(t) = (2 - 3t, -2 + 4t, 2 + 2t)$ in the form of a line passing through the point $(2, -2, 2)$.

3.1.5 Let a vector-valued function of a vector be defined as

$$y_1 = x_1^2 + x_2^2$$
$$y_2 = x_1 - x_2.$$

What is the image of the point (1, 1)? What is the image of the line $x_2 = c$? What is the parametric expression resulting when x_1 is eliminated?

3.1.6. Given the vector-valued function of a vector

$$y_1 = x_1 + 2x_2$$
$$y_2 = x_1 - x_2,$$
$$y_3 = x_1^2 + 2$$

find the image of the point (1, 2). Determine the image of the line $x_2 = x_1$.

3.2 LIMITS AND CONTINUITY OF TRANSFORMATIONS

We have already discussed limits and continuity in the case of single-valued functions in Sections 2.2 and 2.6. We shall now focus on the general case of transformations $\boldsymbol{f}$ from a domain $D \subset R^n$ to R^m, with $m > 1$. But, first some preliminaries.

For a point $x_o \in R$, a (suitably restricted) **δ – neighborhood of $\boldsymbol{x}_o$** will be written as $\delta(x_o) = \{x \in R \mid |x - x_o| < \delta\}$. Thus, $\delta(x_o)$ is a symmetrical open interval about x_o, $x_o - \delta < x < x_o + \delta$, with δ *small and positive*. (Here, δ is taken to be as small as needed.) For $\boldsymbol{x} \in R^n$, a **spherical δ – neighborhood of $\boldsymbol{x}_o$** or **hypersphere with center $\boldsymbol{x}_o$** and radius $\delta > 0$ is the set of points $\delta(\boldsymbol{x}_o) = \{\boldsymbol{x} \in R^n \mid \|\boldsymbol{x} - \boldsymbol{x}_o\| < \delta\}$. (For $n = 2$, $\delta(\boldsymbol{x}_o)$ is called a **circular δ – neighborhood at $\boldsymbol{x}_o$**.) Thus, the distance between $\boldsymbol{x}_o$ any other point $\boldsymbol{x}$ within the hypersphere is strictly less than δ, i.e., $d(\boldsymbol{x}, \boldsymbol{x}_o) = [(\boldsymbol{x} - \boldsymbol{x}_o)^T(\boldsymbol{x} - \boldsymbol{x}_o)]^{1/2} < \delta$ so that $\delta(\boldsymbol{x}_o)$ is an open set. Finally, a deleted **δ – neighborhood of $\boldsymbol{x}_o$** is a δ – neighbohood with the center $\boldsymbol{x}_o$ removed.

Definition 3.1: Let f be a transformation from a domain $D \subset R^n$ to R^m or $f: D \subset R^n \to R^m$, with $m > 1$.

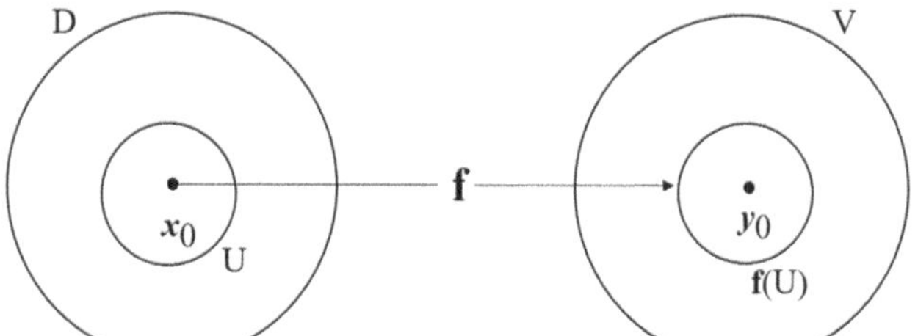

FIGURE 3.1 Point y_o is the limit of the transformation $\boldsymbol{f}$ as $\boldsymbol{x} \to \boldsymbol{x}_o$.

Suppose D is an open set that contains a deleted δ-neighbohood about $\boldsymbol{x}_o$. If for every neighborhood V of y_o there is a deleted δ-neighbohood $U \subset D$ of $\boldsymbol{x}_o$ such that $f(U) \subset V$, then y_o is said to be the limit of the transformation $\boldsymbol{f}$ at $\boldsymbol{x}_o$ or

$$\lim_{\boldsymbol{x} \to \boldsymbol{x}_o} \boldsymbol{f}(\boldsymbol{x}) = y_o. \tag{3.1}$$

(See Figure 3.1.) In terms of limits of vector sequences, $\lim_{\boldsymbol{x} \to \boldsymbol{x}_o} \boldsymbol{f}(\boldsymbol{x}) = y_o$. if and only if $\lim_{k \to \infty} \{\boldsymbol{f}(\boldsymbol{x}_k)\} = y_o$ whenever $\lim_{k \to \infty} \{\boldsymbol{x}_k\} = \boldsymbol{x}_o$, where $\boldsymbol{x}_k \neq \boldsymbol{x}_o$ holds for all $k = 1, 2, \ldots$. Note the following: (1) $\boldsymbol{x}_o$ need not be in D; if $\boldsymbol{x}_o \in D$, $\boldsymbol{f}$ need not be defined at $\boldsymbol{x}_o$ or $\boldsymbol{f}(\boldsymbol{x}_o)$ need not exist; (2) the radius of U is as small as needed in order for $U \subset D$; and (3) this definition requires that $\boldsymbol{f}$ be defined at points arbitrarily close to $\boldsymbol{x}_o$ or D contains points arbitrarily close to $\boldsymbol{x}_o$ so that $\boldsymbol{x}_o$ is actually a limit point.

Equivalently, given a transformation $f: D \subset R^n \to R^m$ and a limit point $\boldsymbol{x}_o \in D$, the point $y_o \in R^m$ is termed the limit of $\boldsymbol{f}$ at $\boldsymbol{x}_o$ or $\lim_{\boldsymbol{x} \to \boldsymbol{x}_o} \boldsymbol{f}(\boldsymbol{x}) = y_o$ if and only if, given an $\varepsilon > 0$, there exists a $\delta > 0$ such that $\boldsymbol{x} \in D$ and $d(\boldsymbol{x}, \boldsymbol{x}_o) < \delta$ imply that $d(f(\boldsymbol{x}), y_o) < \varepsilon$. Again, $f(\boldsymbol{x})$ can be made arbitrarily close to y_o by choosing $\boldsymbol{x}$ sufficiently close to $\boldsymbol{x}_o$. Note that: (1) *sufficiently close* means that δ depends upon the ε selected; and (2) while $\boldsymbol{x}$ is close to $\boldsymbol{x}_o$, it is required that $\boldsymbol{x} \neq \boldsymbol{x}_o$. Looked at in terms of neighborhoods, this definition states that given a neighborhood $\varepsilon(y_o) = \{f(\boldsymbol{x}) \mid d(f(\boldsymbol{x}), y_o) < \varepsilon, \boldsymbol{x} \in D\}$, there exists a neighborhood $\delta(\boldsymbol{x}_o) = \{x \in D \mid d(\boldsymbol{x}, \boldsymbol{x}_o) < \delta\}$ whose intersection with $D \backslash \{\boldsymbol{x}_o\}$ is mapped by $\boldsymbol{f}$ into $\varepsilon(y_o)$.

Since $f(\mathbf{x}) = (f^1(\mathbf{x}), \ldots, f^m(\mathbf{x})), \mathbf{x} \in D$, and f^i, $i = 1, \ldots, m$, is the ith component of f, it follows that the limit of f can be evaluated component-wise, i.e., the convergence of $f(\mathbf{x})$ to y_o is equivalent to the convergence of each component function $f^i(\mathbf{x})$ to the corresponding component of y_o. That is, $y_o = \lim_{\mathbf{x} \to \mathbf{x}_o} f(\mathbf{x})$ if and only if $y_o^i = \lim_{\mathbf{x} \to \mathbf{x}_o} f^i(\mathbf{x}), i = 1, \ldots m$.

We next look to the continuity of f at a specific point. To elaborate, we have

Definition 3.2: Suppose $\mathbf{x}_o$ is an interior point of the domain $D \subset R^n$ of the transformation $f: D \to R^m$. Then f is **continuous** at $\mathbf{x}_o$ if

$$\lim_{\mathbf{x} \to \mathbf{x}_o} f(\mathbf{x}) = f(\mathbf{x}_o). \tag{3.2}$$

Equivalently: (1) f is continuous at $\mathbf{x}_o$ if for every neighborhood V of $f(\mathrm{x}_o)$ there is a neighborhood U of $\mathbf{x}_o$ such that $f(U) \subset V$ or; (2) for every $\varepsilon > 0$ there exists a $\delta > 0$ such that $\|f(\mathbf{x}) - f(\mathbf{x}_o)\| < \varepsilon$ whenever $\|\mathbf{x} - \mathbf{x}_o\| < \delta, \mathbf{x} \in D$; or (3) the transformation $f: R^n \to R^m$ is continuous on D if and only if given an open (resp., closed) set $A \subset R^m$, the inverse image $f^{-1}(A) = \{\mathbf{x} \in R^n | f(\mathbf{x}) \in A\}$ is open (resp., closed) in R^n. Note that: (1) f must be defined at $\mathbf{x}_o$ so that we work with a δ – neighborhood ***of*** x_o and *not* a deleted δ – neighborhood ***of*** x_o; (2) if f is continuous at every point $\mathbf{x}_o \in D \subset R^n$, then f is said to be *continuous on* D; and (3) if $\mathbf{x}_o \in D$ is not a limit point of D, then the limit in (3.2) does not exist. And since $f = (f^1, \ldots, f^m)$, we can state that the transformation function $f: D \subset R^n \to R^m$ is continuous at $\mathbf{x}_o \in D$ if and only if each component function $f^i, i = 1, \ldots, m$, is continuous at $\mathbf{x}_o$.

Besides possessing a limit y_o or being continuous at a point $\mathbf{x}_o$, an additional important property of a transformation function f is that it may be bounded. Specifically, a transformation $f: R^n \to R^m$ is **bounded** on a set $A \subset R^n$ if there exists a scalar M such that $\|f(\mathbf{x})\| \leq M$ for every $\mathbf{x} \in A$. In addition, if f has a limit at $\mathbf{x}_o$, then f is bounded on some deleted δ – neighborhood U_o of $\mathbf{x}_o$.

We finish with

Theorem 3.3: If $f: A \to R^m$ is continuous on a compact (closed and bounded) set $A \subset R^n$, then its image $f(A)$ is also a compact set.

This important theorem will be revisited (and specialized) when we discuss the existence of maximum and minimum values for a real-valued function.

Section 3.2 Exercises:

3.2.1 Does the real-valued function

$$y = f(\mathbf{x}) = f(x_1, x_2) = \frac{x_1^2 - x_2^2}{x_1^2 + x_2^2}$$

have a limit as $\mathbf{x} \to \mathbf{O}$?

Note: (a) f is not defined at $\mathbf{x} = \mathbf{O}$; and (2) the limit must exist and be the same no matter how $\mathbf{x} \to \mathbf{O}$. (Hint: let $\mathbf{x} \to \mathbf{O}$ along the x_1 – axis and then along the x_2 – axis.)

3.2.2 Verify that

$$\lim_{x \to O} f(\mathbf{x}) = \lim_{x \to O} \frac{x_1 x_2}{(x_1^2 + x_2^2)^{1/2}} = 0.$$

Note: f is not defined at $\mathbf{x} = \mathbf{O}$. Use an ε, δ–argument for the verification.

3.2.3 Verify that the real-valued function

$$y = f(\mathbf{x}) = x_1 x_2 \left(\frac{x_1^2 - x_2^2}{x_1^2 + x_2^2} \right)$$

is continuous at $\mathbf{x} = \mathbf{O}$.

3.3 DERIVATIVE OF A SINGLE-VALUED FUNCTION

Sections 3.3–3.5 are concerned with the concept of the derivative of a real-valued function, whether that function has a single argument or multiple arguments. In the instance of, say, a point-to-point mapping $y = f(x), x \in R$, the definition of the derivative hinges upon the existence of a tangent line to a curve; and for the case involving functions of many independent variables, e.g., $y = f(\mathbf{x}), \mathbf{x} \in R^n$, the derivative concept is fundamentally related to the existence of tangent planes to surfaces. In fact, as we shall see in the following discussion, for the general case involving a mapping $\boldsymbol{f}$ from an open domain $D \subset R^n$ to R^m, the differential of $\boldsymbol{f}$ at a point $\boldsymbol{x}_o \in D$ is a linear transformation whose matrix is the derivative.

If the real-valued function $y = f(x)$, with $f: D = R \to R$, is differentiable at a point $x_o \in D$, then the tangent line l to the graph of f at $(x_o, f(x_o))$ has the equation

$$y - f(x_o) = f'(x_o)(x - x_o). \tag{3.3}$$

Here, (3.3) is a linear approximation to the *actual change in* f, $\Delta f = f(x) - f(x_o)$, between points x_o and x. If we set $h = x - x_o$, then $\Delta f(h) = f(x_o + h) - f(x_o)$ and the *linear change in* f is $df(h) = f'(x_o)h$ (Figure 3.2).

It is important to note that we have just approximated a non-linear mapping by a linear one. In fact, as just mentioned, this is the basic idea underlying differential calculus. If we harken back to our discussion of linear transformations (Section 1.10), then we can readily see that the **differential of** $\boldsymbol{f}$ **at** x_o, $df(h) = f'(x_o)h$, is the linear transformation $df \in \mathcal{L}(R, R)$ whose matrix is the **derivative of** $\boldsymbol{f}$ **at** x_o, $f'(x_o)$. (In the current case of a single independent variable x, the said *matrix* is of order (1×1) – a scalar.)

When h is small, the linear change in f, $df(h)$, is a good approximation to the actual change, $\Delta f(h)$, in the sense that

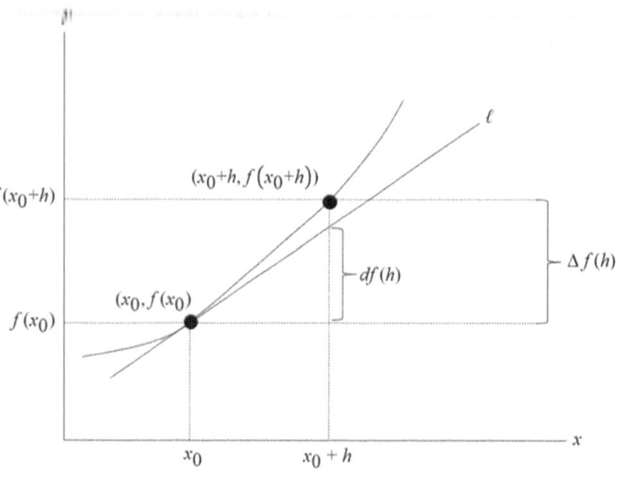

FIGURE 3.2 Actual change in f, $\Delta f(h)$, vs. linear change in f, $df(h)$.

$$\lim_{h \to 0} \frac{\Delta f(h) - df(h)}{h} = \lim_{h \to 0} \frac{f(x_o + h) - f(x_o) - f'(x_o)h}{h} = 0$$

or

$$\lim_{h \to 0} \frac{f(x_o + h) - f(x_o)}{h} = f'(x_o), \tag{3.4}$$

the **derivative of** $\boldsymbol{f}$ **at** x_o (provided, of course, that this limit exists).

We next have the *law of the mean.*

Theorem 3.4: Mean Value Theorem for Derivatives. Suppose a function $y = f(x)$ is continuous over $[a, b] = \{x | a \leq x \leq b\}$ and differentiable over $(a, b) = \{x | a < x < b\}$. Then, there is a point $x = x_o \in (a, b)$ such that

$$f(b) - f(a) = f'(x_o)(b - a). \tag{3.5}$$

Let us now review some important rules of differentiation. Specifically, if $f(x): R \to R$ and $g(x): R \to R$, then:

1. $\frac{d}{dx}(f \pm g) = \frac{df}{dx} \pm \frac{dg}{dx}.$

2. $\frac{d}{dx}(f \cdot g) = \left(\frac{df}{dx}\right)g + \left(\frac{dg}{dx}\right)f.$

3. $\frac{d}{dx}\left(\frac{f}{g}\right) = \frac{g\left(\frac{df}{dx}\right) - f\left(\frac{dg}{dx}\right)}{g^2}, \ g \neq 0.$

4. ***if*** $y = f(x)$ ***and*** $x = g(t)$, ***then, via the chain rule,*** $\frac{dy}{dt} = \frac{df}{dx}\frac{dg}{dt}.$

5. $\frac{d}{dx}(f^n) = nf^{n-1}\frac{df}{dx}.$

6. $\frac{d}{dx}(f^n g^m) = f^{n-1}g^{m-1}\left(ng\frac{df}{dx} + mf\frac{dg}{dx}\right).$

7. $\frac{d}{dx}\left(\frac{f^n}{g^m}\right) = \frac{f^{n-1}}{g^{m+1}}\left(ng\frac{df}{dx} - mf\frac{dg}{dx}\right).$

8. $\frac{d}{dx}(e^{f(x)}) = \frac{df}{dx}e^{f(x)}.$

9. $\frac{d}{dx}(\log_e f(x)) = \frac{1}{f}\frac{df}{dx}, f \neq 0.$

10. $\frac{d}{dx}(\log_a f(x)) = (\log_a e)\frac{1}{f}\frac{df}{dx}.$

11. $\frac{d}{dx}(a^{f(x)}) = a^{f(x)}(\log_e a)\left(\frac{df}{dx}\right).$

12. $\frac{d}{dx}(f(x)^{g(x)}) = gf^{g-1}\frac{df}{dx} + (\log_e f)f^g\frac{dg}{dx}.$

Next, suppose $y = f(\boldsymbol{x})$: $D \subset R^n \rightarrow R$, where $\boldsymbol{x} \in R^n$ and D is an open subset of R^n. For $\boldsymbol{x}_o$ a point interior to D, the *actual change in* f is the increment $\Delta f(\boldsymbol{h}) = f(\boldsymbol{x}_o + \boldsymbol{h}) - f(\boldsymbol{x}_o)$, where $\boldsymbol{h} = \boldsymbol{x} - \boldsymbol{x}_o \in R^n$ with $\|\boldsymbol{h}\|$ sufficiently small. For some vector $\boldsymbol{a} \in R^n$, if the *linear change in f*, $df = \boldsymbol{a}\,\boldsymbol{h}$, approximates $\Delta f(\boldsymbol{h})$ closely so that

$$\lim_{h \to O} \frac{\Delta f(\boldsymbol{h}) - \boldsymbol{a}\,\boldsymbol{h}}{\|\boldsymbol{h}\|} = 0,$$

then we say that **f is differentiable at $\boldsymbol{x}_o$**, and the **differential of f at x_o**, $df(\boldsymbol{h}) = \boldsymbol{a}\,\boldsymbol{h}$, is the linear transformation $df \subset \mathcal{L}(R^n, R)$ whose $(1 \times n)$ matrix $\boldsymbol{a}$ is the **derivative of f** at $\boldsymbol{x}_o$. (As usual, this derivative is uniquely determined if the preceding limit exists.) How is $\boldsymbol{a}$ chosen? In general, if the real-valued function $y = f(\boldsymbol{x})$ has a differential at $\boldsymbol{x}_o$, then

$$\begin{aligned} df(\boldsymbol{h}) &= f_1(\boldsymbol{x}_o)h_1 + f_2(\boldsymbol{x}_o)h_2 + \cdots + f_n(\boldsymbol{x}_o)h_n \\ &= \nabla f(\boldsymbol{x}_o)^T\boldsymbol{h}, \end{aligned} \tag{3.6}$$

where the vector-valued function

$$\nabla f(\boldsymbol{x}_o) = \begin{bmatrix} f_1(\boldsymbol{x}_o) \\ f_2(\boldsymbol{x}_o) \\ \vdots \\ f_n(\boldsymbol{x}_o) \end{bmatrix}$$

is the **gradient vector of f at $\boldsymbol{x}_o$**. Hence, $\boldsymbol{a}$ is $\nabla f(\boldsymbol{x}_o)^T$.

Under what conditions does the differential (3.6) exist? To answer this question, we state: for $f(\boldsymbol{x})$: $D \subset R^n \rightarrow R$, with D open, assume that the partial derivatives $f_i(\boldsymbol{x}_o)$, $i = 1, \ldots, n$, exist throughout D and are continuous at a point $\boldsymbol{x}_o \in D$. Then, f is differentiable at $\boldsymbol{x}_o$, and the differential of f at $\boldsymbol{x}_o$ is the linear transformation $df \in \mathcal{L}(R^n, R)$ defined by $df(\boldsymbol{h}) = \nabla f(\boldsymbol{x}_o)^T\boldsymbol{h}$. (We mention briefly that if f is differentiable at every $\boldsymbol{x} \in D$, then f is said to be **differentiable on D** and is termed a **differentiable function.**) To summarize: the real-valued function $y = f(\boldsymbol{x})$: $D \subset R^n \rightarrow R$, D open, is differentiable at $\boldsymbol{x}_o$ if there exists a linear transformation $\boldsymbol{a} \in \mathcal{L}(R^n, R)$ such that

$$\lim_{h \to O} \frac{f(\boldsymbol{x}_o + \boldsymbol{h}) - f(\boldsymbol{x}_o) - \boldsymbol{a}\,\boldsymbol{h}}{\|\boldsymbol{h}\|} = 0. \tag{3.7}$$

The linear transformation $\boldsymbol{a} \in \mathcal{L}(R^n, R)$ is the **derivative of f at $\boldsymbol{x}_o$** and denoted $\nabla f(\boldsymbol{x}_o)^T$.

Suppose we have a vector $\boldsymbol{u} \in R^n$, with $\|\boldsymbol{u}\| = 1$.[1] Then, for $\boldsymbol{x}_o$ fixed, the equation of the line through the points $\boldsymbol{x}_o$ and $\boldsymbol{x}_o + \boldsymbol{u}$ has the from $\boldsymbol{x}_o + t\boldsymbol{u}$, $0 < t < +\infty$. Let $f(\boldsymbol{x}_o + t\boldsymbol{u})$ represent the image of $\boldsymbol{x}_o + t\boldsymbol{u}$ on this line. Then, we may express the **directional derivative with respect to $\boldsymbol{u}$ of f at $\boldsymbol{x}_o$** as

$$D_{\boldsymbol{u}}f(\boldsymbol{x}_o) = \lim_{t \to 0} \frac{f(\boldsymbol{x}_o + t\boldsymbol{u}) - f(\boldsymbol{x}_o)}{t}, \tag{3.8}$$

provided that the limit exists. How is the directional derivative calculated? To answer this question, set $\boldsymbol{h} = t\boldsymbol{u}$ in (3.7) so that

$$\begin{aligned} &\lim_{t \to 0} \frac{f(\boldsymbol{x}_o + t\boldsymbol{u}) - f(\boldsymbol{x}_o) - \boldsymbol{a}(t\boldsymbol{u})}{\|t\boldsymbol{u}\|} \\ &= \lim_{t \to 0} \frac{f(\boldsymbol{x}_o + t\boldsymbol{u}) - f(\boldsymbol{x}_o)}{t} - \boldsymbol{a}\boldsymbol{u} = 0 \end{aligned}$$

or

$$D_{\boldsymbol{u}}f(\boldsymbol{x}_o) = \boldsymbol{a}\boldsymbol{u} = \nabla f(\boldsymbol{x}_o)^T\boldsymbol{u}. \tag{3.9}$$

Hence, the directional derivative of f, the instantaneous rate of change of f at x_o, measured in units of change in f per unit distance in the u-direction, may be expressed as the scalar product of the gradient of f at x_o and a unit vector specifying direction.

Given (3.9), let us examine some of the important properties of the gradient $\nabla f(x_o)$. Since

$$D_u f(x_o) = \nabla f(x_o)^T u = \|\nabla f(x_o)\| \|u\| \cos\theta$$
$$= \|\nabla f(x_o)\| \cos\theta,$$

if $\cos\theta = 1 (\theta = 0^0)$, then $D_u f(x_o) = \|\nabla f(x_o)\|$. Hence, the directional derivative attains its maximum in the direction of the gradient (via the Cauchy-Schwarz inequality, i.e., $|D_u f(x_o)| \leq \|\nabla f(x_o)\| \|u\|$). Thus, $\nabla f (\neq O)$ is a vector: (1) which points locally in the direction of maximum increase of f; (2) whose magnitude $\|\nabla f(x_o)\|$ is the maximum rate of increase in f in the $\nabla f-$ direction (the maximum value of $D_u f$); and (3) is normal to the contour surface $f(x_o) = f^o$ = constant. If u is tangent to $f(x_o) = f^o$ at x_o, then the directional derivative of f at x_o in the u-direction is $D_u f(x_o) = 0$, i.e., if $\nabla f(x_o) \neq 0$, then $D_u f(x_o) = \|\nabla f(x_o)\| \cos\theta = 0$ only if $\cos\theta = 0 (\theta = \pi/2)$ or $\nabla f(x_o)$ and u are orthogonal.

Next, suppose that we now set $h = te_j$, $j = 1, \ldots, n$, $0 < t < +\infty$, where e_j is the jth unit column vector (the jth element in the set of standard basis vectors $\{e_1, \ldots, e_n\}$ for R^n). Then from (3.8) and (3.9), we write the **partial derivative of f with respect to x_j at x_o** as

$$D_{e_j} f(x_o) = \lim_{t \to 0} \frac{f(x_o + te_j) - f(x_o)}{t}$$
$$= a\, e_j = \nabla f(x_o)^T e_j = \frac{\partial f(x_o)}{\partial x_j}, \; j = 1, \ldots, n.$$

Hence, the directional derivatives of f with respect to the standard basis vectors for R^n are the partial derivatives of f at x_o. Under this result, we may consequently express $\nabla f(x_o)^T$ as

$$\nabla f(x_o)^T = (D_1 f(x_o), \ldots, D_n f(x_o)),$$

where, for the sake of simplicity, we have $D_j f(x_o) = D_{e_j} f(x_o)$.

It is important to note that the existence of derivatives in the directions of the standard basis vectors does not imply that derivatives exist in other directions. If f is differentiable, then f has a directional derivative in every direction at x_o, but not conversely – a function may possess directional derivatives in *all* directions, yet still fail to be differentiable.

What is the geometric interpretation of differentiability in R^n? Let's express the concept of a **surface** $S \subset R^{n+1}$ as $S = \{(x, y) | y = f(x)\}$. Given S, and for f differentiable at $x_o \in D \subset R^n$, D open, there is a **tangent hyperplane** $\mathcal{H} = \{(x, y) | y = f(x_o) + f'(x_o)h, \; h \in R^n\} \subset R^{n+1}$.

Then, f is differentiable at x_o if there exists a hyperplane $\mathcal{H}$ tangent to S at $(x_o, f(x_o))$.

A well-known result from basic calculus is that if a function is differentiable at x_o, then it is also continuous there. We may exploit the notion of continuity in order to state *a sufficient condition for differentiability*: let $f(x)$: $D \subset R^n \to R$, D open, be continuous on D and let the partial derivatives $\partial f / \partial x_j$, $j = 1, \ldots, n$, exist for all $x \in D$ and let $\partial f / \partial x_j$ be continuous on D. Then f is differentiable on D.

Two additional points merit our attention. We first have the

Theorem 3.5: Mean Value Theorem. If $f(x)$: $D \subset R^n \to R$, D open, is differentiable on the line segment connecting x_o and $x_o + h$, then $f(x_o + h) = f(x_o) + \nabla f(\eta)^T h$ for some $\eta = x_o + t\,h$, $0 < t < 1$.

Next, suppose $f(x)$: $D \subset R^n \to R$, with D open. If f is continuous on D, then f is said to be a **function of class $C^{(0)}$**. If the partial derivatives $f_j(x)$, $j = 1, \ldots, n$, exist for every $x \in D$ and are continuous on D, then f is a **function of class $C^{(1)}$**. And if all of the partial derivatives $f_{ij}(x)$, $i, j = 1, \ldots, n$, exist for every $x \in D$ and each f_{ij} is continuous on D, then f is termed a **function of class $C^{(2)}$**. In general, if all of the qth order derivatives of f exist for every $x \in D$ and each $f_{j_1, \ldots, j_q}$ is continuous on D, then f is a **function of class$C^{(q)}$**. (A specialization of this definition is: a function $f(x)$: $D \subset R \to R$ is said to be of class $C^{(q)}$ on D

if $f^{(q)}(x)$ exists and is continuous for all $x \in D$.) In the light of this discussion, the *sufficient condition for differentiability* given above can be restated as: if $f(x)$: $D \subset R^n \to R$ is a function of class $C^{(1)}$, then f is a differentiable function. A result which will be utilized quite frequently in our optimization analysis is

Theorem 3.6: Young's Theorem. If $f(x)$: $D \subset R^n \to R$ is of class $C^{(1)}$ and both f_{ij} and f_{ji} are continuous for $i \neq j$ then $f_{ij} = f_{ji}$.

Let f be differentiable at $x_o \in D$. Under what circumstances will f be twice – differentiable there? To answer this question, suppose $f(x)$: $D \subset R^n \to R$, D open, is differentiable on D.

Then f is twice-differentiable at $x_o \in D$ if there exists a linear transformation $b \in \mathcal{L}(R^n, \mathcal{L}(R^n, n))$, such that

$$\lim_{h \to O} \frac{f'(x_o + h) - f'(x_o) - b\,h}{\|h\|} = O.$$

Here, b is termed the **second derivative of f at x_o** and denoted $f''(x_o)$, i.e., $f''(x_o)$ is an nth-order matrix of second-order partial and cross partial derivatives of f at x_o or

$$f''(x_o) = \begin{bmatrix} f_{11}(x_o) & f_{21}(x_o) & \cdots & f_{n1}(x_o) \\ f_{12}(x_o) & f_{22}(x_o) & \cdots & f_{n2}(x_o) \\ \vdots & \vdots & & \vdots \\ f_{1n}(x_o) & f_{2n}(x_o) & \cdots & f_{nn}(x_o) \end{bmatrix}.$$

Moreover, $f''(x_o)^T$ is called the **Hessian matrix of f at** x_o and denoted

$$H_f(x_o) = \begin{bmatrix} f_{11}(x_o) & f_{12}(x_o) & \cdots & f_{1n}(x_o) \\ f_{21}(x_o) & f_{22}(x_o) & \cdots & f_{2n}(x_o) \\ \vdots & \vdots & & \vdots \\ f_{n1}(x_o) & f_{n2}(x_o) & \cdots & f_{nn}(x_o) \end{bmatrix}. \quad (3.10)$$

If f is twice-differentiable for all $x \in D$, then f is said to be twice-differentiable on D.

Let us examine the Hessian of f in a slightly different light. Let the real-valued functions $f^i(x)$, $x \in R^n$, $i = 1, \cdots, m$, be defined throughout an open region $D \subset R^n$. Then **Jacobian matrix** of the m-component vector-valued function $f^T = (f^1, \ldots f^m)$ is represented, at those points x where the partial derivatives of f^i exist, as the $(m \times n)$ matrix

$$J_f(x) = \frac{\partial(f^1, \ldots, f^m)}{\partial(x_1, \ldots, x_n)} = \begin{bmatrix} f_1^1 & f_2^1 & \cdots & f_n^1 \\ f_1^2 & f_2^2 & \cdots & f_n^2 \\ \vdots & \vdots & & \vdots \\ f_1^m & f_2^m & \cdots & f_n^m \end{bmatrix}. \quad (3.11)$$

If we calculate the $(n \times m)$ matrix $\nabla f = \partial f / \partial x$ or

$$\frac{\partial f}{\partial x} = [\nabla f^1, \ldots, \nabla f^m] = \begin{bmatrix} f_1^1 & f_1^2 & \cdots & f_1^m \\ f_2^1 & f_2^2 & \cdots & f_2^m \\ \vdots & \vdots & & \vdots \\ f_n^1 & f_n^2 & \cdots & f_n^m \end{bmatrix},$$

then the Jacobian matrix appears as $J_f(x) = [\partial f / \partial x]^T$. Now, let the real-valued function $y = f(x)$, $x \in R^n$, be defined over an open region $D \subset R^n$. Then, the nth-order Hessian matrix of f is represented, at those points x where f is twice differentiable, as the *nth order Jacobian matrix of the first partial derivative of f* or

$$H_f(x) = \frac{\partial(f_1, \ldots, f_n)}{\partial(x_1, \ldots, x_n)} = \begin{bmatrix} f_{11} & f_{12} & \cdots & f_{1n} \\ f_{21} & f_{22} & \cdots & f_{2n} \\ \vdots & \vdots & & \vdots \\ f_{n1} & f_{n2} & \cdots & f_{nn} \end{bmatrix}. \quad (3.10.1)$$

Clearly, $H_f(x) = H_f(x)^T$ since $f_{ij} = f_{ji}$, $i \neq j$. And since $\nabla f = \partial f / \partial x$, we may alternatively view the Hessian of f as

$$\nabla^2 f = \partial \nabla f / \partial x = [\nabla f_1, \ldots, \nabla f_n] = H_f$$

by symmetry.

One additional point merits our attention. Let $y = f(u, v)$, $u^T = (u_1, \ldots, u_n) \in R^n$, $v^T = (v_1, \ldots, v_m) \in R^m$ be a real-valued function whose first partial derivatives exist over an open region $D \subset R^{n+m}$. The partial gradient of f is a vector-valued function whose elements are the first partial derivatives of f with respect to some subset of variables. The **partial gradients of f** with respect to u, v are, respectively, the $(n \times 1)$ and $(m \times 1)$ vectors

$$\nabla_{u} f = \partial f / \partial \boldsymbol{u} = \begin{bmatrix} f_{u_1} \\ \vdots \\ f_{u_n} \end{bmatrix}, \quad \nabla_{v} f = \partial f / \partial \boldsymbol{v} = \begin{bmatrix} f_{v_1} \\ \vdots \\ f_{v_m} \end{bmatrix}.$$

Additionally, at those points $\boldsymbol{u}$, $\boldsymbol{v}$ where f is twice-differentiable:

$$\nabla^2_{\boldsymbol{uu}} f = \left[\nabla_{\boldsymbol{u}} f_{u_1}, \nabla_{\boldsymbol{u}} f_{u_2} \right], \quad \nabla^2_{\boldsymbol{vv}} f = \left[\nabla_{\boldsymbol{v}} f_{v_1}, \nabla_{\boldsymbol{v}} f_{v_2} \right],$$

$$\nabla^2_{\boldsymbol{uv}} f = \left[\nabla_{\boldsymbol{v}} f_{u_1}, \nabla_{\boldsymbol{v}} f_{u_2} \right] = \nabla^2_{\boldsymbol{vu}} f.$$

For instance, if $f = 2u_1^3 v_1^2 + 3u_2^2 v_2^2$, then

$$\nabla^2_{\boldsymbol{uu}} f = \begin{bmatrix} 12u_1 v_1^2 & 0 \\ 0 & 6v_2^2 \end{bmatrix}, \quad \nabla^2_{\boldsymbol{vv}} f = \begin{bmatrix} 4u_1^3 & 0 \\ 0 & 6u_2^2 \end{bmatrix},$$

$$\nabla^2_{\boldsymbol{uv}} f = \begin{bmatrix} 12u_1^2 v_2 & 0 \\ 0 & 12u_2 v_2 \end{bmatrix}.$$

Section 3.3 Exercises:

3.3.1 Find $f'(x_o)$ using equation (3.4) when $f(x) = 3x^2 + 6x - 10$ and $x_o = 2$, $x \in R$.

3.3.2 Verify that if the function $y = f(x)$, $x \in R$, is differentiable at a point x_o, then it is also continuous there.

3.3.3 Let $f(x) = x^2$, $x \in R$. Is the average of a and b ($a < b$) in equation (3.5) an appropriate value for x_o?

3.3.4 Given $y = f(\boldsymbol{x}) = 3x_1^2 + x_2^3 - 4x_3 + 2x_1x_2x_3$, $\boldsymbol{x} \in R^3$, find (a) $\nabla f(\boldsymbol{x}_1)$; (b) $D_{\boldsymbol{u}} f(\boldsymbol{x}_1)$ in the direction of $\boldsymbol{x}_1^T = (2, 1, 0)$; (c) $\boldsymbol{H}_f(\boldsymbol{x}_o)$; and (d) the maximum value of $D_{\boldsymbol{u}} f(\boldsymbol{x}_o)$, where $\boldsymbol{x}_o^T = (1, 1, 1)$.

3.3.5 Find the Jacobian matrix of $\boldsymbol{f}^T = (f^1, f^2, f^3)$ when

$$f^1(x_1, x_2) = x_1^2 + x_2^2$$
$$f^2(x_1, x_2) = x_1 + 3x_2^2$$
$$f^3(x_1, x_2) = 2x_1 + x_1x_2 + 4x_2.$$

3.3.6 Given the real-valued function $y = f(\boldsymbol{u}, \boldsymbol{v}) = 3u_1 v_1^2 v_2 + 4u_2 v_1^{-1} + u_1^2 u_2 v_1$, find $\nabla_{\boldsymbol{u}} f$ and $\nabla_{\boldsymbol{v}} f$. Also, determine $\nabla^2_{\boldsymbol{uu}} f$, $\nabla^2_{\boldsymbol{vv}} f$, and $\nabla^2_{\boldsymbol{uv}} f$.

3.3.7 Let S be the surface $y = f(x_1, x_2)$ and let (x_1^0, x_2^0, y^0) be a point on S. (Obviously $y^o = f(x_1^0, x_2^0)$.) Suppose there is a plane tangent to S at (x_1^0, x_2^0, y^0) not parallel to the y-axis. It can be demonstrated that the equation of the **plane tangent to S** at (x_1^0, x_2^0, y^0) appears as

$$y - y^o = A(x_1 - x_1^o) + B(x_2 - x_2^o) \text{ or}$$
$$f = f^o + A(x_1 - x_1^o) + B(x_2 - x_2^o).$$

How should A, B be chosen? Let $f(x_1, x_1) = 3x_1^2 + x_2^2$, with $\boldsymbol{x}_o^T = (1, 1)$. What is the equation of the plane tangent to f at $\boldsymbol{x}_o$? In addition, the vector normal to S at $(\boldsymbol{x}_o y_o)$ has the form $\boldsymbol{N}_{\boldsymbol{x}}^T = (A, B, -1)$. Find $\boldsymbol{N}_{\boldsymbol{x}_o}$.

3.4 DERIVATIVES OF VECTOR-VALUED FUNCTIONS

Consider the mapping $y = \boldsymbol{f}(x)$: $R \to R^m$, where $\boldsymbol{f}(x)^T = (f^1(x), \ldots, f^m(x))$. The $(m \times 1)$ **derivative vector of $\boldsymbol{f}$ at x_o**, $\boldsymbol{f}'(x_o)$, will be defined as

$$\boldsymbol{f}'(x_o) = \lim_{h \to 0} \frac{\boldsymbol{f}(x_o + h) - \boldsymbol{f}(x_o)}{h}, \tag{3.12}$$

provided that this limit exists. Since limits in R^m are taken component-wise, we will say that **$\boldsymbol{f}$ is differentiable at x_o** if and only if each of the coordinate functions f^i, $i = 1, \ldots, m$, is differentiable at x_o so that

$$\boldsymbol{f}'(x_o) = \begin{bmatrix} df^1(x_o)/dx \\ \vdots \\ df^m(x_o)/dx \end{bmatrix}. \tag{3.13}$$

To see how (3.12) is rationalized, let us write the *actual change in* $\boldsymbol{f}$ from x_o to $x_o + h$ as $\Delta\boldsymbol{f}(h) = \boldsymbol{f}(x_o + h) - \boldsymbol{f}(x_o)$, and the *linear change in* $\boldsymbol{f}$ as $d\boldsymbol{f}(h) = \boldsymbol{f}'(x_o)h$, $h = x - x_o$. Remember that $d\boldsymbol{f}(h)$ is the change in $\boldsymbol{f}$ along the tangent line to $\boldsymbol{f}$at $(x_o, \boldsymbol{f}(x_o))$, $\boldsymbol{f}(x) = \boldsymbol{f}(x_o) + \boldsymbol{f}'(x_o)(x - x_o)$. For h sufficiently small, $\Delta f(h) - df(h)$ goes to zero faster than h does so that

$$\lim_{h\to 0}\frac{\Delta f(h)-df(h)}{h}=\lim_{h\to 0}\frac{f(x_o+h)-f(x_o)-f'(x_o)h}{h}$$
$$=\lim_{h\to 0}\frac{f(x_o+h)-f(x_o)}{h}-f'(x_o)=O.$$

In terms of linear transformations, the function $f: D\subset R\to R^m$ is **differentiable at $x_o\in D$** if and only if there exists a linear transformation $df\in\mathcal{L}(R, R^m)$, defined by $df(h)=ch$, such that

$$\lim_{h\to 0}\frac{f(x_o+h)-f(x_o)-ch}{h}=0.$$

So if $f(x): R\to R^m$ is differentiable at x_o, then the linear transformation $df\in\mathcal{L}(R, R^m)$, with $df(h)=ch$, is termed the **differential of f at x_0**. The derivative vector c, the matrix of the linear transformation $df(h)$, is the $(m\times 1)$ vector $f'(x_o)$ so that

$$df(h)=f'(x_o)h=\begin{bmatrix} df^1(x_o)/dx \\ \vdots \\ df^m(x_o)/dx\end{bmatrix}h. \qquad (3.14)$$

Let us now look to the general case involving the transformation function $f(x): D\subset R^n\to R^m$, with D an open subset of R^n. At any point $x\in D$, $f(x)^T=(f^1(x),\ldots,f^m(x))\in R^m$, where $f^i(x)$, $i=1,\ldots,m$, is a real-valued function on D.

To obtain the derivative of f at a point $x_o\in D$, let us express the *actual change in* f from x_o to x_o+h as $\Delta f(h)=f(x_o+h)-f(x_o)$ and the *linear (approximate) change in* f or the **differential of f at x_o** as $df(h)=f'(x_o)h$, $h=x-x_o$. Here, $f'(x_o)$, the **derivative of f at x_o**, is the unique $(m\times n)$ matrix associated with the linear transformation $df\in\mathcal{L}(R^n, R^m)$ or

$$df(h)=f'(x_o)h$$
$$=\left[\frac{\partial(f^1,\ldots,f^m)}{\partial(x_1,\ldots,x_n)}\right]_{x_o}h$$
$$=\begin{bmatrix}\frac{\partial f^1(x_0)}{\partial x_1} & \cdots & \frac{\partial f^1(x_0)}{\partial x_n}\\ \vdots & & \vdots\\ \frac{\partial f^m(x_0)}{\partial x_1} & \cdots & \frac{\partial f^m(x_0)}{\partial x_n}\end{bmatrix}h. \qquad (3.15)$$

The matrix $f'(x_o)$ is the Jacobian of f at x_o. Let us rationalize this result. For $\|h\|$ sufficiently small, $\Delta f(h)-df(h)$ goes to zero faster than h does as $h\to O$ or

$$\lim_{h\to O}\frac{\Delta f(h)-df(h)}{\|h\|}=\lim_{h\to O}\frac{f(x_o+h)-f(x_o)-f'(x_o)h}{\|h\|}=O.$$

Hence, the transformation function $f: D\subset R^n\to R^m$, D open, is **differentiable at $x_o\in D$** if and only if there exists a linear transformation $df\in\mathcal{L}(R^n, R^m)$, defined by $df=dh$, such that

$$\lim_{h\to O}\frac{f(x_o+h)-f(x_o)-dh}{\|h\|}=O, \qquad (3.16)$$

where $d=f'(x_o)$ is the $(m\times n)$ Jacobian matrix given in (3.15).

Section 3.4 Exercises:

3.4.1 Let $f(x)^T=(f^1(x), f^2(x), f^3(x)): R\to R^3$, with

$$f^1(x)=6x^2-x^{-1}$$
$$f^2(x)=x+6$$
$$f^3(x)=2x^3.$$

Find the linear transformation $df(h)$ at $x_o=2$.

3.4.2 Let $f(x)^T=(f^1(x), f^2(x)): D\subset R^3\to R^2$, with

$$f^1(x)=x_1^2x_2+x_1x_3-x_2x_3^2$$
$$f^2(x)=x_1x_2x_3.$$

Find the linear transformations df at $x_o^T=(1, 3, 1)$

3.5 DERIVATIVES OF QUADRATIC FUNCTIONS

Let us express a general quadratic function in x as

$$f(x)=a+\sum_{i=1}^{n}b_ix_i+\sum_{i=i}^{n}\sum_{j=1}^{n}q_{ij}x_ix_j$$
$$=a+b^Tx+x^TQx,$$

where b is of order $(n\times 1)$ and Q is an nth-order symmetric matrix. How do we determine the first and second derivatives of f with respect to x? Working first with the term b^Tx it is easily seen that since

$$\frac{\partial(\boldsymbol{b}^T\boldsymbol{x})}{\partial x_k} = b_k$$

is the kth component of $\boldsymbol{b}$,

$$\nabla(\boldsymbol{b}^T\boldsymbol{x}) = \frac{\partial(\boldsymbol{b}^T\boldsymbol{x})}{\partial \boldsymbol{x}} = \boldsymbol{b}.$$

If we next consider the term $\boldsymbol{x}^T\boldsymbol{Q}\boldsymbol{x}$,

$$\begin{aligned}\frac{\partial(\boldsymbol{x}^T\boldsymbol{Q}\boldsymbol{x})}{\partial x_k} &= 2q_{kk}x_k + \sum_{j\neq k} q_{kj}x_j + \sum_{i\neq k} q_{ik}x_i \\ &= \sum_{j=1}^{n} q_{kj}x_j + \sum_{i=1}^{n} q_{ik}x_i \\ &= \boldsymbol{q}_k\boldsymbol{x} + \boldsymbol{q}_k^T\boldsymbol{x},\end{aligned}$$

where $\boldsymbol{q}_k$ is the kth row of $\boldsymbol{Q}$. Since $\partial(\boldsymbol{x}^T\boldsymbol{Q}\boldsymbol{x})/\partial x_k$ is the kth component of $\boldsymbol{Q}\boldsymbol{x} + \boldsymbol{Q}^T\boldsymbol{x}$, it follows that, with $\boldsymbol{Q}$ symmetric,

$$\nabla(\boldsymbol{x}^T\boldsymbol{Q}\boldsymbol{x}) = \frac{\partial(\boldsymbol{x}^T\boldsymbol{Q}\boldsymbol{x})}{\partial \boldsymbol{x}} = \boldsymbol{Q}\boldsymbol{x} + \boldsymbol{Q}^T\boldsymbol{x} = 2\boldsymbol{Q}\boldsymbol{x}.$$

In the light of these results, the first and second derivatives of f with respect to $\boldsymbol{x}$ appear, respectively, as

$$\begin{aligned}\nabla f(\boldsymbol{x}) &= \frac{\partial f(\boldsymbol{x})}{\partial \boldsymbol{x}} = b + 2\boldsymbol{Q}\boldsymbol{x}, \\ \nabla^2 f(\boldsymbol{x}) &= \frac{\partial \nabla f(\boldsymbol{x})}{\partial \boldsymbol{x}} = 2\boldsymbol{Q}.\end{aligned}$$

Section 3.5 Exercises:

3.5.1 Let $\boldsymbol{x}^T\boldsymbol{Q}\boldsymbol{x} = x_1^2 + x_2^2 + x_3^2 + x_1x_2 + 2x_1x_3 + 4x_2x_3$. Find $\nabla(\boldsymbol{x}^T\boldsymbol{Q}\boldsymbol{x})$.

3.5.2 Suppose $f(\boldsymbol{x}) = 6 + 4x_1 + 2x_2 - 6x_3 + x_1^2 + x_2^2 + x_3^2 + 4x_1x_2 - 2x_1x_3 + 4x_2x_3$. Find $\nabla f(\boldsymbol{x})$ and $\nabla^2 f(\boldsymbol{x})$.

3.6 TAYLOR'S FORMULA

3.6.1 A Single Independent Variable

We begin with

Theorem 3.7: Taylor's Theorem. Let the real-valued function $y = f(x)$ and its first $n + 1$ $(n \geq 0)$ derivatives be continuous throughout a closed interval $[a, b]$ containing the point x_o. Then the value of f at any point x *near* x_o is

$$\begin{aligned}f(x) &= f(x_o) + f'(x_o)(x - x_o) + \tfrac{1}{2!}f''(x_o)(x - x_o)^2 \\ &\quad + \cdots + \tfrac{1}{n!}f^{(n)}(x_o)(x - x_o)^n + R_{n+1},\end{aligned} \tag{3.17}$$

where the remainder term R_{n+1} has the form

$$R_{n+1} = \frac{1}{n!}\int_{x_o}^{x} (x - t)^n f^{(n+1)}(t)\,dt. \tag{3.18}$$

Equation (3.17) is known as **Taylor's formula with integral remainder**. While the remainder term can assume various forms, we shall employ **Lagrange's form of the remainder**, or

$$R_{n+1} = \frac{1}{(n+1)!}f^{(n+1)}(\xi)(x - x_o)^{(n+1)},\ x_o < \xi < x. \tag{3.19}$$

Then[2] combining (3.17) and (3.19) renders **Taylor's formula with Lagrange's form of the remainder**, or

$$\begin{aligned}f(x) &= f(x_o) + f'(x_o)(x - x_o) + \tfrac{1}{2!}f''(x_o)(x - x_o)^2 \\ &\quad + \cdots + \tfrac{1}{n!}f^{(n)}(x_o)(x - x_o)^n + \tfrac{1}{(n+1)!}f^{(n+1)}(\xi)(x - x_o)^{n+1},\ x_o < \xi < x.\end{aligned} \tag{3.20}$$

(Note: Taylor's formula with Lagrange's form of the remainder is an extension of the mean value theorem for derivatives, that is, (3.20) coincides with (3.5) if $n = 0$.) Suppose we set $x = x_o + h$. Then, $x_o < \xi < x_o + h$ can be expressed as $\xi = x_o + \theta h$, $0 < \theta < 1$, so that (3.20) becomes

$$\begin{aligned}f(x) &= f(x_o) + f'(x_o)(x - x_o) + \tfrac{1}{2!}f''(x_o)(x - x_o)^2 \\ &\quad + \cdots + \tfrac{1}{n!}f^{(n)}(x_o)(x - x_o)^n + \tfrac{1}{(n+1)!}f^{(n+1)}(x_o + \theta h)^{n+1},\ 0 < \theta < 1.\end{aligned} \tag{3.21}$$

3.6.2 Generalized Taylor's Formula with Remainder

Suppose $\boldsymbol{h} = \boldsymbol{x} - \boldsymbol{x}_o$ is an $(n \times 1)$ vector with components $h_i = x_i - x_i^0$, $i = 1, \ldots, n$. Then the **jth-order differential** of the real-valued function $y = f(\boldsymbol{x})$ at point $\boldsymbol{x}_o$ interior to a closed region $\mathcal{K} \subset R^n$ may be written as

$$d^j f(\boldsymbol{x}_o, \boldsymbol{h}) = \left(h_1\frac{\partial}{\partial x_1} + \cdots + h_n\frac{\partial}{\partial x_n}\right)_{\boldsymbol{x}_o}^{j} f,\ j = 0, 1, \ldots, n.$$

For instance,

$$d^0 f(\boldsymbol{x}_o, \boldsymbol{h}) = f(\boldsymbol{x}_o),$$
$$d^1 f(\boldsymbol{x}_o, \boldsymbol{h}) = \Sigma_{r=1}^{n} h_r \frac{\partial f(\boldsymbol{x}_o)}{\partial x_r},$$
$$d^2 f(\boldsymbol{x}_o, \boldsymbol{h}) = \Sigma_{r=1}^{n} \Sigma_{s=1}^{n} h_r h_s \frac{\partial^2 f(\boldsymbol{x}_o)}{\partial x_r x_s}, \ \ldots.$$

In this regard, we have

Theorem 3.8: Generalized Taylor's Formula with Remainder. Let the real-valued function $y = f(\boldsymbol{x}), \boldsymbol{x} \in R^n$, and its first $n + 1$ ($n \geq 0$) derivatives be continuous throughout a closed region $\mathcal{K}$ of R^n containing the point $\boldsymbol{x}_0$. Then the value of f at any point $\boldsymbol{x}$ *near* $\boldsymbol{x}_o$ is

$$f(\boldsymbol{x}) = f(\boldsymbol{x}_o) + df(\boldsymbol{x}_o, \boldsymbol{h}) + \frac{1}{2!} d^2 f(\boldsymbol{x}_o, \boldsymbol{h}) + \cdots + \frac{1}{n!} d^n f(\boldsymbol{x}_o, \boldsymbol{h}) + R_{n+1}, \tag{3.22}$$

where R_{n+1} is a remainder term of the form

$$R_{n+1} = \frac{1}{(n+1)!} d^{n+1} f(\boldsymbol{x}_o + \theta \boldsymbol{h}, \boldsymbol{h}),\ 0 < \theta < 1. \tag{3.23}$$

An important special case of (3.22) occurs when $\boldsymbol{x} \in R^2$, that is

$$f(x_1, x_2) = \sum_{j=0}^{n} \frac{1}{j!} \left(h_1 \frac{\partial}{\partial x_1} + h_2 \frac{\partial}{\partial x_2} \right)^j_{(x_1^0, x_2^0)} f + R_{n+1},\ 0 < \theta < 1.$$

Then for $n = 2$,

$$\begin{aligned} f(x_1, x_2) &= \Sigma_{j=0}^{2} \frac{1}{j!} \left(h_1 \frac{\partial}{\partial x_1} + h_2 \frac{\partial}{\partial x_2} \right)^j_{(x_1^0, x_2^0)} f \\ &+ \frac{1}{3!} \left(h_1 \frac{\partial}{\partial x_1} + h_2 \frac{\partial}{\partial x_2} \right)^3_{(x_1^0 + \theta h_1, x_2^0 + \theta h_2)} f \\ &= f(x_1^0, x_2^0) + f_1(x_1^0, x_2^0) h_1 + f_2(x_1^0, x_2^0) h_2 \\ &+ \frac{1}{2!} [f_{11}(x_1^0, x_2^0) h_1^2 + 2 f_{12}(x_1^0, x_2^0) h_1 h_2 \\ &+ f_{22}(x_1^0, x_2^0) h_2^2] \\ &+ \frac{1}{3!} \left(h_1 \frac{\partial}{\partial x_1} + h_2 \frac{\partial}{\partial x_2} \right)^3_{(x_1^0 + \theta h_1, x_2^0 + \theta h_2)} f,\ 0 < \theta < 1. \end{aligned} \tag{3.24}$$

Note: the differential operator

$$\left(h_1 \frac{\partial}{\partial x_1} + h_2 \frac{\partial}{\partial x_2} \right)^j$$

is applied to f via the binomial expansion

$$\left(h_1 \frac{\partial}{\partial x_1} + h_2 \frac{\partial}{\partial x_2} \right)^j f = \sum_{i=0}^{j} \binom{j}{i}^j h_1^i h_2^{j-i} \left(\frac{\partial^j f}{\partial x_1^i \partial x_2^{j-i}} \right).$$

Quite often we shall rely upon the second-order expansion

$$\begin{aligned} f(\boldsymbol{x}_o + \boldsymbol{h}) &= f(\boldsymbol{x}_o) + \nabla f(\boldsymbol{x}_o)^T \boldsymbol{h} + \frac{1}{2!} \boldsymbol{h}^T \frac{\partial \nabla f(\boldsymbol{x}_o + \theta \boldsymbol{h})}{\partial \boldsymbol{x}} \boldsymbol{h} \\ &= f(\boldsymbol{x}_o) + \nabla f(\boldsymbol{x}_o)^T \boldsymbol{h} + \frac{1}{2!} \boldsymbol{h}^T \boldsymbol{H}_f(\boldsymbol{x}_o + \theta \boldsymbol{h}) \boldsymbol{h},\ 0 < \theta < 1. \end{aligned} \tag{3.25}$$

Example 3.1: Using (3.20), determine the third-order Taylor expansion of $y = f(x) = e^{2x}$ about the point $x_o = 0$. Here

$$f'(x) = 2e^{2x},\ f''(x) = 4e^{2x},\ \text{and}\ f'''(x) = 8e^{2x}.$$

Also, $f(0) = 1$ with

$$f'(0) = 2,\ f''(0) = 4,\ \text{and}\ f'''(\xi) = 8e^{2\xi},\ 0 < \xi < x,$$

so that

$$\begin{aligned} f(x) &= 1 + 2x + \frac{1}{2!}(4x^2) + \frac{1}{3!} 8(4e^{2\xi} x^3) \\ &= 1 + 2x + 2x^2 + \frac{4}{3} e^{2\xi} x^3,\ 0 < \xi < x. \end{aligned}$$

If (3.21) is employed, the third-order Taylor expansion is

$$\begin{aligned} f(h) &= 1 + 2x + \frac{1}{2!}(4x^2) + \frac{1}{3!}(8e^{2(\theta h)} x^3) \\ &= 1 + 2x + 2x^2 + \frac{4}{3} e^{2\theta h},\ 0 < \theta < 1. \end{aligned}$$

■

Example 3.2: Using (3.24), compute the third-order Taylor expansion of $y = f(x_1, x_2) = e^{2x_1 + 3x_2}$ about the point $(x_1^0, x_2^0) = (0, 0)$. Here, we are interested in calculating

$$\begin{aligned} f(x_1, x_2) &= \Sigma_{j=0}^{2} \frac{1}{j!} \left(h_1 \frac{\partial}{\partial x_1} + h_2 \frac{\partial}{\partial x_2} \right)^j_{(0,0)} f \\ &+ \frac{1}{3!} \left(h_1 \frac{\partial}{\partial x_1} + h_2 \frac{\partial}{\partial x_2} \right)^3_{(\theta h_1, \theta h_2)} f,\ 0 < \theta < 1. \end{aligned}$$

The reader can readily determine that

$$\begin{aligned} f_1(x_1, x_2) &= 2f & f_{111}(x_1, x_2) &= 8f \\ f_2(x_1, x_2) &= 3f & f_{112}(x_1, x_2) &= 12f \\ f_{11}(x_1, x_2) &= 4f & f_{122}(x_1, x_2) &= 18f \\ f_{12}(x_1, x_2) &= 6f & f_{222}(x_1, x_2) &= 27f. \\ f_{22}(x_1, x_2) &= 9f \end{aligned}$$

Evaluating $f, f_1, f_2, f_{11}, f_{12},$ and f_{22} at (0,0) and $f_{111}, f_{112}, f_{122},$ and f_{222} at $(\theta h_1, \theta h_2)$ renders

$$\begin{aligned} f(0,0) &= 1 & f_{111}(\theta h_1, \theta h_2) &= 8e^{(2\theta h_1 + 3\theta h_2)} \\ f_1(0,0) &= 2 & f_{112}(\theta h_1, \theta h_2) &= 12e^{(2\theta h_1 + 3\theta h_2)} \\ f_2(0,0) &= 3 & f_{122}(\theta h_1, \theta h_2) &= 18e^{(2\theta h_1 + 3\theta h_2)} \\ f_{11}(0,0) &= 4 & f_{222}(\theta h_1, \theta h_2) &= 27e^{(2\theta h_1 + 3\theta h_2)}. \\ f_{12}(0,0) &= 6 \\ f_{22}(0,0) &= 9 \end{aligned}$$

Then, from (3.24)

$$f(x_1, x_2) = 1 + 2x_1 + 3x_2 + 2x_1^2 + 6x_1, x_2 + \frac{9}{2}x_2^2 + \left(\frac{4}{3}x_1^3 + 6x_1^2x_2 + 9x_1x_2^2 + \frac{9}{2}x_2^3\right)e^{(\theta(2h_1+3h_2))}, \ 0 < \theta < 1.$$

■

Section 3.6 Exercises:

3.6.1 Use Equation (3.20) to expand $y = f(x) = x^{-1}$, $x \neq 0$, $x \in R$, about the point $x = 1$ for $n = 3$.

3.6.2 For $y = f(\boldsymbol{x}) = 2x_1^2 + x_1x_2 - 3x_2^3$, $\boldsymbol{x} \in R^2$, determine $d^0f(\boldsymbol{x}_o, h)$, $d^1f(\boldsymbol{x}_o, h)$, and $d^2f(\boldsymbol{x}_o, h)$ for $\boldsymbol{x}_o^T = (1, 1)$.

3.6.3 Use (3.24) to determine the Taylor expansion of $y = f(\boldsymbol{x}) = 4x_1^4 + x_2^4$ near the point $\boldsymbol{x}_o^T = (2, 1)$.

NOTES

1 Actually, for any vector $\boldsymbol{x}_1 \in R^n$, $||\boldsymbol{x}_1|| \neq 0$, we may select $\boldsymbol{u}$ as $\boldsymbol{x}_1/||\boldsymbol{x}_1||$, where $\boldsymbol{u}$ is termed the **direction of** $\boldsymbol{x}_1$, i.e., $\boldsymbol{u}$ is a unit vector that points in the $\boldsymbol{x}_1$-direction.

2 This result is a consequence of the **Mean Value Theorem for Integrals**: let $g(t)$, $h(t)$ be continuous on $[\alpha, \beta]$ with $h(t) \geq 0$ for all $t \in [\alpha, \beta]$. Then, for some, $t = \xi$ such that $\alpha < \xi < \beta$, $\int_\alpha^\beta g(t)\,h(t)dt = g(\xi)\int_\alpha^\beta h(t)\,dt$.

Mathematical Foundations 4

4.1 IMPLICIT FUNCTION THEOREMS

Implicit function theorems are theorems that specify conditions that are sufficient for an equation, or a system of equations, to define certain variables as functions of the remaining variables. In this regard, given $F(x_1, x_2) = 0$, under what conditions does F give x_2 as a function f of x_1 (or x_1 as a function of x_2)? That is, can we eliminate x_2 so that the resulting expression involves x_1 alone? The answer is provided by the following existence theorem.

Theorem 4.1: Implicit Function Theorem. Let the real-valued function $F(x_1, x_2) = 0$ be continuously differentiable over an open region $X \subset R^2$ with $(x_1^0, x_2^0) \in X$. Let $F(x_1^0, x_2^0) = 0$ with $F_2(x_1^0, x_2^0) \neq 0$. Then, there exist neighborhoods $\delta(x_1^0)$ and $\varepsilon(x_2^0)$, with $\delta(x_1^0) \times \varepsilon(x_2^0) \subset X$, and a unique continuously differentiable function $f: \delta(x_1^0) \rightarrow \varepsilon(x_2^0)$, such that:

(a) $x_2 = f(x_1)$ for $x_1 \in \delta(x_1^0)$ (the expression holds locally);
(b) $x_2^0 = f(x_1^0)$; and
(c) $F(x_1, f(x_1)) \equiv 0$ for all $x_1 \in \delta(x_1^0)$ (Figure 4.1). (4.1)

Furthermore,

$$\frac{\partial F}{\partial x_1} + \frac{\partial F}{\partial x_2}\frac{df}{dx_1} = 0$$

and thus

$$\frac{df}{dx_1} = -\frac{\partial F/\partial x_1}{\partial F/\partial x_2} = -\frac{F_1}{F_2}, \; F_2 \neq 0 \tag{4.2}$$

(here, df/dx_1 represent the slope of a contour of F); and

$$\frac{d^2f}{dx_1^2} = -\left[\frac{\partial}{\partial x_1}\left(\frac{F_1}{F_2}\right) + \frac{\partial}{\partial x_2}\left(\frac{F_1}{F_2}\right)\frac{dx_2}{dx_1}\right]$$

or

$$\frac{d^2f}{dx_1^2} = -\frac{F_2^2F_{11} - 2F_1F_2F_{12} + F_1^2F_{22}}{F_2^3}, \; F_2 \neq 0. \tag{4.3}$$

Given $F(x_1, x_2) = 0$, Theorem 4.1 provides the conditions under which this equation *implicitly* defines a function mapping x_1 onto x_2, i.e., $F(x_1, x_2) = 0$ implicitly defines (locally) x_2 in terms of x_1 so that we can, in principle, solve for $x_2 = f(x_1)$. Moreover, Equation (4.2) enables us to obtain the derivative of f. In fact, we do not have to actually solve for f – we simply take $df/dx_1 = -F_1/F_2$, $F_2 \neq 0$.

For instance, suppose $U = u(x_1, x_2)$ is a utility function with x_1, x_2 serving as the quantities of two commodities. Setting $u(x_1, x_2) = \bar{U} =$ constant specifies an indifference curve – defined as the locus of commodity combinations (x_1, x_2) which yield the same level of utility $\bar{U}$ (Figure 4.2). Thus there exists a mapping from x_1 to x_2 *induced* by the indifference curve; it indicates, for each x_1, the x_2 required to make $u(x_1, x_2)$

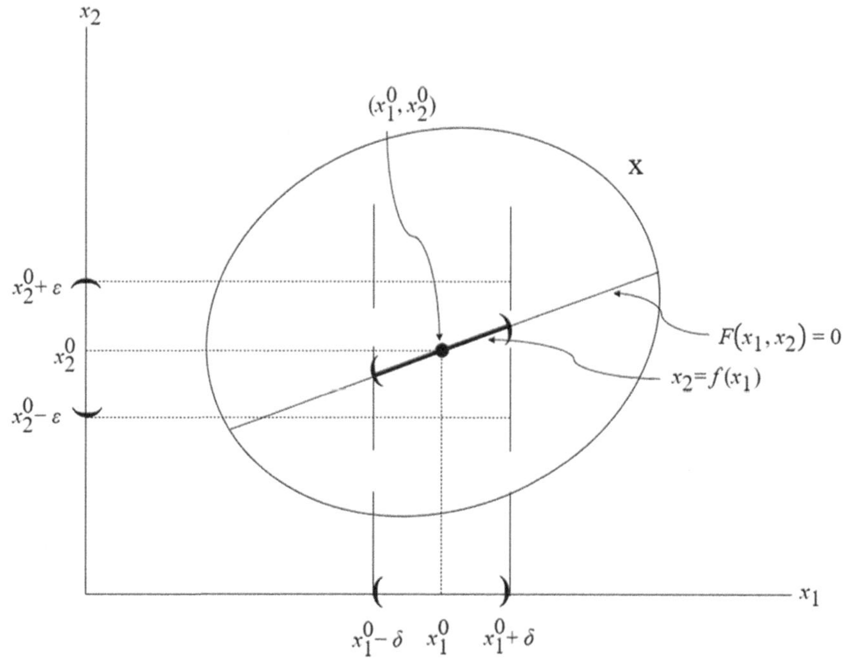

FIGURE 4.1 $F(x_1, x_2) = 0$ implicitly defines x_2 in terms of x_1, $x_2 = f(x_1)$.

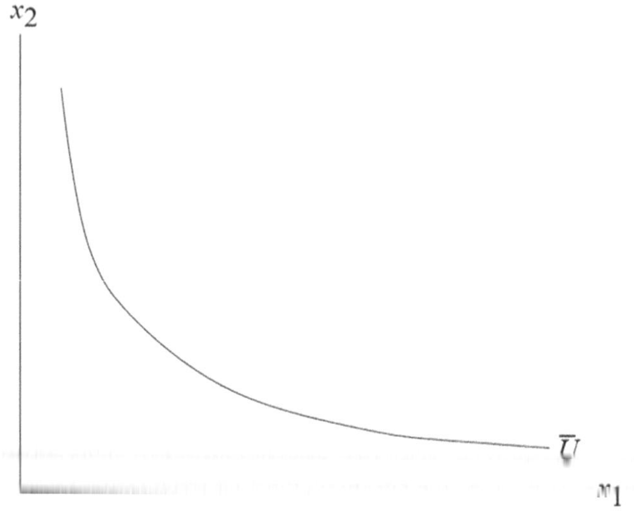

FIGURE 4.2 Indifference curve $\overline{U}$.

equal to $\overline{U}$, i.e., $u(x_1, x_2) = \overline{U}$ implicitly defines a function mapping x_1 onto x_2. Let $u(x_1, x_2) - U = F(x_1, x_2) = 0$. Then the slope of the indifference curve is obtained from (4.2) as

$$\frac{dx_2}{dx_1} = -F_1/F_2 = -u_1/u_2 = -MU_1/MU_2,\ MU_2 \neq 0,$$

where MU_1 (resp., MU_2) is the marginal utility of commodity 1 (resp., commodity 2). Then

$$MU_1/MU_2 = -dx_2/dx_1 = MRS_{1,2},$$

the marginal rate of substitution between commodities 1 and 2 or the rate at which x_1 must be substituted for x_2 in order to keep utility constant at the level $\overline{U}$. Looking again to Figure 4.2 we see that $dMRS_{1,2}/dx_1 < 0$, i.e., we have a diminishing marginal rate of substitution between commodities x_1 and x_2 -- the indifference curve $\overline{U}$ is convex to the origin (more on the convexity notion later on).

Example 4.1: Let $F(x_1, x_2) = x_1^2 + x_2^2 - 10 = 0$, with real $x_1, x_2 > 0$. Let us find dx_2/dx_1. From (4.2),

$$\frac{dx_2}{dx_1} = -\frac{F_1}{F_2} = -\frac{2x_1}{2x_2} = -x_1/x_2,\ x_2 \neq 0.$$

As a check, find $x_2^2 = 10 - x_1^2$ from which we obtain $x_2 = \pm(10 - x_1^2)^{1/2}$. With $x_2 > 0$, $x_2 = (10 - x_1^2)^{1/2} = f(x_1)$, $x_1^2 < 10$. Then

$$f' = \frac{dx_2}{dx_1} = \frac{1}{2}(10 - x_1^2)^{-1/2}(-2x_1) = -x_1/x_2,\ x_2 \neq 0,$$

as expected. ∎

A slight generalization of the preceding existence theorem is provided by Theorem 4.2 which follows. Given $F(\boldsymbol{x}) = 0$, $\boldsymbol{x} \in R^n$, under what conditions does F implicitly define x_n as a function f of $x_1, \ldots, x_{n-1}$? That is, can we eliminate x_n so that the resulting expression involves $x_1, \ldots, x_{n-1}$ only? Let

$$\boldsymbol{x} = \begin{bmatrix} x_1 \\ \vdots \\ x_n \end{bmatrix}, \hat{\boldsymbol{x}} = \begin{bmatrix} x_1 \\ \vdots \\ x_{n-1} \end{bmatrix} \text{ so that } \boldsymbol{x} = \begin{bmatrix} \hat{\boldsymbol{x}} \\ x_n \end{bmatrix}.$$

We then have

Theorem 4.2: Implicit Function Theorem. Let the real-valued function $F(\boldsymbol{x}) = 0$ be continuously differentiable over an open region $X \subset R^n$ with $\boldsymbol{x}_o \in X$. Let $F(\boldsymbol{x}_o) = 0$ with $F_n(\boldsymbol{x}_o) \neq 0$. Then, for $\hat{\boldsymbol{x}}^T = (x_1, \ldots, x_{n-1})$, there exist neighborhoods $\delta(\hat{\boldsymbol{x}}_o)$ and $\varepsilon(x_n^o)$, with $\delta(\hat{\boldsymbol{x}}_o) \times \varepsilon(x_n^o) \subset X$, and a unique continuously differentiable function $f\colon \delta(\hat{\boldsymbol{x}}_o) \to \varepsilon(x_n^o)$, such that:

(a) $x_n = f(\hat{x})$ for $\hat{x} \in \delta(\hat{x}_o)$;

(b) $x_n^o = f(\hat{x}_o)$; and

(c) $F(\hat{x}, f(\hat{x})) \equiv 0$ for all $\hat{x} \in \delta(\hat{x}_o)$. (4.4)

Furthermore,

$$\frac{\partial F}{\partial x_i} + \frac{\partial F}{\partial x_n}\frac{\partial f}{\partial x_i} = 0$$

and thus

$$\frac{\partial f}{\partial x_i} = -\frac{\partial F/\partial x_i}{\partial F/\partial x_n} = -\frac{F_i}{F_n}, \quad F_n \neq 0,\ i = 1, \ldots, n-1 \tag{4.5}$$

(here $\partial f/\partial x_i$ represents the slope of a contour of F in x_i, x_n – space); and

$$\frac{\partial^2 f}{\partial x_i^2} = -\frac{F_n^2 F_{ii} - 2F_i F_n F_{in} + F_i^2 F_{nn}}{F_n^3}, \quad F_n \neq 0,\ i = 1, \ldots, n-1. \tag{4.6}$$

Example 4.2: Let $F(x_1, x_2, x_3) = x_1^2 e^{2x_2} + x_3^3 e^{x_1} - 100 = 0$, with real $x_1, x_2, x_3 > 0$. Suppose $F_3 \neq 0$ so that $x_3 = f(x_1, x_2)$ exists. Let us determine $\partial f/\partial x_1$ and $\partial f/\partial x_2$. From (4.5) we have

$$\frac{\partial f}{\partial x_1} = -\frac{F_1}{F_3} = -\frac{2x_1 e^{2x_2} + x_3^3 e^{x_1}}{3x_3^2 e^{x_1}};$$

$$\frac{\partial f}{\partial x_2} = -\frac{F_2}{F_3} = -\frac{2x_1^2 e^{2x_2}}{3x_3^2 e^{x_1}}.$$ ■

An even broader generalization of Theorem 4.1 is provided by Theorem 4.3 stated below. To set the stage for this theorem, let $f^j(\boldsymbol{x}, \boldsymbol{\alpha}) = 0$, $j = 1, \ldots, m$. Under what conditions can we solve these equations for $\boldsymbol{x}$ in terms of $\boldsymbol{a}$? That is, can we eliminate the m variables x_j, $j = 1, \ldots, m$, by expressing each as a function ϕ^j of the s variables α_l, $l = 1, \ldots, s$. Let

$$\boldsymbol{F} = \begin{bmatrix} f^1 \\ \vdots \\ f^m \end{bmatrix}, \boldsymbol{x} = \begin{bmatrix} x_1 \\ \vdots \\ x_m \end{bmatrix}, \boldsymbol{\alpha} = \begin{bmatrix} \alpha_1 \\ \vdots \\ \alpha_s \end{bmatrix}, \text{ and } \boldsymbol{\phi} = \begin{bmatrix} \phi^1 \\ \vdots \\ \phi^m \end{bmatrix}.$$

We then have

Theorem 4.3: Implicit Function Theorem. Let the real-valued functions $f^j(\boldsymbol{x}, \boldsymbol{\alpha}) = 0$, $j = 1, \ldots, m$, be continuously differentiable over an open region $X \subset R^{m+s}$ with $(\boldsymbol{x}_o, \boldsymbol{\alpha}_o) \in X$. Let $f(\boldsymbol{x}_o, \boldsymbol{\alpha}_o) = 0$ with

$$|J_F(x_o)| \quad \det\left[\frac{\partial(f^1, \ldots, f_m)}{\partial(x_1, \ldots, x_n)}\right]_{(x_o, \boldsymbol{\alpha}_o)} \neq 0.$$

(the $m \times n$ Jacobian determinant of $\boldsymbol{F}$ at $(\boldsymbol{x}_o, \boldsymbol{\alpha}_o)$ has rank m).

Then there exist neighborhoods $\varepsilon(\boldsymbol{x}_o)$ and $\delta(\boldsymbol{\alpha}_o)$, with $\varepsilon(\boldsymbol{x}_o) \times \delta(\boldsymbol{\alpha}_o) \subset X$, and a unique continuously differentiable vector-valued function $\boldsymbol{\phi}\colon \delta(\boldsymbol{\alpha}_o) \to \varepsilon(\boldsymbol{x}_o)$ such that:

(a) $\boldsymbol{x} = \boldsymbol{\phi}(\boldsymbol{\alpha})$ for $\boldsymbol{\alpha} \in \delta(\boldsymbol{\alpha}_o)$;

(b) $\boldsymbol{x}_o = \boldsymbol{\phi}(\boldsymbol{\alpha}_o)$; and

(c) $\boldsymbol{F}(\boldsymbol{\phi}(\boldsymbol{\alpha}), \boldsymbol{\alpha}) \equiv \boldsymbol{O}$ for all $\boldsymbol{\alpha} \in \delta(\boldsymbol{\alpha}_o)$.

Furthermore,

$$\frac{\partial}{\partial \alpha_k} F(\phi(\alpha), \alpha) = \frac{\partial F}{\partial \phi}\frac{\partial \phi}{\partial \alpha_k} + \frac{\partial F}{\partial \alpha_k} = O$$

or

$$J_F(x)\frac{\partial x}{\partial \alpha_k} + \frac{\partial F}{\partial \alpha_k} = O$$

and thus

$$\frac{\partial x}{\partial \alpha_k} = -[J_F(x)]^{-1}\frac{\partial F}{\partial \alpha_k}, \tag{4.7}$$

where

$$\frac{\partial x}{\partial \alpha_k} = \begin{bmatrix} \partial \phi^1/\partial \alpha_k \\ \vdots \\ \partial \phi^m/\partial \alpha_k \end{bmatrix}, \quad \frac{\partial F}{\partial \alpha_k} = \begin{bmatrix} \partial f^1/\partial \alpha_k \\ \vdots \\ \partial f^m/\partial \alpha_k \end{bmatrix}.$$

Example 4.3: Suppose

$$F = \begin{bmatrix} f^1(x, \alpha) \\ f^2(x, \alpha) \end{bmatrix} = \begin{bmatrix} 2x_1 + x_2 + \alpha \\ x_1 x_2 - \alpha \end{bmatrix}$$

with $x_o^T = (1, 1)$. Then

$$|J_F(x)| = \left|\frac{\partial(f^1, f^2)}{\partial(x_1, x_2)}\right| = \begin{vmatrix} \frac{\partial f^1}{\partial x_1} & \frac{\partial f^1}{\partial x_2} \\ \frac{\partial f^2}{\partial x_1} & \frac{\partial f^2}{\partial x_2} \end{vmatrix}$$

$$= \begin{vmatrix} 2 & 1 \\ x_2 & x_1 \end{vmatrix} = 2x_1 - x_2$$

so that $|J_F(x_o)| = 1 \neq 0$. Then, from (4.7),

$$\frac{\partial x}{\partial \alpha} = -[J_F(x)]^{-1}\frac{\partial F}{\partial \alpha}$$

$$= -\begin{bmatrix} \frac{x_1}{2x_1 - x_2} & \frac{-1}{2x_1 - x_2} \\ \frac{-x_2}{2x_1 - x_2} & \frac{2}{2x_1 - x_2} \end{bmatrix}\begin{bmatrix} 1 \\ -1 \end{bmatrix} = \begin{bmatrix} -\frac{x_1 + 1}{2x_1 - x_2} \\ \frac{x_2 + 2}{2x_1 - x_2} \end{bmatrix}.$$

Section 4.1 Exercises:

4.1.1 Suppose $F(x_1 x_2) = a^2 x_1^2 + b^2 x_2^2 - a^2 b^2 = 0$. Find dx_2/dx_1 and d^2x_2/dx_1^2 using (4.2) and (4.3), respectively.

4.1.2 Let $F(x_1 x_2) = x_1^3 + x_2^3 - 3x_1 x_2 = 0$. Determine dx_2/dx_1 and d^2x_2/dx_1^2 via (4.2) and (4.3), respectively.

4.1.3 Let $F(x_1, x_2, x_3) = x_1 + x_2 - x_3 + x_1 x_2 x_3 = 0$. Without explicitly solving for $x_3 = f(x_1, x_2)$, find $\partial f/\partial x_1$, $\partial f/\partial x_2$, and $\partial^2 f/\partial x_1^2$.

4.1.4 Suppose $F(x_1, x_2, x_3) = x_1^2 - x_3^2 + (x_3 - x_2)(x_3 + x_2) = 0$. Find the slope of a contour of F in x_1, x_3 − space.

4.1.5 Let

$$F = \begin{bmatrix} f^1(x, \alpha) \\ f^2(x, \alpha) \end{bmatrix} = \begin{bmatrix} x_1 - x_2 + \alpha_1 \alpha_2 \\ x_1 x_2 - \alpha_1 + \alpha_2 \end{bmatrix}$$

with $x_o^T = (2, -1)$. Find $\partial x/\partial a_1$ and $\partial x/\partial a_2$.

4.2 CHAIN OR COMPOSITE FUNCTION RULES

A given function may often be considered as being "built up" by *composition* from a number of other functions. For instance, suppose

$$y = f(x_1, x_1) = x_1^2 + x_2^2$$
$$x_1 = g(t) = 3t^{-1}$$
$$x_2 = h(t) = 2t^3.$$

Then, $y = F(t) = f(g(t), h(t)) = 9t^{-2} + 4t^6$. Here, f is a **composite function** of t; it expresses y in terms of t **indirectly** through the intermediate variables x_1 and x_2. Our objective will now be to obtain, in a systematic fashion, derivatives of selected composite functions.

We start first

Theorem 4.4: Chain Rule. Let the real-valued function $y = f(x)$, $x \in R^n$, be continuously differentiable on a region $X \subset R^n$ and let the $x_i = g^i(t)$, $t \in R$, be continuously differentiable real-valued functions on a region $\overline{X} \subset R$. (It is assumed that if $t \in \overline{X}$, then $x \in X$.) For $y = f(g^1(t), \ldots, g^n(t)) = F(t)$,

$$\frac{dF}{dt} = \sum_{i=1}^{n} \frac{\partial f}{\partial x_i}\frac{dx_i}{dt} = \nabla f^T \left(\frac{dx}{dt}\right). \tag{4.8}$$

For instance, in the preceding definition of a composite function we determined that $F(t) = 9t^{-2} + 4t^6$ so that

$$F'(t) = -18t^{-3} + 24t^5$$

or, via (4.8),

$$\begin{aligned} F'(t) &= \frac{\partial f}{\partial x_1}\frac{dx_1}{dt} + \frac{\partial f}{\partial x_2}\frac{dx_2}{dt} \\ &= 2x_1(-3t^{-2}) + 2x_2(6t^2) \\ &= 2(3t^{-1})(-3t^{-2}) + 2(2t^3)(6t^2) \\ &= 18t^{-3} + 24t^5. \end{aligned}$$

We next have

Theorem 4.5: Chain Rule. Let the real-valued function $y = f(\mathbf{x})$, $\mathbf{x} \in R^n$, be continuously differentiable on a region $X \subset R^n$ and let the $x_i = g^i(\mathbf{t})$, $\mathbf{t} \in R^m$, $i = 1, \ldots, n$, be continuously differentiable real-valued functions on a region $\overline{X} \subset R$. (It is assumed that if $\mathbf{t} \in \overline{X}$, then $\mathbf{x} \in X$.) For $y = f(g^1(\mathbf{t}), \ldots, g^n(\mathbf{t})) = F(\mathbf{t})$,

$$\frac{\partial F}{\partial t_j} = \sum_{i=1}^{n} \frac{\partial f}{\partial x_i}\frac{\partial x_i}{dt_j}, \; j = 1, \ldots, m. \tag{4.9}$$

Alternatively, for $\mathbf{g}(\mathbf{t})^T = (g^1(\mathbf{t}), \ldots, g^n(\mathbf{t}))$ a differentiable vector-valued function, $y = f(\mathbf{x}) = f(\mathbf{g}(\mathbf{t})) = F(\mathbf{t})$ and thus

$$\frac{\partial F}{\partial t_j} = \nabla f^T \frac{\partial \mathbf{g}}{\partial t_j}, \; j = 1, \ldots, m, \tag{4.10}$$

where

$$\frac{\partial \mathbf{g}}{\partial t_j} = \begin{bmatrix} \partial g^1/\partial t_j \\ \vdots \\ \partial g^n/\partial t_j \end{bmatrix}.$$

Furthermore,

$$\underset{(1\times m)}{\nabla F} = \nabla f^T \frac{\partial \mathbf{x}}{\partial \mathbf{t}} = \nabla f^T \mathbf{J_g}(\mathbf{t}), \tag{4.11}$$

where $\mathbf{J_g}(\mathbf{t})$ is the $(n \times m)$ Jacobian matrix

$$\frac{\partial(g^1, \ldots, g^n)}{\partial(t_1, \ldots, t_m)} = \begin{bmatrix} \partial g^1/\partial t_1 & \partial g^1/\partial t_2 & \cdots & \partial g^1/\partial t_m \\ \partial g^2/\partial t_1 & \partial g^2/\partial t_2 & \cdots & \partial g^2/\partial t_m \\ \vdots & \vdots & & \vdots \\ \partial g^n/\partial t_1 & \partial g^n/\partial t_2 & \cdots & \partial g^n/\partial t_m \end{bmatrix}.$$

(Note that the jth component of ∇F is $\partial F/\partial t_j$.) If the second-order derivatives of f are continuous on X and the g^i, $i = 1, \ldots, n$, have continuous second-order derivatives on $\overline{X}$, then

$$\begin{aligned} \frac{\partial^2 F}{\partial t_j^2} &= \frac{\partial}{\partial t_j}\left(\sum_{i=1}^{n} \frac{\partial f}{\partial x_i}\frac{\partial x_i}{\partial t_j}\right) = \sum_{i=1}^{n} \left[\frac{\partial}{\partial t_j}\left(\frac{\partial f}{\partial x_i}\right)\right]\frac{\partial x_i}{\partial t_j} \\ &\quad + \sum_{i=1}^{n} \frac{\partial f}{\partial x_i}\frac{\partial^2 x_i}{\partial t_j^2}. \end{aligned} \tag{4.12}$$

(Note: $\left[\frac{\partial}{\partial t_j}\left(\frac{\partial f}{\partial x_1}\right)\right] = \frac{\partial^2 f}{\partial x_1^2}\frac{\partial x_i}{\partial t_j} + \frac{\partial^2 f}{\partial x_2 x_1}\frac{\partial x_2}{\partial t_j} + \cdots + \frac{\partial^2 f}{\partial x_n \partial x_1}\frac{\partial x_n}{\partial t_j}$, etc. This is because $\partial f/\partial x_1$ is a function of $x_1, \ldots, x_n$, with each x_i in turn dependent upon $t_1, \ldots, t_m$, $i = 1, \ldots, n$.) In this regard, for $(\mathbf{H}_f)_i$ the ith column of the Hessian matrix of f,

$$\mathbf{H}_f = \frac{\partial(f_1, \ldots, f_n)}{\partial(x_1, \ldots, x_n)},$$

we have

$$\begin{aligned} \frac{\partial^2 F}{\partial t_j^2} &= \sum_{i=1}^{n} \left[(\mathbf{H}_f)_i^T \frac{\partial \mathbf{g}}{\partial t_j}\right]\frac{\partial x_i}{\partial t_j} + \sum_{i=1}^{n} \frac{\partial f}{\partial x_i}\frac{\partial^2 x_i}{\partial t_j^2} \\ &= \sum_{i=1}^{n}\sum_{k=1}^{n} f_{ik}\frac{\partial x_i}{\partial t_j}\frac{\partial x_k}{\partial t_j} + \sum_{i=1}^{n} \frac{\partial f}{\partial x_i}\frac{\partial^2 x_i}{\partial t_j^2} \\ &- \left(\frac{\partial \mathbf{g}}{\partial t_j}\right)^T \mathbf{H}_f \left(\frac{\partial \mathbf{g}}{\partial t_j}\right) + \sum_{i=1}^{n} \frac{\partial f}{\partial x_i}\frac{\partial^2 x_i}{\partial t_j^2} \end{aligned} \tag{4.13}$$

Section 4.2 Exercises:

4.2.1. Let $y = f(x_1, x_2, x_3) = x_1x_2 - x_2x_3 + x_1 + x_3$ with

$$\begin{aligned} x_1 &= g^1(t) = 2t + t^2 \\ x_2 &= g^2(t) = t - t^3 \\ x_3 &= g^3(t) = 6t. \end{aligned}$$

Using (4.8), find dF/dt for $f = (g^1(t), g^2(t), g^3(t))$.

4.2.2. Suppose $y = f(x_1, x_2) = 2x_1 - 6x_2^2$ with

$$x_1 = g^1(\mathbf{t}) = 3t_1 - 4t_2$$
$$x_2 = g^2(\mathbf{t}) = t_1^2 + 2t_2^{-1}.$$

For $\mathbf{g}(\mathbf{t})^T = (g^1(\mathbf{t}), g^2(\mathbf{t}))$, find $\partial F/\partial t_2$ via (4.10) and ∇F using (4.11). Use (4.13) to determine $\partial^2 F/\partial t_2^2$.

4.3 FUNCTIONAL DEPENDENCE

The notion of **functional dependence** considers the circumstances under which a functional relationship exists among a set of functions. More specifically, we have

Definition 4.1: Let $y_1 = f^1(\mathbf{x}), \ldots, y_m = f^m(\mathbf{x}), \mathbf{x} \in R^n, m \geq n$, be a continuously differentiable mapping of a region $X \subset R^n$ *into* R^m. Let $\overline{X}$ be a region of R^m which contains the image of X. If there exists a continuously differentiable real-valued function F on $\overline{X}$ such that $F(y_1, \ldots, y_m) \not\equiv 0$ in a neighborhood of any point in $\overline{X}$, and if $F(f^1(\mathbf{x}), \ldots, f^m(\mathbf{x})) = 0$ for all $\mathbf{x} \in X$, then the $f^j, j = 1, \ldots, m$, are **functionally dependent** on X.

Quite trivially, the functions $y_1 = f^1(x_1, x_2) = x_1 + x_2$ and $y_2 = f^2(x_1, x_2) = x_1^2 + x_2^2 + 2x_1x_2$ are functionally dependent since $F(y_1, y_2) = y_2 - y_1^2 = 0$ for all $\mathbf{x} \in R^2$. How do we check for functional dependence among a set of functions? The next two theorems provide the answer.

Theorem 4.6: Let the real-valued functions $y_i = f^i(x)$, $x \in R^n$, $i = 1, \ldots, n$, be continuously differentiable over a region $X \subset R^n$. A necessary and sufficient condition that the f^i, $i = 1, \ldots, n$, satisfy a functional relationship $F(f^1(\mathbf{x}), \ldots, f^n(\mathbf{x})) = 0$ which does not explicitly contain $\mathbf{x}$ is that

$$|J_F(\mathbf{x})| = \left|\frac{\partial(f^1, \ldots, f^n)}{\partial(x_1, \ldots, x_n)}\right| \equiv 0 \qquad (4.14)$$

for all $\mathbf{x} \in X$. If $\partial(f^1, \ldots, f^n)/\partial(x_1, \ldots, x_n)$ is such that every minor of order greater than r vanishes identically while at least one minor of order r does not vanish, then there will be exactly n-r independent functional relations connecting the variables $y_1, \ldots, y_n$.

Example 4.4: Check that the two functions given above, namely $y_1 = f^1(x_1, x_2) = x_1 + x_2$ and $y_2 = f^2(x_1, x_2) = x_1^2 + x_2^2 + x_1x_2$, are functionally dependent. From (4.14)

$$\left|\frac{\partial(f^1, f^2)}{\partial(x_1, x_2)}\right| = \begin{vmatrix} f_1^1 & f_2^1 \\ f_1^2 & f_2^2 \end{vmatrix} = \begin{vmatrix} 1 & 1 \\ 2x_1 + 2x_2 & 2x_2 + 2x_1 \end{vmatrix}$$
$$= 2x_2 + 2x_1 - (2x_1 + 2x_2) \equiv 0$$

for all $\mathbf{x} \in R^2$. Hence, these expressions are functionally dependent. But if we now replace f^2 by $f^2 = x_1^2 + x_2^2$, then

$$\begin{vmatrix} f_1^1 & f_2^1 \\ f_1^2 & f_2^2 \end{vmatrix} = \begin{vmatrix} 1 & 1 \\ 2x_1 & 2x_2 \end{vmatrix} = 2x_2 - 2x_1 \not\equiv 0$$

for all $\mathbf{x} \in R^2$. Clearly functional dependence no longer holds.

Example 4.5: Suppose

$$y_1 = f^1(x_1, x_2, x_3) = x_1 + x_2 + x_3$$
$$y_2 = f^2(x_1, x_2, x_3) = x_1x_2 + x_1x_3 + x_2x_3$$
$$y_3 = f^3(x_1, x_2, x_3) = x_1^2 + x_2^2 + x_3^2.$$

Are these expressions functionally dependent? The reader can readily check that

$$\left|\frac{\partial(f^1, f^2, f^3)}{\partial(x_1, x_2, x_3)}\right| = \begin{vmatrix} 1 & 1 & 1 \\ x_2 + x_3 & x_1 + x_3 & x_1 + x_2 \\ 2x_1 & 2x_2 & 2x_3 \end{vmatrix} \equiv 0$$

for all $\mathbf{x} \in R^3$. Hence these relationships are functionally dependent with

$$F(y_1, y_2, y_3) = y_3 - y_1^2 - 2y_2 = 0$$

for all $\mathbf{x} \in R^3$.

We next have

Theorem 4.7: A necessary and sufficient condition for the real-valued functions $y_j = f^j(\mathbf{x}), \mathbf{x} \in R^n, j = 1, \ldots, m < n$, to be functionally dependent in a region $X \subset R^n$ is that *every* mth order Jacobian determinant of these m functions with respect to m of the n variables vanishes identically in X, i.e., the $f^j, j = 1, \ldots, m$, satisfy $F(f^1(\boldsymbol{x}),\ldots,f^m(\boldsymbol{x})) = 0$ if

$$\left|\frac{\partial(f^1,\ldots,f^m)}{\partial(x_{k_1}, x_{k_2},\ldots,x_{k_m})}\right| \equiv 0$$

for all choices of m distinct indices $k_1, \ldots, k_m$ from among the n numbers $1, \ldots .n$. (Alternatively, if the real-valued functions $f^1, \ldots, f^m$, $m \leq n$, are functionally dependent on $X \subset R^n$, then the rank of $\partial(f^1,\ldots,f^m)/\partial(x_1,\ldots,x_n) < m$ at each interior point of X. If $m = n$, $|\partial(f^1,\ldots,f^n)/\partial(x_1,\ldots,x_n)| \equiv 0$ on X.)

Section 4.3 Exercises:

4.3.1. Are the expressions

$$y_1 = f^1(\boldsymbol{x}) = 3x_1^2 - x_2^2$$
$$y_2 = f^2(\boldsymbol{x}) = 2x_1x_2$$

functionally related?

4.3.2. Are the equations

$$y_1 = f^1(\boldsymbol{x}) = In(x_1 + x_2)$$
$$y_2 = f^2(\boldsymbol{x}) = x_1^2 + x_2^2 + 2x_1x_2 + 1$$

functionally dependent?

4.3.3. Let

$$y_1 = f^1(x_1, x_2, x_3) = x_1 + x_2 + x_3$$
$$y_2 = f^2(x_1, x_2, x_3) = x_1x_2 + x_2x_3 + \frac{1}{2}x_3^2.$$

Are y_1, y_2 functionally dependent?

Global and Local Extrema of Real-Valued Functions

5.1 CLASSIFICATION OF EXTREMA

Our objective in this chapter is to describe the varieties of extreme values which a function $y = f(\boldsymbol{x}) : D \subset R^n \to R$ may attain over some domain D. Such values of f are generally referred to as **extrema**, a term which is taken to depict either a maximum or minimum value of f over D.

Extreme values of f are defined relative to domain D or to some spherical δ-neighborhood about a point $\boldsymbol{x} \in D$, $\delta(\boldsymbol{x})$. In this regard, f may possess an extremum in either a global or local sense, i.e., a **global extremum** is defined relative to a specific region D *given in advance* while a **local extremum** is defined relative to only $\delta(\boldsymbol{x}) \subset D$. More specifically, we have

Definition 5.1: (a) $f(\boldsymbol{x}) : D \subset R^n \to R$ has a **global extremum** at a point $x_o \in D$ if $\boldsymbol{x}_o$ yields an extremum relative to all other points contained within the limits of a predetermined domain; and (b) $f(\boldsymbol{x}) : D \subset R^n \to R$ has a **local extremum** at a point $\boldsymbol{x}_o$ if one can choose a spherical δ-neighborhood about $\boldsymbol{x}_o$, $\delta(\boldsymbol{x}_o) = \{\boldsymbol{x} \mid \|\boldsymbol{x} - \boldsymbol{x}_o\| < \delta, \boldsymbol{x} \in R^n\}$, such that $\boldsymbol{x}_o$ yields an extremum relative to all other points within $\delta(\boldsymbol{x}_o) \subset D$.

By convention, a global extremum will be defined over a closed domain D, but a local extremum will be specified over an open domain.

5.2 GLOBAL EXTREMA

Suppose $y = f(\boldsymbol{x})$: $D \subset R^n \to R$ is a function defined on a *closed* domain D and let the points $\boldsymbol{x}_o, \boldsymbol{x}_1 \in D$. Then

Definition 5.2: (a) $f(\boldsymbol{x})$ has a **global maximum** at $\boldsymbol{x}_o \in D$ if $f(\boldsymbol{x}_o) \geq f(\boldsymbol{x})$ for all points $\boldsymbol{x}(\neq \boldsymbol{x}_o) \in D$; and (b) $f(\boldsymbol{x})$ has a **global minimum** at $\boldsymbol{x}_1 \in D$ if $f(\boldsymbol{x}_1) \leq f(\boldsymbol{x})$ for all $\boldsymbol{x}(\neq \boldsymbol{x}_1) \in D$.

It should be obvious from the preceding definition of a global maximum that at no point $\boldsymbol{x} \in D$ is $f(\boldsymbol{x}) > f(\boldsymbol{x}_o)$. However, it may be true that, for some point(s) $\boldsymbol{x}(\neq \boldsymbol{x}_o) \in D$, $f(\boldsymbol{x}) = f(\boldsymbol{x}_o)$. When equality holds between $f(\boldsymbol{x}_o)$ and $f(\boldsymbol{x})$ for at least one $\boldsymbol{x}(\neq \boldsymbol{x}_o) \in D$, we say that f attains a **weak global maximum** at $\boldsymbol{x}_o$. Similarly, f attains a **weak global minimum** at $\boldsymbol{x}_1$ if nowhere within D is $f(\boldsymbol{x}_1) > f(\boldsymbol{x})$ while $f(\boldsymbol{x}) = f(\boldsymbol{x}_1)$ for at least one point $\boldsymbol{x}(\neq \boldsymbol{x}_1) \in D$.

We next note that f attains a **strong global maximum** at $\boldsymbol{x}_o$ if $f(\boldsymbol{x}_o)$ is absolutely greater than the value of f at any other point $\boldsymbol{x} \in D$, i.e. $f(\boldsymbol{x}_o) > f(\boldsymbol{x})$ for all $\boldsymbol{x}(\neq \boldsymbol{x}_o) \in D$. And f has a **strong global minimum** at $\boldsymbol{x}_1$ if $f(\boldsymbol{x}_1) < f(\boldsymbol{x})$ for all $\boldsymbol{x}(\neq \boldsymbol{x}_1) \subset D$.

As Figure 5.1 reveals, for $D = [a, b] \subset R$, f attains a strong global maximum at $\boldsymbol{x}_o(f(x_o) > f(x), x \in D)$ and a strong global minimum at $x_1(f(x_1) < f(x), x \in D)$. If the domain of f is redefined as $D = [a', b'] \subset R$, then f has a weak global minimum at x_2 $(f(x_2) \leq f(x)$ for $x \in [a', b']$ and there are x's where $f(x) = f(x_2))$ and a weak global maximum at $x_3(f(x_3) \geq f(x)$ for $x \in [a', b']$ and there are x's where $f(x) = f(x_3))$. Note that if now $D = [a', a'']$, then for each point $x \in D$, f simultaneously attains a weak global maximum and a weak global minimum. There are no strong global extrema of f anywhere over $[a', a'']$.

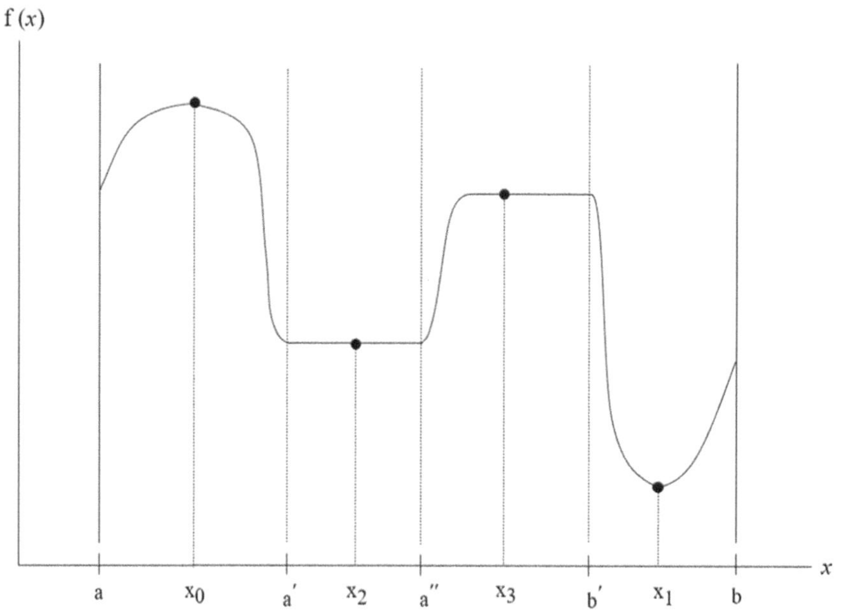

FIGURE 5.1 Varieties of global extrema.

Section 5.2 Exercises:

5.2.1 Suppose $y = f(x) = x^3 + 4$. (a) Does f attain a global extremum at $x = 0$ for $x \in D = (-\infty, +\infty)$? (b) For $D = [0, +\infty)$, does f attain a global extremum? (c) For $D = (0, +\infty)$, does f have a global extremum?

5.2.2 Does $f(x) = x^2$ attain a global maximum over $D = [-1, 1]$? Over $(-1, 1)$? Does f attain a global minimum over $[0, 3]$? Over $(0, 3)$?

5.3 LOCAL EXTREMA

Let $y = f(\boldsymbol{x}): D \subset R^n \rightarrow R$ be defined over the *open* domain D with points $\boldsymbol{x}_0, \boldsymbol{x}_1 \in D$.

Definition 5.3: (a) $f(\boldsymbol{x})$ has a **local maximum** at $\boldsymbol{x}_0 \in D$ if there is a spherical δ-neighborhood about $\boldsymbol{x}_o$, $\delta(\boldsymbol{x}_o)$, such that $f(\boldsymbol{x}) \leq f(\boldsymbol{x}_o)$ for all $\boldsymbol{x} \in \delta(\boldsymbol{x}_o)$; and (b) f has a **local minimum** at $\boldsymbol{x}_1 \in D$ if there is a spherical δ-neighborhood about $\boldsymbol{x}_1$, $\delta(\boldsymbol{x}_1)$, such that $f(\boldsymbol{x}) \geq f(\boldsymbol{x}_1)$ for all $\boldsymbol{x} \in \delta(\boldsymbol{x}_1)$.

Clearly, $f(\boldsymbol{x}) \leq f(\boldsymbol{x}_o)$ means that $f(\boldsymbol{x}_o)$ is at least as large as $f(\boldsymbol{x})$ for all points $\boldsymbol{x}\,(\neq \boldsymbol{x}_o)$ for some distance δ in any direction from $\boldsymbol{x}_o$. Let us express the condition $f(\boldsymbol{x}) \leq f(\boldsymbol{x}_o)$ for all $\boldsymbol{x}\,(\neq \boldsymbol{x}_o) \in \delta(\boldsymbol{x}_o)$ in a slightly different way. Suppose we set

$$\boldsymbol{x} = \boldsymbol{x}_o + \boldsymbol{h} = \begin{bmatrix} x_1^o + h_1 \\ \vdots \\ x_n^o + h_n \end{bmatrix}.$$

Then f will have a local maximum at $\boldsymbol{x}_o$ if, for $||\boldsymbol{h}|| < \delta$ sufficiently small,

$$f(\boldsymbol{x}_o + \boldsymbol{h}) - f(\boldsymbol{x}_o) \leq 0. \tag{5.1}$$

[f has a local maximum at $\boldsymbol{x}_o$]

Similarly, $f(\boldsymbol{x}) \geq f(\boldsymbol{x}_1)$ means that $f(\boldsymbol{x})$ is at least as large as $f(\boldsymbol{x}_1)$ for all points $\boldsymbol{x}\,(\neq \boldsymbol{x}_1)$ for a distance δ in any direction from $\boldsymbol{x}_1$. Expressed in an alternative fashion, the local minimum condition $f(\boldsymbol{x}) \geq f(\boldsymbol{x}_1)$ for all $\boldsymbol{x}\,(\neq \boldsymbol{x}_1) \in \delta(\boldsymbol{x}_1)$ becomes

$$f(\boldsymbol{x}_1 + \boldsymbol{h}) - f(\boldsymbol{x}_1) \geq 0 \tag{5.2}$$

[f has a local minimum at $\boldsymbol{x}_1$]

for $||\boldsymbol{h}|| < \delta$ sufficiently small, where

$$\boldsymbol{x} = \boldsymbol{x}_1 + \boldsymbol{h} = \begin{bmatrix} x_1^1 + h_1 \\ \vdots \\ x_n^1 + h_n \end{bmatrix}.$$

With reference to Definition 5.3a, if $f(\boldsymbol{x}) \leq f(\boldsymbol{x}_o)$ for all $\boldsymbol{x}\,(\neq \boldsymbol{x}_o) \in \delta(\boldsymbol{x}_o)$ and $f(\boldsymbol{x}_o) = f(\boldsymbol{x})$ for at least one point $\boldsymbol{x}\,(\neq \boldsymbol{x}_0) \in \delta(\boldsymbol{x}_o)$, then f has a **weak local maximum** at $\boldsymbol{x}_o$. But if$f(\boldsymbol{x}_o) > f(\boldsymbol{x})$ for all $\boldsymbol{x}\,(\neq \boldsymbol{x}_0) \in \delta(\boldsymbol{x}_o)$, then f has a **strong local maximum** at $\boldsymbol{x}_o$. Turning to Definition 5.3b, f has a **weak local minimum** at $\boldsymbol{x}_1$ if $f(\boldsymbol{x}_1) = f(\boldsymbol{x})$ for at least one other point $\boldsymbol{x}\,(\neq \boldsymbol{x}_1) \in \delta(\boldsymbol{x}_1)$. A **strong local minimum** is attained at $\boldsymbol{x}_1$ if $f(\boldsymbol{x}_1) < f(\boldsymbol{x})$ for all $\boldsymbol{x}\,(\neq \boldsymbol{x}_1) \in \delta(\boldsymbol{x}_1)$. As Figure 5.2 reveals, for

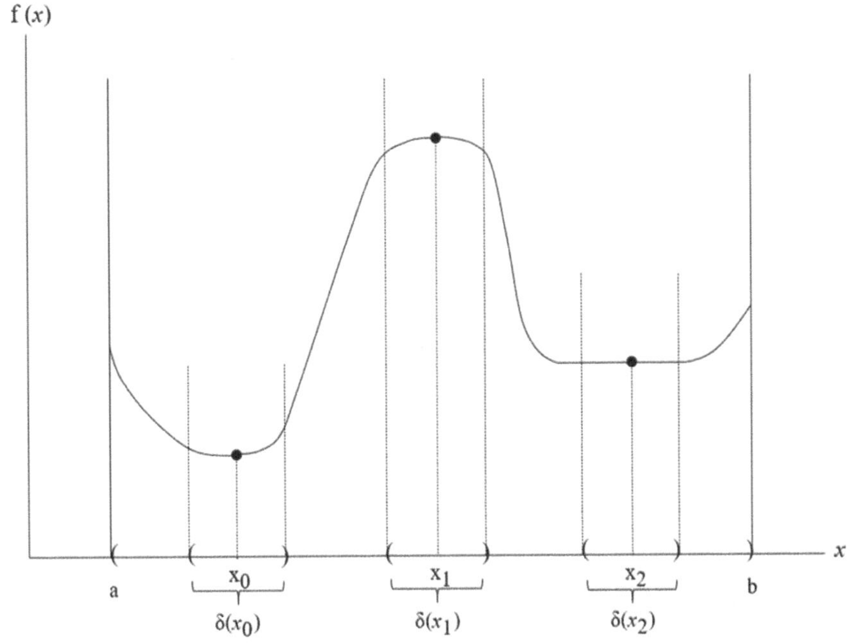

FIGURE 5.2 Varieties of local extrema.

$D = (a, b)$, $f(\boldsymbol{x})$ attains a strong local minimum at x_o (since $f(x_o) < f(x)$ for all $x(\neq x_o) \in \delta(x_o)$), a strong local maximum at x_1 (given that $f(x_1) > f(x)$ for all $x(\neq x_1) \in \delta(x_1)$), and a weak local minimum at x_2 (since $f(x_2) \leq f(x)$ and $f(x_2) = f(x)$ for all $x(\neq x_2) \in \delta(x)$). Can x_2 also represent a weak local maximum? Why?

Example 5.1: Let $y = f(x) = x^2$ for all real $x \in D = (-\infty, +\infty)$. Verify that f has a strong local minimum at $x_o = 0$. From Equation (5.2)

$$\begin{aligned} f(x_o + h) - f(x_o) &= f(h) - f(0) \\ &= h^2 > 0 \text{ for all } |h| < \delta. \end{aligned}$$

Does f attain an extremum at $x_1 = 2$? Now (5.2) yields

$$\begin{aligned} f(2 + h) - f(2) &= (2 + h)^2 - 4 \\ &= 4h + h^2 > 0(<0) \text{ for } h > 0(<0), \\ & \quad |h| < \delta. \end{aligned}$$

Since $f(2 + h) - f(2)$ changes sign when the sign of h changes, f does not attain a local extremum at $x_1 = 2$. ■

Example 5.2: Let $y = f(x) = 10 - 2(x - 3)^2$ for all real $x \in D = (1, 5)$. Show that f has a strong local maximum at $x_o = 3$. From (5.1),

$$\begin{aligned} f(x_o + h) - f(x_o) &= f(3 + h) - f(3) \\ &= -2h^2 < 0 \text{ for all } |h| < \delta. \end{aligned}$$

Does f attain an extremum at $x_1 = 2$? Again looking to (5.1) we have

$$f(2 + h) - f(2) = -2h^2 + 4h > 0(<0) \text{ for } h > 0(<0), \quad |h| < \delta.$$

Since the sign of $f(2 + h) - f(2)$ changes when the sign of h does, it is evident that f does not attain a local extremum at $x_1 = 2$. ■

Example 5.3: Let $y = f(x_1, x_2) = x_1^2 + x_2^2 + 2x_1 - x_2 - 3$ be defined for all real $\boldsymbol{x}^T = (x_1, x_2) \in R^2$. Demonstrate that f attains a strong local minimum at $\boldsymbol{x}_o^T = \left(-1, \frac{1}{2}\right)$. For

$$\boldsymbol{x} = \boldsymbol{x}_o + \boldsymbol{h} = \begin{bmatrix} x_1^0 + h_1 \\ x_2^0 + h_2 \end{bmatrix} = \begin{bmatrix} -1 + h_1 \\ \frac{1}{2} + h_2 \end{bmatrix},$$

we have

$$\begin{aligned} f(\boldsymbol{x}_o + \boldsymbol{h}) &= (-1 + h_1)^2 + \left(\tfrac{1}{2} + h_2\right)^2 + 2(-1 + h_1) \\ &\quad - \left(\tfrac{1}{2} + h_2\right) - 3 \\ &= h_1^2 + h_2^2 - \tfrac{17}{4}. \end{aligned}$$

And with $f(\boldsymbol{x}_o) = -\frac{17}{4}$, (1.2) yields

$$f(\boldsymbol{x}_o + \boldsymbol{h}) - f(\boldsymbol{x}_o) = \boldsymbol{h}_1^2 + \boldsymbol{h}_2^2.$$

Since this last expression is strictly positive for all $h_1^2 + h_2^2 < \delta^2$ sufficiently small, we see that f indeed has a strong local minimum at $\boldsymbol{x}_o$.

Does f attain an extremum at $\boldsymbol{x}_1^T = (1, 1)$? Again looking to Equation (5.2)

$$\begin{aligned} f(\boldsymbol{x}_1 + \boldsymbol{h}) - f(\boldsymbol{x}_1) &= (1 + h_1)^2 + (1 + h_2^2) + 2(1 + h_1) \\ &\quad - (1 + h_2) - 3 \\ &= h_1^2 + h_2^2 + 4h_1 + h_2. \end{aligned}$$

Since the sign of the difference on the left-hand side of this equation changes for changes in the sign of h_1 and h_2 on the right-hand side, with h_1, h_2 sufficiently small or $||\boldsymbol{h}|| < \delta$ it follows that f does not attain an extremum at $\boldsymbol{x}_1^T = (1, 1)$.

It should by now be apparent that the categories of global and local extrema are not mutually exclusive; an extremum of f can simultaneously be classified as global and local. Specifically, if a global extremum occurs at an interior point of a closed domain $D \subset R^n$, then it must also be a local extremum, although the converse of this statement need not be true. But if f attains a global extremum at a boundary point of a closed domain D, then this extremum cannot also be of the local variety. Hence any global extremum must be either a local extremum or occur at a boundary point of its closed domain.

Section 5.3 Exercises:

5.3.1 Does $f(x) = x^2 - 5x + 8$, $x \in R$, attain a local extremum at $x_o = 1$? At $x_1 = 2.5$?

5.3.2 Does $f(x) = (1 + x_1)(1 + x_2)$, $x \in R^2$, have a local extremum at the point $\boldsymbol{x}_o^T = (1, 2)$? At $\boldsymbol{x}_1^T = (-1, -1)$?

5.3.3 Does $f(x) = 2x_1^2 + x_2^2 - x_1 - 3x_2 + 4$ have a local minimum at $\boldsymbol{x}_o^T = \left(\frac{1}{4}, \frac{3}{2}\right)$?

Global Extrema of Real-Valued Functions

6.1 EXISTENCE OF GLOBAL EXTREMA

In the discussion which follows, we shall assume that the domain D of a continuous function $y = f(\boldsymbol{x}) : D \subset R^n \rightarrow R$ is a closed, bounded region. To set the stage for the developments herein, we shall offer a brief review of some key properties of bounded sets and functions.

Definition 6.1: A set $S \subset R^n$ is **bounded** if and only if there exists a real number $N > 0$ such that for all points $\boldsymbol{x} \in S$, $||\boldsymbol{x}|| \leq N$. (If $S \subset R$, then $|x| \leq N$ for every $x \in S$.)

For $x \in R$, we have the following sequence of definitions.

Definition 6.2: (a) A set $S \subset R$ is said to be **bounded from above** if and only if there exists a real number M such that, for every $x \in S$, $x \leq M$. Here M is termed an **upper bound** for S, and (b) A set $S \subset R$ is **bounded from below** if and only if there exists a real number m such that, for every $x \in S$, $m \leq x$. Thus, m is said to be a **lower bound** for S.

In this regard, we have

Definition 6.3: A set $S \subset R$ which is bounded from both above and below is said to be **bounded**.

Thus, a set $S \subset R$ is termed **unbounded** if either it fails to possess an upper bound or a lower bound or if it possesses neither an upper nor a lower bound.

Upper and lower bounds for a set $S \subset R$ are not unique. For instance, if M is an upper bound for S and $M' > M$, then M' is also an upper bound for S; and if m is a lower bound for S and $m' < m$, then m' is also a lower bound for S. Additionally, an upper or lower bound for S may not be a member of S, e.g., if $S = \{x|a \leq x \leq b\}$, then a and b are, respectively, lower and upper bounds for S with both $a, b \in S$. Furthermore, for $S' = \{x|a < < b\}$, a and b are, respectively, lower and upper bounds for S', with $a, b \notin S'$. (Note that, for both S and S', any number less than a is a lower bound; and any number greater than b is an upper bound.)

Let us now consider the following questions. Of all the upper bounds for a set $S \subset R$, can we find one which is smaller than any other upper bound for S? Similarly, of all the lower bounds for S, is there one that is larger than any other lower bound for S? That is, can we find the *least upper bound* or **supremum** of S (denoted sup (S)); and can we determine the *greatest lower bound* or **infimum** of S (denoted inf (S))? Before answering these questions, let us state the qualifications that sup (S) and inf (S) must possess. Specifically,

Definition 6.4:

a. L is a supremum for a set S if and only if:

i. L is an upper bound for S;

ii. No number less than L is an upper bound for S.

b. l is an infimum for S if and only if:

i. l is a lower bound for S;

ii. no number greater than l is a lower bound for S.

Given these characteristics of a supremum (infimum), the answer to the preceding questions are provided by

Theorem 6.1: (**Completeness Axiom**). If $S \subset R$ is a bounded, non-empty set, then there exist real numbers l, L such that $l = \inf(S) \leq L = \sup(S)$.

As this theorem indicates, every non-empty set of real numbers which has an upper bound has a supremum; and every non-empty set of real numbers which has a lower bound has an infimum. Moreover, for any bounded set S: (a) l and L are unique; and (b) l and L may not be members of S. (For instance, if $S = \{x | a \leq x \leq b\}$, then $a = \inf(S)$, $b = \sup(S)$, and both $\inf(S), \sup(S) \in S$. And if $S' = \{x | a < x < b\}$, then $a = \inf(S')$, $b = \sup(S')$ but $\inf(S'), \sup(S') \notin S'$.)

Under what circumstances will a set $S \subset R$ possess *extreme elements*? For S a bounded, non-empty set, we know that both $L = \sup(S)$ and $l = \inf(S)$ exist. If $L \in S$, then S is said to have a *greatest element*, namely L; and if $l \in S$, then S possesses l as its *least element*. What type of set must S be in order for it to possess l as its least element and L as its greatest element? The answer is that S must be compact, i.e., In general, a set $S \subset R^n$ is **compact** if it is closed and bounded. For $S \subset R$, the closed interval $S = \{x | a \leq x \leq b\}$ does nicely – it is a bounded set of points such that $b = L = \sup(S)$ is its greatest element while $a = l = \inf(S)$ is its least element.

As mentioned earlier, we also need to consider the notion of a function that is bounded over its domain. To this end, we have

Definition 6.5: A function $y = f(\boldsymbol{x}) : D \subset R^n \rightarrow R$ is said to be **bounded** over D if and only if there exists a real number $N > 0$ such that $|f(\boldsymbol{x})| \leq N$ for all $\boldsymbol{x} \in D$.

What about the existence of the least and greatest values of a real-valued function? The answer is provided by

Theorem 6.2: (Weierstrass's Theorem). A function $y = f(\boldsymbol{x}) : D \subset R^n \rightarrow R$ continuous over a non-empty, compact (closed and bounded) region D takes on a least *and* a greatest value at least once over D.

Hence, this theorem informs us that if the domain of a continuous function f is compact, *then so is its range* (denoted R_f), i.e., the range of f is also a closed, bounded set, and thus, by the completeness axiom, the values of f have a supremum as well as an infimum. Moreover, the said supremum and infimum are themselves elements of R_f and consequently are, respectively, the greatest and least elements of R_f. Thus, the supremum of R_f is the global maximum of f; and the infimum of R_f is the global minimum of f. To summarize: Weierstrass's Theorem establishes the fact that a real-valued function that is continuous over a compact domain D attains a global maximum *and* a global minimum value at some point(s) of that domain.

We just found that Weierstrass's theorem guarantees the existence of an $N > 0$ such that $|f(\boldsymbol{x})| \leq N$ for all $\boldsymbol{x} \in D$. This result is accommodated if we choose N to be the larger of the global maximum and the global minimum value of f. Hence, the existence of global extrema provides us with a sufficient condition for f to be bounded on D. However, it is not a necessary condition since f may be bounded even if its range does not possess a least and a greatest element. Conversely, if f is bounded over D, there must exist numbers $\inf(R_f)$ and $\sup(R_f)$ such that $\inf(R_f) \leq f(\boldsymbol{x}) \leq \sup(R_f)$ for all $\boldsymbol{x} \in D$. However, $\inf(R_f)$ and $\sup(R_f)$ need not be members of R_f. Hence the boundedness of the range of f is not a sufficient condition for f to attain a least and a greatest element over its domain. So, if f is bounded we have a necessary condition for the existence of global extrema.

Weierstrass's theorem guarantees the existence of a global maximum and a global minimum value over the domain D of f. However, it did not explicitly associate any specific value(s) of $\boldsymbol{x} \in D$ with $\sup(R_f)$ and $\inf(R_f)$. How then can we be sure that we will be able to find at least one $\boldsymbol{x} \in D$ (call it $\boldsymbol{x}_o$) such that f maps $\boldsymbol{x}_o$ into $\sup(R_f)$? Similarly, is there at least one $\boldsymbol{x} \in D$ (say, $\boldsymbol{x}_1$) such that f maps $\mathbf{x}_1$ into inf (R_f)? As we shall now see, the answer to both of these questions is *yes*. The justification for an affirmative response is provided by

Theorem 6.3: **(Intermediate-Value Theorem)**. Let $a,b \in R_f$ with $a < b$. In addition, let $y = f(x)$: $[a, b] \subset R \rightarrow R$ be a continuous function with $f(a) \neq f(b)$. Then if c is any number between $f(a)$ and $f(b)$, there exists at least one number $\bar{x}$ between a and b such that $f(\bar{x}) = c$.

So as x varies from a to b, $f(x)$ takes on every value between $f(a)$ and $f(b)$. And since sup (R_f) and inf (R_f) are both elements of R_f, we may be certain that there exists at least one point $x_o \in [a, b]$ where $f(x_o)$ = sup (R_f) = global maximum of f over $[a, b]$; and at least one point $x_1 \in [a, b]$ where $f(x_1)$ = inf (R_f) = global minimum of f over $[a, b]$.

One additional point is in order. Under a suitable transformation, we may express the global minimum of a function in terms of its global maximum. To see this, let R_f be a closed, bounded set of real numbers, say, $R_f = \{r | r_1 \leq r \leq r_2, r \in R_f\}$. Then, inf $(R_f) = -\sup\{-r | r \in R_f\}$. As this relationship indicates, if we desire to obtain the global minimum of a continuous real-valued function over a compact set, then we may do so by taking the negative of the global maximum of $-f$ or

$$\text{global min}\{f(\mathbf{x})\} = -\text{global max}\{-f(\mathbf{x})\}, \mathbf{x} \in R^n. \quad (6.1)$$

Example 6.1: Let $y = f(x) = x^2$ for all real $x \in [a, b]$, $x > 0$ (Figure 6.1a). Since $R_f = [a^2, b^2]$, R_f is a bounded set with a^2 = inf (R_f) and b^2 = sup (R_f) both members of R_f. Thus, a^2 and b^2 are, respectively, the least and greatest elements in R_f so that a^2 is the strong global minimum of f and b^2 is the strong global maximum of f over $[a, b]$. If we redefine the domain of f as the open interval (a, b) (Figure 6.1b), then $R_f = (a^2, b^2)$ is also an open interval, and thus, even though R_f is a bounded set, a^2 = inf (R_f) and b^2 = sup (R_f) are not members of R_f. Thus, R_f possesses neither a least nor greatest element and thus f has no global maximum or global minimum over (a, b). ■

Example 6.2: Let the real-valued function $y = f(x)$ be defined as

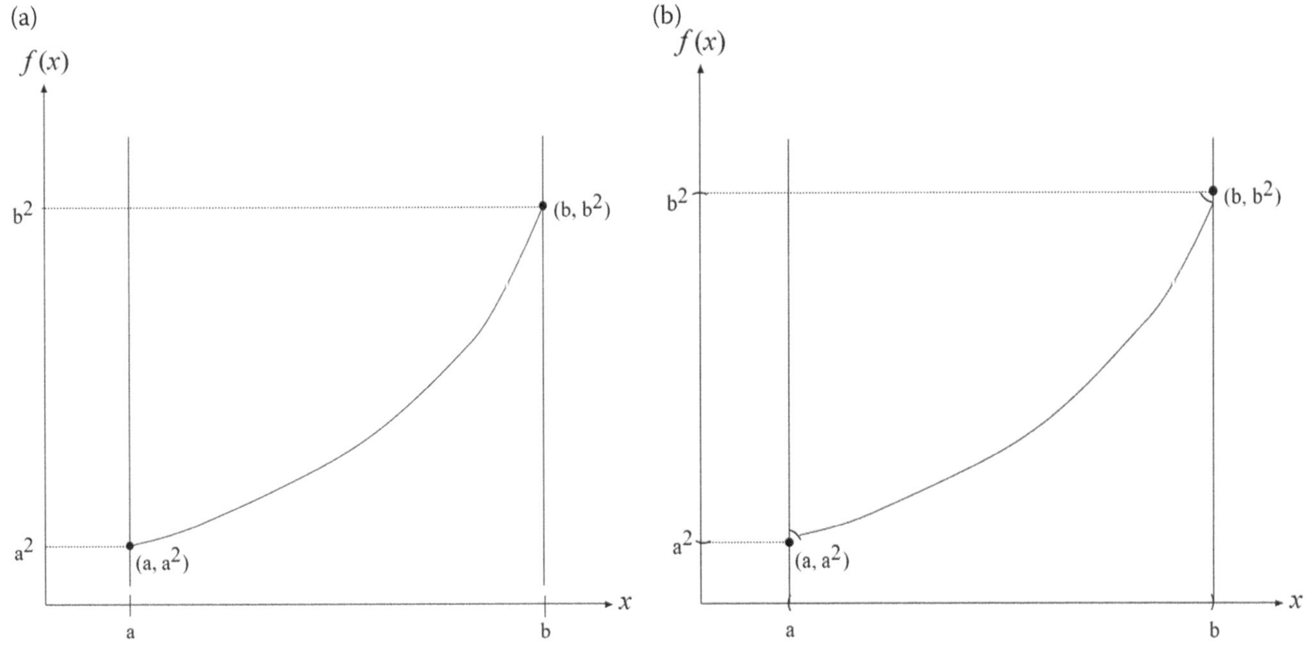

FIGURE 6.1 (a) f attains a strong global maximum and a strong global minimum; (b) f attains neither a global maximum nor a global minimum.

$$f(x) = \begin{cases} 3/4 \text{ when } x = 1; \\ x^2 \text{ for } 1 < x < 6; \\ 44 \text{ when } x = 6. \end{cases}$$

Here, the domain of f is $D_f = [1, 6]$ and the range of f is the bounded set $R_f = \{3/4\} \cup \{1 < y < 36\} \cup \{44\}$ (Figure 6.2a). Although f is discontinuous at $x = 1$ and $x = 6$, its range possesses a least and a greatest element since $3/4 = \inf(R_f)$ and $44 = \sup(R_f)$ are both members of R_f. Hence, 3/4 and 44 are, respectively, the strong global minimum and strong global maximum of f on [1, 6]. By redefining $f(x)$ as 2 when $x = 1$ and 18 when $x = 6$ (Figure 6.1b), we see that f possesses neither a global maximum nor a global minimum over [1, 6] since R_f, while remaining a bounded set, has neither a least nor a greatest element.

To summarize: the discussion just offered pertaining to the existence of global extrema indicates that such extrema *may not exist* either because: (1) domain D is bounded but not closed (Example 6.1); (2) f is discontinuous at some point(s) on D (Example 6.2); or (3) D is an unbounded or infinite set (e.g., if $y = f(x) = x^2$ and $x \in [a,+\infty)$, $x > 0$, then the domain of f is unbounded and, consequently $R_f = \{y | a^2 \leq y < +\infty\}$ is also unbounded and thus f only has a least element, a^2, which corresponds to the strong global minimum of f). Thus, Weierstrass's theorem is only a sufficient condition for a global extremum; this theorem makes no mention of the existence or non-existence of global extrema if at least one of its assumptions does not hold. However, one fact is absolutely certain – we can be sure that a function admits both a global maximum and a global minimum value if it is continuous over a compact domain.

Section 6.1. Exercises:

6.1.1 Let $y = f(x) = x^{-1}$, $x \in [-1, 1]$, $x \neq 0$, $x \in R$. Since $f(-1) = -1$ and $f(1) = 1$, does f take on all values between -1 and 1? Is f bounded?

6.1.2 Does the real-valued function $y = f(x) = |x|$, $x \in [-1, 1]$ have a maximum? A minimum? Why?

6.1.3 Can a function be bounded on a set without assuming a maximum or minimum on the set? Is this statement true for

$$y = f(x) = \begin{cases} x, & x \in (0, 1); \\ \frac{1}{2}, & x = 0 \text{ or } 1? \end{cases}$$

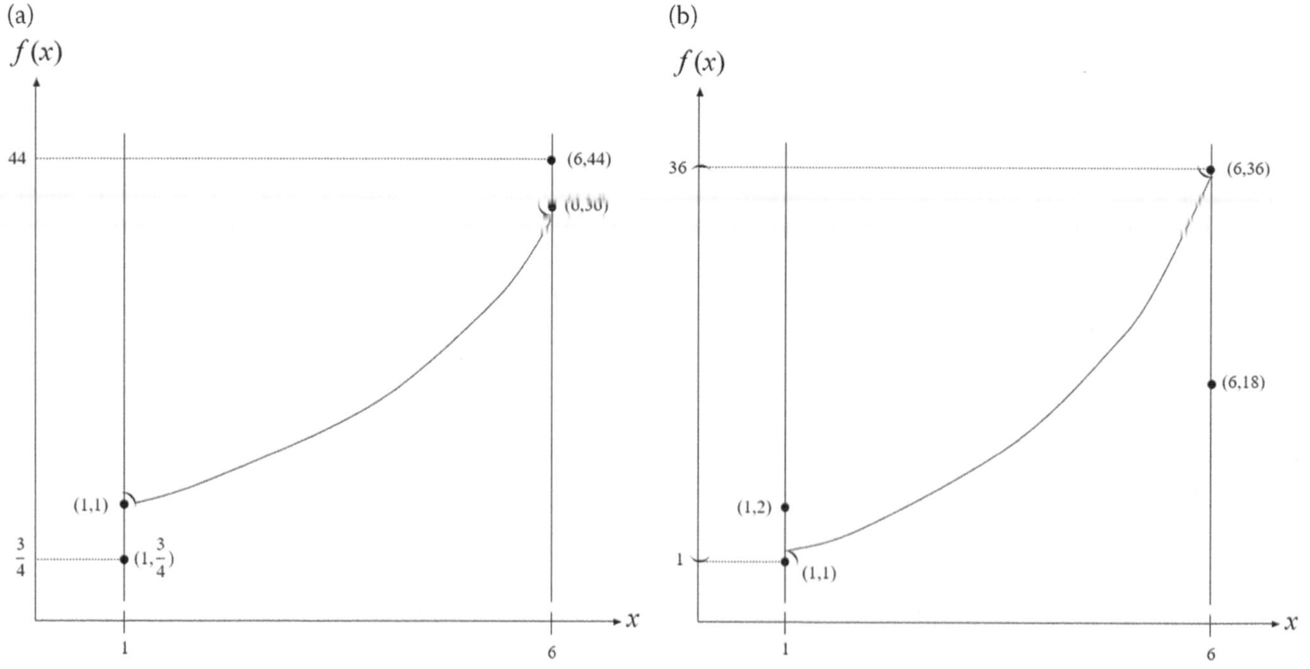

FIGURE 6.2 (a) f attains a strong global maximum and a strong global minimum; (b) f attains neither a global maximum nor a global minimum.

Is this the case if we redefine the first portion of the function as $f(x) = x,\ x \in [0, 1]$?

6.1.4 Does the real-valued function $y = f(x) = x^2$ have a maximum for $x \in (0, 1)$? A minimum?

6.1.5 Suppose $y = f(x) = \frac{1}{x_1 + x_2}: D \subset R^2 \rightarrow R,\ D = \{x | 0 < x_1 < 1,\ 0 < x_2 < 1\}$. Does f attain an upper bound on D?

6.2 EXISTENCE OF GLOBAL EXTREMA: ANOTHER LOOK

One of the assumptions made in the statement of Weierstrass's theorem is that the function $y = f(\mathbf{x}) : D \subset R^n \rightarrow R$ is *continuous*. Specifically, from the definition of continuity at a point $\mathbf{x}_o$ we have: given an $\varepsilon > 0$ there exists a $\delta > 0$ such that $|f(\mathbf{x}) - f(\mathbf{x}_o)| < \varepsilon$ whenever $\mathbf{x} \in \delta(\mathbf{x}_0) = \{\mathbf{x}|\ ||\mathbf{x} - \mathbf{x}_o||\} < \delta$ i.e.,

$$\mathbf{x}_o - \delta < \mathbf{x} < \mathbf{x}_o + \delta \text{ implies } f(\mathbf{x}_o) - \varepsilon < f(\mathbf{x}) < f(\mathbf{x}_o) + \varepsilon,\ \mathbf{x} \in D.$$

Let us now *weaken* the assumption of continuity by assuming that f is semi-continuous. To this end we state

Definition 6.6: A function $y = f(\mathbf{x}) : D \subset R^n \rightarrow R$ is said to be:

a. upper semi-continuous at $\mathbf{x}_o \in D$ if for each $\varepsilon > 0$ there exists a $\delta > 0$ such that $||\mathbf{x} - \mathbf{x}_o|| < \delta,\ \mathbf{x} \in D$, implies $f(\mathbf{x}) < f(\mathbf{x}_o) + \varepsilon$ (Figure 6.3a);

b. lower semi-continuous at $\mathbf{x}_o \in D$ if for each $\varepsilon > 0$ there exists a $\delta > 0$ such that $||\mathbf{x} - \mathbf{x}_o|| < \delta,\ \mathbf{x} \in D$, implies $f(\mathbf{x}_o) - \varepsilon < f(\mathbf{x})$ (Figure 6.3b).

Example 6.3: The real-valued function

$$f(x) = \begin{cases} 1/2, & x = 1; \\ x + 1, & x \neq 1 \end{cases}$$

is lower semi-continuous (Figure 6.4a) while the real-valued function

$$f(x) = \begin{cases} 1/2, & x = 0; \\ x^3, & x \neq 0 \end{cases}$$

is upper semi-continuous (Figure 6.4b). ■

Additionally, the function $y = f(\mathbf{x}) : D \subset R^n \rightarrow R$ is upper semi-continuous on D if it is upper semi-continuous at each point of D or, equivalently, if either of the following conditions hold:

i. the **upper level set** $U_f(\alpha) = \{\mathbf{x} | f(\mathbf{x}) \geq \alpha,\ \mathbf{x} \in D\}$ is closed relative to D for each $\alpha \in R$.

ii. the **hypograph of** ***f***, $hyp(f) = \{(\mathbf{x},\alpha) | f(\mathbf{x}) \geq \alpha,\ \mathbf{x} \in D,\ \alpha \in R\}$ is closed relative to $D \times R \subset R^{n+1}$ Figure 6.5).

Similarly, the function $y = f(\mathbf{x}) : D \subset R^n \rightarrow R$ is lower semi-continuous on D if it is lower semi-continuous at each point of D, or equivalently, if either of the following conditions hold:

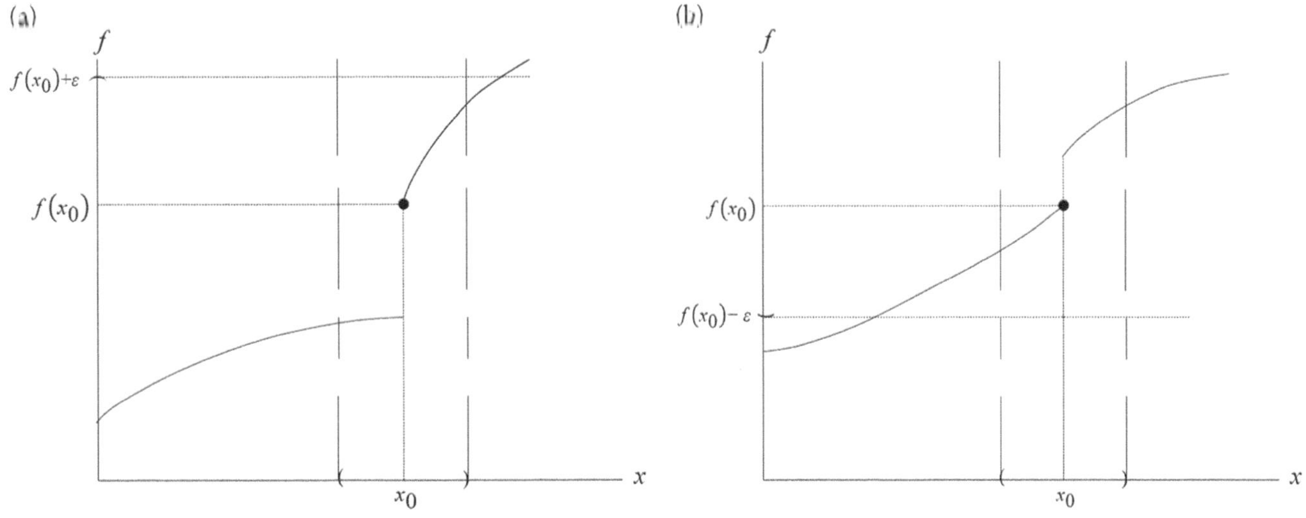

FIGURE 6.3 (a) f is upper semi-continuous at x_o; (b) f is lower semi-continuous at x_o.

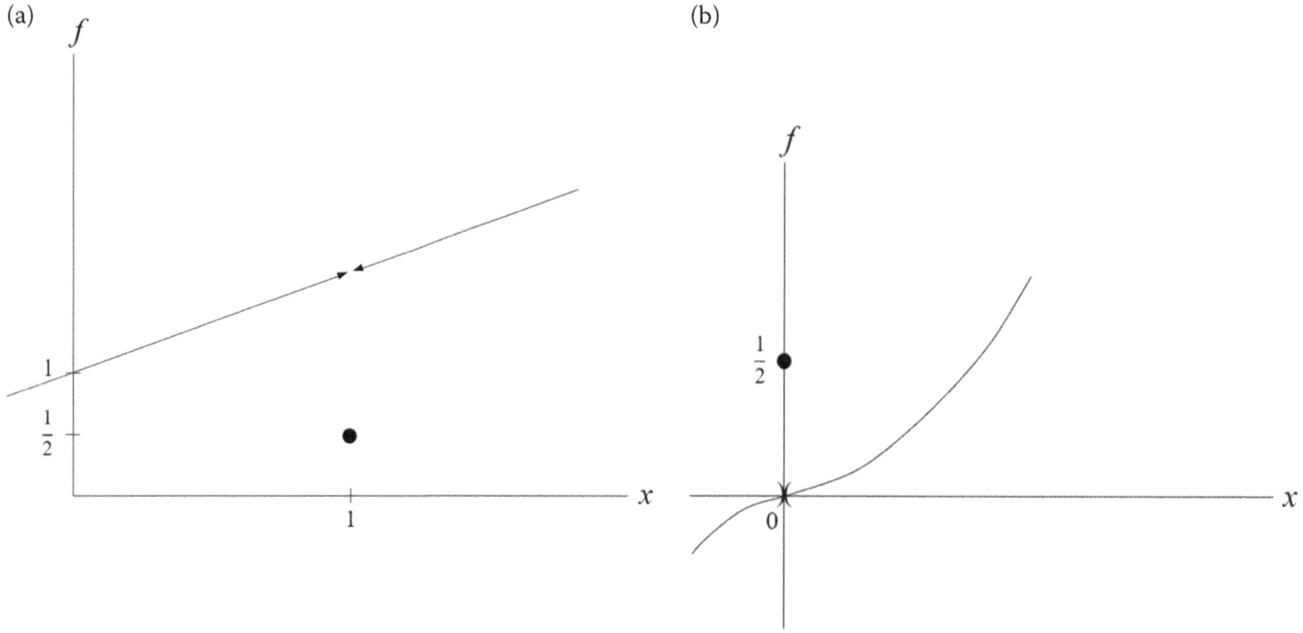

FIGURE 6.4 (a) f is lower semi-continuous at $x = 1$; (b) f is upper semi-continuous at $x = 0$.

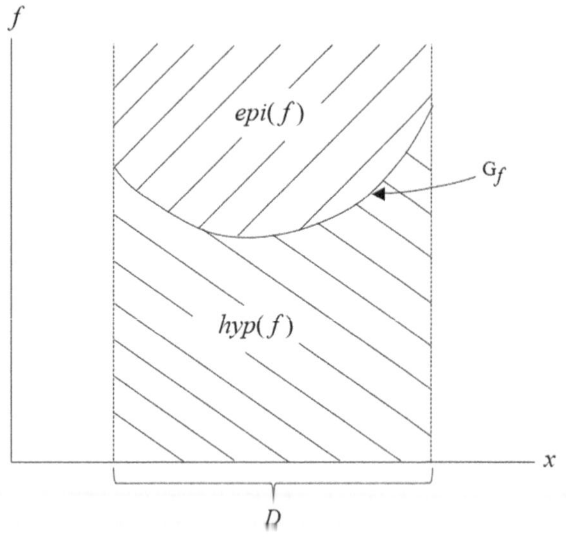

FIGURE 6.5 The graph, epigraph, and hypograph of f over $D \subset R$.

iii. the **lower level set** $L_f(\alpha) = \{\boldsymbol{x} | f(\boldsymbol{x}) \leq \alpha, \boldsymbol{x} \in D\}$ is closed relative to D for each $\alpha \in R$.

iv. the **epigraph of** $\boldsymbol{f}$, $epi\ (f) = \{(\boldsymbol{x},\alpha) | f(\boldsymbol{x}) \leq \alpha, \boldsymbol{x} \in D, \alpha \in R\}$ is closed relative to $D \times R \subset R^{n+1}$ (Figure 6.5).

As Figure 6.5 reveals, if we write the **graph of** $\boldsymbol{f}$ as $G_f = \{(x, f(x)) | x \in D\} \subset R^2$, then hyp (f) and epi (f) are two sets related to G_f in that hyp (f) consists of points on or below the graph of f while epi (f) consists of points on or above the graph of f.

We note briefly that f is lower semi-continuous at $\boldsymbol{x}_o \in D$ if and only if $-f$ is upper semi-continuous at $\boldsymbol{x}_o \in D$. Moreover, if f is both upper and lower semi-continuous at $\boldsymbol{x}_o \in D$, then f is continuous at $\boldsymbol{x}_o \in D$.

Let us harken back to Weierstrass's theorem (Theorem 6.2), which states essentially that a function $y = f(\boldsymbol{x}) : D \subset R^n \rightarrow R$ continuous over a compact domain D takes on a least and a greatest value (a global minimum and a global maximum) at least once over D. We also noted earlier that Weierstrass's theorem serves as a sufficient condition for the existence of a global maximum and a global minimum on D. If we now substitute semi-continuity for the continuity of f on D, then we have a weakened or less restrictive sufficient condition for the existence of global extrema. Specifically, we can now state an extended or revised version of Weierstrass's theorem.

Theorem 6.4: (Modified Weierstrass's Theorem)

a. An upper semi-continuous function $y = f(\boldsymbol{x})$: $D \subset R^n \to R$, with D compact, is bounded from above and attains in D the value

$$\alpha = \sup_{\boldsymbol{x} \in D} f(\boldsymbol{x}) = \max_{\boldsymbol{x} \in D} f(\boldsymbol{x});$$

b. A lower semi-continuous function $y = f(\boldsymbol{x})$: $D \subset R^n \to R$, with D compact, is bounded from below and attains in D the value

$$\beta = \inf_{\boldsymbol{x} \in D} f(\boldsymbol{x}) = \min_{\boldsymbol{x} \in D} f(\boldsymbol{x}).$$

For instance, the implication of part (a) is that there exists an $\boldsymbol{x}_o \in D$ such that, for all $\boldsymbol{x}\ (\neq \boldsymbol{x}_o) \in D, f(\boldsymbol{x}) \leq f(\boldsymbol{x}_o)$. This is because a is finite over D and there exists an $\boldsymbol{x}_o \in D$ such that $f(\boldsymbol{x}_o) = \alpha$.

Section 6.2 Exercises:

6.2.1 Is the real-valued function

$$y = f(x) = \begin{cases} 1, & x \geq 0; \\ -1, & x < 0 \end{cases}$$

upper or lower semi-continuous?

6.2.2 Is the real-valued function

$$y = f(x) = \begin{cases} 0.5, & x < 1; \\ 1, & x = 1; \\ 0.25, & x > 1 \end{cases}$$

upper or lower semi-continuous?

6.2.3 Are the following statements true or false?

1. A real-valued function $y = f(\boldsymbol{x}) : D \to R$ is upper semi-continuous if and only if the set $\{x | f(x) < \alpha, x \in D, \alpha \in R\}$ is open.
2. A real valued function $y = f(\boldsymbol{x}) : D \to R$ is lower semi-continuous if and only if the set $\{x | f(x) > \alpha, x \in D, \alpha \in R\}$ is open.

   For D a metric space:

3. The real-valued function $y = f(\boldsymbol{x}) : D \to R$ is upper semi-continuous at $\boldsymbol{x}_o \in D$ if

$$\limsup_{x \to x_o} f(x) \leq f(x_o).$$

4. The real-valued function $y = f(\boldsymbol{x}) : D \to R$ is lower semi-continuous at $\boldsymbol{x}_o \in D$ if

$$\liminf_{x \to x_o} f(x) \geq f(x_o).$$

6.2.4 Sketch the epigraph of $y = f(x) = e^x$, $x \in [0, a] \subset R$.

6.2.5 Sketch the epigraph of $y = f(x) = |x|$, $x \in [-1, 1] \subset R$.

6.2.6 Sketch the hypograph of $y = f(x) = x^a$, $0 \leq a \leq 1$, $a \in R$, $x \in R$.

6.2.7 Sketch the hypograph of $y = f(x) = \log x$ for $x > 0$, $x \in R$.

CHAPTER 7

Local Extrema of Real-Valued Functions

7.1 FUNCTIONS OF A SINGLE INDEPENDENT VARIABLE

In Chapter 5, we specified the notion of a *local* extreme value of a real-valued function. So for a function such as $y = f(x): D \subset R \to R$, it is imperative to examine the conditions under which local extrema may be recognized and located. As an aid in this endeavor, let us first consider

Definition 7.1: Let $y = f(x): [a, b] \subset R \to R$. Then f is termed an **increasing (decreasing) function** over $[a, b]$ if and only if $f(x_1) \le f(x_2)$ $(f(x_1) \ge f(x_2))$ when $a \le x_1 < x_2 \le b$.

If f is increasing or decreasing (but not both) over $[a, b]$, it is said to be **monotonic**. And if strict inequality holds in Definition 7.1, then f is termed **strictly increasing** (**strictly decreasing**) over $[a, b]$ and thus is said to be **strictly monotonic**.

Rather than work with function values, let us frame the preceding definition in terms of the slope of f. Specifically, we have

Theorem 7.1: Let $y = f(x): [a, b] \subset R \to R$ be continuous over $[a, b]$ and differentiable over (a, b). Then f is an increasing (decreasing) function over (a, b) if and only if $f'(x) \ge 0 (f'(x) \le 0)$ for all $x \in (a, b)$.

Here, too, strict inequality implies that f is strictly increasing (strictly decreasing) over (a, b). If $f'(x) = 0$ for all $x \in (a, b)$, then f is said to be constant over (a, b) and thus may be characterized as both increasing *and* decreasing (though not strictly) over (a, b).

Let us sharpen the focus of Theorem 7.1 by considering the behavior of f and f' at a specific value of x, namely $\bar{x}$. We thus state

Theorem 7.2: Let $y = f(x): [a, b] \subset R \to R$ be continuous over $[a, b]$ and differentiable over (a, b), with $\bar{x}$ taken to be an interior point of (a, b). Then, if $f'(\bar{x}) > 0 (f'(\bar{x}) < 0)$:

a. f is strictly increasing (strictly decreasing) at $\bar{x}$; and

b. in any δ – neighborhood of $\bar{x}$, $\delta(\bar{x})$,f takes on values both smaller and larger than $f(\bar{x})$.

Furthermore, if $f'$is continuous at $\bar{x}$, then f' exhibits *persistence of sign*, i.e.,

c. with $f'(\bar{x}) \neq 0, f'$ has the *same sign* as $f'(\bar{x})$ for *all* $x \in \delta(\bar{x})$.

The importance of part (c) of this theorem cannot be overstated. In fact, it behooves us to examine its plausibility. The continuity of f at $\bar{x}$ means that for any $\varepsilon > 0$ there exists a $\delta > 0$ such that $|f'(x) - f'(\bar{x})| < \varepsilon$ whenever $|x - \bar{x}| < \delta$ or

$$f'(\bar{x}) - \varepsilon < f'(x) < f'(\bar{x}) + \varepsilon$$
$$\text{when } x \in \delta(\bar{x}). \tag{7.1}$$

Take $\varepsilon = |f'(\bar{x})|$. Hence, (7.1) becomes

$$f'(\bar{x}) - |f'(\bar{x})| < f'(x) < f'(\bar{x}) + |f'(\bar{x})|$$
$$\text{when } x \in \delta(\bar{x}). \tag{7.1.1}$$

When $f'(\bar{x}) > 0$, (7.1.1) becomes

$$0 < f'(x) < 2f'(\bar{x}) \text{ for all } x \in \delta(\bar{x}); \tag{7.2}$$

and when $f'(\bar{x}) < 0$, (7.1.1) yields

$$-2f'(\bar{x}) < f'(x) < 0 \text{ for all } x \in \delta(\bar{x}). \tag{7.3}$$

From part (a) of Theorem 7.1, $f'(\bar{x}) > 0 (f'(\bar{x}) < 0)$ is sufficient but not necessary for f to be strictly increasing (strictly decreasing) at $\bar{x}$, i.e., f may be a strictly increasing (strictly decreasing) function at $\bar{x}$even though $f'(\bar{x}) = 0$. For example, consider $y = f(x) = x^3$, $-\infty < x < +\infty$. This function is strictly increasing at $\bar{x} = 0$, yet $f'(0) = 0$.

Let us reexamine the *persistence of sign* attribute of continuous functions offered earlier. The following generalization will be employed quite frequently in our search for local extrema. Specifically, let $y = f(x)$: $[a, b] \subset R \to R$. Then:

1. if f is continuous at $\bar{x}$ with $f(\bar{x}) \neq 0$, then $f(\bar{x} + \theta h)$, $0 < \theta < 1$, has the same sign as $f(\bar{x})$, $|h| < \delta$;
2. if f' is continuous at $\bar{x}$ with $f'(\bar{x}) \neq 0$, then $f'(\bar{x} + \theta h)$, $0 < \theta < 1$, has the same sign as $f'(\bar{x})$, $|h| < \delta$; and
3. in general, if $f^{(i)}$ is continuous at $\bar{x}$ with $f^{(i)}(\bar{x}) \neq 0$, then $f^{(i)}(\bar{x} + \theta h)$, $0 < \theta < 1$, has the same sign as $f^{(i)}(\bar{x})$, $|h| < \delta$, $i = 0, 1, \ldots, n$ (provided, of course, that these higher-order derivatives exist at $\bar{x}$).

If it is known that f attains a local maximum at a point $x_o \in (a, b)$, then f must be increasing for $x_o - \delta < x < x_o$ and decreasing for $x_o < x < x_o + \delta$. But this is equivalent to the condition that $f'(x) \geq 0$ for all $x \in (x_o - \delta, x_o)$ and $f'(x) \leq 0$ for all $x \in (x_o, x_o + \delta)$ (by Theorem 7.1). Now, if $f'(x)$ is non-negative for $x_o - \delta < x < x_o$ and non-positive for $x_o < x < x_o + \delta$, then f' should equal zero somewhere within $\delta(x_o)$. This observation can be formalized by the following theorem.

Theorem 7.3: (Existence Theorem). If a real-valued function, continuous over $[a, b]$, differs in sign at its endpoints, then it must *vanish* (it equals zero) at some interior point(s) of $[a, b]$.

Let us apply this theorem to $f'(x)$ (assumed continuous on $[a, b]$). If $f'(x)$ is non-negative for $x \in (x_o - \delta, x)$ and non-positive for $x \in (x_o, x_o + \delta)$, then clearly $f'(x)$ must vanish for at least one $x \in \delta(x_o)$. For which values of $x \in \delta(x_o)$ does $f'(x)$ vanish? If f has a *strong local maximum* on (a, b), then $f'(x)$vanishes for only one value of $x \in \delta(x_o)$, namely $x = x_o$. Hence $f'(x_o) = 0$. (A similar argument holds for f attaining a strong local minimum at, say, a point $x_1 \in (a, b)$, i.e., we require $f'(x_1) = 0$.) If f has a *weak local maximum* at $x_o \in (a, b)$, not only does $f'(x_o) = 0$, but $f'(x)$ vanishes for all other values of $x \in \delta(x_o)$. (Note that f can assume, say, a local extremum at x_o without f' vanishing there. This case emerges if f is not differentiable at x_o, i.e., f' has a finite discontinuity at $x_o$and does not exist there. For example, consider $y = f(x) = |x|$, $-\infty < x < +\infty$.)

Let us now consider some additional terminology.

Definition 7.2: The point $\bar{x}$is termed a **critical value** of $x \in (a, b)$ if: (a) $f'(\bar{x}) = 0$, or (b) $f'(\bar{x})$ fails to exist.

When $f'(\bar{x}) = 0$,f is neither increasing nor decreasing; the value of f at $\bar{x}$ is *stationary* in that the tangent line to f at $\bar{x}$ is horizontal. In this regard we have

Definition 7.3: Any point $(\bar{x}, f(\bar{x}))$ where f' vanishes is called a **stationary point**, with $f(\bar{x})$ termed a **stationary value** of f.

So if $\bar{x}$ is a critical value of $x \in (a, b)$, then either $f'(\bar{x}) = 0$ (and thus $f(\bar{x})$ is a stationary value of f on (a, b)), or $f'(\bar{x})$ does not exist.

Can f possess a stationary point that is not an extremum? The answer is in the affirmative in that f may have a **horizontal point of inflection** for some critical value of $x = x' \in (a, b)$ (see, for instance, point A of Figure 7.1a). In general, we have

Definition 7.4: A point where a curve crosses over its tangent line (horizontal or not) and changes the direction of its concavity from upwards to downwards (or *vice versa*) is a **point of inflection**.

So if a function with a continuous first derivative over (a, b) possesses a local extremum, then this extreme point is also a stationary point – but not every stationary point is an extreme point.

7.2 A NECESSARY CONDITION FOR A LOCAL EXTREMUM

It should be evident that if we are to determine any extreme values of a function with a continuous first

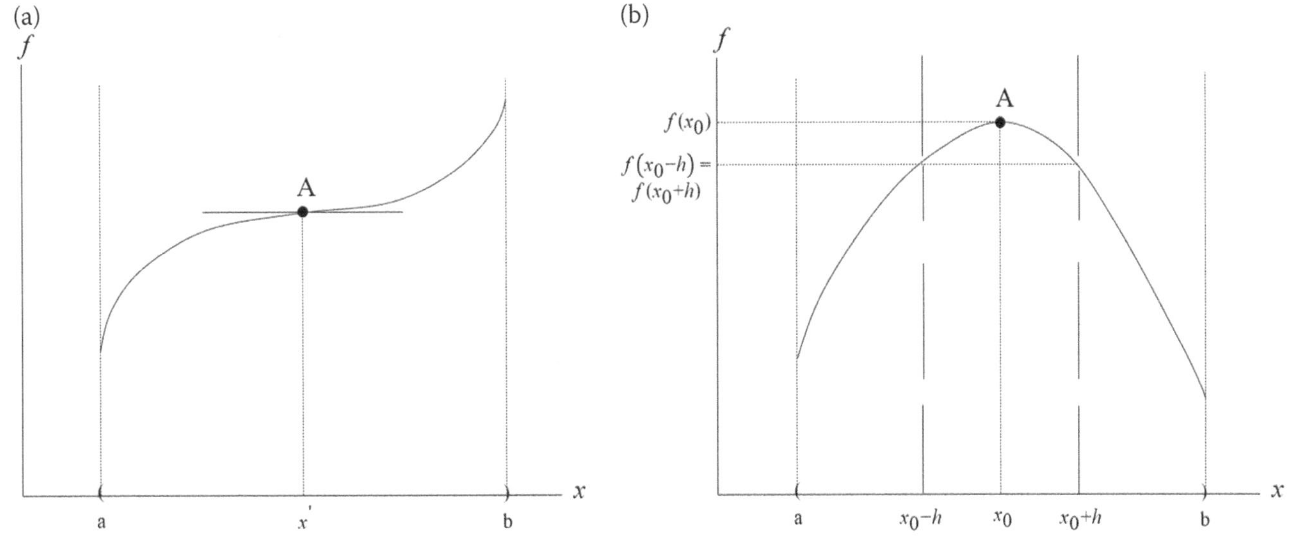

FIGURE 7.1 (a) f has a horizontal point of inflection at x'; (b) f has a local maximum at x_o.

derivative, our search must begin with an examination of its set of stationary points. To this end, we shall consider necessary, sufficient, and necessary and sufficient conditions for isolating local extreme values. We begin with

Theorem 7.4: (A Necessary Condition). Let $y = f(x)$ be a real-valued function of class $C^{(1)}$ throughout a δ-neighborhood of the point $x_o \in (a, b)$. If f has a local extremum at x_o, then $f'(x_o) = 0$.

Proof: Since this condition is prerequisite for a local extremum of f, we shall demonstrate that the said extremum cannot occur at $x_o$if $f'(x_o) \neq 0$. Consider the following two cases.

1. If $f'(x_o) > 0$, then Taylor's formula with remainder of order 1 can be written as

$$f(x_o + h) - f(x_o) = f'(x_o + \theta h)h, \quad 0 < \theta < 1, \ |h| < \delta. \tag{7.3}$$

According to the *persistence of sign* attribute of a continuous derivative, $f'(x_o) > 0$ implies that $f'(x_o + \theta h) > 0$ for all $|h| < \delta$. For $h > 0, f'(x_o + \theta h)h > 0$ and thus $f(x_o + h) - f(x_o) > 0$; and for $h < 0, f'(x_o + \theta h)h < 0$ so that now $f(x_o + h) - f(x_o) < 0$. Hence, $f(x_o + h) - f(x_o)$ is negative for x close to x_o, but smaller; it is positive for x close to x_o, but larger. However, this cannot happen if a local extremum occurs at x_o (see Figure 7.1b).

2. If $f'(x_o) < 0$, the *persistence of sign* property informs us that $f'(x_o + \theta h) < 0$ for$|h| < \delta$ and thus $f'(x_o + \theta h)h$ is negative for $h > 0$ and positive for $h < 0$. Hence $f(x_o)$ is not an extreme value of f on (a, b) since the sign of $f(x_o + h) - f(x_o)$ changes when the sign of h does. Q. E. D.

Example 7.1: Does the real-valued function $y = f(x) = 3 - 2x + x^2, x \in (-1, 4)$, possess a local extremum? Setting $f' = -2 + 2x = 0$ yields the critical value $x = 1$. Hence the stationary value of f over $(-1\ 4)$ is $f(1) = 2$. Calculating $f(x_o + h) - f(x_o)$ we find that

$$f(1 + h) - f(1) = h^2 > 0, \ |h| < \delta.$$

Hence f assumes a strong local minimum at $x = 1$.

Example 7.2: Determine if the real-valued function $y = f(x) = 5 - (x - 3)^3, x \in (1, 5)$, attains a local extremum. Setting $f' = -3(x - 3)^2 = 0$ we see that f is stationary at the critical value $x = 3$. Is $f(3) = 5$ a local extremum of f over $(1, 5)$? Since

$$f(3 + h) - f(3) = -h^3 > 0 \text{ for } h < 0$$

and

$$f(3 + h) - f(3) = -h^3 < 0 \text{ for } h > 0,$$

it is obvious that f does not attain a local extremum over $(1, 5)$. (Remember that, for a local extremum, the sign of $f(x_o + h) - f(x_o)$ cannot change when the sign of h changes, $|h| < \delta$.)

While Theorem 7.4 offers an operational procedure for determining the points at which f is stationary over (a, b) (simply set $f'(x) = 0$ and solve for the value(s) of x which makes f' vanish), it provides no insight into whether any particular (stationary) point corresponds to a maximum, minimum, or horizontal point of inflection of f. To resolve this issue, we shall examine conditions which are either sufficient or necessary and sufficient to identify a local extremum. As explained below, a sufficient condition allows us to distinguish a local maximum from a local minimum. But a necessary and sufficient condition goes a step further - - it enables us to distinguish a local maximum from a local minimum from a stationary inflectional value. Thus a complete or exhaustive classification of stationary points is provided by a necessary and sufficient condition.

7.3 A SUFFICIENT CONDITION FOR A LOCAL EXTREMUM

Let us utilize the first as well as the second derivative of f, f'', to develop a sufficient condition for a local extremum. We thus have

Theorem 7.5: (A Sufficient Condition). Let $y = f(x)$ be a real-valued function of class $C^{(2)}$ throughout a δ - neighborhood of the point $x_o \in (a, b)$. If $f'(x_o) = 0$ and $f''(x_o) \neq 0$, then f has a strong local extremum at x_o. Moreover: (a) if $f''(x_o) < 0$, f has a strong local maximum at x_o; and (b) if $f''(x_o) > 0$, f has a strong local minimum at x_o.

Proof: Let us specify the second-order Taylor expansion of f near x_o (given that, by hypothesis, $f'(x_o) = 0$ and $f''(x_o) \neq 0$) as

$$f(x_o + h) - f(x_o) = \frac{1}{2!} f''(x_o + \theta h) h^2,$$

$$0 < \theta < 1,\ |h| < \delta. \qquad (7.4)$$

Since f'' is continuous and non-vanishing at x_o, the *persistence of sign* property of f'' informs us that we can find an arbitrarily small interval about x_o, $\delta(x_o)$, such that f'' has the same sign as $f''(x_o)$ for all $x \in \delta(x_o)$. Consider the following cases:

1. If $f''(x_o) < 0$, then $\frac{1}{2!} f''(x_o + \theta h) h^2 < 0$ when x, and thus $x_o + \theta h$, is close to x_o, $|h| < \delta$. Hence the right-hand side of (7.4) is negative and thus $f(x_o + h) - f(x_o) < 0$ for all $x \in \delta(x_o)$ and thus f assumes a strong local maximum at x_o.

2. For $f''(x_o) > 0$, it follows that $\frac{1}{2!} f''(x_o + \theta h)$ $h^2 > 0$ when x, and thus $x_o + \theta h$, is close to x_o, $|h| < \delta$. Hence the right-hand side of (7.4) is positive and thus $f(x_o + h) - f(x_o) > 0$ for all $x \in \delta(x_o)$ so that f has a strong local minimum at x_o. Q. E. D.

The preceding discussion has indicated that the sign of f'' enables us to distinguish a local maximum from a local minimum at a point were $f' = 0$. But the case where f'' also equals zero at this point is left undecided. Hence Theorem 7.5 represents a condition which is only sufficient (and thus incomplete since local extrema may still exist at a point where $f'' = 0$) and not necessary and sufficient to identify a strong local extremum.

Example 7.3: Let the real-valued function $y = f(x) = x^3 - 6x^2 + 15$ be defined for all $x \in (-2, 7)$. Isolate any local extreme points of f. Setting $f' = 0$ and solving for the critical values of x yields $x = 0,4$. From $f'' = 6x - 12$ we have, for $x = 0$, $f''(0) = -12 < 0$; and for $x = 4$, $f''(4) = 12 > 0$. Hence f has a strong local maximum at $x = 0$ and a strong local minimum at $x = 4$.

7.4 A NECESSARY AND SUFFICIENT CONDITION FOR A LOCAL EXTREMUM

The preceding sufficient condition for a local extremum was inconclusive when both f' and f'' vanished at a critical $x = x_o \in (a, b)$. A generalization of Theorem 7.5 will handle this situation.

Theorem 7.6: (A Necessary and Sufficient Condition). Let $y = f(x)$ be a real-valued function of class $C^{(n)}$ throughout a δ - neighborhood of the point $x_o \in (a, b)$. If $f'(x_o) = f''(x_o) = \ldots = f^{n-1}(x_o) = 0$ and $f^{(n)}(x_o) \neq 0$, then:

1. f attains a strong local extremum at x_o if n is even. Moreover:

   a. if $f^{(n)}(x_o) < 0$, f has a strong local maximum at x_o; and

b. if $f^{(n)}(x_o) > 0$, f has a strong local minimum at x_o.

2. f attains a horizontal point of inflection at x_o if n is odd.

Proof: Consider the nth order Taylor expansion of f near x_o(under the hypothesis that $f^{(i)}(x_o) = 0$, $i = 1, \ldots, n-1$ and $f^{(n)}(x_o) \neq 0$) or

$$f(x_o + h) - f(x_o) = \frac{1}{n!} f^{(n)}(x_o + \theta h) h^n,$$
$$0 < \theta < 1, \ |h| < \delta. \quad (7.5)$$

1. Suppose n is even. Since $f^{(n)}$ is continuous and non-vanishing at x_o, we can find an arbitrarily small δ - neighborhood about x_o, $\delta(x_o)$, such that, by the *persistence of sign* property, $\frac{1}{n!} f^{(n)}(x_o + \theta h) h^n$ has the same sign as $f^{(n)}(x_o)$ when x, and thus $x_o + \theta h$, is close to x_o, $|h| < \delta$. If $f^{(n)}(x_o) < 0$, then the right-hand side of (7.5) is negative and thus $f(x_o + h) - f(x_o) < 0$ for all $|h| < \delta$. Hence f has a strong local maximum at x_o. When $f^{(n)}(x_o) > 0$, the right-hand side of (7.5) is positive and thus $f(x_o + h) - f(x_o) > 0$ for all $|h| < \delta$ so that, in this instance, f has a strong local minimum at x_o.
2. If n is odd, h^n is negative for $x_o - \delta < x < x_o$ and positive for $x_o < x < x_o + h$, while, again, due to the *persistence of sign* property, $f^{(n)}(x_o + \theta h)$ is of unchanging sign for $|h| < \delta$. However, the sign of the right-hand side of (7.5) changes as x passes through x_o from left to right since the sign of h^n does. That is, when $f^{(n)}(x_o) < 0$, it follows that $f^{(n)}(x_o + \theta h) < 0$ but $\frac{1}{n!} f^{(n)}(x_o + \theta h) h^n$ is positive for $h < 0$ and negative for $h > 0$. This implies that the sign of $f(x_o + h) - f(x_o)$ changes when the sign of h does. Similarly, for $f^{(n)}(x_o) > 0$, we may again invoke the *persistence of sign* property of $f^{(n)}(x_o + \theta h)$ and conclude that $f(x_o + h) - f(x_o)$ is negative when $h < 0$ and positive when $h > 0$, $|h| < \delta$. Clearly, in this circumstance, f cannot have a strong local extremum at x_o - - the stationary point attained is a horizontal point of inflection at x_o (it is horizontal since $f'(x_o) = 0$). Q. E. D.

Example 7.4: Suppose the real-valued function $y = f(x) = (x-3)^4 + 10$ is defined for all $x \in (0, 6)$. Does f possess a local extremum over this interval? Let us first find

$$f'(x) = 4(x-3)^3, f''(x) = 12(x-3)^2,$$
$$f'''(x) = 24(x-3), \text{ and } f^{(iv)}(x) = 24,$$

where $f'(3) = f''(3) = f'''(3) = 0$ and $f^{(iv)}(3) \neq 0$. Since the order of the first non-vanishing derivative at $x = 3$ is even, part (1) of Theorem 7.6 indicates that f has a strong local extremum at $x = 3$. And with $f^{(iv)}(3) = 24 > 0$, part (1.b) of this theorem informs us that f has a strong local minimum at $x = 3$.

Example 7.5: Does the real-valued function $y = f(x) = 5 - (x-3)^3$, $x \in (1, 5)$, attain a local extremum over its domain? From

$$f'(x) = -3(x-3)^2, f''(x) = -6(x-3),$$
$$\text{and } f'''(x) = -6,$$

with $f'(3) = f''(3) = 0$ and $f'''(3) \neq 0$, it follows that since the order of the first non-vanishing derivative at $x = 3$ is odd, the function attains a horizontal point of inflection at $x = 3$.

Section 7.4 Exercises:

7.4.1 Does the real-valued function $y = f(x) = 2x^3 - 6x^2 + \frac{11}{6}x + 8$ attain a local extremum? Does f have a point of inflection?

7.4.2 Does the real-valued function $y = f(x) = x + x^{-1}$, $x \neq 0$, attain a local extremum? Does f have a point of inflection?

7.4.3 Does the real-value function $y = f(x) = \frac{1}{3}x^3 - 2x^2 + 3x + 10$ attain a local extremum? Does f have a point of inflection?

7.5 FUNCTIONS OF N INDEPENDENT VARIABLES

We begin with

Theorem 7.7: (A Necessary Condition). Let $y = f(\boldsymbol{x}) : D \subset R^n \to R$, D an open domain, be a function of class $C^{(1)}$ throughout a spherical δ - neighborhood $\delta(\boldsymbol{x}_o) \subset D$. If f has a local extremum at $\boldsymbol{x}_o \in D$, then $\nabla f(\boldsymbol{x}_o) = \boldsymbol{O}$.

Proof: Since the vanishing of ∇f at $\boldsymbol{x}_o$ is a prerequisite for the occurrence of a local extremum, we shall argue that a local extremum cannot occur at $\boldsymbol{x}_o$ unless $\nabla f = \boldsymbol{O}$.

there. Consider the first-order Taylor expansion of f at $\boldsymbol{x}_o$ (under the assumption that $\nabla f(\boldsymbol{x}_o) \neq \boldsymbol{O}$) or

$$f(\boldsymbol{x}_o + \boldsymbol{h}) - f(\boldsymbol{x}_o) = \nabla f(\boldsymbol{x}_o + \theta \boldsymbol{h})^T \boldsymbol{h},$$
$$0 < \theta < 1, \ \|\boldsymbol{h}\| < \delta. \quad (7.6)$$

Assume that $f_k(\boldsymbol{x}_o) \neq 0$ but all other partial derivatives of f vanish at $\boldsymbol{x}_o$. By the *persistence of sign* property, $f_k(\boldsymbol{x}_o + \theta \boldsymbol{h})$ has the same sign as $f_k(\boldsymbol{x}_o)$ for all $\boldsymbol{x}$ near $\boldsymbol{x}_o$ with h_k sufficiently small. Then

$$\nabla f(\boldsymbol{x}_o + \theta \boldsymbol{h})^T \boldsymbol{h} = f_k(\boldsymbol{x}_o + \theta \boldsymbol{h}) h_k \neq 0, \ |h_k| < \delta.$$

If $f_k(\boldsymbol{x}_o) > 0$, then $f_k(\boldsymbol{x}_o + \theta \boldsymbol{h}) > 0$ also and thus $f(\boldsymbol{x}_o + \boldsymbol{h}) - f(\boldsymbol{x}_o) = f_k(\boldsymbol{x}_o + \theta \boldsymbol{h}) h_k > 0 \ (< 0)$ as $h_k > 0 \ (< 0)$, $0 < \theta < 1$, $|h_k| < \delta$. Hence f cannot attain a local extremum at $\boldsymbol{x}_o$ since the sign of $f(\boldsymbol{x}_o + \boldsymbol{h}) - f(\boldsymbol{x}_o)$ changes whenever the sign of h_k does.

If $f_k(\boldsymbol{x}_o) < 0$, then $f_k(\boldsymbol{x}_o + \theta \boldsymbol{h}) < 0$ and $f(\boldsymbol{x}_o + \boldsymbol{h}) - f(\boldsymbol{x}_o) = f_k(\boldsymbol{x}_o + \theta \boldsymbol{h}) h_k < 0 \ (> 0)$ as $h_k > 0 \ (< 0)$, $0 < \theta < 1$, $|h_k| < \delta$. Again, a contradiction emerges. Thus f cannot attain a local extremum at $\boldsymbol{x}_o$ unless $f_i = 0, i = 1, \ldots, n$. Q. E. D.

When ∇f vanishes at an interior point of $D \subset R^n$, the value of f at this point is said to be stationary. In this regard,

Definition 7.5: Any point $(\bar{\boldsymbol{x}}, \bar{f}) \in R^{n+1}$ where ∇f vanishes is termed a **stationary point** while $\bar{f}$ is called a **stationary value** of f.

Also,

Definition 7.6: The point $\bar{\boldsymbol{x}} \in D \subset R^n$ is called a critical point of D if (a) $\nabla f(\bar{\boldsymbol{x}}) = \boldsymbol{O}$, or (b) $\nabla f(\bar{\boldsymbol{x}})$ fails to exist.

So if $\bar{\boldsymbol{x}} \in D$ is a critical point where ∇f exists, it follows that $\nabla f(\bar{\boldsymbol{x}}) = \boldsymbol{O}$ and $\bar{f}$ is a stationary value of f over D. Hence every extreme point where f is differentiable is a stationary point, but not conversely, i.e., f may attain a saddle point at $\bar{\boldsymbol{x}}$. Specifically,

Definition 7.7: A stationary point $(\bar{\boldsymbol{x}}, \bar{f})$, $\boldsymbol{x} \in R^2$, is termed a **saddle point** if $\bar{f}$ is simultaneously a strong local maximum in one direction and a strong local minimum in another. Here $\bar{f}$ is called a **saddle value** or **minimax value** of f on $D \subset R^2$.

So while $\nabla f(\bar{\boldsymbol{x}}) = \boldsymbol{O}$, it is not the case that $f(\boldsymbol{x}_o + \boldsymbol{h}) - f(\boldsymbol{x}_o) \leq 0 (\geq 0)$, $\|\boldsymbol{h}\| < \delta$.

Example 7.6: Let the real-valued function $y = f(\boldsymbol{x}) = -2x_1^2 - 3x_2^2 + 3x_1x_2 - 2x_1 - x_2 - 10$ be defined throughout the entirety of $D = R^2$. Does f attain a stationary value over D? From our necessary condition for a local extremum we set $\nabla f = \boldsymbol{O}$ to obtain

$$-4x_1 + 3x_2 = 2$$
$$3x_1 - 6x_2 = 1.$$

Solving this system simultaneously yields $\boldsymbol{x}_o^T = \left(-1, -\frac{2}{3}\right)$. Clearly f is stationary at this point.

Example 7.7: Does the real-valued function $y = f(\boldsymbol{x}) = -4x_1x_2 - 2x_1 - 2x_2 + 8$ attain a local extremum over $D = R^2$? Setting $\nabla f = \boldsymbol{O}$ renders the critical point $\boldsymbol{x}_o^T = \left(\frac{1}{2}, \frac{1}{2}\right)$. For $\boldsymbol{x} = \boldsymbol{x}_o + \boldsymbol{h}$, $f(\boldsymbol{x}_o + \boldsymbol{h}) = 4h_1h_2 + 7$. Since $f(\boldsymbol{x}_o) = 7$, we find that $f(\boldsymbol{x}_o + \boldsymbol{h}) - f(\boldsymbol{x}_o) = 4h_1h_2$, $\|\boldsymbol{h}\| < \delta$. Clearly this difference may be positive, negative, or zero, depending on the values given to h_1, h_2. Thus f does not attain a local extremum over D. In fact, f has a saddle point at $(\boldsymbol{x}_o, f(\boldsymbol{x}_o)) \in R^3$.

Suppose we have isolated a stationary point of a continuously differentiable function over an open domain $D \subset R^n$. How do we distinguish a local maximum from a local minimum at the said point? To answer this question, we look to

Theorem 7.8: (A Sufficient Condition). Let the real-valued function $y = f(\boldsymbol{x})$, $\boldsymbol{x} \in D \subset R^n$, D an open domain, be a function of class $C^{(2)}$ throughout a spherical δ-neighborhood of the point $\boldsymbol{x}_o$,$\delta(\boldsymbol{x}_o)$. If $\nabla f(\boldsymbol{x}_o) = \boldsymbol{O}$ and the quadratic form $Q(\boldsymbol{h}) = \boldsymbol{h}^T \boldsymbol{H}_f(\boldsymbol{x}_o) \boldsymbol{h}$ is negative (positive) definite for all $\boldsymbol{h} \neq \boldsymbol{O}$, then f attains a strong local extremum at $\boldsymbol{x}_o$. That is, if $\nabla f(\boldsymbol{x}_o) = \boldsymbol{O}$ and

(a) $f_{11}^o < 0$, $\begin{vmatrix} f_{11}^o & f_{12}^o \\ f_{21}^o & f_{22}^o \end{vmatrix} > 0, \ldots, (-1)^n |\boldsymbol{H}_f(\boldsymbol{x}_o)| > 0$, then

[$Q(\boldsymbol{h})$ is negative definite]

f has a strong local maximum at $\boldsymbol{x}_o$; while if

(b) $f_{11}^o > 0$, $\begin{vmatrix} f_{11}^o & f_{12}^o \\ f_{21}^o & f_{22}^o \end{vmatrix} > 0, \ldots, |\boldsymbol{H}_f(\boldsymbol{x}_o)| > 0$, then

[$Q(\boldsymbol{h})$ is positive definite]

f has a strong local minimum at $\boldsymbol{x}_o$.

Proof: From a second-order Taylor expansion of f, given that $\nabla f(\boldsymbol{x}_o) = \boldsymbol{O}$ and $Q(\boldsymbol{h}) > 0(<0)$ for $\boldsymbol{h} \neq \boldsymbol{O}$, we have

$$f(\boldsymbol{x}_o + \boldsymbol{h}) - f(\boldsymbol{x}_o) = \frac{1}{2!}\boldsymbol{h}^T \boldsymbol{H}_f(\boldsymbol{x}_o + \theta\boldsymbol{h})\boldsymbol{h},$$
$$0 < \theta < 1, \ \|\boldsymbol{h}\| < \delta.$$

With the second-order partial derivatives of f continuous at $\boldsymbol{x}_o$, $\boldsymbol{h}^T\boldsymbol{H}_f(\boldsymbol{x}_o + \theta\boldsymbol{h})\boldsymbol{h}$ has the same sign as $\boldsymbol{h}^T\boldsymbol{H}_f(\boldsymbol{x}_o)\boldsymbol{h}$ when $\boldsymbol{x}$, and thus $\boldsymbol{x}_o + \theta\boldsymbol{h}$, is near $\boldsymbol{x}_o$. Hence $f(\boldsymbol{x}_o + \boldsymbol{h}) - f(\boldsymbol{x}_o)$ has the same sign as $\boldsymbol{h}^T\boldsymbol{H}_f(\boldsymbol{x}_o)\boldsymbol{h}$, $\|\boldsymbol{h}\| < \delta$.

When $\boldsymbol{h}^T\boldsymbol{H}_f(\boldsymbol{x}_o)\boldsymbol{h} < 0$, $\boldsymbol{h} \neq \boldsymbol{O}$(negative definiteness), $f(\boldsymbol{x}_o + \boldsymbol{h}) - f(\boldsymbol{x}_o) < 0$ for $\|\boldsymbol{h}\| < \delta$ so that f has a strong local maximum at $\boldsymbol{x}_o$. If $\boldsymbol{h}^T\boldsymbol{H}_f(\boldsymbol{x}_o)\boldsymbol{h} > 0$, $\boldsymbol{h} \neq \boldsymbol{O}$ (positive definiteness), $f(\boldsymbol{x}_o + \boldsymbol{h}) - f(\boldsymbol{x}_o) > 0$ for small $\|\boldsymbol{h}\|$. In this instance f has a strong local minimum at $\boldsymbol{x}_o$. Q. E. D.

What about the cases in which $\nabla f(\boldsymbol{x}_o) = \boldsymbol{O}$ and $Q(\boldsymbol{h})$ is indefinite or semi-definite? For $Q(\boldsymbol{h})$ indefinite (i.e., positive for some vectors $\boldsymbol{h}$ and negative for others), $f(\boldsymbol{x}_o + \boldsymbol{h}) - f(\boldsymbol{x}_o)$ is correspondingly positive for suitably restricted variations in $\boldsymbol{h}$ from $\boldsymbol{x}_o$ in one direction and negative for like variations in another. Hence f does not attain a strong local extremum at $\boldsymbol{x}_o$. In fact, for $\boldsymbol{x} \in R^2$, f has a saddle point at $\boldsymbol{x}_o$.

If $Q(\boldsymbol{h})$ is semi-definite, no conclusion about extreme values of f can be reached. In this instance our classification scheme is incomplete - - Theorem 7.8 provides only a sufficient condition for a strong local extremum and not one which is both necessary and sufficient.

Example 7.8: Suppose the real-valued function $y = f(\boldsymbol{x}) = -2x_1^2 - 3x_2^2 + x_1x_2 + x_1 + x_2$ is defined for all $\boldsymbol{x} \in R^2$. Does f attain an extremum over its domain? Setting $\nabla f = \boldsymbol{O}$ enables us to find $\boldsymbol{x}_o^T = \left(\frac{7}{23}, \frac{5}{23}\right)$. And with

$$\boldsymbol{H}_f(\boldsymbol{x}_o) = \begin{bmatrix} -4 & 1 \\ 1 & -6 \end{bmatrix},$$

we see that $M_1 = -4 < 0$ and $M_2 = |\boldsymbol{H}_f(\boldsymbol{x}_o)| = 23 > 0$. Hence $Q(\boldsymbol{h})$ is negative definite and thus f attains a strong local maximum at $\boldsymbol{x}_o$.

Example 7.9: Let the real-valued function $y = f(\boldsymbol{x}) = x_1^3 - x_2^2 + x_1x_2 - x_1 - x_2$ be defined everywhere on R^2. Upon setting $\nabla f = \boldsymbol{O}$ we can solve the resulting (nonlinear) equation system simultaneously so as to obtain $x_1 = \left(\frac{1}{2} \pm \frac{5}{2}\right)/6$. Hence the critical points which emerge are $\boldsymbol{x}_o^T = \left(\frac{1}{2}, \frac{1}{4}\right)$ and $\boldsymbol{x}_1^T = \left(-\frac{1}{3}, \frac{2}{3}\right)$. With

$$\boldsymbol{H}_f(\boldsymbol{x}_o) = \begin{bmatrix} 3 & 1 \\ 1 & 2 \end{bmatrix},$$

it is easily see that $M_1 = 3$ and $M_2 = |\boldsymbol{H}_f(\boldsymbol{x}_o)| = 5$ so that $\boldsymbol{H}_f(\boldsymbol{x}_o)$ is a positive definite matrix and thus f has a strong local minimum at $\boldsymbol{x}_o$. For

$$\boldsymbol{H}_f(\boldsymbol{x}_1) = \begin{bmatrix} -2 & 1 \\ 1 & 2 \end{bmatrix},$$

$M_1 = -2$ and $M_2 = -5$. In this instance f attains a saddle point at $\boldsymbol{x}_1$ since $|\boldsymbol{H}_f(\boldsymbol{x}_1)| < 0$.

Section 7.5 Exercises:

7.5.1 Does the real-valued function $y = f(\boldsymbol{x}) = x_1^3 + x_2^3 - x_1 - x_2$ attain a local extremum over the positive orthant?

7.5.2 Does the real-valued function $y = f(\boldsymbol{x}) = -5(2 - x_1)^2 - 2(3 - x_2)^2$ attain a local extremum?

7.5.3 Does the real-valued function $y = f(\boldsymbol{x}) = -x_1^2 - x_2^2 + 6x_1x_2 - 2x_1 - 3x_2$ attain a local extremum over the non-negative orthant?

7.5.4 Does the real-valued function $y = f(\boldsymbol{x}) = x_1^2 + x_2^2 + x_3^2 + x_1x_2 + x_2x_3$ attain a local extremum in R^4?

7.5.5 Determine if the real-valued function $y = f(\boldsymbol{x}) = (x_1 - 2x_2^2)(x_1 - x_2^2)$ has a local extremum.

7.5.6 Does the real-valued function $y = f(\boldsymbol{x}) = x_1 + 3ex_2 - e^{x_1} - e^{3x_2}$, where e is the base of the natural logarithm, have a local extremum?

7.5.7 Let a sample point in the X, Y – plane be indicated as (X_i, Y_i), $i = 1, \ldots, n$, and let the sample regression line appear as $\hat{Y} = \hat{\beta}_o + \hat{\beta}_1X$. Also, let the ith residual or deviation from this line appear as $e_i = Y_i - \hat{Y}_i$. Determine $\hat{\beta}_o$ and $\hat{\beta}_1$ via the *Principle of Least Squares*: to obtain the line of best fit through the scatter of points (X_i, Y_i), $i = 1, \ldots, n$, choose $\hat{\beta}_o$ and $\hat{\beta}_1$ so as to minimize $\Sigma_i\, e_i^2 = \Sigma_i\,(Y - \hat{Y}_i)^2 = \Sigma_i\,(Y_i - \hat{\beta}_o - \hat{\beta}_1X_i)^2 = f(\hat{\beta}_o, \hat{\beta}_1)$.

7.6 ECONOMIC APPLICATIONS

7.6.1 Elasticity of Demand

Let us express the own-price demand function g for a product as

$$q = g(p), \quad \frac{dg}{dp} = \frac{dq}{dp} < 0 \text{ (the law of demand)}, \tag{7.7}$$

where q is quantity demanded and p is the product's own price. If g is continuous and single-valued, then its inverse g^{-1}, the **inverse own-price demand function**, exists with

$$p = g^{-1}(q), \; dg^{-1}/dq = \frac{dp}{dq} = \frac{1}{dg/dp}. \tag{7.8}$$

Suppose we write the own-price point elasticity of demand coefficient as

$$\begin{aligned} \in_D &= \frac{\text{\% change in quantity demanded}}{\text{\% change in price}} \\ &= \frac{dq}{dp}\frac{p}{q} = \frac{d \ln q}{d \ln p} < 0 \end{aligned} \tag{7.9}$$

since $dq/dp < 0$. Then:

a. $| \in_D | > 1$ - - *demand is elastic*;

b. $0 < | \in_D | < 1$ -- *demand is inelastic*; and

c. $| \in_D | = 1$ - - *demand is of unitary elasticity.*

Looking to the graphical interpretation of (7.9), let us work with the inverse own-price demand function (Figure 7.2).

$$| \in_D | = \left| \frac{1}{dp/dq}\frac{p}{q} \right| = \frac{1}{AC/CE}\frac{AC}{OC} = \frac{CE}{OC}$$

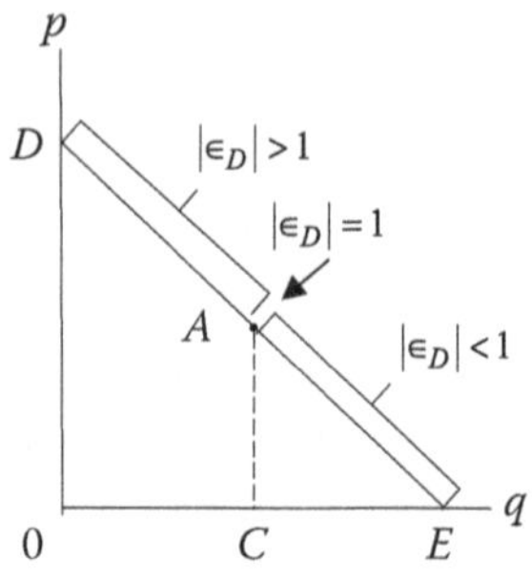

FIGURE 7.2 Point elasticity of demand.

At point D, $| \in_D | \to \infty$ as $q \to 0$.
At point A, $| \in_D | = 1$ with $CE = OC$.
At point E, $| \in_D | \to 0$ as $p \to 0$.

Let us next consider the concept of the cross-price elasticity of demand for a product. Here we consider two individual goods, q_1 and q_2 and their respective prices, p_1 and p_2. Additionally, let M represent money income. Then the demand function for q_1 is $q_1 = g^1(p_1, p_2, M)$ while the demand function for q_2 is $q_2 = g^2(p_1, p_2, M)$. In general, let $\in_{ij}$ denote the **cross-price elasticity of good*i*with respect to the price of commodity*j***. (If $i = j$, $\in_{ij}$ is the own-price elasticity of commodity i.)

By definition,

$$\in_{12} = \frac{\partial q_1}{\partial p_2}\frac{p_2}{q_1} = \frac{\partial(\ln q_1)}{\partial(\ln p_2)}. \tag{7.10}$$

Then:

a. $\in_{12} > 0$ - - q_1 and q_2 are **gross substitutes**. So as p_2 increases, the quantity demanded of q_1 increases. (Note that $\in_{12}$ involves both income and substitution effects; if the income effect is eliminated, they are *net* substitutes).

b. $\in_{12} < 0$- - q_1 and q_2 are **gross complements.** As p_2 increases, the quantity demand of q_1 decreases, (Here, too, if the income effect is eliminated, they are *net* complements.)

c. $\in_{12} = 0$- - q_1 and q_2 are **independent**. As p_2 increases, the quantity demanded of q_1 is not affected.

We next consider the **income elasticity of demand for good*i*** or

$$\in_{iM} = \frac{\partial q_1}{\partial M}\frac{M}{q_i} = \frac{\partial(\ln q_i)}{\partial(\ln M)}. \tag{7.11}$$

In this regard:

a. $\in_{iM} > 0$- - q_i is a **superior** or **normal good.** As M increases, the quantity demanded of q_i also increases.

i. $0 < \in_{iM} < 1$ - - q_i is a **necessity**. As M increases, the quantity demanded of q_i also increases, but less than proportionately.

ii. $\in_{iM} > 1$- - q_i is a **luxury**. As M increases, the quantity demanded of q_i also increases, but more than proportionately.

b $\in_{iM} < 0$- -q_i is an **inferior good.** As M increases, the quantity demanded of q_i decreases with a shift to more attractive alternatives. (If q_i displays *extreme inferiority*, it is termed a **Giffen good.**)

Section 7.6.1 Exercises:

7.6.1.1 The own-price demand function for q is $q = g(p) = 100 - 10p$. Is demand elastic at $(p, q) = (1.50, 85)$?

7.6.1.2 The own-price demand function for q is $q = g(p) = cp^{-1.2}$, c= constant. Is demand elastic?

7.6.1.3 The demand function for q_1 is $q_1 = g^1(p_1, p_2, M) = 1000 - \frac{3p_1M}{p_2}$. Find: $|\in_{11}|$, $\in_{12}$ and $\in_{1M}$ when $(p_1, p_2, M, q_1) = (10, 5, 600, 280)$.

76.1.4 Suppose the demand function for q_1 is $q_1 = 100\, p_1^{-1.3} p_2^{2.8} p_3^{-3.4} p_4^{0.001} M^{0.617}$. Find: $|\in_{11}|$, $\in_{12}$, $\in_{13}$, $\in_{14}$, and $\in_{1M}$.

7.6.1.5 Suppose

$$-1.2 = \frac{dq}{dp}\frac{p}{q}.$$

Verify that the associated own-price demand function is the one given in Exercise 7.6.1.2.

7.6.2 Production and Cost

We may view a production function as a technical relationship indicating the maximum quantity of output forthcoming from various combinations of inputs. For instance, we may express a production function by the equation

$$q = F(l, k), \tag{7.12}$$

where q represents a flow of output determined from a flow of labor (l) and capital (k) services in production.

In the short run, k (plant and equipment) will be held fixed at $k = \bar{k}$ so that (7.12) can be rewritten as

$$TP = q = F(l, \bar{k}) = f(l), \tag{7.13}$$

where f is the **total product (TP) of labor function**. It is assumed that f is continuous and single-valued with $f(0) = 0$. Additionally, we may define the **average product (AP) of labor function** as

$$AP = f(l)/l \tag{7.14}$$

(an indicator of the *efficiency* of the production process) and the **marginal product (MP) of labor function** as

$$MP = df/dl \tag{7.15}$$

(Figure 7.3. Relative to this figure we can note the following:

1. for $0 < l < l'$ we have **increasing returns to labor** ($f' > 0, f'' > 0$);
2. for $l' < l$, we have a range of **decreasing returns to labor** ($f' > 0, f'' < 0$);
3. point A depicts a point of diminishing returns to labor ($f' > 0, f'' = 0$)with MP at a maximum (point C);
4. AP corresponds graphically to the slope of a ray from the origin to the TP curve;
5. when AP is increasing, MP exceeds AP; when AP is at a maximum (points B, D)

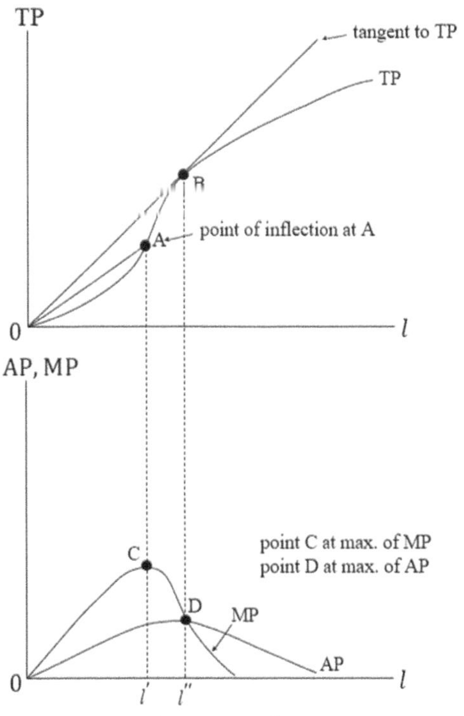

FIGURE 7.3 TP, MP, and AP curves.

$MP = AP$; and when AP is decreasing, MP is less than AP.

Consider next the **output elasticity of labor** - - an index of the output response to an increase in labor or

$$\eta = \frac{\text{\% change in output}}{\text{\% change in labor}} = \frac{dq}{dl}\frac{l}{q} = MP/AP. \quad (7.16)$$

Since η measures increasing or decreasing average returns to labor, it follows that (from Figure 7.3):

a. when AP is increasing, MP exceeds AP and thus $\eta > 1$;

b. when AP is at a maximum, $MP = AP$ and thus $\eta = 1$; and

c. when AP is decreasing, MP is less than AP and thus $\eta < 1$.

Turning now to cost in the short run, remember that economic costs are either *explicit* or *implicit*. That is, **explicit (external) costs** are essentially cash disbursements to suppliers of resources operating outside the firm and thus constitute **accounting costs** (fees for equipment rental, direct wages and benefits, payments to vendors of raw materials, etc.) **Implicit (internal) costs** involve no direct monetary payments for resources. Such costs are *imputed* or assigned on an **opportunity cost** basis, i.e., foregone rewards of entrepreneurship; rewards to self-employed resources/services; rewards for risk-taking, etc. If these foregone rewards are adequately compensated, the firm (organization) earns **a normal profit**. In sum:

economic cost = **opportunity cost**
= explicit cost + implicit cost
= accounting cost + normal profit.

Let us define **total variable cost** (TVC) as a *direct cost* (it varies directly with output) which includes all explicit input charges which depend on the rate of output. **Total fixed cost** (TFC) is *overhead cost* (e.g., long-term contractual payments such as interest, depreciation, etc.) which is incurred even if output is zero and involves all charges which are independent of the rate of output. Thus, TVC is *avoidable*; TFC is *unavoidable*.

Given these considerations, we can write **total cost** (TC) as

$$TC = TVC + TFC = TVC + \bar{C}, \quad (7.17)$$

where $\bar{C}$= constant. Then **average cost** (AC) is the sum of **average variable cost** (AVC) and **average fixed cost** (AFC) or

$$AC = \frac{TC}{q} = \frac{TVC}{q} + \frac{TFC}{q} = AVC + ATC \quad (7.18)$$

while **marginal cost** (MC) equals

$$MC = \frac{dTC}{dq} = \frac{dTVC}{dq}. \quad (7.19)$$

As we shall see, the shape of the TC function is determined by the shape of the underlying (short-run) production function $q = f(l)$. To see this, let us assume that f is continuous and single-valued. Then its inverse $l = f^{-1}(q)$ exists. Then, since capital is fixed at $\bar{k}$ = constant, total variable cost is labor cost wl, where w is the (constant) money wage rate. Given this setting,

$$TC = wl + \bar{C} = wf^{-1}(q) + \bar{C}, \quad (7.20)$$

which is a function of q alone. Then **average variable cost** (AVC) is

$$AVC = \frac{wf^{-1}(q)}{q}; \quad (7.21)$$

average fixed cost (AFC) is

$$AFC = \bar{C}/q, \quad (7.22)$$

average cost (AC) is

$$AC = AVC + AFC = \frac{wf^{-1}(q) + \bar{C}}{q}; \quad (7.23)$$

and **marginal cost** (MC) is

$$MC = \frac{dTC}{dq} = \frac{dTVC}{dq} = w\frac{df^{-1}(q)}{dq}. \quad (7.24)$$

All of this discussion is summarized in Figure 7.4. Note the following:

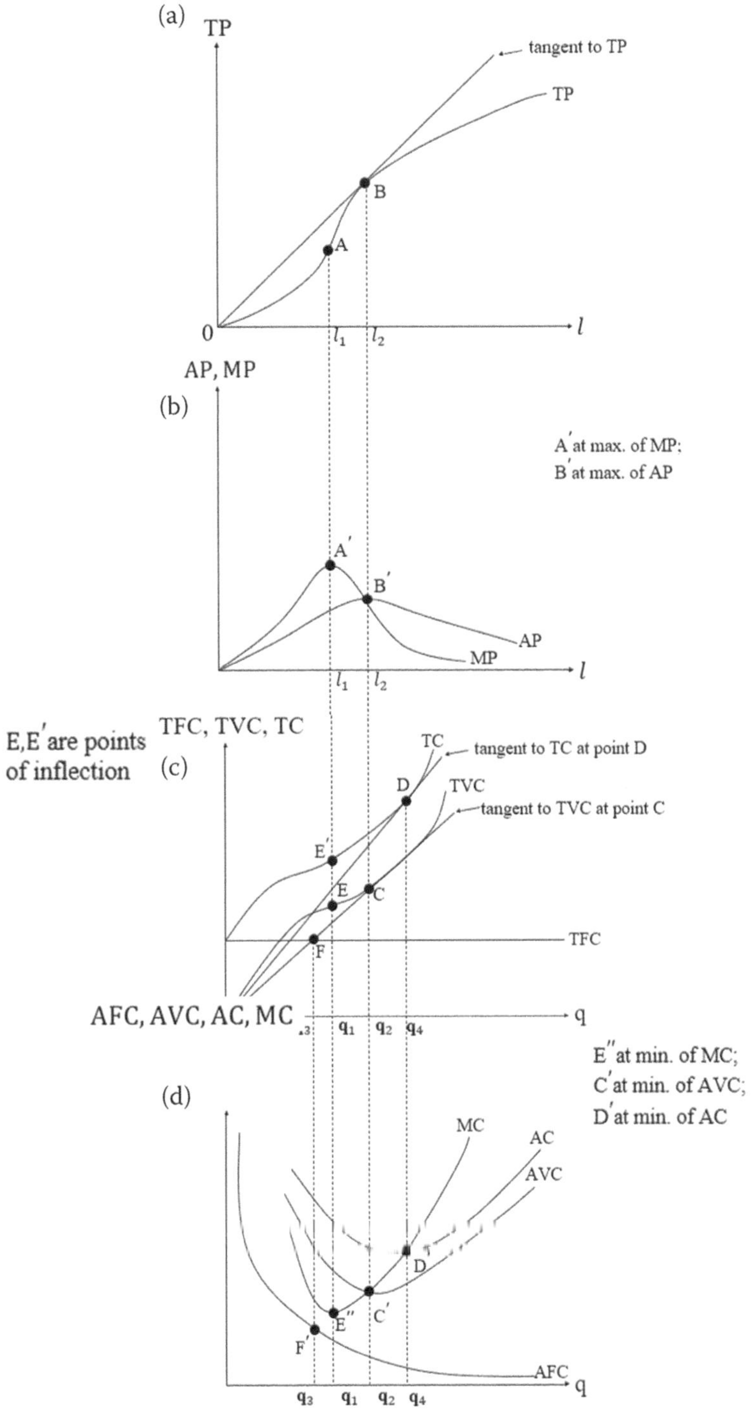

FIGURE 7.4 (a) Total product, (b) average and marginal product, (c) total cost components, (d) average cost components.

1. Panels (a) and (c): when *TP* exhibits a point of inflection (diminishing returns to labor sets in), then *TC* and *TVC* also display a point of inflection.

2. Panels (b) and (d): when *MP* attains a maximum, the slope of *TC* as well as the slope of *TVC* reaches a minimum, i.e., *MC* is at a minimum. And when *AP* reaches a maximum, *AVC* is at a minimum.

3. The output level (q_4) at which *AC* is at a minimum is termed **capacity output**; it is the largest producible without encountering an increase in the average cost of production.

Another way of looking at the relationship between MC and AC is via the **elasticity of total cost** – an index of the responsiveness of total cost to changes in output (it measures increasing or decreasing average cost) or

$$\tau = \frac{\text{\% change in total cost}}{\text{\% change in output}} = \frac{dC}{dq}\frac{q}{C} = \frac{MC}{AC}. \quad (7.25)$$

In this regard:

1. when AC is decreasing, $MC < AC$ and thus $\tau < 1$;
2. when AC is increasing $MC > AC$ and thus $\tau > 1$; and
3. when AC is at a minimum, $MC = AC$ and thus $\tau = 1$.

Section 7.6.2 Exercises:

7.6.2.1 Point A in Figure 7.3 was termed a point of diminishing returns to the labor input. Can you state the *law of diminishing returns*? Why is it termed an *empirical generalization*?

7.6.2.2 Verify that when AP is at a maximum, $MP = AP$.

7.6.2.3 Suppose $q = F(l, k) = 1.5l^{1/2}k^{1/2}$ with $k = \bar{k} = 100$. Find the short-run TP, MP, and AP functions. Is there a point of diminishing returns? Determine η.

7.6.2.4 Let $q = F(l) = 5\,l + 8\,l^2 - 6\,l^3$. Find the short-run AP and MP functions. Also, determine the points where AP and MP are at a maximum. Discuss the behavior of η.

7.6.2.5 Verify that when AC is at a minimum, $MC = AC$.

7.6.2.6 Suppose $TC = 1000 + 100q -$ [illegible]$0q^2 + 1q^3$.

a. Identify TVC and TFC.

b. Find: AC, AVC, AFC, and MC.

c. Find the output levels for minimum AC, minimum AVC, and minimum MC.

7.6.2.7 Suppose $q = F(l) = 15\ l^{1/2}$, $TFC = 1000$, and $w =$ \$8. Find the TC, AC, AVC, and MC functions. Also find τ at minimum AC.

7.6.3 Elasticity and Total Revenue

Let the own-price demand function for commodity $q, q = g(p)$, be continuous and single-valued. Then the inverse own-price demand function exists and appears as $p = g^{-1}(q)$. Thus total revenue (TR), average revenue (AR), and marginal revenue (MR) can be expressed, respectively, as

$$\begin{aligned} &(a)\quad TR = pq = g^{-1}(q)q; \\ &(b)\quad AR = \tfrac{TR}{q} = p = g^{-1}(q);\ \textbf{and} \\ &(c)\quad MR = \tfrac{dTR}{dq} = p + q\tfrac{dp}{dq} < p \end{aligned} \quad (7.26)$$

(the MR function lies below th AR function).

How do we interpret (7.26c)? Looking to Figure 7.5, let price decrease from p_1 to p_2. Hence quantity demanded increases from q_1 to q_2. The impact on TR is:

TR decreases by $dp \cdot q_1$;

TR increases by $dq \cdot p_2$;

Net change in TR = gain in revenue + loss in revenue or

$dTR = dq \cdot p_2 + dp \cdot q_1$ or

$$\frac{dTR}{dq} = p_2 + q_1\frac{dp}{dq} = MR$$

and thus

$$MR = p + q\frac{dp}{dq}, \quad (7.27)$$

where, for a *small* decrease in price, p is the *new price* and q is the *old quantity demanded*, and dp/dq is the change in price per unit change in quantity demanded. Let us write, from (7.9),

$$\epsilon_D = -\frac{dq}{dp}\frac{p}{q} > 0. \quad (7.28)$$

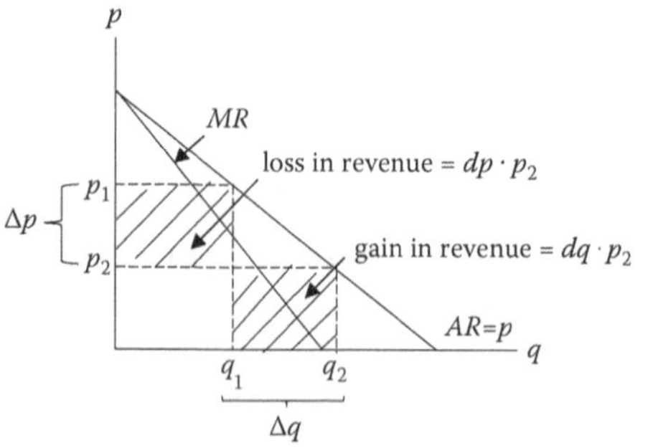

FIGURE 7.5 Average revenue and Marginal revenue.

Then, from (7.27),

$$MR = p + q\frac{dp}{dq} = p\left(1 + \frac{q}{p}\frac{dp}{dq}\right) = p\left(1 - \frac{1}{\in_D}\right). \quad (7.29)$$

(Note that $\lim_{\in_D \to \infty} p\left(1 - \frac{1}{\in_D}\right) = p$.)

Section 7.6.3 Exercises:

7.6.3.1 Verify the following proposition:
When the inverse own-price demand curve is linear and downward sloping, the MR curve is linear and downward sloping; moreover, the MR curve is twice as steep as the former curve.

7.6.3.2 Given Equation (7.29), verify that when total revenue is at a maximum, $\in_D = 1$. Also, given the preceding proposition (Exercise 7.6.3.1), graphical illustrate the notion that when demand is elastic (inelastic) price and TR move in opposite directions (in the same direction).

7.6.3.3 Use Equation (7.28) and the proposition given in Exercise 7.6.3.1 to demonstrate that when $\in_D = 1$, output is $q = a/2b$.

7.6.4 Profit Maximization

Broadly defined, **total profit** (P) equals sales revenue (TR) less total cost (TC) or $P = TR - TC$. Consider the notion of

Accounting profit = sales revenue less *all costs* chargeable against the product sold = sales revenue less *direct costs* (materials, labor, utilities, etc.) chargeable against the product sold less *indirect costs* (overhead, managerial salaries, depreciation, etc.) chargeable against the product sold.

Additionally consider

Normal profit = charges *imputed* against the capital owned by the firm for risk taking.

Thus, normal profit is assigned on an *opportunity cost* basis, i.e., it reflects how much the resources would earn in their next best alternative employment. Hence this cost element reflects a return to entrepreneurship. Give the preceding considerations, we can define

Economic (pure) profit = accounting profit less normal profit.

Thus, this concept involves total revenue less *all* costs of production. So, when economic profit is positive, firms earn in excess of what it would take to keep resources in the industry. This signals resources to enter the industry. If long-run economic profit is zero, there is no incentive to move resources into or out of the industry - - resources have no better alternative employment opportunity.

In what follows we shall consider a set of short-run profit maximization models under a variety of market condition. In all models considered the flow of capital services in production is assumed constant at $k = \bar{k}$ and, in all cases, the decision variable will be either the flow of output (q) or the flow of labor services (l).

Regarding the product market, it will be assumed that the firm is either a **perfect competitor** (the firm is a *price-taker* and faces an infinitely elastic demand curve for its *homogeneous product* so that product price (p) is constant and *exogenously* determined) or a **monopolist** (the firm is the industry and faces a *downward sloping demand curve* for its *highly differentiated product*). If we assume perfect competition in the labor market, then the money wage rate (w) is taken to be constant (there is an infinitely elastic labor supply curve). If **monopsony** exists in the labor market (there is a single employer of labor) then average labor cost is an increasing function of labor.

7.6.4.1 Profit Maximization Under Perfect Competition in the Product and Factor (Labor) Markets

The profit function can be written as

$$\begin{aligned} P = TR - TC &= pq - TVC - TFC \\ &= pq - wf^{-1}(q) - TFC, \end{aligned} \quad (7.30)$$

where, under the given market conditions, p = constant and w = constant. From the first-derivative condition for profit maximization we have

$$\frac{dP}{dq} = \frac{dTR}{dq} - \frac{dTC}{dq} = MR - MC = 0$$

or $p = MR = MC$ at the two output levels q' and q^* (Figure 7.6a). Should the firm produce at q' or q^*? To answer this question, we need to examine the second-derivative condition for a maximum:

$$\frac{d^2P}{dq^2} = \frac{d^2TR}{dq^2} - \frac{d^2TC}{dq^2} = \frac{d^2TC}{dq^2} < 0$$

or $dMC/dq > 0$ at $q^* > q'$. That is, to maximize profit MC must be an increasing function of output at q^* or MC

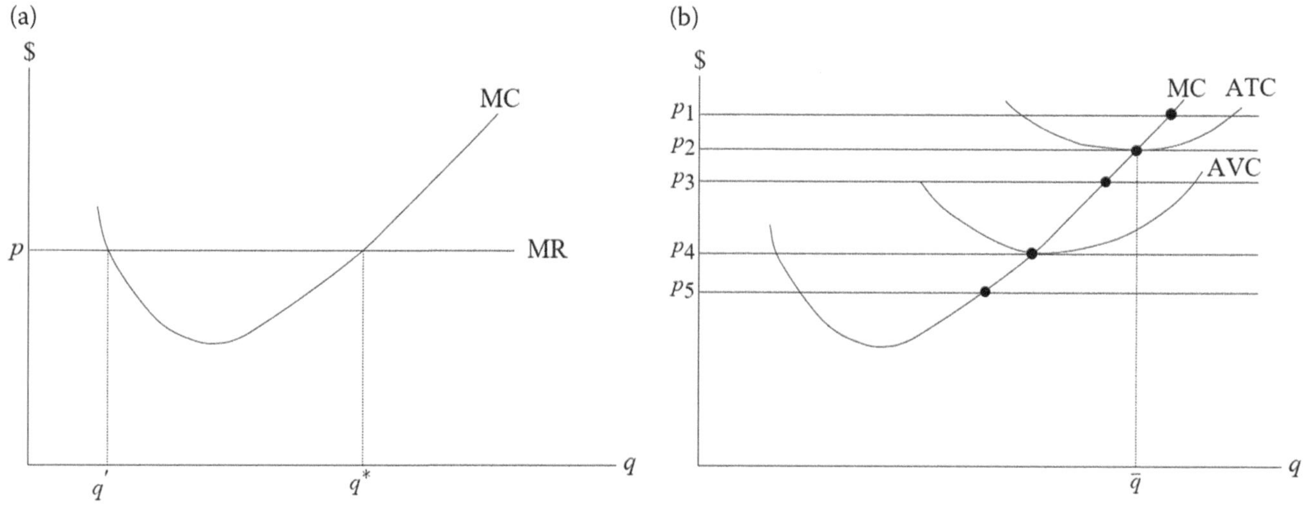

FIGURE 7.6 (a) Determining profit-maximizing output q^*; (b) profit maximization and product price variation.

must cut MR from below. Stated alternatively, from Equation (7.30), we must have $w(d^2f^{-1}/dq^2) > 0$ at q^*, i.e., q^* must be found beyond the point of diminishing returns in production.

Let us consider the implications for profit maximization of an assortment of product prices (Figure 7.6b):

1. for $p = p_1 > ATC$, $TR > TC$ and the firm makes a positive economic or pure profit.
2. for $p = p_2 = ATC$, $TR = TC$ and total economic profit is zero. In fact, break-even output $\bar{q}$ occurs when $p = \min ATC$.
3. for $p = p_3$, $AVC < p < ATC$ so that, even though $TC > TR$ and a loss is incurred, some of the fixed cost is covered; the firm continues production.
4. for $p = p_4 = AVC$, TC exceeds TR by TFC (the loss is $-TFC$). In this instance it makes no difference whether the firm operates or not.
5. for $p = p_5 < AVC$, TCexceeds TR by more than TFC. The firm should shut down.

Given this discussion, we can now state the firm's *optimal decision rule*: if $p \geq AVC$, produce the level of output at which $p = MR = MC$; if $p < AVC$, produce nothing. This rule leads us to define the firms **short-run supply function** – that portion of the MC curve at and above min AVC (the bold portions in Figure 7.7).

It is instructive to respecify the first - and second - order conditions for profit maximization in terms of factor market equilibrium. For this purpose, the profit function appears as

$$P = TR - TC = pq - wl - TFC = pf(l) - wl - TFC. \tag{7.31}$$

Given this expression, the first-order condition for profit maximization is

$$\frac{dP}{dq} = p\frac{df}{dl} - w = 0 \tag{7.32}$$

or $p(df/dl) = w$ at l^*, i.e., at l^* the **value of the marginal product of labor** (VMP_l) must equal the money wage rate. Thus the cost to the firm of an additional unit of labor should equal the value of what that unit produces.

Looking to the second-order condition for profit maximization we have, from Equation (7.32),

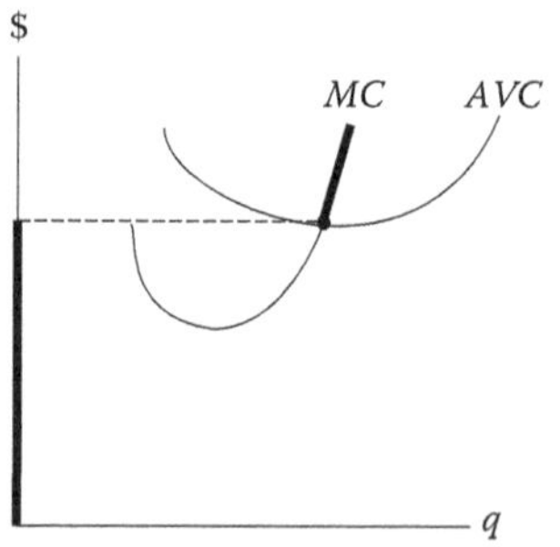

FIGURE 7.7 Short-run supply function.

$$\frac{d^2P}{dl^2} = p\frac{d^2f}{dl^2} = p\frac{dVMP_l}{dl} < 0 \text{ at } l^* \qquad (7.33)$$

or the value of the marginal product of labor curve must be a decreasing function of labor at l^* (i.e., l^* is located beyond the point of diminishing returns in production).

At this point in our discussion, we need some additional definitional concepts. We already know that

$$\text{total revenue} = pq = pf(l); \text{ and } VMP_l = p\frac{df}{dl}.$$

Building upon these notions we have:

average revenue product of labor

$$(ARP_l) = \frac{pf(l)}{l} = p\left(\frac{f(l)}{l}\right), \qquad (7.34)$$

where $f(l)/l$ is the average product of labor; and

average net revenue product of labor $(ANRP_l)$ = ARP_l− fixed capital cost per worker,

i.e., since $P = pf(l) - wl - r\bar{k}$ (r is the price of a unit of capital services in production), it follows that

$$\frac{P}{l} = p\frac{f(l)}{l} - w - \frac{r\bar{k}}{l}$$

or

$$p\frac{f(l)}{l} - \frac{r\bar{k}}{l} = \frac{P}{l} + w$$

and thus

$$ANRP_l = ARP_l - \frac{r\bar{k}}{l} = \frac{P}{l} + w, \qquad (7.35)$$

where $r\bar{k}/l$ is the *fixed capital cost per worker.*

Given these additional concepts, we can now examine Figure 7.8. Note that since we have perfect competition in the labor market, w represents the infinitely elastic labor supply curve. Additionally, the optimal output level q^* is determined from the production function as $q^* = f(l^*)$ while the **normal profit point** (the level of profit which makes the firm competitive in the long run - - and includes opportunity cost) is at max $ANRP_l$. Hence the back-slashed area P depicts **pure profit**.

Given the preceding discussion, it should be apparent that the firm's *optimal decision rule* for labor employment can be framed as: if $w \leq ARP_l$, hire labor up to the point at which $VMP_l = w$; if $w > ARP_1$, hire none. Hence the firm's short-run demand curve for labor appears in Figure 7.9.

7.6.4.2 Monopoly in the Product Market and Perfect Competition in the Factor (Labor) Market

As usual, we may define the profit function as $P = TR - TC$. Then the first-order condition for profit maximization appears as

$$\frac{dP}{dq} = \frac{dTR}{dq} - \frac{dTC}{dq} = 0$$

or

$$p \neq MR = MC \text{ at } q^* \qquad (7.36)$$

since now $p \neq$ constant, i.e., the monopolist faces a downward sloping demand curve for its product. In

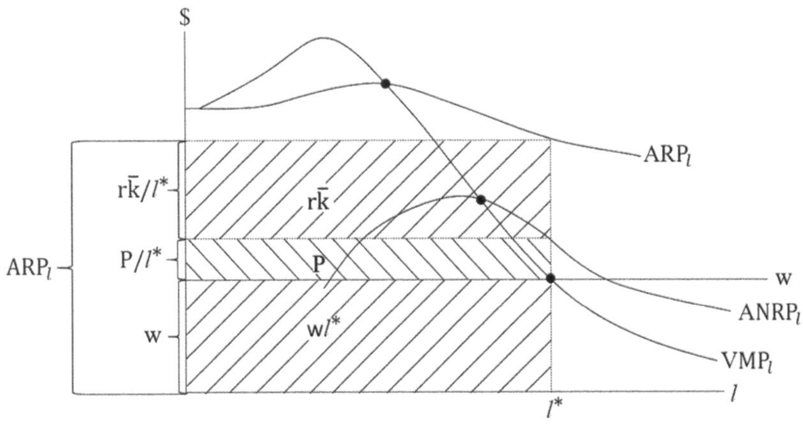

FIGURE 7.8 Determining the optimal labor input level l^*.

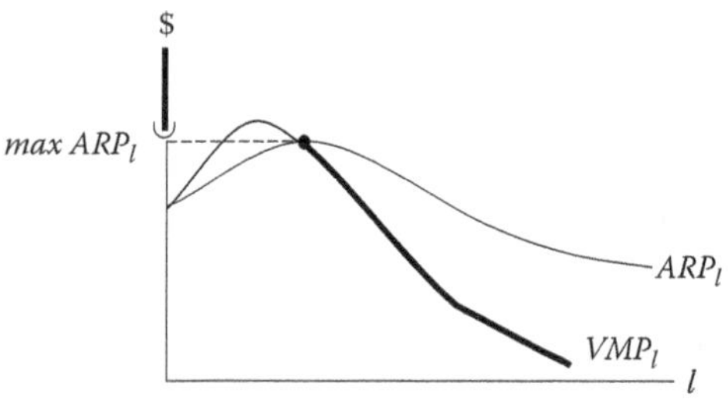

FIGURE 7.9 Short-run demand curve for labor.

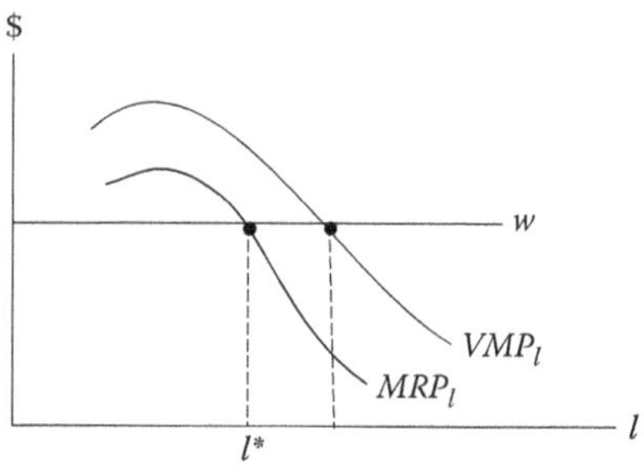

FIGURE 7.11 Since $MR < p$, $MRP_l < VMP_l$.

fact, under monopoly in the product market, $p > MR$. Looking to the second-order condition for profit maximization, we require that

$$\frac{d^2P}{dq^2} = \frac{dMR}{dq} - \frac{dMC}{dq} < 0$$

at q^* or $dMR/dq < dMC/dq$, i.e., the slope of the MR curve must be less than the slope of the MC curve or MC must cut MR from below (Figure 7.10).

To consider profit-maximization from the labor market perspective, let us express the profit function in terms of l as

$$\begin{aligned} P = TR - TC &= pq - wl - TFC \\ &= g^{-1}(q)q - wl - TFC \qquad (7.37) \\ &= g^{-1}(f(l))f(l) - wl - TFC. \end{aligned}$$

Then the first-order condition for maximizing profit is

$$\begin{aligned} \frac{dP}{dl} &= g^{-1}\frac{df}{dl} + f(l)\frac{dg^{-1}}{dq}\frac{df}{dl} - w \\ &= \left(p + q\frac{dp}{dq}\right)\frac{df}{dl} - w = p\left(1 - \frac{1}{\epsilon_D}\right)\frac{df}{dl} - w = 0 \end{aligned}$$

or

$$p\left(1 - \frac{1}{\epsilon_D}\right)\frac{df}{dl} = MR{\cdot}MP_l = MRP_l = w \text{ at } l^*, \quad (7.38)$$

i.e., the firm should hire labor up to the point l^* at which the **marginal revenue product of labor** (MRP_l) equals the money wage rate. Since $MR < p$, the MRP_l curve lies below the VMP_l curve so that, for a given money wage rate, the monopolist will hire less labor than the perfect competitor in the product market (Figure 7.11).

Looking to the second-order condition for profit maximization, given $dP/dl = MR{\cdot}MP_l - w$,

$$\begin{aligned} \frac{d^2P}{dl^2} &= \frac{dMRP_l}{dl} = MR\frac{dMP_l}{dl} + MP_l\frac{dMR}{dq}\frac{dq}{dl} \\ &= MR\frac{d^2f}{dl^2} + \frac{dMR}{dq}\left(\frac{df}{dl}\right)^2 < 0 \\ &\quad (>0)(<0) \quad (<0)\ (>0) \end{aligned}$$

at l^* since $d^2f/dl^2 < 0$ (we are beyond the point of diminishing returns in production). Thus, at l^*, the MRP_l curve must be a decreasing function of labor.

We can next define

$$TR = pq = g^{-1}(q)q = g^{-1}(f(l))f(l);$$

$ARP_l = \frac{TR}{l} = g^{-1}(f(l))\frac{f(l)}{l} = AR{\cdot}AP_l$, where AR is average revenue and AP_l is the average product of labor;

$MRP_l = p\left(1 - \frac{1}{\epsilon_D}\right)\frac{df}{dl} = MR{\cdot}MP_l$; and $ANRP_l = ARP_l -$ fixed capital cost per worker $= ARP_l - \frac{r\bar{k}}{l} = \frac{P}{l} + w$.

Then, from Figure 7.12, l^* is determined where $MRP_l = w$ and thus $q^* = f(l^*)$. Then the firm's *optimal decision rule* appears as: if $w \leq ARP_l$, hire labor up to the point at which $MRP_l = w$; if $w > ARP_l$, hire none.

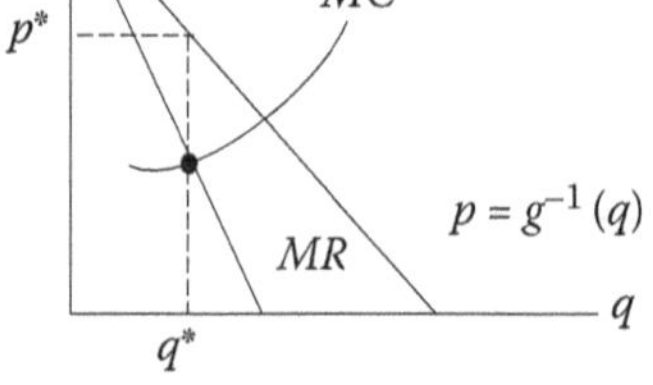

FIGURE 7.10 Determining the profit-maximizing price-quantity combination (q^*, p^*).

7.6.4.3 Perfect Competition in the Product Market and Monopsony in the Factor (Labor) Market

Under a regime of **monopsony** in the labor market there exists a single buyer of labor services. Hence

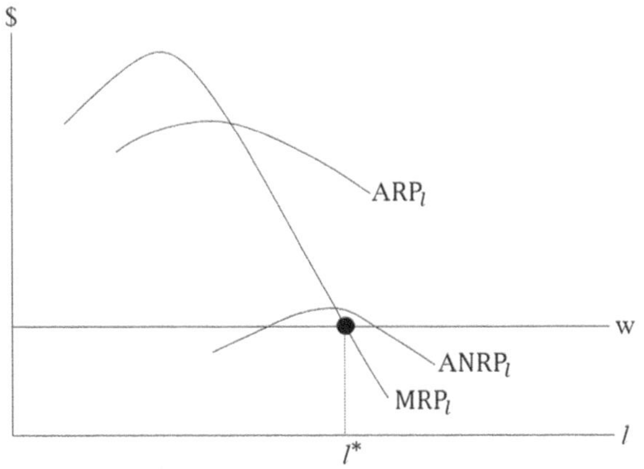

FIGURE 7.12 Determining the optimal labor input level l^*.

factor cost per unit increases as the number of units of labor purchased increases. Thus, w is no longer constant but expressible as an increasing function of labor or

$$w = h(l), \quad \frac{dw}{dl} > 0, \tag{7.39}$$

i.e., **average labor cost** is an increasing function of l.

We can now write

$$TC = wl + TFC = h(l)l + TFC$$

and thus **marginal factor cost** (MC_l) is

$$\begin{aligned} MC_l = \frac{dTC}{dl} &= h(l) + l\frac{dh}{dl} \\ &= w + l\frac{dw}{dl} > w \end{aligned} \tag{7.40}$$

(Figure 7.13). Thus, MC_l equals average labor cost plus the increase in average labor cost necessary to induce an increase in the quantity of labor supplied times the number of units for which it must be paid.

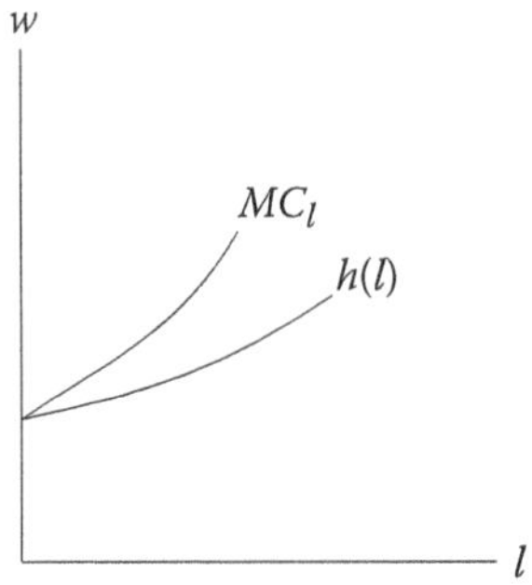

FIGURE 7.13 The market supply curve of labor is $h(l) < MC_l$.

Using (7.40) we can rewrite MC_l as

$$MC_l = w\left(1 + \frac{l}{w}\frac{dw}{dl}\right) = w\left(1 + \frac{1}{\mu}\right), \tag{7.41}$$

where

$$\mu = \frac{dl}{dw}\frac{w}{l} \tag{7.42}$$

is the **point elasticity of labor supply coefficient** - - it represents an index of the responsiveness of the quantity of labor supplied to a change in the wage rate (Figure 7.14). (Note that $\lim_{\mu\to\infty} MC_l = \lim_{\mu\to\infty} w\left(1 + \frac{1}{\mu}\right) = w$.)

Looking next to the profit function we have

$$\begin{aligned} P = TR - TC &= pq - h(l)l - TFC \\ &= pf(l) - h(l)l - TFC. \end{aligned} \tag{7.43}$$

Then the first-derivative condition for maximizing P is

$$\frac{dP}{dl} = p\frac{df}{dl} - h(l) - l\frac{dh}{dl} = p\frac{df}{dl} - w\left(1 + \frac{1}{\mu}\right) = 0$$

or $VMP_l = MC_l$ at l^*. For the second-order condition for a maximum we have

$$\frac{d^2P}{dl^2} = p\frac{d^2f}{dl^2} - \frac{dMC_l}{dl} < 0$$

or

$$p\frac{d^2f}{dl^2} = \frac{dVMP_l}{dl} < \frac{dMC_l}{dl} \text{ at } l^*.$$

As this condition reveals, the slope of the VMP_l curve must be less than the slope of the MC_l curve at l^* or the MC_l curve must cut the VMP_l curve from below (point A of Figure 7.15). The optimal money wage rate w^* can now be determined from the optimal labor employment rate l^* as $w^* = h(l^*)$. Additionally, the optimal output level is $q^* = f(l^*)$. Interestingly enough, the short-run demand curve for labor is simply the single point A. Given the preceding discussion, the firm's *optimal decision rule* reads as: if $TR > TVC$ (the firm covers some fixed cost), the firm should hire labor up

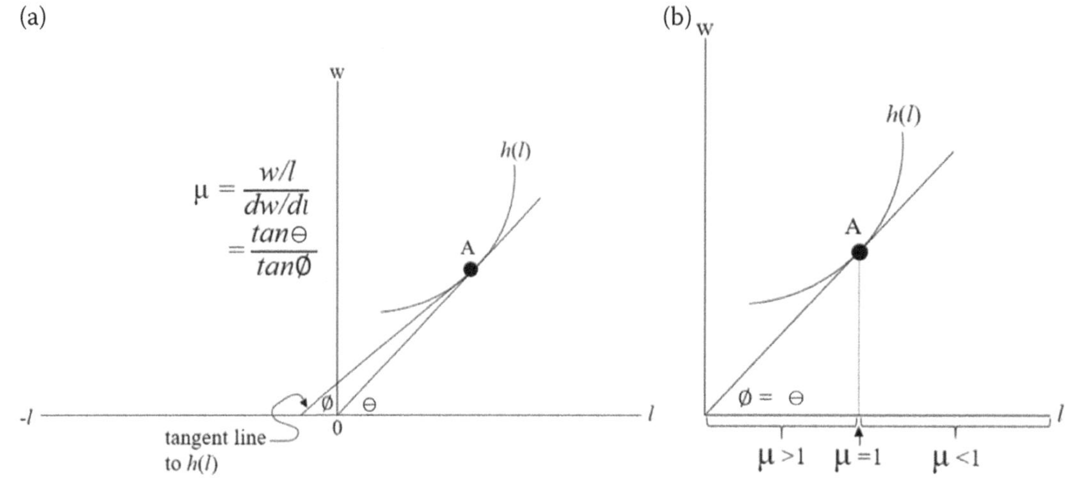

FIGURE 7.14 (a) μ defined graphically; (b) the range of μ values.

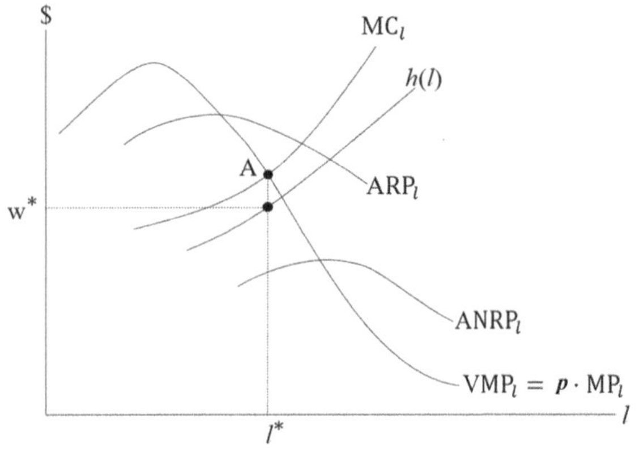

FIGURE 7.15 Determining the optimal input level l^* and the optimal wage rate w^*.

to the point at which $VMP_l = MC_l$; if $TR < TVC$, hire none.

7.6.4.4 *Monopoly in the Product Market and Monopsony in the Factor (Labor) Market*

Given $p = g^{-1}(q)$ and $q = f(l)$, we can write $p = g^{-1}(f(l))$. And with $w = h(l)$, the profit function is

$$\begin{aligned} P = TR - TC &= pq - wl - TFC \\ &= g^{-1}(f(l)) - h(l)l - TFC. \end{aligned} \tag{7.44}$$

The first-order condition for maximizing profit is

$$\frac{dP}{dl} = p\left(1 - \frac{1}{\epsilon_D}\right)\frac{df}{dl} - w\left(1 + \frac{1}{\mu}\right) = 0$$

or $MRP_l = MC_l$ at l^*. And if we express $dP/dl = MR{\cdot}MP_l - MC_l = MR\frac{df}{dl} - MC_l$, then we can write the second-order condition for profit maximization as

$$\begin{aligned} \frac{d^2P}{dl^2} &= MR\frac{d^2f}{dl^2} + \frac{df}{dl}\frac{dMR}{dq}\frac{dq}{dl} - \frac{dMC_l}{dl} \\ &= MR\frac{d^2f}{dl^2} + \frac{dMR}{dq}\left(\frac{df}{dl}\right)^2 - \frac{dMC_l}{dl} \\ &= \frac{dMRP_l}{dl} - \frac{dMC_l}{dl} < 0 \text{ at } l^*, \end{aligned}$$

i.e., the slope of the MRP_l curve must be less than the slope of the MC_l at l^* or MC_l must cut MRP_l from below (point A of Figure 7.16). Hence the firm's *optimal decision rule* can be expressed as: if $TR > TVC$ (some fixed cost is covered), hire labor up to the point at which $MRP_l = MC_l$; if $TR < TVC$, hire none. In the light of our preceding discussion(s) it should be evident that the firm's short-run demand curve for labor is that portion of the MRP_l curve below maximum

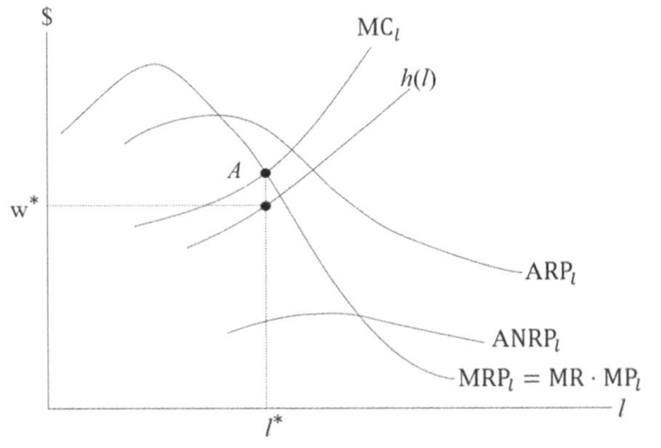

FIGURE 7.16 Determining the optimal labor input level l^* and the optimal wage rate w^*.

ARP_l, Given l^*, we can readily determine $q^* = f(l^*)$ and $w^* = h(l^*)$.

Section 7.6.4 Exercises:

7.6.4.1 Let $p = \$20$ with $TC = 100 + 10q - 0.02q^2 + 0.001q^3$. Find the profit-maximizing level of output q^*. Also, find the level of q at which AVC and MC are at a minimum. Explain the behavior of average fixed cost (AFC).

7.6.4.2 Let $p = \$5$, $w = \$4$, with $q = 10l^{1/2}k^{1/2}$. Assume $k = \bar{k} = 100$. Express total profit as a function of output and then determine the profit-maximizing q and l values.

7.6.4.3 Given the information provided in the preceding problem, express total profit as a function of labor and then determine the profit-maximizing q and l values.

7.6.4.4 Let $p = g^{-1}(q) = 100 - 4q$ with $TC = 50 + 20q$.

a. Determine the profit-maximizing price-quantity combination (q^*, p^*).

b. Suppose the government imposes a specific sales tax of $\$8$/unit on the output of the firm. Maximize post-tax profit and determine

   i. post-tax q.

   ii. post-tax p.

   iii. amount of tax revenue collected by the government.

   Also, verify that: product price increases by less than the unit tax but profit decreases by more than the amount of tax revenue going to the government.

c. Why couldn't the firm *shift* all of the tax/unit to consumers in the form of a higher price? Think elasticity! That is, the more inelastic the demand for the product, the larger the portion of the tax *shifted forward* to consumers. Determine if demand is elastic at the profit-maximizing combination (q^*, p^*).

7.6.4.5 Given the following information, determine the optimal labor input level. Also find p^*, q^* and w^*. Note: solve this problem two ways – first, by directly maximizing the profit function (in terms of l); and second, by employing the equality $MRP_l = MC_l$.

$$p = 100 - q,$$

$$q = 5lk \text{ with } k = \bar{k} = 5,$$

$$w = 15 + \frac{1}{2}l.$$

7.6.4.6 A perfectly competitive firm has the total cost function $TC = 0.1q^3 - 2q^2 + 15q + 100$. Determine the firm's short-run supply function.

7.6.4.7 Determine the optimal input combination (l^*, k^*) of a firm if the production function is $q = F(l, k) = 20 - l^2 + 10l - 2k^2 + 5k$, the (constant) prices of the inputs l and k are, respectively, $w = \$2$ and $r = \$1$ and the price of output is $\$5$.

7.6.4.8 The inverse demand functions for products x and y are, respectively, $p_x = 36 - 3x$ and $p_y = 40 - 5y$. The joint cost function is $TC(x, y) = x^2 + 2xy + 3y^2$. Find the profit-maximizing price-quantity combinations for the two products.

7.6.4.9 A perfectly competitive firm can produce its output q in either of two plants at a unit price of $p = \$216$. The plant cost functions are, respectively,
$TC_1 = 20q_1^3 - 24q_1 + 5$ and $TC_2 = 27q^2 + 10$, where $q_1 (q_2)$ is the amount of the single output produced in plant 1 (2). What is the amount of

output produced in each plant and the total amount sold to maximize profit? (Assume that all output produced is sold.)

7.6.4.10 Suppose a firm operates in two distinct markets, where the inverse demand functions for markets 1 and 2 are, respectively, $p_1 = 80 - 5q_1$ and $p_2 = 180 - 20q_2$, with $q = q_1 + q_2$ representing output as a whole and $TC = 50 + 20q$. Find the profit-maximizing price and quantity combinations for each market along with q. Verify that the price will be lower in the market with the greater demand elasticity.

7.6.4.11 Suppose a firm maximizes sales revenue (TR) subject to a minimum profit constraint ($P = TR - T \geq P_o$). Given that $P_o = \$334$, $p = 100 - 4q$, and $TC = 50 + 20q$, determine the sales-maximizing output level given that the decision process involves the following steps:

1. find the q (call it q_s) that provides an unconstrained maximum for TR. This is the appropriate solution if $P\ (q_s) \geq P_o$.
2. if $P\ (q_s) < P_o$, solve $P\ (q) = P_o$ for q.

Convex and Concave Real-Valued Functions

8.1 CONVEX SETS

The convexity or concavity of real-valued functions is an important property when it comes to determining their extreme values (i.e., minimum or maximum values) over a convex domain. In this regard, let us examine the properties of the domain of a convex or concave function. To this end we have

Definition 8.1: A region on $\mathcal{K} \subset R^n$ is **convex** if for any two points $\boldsymbol{x}_1, \boldsymbol{x}_2 \in \mathcal{K}$, the line segment joining them is also in $\mathcal{K}$, i.e., the **internal average** or **convex combination** of $\boldsymbol{x}_1$ and $\boldsymbol{x}_2$,

$$\boldsymbol{x}_c = \theta \boldsymbol{x}_2 + (1 - \theta)\boldsymbol{x}_1,\ 0 \leq \theta \leq 1,$$

is also a member of $\mathcal{K}$, where

$$\boldsymbol{x}_c^T = (x_1^c, \ldots, x_n^c) = (\theta x_1^2 + (1 - \theta)x_1^1, \ldots, \theta x_n^2 + (1 - \theta)x_n^1).$$

Clearly, the components of $\boldsymbol{x}_c$ are themselves convex combinations of the components of $\boldsymbol{x}_1$, $\boldsymbol{x}_2$. Here, $\boldsymbol{x}_c$ represents, for a given value of θ, a point on the line segment connecting $\boldsymbol{x}_1$, $\boldsymbol{x}_2$ (see Figure 8.1a for $\boldsymbol{x} \in R^2$). So for $\theta = 0$, $\boldsymbol{x}_c = \boldsymbol{x}_1$; and as θ increases in value from 0 to 1, the point $\boldsymbol{x}_c$ travels along the line segment joining $\boldsymbol{x}_1$ and $\boldsymbol{x}_2$ from $\boldsymbol{x}_1$ to $\boldsymbol{x}_2$, at which point $\theta = 1$ and thus $\boldsymbol{x}_c = \boldsymbol{x}_2$.

Often a convex set $\mathcal{K}$ is depicted as one which is *non-reentrant*, i.e., $\mathcal{K}$ is not convex if the variable point $\boldsymbol{x}_c$ on the line segment between $\boldsymbol{x}_1$ and $\boldsymbol{x}_2$ leaves and then reenters $\mathcal{K}$ as θ steadily increases in value from 0 to 1 (Figure 8.1b). A moments reflection reveals that each of the following items is a convex set:

a. a **plane** $\{\boldsymbol{x} | C^T\boldsymbol{x} = \alpha, \boldsymbol{x} \in R^n\}$ as well as the **open and closed half-planes** $\{\boldsymbol{x} | C^T\boldsymbol{x} < \alpha (>\alpha), \boldsymbol{x} \in R^n\}$ and $\{\boldsymbol{x} | C^T\boldsymbol{x} \leq \alpha (\geq \alpha), \boldsymbol{x} \in R^n\}$, respectively.

b. a **spherical δ-neighborhood** about $\boldsymbol{x}_o$, $\delta(\boldsymbol{x}_o) = \{\boldsymbol{x} | \|\boldsymbol{x} - \boldsymbol{x}_o\| < \delta, \boldsymbol{x} \in R^n\}$;

c. the **singleton** $\{\boldsymbol{x}\} \in R^n$;

d. a subspace of R^n defined as $\mathcal{S} = \{\boldsymbol{x} | \boldsymbol{x} = \alpha_1\boldsymbol{x}_1 + \alpha_2\boldsymbol{x}_2$, with $\boldsymbol{x}_1, \boldsymbol{x}_2 \in \mathcal{S}$ and $\alpha_1, \alpha_2 \in R\}$;

e. A **ray** $\{\boldsymbol{x} | \boldsymbol{x} = \boldsymbol{x}_o + \lambda \boldsymbol{d}, \boldsymbol{x} \in R^n, 0 \leq \lambda \in R\}$, where $\boldsymbol{d} \in R^n$ is a *direction* and $\boldsymbol{x}_o$ is a *fixed point*;

f. The interior and closure of a convex set $\mathcal{S} \subset R^n$;

g. The sets $S = S_1 \cap S_2$ and $T = S_1 \pm S_2 = \{\boldsymbol{x}_1 \pm \boldsymbol{x}_2 | \boldsymbol{x}_1 \in S_1, \boldsymbol{x}_2 \in S_2\}$ if $S_1, S_2 \subset R^n$ are convex; and

h. The **convex hull** of an arbitrary set $A \subset R^n$, denoted co (A), formed as the intersection of all convex sets containing A. (Here co (A) is the

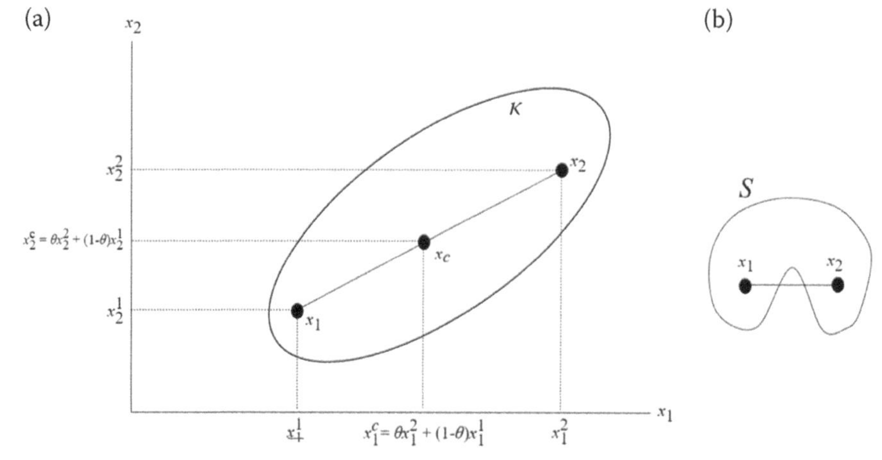

FIGURE 8.1 a. $\mathcal{K} \subset R^2$ is a convex set; b. is not a convex set.

smallest convex set in R^n containing A.) Alternatively, co (A) may be defined as the set of all convex combinations of points from A or co $(A) = \{\boldsymbol{x} | \boldsymbol{x} = \sum_{i=1}^{m} \theta_i \boldsymbol{x}_i, \boldsymbol{x}_i \in A, \sum_{i=1}^{m} \theta_i = 1, 0 \le \theta_i \in R$ for all i and m is an arbitrary positive integer$\}$.

Section 8.1 Exercises:

8.1.1 Verify that the hyperplane $\{x | C^T x = \alpha, C \neq O, C \in R^n, \alpha \in R\}$ in R^n is a convex set.

8.1.2 Verify that the ray $\{x | x = x_o + \lambda d, x \in R^n, 0 \le \lambda \in R\}$ is a convex set.

8.1.3 Suppose S_1, S_2 are convex sets in R^n. Verify that their intersection $S = S_1 \cap S_2$ is a convex set.

8.2 CONVEX AND CONCAVE REAL-VALUED FUNCTIONS

We now consider functions that may be classified as convex or concave over their domain. Specifically,

Definition 8.2: Let $y = f(\boldsymbol{x})$: $\mathcal{D} \subset R^n \to R$, where $\mathcal{D}$ is a convex set. The function f is **convex** (**concave**) on $\mathcal{D}$ if linear interpolation between any two points on f never underestimates (never overestimates) the value of f at the point of interpolation.

So if f is defined throughout the entirety of a convex region $\mathcal{D} \subset R^n$, then f is convex on $\mathcal{D}$ if and only if, for any two points $\boldsymbol{x}_1, \boldsymbol{x}_2 \in \mathcal{D}$,

$$f(\boldsymbol{x}_c) \le \theta\, f(\boldsymbol{x}_2) + (1-\theta) f(\boldsymbol{x}_1),\ 0 \le \theta \le 1. \quad (8.2)$$

And if f is concave over $\mathcal{D} \subset R^n$, the inequality in (8.2) is reversed. Hence, f is concave over $\mathcal{D}$ if and only if, for any two points $\boldsymbol{x}_1, \boldsymbol{x}_2 \in \mathcal{D}$,

$$f(\boldsymbol{x}_c) \ge \theta\, f(\boldsymbol{x}_2) + (1-\theta) f(\boldsymbol{x}_1),\ 0 \le \theta \le 1. \quad (8.3)$$

We shall term f **strictly convex** (**strictly concave**) over $\mathcal{D}$ if and only if, for any two points $\boldsymbol{x}_1, \boldsymbol{x}_2 \in \mathcal{D}$, with $\boldsymbol{x}_1 \neq \boldsymbol{x}_2$ and $0 < \theta < 1$, strict inequality holds in (8.2) ((8.3)).

To illustrate the notion of convexity or concavity geometrically, let us take $\mathcal{D} = [a, b] \subset R$, $0 < a < b$. Let our **point of interpolation** be $x_c = \theta x_2 + (1-\theta) x_1$, $0 \le \theta \le 1$. Hence, **linear interpolation of** f at x_c is represented as $\theta\, f(x_2) + (1-\theta) f(x_1)$. Thus linear interpolation of f between x_1 and x_2 is represented as a point on the straight line between $(x_1, f(x_1))$ and $(x_2, f(x_2))$ while the value of f at x_c is $f(\theta x_2 + (1-\theta) x_1)$. So if (8.2) is to hold so that f is convex, then the line segment joining $(x_1, f(x_1))$ and $(x_2, f(x_2))$ must lie on or above the function (Figure 8.2a). And if (8.3) holds for f concave, the line segment connecting $(x_1, f(x_1))$ and $(x_2, f(x_2))$ must lie on or below the function (Figure 8.2b). If f is strictly concave (strictly convex) over $\mathcal{D}$, the line segment joining $(x_1, f(x_1))$ and $(x_2, f(x_2))$ lies totally below (totally above) for all values of $x \in (x_1, x_2) \subset [a, b]$ since, in this instance, $0 < \theta < 1$. Note that Equations (8.2), (8.3) are equivalent if f is linear over $\mathcal{D}$ since, in this instance $f(\boldsymbol{x}_c) = \theta\, f(\boldsymbol{x}_2) + (1-\theta) f(\boldsymbol{x}_1)$ for all $\boldsymbol{x}_1, \boldsymbol{x}_2 \in \mathcal{D}$.

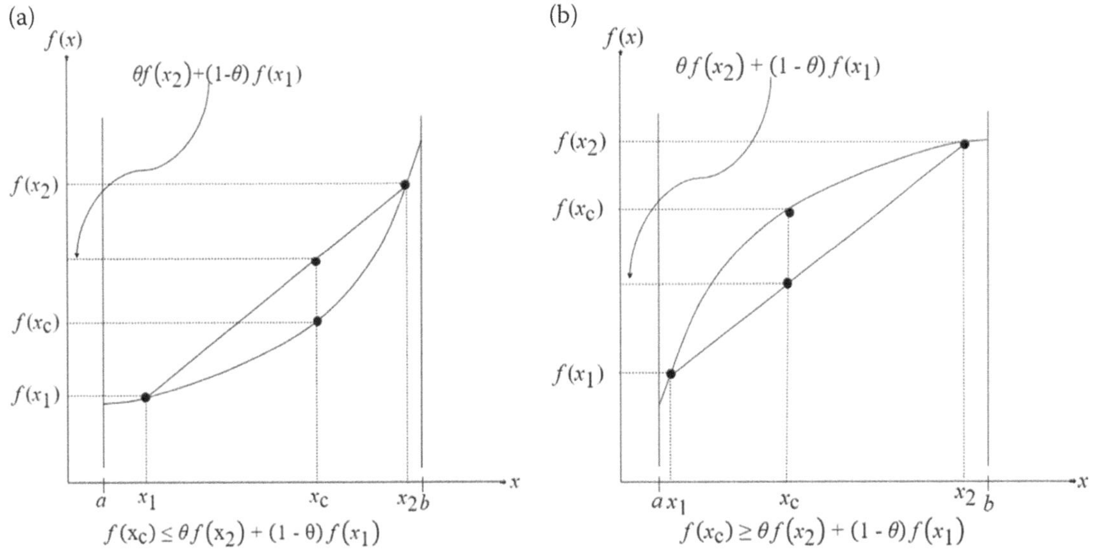

FIGURE 8.2 a. f is convex over $\mathscr{D} = [a, b]$; b. f is concave over $\mathscr{D} = [a, b]$.

Example 8.1: Suppose $y = f(x) = x^2$ is defined for all $x \in \mathcal{D} = [a, b]$. Demonstrate that f is convex on $\mathcal{D}$ via Equation (8.2). Here, we must show that $f(x_c) - \theta f(x_2) - (1-\theta) f(x_1) \leq 0$, $0 \leq \theta \leq 1$. For $x_1, x_2 \in \mathcal{D}$, $f(x_1) = x_1^2$, $f(x_2) = x_2^2$, and, for $x_c = \theta x_2 + (1-\theta)x_1$,

$$f(x_c) = (\theta x_2 + (1-\theta)x_1)^2 = \theta^2 x_2^2 + 2\theta(1-\theta)x_1 x_2 + (1-\theta)^2 x_1^2.$$

Then from (8.2)

$$\theta^2 x_2^2 + 2\theta(1-\theta)x_1 x_2 + (1-\theta)^2 x_1^2 - \theta x_2^2 - (1-\theta)x_1^2 \leq 0$$

or

$$\theta(\theta-1)x_2^2 + \theta(\theta-1)x_1^2 - 2\theta(\theta-1)x_1 x_2 \leq 0, \; 0 \leq \theta \leq 1.$$

For $\theta = 0$ or 1, this expression reduces (trivially) to zero, while for $0 < \theta < 1$, this inequality simplifies to $(x_1 - x_2)^2 \geq 0$. Thus (8.2) holds and thus f is convex on $\mathcal{D}$. If $x_1 \neq x_2$, then $(x_1 - x_2)^2 > 0$ and thus f is strictly convex over $\mathcal{D}$. ■

Some important properties of convex (concave) functions are:

1. If $f(\boldsymbol{x}): \mathcal{D} \subset R^n \rightarrow R$, $\mathcal{D}$ a convex set, is convex, then $-f(\boldsymbol{x})$ is concave on $\mathcal{D}$, and conversely.
2. Let $f^1(\boldsymbol{x}), \ldots, f^m(\boldsymbol{x}): \mathcal{D} \subset R^n \rightarrow R$ be convex (concave) functions. Then the weighted sum of the functions f^i, $i = 1, \ldots, m$,

$$f(\boldsymbol{x}) = \sum_{i=1}^{m} \alpha_i f^i(\boldsymbol{x}), \; \alpha_i > 0, \; i = 1, \ldots, m,$$

is a convex (concave) function.

3. Let $g(\boldsymbol{x}): \mathcal{D} \subset R^m \rightarrow R$, be a concave function, where $\mathcal{D} = \{\boldsymbol{x} | g(\boldsymbol{x}) > 0\}$. Define $h(\boldsymbol{x}): \mathcal{D} \rightarrow R$ as $h(\boldsymbol{x}) = g(\boldsymbol{x})^{-1}$. Then $h(\boldsymbol{x})$ is convex on $\mathcal{D}$.
4. Suppose $f(\boldsymbol{x}): \mathcal{D} \subset R^n \rightarrow R$ with $\mathcal{D}$ convex and $g: R \rightarrow R$, with the range of g containing $f(\boldsymbol{x})$, $\boldsymbol{x} \in \mathcal{D}$. Also, let h be the composition of f and g or $h(\boldsymbol{x}) = g \circ f = g(f(\boldsymbol{x})): R^n \rightarrow R$. Then:
   a. if g is concave and increasing and f is concave, then $h = g(f)$ is concave;
   b. if f is convex and increasing and f is convex, then $h = g(f)$ is convex;
   c. if g is concave and decreasing and f is convex, then $h = g(f)$ is concave; and
   d. if g is convex and decreasing and f is concave, then $h = g(f)$ is convex.

In general, if $f^i(\boldsymbol{x}): \mathcal{D} \subset R^n \rightarrow R$, $i = 1, \ldots, m$, with $\mathcal{D}$ convex and $g(\boldsymbol{x}) = g(f^1(\boldsymbol{x}), \ldots, f^m(\boldsymbol{x})): R^m \rightarrow R$, $\boldsymbol{x} \in \mathcal{D}$, then:

> if g is concave (convex) and increasing in each argument f^i and each f^i is concave (convex), then $g(f^1(\boldsymbol{x}), \ldots, f^m(\boldsymbol{x}))$ is concave (convex).

5. For a convex function $f(\mathbf{x}): \mathcal{D} \subset R^n \to R$, with $\mathcal{D}$ a convex set, let $L_f(\alpha) = \{\mathbf{x} \in \mathcal{D} | f(\mathbf{x}) \leq \alpha, \alpha \in R\}$ denote the **lower level set** of f. If f is concave, its **upper level set** is structured as $U_f(\alpha) = \{\mathbf{x} \in \mathcal{D} | f(\mathbf{x}) \geq \alpha, \alpha \in R\}$. Then $L_f(\alpha)(U_f(\alpha))$ is a convex set for each α.

6. Let $f(\mathbf{x}): \mathcal{D} \subset R^n \to R$ be a convex (concave) function, with $\mathcal{D}$ a convex set. Then f is continuous on the interior of $\mathcal{D}$ (with any points of discontinuity occurring only on the boundary of $\mathcal{D}$). Moreover, if $\mathcal{D}$ is open and f is convex (concave), then f is continuous on $\mathcal{D}$.

Section 8.2 Exercises:

8.2.1 Suppose $y = f(\mathbf{x})$, $\mathbf{x} \in R^n$, is a concave function. Demonstrate that $-f(\mathbf{x})$ is convex.

8.2.2 Let $f^1(\mathbf{x}), f^2(\mathbf{x})$, with $\mathbf{x} \in R^n$, be concave functions. Verify that their sum $f(\mathbf{x}) = f^1(\mathbf{x}) + f^2(\mathbf{x})$ is also concave.

8.2.3 Let $y = f(\mathbf{x}) = x_1^2 + x_2^2$, $\mathbf{x} \in R^2$. Is f convex? Strictly convex?

8.2.4 Verify 4.a from properties of concave (convex) functions in Section 8.2.

8.2.5 a. Verify 4.a from properties of concave (convex) functions in Section 8.2 for $x \in R$ with both g and f twice continuously differentiable and $h(x) = g \circ f = g(f(x))$.

b. Does this result hold for $f = x^{3/4}$, $x > 0$, and $g = -f(x)^2$?

8.2.6 Verify that if $y = f(\mathbf{x}): \mathcal{D} \subset R^n \to R$ is concave, then its upper level set $U_f(\beta) = \{\mathbf{x} \in \mathcal{D} | f(\mathbf{x}) \geq \beta, \beta \in R\}$ is convex. (See property 5 of Section 8.2.)

8.3 SUPERGRADIENTS OF CONCAVE AND SUBGRADIENTS OF CONVEX FUNCTIONS

We noted in Chapter 6 that there are two sets associated with the graph of a real-valued function $y = f(\mathbf{x}): \mathcal{D} \subset R^n \to R$, namely the **hypograph of** f, $\text{hyp}(f) = \{(\mathbf{x}, \alpha) | \alpha \leq f(\mathbf{x}), \mathbf{x} \in \mathcal{D}\} \subset R^{n+1}$ and the **epigraph of** f, $\text{epi}(f) = \{(\mathbf{x}, \alpha) | \alpha \geq f(\mathbf{x}), \mathbf{x} \in \mathcal{D}\} \subset R^{n+1}$. As indicated in Figures 8.3a, b, epi (f) and hyp (f) are convex sets. (A moments reflection reveals that the upper (lower) level set of f is the projection of hyp (f) (epi (f)) on $\mathcal{D}$.)

This observation leads us to

Theorem 8.1: Let $y = f(\mathbf{x}): \mathcal{D} \subset R^n \to R$, with $\mathcal{D}$ a convex set. Then f is concave (convex) if and only if hyp (f) (epi (f)) is a convex set.Additionally, a function $f(\mathbf{x}): \mathcal{D} \subset R^n \to R$, $\mathcal{D}$ a convex set, is termed a **closed concave function** if hyp (f) is a closed convex subset of R^{n+1}. In this regard, if f is a closed concave function, then its upper level sets are closed convex subsets of R^n. For a **closed convex function** f, epi (f) is closed and convex as are the lower level sets of f.

A glance back at Figures 8.3a, b reveals that since the epigraph of a convex function and the hypograph of a

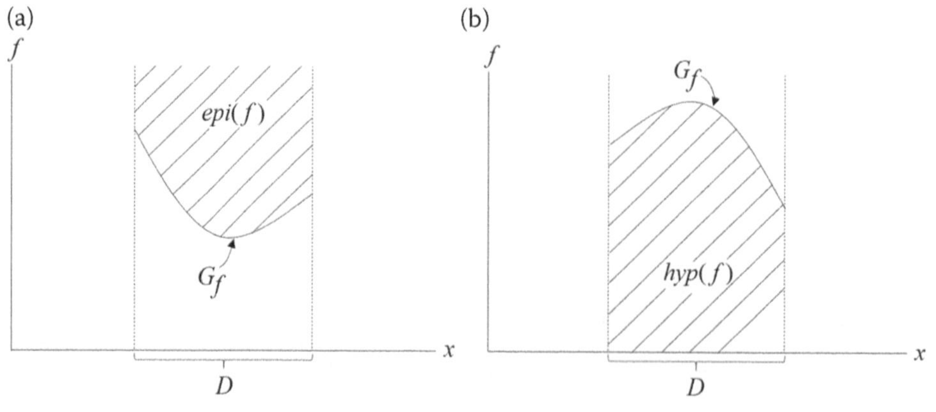

FIGURE 8.3 The graph of f, $G_f = \{(\mathbf{x}, f(\mathbf{x})) | \mathbf{x} \in \mathscr{D}\} \subset R^2$, and: a. the convex epigraph of f, epi(f), for f a convex function; b. the convex hypograph of f, hyp (f), for f a concave function.

concave function are convex sets, they admit tangent or supporting hyperplanes at their boundary points. These supporting hyperplanes are called supergradients (subgradients). More formally,

Definition 8.3: Let $y = f(\boldsymbol{x})$: $\mathcal{D} \subset R^n \rightarrow R$, with $\mathcal{D}$ a convex set. If f is concave on $\mathcal{D}$, then the vector ξ is a **supergradient of f** at $\boldsymbol{x}_o \in \mathcal{D}$ if

$$f(\boldsymbol{x}) \leq f(\boldsymbol{x}_o) + \xi^T(\boldsymbol{x} - \boldsymbol{x}_o), \ \boldsymbol{x} \in \mathcal{D}; \qquad (8.4)$$

if f is convex on $\mathcal{D}$, then the vector ξ is a **subgradient of f** at $\boldsymbol{x}_o \in \mathcal{D}$ if

$$f(\boldsymbol{x}) \geq f(\boldsymbol{x}_o) + \xi^T(\boldsymbol{x} - \boldsymbol{x}_o), \ \boldsymbol{x} \in \mathcal{D}. \qquad (8.5)$$

Note that the expression on the right-hand side of each of the inequalities (8.4), (8.5) is a supporting hyperplane of f at $\boldsymbol{x}_o$ while ξ is the slope of the supporting hyperplane (Figure 8.4a,b). (For instance, in Figure 8.4a, $\tan\theta = \xi = \text{rise/run} = \text{rise}/(x - x_o)$ or rise $= \xi(x - x_o)$.)

Let us examine, say, the subgradient of f from another viewpoint. For instance, suppose f is convex at the boundary point $(\boldsymbol{x}_o, f(\boldsymbol{x}_o)) \in$ epi (f) (Figure 8.5). At $(\boldsymbol{x}_o, f(\boldsymbol{x}_o))$ the vector normal to the supporting hyperplane $\mathcal{H}$ of f is $(\xi^T, -1)$ so that

$$\begin{pmatrix} \xi \\ -1 \end{pmatrix}^T \begin{pmatrix} \boldsymbol{x} - \boldsymbol{x}_o \\ f(\boldsymbol{x}) - f(\boldsymbol{x}_o) \end{pmatrix} \leq 0$$

or Equation (8.5) obtains.

A real-valued function f is termed **subdifferentiable at $\boldsymbol{x}_o$** if there exists at least one subgradient of f at $\boldsymbol{x}_o$; and f is said to be **subdifferentiable** if it is subdifferentiable for all $\boldsymbol{x} \in \mathcal{D}$. Similarly, f is termed **superdifferentiable at $\boldsymbol{x}_o$** if there exists at least one supergradient of f at $\boldsymbol{x}_o$; and f is said to be **superdifferentiable** if it is super-differentiable for all $\boldsymbol{x} \in \mathcal{D}$.

Definition 8.4: Suppose $y = f(\boldsymbol{x})$: $\mathcal{D} \subset R^n \rightarrow R$ is concave. The **super-differential of f at $\boldsymbol{x}_o$** is the set of all supergradients of f at $\boldsymbol{x}_o$ and denoted

$$\partial f(\boldsymbol{x}_o) = \{\xi | \xi^T(\boldsymbol{x} - \boldsymbol{x}_o) \geq f(\boldsymbol{x}) - f(\boldsymbol{x}_o), \ \boldsymbol{x} \in \mathcal{D}\}. \qquad (8.6)$$

For f convex, the **subdifferential of f at $\boldsymbol{x}_o$** is the set of all subgradients of f at $\boldsymbol{x}_o$ and written

$$\partial f(\boldsymbol{x}_o) = \{\xi | \xi^T(\boldsymbol{x} - \boldsymbol{x}_o) \leq f(\boldsymbol{x}) - f(\boldsymbol{x}_o), \ \boldsymbol{x} \in \mathcal{D}\}. \qquad (8.7)$$

Looking to the properties of the subdifferential of f at $\boldsymbol{x}_o$ we see that $\partial f(\boldsymbol{x}_o)$ is always a closed, convex set (since it is the intersection of an infinite number of closed half-planes or $\partial f(x_o) = \bigcap_{x \in D} \{g | f(x) \geq f(x_o) + g^T(x - x_o)\}$, with the elements $\boldsymbol{g} \in \partial f(\boldsymbol{x}_o)$ the subgradients of f at $\boldsymbol{x}_o$) and, if $\boldsymbol{x}_o$ is an interior point of $\mathcal{D}$, then $\partial f(\boldsymbol{x}_o)$ is nonempty and bounded. Similarly, the superdifferentiable of f at $\boldsymbol{x}_o$, $\partial f(\boldsymbol{x}_o)$, is always a closed, convex set given that it is, likewise, the intersection of an infinite number of closed half-planes or $\partial f(x_o) = \bigcap_{x \in \mathcal{D}} \{g | f(x) \leq f(x_o) + g^T(x - x_o)\}$, $\boldsymbol{g} \in \partial f(\boldsymbol{x}_o)$. And if $\boldsymbol{x}_o \in \text{int}(\mathcal{D})$, then $\partial f(\boldsymbol{x}_o)$ is nonempty and bounded.

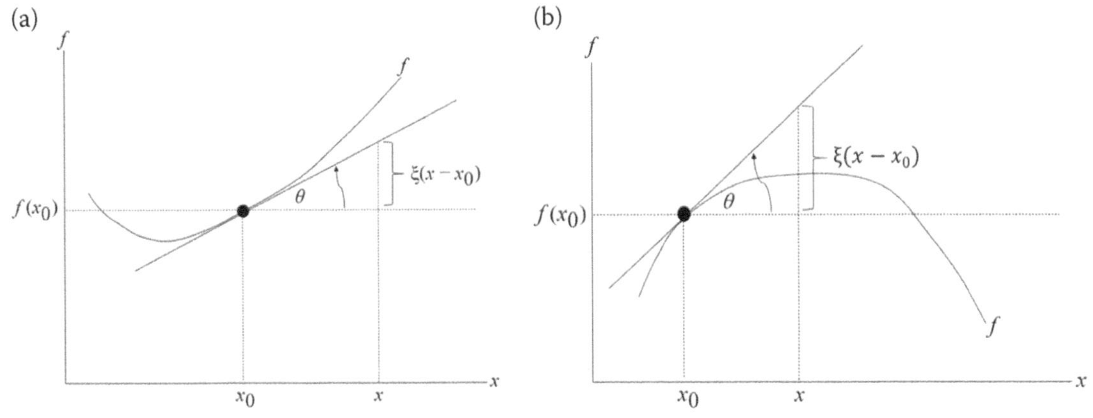

FIGURE 8.4 a. Subgradient of a convex function; b. supergradient of a concave function.

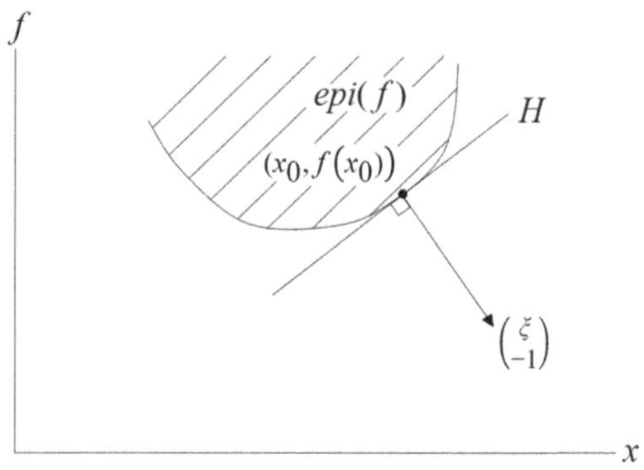

FIGURE 8.5 The vector $(\xi^T, -1)$ is normal to the supporting hyperplane $\mathcal{H}$.

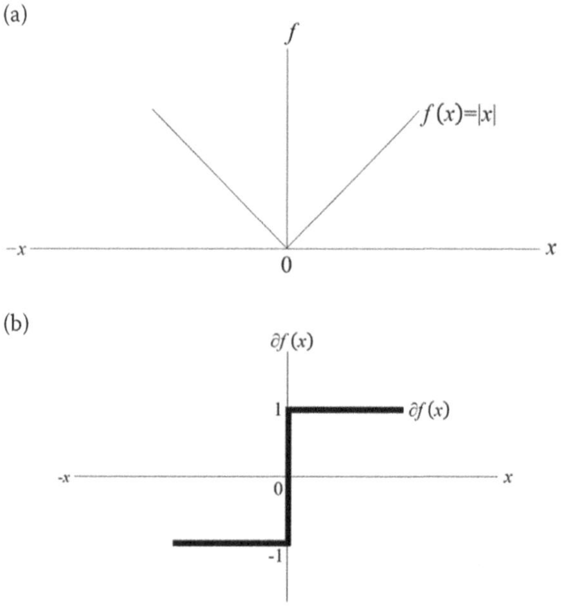

FIGURE 8.6 a. Graph of $f(x) = |x|$; b. graph of $\partial f(x)$.

Example 8.2: Given the real-valued function $y = f(x) = |x|$ (Figure 8.6a), let us determine the subdifferential of f. Looking to Figure 8.6b we see that: (a) for $x < 0$, $\partial f(x) = \{-1\}$; (b) for $x > 0$ $\partial f(x) = \{1\}$; and (c) for $x = 0$, $\partial f(0) = [-1, 1]$ since ∂f is defined by $|x| \geq \xi x$ for all x and $\xi \in [-1, 1]$ (see Boyd and Vandenberger [2008]). ■

The next theorem states that every concave (convex) real-valued function possesses at least one supergradient (subgradient) at points within the interior of its convex domain. Specifically, we have

Theorem 8.4: Suppose $\mathcal{D}$ is a convex set in R^n and $y = f(\boldsymbol{x})$: $\mathcal{D} \subset R^n \to R$ is concave. Then for $\boldsymbol{x}_o \in \text{int}(\mathcal{D})$, there exists a vector ξ such that the hyperplane $\mathcal{H} = \{(\boldsymbol{x}, y) | y = f(\boldsymbol{x}_o) + \xi^T(\boldsymbol{x} - \boldsymbol{x}_o)\}$ supports hyp (f) at the point $(\boldsymbol{x}_o, f(\boldsymbol{x}_o))$. That is, the vector ξ is a supergradient of f at $(\boldsymbol{x}_o, f(\boldsymbol{x}_o))$ with

$$f(\boldsymbol{x}) \leq f(\boldsymbol{x}_o) + \xi^T(\boldsymbol{x} - \boldsymbol{x}_o), \ \boldsymbol{x} \in \mathcal{D}. \tag{8.8}$$

But if f is taken to be a convex on $\mathcal{D}$, then obviously the inequality in (8.8) must be reversed or

$$f(\boldsymbol{x}) \geq f(\boldsymbol{x}_o) + \xi^T(\boldsymbol{x} - \boldsymbol{x}_o), \ \boldsymbol{x} \in \mathcal{D},$$

where ξ is now a subgradient of f at $(\boldsymbol{x}_o, f(\boldsymbol{x}_o))$ with $\mathcal{H}$ a support for epi (f).

It should be evident that if f is assumed to be *strictly concave* (*strictly convex*) then (8.8) and (8.9) hold with strict inequality. (Note that the converse of this theorem might not always be true, i.e., if for each point $x \in \text{int}(\mathcal{D})$ there exists a supergradient (subgradient) supporting hyp (f) (epi (f)), then f is not necessarily a concave (convex) function.) However, let us consider the circumstances under which the converse of Theorem 8.4 is valid. Specifically, let us restrict our discussion to the set of interior points of $\mathcal{D}$.

Theorem 8.5: Let $y = f(\boldsymbol{x}): \mathcal{D} \subset R^n \to R$ with $\mathcal{D}$ a convex set. Suppose that for each point $\boldsymbol{x} \in \text{int}(\mathcal{D})$ there exists a supergradient vector ξ such that

$$f(\boldsymbol{x}) \leq f(\boldsymbol{x}_o) + \xi^T(\boldsymbol{x} - \boldsymbol{x}_o),\ \boldsymbol{x} \in \mathcal{D}.$$

Then f is concave on int $(\mathcal{D})$. If for each point $\boldsymbol{x} \in \text{int}(\mathcal{D})$ there exists a subgradient vector ξ such that

$$f(\boldsymbol{x}) \geq f(\boldsymbol{x}_o) + \xi^T(\boldsymbol{x} - \boldsymbol{x}_o),\ \boldsymbol{x} \in \mathcal{D},$$

then f is convex on int $(\mathcal{D})$.

The preceding discussion of the supergradient (subgradient) vector begs the following question. Specifically, what is the importance or role of the supergradient (subgradient) of a function $f(\boldsymbol{x}): \mathcal{D} \subset R^n \to R$, $\mathcal{D}$ convex? As we shall soon see (Section 8.4), if, say, a *convex differentiable* function f has a gradient vector $\nabla f(\boldsymbol{x}_o)$ at every point $\boldsymbol{x}_o$, then f satisfies the inequality $f(\boldsymbol{x}) \geq f(\boldsymbol{x}_o) + \nabla f(\boldsymbol{x}_o)^T(\boldsymbol{x} - \boldsymbol{x}_o)$. But if f is *convex and not differentiable* over $\mathcal{D}$, then the subgradient serves as a *generalization of the gradient*. Consistent with this remark is the result that the subdifferential (superdifferential) of a convex (concave) function f, $\partial f(\boldsymbol{x}_o)$, is the singleton $\{\xi\} \in R^n$ if and only if f is differentiable at $\boldsymbol{x}_o$, in which case $\xi = \nabla f(\boldsymbol{x}_o)$ or $\xi_j = (\partial f / \partial x_j)_{\boldsymbol{x}_o}$, $j = 1, \ldots, n$. Moreover, the subgradients (supergradients) can be characterized by directional derivatives. To see this, let us first review some notation (Section 3.3). Let the directional derivative of f at $\boldsymbol{x}_o$ in the direction $\boldsymbol{u}$ be expressed as

$$\mathcal{D}_{\boldsymbol{u}} f(\boldsymbol{x}_o) = \lim_{t \to o} \frac{f(\boldsymbol{x}_o + t\boldsymbol{u}) - f(\boldsymbol{x}_o)}{t}, \quad (8.10)$$

provided, of course, that the indicated limit exists. If this limit exists for all $u \in \mathcal{D}$, then f is termed **directionally differentiable** at $\boldsymbol{x}_o$. In fact, every convex (concave) function is directionally differentiable. That is,

Theorem 8.6: Let $y = f(\boldsymbol{x}): \mathcal{D} \subset R^n \to R$ with f convex (concave). Then at every point $\boldsymbol{x}_o \in \mathcal{D}$ and in every direction $\boldsymbol{u}$, the directional derivative $\mathcal{D}_{\boldsymbol{u}} f(\boldsymbol{x}_o)$ exists.Given (8.10), we can now establish the connection between the directional derivative and the subgradient (supergradient) of a convex (concave) function.

Theorem 8.7: Let $y = f(\boldsymbol{x}): \mathcal{D} \subset R^n \to R$ be convex (concave) with f finite at $\boldsymbol{x}_o \in \mathcal{D}$. A vector $\xi \in R^n$ is a subgradient (supergradient) of f at $\boldsymbol{x}_o \in \mathcal{D}$ if and only if $\mathcal{D}^+_{\boldsymbol{u}} f(\boldsymbol{x}_o) \geq \xi^T\boldsymbol{u}\ (\leq \xi^T\boldsymbol{u})$ for all $\xi \in \partial f(\boldsymbol{x}_o) \neq \phi$ and every direction $\boldsymbol{u} \in \mathcal{D}$, where $\mathcal{D}^+_{\boldsymbol{u}} f(\boldsymbol{x}_o)$ is the right-sided (meaning that $t \to 0+$) directional derivative of f at $\boldsymbol{x}_o$ in the $\boldsymbol{u}$-direction.These inequalities hold as equalities for at least one subgradient (supergradient) in $\partial f(\boldsymbol{x}_o)$. That is, if int $(\mathcal{D}) \neq \varnothing$, then for all $\boldsymbol{x}_o \in \text{int}(\mathcal{D})$, $\mathcal{D}^+_{\boldsymbol{u}} f(\boldsymbol{x}_o)$ exists and

$$\mathcal{D}^+_{\boldsymbol{u}} f(\boldsymbol{x}_o) = \sup_{\xi \in \partial f(\boldsymbol{x}_o)} \{\xi^T\boldsymbol{u}\}.$$

Section 8.3 Exercises:

8.3.1 Prove Theorem 8.1 for f concave. (Hint: (a) pick two arbitrary points $(\boldsymbol{x}', y')$ and $(\boldsymbol{x}'', y'')$ within hyp (f) and show that their convex combination satisfies $y_c \leq f(\boldsymbol{x}_c)$ and that $(\boldsymbol{x}_c, y_c)$ is an element of hyp (f) and thus hyp (f) is convex. (b) Pick arbitrary $\boldsymbol{x}', \boldsymbol{x}'' \in \mathcal{D}$ with $(\boldsymbol{x}', f(\boldsymbol{x}'))$ and $(\boldsymbol{x}'', f(\boldsymbol{x}''))$ elements of hyp (f). Under the convexity of hyp (f) demonstrate that f is concave).

8.3.2 Suppose $y = f(x)$, $x \in R$, is defined as

$$f(x) = \begin{cases} 6 - x, & 0.38 \leq x \leq 2.62; \\ 6 - (x - 1)^2, & \text{elsewhere.} \end{cases}$$

a. What are the supergradients of f at $x = 0.38$ and $x = 2.62$?

b. What is the supergradient of f over the interval $0.38 \leq x \leq 2.62$?

c. What are the supergradients of f for $x < 0.38$ and $x > 2.62$?

8.3.3 Given a convex set $X = \{\boldsymbol{x} | x_1^2 + x_2^2 \leq 4, \boldsymbol{x} \in R^2\}$, find a supporting hyperplane to X at $\boldsymbol{x}_o^T = (\sqrt{2}, \sqrt{2})$.

8.3.4 Determine the subdifferential of the convex function

$$f(x) = \begin{cases} 1 - 2x, & x < 1; \\ -3 + 2x, & x \geq 1 \end{cases}$$

at $x_o = 1, x \in R$.

8.3.5 Determine the superdifferential of the concave function

$$f(x) = \begin{cases} x, & x < 0; \\ -x^2, & x \geq 0 \end{cases}$$

at the points -1, 0, 1, and 2, $x \in R$.

8.3.6 Find the directional derivative of the real-valued function $y = f(\boldsymbol{x}) = 2x_1^2 - x_1x_2, \boldsymbol{x} \in R^2$, at $\boldsymbol{x}_o^T = (1, 1)$ in the $\boldsymbol{u}^T = (2, 1)$ direction.

8.4 DIFFERENTIABLE CONVEX AND CONCAVE REAL-VALUED FUNCTIONS

In this section we consider convex and concave real-valued functions which are differentiable and twice-differentiable over a convex domain. If f is differentiable over an open convex region $\mathcal{D} \subset R^n$, then a necessary condition for any such function to be concave (convex) on $\mathcal{D}$ is

Theorem 8.8: Let $y = f(\boldsymbol{x})$: $\mathcal{D} \subset R^n \rightarrow R$ be a function of class $C^{(1)}$ at the point $\boldsymbol{x}_o \in \mathcal{D}$, with $\mathcal{D}$ open and convex. If f is concave at $\boldsymbol{x}_o$, then

$$f(\boldsymbol{x}) \leq f(\boldsymbol{x}_o) + \nabla f(\boldsymbol{x}_o)^T(\boldsymbol{x} - \boldsymbol{x}_o), \boldsymbol{x} \in \mathcal{D}. \quad (8.11)$$

And if f is convex at $\boldsymbol{x}_o$, then

$$f(\boldsymbol{x}) \geq f(\boldsymbol{x}_o) + \nabla f(\boldsymbol{x}_o)^T(\boldsymbol{x} - \boldsymbol{x}_o), \quad \boldsymbol{x} \in \mathcal{D}. \quad (8.12)$$

Here (8.11) and (8.12) depict the **linearization of f at $\boldsymbol{x}_o$**.[1]

Proof: Let f be concave at $\boldsymbol{x}_o$ (a similar proof holds for f convex at $\boldsymbol{x}_o$). Then

$$f(\theta \ \boldsymbol{x} + (1 - \theta)\boldsymbol{x}_o) \geq \theta f(\boldsymbol{x}) + (1 - \theta) f(\boldsymbol{x}_o), 0 \leq \theta \leq 1,$$

and, for $\boldsymbol{h} = \boldsymbol{x} - \boldsymbol{x}_o$,

$$\frac{f(\boldsymbol{x}_o + \theta \boldsymbol{h}) - f(\boldsymbol{x}_o)}{\theta} \geq f(\boldsymbol{x}) - f(\boldsymbol{x}_o).$$

Using a first-order Taylor expansion of f at $\boldsymbol{x}_o$ we have

$$f(\boldsymbol{x}_o + \theta \boldsymbol{h}) = f(\boldsymbol{x}_o) + \theta \nabla f(\boldsymbol{x}_o + \tau \theta \boldsymbol{h})^T \boldsymbol{h}, \quad 0 < \tau < 1.$$

A substitution of this expression into the preceding inequality yields

$$\nabla f(\boldsymbol{x}_o + \tau \theta \boldsymbol{h})^T \boldsymbol{h} \geq f(\boldsymbol{x}) - f(\boldsymbol{x}_o).$$

Then

$$\lim_{\theta \to 0} \nabla f(\boldsymbol{x}_o + \tau \theta \boldsymbol{h})^T \boldsymbol{h} = \nabla f(\boldsymbol{x}_o)^T \boldsymbol{h} \geq f(\boldsymbol{x}) - f(\boldsymbol{x}_o)$$

or

$$f(\boldsymbol{x}) \leq f(\boldsymbol{x}_o) + \nabla f(\boldsymbol{x}_o)^T(\boldsymbol{x} - \boldsymbol{x}_o), \boldsymbol{x} \in \mathcal{D}. \quad \textbf{Q. E. D.}$$

If f is strictly concave (strictly convex) at $\boldsymbol{x}_o \in \mathcal{D}$, then (8.11) ((8.12)) holds with strictly inequality for $\boldsymbol{x}(\neq \boldsymbol{x}_o) \in \mathcal{D}$.

Equations (8.11) and (8.12), respectively, stipulate that if f is concave (convex) on $\mathcal{D}$, then a linear approximation to f at $\boldsymbol{x}_o$ always overestimates (underestimates) the function, i.e., for f concave (convex), f lies everywhere below (above) its tangent hyperplane.

A necessary and sufficient condition for a differentiable function to be convex or concave over an open convex domain $\mathcal{D} \subset R^n$ is

Theorem 8.9: Let $y = f(\boldsymbol{x})$: $\mathcal{D} \subset R^n \rightarrow R$ be a function of class $C^{(1)}$ over $\mathcal{D}$, with $\mathcal{D}$ open and convex. Then f is convex over $\mathcal{D}$ if and only if

$$f(\boldsymbol{x}_2) - f(\boldsymbol{x}_1) \geq \nabla f(\boldsymbol{x}_1)^T(\boldsymbol{x}_2 - \boldsymbol{x}_1) \text{ for any } \boldsymbol{x}_1, \boldsymbol{x}_2 \in \mathcal{D}; \quad (8.13)$$

f is concave over $\mathcal{D}$ if and only if

$$f(\boldsymbol{x}_2) - f(\boldsymbol{x}_1) \leq \nabla f(\boldsymbol{x}_1)^T(\boldsymbol{x}_2 - \boldsymbol{x}_1) \text{ for any } \boldsymbol{x}_1, \boldsymbol{x}_2 \in \mathcal{D}. \quad (8.14)$$

Proof: (Necessity) If f is concave (convex) at each $\boldsymbol{x}_1 \in \mathcal{D}$, this portion of the proof proceeds as in Theorem 8.8.

(Sufficiency) For $\boldsymbol{x}_1, \boldsymbol{x}_2 \in \mathcal{D}$, $\boldsymbol{x}_c = \theta \boldsymbol{x}_2 + (1 - \theta)\boldsymbol{x}_1 \in \mathcal{D}$, $0 \leq \theta \leq 1$.

From Equation (8.14),

$$f(\boldsymbol{x}_1) - f(\boldsymbol{x}_c) \leq \theta \nabla f(\boldsymbol{x}_c)^T (\boldsymbol{x}_1 - \boldsymbol{x}_c),$$

$$f(\boldsymbol{x}_2) - f(\boldsymbol{x}_c) \leq -(1-\theta) \nabla f(\boldsymbol{x}_c)^T (\boldsymbol{x}_2 - \boldsymbol{x}_c).$$

Weighting the first inequality by $1-\theta$, and the second by θ, and summing yields

$$\theta f(\boldsymbol{x}_2) + (1-\theta) f(\boldsymbol{x}_1) - f(\boldsymbol{x}_c) \leq 0.$$

Hence, f is concave on $\mathcal{D}$. (A similar proof utilizing Equation (8.13) indicates that f is convex over $\mathcal{D}$.) Q. E. D.

Additionally, f is strictly convex (strictly concave) over $\mathcal{D}$ if and only if Equation (8.13) (Equation (8.14)) holds with strict inequality for any $\boldsymbol{x}_1, \boldsymbol{x}_2 \in \mathcal{D}$, $\boldsymbol{x}_1 \neq \boldsymbol{x}_2$.

We next turn to a necessary and sufficient condition for a twice-differentiable function to be convex or concave over an open convex region $\mathcal{D} \subset R^n$.

Theorem 8.10: Let $y = f(\boldsymbol{x})$: $\mathcal{D} \subset R^n \rightarrow R$ be a function of class $C^{(2)}$, with $\mathcal{D}$ open and convex. Then f is concave (convex) over $\mathcal{D}$ if and only if the quadratic form

$$Q(\boldsymbol{h}) = \sum_{r=1}^{n} \sum_{s=1}^{n} h_r h_s \frac{\partial^2 f(\boldsymbol{x})}{\partial x_r \partial x_s} = \boldsymbol{h}^T \boldsymbol{H}_f(\boldsymbol{x}) \boldsymbol{h}$$

is negative semi-definite (positive semi-definite) for all $\boldsymbol{h}$ and all points $\boldsymbol{x} \in \mathcal{D}$. Here, $\boldsymbol{h} \in R^n$ and $\boldsymbol{H}_f(\boldsymbol{x})$ is the Hessian matrix of f at $\boldsymbol{x}$.

Proof: A second-order Taylor expansion of f near $\boldsymbol{x}_c = \theta \boldsymbol{x}_1 + (1-\theta)\boldsymbol{x}_2 \in \mathcal{D}$, $0 \leq \theta \leq 1$, can be expressed as

$$f(\boldsymbol{x}) = f(\boldsymbol{x}_c) + \nabla f(\boldsymbol{x}_c)^T \boldsymbol{h} + \frac{1}{2!} \boldsymbol{h}^T \boldsymbol{H}_f(\boldsymbol{x}_c + \eta \boldsymbol{h}) \boldsymbol{h}, \ 0 < \eta < 1. \tag{8.15}$$

(Necessity) Let f be concave (convex). From (8.15),

$$f(\boldsymbol{x}) - f(\boldsymbol{x}_c) - \nabla f(\boldsymbol{x}_c)^T \boldsymbol{h} = \frac{1}{2!} \boldsymbol{h}^T \boldsymbol{H}_f(\boldsymbol{x}_c + \eta \boldsymbol{h}) \boldsymbol{h}, \ 0 < \eta < 1. \tag{8.15.1}$$

With f concave (convex) over $\mathcal{D}$, the left-hand side of (8.15.1), and thus its right-hand side, is non-positive (non-negative). Hence, $\boldsymbol{h}^T \boldsymbol{H}_f(\boldsymbol{x}_c + \eta \boldsymbol{h}) \boldsymbol{h} \leq 0 (\geq 0)$.

(Sufficiency) Upon substituting $\boldsymbol{x} = \boldsymbol{x}_1, \boldsymbol{x}_2 \in \mathcal{D}$ into Equation (8.15) we obtain

$$f(\boldsymbol{x}_j) = f(\boldsymbol{x}_c) + \nabla f(\boldsymbol{x}_c)^T \boldsymbol{h}_j + \frac{1}{2!} \boldsymbol{h}_j^T \boldsymbol{H}_f(\boldsymbol{x}_c + \eta_j \boldsymbol{h}_j) \boldsymbol{h}_j, \ 0 < \eta_j < 1, \ j = 1, 2,$$

where $\boldsymbol{h}_j$, $j = 1, 2$, denotes an $(n \times 1)$ vector with components $h_1^j, \ldots, h_n^j$. Weighting $f(\boldsymbol{x}_2)$ by θ and $f(\boldsymbol{x}_1)$ by $1-\theta$, $0 \leq \theta \leq 1$, and summing, yields

$$\begin{aligned} \theta f(\boldsymbol{x}_2) + (1-\theta) f(\boldsymbol{x}_1) &= f(\boldsymbol{x}_c) \\ &+ \theta \nabla f(\boldsymbol{x}_c)^T \boldsymbol{h}_2 + (1-\theta) \nabla f(\boldsymbol{x}_c)^T \boldsymbol{h}_1 \\ &+ \tfrac{\theta}{2!} \boldsymbol{h}_2^T \boldsymbol{H}_f(\boldsymbol{x}_c + \eta_2 \boldsymbol{h}_2) \boldsymbol{h}_2 \\ &+ \tfrac{(1-\theta)}{2!} \boldsymbol{h}_1^T \boldsymbol{H}_f(\boldsymbol{x}_c + \eta_1 \boldsymbol{h}_1) \boldsymbol{h}_1, \\ & 0 < \eta_j < 1, \ j = 1, 2. \end{aligned} \tag{8.16}$$

Since $\nabla f(\boldsymbol{x}_c)^T [\theta \boldsymbol{h}_2 + (1-\theta) \boldsymbol{h}_1] = 0$, (8.16) becomes

$$\begin{aligned} & \theta f(\boldsymbol{x}_2) + (1-\theta) f(\boldsymbol{x}_1) - f(\boldsymbol{x}_c) \\ &= \tfrac{\theta}{2!} \boldsymbol{h}_2^T \boldsymbol{H}_f(\boldsymbol{x}_c + \eta_2 \boldsymbol{h}_2) \boldsymbol{h}_2 \\ &+ \tfrac{(1-\theta)}{2!} \boldsymbol{h}_1^T \boldsymbol{H}_f(\mathrm{x}_c + \eta_1 \boldsymbol{h}_1) \boldsymbol{h}_1, \\ & 0 < \eta_j < 1, \ j = 1, 2. \end{aligned} \tag{8.16.1}$$

Now, if the quadratic forms $\boldsymbol{h}_j^T \boldsymbol{H}_f(\boldsymbol{x}_c + \eta_j \boldsymbol{h}_j) \boldsymbol{h}_j \leq 0 (\geq 0)$, i.e., are negative semi-definite (positive semi-definite), $j = 1, 2$, then $\theta f(\boldsymbol{x}_2) + (1-\theta) f(\boldsymbol{x}_1) \leq f(\boldsymbol{x}_c) (\geq f(\boldsymbol{x}_c))$. Thus f is concave (convex) over $\mathcal{D}$. Q. E. D. (An alternative and important methodology for executing the proofs of Theorems 8.9 and 8.10 appears in Appendix 8.A.)

If $Q(\boldsymbol{h}) = \boldsymbol{h}^T \boldsymbol{H}_f(\boldsymbol{x}) \boldsymbol{h} < 0 (>0)$, i.e., is negative (positive) definite for $\boldsymbol{h} \neq \boldsymbol{O}$ and all points $\boldsymbol{x} \in \mathcal{D}$, then f is strictly concave (strictly convex) over $\mathcal{D}$. But if f is strictly concave (strictly convex) on $\mathcal{D}$, then $\boldsymbol{h}_j^T \boldsymbol{H}_f(\boldsymbol{x}) \boldsymbol{h} \leq 0 (\geq 0)$ for all $\boldsymbol{x} \in \mathcal{D}$. Hence, we may not legitimately conclude that $\boldsymbol{h}_j^T \boldsymbol{H}_f(\boldsymbol{x}) \boldsymbol{h} < 0 (>0)$, $\boldsymbol{x} \in \mathcal{D}$.

Example 8.3: A specialization of Theorem 8.10 to $x \in R$ reveals that $y = f(x)$: $\mathcal{D} = (a, b) \subset R \rightarrow R$ is convex (concave) over $\mathcal{D}$ if and only if $f''(x) \geq 0 (\leq 0)$. If $f''(x) > 0 (< 0)$ for all $\boldsymbol{x} \in \mathcal{D}$, then f is strictly convex (strictly concave) on $\mathcal{D}$. But if f is strictly convex (strictly concave) over $\mathcal{D}$, it is not necessarily the case that $f''(x) > 0 (< 0)$ for all $\boldsymbol{x} \in \mathcal{D}$, i.e., if f is strictly convex (strictly concave) on $\mathcal{D}$, we may only conclude that $f''(x) \geq 0 (\leq 0)$.

For instance, $y = f(x) = 1 - 3x + x^4$, $\mathcal{D} = (-1, 1)$, is strictly convex over its domain with $f''(x) = 12x^2$. However, $f''(0) = 0$ so that $f''(x) \geq 0$ over $\mathcal{D}$. ■

Example 8.4: Is the real-valued function $y = f(\boldsymbol{x}) = (x_1 x_2)^{\frac{1}{2}}$ concave for all $\boldsymbol{x} > \boldsymbol{O}$? From the Hessian matrix of f,

$$\boldsymbol{H}_f = \begin{bmatrix} f_{11} & f_{12} \\ f_{21} & f_{22} \end{bmatrix} = \begin{bmatrix} -\frac{1}{4}x_1^{-\frac{3}{2}}x_2^{\frac{1}{2}} & \frac{1}{4}x_1^{-\frac{1}{2}}x_2^{-\frac{1}{2}} \\ \frac{1}{4}x_1^{-\frac{1}{2}}x_2^{-\frac{1}{2}} & -\frac{1}{4}x_1^{\frac{1}{2}}x_2^{-\frac{3}{2}} \end{bmatrix},$$

we see that $f_{11}, f_{22} < 0$ and $|\boldsymbol{H}_f| = 0$. Hence, $Q(\boldsymbol{h}) = \boldsymbol{h}^T \boldsymbol{H}_f \boldsymbol{h}$ is negative semi-definite for all $\boldsymbol{x} > \boldsymbol{O}$ so that f is concave over the positive orthant. ■

Example 8.5: Let the real-valued function $y = f(\boldsymbol{x}) = x_1^2 + x_1 x_2 + 2x_2^2$ be defined for all $\boldsymbol{x} \in \mathcal{D} = R^2$. Since

$$f_{11} = 2 > 0, \ |\boldsymbol{H}_f| = \begin{vmatrix} 2 & 1 \\ 1 & 4 \end{vmatrix} = 7 > 0,$$

we may conclude that $Q(\boldsymbol{h}) = \boldsymbol{h}^T \boldsymbol{H}_f(\boldsymbol{x})\boldsymbol{h} > 0$ for all $\boldsymbol{h} \neq \boldsymbol{O}$ and all points $\boldsymbol{x}$. Hence, it follows that $Q(\boldsymbol{h})$ is positive definite and thus f is strictly convex over $\mathcal{D}$. ■

Example 8.6: Let the real-valued function $y = f(x) = 2x_1^2 + x_2^2$ be defined for all $\boldsymbol{x} \in \mathcal{D} = R^2$. Is f convex or concave over $\mathcal{D}$? Since $|\boldsymbol{H}_f| = -8 < 0$, $Q(\boldsymbol{h}) = \boldsymbol{h}^T \boldsymbol{H}_f(\boldsymbol{x})\boldsymbol{h}$ is indefinite so that f is neither convex nor concave. In fact, the surface is saddle-shaped over $\mathcal{D}$. ■

Example 8.7: Is the real-valued function $y = f(\boldsymbol{x}) = -2x_1^{-2} + x_1 x_3 - x_2^2 + x_2 x_3 + x_2 x_4 - x_3^2 - x_4^{-2}$ concave at $\boldsymbol{x}_o^T = (1, 0, 0, 1)$? From

$$\begin{aligned} &f_{11}^o = -12, f_{12}^o = 0, f_{23}^o = 1, f_{34}^o = 0, \\ &f_{22}^o = -2, f_{13}^o = 1, f_{24}^o = 1, \\ &f_{33}^o = -2, f_{14}^o = 0, \\ &f_{44}^o = -6, \end{aligned}$$

we obtain the Hessian matrix

$$\boldsymbol{H}_f(\boldsymbol{x}_o) = \begin{bmatrix} -12 & 0 & 1 & 0 \\ 0 & -2 & 1 & 1 \\ 1 & 1 & -2 & 0 \\ 0 & 1 & 0 & -6 \end{bmatrix}.$$

Since the naturally ordered principal minors of this matrix alternate in sign, starting negative (i.e., they form the sequence of values −12, 24, −34, 93), we see that $\boldsymbol{H}_f(\boldsymbol{x}_o)$, or $Q(\boldsymbol{h}) = \boldsymbol{h}^T \boldsymbol{H}_f(\boldsymbol{x}_o)\boldsymbol{h}$, is negative definite and thus f is strictly concave at $\boldsymbol{x}_o$. ■

We now close this section with a few summary comments pertaining to the interface between concave or convex functions and the sign definiteness or semi-definiteness of the quadratic form $\boldsymbol{x}^T \boldsymbol{H}_f(\boldsymbol{x})\boldsymbol{x}$. Specifically, suppose $y = f(\boldsymbol{x})$: $\mathcal{D} \subset R^n \to R$, $\mathcal{D}$ an open set, is a function of class $C^{(2)}$. Then:

1. if $\boldsymbol{H}_f(\boldsymbol{x})$ is positive definite (its naturally ordered principal minors are all positive, i.e., $M_k > 0$ for all $\boldsymbol{x} \in \mathcal{D}$, $k = 1, \ldots, n$), then f is strictly convex.
2. if $\boldsymbol{H}_f(\boldsymbol{x})$ is negative definite (its naturally ordered principal minors alternate in sign, the first being negative, i.e., $(-1)^k M_k > 0$ for all $\boldsymbol{x} \in \mathcal{D}$, $k = 1, \ldots, n$), then f is strictly concave.
3. f is convex (concave) on $\mathcal{D}$ if and only if $\boldsymbol{H}_f(\boldsymbol{x})$ is positive (negative) semi-definite for all $\boldsymbol{x} \in \mathcal{D}$.

Section 8.4 Exercises:

8.4.1 Given the real-valued function

$$f(x) = \begin{cases} x, & x < 0; \\ -x^2, & x \geq 0, \end{cases}$$

determine the support functions for f at the points $x = -1, 0, 1$, and 2.

8.4.2 Are the real-valued functions $y = f(x)$, $x \in R$, of class $C^{(2)}$ strictly concave or strictly convex at the indicated points (x_o):

a. $f(x) = x^4$, $x_o = 1$.

b. $f(x) = -x^2$, $x_o = 2$.

8.4.3 Determine if the real-valued functions of class $C^{(2)}$ are convex or concave.

a. $f(\boldsymbol{x}) = x_1^2 + x_1x_2 + x_2^2$.

b. $f(\boldsymbol{x}) = x_1^2x_2^2$.

c. $f(\boldsymbol{x}) = 4x_1^2 + x_1x_2 + 2x_2^2 + x_1x_3 + x_3^2$.

d. $f(\boldsymbol{x}) = x_1^2 + x_2^2 + x_3^2 + x_1x_3 + x_2x_3$.

e. $f(\boldsymbol{x}) = 2x_1^2 + x_2^2 + 4x_3^2 - x_1 - 2x_3$.

f. $f(\boldsymbol{x}) = -2x_1^2 + 4x_1x_2 - 2x_2^2 - 2x_3^2$.

8.5 EXTREMA OF CONVEX AND CONCAVE REAL-VALUED FUNCTIONS

It should be intuitively clear that the most important cases involving the attainment of global and local extrema by convex or concave real-valued functions occur when these functions are strictly convex or strictly concave over some convex domain $\mathcal{D}$. The following set of theorems deals exclusively with the strict case.

Theorem 8.11: Let $y = f(\boldsymbol{x})$: $\mathcal{D} \subset R^n \rightarrow R$ be strictly concave, with $\mathcal{D}$ closed and convex. Then f assumes a (strong) global maximum somewhere over $\mathcal{D}$.
Clearly, this extremum must be unique and occurs at an interior point of $\mathcal{D}$ (hence, it is also a strong local maximum) or on the boundary of $\mathcal{D}$. The global minimum of a strictly concave function always occurs on the boundary of $\mathcal{D}$, and may or may not be unique. Hence, a strictly concave function never assumes a local minimum over $\mathcal{D}$.

We next have

Theorem 8.12: Let $y = f(\boldsymbol{x})$: $\mathcal{D} \subset R^n \rightarrow R$ be strictly convex, with $\mathcal{D}$ closed and convex. Then f assumes a (strong) global minimum somewhere over $\mathcal{D}$.
It is also true that this extremum is unique and occurs at an interior point of $\mathcal{D}$ (so that it is also a strong local minimum) or at one of its boundary points. The global maximum of a strictly convex function must occur at a boundary point of $\mathcal{D}$. It may or may not be unique. Thus a strictly convex function never attains a local maximum anywhere over its domain.

To operationalize the preceding two theorems, we have

Theorem 8.13: Let $y = f(\boldsymbol{x})$: $\mathcal{D} \subset R^n \rightarrow R$ be concave (convex) with $\mathcal{D}$ closed and convex. Then:

i. any local maximum (local minimum) is also the global maximum (global minimum) of f over $\mathcal{D}$.

ii. the set $\mathcal{D}^*$ of all maximal (minimal) points $\boldsymbol{x}$ is convex.

iii. $\mathcal{D}^*$ has at most one element if f is strictly concave (strictly convex).

Proof: (i) If $\boldsymbol{x}_o$ is a local maximizer of f, then there exists a $\delta > 0$ such that $f(\boldsymbol{x}_o) \geq f(\boldsymbol{x})$ for all $\boldsymbol{x} \in \delta(\boldsymbol{x}_o) \subset \mathcal{D}$. Let $\boldsymbol{x}_1 \in \mathcal{D}$ and, for $\lambda \in [0, 1]$, $(1-\lambda)\boldsymbol{x}_o + \lambda\boldsymbol{x}_1 \in \mathcal{D}$. For sufficiently small $\lambda_o > 0$, $\|((1-\lambda_o)\boldsymbol{x}_o + \lambda_o\boldsymbol{x}_1) - \boldsymbol{x}_o\| < \delta$. Hence

$$f(\boldsymbol{x}_o) \geq f((1-\lambda_o)\boldsymbol{x}_o + \lambda_o\boldsymbol{x}_1).$$

With f concave,

$$f((1-\lambda_o)\boldsymbol{x}_o + \lambda_o\boldsymbol{x}_1) \geq (1-\lambda_o)f(\boldsymbol{x}_o) + \lambda_o f(\boldsymbol{x}_1).$$

Upon combining the last two inequalities and dividing by λ_o renders $f(\boldsymbol{x}_o) \geq f(\boldsymbol{x}_1)$. Thus $\boldsymbol{x}_o$ yields a global maximum of f over $\mathcal{D}$. (A similar argument holds for f strictly concave on $\mathcal{D}$.)

ii. Let $\boldsymbol{x}_o \in \mathcal{D}^*$ and consider the upper level set $U_f(f(\boldsymbol{x}_o)) = \{\boldsymbol{x} \in \mathcal{D} | f(\boldsymbol{x}) \geq f(\boldsymbol{x}_o)\}$. Hence, $U_f(f(\boldsymbol{x}_o)) = \mathcal{D}^*$ since, if f is concave on $\mathcal{D}$, then the upper level set is convex.

iii. Suppose there exists two points $\boldsymbol{x}_1, \boldsymbol{x}_2 \in \mathcal{D}^*$ for which $f(\boldsymbol{x}_1) = f(\boldsymbol{x}_2)$. Under strict concavity, we have $f(\theta\boldsymbol{x}_1 + (1-\theta)\boldsymbol{x}_2) > \theta f(\boldsymbol{x}_1) + (1-\theta)f(\boldsymbol{x}_2) = f(\boldsymbol{x}_1) = f(\boldsymbol{x}_2)$, $\theta \in (0, 1)$. Clearly, this contradicts the global maximality assumption. Q. E. D.

Let us now examine a necessary and sufficient condition for the existence of a strong global extremum. We start with

Theorem 8.14: Let $y = f(\boldsymbol{x}): \ \mathcal{D} \subset R^n \to R$ be strictly convex (strictly concave), with $\mathcal{D}$ convex. A point $\boldsymbol{x}_o$ yields strong global minimum (strong global maximum) of f if and only if there exists a subgradient (supergradient) vector ξ at $\boldsymbol{x}_o$ such that $\xi^T(\boldsymbol{x} - \boldsymbol{x}_o) > 0(<0), \ \boldsymbol{x} \in \mathcal{D}$. Moreover, if f is of class $C^{(1)}$ at $\boldsymbol{x}_o$, then $\boldsymbol{x}_o$ yields a strong global minimum (strong global maximum) of f if and only if $\nabla f(\boldsymbol{x}_o)^T(\boldsymbol{x} - \boldsymbol{x}_o) > 0(<0), \boldsymbol{x} \in \mathcal{D}$.

Note that if f is $C^{(1)}$ with $\mathcal{D}$ convex and open, then $\boldsymbol{x}_o$ yields a strong global minimum (strong global maximum) if and only if there exists a *zero subgradient* (supergradient) of f at $\boldsymbol{x}_o$. More formally,

Theorem 8.15: Let $y = f(\boldsymbol{x}): \ \mathcal{D} \subset R^n \to R$ be strictly concave (strictly convex) and of class $C^{(1)}$ at a point $\boldsymbol{x}_o \in \text{int}(\mathcal{D})$, with $\mathcal{D}$ open and convex. Then f has a strong global maximum (strong global minimum) at $\boldsymbol{x}_o$ if and only if $\nabla f(\boldsymbol{x}_o) = \boldsymbol{O}$.

Proof: (Sufficiency). For f strictly concave at $\boldsymbol{x}_o$, we know that

$$f(\boldsymbol{x}) < f(\boldsymbol{x}_o) + \nabla f(\boldsymbol{x}_o)^T(\boldsymbol{x} - \boldsymbol{x}_o), \boldsymbol{x}(\neq \boldsymbol{x}_o) \in \mathcal{D}, \quad (8.17)$$

i.e., f lies everywhere below its tangent hyperplane. If $\nabla f(\boldsymbol{x}_o) = \boldsymbol{O}$, (8.17) becomes $f(\boldsymbol{x}) < f(\boldsymbol{x}_o), \boldsymbol{x}(\neq \boldsymbol{x}_o) \in \mathcal{D}$. Hence, f has a strong local maximum which is also the strong global maximum of f at $\boldsymbol{x}_o$. For f strictly convex at $\boldsymbol{x}_o$, the function lies everywhere above its tangent hyperplane so that

$$f(\boldsymbol{x}) > f(\boldsymbol{x}_o) + \nabla f(\boldsymbol{x}_o)^T(\boldsymbol{x} - \boldsymbol{x}_o), \boldsymbol{x}(\neq \boldsymbol{x}_o) \in \mathcal{D}, \quad (8.18)$$

whereupon setting $\nabla f(\boldsymbol{x}_o) = \boldsymbol{O}$ implies $f(\boldsymbol{x}) > f(\boldsymbol{x}_o)$, $\boldsymbol{x}(\neq \boldsymbol{x}_o) \in \mathcal{D}$. In this instance f has a strong local and thus strong global minimum at $\boldsymbol{x}_o$.

(Necessity). With $\nabla f(\boldsymbol{x}_o) \neq \boldsymbol{O}$, the above necessary condition for a local extremum (Theorem 7.7) informs us that $f(\boldsymbol{x}_o)$ is not an extreme value of f over $\mathcal{D}$.

Looked at in an alternative fashion, for f strictly convex (strictly concave), $\boldsymbol{O} \in \partial f(\boldsymbol{x}_o)$ if and only if f attains its strong global minimum (strong global maximum) at $\boldsymbol{x}_o$. Q. E. D.

The gist of this theorem is that if a function is strictly concave (strictly convex) over its domain, then the vanishing of the gradient of f constitutes a *necessary and sufficient condition* for the existence of a strong local extremum (and thus, for a strong global extremum). Thus the property of strict concavity (strict convexity) replaces the usual sufficient condition for a strong local maximum (strong local minimum). Hence, the ordinary sufficient condition for a strong local extremum over $\mathcal{D}$ ($\boldsymbol{h}^T \boldsymbol{H}_f(\boldsymbol{x}_o)\boldsymbol{h}$ is negative or positive definite for all $\boldsymbol{h} \neq \boldsymbol{O}, \boldsymbol{x} \in \mathcal{D}$) and the notions of strict concavity or strict concavity of f over this region are one and the same.

Section 8.5 Exercises:

8.5.1 Do the following functions attain a local extremum for $\boldsymbol{x} \in R^2$?

a. $f(\boldsymbol{x}) = x_1^3 + 4x_1x_2 - x_2^2$.

b. $f(\boldsymbol{x}) = 2x_1^2 + x_2^{-1} + x_2$.

c. $f(\boldsymbol{x}) = (x_1 - 2)^2 + (x_2 - 3)^2$.

d. $f(\boldsymbol{x}) = 6x_1^3 + 2x_1x_2 - 2x_1^2 + x_2^2 + 10$.

8.5.2 Do the following functions attain a local extremum for $\boldsymbol{x} \in R^3$?

a. $f(\boldsymbol{x}) = -x_1^2 - 4x_2^2 - x_3^2 + 2x_1x_2 + x_2x_3$.

b. $f(\boldsymbol{x}) = x_1^2 + 3x_2^2 - 2x_3^2 + x_1x_2 - x_1x_3 - x_2x_3 - x_1 - 2x_2 - x_3$.

8.6 STRONGLY A-CONCAVE AND STRONGLY A-CONVEX FUNCTIONS [AVRIEL *ET AL.* (2010); VIAL (1982, 1983)]

A subclass of strictly concave or strictly convex functions are those that may be characterized as strongly α-concave or strongly α-convex over a convex domain. Specifically, we have

Definition 8.5: A function of class $C^{(2)}$, $y = f(\boldsymbol{x}): \ \mathcal{D} \subset R^n \to R$, $\mathcal{D}$ open and convex, is termed **stronglyα-concave** on $\mathcal{D}$ if and only if there exists an $\alpha > 0$, $\alpha \in R$, such that for every $\boldsymbol{x} \in \mathcal{D}$,

$$\nabla^2 f(\boldsymbol{x}) + \alpha\, \boldsymbol{I}_n \leq \boldsymbol{O}, \quad \text{[negative semidefinite]} \tag{8.19}$$

where $\nabla^2 f(\boldsymbol{x})$ is the Hessian matrix of f at $\boldsymbol{x}$. The function f is termed **strongly α-convex** if and only if there exists an $\alpha > 0$, $\alpha \in R$, such that for every $\boldsymbol{x} \in \mathcal{D}$,

$$\nabla^2 f(\boldsymbol{x}) - \alpha\, \boldsymbol{I}_n \geq \boldsymbol{O}. \quad \text{[positive semidefinite]} \tag{8.20}$$

An alternative to this definition is provided by

Definition 8.5.1: Let $y = f(\boldsymbol{x})$: $\mathcal{D} \subset R^n \rightarrow R$, $\mathcal{D}$ open and convex. Then:

i. $f(\boldsymbol{x})$ is strongly α-concave if it can be expressed as

$$f(\boldsymbol{x}) = \varnothing(\boldsymbol{x}) - \frac{\alpha}{2}\boldsymbol{x}^T\boldsymbol{x}, \tag{8.19.1}$$

where $\varnothing$ is a concave function defined on $\mathcal{D}$ and $\alpha > 0$, $\alpha \in R$. Similarly, $f(\boldsymbol{x})$ is strongly α-convex if it can be written as

$$f(\boldsymbol{x}) = \psi(\boldsymbol{x}) + \frac{\alpha}{2}\boldsymbol{x}^T\boldsymbol{x}, \tag{8.20.1}$$

where ψ is convex on $\mathcal{D}$; or

ii. $f(\boldsymbol{x})$ is strongly α-concave if and only if the function

$$g(\boldsymbol{x}) = f(\boldsymbol{x}) + \frac{\alpha}{2}\boldsymbol{x}^T\boldsymbol{x} \tag{8.19.2}$$

is concave. Similarly, $f(\boldsymbol{x})$ is strongly α-convex if and only if the function

$$g(\boldsymbol{x}) = f(\boldsymbol{x}) - \frac{\alpha}{2}\boldsymbol{x}^T\boldsymbol{x} \tag{8.20.2}$$

is convex.

For $\boldsymbol{x}_1, \boldsymbol{x}_2 \in \mathcal{D}$, we have, from Taylor's Theorem ((3.6.8)),

$$f(\boldsymbol{x}_2) = f(\boldsymbol{x}_1) + \nabla f(\boldsymbol{x}_1)^T(\boldsymbol{x}_2 - \boldsymbol{x}_1) + \frac{1}{2}(\boldsymbol{x}_2 - \boldsymbol{x}_1)^T\nabla^2 f(\boldsymbol{z})(\boldsymbol{x}_2 - \boldsymbol{x}_1) \tag{8.21}$$

where $\boldsymbol{z} = \boldsymbol{x}_1 + \theta(\boldsymbol{x}_2 - \boldsymbol{x}_1)$, $\theta \in [0, 1]$. By the strongly α-convexity assumption (8.19), the last term on the right-hand side of (8.21) will not exceed $-\frac{\alpha}{2}\|\boldsymbol{x}_2 - \boldsymbol{x}_1\|^2$. Hence

$$f(\boldsymbol{x}_2) \leq f(\boldsymbol{x}_1) + \nabla f(\boldsymbol{x}_1)^T(\boldsymbol{x}_2 - \boldsymbol{x}_1) - \frac{\alpha}{2}\|\boldsymbol{x}_2 - \boldsymbol{x}_1\|^2. \tag{8.22}$$

(Note that if $\alpha = 0$, we get Equation (8.11) - - a basic characteristic of concave functions.) For f strongly α-convex on $\mathcal{D}$, (8.20) enables us to write

$$f(\boldsymbol{x}_2) \geq f(\boldsymbol{x}_1) + \nabla f(\boldsymbol{x}_1)^T(\boldsymbol{x}_2 - \boldsymbol{x}_1) + \frac{\alpha}{2}\|\boldsymbol{x}_2 - \boldsymbol{x}_1\|^2, \tag{8.23}$$

for $\boldsymbol{x}_1, \boldsymbol{x}_2 \in \mathcal{D}$. If $\alpha = 0$, then Equation (8.12) obtains - - a key property of convex functions. Our interpretation of (8.22) ((8.23)) is that, for any point $\boldsymbol{x}_2 \in \mathcal{D}$, there exists a concave (convex) quadratic upper (lower) bound to $f(\boldsymbol{x}_2)$.

Suppose that $f(\boldsymbol{x})$ is *not* differentiable on a convex domain $\mathcal{D} \subset R^n$. Then f is strongly α-concave if and only if there exists an $\alpha > 0$, $\alpha \in R$, such that, for $\boldsymbol{x}_1, \boldsymbol{x}_2 \in \mathcal{D}$,

$$f(\lambda\boldsymbol{x}_1 + (1-\lambda)\boldsymbol{x}_2) \geq \lambda f(\boldsymbol{x}_1) + (1-\lambda)f(\boldsymbol{x}_2) + \frac{\alpha}{2}\lambda(1-\lambda)\|\boldsymbol{x}_2 - \boldsymbol{x}_1\|^2, \tag{8.24}$$

$\lambda \in [0, 1]$. (Note that (8.24) is identical to the definition of a concave function for $\alpha = 0$.) For f strongly α-convex,

$$f(\lambda\boldsymbol{x}_1 + (1-\lambda)\boldsymbol{x}_2) \leq \lambda f(\boldsymbol{x}_1) + (1-\lambda)f(\boldsymbol{x}_2) - \frac{\alpha}{2}\lambda(1-\lambda)\|\boldsymbol{x}_2 - \boldsymbol{x}_1\|^2, \tag{8.25}$$

$\lambda \in [0, 1]$. Additionally, if we do not admit differentiability of f, then we can replace $\nabla f(\boldsymbol{x}_1)$ by $\xi \in \partial f(\boldsymbol{x}_1)$. Thus f is strongly α-concave if it is superdifferentiable and, for $\boldsymbol{x}_1, \boldsymbol{x}_2 \in \mathcal{D}$,

$$f(\boldsymbol{x}_2) - f(\boldsymbol{x}_1) \leq \xi^T(\boldsymbol{x}_2 - \boldsymbol{x}_1) - \frac{\alpha}{2}\|\boldsymbol{x}_2 - \boldsymbol{x}_1\|^2. \tag{8.26}$$

And if f is subdifferentiable for $\boldsymbol{x}_1, \boldsymbol{x}_2 \in \mathcal{D}$,

$$f(\mathbf{x}_2) - f(\mathbf{x}_1) \geq \xi^T(\mathbf{x}_2 - \mathbf{x}_1) + \frac{\alpha}{2}\|\mathbf{x}_2 - \mathbf{x}_1\|^2. \quad (8.27)$$

We noted above that, say, strongly α-concave functions are a subclass of strictly concave functions. Hence, the following chain of implications holds:

f is strongly α-concave $\rightarrow f$ is strictly concave $\rightarrow f$ is concave.

To verify this, we need only demonstrate that the strongly α-concavity of f implies its strict concavity. To this end let

$$f(\lambda\mathbf{x}_1 + (1-\lambda)\mathbf{x}_2) + \tfrac{\alpha}{2}\|\lambda\mathbf{x}_1 + (1-\lambda)\mathbf{x}_2\|^2 \geq \lambda f(\mathbf{x}_1) + (1-\lambda)f(\mathbf{x}_2) + \lambda\tfrac{\alpha}{2}\|\mathbf{x}_1\|^2 + (1-\lambda)\tfrac{\alpha}{2}\|\mathbf{x}_2\|^2.$$

But

$$-\lambda\frac{\alpha}{2}\|\mathbf{x}_1\|^2 - (1-\lambda)\frac{\alpha}{2}\|\mathbf{x}_2\|^2 + \frac{\alpha}{2}\|\lambda\mathbf{x}_1 + (1-\lambda)\mathbf{x}_2\|^2 < 0$$

for all $\mathbf{x}_1, \mathbf{x}_2 (\mathbf{x}_1 \neq \mathbf{x}_2)$ and $\lambda \in (0, 1)$ since $-\|\mathbf{x}_1\|^2$ and $-\|\mathbf{x}_2\|^2$ are strictly concave. Hence,

$$f(\lambda\mathbf{x}_1 + (1-\lambda)\mathbf{x}_2) > \lambda f(\mathbf{x}_1) + (1-\lambda)f(\mathbf{x}_2).$$

Let us consider the special case where $\mathcal{D} = R$ and f is of class $C^{(2)}$. Then:

i. f is concave (convex) if and only if $f''(x) \leq 0 (\geq 0)$ for all x;

ii. f is strictly concave (strictly convex) if $f''(x) < 0 (>0)$ for all x (this is sufficient but not necessary);

iii. f is strongly α-concave (strongly α-convex) if and only if $f''(x) \leq \alpha (\geq \alpha)$ for all x.

What are the implications of, say, the strongly α-concavity of f for finding an extremum of any such function? We have seen that if f is continuous on a compact domain, then it has a maximum and a minimum over that domain. Like a strictly concave function, a strongly

α-concave function has a *unique* maximum on a compact domain. Here, the usual first-order condition for a local maximum of a strongly concave $C^{(1)}$ function at a point $\mathbf{x}_o$ is also sufficient for the existence of a global maximum of f at $\mathbf{x}_o$. This is because, in a neighborhood of $\mathbf{x}_o$, a strongly α-concave $C^{(2)}$ function exhibits the curvature of a negative definite quadratic form.

Example 8.8: Utilize Equation (8.20.1) to demonstrate that $f(x) = x^2$ is strongly α-convex. From $f(x) = \psi(x) - \frac{\alpha}{2}x^2$ we have

$$f'(x) = \psi'(x) - \alpha x,$$
$$f''(x) = \psi''(x) - \alpha.$$

Then $2 = \psi''(x) - \alpha$ ***or*** $\psi''(x) = 2 - \alpha$. Hence, as required, $\psi(x)$ is convex for $\psi''(x) = 2 - \alpha \geq 0$. ■

Section 8.6 Exercises:

8.6.1

a. Is the function $f(x) = -x^4$, $x \in R$, strongly α-concave at $x = 0$?

b. Is the function $f(x) = e^x$, $x \in R$, strongly α-convex?

c. Is the function $f(x) = x^2 + |x|$, $x \in R$, strongly α-convex?

8.6.2 Is the function $f(\mathbf{x}) = x_1^2 + x_2^2$, $\mathbf{x} \in R^2$, strongly α-convex? (Use Equation (8.20) to frame your answer.)

8.6.3 For $\mathbf{x}_1, \mathbf{x}_2 \in \mathcal{D}$, $\mathcal{D}$ convex, and $\lambda \in R$, verify that $\lambda\|\mathbf{x}_1\|^2 + (1-\lambda)\|\mathbf{x}_2\|^2 - \lambda(1-\lambda)\|\mathbf{x}_1 - \mathbf{x}_2\|^2 = \|\lambda\mathbf{x}_1 + (1-\lambda)\mathbf{x}_2\|^2$.

8.7 CONJUGATE FUNCTIONS (ROCKAFELLAR [1970, 1974]; FENCHEL [1949])

8.7.1 Some Preliminary Notions

To set the stage for a discussion of functions conjugate to convex (concave) functions, let us consider some requisite terminology. We start with

Definition 8.6: Suppose $X = Y = R^n$ are two finite linear spaces. A pairing of X and Y is termed a **bilinear form** $\langle \cdot,\cdot \rangle = \langle \mathbf{x}, \mathbf{y} \rangle$ on $X \times Y$, with $\mathbf{x}$ restricted to X and $\mathbf{y}$ restricted to Y. For our purposes, let us take X, Y

to be paired spaces with $< \boldsymbol{x}, \boldsymbol{y} > = \boldsymbol{x}^T\boldsymbol{y}$.

Additionally, we shall focus on convex (concave) functions on all of R^n or on such functions whose values are real or $\pm\infty$ and whose domain is a subset X of R^n, i.e., for X a real linear space, f is actually taken to be the **extended convex (concave) function** $f: X \to [-\infty, +\infty]$ although, for convenience, we shall refer to it simply as convex (concave).

We noted earlier that the function $f: X \to [-\infty, +\infty]$ is convex if its epigraph $\text{epi}(f) = \{(\boldsymbol{x}, \alpha) | f(\boldsymbol{x}) \le \alpha, \boldsymbol{x} \in X, \alpha \in R\}$ is a convex subset of $X \times R \subset R^{n+1}$.

Definition 8.7: The effective **domain** of a convex function $f: X \to [-\infty, +\infty]$ is the set

$$\begin{aligned}\text{dom}(f) &= \{\boldsymbol{x} | f(\boldsymbol{x}) < +\infty, \boldsymbol{x} \in R^n\}\\ &= \{\boldsymbol{x} | \text{ there exists an } \alpha < +\infty \text{ such that } (x, \alpha) \in \text{epi}(f)\}.\end{aligned}$$

Thus dom (f) is the projection of epi (f) onto R^n (see Figure 8.7); it is a convex set if f itself is a convex function (although the converse of this assertion does not hold.)

Definition 8.8: A convex function $f: X \to [-\infty, +\infty]$ is said to be **proper** if $\text{epi}(f) \neq \emptyset$ and contains no vertical lines, i.e., $f(\boldsymbol{x}) < +\infty$ for at least one $\boldsymbol{x}$ and $f(\boldsymbol{x}) > -\infty$ for every $\boldsymbol{x}$.

Next, suppose $f: X \to [-\infty, +\infty]$ is a convex function defined on R^n. For points $\boldsymbol{x} \in R^n$, consider the collection of all **linear affine functions** h of the form $h(\boldsymbol{x}) = \boldsymbol{a}^T\boldsymbol{x} - b$, $b \in R$, such that $h(\boldsymbol{x}) \le f(\boldsymbol{x})$. Then

Definition 8.9: The **lower support set** for f is the set $l(f) = \{(\boldsymbol{a}, b) | h(\boldsymbol{x}) \le f(\boldsymbol{x}), \boldsymbol{a} \in R^n, b \in R, \boldsymbol{x} \in R^n\}$. In this regard we have

Definition 8.10: The **closure** of a convex function $f: X \to [-\infty, +\infty]$, $cl(f)$, is the pointwise supremum of all affine functions h with values $h(\boldsymbol{x}) \le f(\boldsymbol{x})$ or

$$cl(f) = \sup_{(\boldsymbol{a},b) \in l(f)} \{\boldsymbol{a}^T\boldsymbol{x} - b\}.$$

Obviously $cl(f) \le f(\boldsymbol{x})$ for every $\boldsymbol{x} \in R^n$.

Definition 8.11: A convex function $f: X \to [-\infty, +\infty]$ is **closed** if $cl(f) = f(\boldsymbol{x})$. A proper convex function f is closed if and only if the (convex) lower level set $L_f(\alpha) = \{\boldsymbol{x} | f(\boldsymbol{x}) \le \alpha, \boldsymbol{x} \in R^n, \alpha \in R\}$ is closed for *every* $\alpha \in R$.

It is instructive to view the concept of a closed convex function in a slightly different light. We know that a function $f: X \to [-\infty, +\infty]$ is lower semicontinuous if the lower level set $L_f(\alpha)$ is closed for *all* $\alpha \in R$ or, equivalently, if epi (f) is closed as a subset of $X \times R$. In this regard we have

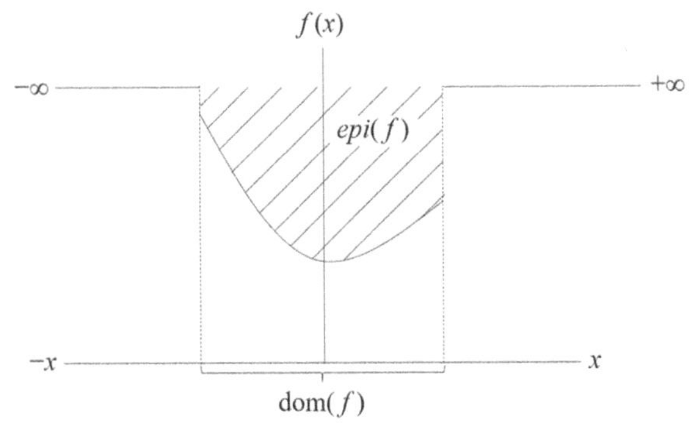

FIGURE 8.7 The effective domain of f, dom (f).

Definition 8.12: The **lower semicontinuous hull** of *f*, lsc (*f*), is the greatest lower semicontinuous function less than α equal to *f*. It is convex.

Then

Definition 8.13: The **closure** of f is defined as the convex set

$$cl(f) = \begin{cases} lsc(f) \text{ for all } \boldsymbol{x} \text{ if } lsc(f) > -\infty \text{ for all } \boldsymbol{x}; \\ -\infty \text{ for all } \boldsymbol{x} \text{ if } lsc(f) = -\infty \text{ for some } \boldsymbol{x}. \end{cases}$$

Hence, we can state that a convex function *f* is closed if $f(\boldsymbol{x}) = lsc(f)$, i.e., if f is a lower semicontinuous function which nowhere takes on the value $-\infty$. If *lsc* (*f*) has a finite value at some point $\boldsymbol{x}$, then *lsc* (*f*) and *f* are proper and $lsc(f) = cl(f)$.

The epigraph of the closure of *f* is the closure of the epigraph of f or $\text{epi}(cl(f)) = \overline{\text{epi}(f)}$. In fact, this last equality serves as the definition of the closure of a proper convex function, i.e., for a proper convex function, we need only close epi (*f*) to obtain its closure. So for proper convex functions, the notion of closedness and lower semicontinuity are identical and thus interchangeable.

Having just dealt with convex functions, let us now turn to a parallel discussion involving concave functions. This presentation will be offered in a condensed form.

For a concave function $g: X \to [-\infty, +\infty]$, let

$$\text{hyp}(g) = \{(\boldsymbol{x}, \alpha) | \alpha \le g(\boldsymbol{x}), \boldsymbol{x} \in R^n, \alpha \in R\};$$

$$\text{dom}(g) = \{\boldsymbol{x} | g(\boldsymbol{x}) > -\infty, \boldsymbol{x} \in R^n\}.$$

Also, g is proper if $g(\boldsymbol{x}) > -\infty$ for at least one $\boldsymbol{x}$ and $g(\boldsymbol{x}) < +\infty$ for every $\boldsymbol{x}$, i.e., if $-g$ is proper. (Remember that g is concave if $-g$ is convex.) The closure of g, $cl(g)$, is the pointwise infimum of all the linear affine functions such that $h(\boldsymbol{x}) \ge g(\boldsymbol{x})$ or

$$cl(g) = \inf_{(\boldsymbol{a}, b) \in u(f)} \{\boldsymbol{a}^T \boldsymbol{x} - b\},$$

where $u(f) = \{(\boldsymbol{a}, b) | h(\boldsymbol{x}) \ge f(\boldsymbol{x}), \boldsymbol{a} \in R^n, b \in R^n\}$ is the **upper support set** for *f*. Additionally, we can write $cl(g) = -(cl(-g))$. The concave function $g: X \to [-\infty, +\infty]$ is closed if $cl(g) = g$ or if $-g$ is closed.

If g is a proper concave function, then g is closed if and only if it is upper semicontinuous, i.e., if and only if the (convex) **upper level sets** $U_f(\alpha) = \{\boldsymbol{x} | g(\boldsymbol{x}) \ge \alpha, \boldsymbol{x} \in R^n, \alpha \in R\}$ are all closed for each α. The **upper semicontinuous hull** of g, usc (g), is the lowest upper semicontinuous function greater than or equal to g, and the closure of g is

$$cl(g) = \begin{cases} usc(g) \text{ for all } \boldsymbol{x} \text{ if } usc(g) < +\infty \text{ for all } \boldsymbol{x}; \\ +\infty \text{ for all } \boldsymbol{x} \text{ if } usc(g) = +\infty \text{ for all } \boldsymbol{x}. \end{cases}$$

Hence, the class of upper semicontinuous concave functions g which are proper is equivalent to the class of closed concave functions.

8.7.2. Conjugacy Defined

For $X = R^n$ a finite linear space, let $Y = R^n$ serve as its (finite linear) *dual space*. Then the dual pairing for these spaces, as indicated above, can be represented as $\langle \cdot, \cdot \rangle = \langle \boldsymbol{x}, \boldsymbol{y} \rangle = \boldsymbol{x}^T \boldsymbol{y}: X \times Y \to R$. For closed convex $f: X \to [-\infty, +\infty]$, let the convex conjugate of f be denoted as $f^*: Y \to [-\infty, +\infty]$. How shall f^* be structured? To answer this question, let us harken back to Definition 8.10 (*cl* (*f*) on R^n with values $h(\boldsymbol{x}) \le f(\boldsymbol{x})$). Let $h(\boldsymbol{x}) = \boldsymbol{y}^T \boldsymbol{x} - b$, $b \in R$. Given $\boldsymbol{y} \in Y$, for what values of b will $h(\boldsymbol{x}) \le f(\boldsymbol{x})$, $\boldsymbol{x} \in X$, be satisfied? Clearly, $h(\boldsymbol{x}) \le f(\boldsymbol{x})$ for every $\boldsymbol{x}$ if and only if

$$b \ge \boldsymbol{y}^T \boldsymbol{x} - f(\boldsymbol{x}) \text{ or } b \ge \sup\{\boldsymbol{y}^T \boldsymbol{x} - f(\boldsymbol{x})\}.$$

The greatest lower bound of b satisfying this inequality is a function of $\boldsymbol{y}$ and will be termed the **conjugate function of *f***, f^*, and expressed as

$$f^*(\boldsymbol{y}) = \sup_{\boldsymbol{x}} \{\boldsymbol{y}^T \boldsymbol{x} - f(\boldsymbol{x})\}, \ \boldsymbol{x} \in R^n. \tag{8.28}$$

Here the domain of f^* consists of those $\boldsymbol{y} \in Y$ for which the supremum in (8.28) is finite or for which the difference $\boldsymbol{y}^T \boldsymbol{x} - f(\boldsymbol{x})$ is bounded above on X.

What is the geometric interpretation of this equation? Figure 8.8 tells the story. Suppose $y \in R$ and $f: R \to [-\infty, +\infty]$. Given $f(x)$ and given a

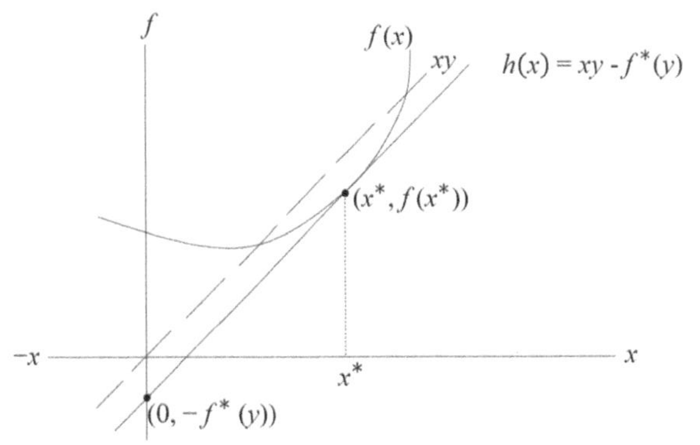

FIGURE 8.8 The conjugate function f^*.

slope coefficient y, we search along the x-axis to locate the x value, x^*, that maximizes the difference between the line yx and $f(x)$ (assuming, of course, that the supremum is finite). Once x^* is determined, we can define an affine function $h(x)$ with slope y passing through the point $(x^*, f(x^*))$ and having a vertical intercept equal to $-f^*(y)$. That is, since at $(x^*, f(x^*))$ we have $f(x) = h(x)$, it follows from Equation (8.28) that $h(x) = yx - f^*(y)$ is the affine function with slope y lying below f. If f is $C^{(1)}$, this occurs at the x value, x^*, for which $y = f'(x^*)$.

Looking to some of the properties of conjugate functions we have:

1. $\text{epi}(f^*) = \{(y, b) | y \in R^n, b \in R, y^T x - b \leq f(x), x \in R^n\}$,

i.e., $\text{epi}(f^*)$ equals the lower support set $l(f)$ and thus $\text{epi}(f^*)$ is a convex set with f^* a convex function.

2. The **conjugate function of f^*** is

$$f^{**}(x) = \sup_y \{y^T x - f^*(y)\}, \; y \in R^n. \quad (8.29)$$

If f is a closed convex function, then $f^{**} = f$, i.e., the conjugate of the conjugate of f is the original function f itself. In this circumstance the conjugate of f^* is a closed convex function and proper if and only if f is proper.

3. A moments reflection reveals that $f^{**} = cl(f)$. Additionally $l(f^*) = epi(f^{**}) = epi(cl(f))$ and $\boldsymbol{cl}(f^*) = (cl(f))^*$. It is also the case that $f^{**}(x) \leq f(x)$.

4. **Fenchel's inequality** $f(x) + f^*(y) \geq x^T y$ holds for any vectors $x, y \in R^n$.

5. A function $f: R^n \to R$ is **separable** if $f(x) = \sum_{i=1}^n f^i(x_i)$. Then the conjugate of f, f^*, is also separable.

6. For f convex on R^n and $a \in R$:

   i. if $j(x) = f(x) + a$, then $j^*(y) = f^*(y) - a$;

   ii. if $j(x) = f(x + z)$, then $j^*(y) = f^*(y) - y^T z$, $z \in R^n$;

   iii. if $j(x) = f(ax)$, $a \neq 0$, then $j^*(y) = f^*(y/a)$; and

   iv. if $j(x) = af(x)$, $a > 0$, then $j^*(y) = af^*(y/a)$.

7. For f convex on R^n, the subgradient $\xi \in \partial f(x)$ if and only if $f^*(\xi) = \xi^T x - f(x)$, where $\partial f(x)$ is the subdifferential of f at x. (To see this, suppose ξ satisfies (8.5) or $f(y) \geq f(x) + \xi^T(y - x)$, $y \in R^n$. Then $\xi \in \partial f(x)$ if and only if $\xi^T x - f(x) \geq \xi^T y - f(y)$ and thus $\xi^T x - f(x) = \sup_y \{\xi^T y - f(y)\} = f^*(\xi)$.)

Example 8.9: Given the following assortment of functions, determine f^* and f^{**}.

a. f is affine or $f(x) = ax + b$, $b \in R$. Then in (8.28), $yx - f = yx - ax - b$ is bounded for $y = a(=\text{constant})$. Hence, the domain of f^* is the singleton $\{a\}$ and $f^*(y) = f^*(a) = -b$. Additionally, from (8.29), $f^{**}(x) = \sup_y \{yx - f^*(y)\} = \sup_y \{yx - (-b)\} = ax + b$ for y a member of the domain of f^* (which is obviously bounded so that the supremum is finite).

b. f is the exponential function $f(x) = e^x$. Clearly, $xy - e^x$ reaches an attained supremum or maximum when $y - e^x = 0$ or $x = \ln y$, $y > 0$.

Hence, $f^*(y) = y \ln y - y$, where the domain of f^* consists of positive real y values. Moreover $f^{**}(x) = \sup_y\{y \ln y - (y \ln y - y)\} = \sup_y\{y\} = e^x$.

c. Let $f = x^{-1}$, $x > 0$. Then $yx - x^{-1}$ does not have a finite supremum for $y > 0$. Also, the supremum in (8.28) is zero for $y = 0$ and the attained supremum or maximum of this expression occurs for $y + x^{-2} = 0$ or $x = (-y)^{-1/2}$ for $y < 0$. Thus $f^*(y) = y(-y)^{-1/2} - (-y)^{1/2} = -2(-y)^{1/2}, y \leq 0$. Additionally, $f^{**}(x) = \sup_y\{y(-y)^{-1/2} - (-2(-y)^{1/2})\} = x^{-1}$.

d. f is the quadratic form $f(\boldsymbol{x}) = \frac{1}{2}\boldsymbol{x}^T\boldsymbol{Q}\boldsymbol{x}$. With $\boldsymbol{Q}$ positive definite, f is a (strictly) convex function of $\boldsymbol{x}$. The difference $\boldsymbol{y}^T\boldsymbol{x} - \frac{1}{2}\boldsymbol{x}^T\boldsymbol{Q}\boldsymbol{x}$ as a function of $\boldsymbol{x}$ is bounded above for all vectors $\boldsymbol{y}$. Hence, the attained supremum or maximum of this difference occurs when $\boldsymbol{y} - \boldsymbol{Q}\boldsymbol{x} = \boldsymbol{O}$ or $\boldsymbol{x} = \boldsymbol{Q}^{-1}\boldsymbol{y}$, with $f^*(y) = \boldsymbol{y}^T\boldsymbol{Q}^{-1}\boldsymbol{y} - \frac{1}{2}(\boldsymbol{Q}^{-1}\boldsymbol{y})^T\boldsymbol{Q}(\boldsymbol{Q}^{-1}\boldsymbol{y}) = \frac{1}{2}\boldsymbol{y}^T\boldsymbol{Q}^{-1}\boldsymbol{y}$ (since $\boldsymbol{Q}$ is taken to be a symmetric matrix). Moreover, $f^{**}(x) = \sup_y\left\{\boldsymbol{y}^T\boldsymbol{Q}^{-1}\boldsymbol{y} - \frac{1}{2}\boldsymbol{y}^T\boldsymbol{Q}^{-1}\boldsymbol{y}\right\} = \frac{1}{2}\boldsymbol{y}^T\boldsymbol{Q}^{-1}\boldsymbol{y} = \frac{1}{2}(\boldsymbol{Q}\boldsymbol{x})^T\boldsymbol{Q}^{-1}\boldsymbol{Q}\boldsymbol{x} = \frac{1}{2}\boldsymbol{x}^T\boldsymbol{Q}\boldsymbol{x}$. ■

For a concave function $g: X \to [-\infty, +\infty]$, its conjugate will be specified as

$$g^*(\boldsymbol{y}) = \inf_{\boldsymbol{x}}\{\boldsymbol{y}^T\boldsymbol{x} - g(\boldsymbol{x})\}, \boldsymbol{x} \in R^n. \quad (8.30)$$

Here g^* is a closed concave function with $g^{**} = cl(g)$, where

$$g^{**}(\boldsymbol{x}) = \inf_{\boldsymbol{y}}\{\boldsymbol{y}^T\boldsymbol{x} - g^*(\boldsymbol{y})\}, \boldsymbol{y} \in R^n, \quad (8.31)$$

and $cl(g)$ was defined in the preceding section.

What is the connection between g^* and f^*? This relationship can be depicted as follows. By definition (remember that $g = -f$, f convex),

$$\begin{aligned} f^*(\boldsymbol{y}) &= \sup_{\boldsymbol{x}}\{\boldsymbol{y}^T\boldsymbol{x} - f(\boldsymbol{x})\} \\ &= \sup_{\boldsymbol{x}}\{-[\boldsymbol{y}^T\boldsymbol{x} - g(\boldsymbol{x})]\} \\ &= -\inf_{\boldsymbol{x}}\{(-\boldsymbol{y}^T)\boldsymbol{x} - g(\boldsymbol{x})\} = -g^*(-\boldsymbol{y}). \end{aligned} \quad (8.32)$$

Here, too, we can establish a connection between supergradients and conjugate functions (using Equation (8.4)). Specifically, for g a concave function on R^n, $\xi \in \partial g(\boldsymbol{x})$ if $g^*(\xi) = \xi^T\boldsymbol{x} - g(\boldsymbol{x})$, where ξ is a supergradient and $\partial g(\boldsymbol{x})$ is the superdifferential of f at $\boldsymbol{x}$.

Section 8.7 Exercises:

8.7.1 Suppose $S \subset R^n$ is a convex set. The **indicator function** of S is defined as

$$f(\boldsymbol{x}) = \begin{cases} 0, & \boldsymbol{x} \in S \\ +\infty & \text{otherwise.} \end{cases}$$

What is the conjugate function of S?

8.7.2 Determine the conjugate function of $f(x) = x \log x$.

8.7.3 Given that $g(x) = \ln x$, find $g^*(y)$ and $g^{**}(x)$.

8.8 CONJUGATE FUNCTIONS IN ECONOMICS (LAU[1978]; DIEWERT [1973,1982]; BLUME [2008]; JORGENSEN AND LAU [1974]; BECKMAN AND KAPUR [1977])

Let a production function be of the form $q = F(l, k)$, where $l(k)$ is the flow of labor (capital) services in production. If we have a state of perfect competition in the product and factor markets, then the price of a unit of output (p) and the money price of labor ($\hat{w}$) are constant. Then the firm's short-run production function, with $k = \bar{k} =$ constant, can be expressed as $q = F(l, \bar{k})$. Given these considerations, the short-run profit function is

$$P = TR - TC = pF(l, \bar{k}) - \hat{w}l. \quad (8.33)$$

Upon dividing both sides of Equation (8.33) by p, we get the short-run **normalized profit function**

$$P^o = F(l, k) - wl, \quad (8.34)$$

where $P^o = P/p$ is **normalized profit** and $w = \hat{w}/p$ is the **normalized wage rate**. If l_o denotes the profit-maximizing l value, then the **normalized maximum value of profit** is

$$\pi^o = F(l_o, \bar{k}) - wl_o = G(w, \bar{k}). \quad (8.35)$$

Then it can be shown that the maximum of P is equivalent to the maximum of P^o (the same l_o obtains) and thus there exists a one-to-one correspondence between (8.33) and (8.35).

To facilitate the discussion which follows, we shall assume that the production function $q = F(l, \bar{k})$ has the following properties:

a. F is a finite, non-negative real-valued function defined on $\bar{R}_+^2$, and for each $\bar{k} \in \bar{R}_+$, $F(0, \bar{k}) = 0$.

b. F is continuous on $\bar{R}_+^2$.

c. For each $l \in \bar{R}_+$, F is of class $C^{(1)}$ on R_+.

d. F is non-decreasing in l and $\bar{k}$ on $\bar{R}_+^2$ and strictly increasing in l and $\bar{k}$ on $\bar{R}_+^2$. (8.36)

e. F is concave on $\bar{R}_+$ and strongly concave on R_+.

f. for each $\bar{k} \in \bar{R}_+$,F is of class $C^{(2)}$ on R_+.

g. A bounded and attainable unique solution exists for the normalized profit maximization problem for all $w \in R_+$ and $\bar{k} \in \bar{R}_+$.

In this regard, under assumptions (8.36 a, …, g), for each $w \in R_+$ and $\bar{k} \in \bar{R}_+$, there exists a unique $l_o = l_o(w, \bar{k}) \in R_+$ such that $F(l_o, \bar{k}) - wl_o \geq F(l, \bar{k}) - wl$ for all $l \in \bar{R}_+$. Additionally, $l_o(w, \bar{k})$ is continuous on R_+ for each $\bar{k} \in \bar{R}_+$ and continuous on $\bar{R}_+$ for each $w \in R_+$. And for each $\bar{k} \in \bar{R}_+$,l_o is of class $C^{(1)}$ on R_+.

Given the preceding discussion, we can now conclude that: the normalized profit function $G(w, \bar{k}) = F(l_o(w, \bar{k}), \bar{k}) - wl_o(w, \bar{k})$ is continuous on $R_+ \times \bar{R}_+$, is of class $C^{(2)}$ on R_+ for each $\bar{k} \in \bar{R}_+$, and is of class $C^{(1)}$ on R_+ for each $w \in R_+$. $q_o(w, \bar{k}) = F(l_o(w, \bar{k}), \bar{k})$ is continuous on $R_+ \times \bar{R}_+$ and is of class $C^{(1)}$ on R_+ for each $\bar{k} \in \bar{R}_+$.

We next turn to the notion of **duality** between the production function and the normalized profit function. Suppose $q(=F(l, \bar{k}))$ is given, where q is the maximum amount of output that can be produced by the firm's technology over a short-run time period using l and $\bar{k}$. The firm's maximum profit from producing at least q given p and w is defined by $G(w, \bar{k})$. Under certain regularity conditions, the normalized profit function G also completely describes the technology of the firm. So given G, it can be used to define F. Hence, we have a *duality* between normalized profit and production functions in the sense that either of these functions can describe the technology of the firm equally well in certain circumstances. To spell out these circumstances we must define a set of regularity conditions that G must possess; and given these conditions, we must show that F can be constructed from a given G. More specifically, we need to consider the properties of the class of normalized profit functions which correspond to the class of production functions which satisfy (8.36, a, …, g). Upon doing so we will have established a one-to-one correspondence between the members of these two classes. (But this is not the whole story. As will be explored later on, there also exists a duality between the *production function* and the *indirect production function* as well as between a consumer's *utility function* and the *expenditure function*.)

Let the production function $F(l, \bar{k})$ satisfy properties (8.36, a, …, g). Then we can define the normalized profit function $G(w, \bar{k})$ on $R_+ \times \bar{R}_+$ as the *conjugate* of the production function F, for each $\bar{k} \in \bar{R}_+$, as

$$G(w, k) = F_o(w, \bar{k}) = \inf_l \{wl - F(l, \bar{k})\}. \quad (8.37)$$

Then it can be shown that the normalized profit function $G(w, \bar{k})$ corresponding to a production function $F(l, \bar{k})$ with properties (8.35 a, …, g) possesses the properties:

a. G is a finite positive real-valued function defined on $R_+ \times \bar{R}_+$.

b. G is continuous on $R_+ \times \bar{R}_+$

c. For each $\bar{k} \in \bar{R}_+$, G is of class $C^{(1)}$ and for each $w \in R_+$, F is of class $C^{(1)}$ on R_+.

d. $G(w, \bar{k})$ is non-increasing in w and non-decreasing in $\bar{k}$ on $R_+ \times \bar{R}_+$ and strictly decreasing in q and strictly increasing in $\bar{k}$ on $R_+ \times \bar{R}_+$. (8.38)

e. For each $\bar{k} \in \bar{R}_+$, $G(w, \bar{k})$ is locally strongly convex on R_+.

f. For each $\bar{k} \in \bar{R}_+$, $G(w, \bar{k})$ is of class $C^{(2)}$ on R_+.

g. For each $\bar{k} \in \bar{R}_+$, the range of $G_w(w, \bar{k})$ is all of R_+.

Given properties (8.38 a, ..., g), $G(w, \bar{k})$ is a closed proper convex function on $R_+ \times \bar{R}_+$ with unique conjugate function

$$F(l, \bar{k}) = G_o(w, \bar{k}) = \sup_{w}\{wl - G(w, \bar{k})\}. \quad (8.39)$$

So if $G(w, \bar{k})$ has properties (8.38 a, ..., g), then $F(l, \bar{k})$ has properties (8.36).

Example 8.10: Given the (Cobb-Douglas) short-run production function with $\alpha > 0$ and $\beta > 0$, find the normalized profit function $G(w, \bar{k})$. From (8.37)

$$G(w, k) = \inf_{l}\{wl - Al^{\alpha}\bar{k}^{\beta}\}.$$

The expression within the braces attains a minimum with respect to l when $w - \alpha A l^{\alpha-1}\bar{k}^{\beta} = 0$ or for

$$l_o = \left(\frac{w}{\alpha A}\right)^{\frac{1}{\alpha-1}} \bar{k}^{\frac{-\beta}{\alpha-1}}.$$

Then

$$\begin{aligned} G(w, \bar{k}) &= wl_o - F(l_o, \bar{k}) \\ &= cw^{\frac{\alpha}{\alpha-1}}\bar{k}^{\frac{-\beta}{\alpha-1}}, \end{aligned}$$

where $c = A^{-\frac{1}{\alpha-1}}\ \alpha^{-\frac{1}{\alpha-1}}\left(\frac{\alpha-1}{\alpha}\right)$ = constant. ■

Example 8.11: Given the (long-run) production function $q = F(l, k) = l^{\alpha}k^{\beta}$, $\alpha + \beta < 1$, with $\alpha > 0$ and $\beta > 0$, determine the normalized maximum value of profit $G(w, r)$. Let us write long-run profit as

$$P = TR - TC = pF(l, k) - \hat{w}l - \hat{r}k.$$

Then the long-run **normalized profit function** is

$$P^o = F(l, k) - wl - rk,$$

where $P^o = P/p$, $w = \hat{w}/p$, and $r = \hat{r}/p$. Additionally, the **normalized maximum value of profit** is

$$\pi^o = F(l_o, k_o) - wl_o - rk_o = G(w, r).$$

Looking to the conjugate function we have

$$G(w, r) = \inf_{l.k}(wl + rk - l^{\alpha}k^{\beta}). \quad (*)$$

The necessary conditions for a minimum (the attained infimum) at (l_o, k_o) are

$$\begin{aligned} w - \alpha\ l_o^{\alpha-1}\ k_o^{\beta} &= 0, \\ r - \beta\ l_o^{\alpha}\ k_o^{\beta-1} &= 0. \end{aligned}$$

If we now solve this system for l_o and k_o in terms of w and r we obtain

$$\begin{aligned} l_o &= w^{(1-\beta)/(\alpha+\beta-1)}r^{\beta/(\alpha+\beta-1)}\alpha^{(\beta-1)/(\alpha+\beta-1)} \times \beta^{-\beta/(\alpha+\beta-1)} \\ k_o &= w^{\alpha/(\alpha+\beta-1)}r^{(1-\alpha)/(\alpha+\beta-1)}\alpha^{-\alpha/(\alpha+\beta-1)} \times \beta^{(\alpha-1)/(\alpha+\beta-1)}. \end{aligned}$$

Then substituting l_o and k_o into (*) gives $G(w, r) = (\alpha^{-\alpha/(\alpha+\beta-1)}\beta^{-\beta(\alpha+\beta-1)})\,w^{\alpha/(\alpha+\beta-1)}r^{\beta/(\alpha+\beta-1)}$. ■

Example 8.12: For the (long-run) CES production function $q = F(l, k) = [\delta l^{-\rho} + (1 - \delta)k^{-\rho}]^{-h/\rho}$, where $h(\neq 1)$ is the homogeneity parameter, δ and $1 - \delta$ are distribution parameters ($\delta > 0$), and $\rho = \frac{1}{\sigma} - 1$ is the substitution parameter (a transform of the elasticity of substitution σ). Determine the range of values for h and ρ which will guarantee that F is concave.

We know that F is concave for all $(l, k) \in \bar{R}_+^2$ if $\boldsymbol{H}_F(l, k)$ is negative semi-definite. And this will occur if $0 < h < 1$, $\rho \geq -1$.

Next, given that F is concave, find the normalized maximum value of profit $G(w, r)$. As in the preceding example problem, let

$$\pi^o = F(l_o, k_o) - wl_o - rk_o = G(w, r)$$

and

$$G(w, r) = \inf_{l.k}(wl + rk - [\delta l^{-\rho} + (1 - \delta)k^{-\rho}]^{-h/\rho}). \quad (*)$$

The first-order conditions for a minimum (the attained infimum) at (l_o, k_o) are

$$\begin{aligned} w &= h\ \delta\ l_o^{-\rho-1}[\delta\ l^{-\rho} + (1 - \delta)k_o^{-\rho}]^{-\frac{h}{\rho}-1}, \\ r &= h\,(1 - \delta)\ k_o^{-\rho-1}[\delta\ l_o^{-\rho} + (1 - \delta)k_o^{-\rho}]^{-\frac{h}{\rho}-1}. \quad (**) \end{aligned}$$

From system (**) we find that

$$l_o^{-\rho} = \left[\left(\frac{w}{h\delta}\right)F(l_o, k_o)^{-(h+\rho)/h}\right]^{\rho/(1+\rho)},$$

$$k_o^{-\rho} = \left[\left(\frac{r}{h(1-\delta)}\right)F(l_o, k_o)^{-(h+\rho)/\rho}\right]^{\rho/(1+\rho)}.$$

A substitution of these values into $q = F(l, k)$ yields

$$F(l_o, k_o) = \left[\left(\frac{h\delta}{w}\right)^{-\rho/(1+\rho)} + \left(\frac{h(1-\delta)}{r}\right)^{-\rho/(1+\rho)}\right]^{-\frac{h}{\rho}\left(\frac{1+\rho}{1-h}\right)}. \quad (\S)$$

Now, it can be shown that substituting system (**) into $G(w, r)$ gives $G(w, r) = (h - 1)F(l_o, k_o)$. Hence, inserting (§) into this latter expression gives us $G(w, r) =$

$$(h-1)h^{h/(1-h)}\left[\left(\frac{w}{\delta}\right)^{\rho/(1+\rho)} + \left(\frac{r}{(1-\delta)}\right)^{\rho/(1+\rho)}\right]^{\frac{(1+\rho)}{\rho}\frac{h}{(h-1)}}. \blacksquare$$

Section 8.8 Exercises:

8.8.1 Given $G(w, \bar{k})$ from Example 8.10, find its conjugate function using Equation (8.39).

8.8.2 Let l_1^* and l_2^* be profit-maximizing quantities of labor for w_1 and w_2, respectively. Verify the *monotonicity* property (8.38d).

8.8.3 An alternative representation of Equation (8.39) is

$$F(l, \bar{k}) = \inf\{wl + G(w, \bar{k})\}.$$

Verify the legitimacy of this expression.

APPENDIX 8.A ALTERNATIVE PROOFS OF THEOREMS 8.9 AND 8.10

This appendix offers an alternative methodology for proving theorems such as 8.9 and 8.10. In particular, it enables us to transform a function of variables in R^n to a function displaying a single argument in R, namely θ. We start with

Definition 8.A.1: Let $I = [0, 1]$ denote an interval in R. Given the function $y = f(\boldsymbol{x}): \mathcal{D} \subseteq R^n \rightarrow R$, $\mathcal{D}$ a convex set, and a vector $\boldsymbol{u} \in R^n$, let

$$\varphi(\theta) = f(\boldsymbol{x} + \theta\boldsymbol{u}), \ \theta \in I, \quad \text{(A.8.1)}$$

be the **restriction of** f on a line segment through $\boldsymbol{x} \in \mathcal{D}$.

What we shall now demonstrate is the relationship between the concavity (convexity) of a function f and the concavity (convexity) of its restriction φ along a line $I \in R$. More specifically, we have

Theorem 8.A.1: (Necessary and Sufficient Condition). A function $y = f(\boldsymbol{x}): \mathcal{D} \subseteq R^n \rightarrow R$, $\mathcal{D}$ a convex set, is concave (convex) if and only if its restriction $\varphi(\theta)$, $\theta \in I = [0, 1]$, to every line segment in $\mathcal{D}$ is a concave (convex) function of θ.

Proof (Sufficiency). Assume that $\varphi(\theta)$, $\theta \in [0, 1]$, is concave in θ and let $\boldsymbol{x}_1$, $\boldsymbol{x}_2$ depict arbitrary points in $\mathcal{D}$. Then the restriction of f to the line segment connecting these points is $\varphi(\theta) = f(\theta\boldsymbol{x}_2 + (1 - \theta)\boldsymbol{x}_1)$. With $\varphi(\theta)$ concave,

$$\begin{aligned} f(\theta\boldsymbol{x}_2 + (1-\theta)\boldsymbol{x}_1) &= \varphi(\theta) \\ &= \varphi(\theta\cdot 1 + (1-\theta)\cdot 0) \\ &\geq \theta\varphi(1) + (1-\theta)\varphi(0) \\ &= \theta f(\boldsymbol{x}_2) + (1-\theta)f(\boldsymbol{x}_1) \end{aligned}$$

and thus f is concave.

(Necessity). Suppose that f is concave. For points $z_1, z_2 \in \mathcal{D}$, with $\theta \in [0, 1]$,

$$\begin{aligned}\varphi(\theta z_2 + (1-\theta)z_1) &= f[(\theta z_2 + (1-\theta)z_1))\boldsymbol{x}_2 \\ &\quad + (1 - (\theta z_2 + (1-\theta)z_1))\boldsymbol{x}_1] \\ &= f[\theta(z_1\boldsymbol{x}_2 + (1-z_1)\boldsymbol{x}_1) \\ &\quad + (1-\theta)(z_2\boldsymbol{x}_2 + (1-z_2)\boldsymbol{x}_1)] \\ &\geq \theta f(z_1\boldsymbol{x}_2 + (1-z_1)\boldsymbol{x}_1) \\ &\quad + (1-\theta)f(z_2\boldsymbol{x}_2 + (1-z_2)\boldsymbol{x}_1) \\ &= \theta\varphi(z_1) + (1-\theta)\varphi(z_2)\end{aligned}$$

so that φ is concave. Q. E. D.

As an input into the proofs that follow, let us consider a function $y = f(x)$ which is $C^{(1)}$ on an interval $I \in R$. Then f is concave on I if and only if

$$f(x_2) - f(x_1) \leq f'(x_1)(x_2 - x_1) \text{ for any } x_1, x_2 \in I; \tag{8.A.2}$$

while f is convex on I if and only if

$$f(x_2) - f(x_1) \geq f'(x_1)(x_2 - x_1) \text{ for any } x_1, x_2 \in I. \tag{8.A.3}$$

Let us utilize this result to obtain an **alternative proof of** Theorem 8.9. To this end, our discussion will involve the restriction of the concave $C^{(1)}$ function f on a line segment within $\mathcal{D}$. So for arbitrary $\boldsymbol{x}_1, \boldsymbol{x}_2 \in \mathcal{D}$,

$$\varphi(\theta) = f(\theta\boldsymbol{x}_2 + (1-\theta)\boldsymbol{x}_1) = f(\boldsymbol{x}_1 + \theta(\boldsymbol{x}_2 - \boldsymbol{x}_1)), \ \theta \in [0, 1].$$

Then

$$\varphi'(\theta) = \nabla f(\boldsymbol{x}_1 + \theta(\boldsymbol{x}_2 - \boldsymbol{x}_1))^T(\boldsymbol{x}_2 - \boldsymbol{x}_1) \tag{8.A.4}$$

and

$$\varphi'(0) = \nabla f(\boldsymbol{x}_1)^T(\boldsymbol{x}_2 - \boldsymbol{x}_1). \tag{8.A.5}$$

For $\theta = 0$, $\varphi(0) = f(x_1)$; and for $\theta = 1$, $\varphi(1) = f(x_2)$. Then from Theorem 8.A.1 (f is concave on $\mathcal{D}$ if and only if $\varphi(\theta)$ is concave), inequality (8.A.2), and (8.A.5),

$$\varphi(1) - \varphi(0) \leq \varphi'(0)(1-0) = \varphi'(0)$$

or

$$f(\boldsymbol{x}_2) - f(\boldsymbol{x}_1) \leq \nabla f(\boldsymbol{x}_1)^T(\boldsymbol{x}_2 - \boldsymbol{x}_1). \quad \textbf{Q. E. D.}$$

Finally, we may use the preceding approach to develop an **alternative proof of** Theorem 8.10. Again, our discussion will be limited to the restriction of the concave $C^{(2)}$ function f on a line segment for $\theta \in I = [0, 1]$. (Sufficiency). Let $\boldsymbol{x}_1, \boldsymbol{x}_2$ be arbitrary points in $\mathcal{D}$ with $\varphi(\theta) = f(\boldsymbol{x}_1 + \theta(\boldsymbol{x}_2 - \boldsymbol{x}_1))$. Since Theorem 8.A.1 is equivalent to $\varphi''(\theta) \leq 0$ (since a $C^{(2)}$ function $\varphi(\theta)$ is concave if $\varphi''(\theta) \leq 0$ for all $\theta \in I$), let us determine

$$\begin{aligned}\varphi''(\theta) &= \frac{d}{d\theta}[(\boldsymbol{x}_2 - \boldsymbol{x}_1)^T \nabla f(\boldsymbol{x}_1 + \theta(\boldsymbol{x}_2 - \boldsymbol{x}_1))] \\ &= (\boldsymbol{x}_2 - \boldsymbol{x}_1)^T \nabla^2 f(\boldsymbol{x}_1 + \theta(\boldsymbol{x}_2 - \boldsymbol{x}_1))(\boldsymbol{x}_2 - \boldsymbol{x}_1) \\ &= (\boldsymbol{x}_2 - \boldsymbol{x}_1)^T \boldsymbol{H}_f(\boldsymbol{x}_1 + \theta(\boldsymbol{x}_2 - \boldsymbol{x}_1))(\boldsymbol{x}_2 - \boldsymbol{x}_1) \leq 0 \text{ for all } \theta \in I.\end{aligned}$$

Now, if the Hessian matrix of f, $\boldsymbol{H}_f$, is negative semidefinite, then $\varphi''(\theta) \leq 0$ so that every $\varphi(\theta)$ is concave, and thus f itself is concave.

(Necessity). Let f be concave on $\mathcal{D}$ with $\boldsymbol{y}$ an arbitrary point in $\mathcal{D}$ and $\boldsymbol{u} \in R^n$ a direction. With $\mathcal{D}$ open, there is a $\bar{\theta} > 0$ such that $\boldsymbol{x}_2 = \boldsymbol{y} + \bar{\theta}\boldsymbol{u}$. Since f is concave, φ is concave (via Theorem 8.A.1) and

$$\begin{aligned}0 \geq \varphi''(0) &= (\boldsymbol{x}_2 - \boldsymbol{y})^T \boldsymbol{H}_f(\boldsymbol{y})(\boldsymbol{x}_2 - \boldsymbol{y}) \\ &= (\bar{\theta}\boldsymbol{u})^T \boldsymbol{H}_f(\boldsymbol{y})(\bar{\theta}\mathrm{u}) \\ &= \bar{\theta}^2(\boldsymbol{u}^T \boldsymbol{H}_f(\boldsymbol{y})\boldsymbol{u})\end{aligned}$$

and thus $\boldsymbol{u}^T \boldsymbol{H}_f(\boldsymbol{y})\boldsymbol{u} \leq \boldsymbol{0}$ so this quadratic form is negative semidefinite for all $\boldsymbol{y} \in \mathcal{D}$. Q. E. D.

NOTE

1 Suppose $y = f(\boldsymbol{x})$: $\mathcal{D} \subset R^n \rightarrow R$, $\mathcal{D}$ a convex set, is concave (convex) on $\mathcal{D}$. The **affine linear function** $g(\boldsymbol{x}) = \boldsymbol{C}^T\boldsymbol{x} + \alpha$, $\boldsymbol{C}(\neq \boldsymbol{O}) \in R^n$, $\alpha \in R$, is termed a **support function of** f $at \boldsymbol{x}_o \in \mathcal{D}$ if $g(\mathrm{x})$ is a supporting hyperplane of hyp (f) (epi (f)) at $(\boldsymbol{x}_o, f(xo))$, i.e., if $g(\boldsymbol{x}_o) = f(\boldsymbol{x}_o)$ and $g(\boldsymbol{x}) \geq f(\boldsymbol{x})(\leq f(\boldsymbol{x}))$, $\boldsymbol{x} \in \mathcal{D}$. Since $g(\boldsymbol{x}_o) = f(\boldsymbol{x}_o)$, it follows that $\alpha = f(\boldsymbol{x}_o) - \boldsymbol{C}^T\boldsymbol{x}_o$. Thus the family of support functions of a concave (convex) function f at $\boldsymbol{x}_o$ appears as $g(\boldsymbol{x}) = \boldsymbol{C}^T\boldsymbol{x} + f(\boldsymbol{x}_o) - \boldsymbol{C}^T\boldsymbol{x}_o = f(\boldsymbol{x}_o) + \boldsymbol{C}^T(\boldsymbol{x} - \boldsymbol{x}_o)$, where $\boldsymbol{C}$ satisfies $g(\boldsymbol{x}_o) \geq f(\boldsymbol{x})$ $(\leq f(\boldsymbol{x}))$ or $f(\boldsymbol{x}_o) + \boldsymbol{C}^T(\boldsymbol{x} - \boldsymbol{x}_o) \geq f(\boldsymbol{x})(\leq f(\boldsymbol{x}))$, $\boldsymbol{x} \in \mathcal{D}$. If f is differentiable at $\boldsymbol{x}_o$, then, with $\mathcal{D}$ open, it follows that $\boldsymbol{C} = \nabla g(\boldsymbol{x}) = \nabla f(\boldsymbol{x})$ at $\boldsymbol{x}_o$.

CHAPTER 9

Generalizations of Convexity and Concavity

9.1 INTRODUCTION

In this chapter we shall introduce more general types of functions than those which are characterized as concave or convex. In fact, to establish certain optimality conditions, a less restrictive variety of functions is called for - - functions which display only some of the properties of convex and concave functions and thus constitute a broader class of functions. For instance, we previously defined the upper level set for a function $f(\mathbf{x})$: $\mathcal{D} \subset R^n \rightarrow R$ as

$U_f(\alpha) = \{\mathbf{x} \in \mathcal{D} | f(\mathbf{x}) \geq \alpha, \alpha \in R\}$. If f is a concave function and $\mathcal{D}$ is a convex set, then $U_f(\alpha)$ is a convex set for all $\alpha \in R$. Conversely, suppose it is known that, for some function f, $U_f(\alpha)$ is a convex set for all $\alpha \in R$. Can we then conclude that f is a concave function? The answer is *no*; f need not be concave in order for $U_f(\alpha)$ to be a convex set. Hence, our objective is to characterize functions whose level sets are also convex by generalizing (weakening) the notion of concavity (convexity). However, under this weakening process, certain desirable properties of concave functions are lost; the more general functions may have discontinuities at interior points of $\mathcal{D}$ and a critical point which is a local maximum is not necessarily a global maximum (see (i) of Theorem 8.13).

9.2 QUASICONCAVITY AND QUASICONVEXITY

We start with

Definition 9.1: A function $y = f(\mathbf{x})$: $\mathcal{D} \subset R^n \rightarrow R$, with $\mathcal{D}$ a convex set, is **quasiconcave** on $\mathcal{D}$ if and only if for all $\mathbf{x}_1, \mathbf{x}_2 \in \mathcal{D}$,

$$f(\mathbf{x}_1) \leq f(\mathbf{x}_2) \text{ implies } f(\mathbf{x}_1) \leq f(\mathbf{x}_c), \qquad (9.1)$$

where $\mathbf{x}_c = \lambda\mathbf{x}_1 + (1-\lambda)\mathbf{x}_2$, $\lambda \in [0, 1]$, is the convex combination of $\mathbf{x}_1$ and $\mathbf{x}_2$. Equivalently, under the preceding set of assumptions,

$$f(\mathbf{x}_c) \geq \min \{f(\mathbf{x}_1), f(\mathbf{x}_2)\}. \qquad (9.2)$$

Proof: (Necessity). Suppose f is quasiconcave. Let $\mathbf{x}_1, \mathbf{x}_2 \in \mathcal{D}$, $\lambda \in [0, 1]$, with $\alpha = \min \{f(\mathbf{x}_1), f(\mathbf{x}_2)\}$. Then $\mathbf{x}_1, \mathbf{x}_2 \in U_f(\alpha) = \{\mathbf{x} \in \mathcal{D} | f(\mathbf{x}) \geq \alpha, \alpha \in R\}$. Since $U_f(\alpha)$ is convex, $\mathbf{x}_c \in U_f(\alpha)$ so that $f(\mathbf{x}_c) \geq \alpha$ or (9.2) holds.

(Sufficiency). Conversely, suppose (9.2) holds. For an arbitrary α, let $\mathbf{x}_1, \mathbf{x}_2 \in U_f(\alpha)$ so that $f(\mathbf{x}_1) \geq \alpha$, $f(\mathbf{x}_2) \geq \alpha$. And for $\lambda \in [0, 1]$, (9.2) implies that $f(\mathbf{x}_c) \geq$

$\min\{f(\boldsymbol{x}_1), f(\boldsymbol{x}_2)\} \geq \alpha$ so that $\boldsymbol{x}_c \in U_f(\alpha)$. Hence, $U_f(\alpha)$ is a convex set.

Next, suppose $\boldsymbol{x}_1, \boldsymbol{x}_2 \in \mathcal{D}$ with $\lambda \in [0, 1]$. Then (9.2) holds if and only if (9.1) is true. Q.E.D.

As (9.2) requires, f is quasiconcave if the values of the function along a line segment joining any two points in $\mathcal{D}$ are greater than or equal to the minimum of the function values at the end points of the line segment. That is, the restriction of f along a line joining points $\boldsymbol{x}_1, \boldsymbol{x}_2 \in \mathcal{D}$, $\varphi(\lambda)$, can actually lie above at least one of the end points.

Similarly, we have

Definition 9.2: A function $y = f(\boldsymbol{x})$: $\mathcal{D} \subset R^n \rightarrow R$, with $\mathcal{D}$ a convex set, is **quasiconvex** on $\mathcal{D}$ if and only if for all $\boldsymbol{x}_1, \boldsymbol{x}_2 \in \mathcal{D}$,

$$f(\boldsymbol{x}_1) \leq f(\boldsymbol{x}_2) \text{ implies } f(\boldsymbol{x}_c) \leq f(\boldsymbol{x}_2), \quad (9.3)$$

where $\boldsymbol{x}_c = \lambda \boldsymbol{x}_1 + (1-\lambda)\boldsymbol{x}_2$, $\lambda \in [0, 1]$. Equivalently, under the given assumptions,

$$f(\boldsymbol{x}_c) \leq \max\{f(\boldsymbol{x}_1), f(\boldsymbol{x}_2)\}. \quad (9.4)$$

For instance, for $x \in R$, examples of quasiconcave and quasiconvex functions are provided in Figure 9.1a, b, respectively.

Example 9.1: Let $y = f(x) = 4x - \frac{1}{2}x^2$, $x \in [0, 4]$. Use Equation (9.1) to verify that f is quasiconcave over the given domain. Suppose we select $x_1 = 1$, $x_2 = 2$. Then $f(1) = 7/2 < f(2) = 6$. Set $\lambda = 1/2$ so that $x_c = 3/2$ and thus $f(x_c) = 39/8 > f(1)$. Thus $f(x_1) \leq f(x_2)$ implies $f(x_1) \leq f(x_c)$ or (9.1) so that f is quasiconcave. ■

Note that:

1. the function f is quasiconcave (quasiconvex) on $\mathcal{D}$ if $-f$ is quasiconvex (quasiconcave).
2. every convex (concave) function on $\mathcal{D}$ is quasiconvex (quasiconcave), but not conversely. Hence, quasiconcavity (quasiconvexity) is a generalization of concavity (convexity). For instance, from (8.3),

$$f(\boldsymbol{x}_c) \geq \theta f(\boldsymbol{x}_2) + (1-\theta) f(\boldsymbol{x}_1)$$

or

$$f(\boldsymbol{x}_c) \geq f(\boldsymbol{x}_1) + \theta[f(\boldsymbol{x}_2) - f(\boldsymbol{x}_1)], \ \theta \in [0, 1].$$

If $f(\boldsymbol{x}_2) \geq f(\boldsymbol{x}_1)$, then $f(\boldsymbol{x}_c) \geq f(\boldsymbol{x}_1)$ or (9.1). Hence, concavity implies quasiconcavity.

3. in contrast to concave (convex) functions on $\mathcal{D}$, quasiconcave (quasiconvex) functions can admit discontinuities on their domain (Figure 9.1c). However, if f is continuous on the closure of a convex set $\mathcal{D} \subset R^n$, $\bar{\mathcal{D}}$, and quasiconcave (quasiconvex) on int $(\mathcal{D})$ then f is quasiconcave (quasiconvex) on $\bar{\mathcal{D}}$.
4. While a concave function can be characterized by the convexity of its hypograph, a quasiconcave function can be characterized by the convexity of its level sets. Similarly, a convex function can be characterized by the convexity of its epigraph while a quasiconvex function can also be characterized by the convexity of its level sets.

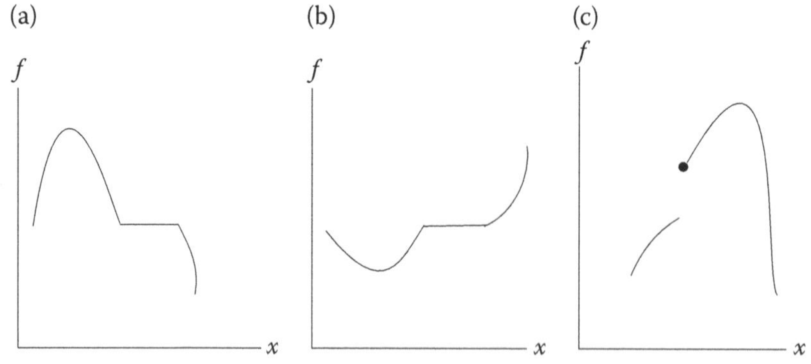

FIGURE 9.1 a. f is quasiconcave; b. f is quasiconvex; c. f is quasiconcave.

5. any increasing transformation of a concave function results in a quasiconcave function.

6. if $\varnothing$: $R \to R$ is an increasing function and f is quasiconcave, then the composition of $\varnothing$ and f, $\varnothing \circ f$, is a quasiconcave function.

7. It was mentioned in the preceding chapter that the sum of concave (convex) functions is also concave (convex). However, this property is not generalizable to quasiconcave (quasiconvex) functions, e.g., the sum of quasiconcave functions is not necessarily quasiconcave.

8. While the concavity (convexity) of a function over a convex domain $\mathcal{D} \subset R^n$ can be extended to concavity (convexity) over the entirety of R^n, such is not the case for quasiconcave (quasiconvex) functions.

Relative to item 4 we have

Theorem 9.1: Let $y = f(\boldsymbol{x})$: $\mathcal{D} \subset R^n \to R$ with $\mathcal{D}$ a convex set. Let the **upper level set** of f be denoted as $U_f(\alpha) = \{\boldsymbol{x} \in \mathcal{D} | f(\boldsymbol{x}) \geq \alpha, \alpha \in R\}$ with its **lower level set** expressed as $L_f(\alpha) = \{\boldsymbol{x} \in \mathcal{D} | f(\boldsymbol{x}) \leq \alpha, \alpha \in R\}$. Then f is quasiconcave (quasiconvex) on $\mathcal{D}$ if and only if $U_f(\alpha)$ $(L_f(\alpha))$ is a convex set for all α.

Proof: (Necessity). Suppose f is quasiconcave with points $\boldsymbol{x}_1, \boldsymbol{x}_2 \in U_f(\alpha)$. Since $f(\boldsymbol{x}_1) \geq \alpha$ and $f(\boldsymbol{x}_2) \geq \alpha$, it follows that $f(\boldsymbol{x}_c) \geq \min\{f(\boldsymbol{x}_1), f(\boldsymbol{x}_2)\} \geq \alpha$ so that $\boldsymbol{x}_c \in U_f(\alpha)$ and thus $U_f(\alpha)$ is convex, where $\boldsymbol{x}_c = \lambda\boldsymbol{x}_1 + (1-\lambda)\boldsymbol{x}_2$, $\lambda \in [0, 1]$.

(Sufficiency). Suppose $f(\boldsymbol{x}_1) = \min\{f(\boldsymbol{x}_1), f(\boldsymbol{x}_2)\}$. Then $U_f(f(\boldsymbol{x}_1)) = \{\boldsymbol{x} \in \mathcal{D} | f(\boldsymbol{x}) \geq f(\boldsymbol{x}_1)\}$. With $\boldsymbol{x}_2 \in U_f(f(\boldsymbol{x}_1))$ and $U_f(f(\boldsymbol{x}_1))$ convex, it follows that $\boldsymbol{x}_c \in U_f(f(\boldsymbol{x}_1))$ for every $\lambda \in [0, 1]$. Hence, $f(\boldsymbol{x}_c) \geq f(\boldsymbol{x}_1) = \min\{f(\boldsymbol{x}_1), f(\boldsymbol{x}_2)\}$ so that f is quasiconcave. Q.E.D.

Section 9.2 Exercises:

9.2.1 Let $y = f(x) = x^2 + 3$, $2 \leq x \leq 10$. Find the lower level set $L_f(\alpha)$ for $\alpha = 15$.

9.2.2 Let $y = f(x) = 2x + 5$, $3 \leq x \leq 6$. Find the upper level set $U_f(\alpha)$ for $\alpha = 15$. What is $L_f(\alpha)$?

9.2.3 Use Equation (9.3) to verify that $f(x) = x^2$, $x \in R$, with $x > 0$, is quasiconvex.

9.2.4 Verify that if f is quasiconcave, then $U_f(\alpha)$ is convex.

9.2.5 Use Equation (9.1) to verify that $f(\boldsymbol{x}) = x_1x_2$, $\boldsymbol{x} \in R^2$, with $\boldsymbol{x} \geq \boldsymbol{O}$, is quasiconcave.

9.3 DIFFERENTIABLE QUASICONCAVE AND QUASICONVEX FUNCTIONS

We start with a necessary and sufficient condition which characterizes the quasiconcavity (quasiconvexity) of a $C^{(1)}$ function in terms of its gradient vector. Specifically,

Theorem 9.2: Let $y = f(\boldsymbol{x})$: $\mathcal{D} \subset R^n \to R$ be $C^{(1)}$ on the open, convex set $\mathcal{D}$. Then:

1. f is quasiconcave on $\mathcal{D}$ if and only if

$$f(\boldsymbol{x}_1) \leq f(\boldsymbol{x}_2) \text{ implies } \nabla f(\boldsymbol{x}_1)^T(\boldsymbol{x}_2 - \boldsymbol{x}_1) \geq 0 \text{ for } \boldsymbol{x}_1, \boldsymbol{x}_2 \in \mathcal{D}; \tag{9.5}$$

2. f is quasiconvex on $\mathcal{D}$ if and only if

$$f(\boldsymbol{x}_1) \geq f(\boldsymbol{x}_2) \text{ implies } \nabla f(\boldsymbol{x}_1)^T(\boldsymbol{x}_2 - \boldsymbol{x}_1) \leq 0 \text{ for } \boldsymbol{x}_1, \boldsymbol{x}_2 \in \mathcal{D}; \tag{9.6}$$

Proof: (Necessity) Suppose f is quasiconcave on $\mathcal{D}$ with $f(\boldsymbol{x}_1) \leq f(\boldsymbol{x}_2)$. Then, for $\lambda \in [0, 1]$, $f(\boldsymbol{x}_c) \geq f(\boldsymbol{x}_1)$ or $f(\boldsymbol{x}_1 + \lambda(\boldsymbol{x}_2 - \boldsymbol{x}_1)) \geq f(\boldsymbol{x}_1)$ so that, for $\lambda \in (0, 1)$,

$$\frac{f(\boldsymbol{x}_1 + \lambda(\boldsymbol{x}_2 - \boldsymbol{x}_1)) - f(\boldsymbol{x}_1)}{\lambda} \geq 0.$$

Then

$$\lim_{\lambda \to 0} \frac{f(\boldsymbol{x}_1 + \lambda(\boldsymbol{x}_2 - \boldsymbol{x}_1)) - f(\boldsymbol{x}_1)}{\lambda} = \nabla f(\boldsymbol{x}_1)^T(\boldsymbol{x}_2 - \boldsymbol{x}_1) \geq 0.$$

Here $\nabla f(\boldsymbol{x}_1)^T(\boldsymbol{x}_2 - \boldsymbol{x}_1)$ represents the directional derivative of f at $\boldsymbol{x}_1$ in the direction $\boldsymbol{x}_2 - \boldsymbol{x}_1 (\neq 0)$. Hence, we see that if f is quasiconcave, then the directional derivative of f at $\boldsymbol{x}_1$ in the $\boldsymbol{x}_2 - \boldsymbol{x}_1$ direction must be non-decreasing.

(Sufficiency). Suppose $f(\boldsymbol{x}_1) \le f(\boldsymbol{x}_2)$ and, for $\boldsymbol{x}_c = \lambda\boldsymbol{x}_2 + (1-\lambda)\boldsymbol{x}_1$, let $f(\boldsymbol{x}_c) < f(\boldsymbol{x}_1)$, $\lambda \in (0, 1)$. By hypothesis,

$$\nabla f(\boldsymbol{x}_c)^T(\boldsymbol{x}_2 - \boldsymbol{x}_c) \ge 0, \quad (*)$$
$$\nabla f(\boldsymbol{x}_c)^T(\boldsymbol{x}_1 - \boldsymbol{x}_c) \ge 0. \quad (**)$$

Substituting $\boldsymbol{x}_c = \lambda\boldsymbol{x}_1 + (1-\lambda)\boldsymbol{x}_2$ into (*) and $\boldsymbol{x}_c = \lambda\boldsymbol{x}_2 + (1-\lambda)\boldsymbol{x}_1$ into (**) and simplifying yields

$$\nabla f(\boldsymbol{x}_c)^T(\boldsymbol{x}_2 - \boldsymbol{x}_1) \ge 0,$$
$$\nabla f(\boldsymbol{x}_c)^T(\boldsymbol{x}_1 - \boldsymbol{x}_2) \ge 0$$

and thus

$$\nabla f(\boldsymbol{x}_c)^T(\boldsymbol{x}_1 - \boldsymbol{x}_2) = 0.$$

Define $S = \{\boldsymbol{x} \in R^n | f(\boldsymbol{x}) \ge f(\boldsymbol{x}_1), \boldsymbol{x} = \theta\boldsymbol{x}_1 + (1-\theta)\boldsymbol{x}_c, \theta \in [0, 1)\}$. Then for $\boldsymbol{x}_o \in S$ close to $\boldsymbol{x}_c$ and an $\bar{\boldsymbol{x}} \in (\boldsymbol{x}_o, \boldsymbol{x}_c)$ such that, via the Mean Value Theorem,

$$\begin{aligned} f(\boldsymbol{x}_c) &= f(\boldsymbol{x}_o) + \nabla f(\bar{\boldsymbol{x}})^T(\boldsymbol{x}_c - \boldsymbol{x}_o) \\ &= f(\boldsymbol{x}_o) + \theta_o \nabla f(\bar{\boldsymbol{x}})^T(\boldsymbol{x}_c - \boldsymbol{x}_1), \ 0 \le \theta_o \le 1, \end{aligned}$$

and thus

$$f(\boldsymbol{x}_c) = f(\boldsymbol{x}_o) + \lambda\ \theta_o \nabla f(\bar{\boldsymbol{x}})^T(\boldsymbol{x}_1 - \boldsymbol{x}_2).$$

But since we can write $\bar{\boldsymbol{x}} = \alpha\boldsymbol{x}_1 + (1-\alpha)\boldsymbol{x}_2$ and $f(\boldsymbol{x}_1) < f(\bar{\boldsymbol{x}})$, we have $\nabla f(\bar{\boldsymbol{x}})^T(\boldsymbol{x}_1 - \boldsymbol{x}_2) = 0$. With $f(\boldsymbol{x}_c) = f(\boldsymbol{x}_o) \ge f(\boldsymbol{x}_1)$, we see that this contradicts our previous inequality that $f(\boldsymbol{x}_c) < f(\boldsymbol{x}_1)$. Hence, (9.1) must hold so that f is quasiconcave. Q.E.D.

An alternative proof of this theorem is provided in Appendix 9.A.

So if f is a $C^{(1)}$ quasiconcave function with $\boldsymbol{x}_1, \boldsymbol{x}_2 \in \mathcal{D}$ and $f(\boldsymbol{x}_1) \le f(\boldsymbol{x}_2)$, then the directional derivative of f at $\boldsymbol{x}_1$ in the $\boldsymbol{x}_2 - \boldsymbol{x}_1$ direction is nonnegative. Thus, the directional derivative of f provides the correct information about the direction of change of the function.

Let us examine the geometry underlying (9.5). Given $U_f(\alpha)$, we know that $\nabla f(\boldsymbol{x}_o)$ must be orthogonal to the level curve $f(\boldsymbol{x}_o) = \alpha$ and thus to the tangent hyperplane $\nabla f(\boldsymbol{x}_o)^T(\boldsymbol{y} - \boldsymbol{x}_o) = 0$ at $\boldsymbol{x}_o$ (Figure 9.2). So for any $\boldsymbol{x} \in U_f(\alpha)$, $f(\boldsymbol{x}) \ge \alpha = f(\boldsymbol{x}_o)$ and thus the angle between $\nabla f(\boldsymbol{x}_o)$ and $(\boldsymbol{x} - \boldsymbol{x}_o)$ must be acute ($\le \pi/2$) or $\nabla f(\boldsymbol{x}_o)^T(\boldsymbol{x} - \boldsymbol{x}_o) \ge 0$. Hence, points $\boldsymbol{x} \in U_f(\alpha)$ all lie above the hyperplane tangent to the level curve $f(\boldsymbol{x}_o) = \alpha$.

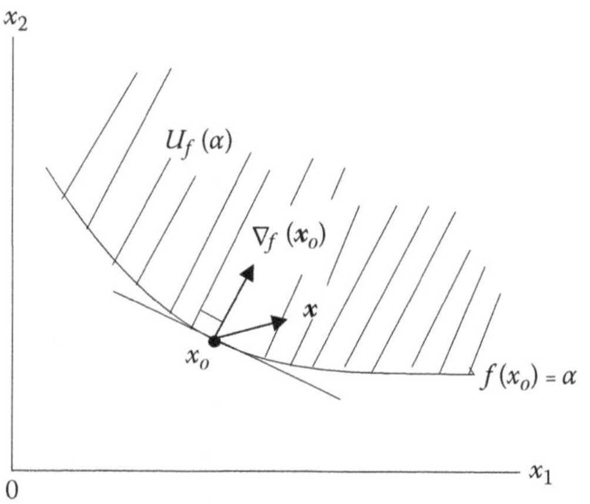

FIGURE 9.2 The angle between $\nabla f(\boldsymbol{x}_o)$ and $\boldsymbol{x} - \boldsymbol{x}_o$ must be acute $\nabla f(\boldsymbol{x}_o)^T(\boldsymbol{x} - \boldsymbol{x}_o) \ge 0$.

Example 9.2: Check for the quasiconcavity of the function given in Example 9.1 using (9.5). For $1 = x_1 < x_2 = 2$ we have $f(x_1) = f(1) = 7/2 < 6 = f(2) = f(x_2)$ while $\nabla f(x_1)^T(x_2 - x_1) = 4 - x_1 \ge 0$ for $x_1 \in [0, 4]$. Clearly, $f(x_1) \le f(x_2)$ implies $f(x_1) \le f(x_c)$ or (9.5) holds so that f is quasiconcave. ■

Next, suppose that f of class $C^{(2)}$ over its domain. Then a test for the quasiconcavity (quasiconvexity) of f using second-order information is provided by Theorem 9.3, a necessary and sufficient condition for f to be quasiconcave (quasiconvex). We start with

Definition 9.3: The $(n+1)$st order **bordered Hessian matrix** of a $C^{(2)}$ function $y = f(\boldsymbol{x})$: $\mathcal{D} \subset R^n_+ \to R$ at a point $\boldsymbol{x} \in \mathcal{D}$, $\mathcal{D}$ open, is denoted as

$$\bar{\boldsymbol{H}} = \boldsymbol{H}_f^f(\boldsymbol{x}) = \begin{bmatrix} 0 & \nabla f(\boldsymbol{x})^T \\ \nabla f(\boldsymbol{x}) & \boldsymbol{H}_f(\boldsymbol{x}) \end{bmatrix} = \begin{bmatrix} 0 & f_1 & \cdots & f_n \\ f_1 & f_{11} & \cdots & f_{1n} \\ \vdots & \vdots & & \vdots \\ f_n & f_{n1} & \cdots & f_{nn} \end{bmatrix}$$

(where the superscript "f" on the Hessian matrix $\boldsymbol{H}_f$ indicates that the first partial derivatives of f are used to border $\boldsymbol{H}_f$). Also, let

$$|\bar{H}_r| = \begin{vmatrix} 0 & \nabla f(\boldsymbol{x})_r^T \\ \nabla f(\boldsymbol{x})_r & \boldsymbol{H}_f(\boldsymbol{x})_{rr} \end{vmatrix} = \begin{vmatrix} 0 & f_1 & \cdots & f_r \\ f_1 & f_{11} & \cdots & f_{1r} \\ \vdots & \vdots & & \vdots \\ f_r & f_{r1} & \cdots & f_{rr} \end{vmatrix},$$

$r = 1, \ldots, n$, denote the rth naturally ordered principal minor of $\bar{\boldsymbol{H}}$.

We then have

Theorem 9.3: Let $y = f(\boldsymbol{x}): \mathcal{D} \subset R_+^n \rightarrow R$ be of class $C^{(2)}$ on the open, convex set $\mathcal{D}$, with int $(\mathcal{D}) \neq \emptyset$. Then:

1. (Necessity)
   a. if f is quasiconcave on $\mathcal{D}$, then for all $\boldsymbol{x} \in \mathcal{D}$, $\boldsymbol{x} \geq \boldsymbol{O}$,

$$(-1)^r |\bar{H}_r| \geq 0,\ r = 1, \ldots, n; \tag{9.7}$$

   b. if f is quasiconvex on $\mathcal{D}$, then for all $\boldsymbol{x} \in \mathcal{D}$, $\boldsymbol{x} \geq \boldsymbol{O}$,

$$|\bar{H}_r| \leq 0,\ r = 1, \ldots, n. \tag{9.8}$$

2. (Sufficiency) Conversely, for $\boldsymbol{x} > \boldsymbol{O}$,
   a. if

$$(-1)^r |\bar{\boldsymbol{H}}_r| > 0,\ r = 1, \ldots, n, \tag{9.9}$$

   then f is quasiconcave on $\mathcal{D}$ (i.e., if the $|\bar{\boldsymbol{H}}_r|$ alternate in sign, beginning negative, then f is quasiconcave on $\mathcal{D}$);
   b. if

$$|\bar{\boldsymbol{H}}_r| < 0,\ r = 1, \ldots, n, \tag{9.10}$$

then f is quasiconvex on $\mathcal{D}$.

(See Appendix 9.A for the rationalization of the use of this theorem.)

Example 9.3: Demonstrate that the function $y = f(\boldsymbol{x}) = x_1 x_2$ is quasiconcave over $\mathcal{D} = \{\boldsymbol{x} | \boldsymbol{x} \geq \boldsymbol{O}\} \subset R^2$. Let us start with the bordered Hessian matrix

$$\bar{\boldsymbol{H}} = \begin{bmatrix} 0 & \nabla f(\boldsymbol{x})^T \\ \nabla f(\boldsymbol{x}) & \boldsymbol{H}_f(\boldsymbol{x}) \end{bmatrix} = \begin{bmatrix} 0 & f_1 & f_2 \\ f_1 & f_{11} & f_{12} \\ f_2 & f_{21} & f_{22} \end{bmatrix} = \begin{bmatrix} 0 & x_2 & x_1 \\ x_2 & 0 & 1 \\ x_1 & 1 & 0 \end{bmatrix}.$$

Then

$$(-1)|\bar{\boldsymbol{H}}_1| = -\begin{vmatrix} 0 & x_2 \\ x_2 & 0 \end{vmatrix} = x_2^2 > 0,$$

$$(-1)^2|\bar{\boldsymbol{H}}_2| = \begin{vmatrix} 0 & x_2 & x_1 \\ x_2 & 0 & 1 \\ x_1 & 1 & 0 \end{vmatrix} = 2x_1x_2 > 0.$$

Thus f is quasiconcave on $\mathcal{D}$.

More generally, we can also demonstrate that $y = f(\boldsymbol{x}) = x_1^\alpha x_2^\beta$, with $\alpha, \beta > 0$, is quasiconcave on $\mathcal{D}$. Here, from (9.9),

$$(-1)|\bar{\boldsymbol{H}}_1| = \alpha^2 x_1^{2(\alpha-1)} x_2^{2\beta} > 0,$$

$$(-1)^2|\bar{\boldsymbol{H}}_2| = \alpha\beta(\alpha + \beta) x_1^{3\alpha-2} x_2^{3\beta-2} > 0.$$

Thus $f = x_1^\alpha x_2^\beta$ is quasiconcave. ■

Example 9.4: Suppose $y = f(x_1, x_2)$ is $C^{(2)}$ on an open convex set $\mathcal{D} \subset R^2$. Additionally, write $f(x_1, x_2) - y = F(x_1, x_2) = 0$ and let F be monotonically increasing on $\mathcal{D}$ so that $F_1, F_2 > 0$. If

$$|\bar{\boldsymbol{H}}| = \begin{vmatrix} 0 & F_1 & F_2 \\ F_1 & F_{11} & F_{12} \\ F_2 & F_{21} & F_{22} \end{vmatrix} = -F_1^2 F_{22} + 2F_1F_2F_{12} - F_2^2 F_{11} > 0,$$

then F is quasiconcave on $\mathcal{D}$. (Conversely, if F is quasiconcave on $\mathcal{D}$, then $|\bar{\boldsymbol{H}}| \geq 0$.) Hence, the upper level set $U_f(\alpha)$ is convex and the level curve $f(x_1, x_2) = \alpha$ (the lower boundary of $U_f(\alpha)$) is convex to the origin.

We previously found (Theorem 4.1.1.) that for $F(x_1, x_2) = 0$ of class $C^{(n)}$ for $\boldsymbol{x} \in \mathcal{D} \subset R^2$ ($\mathcal{D}$ is now assumed open *and* convex), there exists, locally, a function $x_2 = g(x_1)$ (which depicts a contour or level curve of f) such that

$$\frac{dx_2}{dx_1} = -\frac{F_1}{F_2},\ F_2 \neq 0,$$

and

$$\frac{d^2x_2}{dx_1^2} = -\frac{F_1^2F_{22} - 2F_1F_2F_{12} + F_2^2F_{11}}{F_2^3},\ F_2^3 \neq 0,$$

or

$$\frac{d^2x_2}{dx_1^2} = \frac{|\bar{H}|}{F_2^3} \geq 0.$$

Thus the quasiconcavity of F implies the convexity of g (the level curves are convex to the origin) so that $d^2x_2/dx_1^2 \geq 0$. ∎

What are the implications of quasiconcavity or quasiconvexity for determining a local (global) extremum of any such function? Suppose $f\colon \mathcal{D} \subset R^n \to R$, $\mathcal{D}$ convex, is quasiconcave. If $\boldsymbol{x}_o$ provides a local maximum of f, then this extremum need not be a global maximum of f on $\mathcal{D}$. However,

Theorem 9.4: Let $f\colon \mathcal{D} \subset R^n \to R$, $\mathcal{D}$ convex, be quasiconcave (quasiconvex). Then the set of points $\mathcal{D}^*$ at which f attains its global maximum (minimum) is convex.

Proof: Let $\mathcal{D}^* = \{\boldsymbol{x} \in \mathcal{D} | f(x) \leq M\} \neq \phi$, where M is the maximum value of f on $\mathcal{D}$. Then the convexity of $\mathcal{D}^*$ follows from the convexity of the upper level sets of a quasiconcave function. Q. E. D.

Section 9.3 Exercises:

9.3.1 Use Equation (9.5) to demonstrate that $f(\boldsymbol{x}) = x_1x_2$, $\boldsymbol{x} \in R^2$, with $\boldsymbol{x} \geq \boldsymbol{O}$, is quasiconcave.

9.3.2 Let $f(x) = x^3$, $x \in R$. Use Equation (9.6) to determine if f is quasiconvex.

9.3.3 Use Theorem 9.3 to determine the quasiconcavity (quasiconvexity) of the following functions for $\boldsymbol{x} \in R^2$, $\boldsymbol{x} > \boldsymbol{O}$:

a. $f = -x_1x_2$.

b. $f = x_1^3 + x_2^3$.

c. $f = -x_1^2 - x_2^2$.

d. $f = e^{x_1} + e^{x_2}$.

9.4 STRICTLY QUASICONCAVE AND QUASICONVEX FUNCTIONS

We now look to strengthening somewhat the notions of quasiconcavity and quasiconvexity. To this end we have

Definition 9.4: A function $y = f(\boldsymbol{x})\colon\ \mathcal{D} \subset R^n \to R$, with $\mathcal{D}$ a convex set, is **strictly quasiconcave** on $\mathcal{D}$ if and only if for each $\boldsymbol{x}_1, \boldsymbol{x}_2 \in \mathcal{D}$, with $f(\boldsymbol{x}_1) \neq f(\boldsymbol{x}_2)$,

$$f(\boldsymbol{x}_1) < f(\boldsymbol{x}_2) \text{ implies } f(\boldsymbol{x}_1) < f(\boldsymbol{x}_c), \quad (9.11)$$

where $\boldsymbol{x}_c = \lambda\boldsymbol{x}_1 + (1-\lambda)\boldsymbol{x}_2$, $\lambda \in (0, 1)$. Equivalently, under the given assumptions,

$$f(\boldsymbol{x}_c) > \min\{f(\boldsymbol{x}_1), f(\boldsymbol{x}_2)\}. \quad (9.12)$$

Thus f is strictly quasiconcave if and only if for every two distinct points in $\mathcal{D}$, the value of the function at an interior point of the line segment joining the two points is greater than the minimum of the function values at the end points. Similarly,

Definition 9.5: A function $y = f(\boldsymbol{x})\colon\ \mathcal{D} \subset R^n \to R$, with $\mathcal{D}$ a convex set, is **strictly quasiconvex** on $\mathcal{D}$ if and only if for each $\boldsymbol{x}_1, \boldsymbol{x}_2 \in \mathcal{D}$, with $f(\boldsymbol{x}_1) \neq f(\boldsymbol{x}_2)$,

$$f(\boldsymbol{x}_2) < f(\boldsymbol{x}_1) \text{ implies } f(\boldsymbol{x}_c) < f(\boldsymbol{x}_1), \quad (9.13)$$

$\boldsymbol{x}_c = \lambda\boldsymbol{x}_1 + (1-\lambda)\boldsymbol{x}_2$, $\lambda \in (0, 1)$, or, under the stated assumptions,

$$f(\boldsymbol{x}_c) < \max\{f(\boldsymbol{x}_1), f(\boldsymbol{x}_2)\}. \quad (9.14)$$

Figure 9.3 offers some examples of strictly quasiconcave and strictly quasiconvex functions for $x \in R$.

Suppose f is of class $C^{(1)}$ on an open convex set $\mathcal{D} \subset R^n$ and that $\boldsymbol{x}_1$ is not a critical point of f. We then have

Theorem 9.6: A $C^{(1)}$ function $y = f(\boldsymbol{x})\colon\ \mathcal{D} \subset R^n \to R$, $\mathcal{D}$ open and convex with $\nabla f(\boldsymbol{x}) \neq \boldsymbol{O}$ for all $\boldsymbol{x} \in \mathcal{D}$, is strictly quasiconcave on $\mathcal{D}$ if and only if for every $\boldsymbol{x}_1, \boldsymbol{x}_2 \in \mathcal{D}$,

$$f(\boldsymbol{x}_1) \leq f(\boldsymbol{x}_2),\ \nabla f(\boldsymbol{x}_1) \neq \boldsymbol{O} \text{ implies}$$
$$\nabla f(\boldsymbol{x}_1)^T(\boldsymbol{x}_2 - \boldsymbol{x}_1) > 0. \quad (9.15)$$

Also,

Theorem 9.7: A $C^{(1)}$ function $y = f(\boldsymbol{x})\colon\ \mathcal{D} \subset R^n \to R$, $\mathcal{D}$ open and convex with $\nabla f(\boldsymbol{x}) \neq \boldsymbol{O}$ for all $\boldsymbol{x} \in \mathcal{D}$, is strictly quasiconvex on $\mathcal{D}$ if and only if for every $\boldsymbol{x}_1, \boldsymbol{x}_2 \in \mathcal{D}$,

$$f(\boldsymbol{x}_1) \geq f(\boldsymbol{x}_2),\ \nabla f(\boldsymbol{x}_1) \neq \boldsymbol{O} \text{ implies}$$
$$\nabla f(\boldsymbol{x}_1)^T(\boldsymbol{x}_2 - \boldsymbol{x}_1) < 0. \quad (9.16)$$

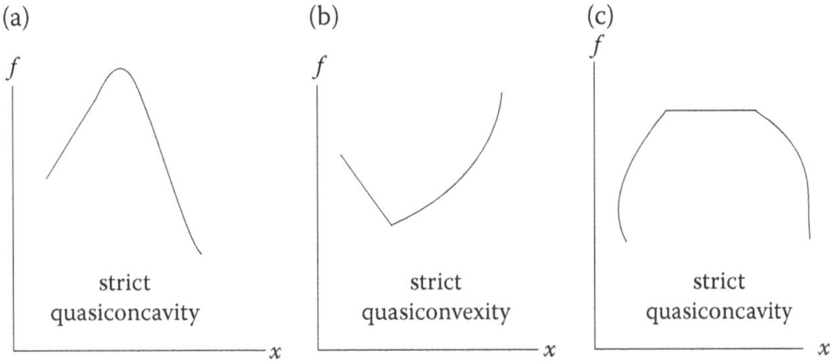

FIGURE 9.3 a. f is strictly quasiconcave; b. f is strictly quasiconvex; and c. f is strictly quasiconcave.

What is the significance of the concept of strict quasiconcavity (strict quasiconvexity)? A glance back at equation (9.5) reveals that for f quasiconcave, the directional derivative of f at $\mathbf{x}_1$ can equal zero even if $f(\mathbf{x}_1) < f(\mathbf{x}_2)$. Thus $\nabla f(\mathbf{x}_1) = \mathbf{O}$ might prompt us to reach the incorrect conclusion that f has a local (global) maximum at $\mathbf{x}_1$. The notion of strict quasiconcavity avoids this problem.

We note briefly that:

1. The function f is strictly quasiconcave (strictly quasiconvex) on if $-f$ is strictly quasiconvex (strictly quasiconcave);
2. Every strictly concave (strictly convex) function is also strictly quasiconcave (strictly quasiconvex);
3. We previously noted that every strictly concave (strictly convex) function is also concave (convex). But a function which is strictly quasiconcave (strictly quasiconvex) is not necessarily quasiconcave (quasiconvex). However, if f is upper semicontinuous (lower semicontinuous)[1] on a convex set $\mathcal{D}$, then strict quasiconcavity (strict quasiconvexity) implies quasiconcavity (quasiconvexity). To verify these assertions, we have

Theorem 9.8: Let $f\colon\ \mathcal{D} \subset R^n \to R$, $\mathcal{D}$ a convex set, be strictly quasiconcave and upper semicontinuous. Then f is quasiconcave.

Proof: Let $\mathbf{x}_1, \mathbf{x}_2 \in \mathcal{D}$ with $f(\mathbf{x}_1) \neq f(\mathbf{x}_2)$. If f is strictly quasiconcave, then, via (9.12), $f(\mathbf{x}_c) > \min\{f(\mathbf{x}_1), f(\mathbf{x}_2)\}$, $\lambda \in (0, 1)$. Suppose $f(\mathbf{x}_1) = f(\mathbf{x}_2)$. If f is quasiconcave, then, from (9.1), $f(\mathbf{x}_c) \geq f(\mathbf{x}_1)$, $\lambda \in (0, 1)$. By contradiction, let $f(\theta\mathbf{x}_1 + (1-\theta)\mathbf{x}_2) < f(\mathbf{x}_1)$ for $\theta \in (0, 1)$. Let $\mathbf{x} = \theta\mathbf{x}_1 + (1-\theta)\mathbf{x}_2$. With f upper semicontinuous, there exists a $\lambda \in (0, 1)$ such that $f(\mathbf{x}) < f(\lambda\mathbf{x}_1 + (1-\lambda)\mathbf{x}) < f(\mathbf{x}_1) = f(\mathbf{x}_2)$. Since $\mathbf{x}$ can be expressed as a convex combination of $\lambda\mathbf{x}_1 + (1-\lambda)\mathbf{x}$ and $\mathbf{x}_2$, and with f strictly quasiconcave and $f(\lambda\mathbf{x}_1 + (1-\lambda)\mathbf{x}) < f(\mathbf{x}_2)$, it follows that $f(\mathbf{x}) > f(\lambda\mathbf{x}_1 + (1-\lambda)\mathbf{x})$. But this contradicts the previous result that $f(\mathbf{x}) < f(\lambda\mathbf{x}_1 + (1-\lambda)\mathbf{x}) < f(\mathbf{x}_1) = f(\mathbf{x}_2)$. Q.E.D.

4. Definitions 9.4 and 9.5 preclude f from displaying any "flat spots" over $\mathcal{D}$ save for extreme points (see Figure 9.3c). Also, the boundary $f(\mathbf{x}) = \alpha$ of the upper level set of f, $U_f(\alpha)$, is precluded from exhibiting any "flat spots" if f is strictly quasiconcave (Figures 9.4a, b).
5. A local maximum (local minimum) of a strictly quasiconcave (strictly quasiconvex) function is also a global maximum (global minimum). More formally,

Theorem 9.9: Let $f\colon\ \mathcal{D} \subset R^n \to R$ be strictly quasiconcave (strictly quasiconvex) on the convex set $\mathcal{D}$ with $\mathbf{x}_o \in \mathcal{D}$. If $f(\mathbf{x}_o)$ is a local maximum (local minimum), then it is also a global maximum (global minimum) of f on $\mathcal{D}$.

It is important to note that a quasiconcave (quasiconvex) function can have a local maximum (local minimum) that is not a global maximum (global minimum).

6. Theorem 9.3 cannot be used as a test for *strict* quasiconcavity (*strict* quasiconvexity). However, the following sufficient condition holds for f of class $C^{(2)}$:

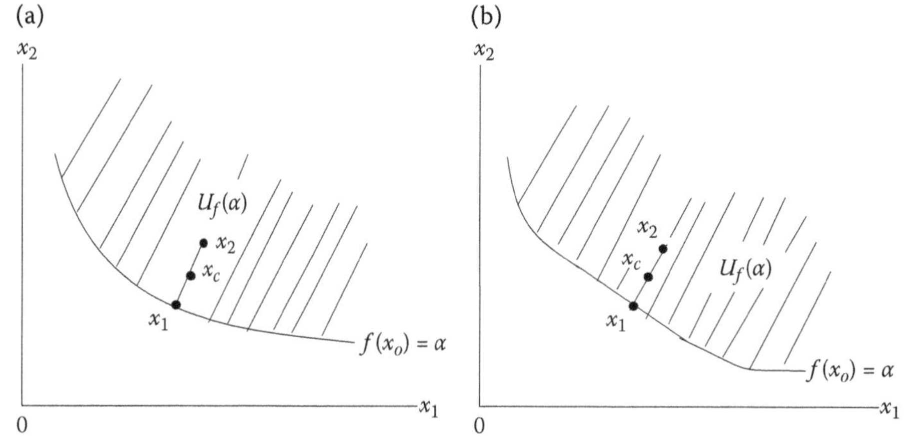

FIGURE 9.4 a. f is strictly quasiconcave; b. f is quasiconcave.

i. if $\mathcal{D} \subset R^n_{++}$, f is monotonically increasing, and $(-1)^r|\bar{\boldsymbol{H}}_r| > 0$, $r = 1, \ldots, n$, for all $\boldsymbol{x} \in \mathcal{D}$, then f is strictly quasiconcave;

ii. if $\mathcal{D} \subset R^n_{++}$, f is monotonically decreasing, and $|\bar{\boldsymbol{H}}_r| < 0$, $r = 1, \ldots, n$, for all $\boldsymbol{x} \in \mathcal{D}$, then f is strictly quasiconvex.

It was mentioned above (note 5) that if, for instance, f is strictly quasiconcave over a convex set $\mathcal{D}$ and attains a local maximum on $\mathcal{D}$, then the said maximum is also a global maximum. However, the global maximum so attained is not necessarily unique (Figure 9.3). In order to assume the uniqueness of the global maximum, when it exists, situations such as the one depicted in Figure 9.3c must be ruled out. In this regard, we must strengthen the concept of quasiconcavity to one of strong quasiconcavity (Section 9.5).

Section 9.4 Exercises:

9.4.1 Are any of the functions appearing in Exercise 9.3.3 strictly quasiconcave (strictly quasiconvex)?

9.4.2 Which of the following statements are correct?

a. Every strictly quasiconcave function is quasiconcave.

b. A strictly quasiconcave function attains its maximum at no more than one point.

c. Strict quasiconcavity is not a proper generalization of concavity, but only of strict concavity.

d. A concave function that is flat over some region cannot be strictly quasiconcave, but every strictly concave function is strictly quasiconcave.

9.4.3 Prove the following proposition. Let $f : \subset R \to R$. If f is strictly quasiconcave, then there is no point $\boldsymbol{x}_o \in \mathcal{D}$ that is a local minimum of f.

9.5 STRONGLY QUASICONCAVE AND STRONGLY QUASICONVEX FUNCTIONS

We first have

Definition 9.6: A function $y = f(\boldsymbol{x})$: $\mathcal{D} \subset R^n \to R$, with $\mathcal{D}$ a convex set, is **strongly quasiconcave** on $\mathcal{D}$ if for each $\boldsymbol{x}_1, \boldsymbol{x}_2 \in \mathcal{D}$, $\boldsymbol{x}_1 \neq \boldsymbol{x}_2$,

$$f(\boldsymbol{x}_1) < f(\boldsymbol{x}_2) \text{ implies } f(\boldsymbol{x}_1) < f(\boldsymbol{x}_c), \quad (9.17)$$

where $\boldsymbol{x}_c = \lambda\boldsymbol{x}_1 + (1-\lambda)\boldsymbol{x}_2$, $\lambda \in (0, 1)$, is the convex combination of $\boldsymbol{x}_1$ and $\boldsymbol{x}_2$. Equivalently, under the given assumptions,

$$f(\boldsymbol{x}_c) > \min\ \{f(\boldsymbol{x}_1), f(\boldsymbol{x}_2)\}. \quad (9.18)$$

Similarly,

Definition 9.7: A function $y = f(\boldsymbol{x})$: $\mathcal{D} \subset R^n \to R$, with $\mathcal{D}$ a convex set, is **strongly quasiconvex** on $\mathcal{D}$ if for each $\boldsymbol{x}_1, \boldsymbol{x}_2 \in \mathcal{D}$, $\boldsymbol{x}_1 \neq \boldsymbol{x}_2$,

$$f(\boldsymbol{x}_2) < f(\boldsymbol{x}_1) \text{ implies } f(\boldsymbol{x}_c) < f(\boldsymbol{x}_1), \quad (9.19)$$

where $\boldsymbol{x}_c = \lambda\boldsymbol{x}_1 + (1-\lambda)\boldsymbol{x}_2$, $\lambda \in (0, 1)$, is the convex combination of x_1 and x_2. Equivalently, under the stated assumptions,

$$f(x_c) < \max\{f(x_1), f(x_2)\}.$$

(The functions appearing in Figures 9.2a, b are, respectively, strongly quasiconcave and strongly quasiconvex.)

Some important considerations regarding these functions are:

1. the function f is strongly quasiconcave (strongly quasiconvex) if $-f$ is strongly quasiconvex (strongly quasiconcave).
2. every strictly concave (strictly convex) function is also strongly quasiconcave (strongly quasiconvex).
3. every strongly quasiconcave (strongly quasiconvex) function is also strictly quasiconcave (strictly quasiconvex). Hence, strict quasiconcavity (strict quasiconvexity) is a weaker condition than strong quasiconcavity (strong quasiconvexity).
4. every strongly quasiconcave (strongly quasiconvex) function is also quasiconcave (quasiconvex), and this is true without the requirement of upper semicontinuity (lower semicontinuity).
5. If f is strongly quasiconcave (strongly quasiconvex) on a convex set $\mathcal{D}$, then its attained maximum (minimum) is unique. In other words, a local maximal (minimal) solution is the unique global maximal (global minimal) solution (see Figure 5.2) or, in the case of, say, strong quasiconcavity, f is a *unimodal function*. To see this we state

Theorem 9.10: Let $f\colon \mathcal{D} \subset R^n \to R$, with f strongly quasiconcave and $\mathcal{D}$ convex. If x_o yields a local maximal solution for f, then x_o is also the unique or strong global maximum for f.

Proof: For x_o a local maximum, there exists a $\delta > 0$ such that, for all $x \in \delta(x_o) \cap \mathcal{D}, f(x_o) \geq f(x)$. Suppose to the contrary that there is a point $\bar{x} \in \mathcal{D}$, $\bar{x} \neq x_o$, such that $f(\bar{x}) \geq f(x_o)$. With f strongly quasiconcave, we have, via (9.18), that $f(\lambda\bar{x} + (1-\lambda)x_o) > \min\{f(\bar{x}), f(x_o)\} = f(x_o)$, $\lambda \in (0, 1)$. For λ sufficiency small, $\lambda\bar{x} + (1-\lambda)x_o \in \delta(x_o) \cap \mathcal{D}$ so that the preceding inequality violates the assumption that x_o is a local maximal solution. Q.E.D.

6. The function $f\colon \mathcal{D} \subset R^n \to R$ is **strongly α-quasiconcave** if there is an $\alpha > 0$, $\alpha \in R$, such that, for $x_c = \lambda x_1 + (1-\lambda)x_2$,

$$f(x_c) \geq \min\{f(x_1), f(x_2)\} + \lambda(1-\lambda)\alpha\|x_2 - x_1\|^2$$

for all $x_1, x_2 \in \mathcal{D}$ and $\lambda \in [0, 1]$; f is **strongly α-quasiconvex** if, under the same conditions,

$$f(x_c) \leq \max\{f(x_1), f(x_2)\} - \lambda(1-\lambda)\alpha\|x_2 - x_1\|^2.$$

If, for instance, f is a differentiable strictly or strongly quasiconcave function of class $C^{(1)}$ on a convex set $\mathcal{D} \subset R^n$, then, for a point $x_o \in \mathcal{D}$, it does not necessarily follow, as was the case for a concave function, that $\nabla f(x_o) = O$ implies that x_o yields a global maximum of f. What is needed is the specification of a class of functions which, like concave functions, guarantee that if $\nabla f(x_o) = O$, then $f(x_o) \geq f(x)$, $x \in \mathcal{D}$. This leads to the definition of pseudoconcave and pseudoconvex functions (Section 9.6).

Section 9.5 Exercises:

9.5.1 Which of the following statements are true?

a. Every strictly concave function is strongly quasiconcave.

b. Every strongly quasiconvex function is strictly quasiconvex.

c. Every strongly quasiconcave function is quasiconcave.

9.5.2 Is the function $f(x) = x^2\colon \mathcal{D} \subset R \to R$ strongly quasiconcave?

9.6 PSEUDOCONCAVE AND PSEUDOCONVEX FUNCTIONS

As will be demonstrated below, the importance of, say, pseudoconcave functions is that they provide a bridge between the families of quasiconcave and concave functions. While exhibiting some of the characteristics of quasiconcave functions, pseudoconcave functions, like concave functions, have the property that if $\nabla f(\boldsymbol{x}_o) = \boldsymbol{O}$, then the critical point $\boldsymbol{x}_o$ gives rise to a global maximum of f. (Such is not the case for quasiconcave or strictly quasiconcave functions.)

We start with

Definition 9.8: A function $y = f(\boldsymbol{x})$: $\mathcal{D} \subset R^n \to R$ of class $C^{(1)}$, with $\mathcal{D}$ an open convex set, is **pseudoconcave** on $\mathcal{D}$ if

$$\nabla f(\boldsymbol{x}_2)^T(\boldsymbol{x}_1 - \boldsymbol{x}_2) \leq 0 \text{ implies}$$
$$f(\boldsymbol{x}_1) \leq f(\boldsymbol{x}_2),\ \boldsymbol{x}_1, \boldsymbol{x}_2 \in \mathcal{D}. \tag{9.20}$$

Equivalently,

$$f(\boldsymbol{x}_1) > f(\boldsymbol{x}_2) \text{ implies}$$
$$\nabla f(\boldsymbol{x}_2)^T(\boldsymbol{x}_1 - \boldsymbol{x}_2) > 0. \tag{9.20.1}$$

As (9.20) indicates, if the directional derivative of f at a point $\boldsymbol{x}_2$ in the $(\boldsymbol{x}_1 - \boldsymbol{x}_2)-$ direction is nonpositive, then f is nonincreasing in that direction. Similarly,

Definition 9.9: A function $y = f(\boldsymbol{x})$: $\mathcal{D} \subset R^n \to R$ of class $C^{(1)}$, with $\mathcal{D}$ an open convex set, is **pseudoconvex** on $\mathcal{D}$ if

$$\nabla f(\boldsymbol{x}_2)^T(\boldsymbol{x}_1 - \boldsymbol{x}_2) \geq 0 \text{ implies}$$
$$f(\boldsymbol{x}_1) \geq f(\boldsymbol{x}_2),\ \boldsymbol{x}_1, \boldsymbol{x}_2 \in \mathcal{D}. \tag{9.21}$$

Equivalently,

$$f(\boldsymbol{x}_1) < f(\boldsymbol{x}_2) \text{ implies } \nabla f(\boldsymbol{x}_2)^T(\boldsymbol{x}_1 - \boldsymbol{x}_2) < 0. \tag{9.21.1}$$

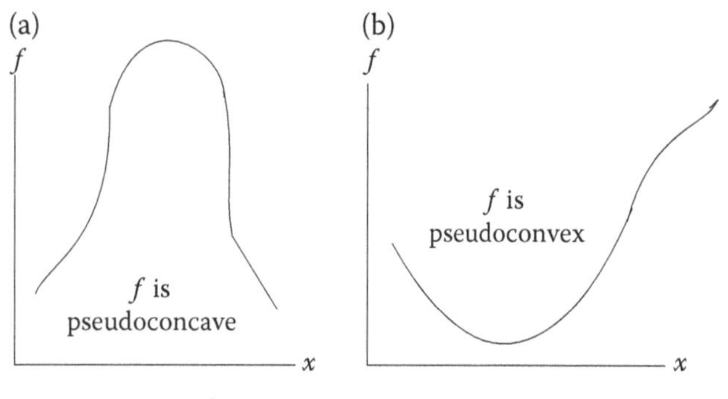

FIGURE 9.5 a. f is pseudoconcave; b. f is pseudoconvex.

(Figure 9.5 exhibits functions which are pseudoconcave and pseudoconvex.) If the *implied inequalities* hold strictly in (9.20) and (9.21) when $\boldsymbol{x}_1 \neq \boldsymbol{x}_2$, then f is **strictly pseudoconcave** and **strictly pseudoconvex**, respectively. A strictly pseudoconcave (strictly pseudoconvex) function is pseudoconcave (pseudoconvex), but not conversely.

We note further that:

1. the function f is pseudoconcave (pseudoconvex) on $\mathcal{D}$ if it is pseudoconcave (pseudoconvex) at each point of $\mathcal{D}$.
2. the function f is pseudoconcave (pseudoconvex) on $\mathcal{D}$ if and only if $-f$ is pseudoconvex (pseudoconcave) at each point of $\mathcal{D}$.
3. A concave function of class $C^{(1)}$ is pseudoconcave, which in turn is strictly quasiconcave. If f is a strictly pseudoconcave function of class $C^{(1)}$, then it is also strongly quasiconcave, which in turn is strictly quasiconcave.
4. The difficulty with a strictly quasiconcave function is that a point such as $\boldsymbol{x}_o \in \mathcal{D}$, $\mathcal{D}$ open and convex, with $\nabla f(\boldsymbol{x}_o) = \boldsymbol{O}$ (f attains a local maximum at $\boldsymbol{x}_o$) may not also give rise to a global maximum. However, if f is pseudoconcave, then any local maximum is also a global maximum. This observation is incorporated in

Theorem 9.11: Let f: $\mathcal{D} \subset R^n \to R$, $\mathcal{D}$ open and convex, be a pseudoconcave function of class $C^{(1)}$ and suppose $\nabla f(\boldsymbol{x}_o) = \boldsymbol{O}$, $\boldsymbol{x}_o \in \mathcal{D}$. Then $\boldsymbol{x}_o$ is a global maximum of f. (If f is strictly pseudoconcave on $\mathcal{D}$, then $\boldsymbol{x}_o$ is a unique or strong global maximum of f)

Proof: From (9.20) we can write $\nabla f(\boldsymbol{x}_0)^T(\boldsymbol{x} - \boldsymbol{x}_0) \leq 0$ implies $f(\boldsymbol{x}) \leq f(\boldsymbol{x}_o)$. If $\nabla f(\boldsymbol{x}_o) = \boldsymbol{O}$, then obviously $f(\boldsymbol{x}) \leq f(\boldsymbol{x}_o)$ so that $\boldsymbol{x}_o$ is a global maximizer for f on $\mathcal{D}$. Q.E.D.

In like fashion we can state

Theorem 9.12: Let $y = f(\boldsymbol{x})$: $\mathcal{D} \subset R^n \to R$, $\mathcal{D}$ open and convex, be a pseudoconvex function of class $C^{(1)}$ with $\nabla f(\boldsymbol{x}_o) = \boldsymbol{O}$, $\boldsymbol{x}_o \in \mathcal{D}$. Then $\boldsymbol{x}_o$ is a global minimum of f. (If f is strictly pseudoconvex on $\mathcal{D}$, then $\boldsymbol{x}_o$ is a unique or strong global minimum of f.)

5. It is important to note that pseudoconcavity requires that, say, (9.20.1) is satisfied for all points in $\mathcal{D}$; quasiconcavity implies that (9.20.1) is satisfied when $\nabla f(\boldsymbol{x}_2) \neq \boldsymbol{O}$. More specifically, we have

Theorem 9.13: Suppose that the function $y = f(\boldsymbol{x})$: $\mathcal{D} \subset R^n \to R$, $\mathcal{D}$ open and convex, is of class $C^{(1)}$ on $\mathcal{D}$.

i. if f is pseudoconcave (pseudoconvex) on $\mathcal{D}$, then f is quasiconcave (quasiconvex) on $\mathcal{D}$.

ii. if $\nabla f(\boldsymbol{x}) \neq \boldsymbol{O}$ for all $\boldsymbol{x} \in \mathcal{D}$, then f is pseudoconcave (pseudoconvex) on $\mathcal{D}$ if and only if f is quasiconcave (quasiconvex) on $\mathcal{D}$.

So if, for instance, f is quasiconcave on $\mathcal{D}$, then f is pseudoconcave at any point, say, $\boldsymbol{x}_o \in \mathcal{D}$ where $\nabla f(\boldsymbol{x}_o) \neq \boldsymbol{O}$ so that the condition $\nabla f(\boldsymbol{x}_o)^T(\boldsymbol{x} - \boldsymbol{x}_o) \leq 0$ implies $f(\boldsymbol{x}) \leq f(\boldsymbol{x}_o)$, $\boldsymbol{x} \in \mathcal{D}$. Thus a quasiconcave function is pseudoconcave when there are not critical points.

Additional considerations involving pseudoconcavity at a critical point are provided in Appendix 9.B.

6. if f is of class $C^{(2)}$, then the sufficient condition for quasiconcavity (Theorem 9.3) is also sufficient for pseudoconcavity [Avriel (2003); Ferland (1971)]. For details on this statement see Appendix 9.B.

A summary of the interrelationships between the various types of concavity, and generalizations thereof, is presented in the following diagram (Figure 9.6), where $\boldsymbol{x}_c = \lambda\boldsymbol{x}_1 + (1 - \lambda)\boldsymbol{x}_2$ [Bazaraa, *et al.* (2006); Mangasarian

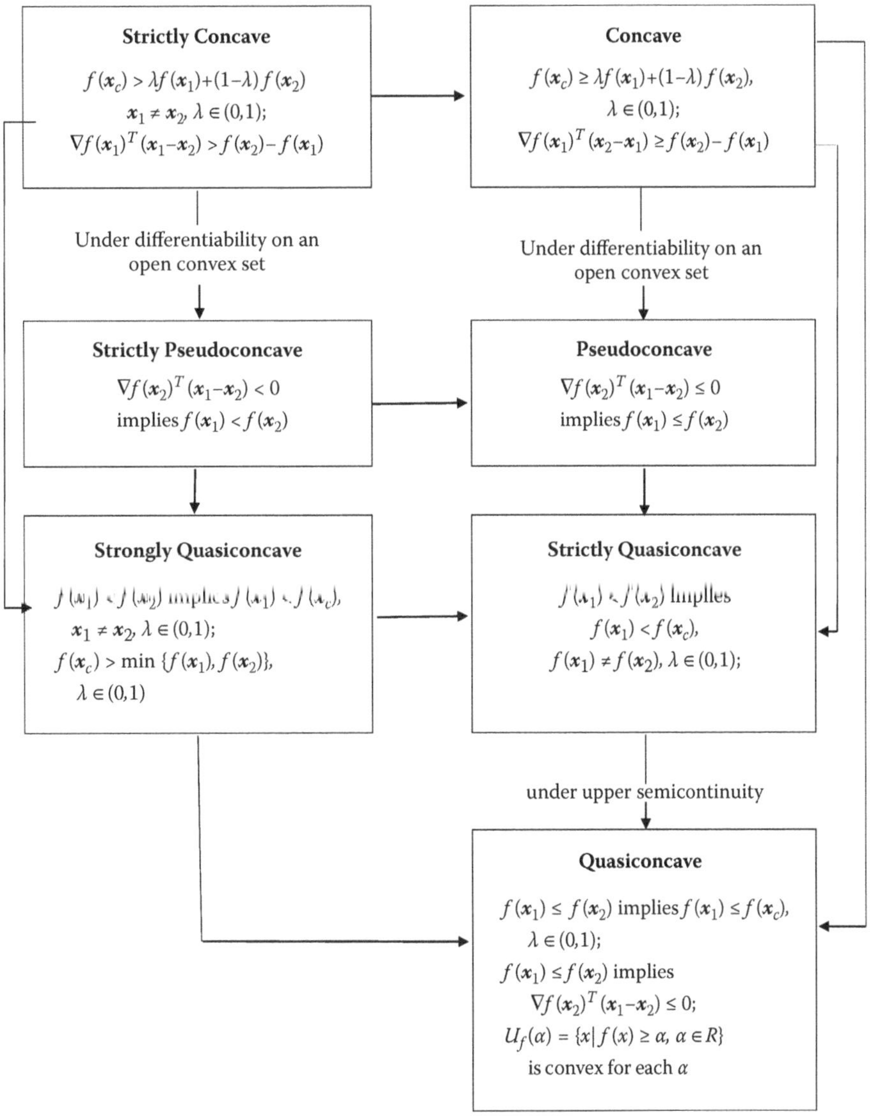

FIGURE 9.6 Types of concavity and its generalizations.

(1969)]. (A similar diagram can be developed for convex functions and their generalizations.)

Section 9.6 Exercises:

9.6.1 Are the following functions pseudoconvex?

a. $f(x) = -x^2 - x, \quad \mathcal{D} = \{x \in R | 0 \leq x \leq 1\}$.

b. $f(x) = x^3 + x, x \in R$. (Hint: use Equation 9.20.1.)

9.6.2 Which of the following statements are true?

a. If a function is pseudoconcave, then it is also strictly quasiconcave.

b. If a function is strictly quasiconcave, it need not be pseudoconcave.

c. A function is strictly quasiconcave if it is of class $C^{(1)}$ and pseudoconcave on an open convex set $\mathcal{D} \subset R^n$.

APPENDIX 9.A **ADDITIONAL THOUGHTS ON** THEOREMS 9.2, 9.3, AND 9.6 (CAMBINI AND MARTEIN [2009]; MANGASARIAN [1969]; BORWEIN AND LEWIS [2000])

As was the case in Appendix 8.A, this appendix also offers an alternative and instructive methodology for proving theorems pertaining to quasiconcave (quasiconvex) functions. Given the definition of the restriction of f (Definition 8.A.1) appearing in Appendix 8.A, we start with

Theorem 9.A.1

Given the $C^{(1)}$ restriction $\varphi(\theta)$ on $I = [0, 1] \subset R$, φ is quasiconcave on I if and only if for $\theta_1, \theta_2 \in I$,

$$\varphi(\theta_1) \leq \varphi(\theta_2) \text{ implies } \varphi'(\theta_1)(\theta_2 - \theta_1) \geq 0. \tag{9.A.1}$$

Given this result, we can now undertake an alternative proof of Theorem 9.2. Specifically,

Theorem 9.A.2

Let $y = f(\boldsymbol{x})$: $\mathcal{D} \subset R^n \rightarrow R$ be of class $C^{(1)}$ on the convex set $\mathcal{D}$. Then f is quasiconcave on $\mathcal{D}$ if and only if, for $\boldsymbol{x}_1, \boldsymbol{x}_2 \in \mathcal{D}$,

$$f(\boldsymbol{x}_1) \leq f(\boldsymbol{x}_2) \text{ implies } \nabla f(\boldsymbol{x}_1)^T(\boldsymbol{x}_2 - \boldsymbol{x}_1) \geq 0. \tag{9.A.2}$$

Proof: (Necessity) For f quasiconcave, let $\boldsymbol{x}_1, \boldsymbol{x}_2 \in \mathcal{D}$ with $f(\boldsymbol{x}_1) \leq f(\boldsymbol{x}_2)$. Let $\varphi(\theta) = f(\boldsymbol{x}_1 + \theta(\boldsymbol{x}_2 - \boldsymbol{x}_1))$, $\theta \in [0, 1]$, so that $\varphi(0) \leq \varphi(1)$. Then, via the preceding theorem, $\varphi'(0) = \nabla f(\boldsymbol{x}_1)^T(\boldsymbol{x}_2 - \boldsymbol{x}_1) \geq 0$.

(Sufficiency) Suppose (9.A.2) holds. If we assume that f is not quasiconcave, then there exists a restriction $\varphi(\theta) = f(\boldsymbol{x}_1 + \theta(\boldsymbol{x}_2 - \boldsymbol{x}_1))$, $\boldsymbol{x}_1, \boldsymbol{x}_2 \in \mathcal{D}$, $\theta \in [0, 1]$ which *is not* quasiconcave and thus, via Theorem 9.A.1, there exists $\theta_1, \theta_2 \in [0, 1]$ such that $\varphi(\theta_1) \leq \varphi(\theta_2)$ and $\varphi'(\theta_1)(\theta_2 - \theta_1) < 0$. Set $\bar{\boldsymbol{x}}_k = \boldsymbol{x}_1 + \theta_k(\boldsymbol{x}_2 - \boldsymbol{x}_1)$, $k = 1, 2$. Then $f(\bar{\boldsymbol{x}}_1) \leq f(\bar{\boldsymbol{x}}_2)$ and $\varphi'(\theta_1) = \nabla f(\bar{\boldsymbol{x}}_1)^T(\boldsymbol{x}_2 - \boldsymbol{x}_1) = (\theta_2 - \theta_1)^{-1}\nabla f(\bar{\boldsymbol{x}}_1)^T(\bar{\boldsymbol{x}}_2 - \bar{\boldsymbol{x}}_1)$ and thus $\varphi'(\theta_1)(\theta_2 - \theta_1) = \nabla f(\bar{\boldsymbol{x}}_1)^T(\bar{\boldsymbol{x}}_2 - \bar{\boldsymbol{x}}_1) < 0$, which contradicts (9.A.2). Q.E.D.

This next theorem renders a proof of Theorem 9.6, which provides a description of a quasiconcave function f which does not possess a critical point. We thus have

Theorem 9.A.3

A function $y = f(\boldsymbol{x})$: $\mathcal{D} \subset R^n \rightarrow R$ of class $C^{(1)}$, $\mathcal{D}$ open and convex with $\nabla f(\boldsymbol{x}) \neq \boldsymbol{O}$ for all $\boldsymbol{x} \in \mathcal{D}$, is strictly quasiconcave on $\mathcal{D}$ if and only if for every $\boldsymbol{x}_1, \boldsymbol{x}_2 \in \mathcal{D}$,

$$f(\mathbf{x}_1) \le f(\mathbf{x}_2),\ \nabla f(\mathbf{x}_1) \ne \mathbf{O} \text{ implies } \nabla f(\mathbf{x}_1)^T(\mathbf{x}_2 - \mathbf{x}_1) > 0. \tag{9.A.3}$$

Proof: (Necessity) Suppose f is strictly quasiconcave. If $f(\mathbf{x}_1) < f(\mathbf{x}_2)$, then $\nabla f(\mathbf{x}_1)^T(\mathbf{x}_2 - \mathbf{x}_1) > 0$. If $f(\mathbf{x}_1) = f(\mathbf{x}_2)$, strict quasiconcavity implies that there exists a $\bar{\theta} \in (0, 1)$ such that $f(\mathbf{x}_1) < f(\bar{\mathbf{x}})$, where $\bar{\mathbf{x}} = \mathbf{x}_1 + \bar{\theta}(\mathbf{x}_2 - \mathbf{x}_1)$. Hence, $\nabla f(\mathbf{x}_1)^T(\bar{\mathbf{x}} - \mathbf{x}_1) = \nabla f(\mathbf{x}_1)^T(\bar{\theta}(\mathbf{x}_2 - \mathbf{x}_1)) > 0$ so that (9.A.3) holds.

(Sufficiency) Suppose (9.A.3) holds. Then f is quasiconcave. If f is not strictly quasiconcave, there exists $\mathbf{x}_1, \mathbf{x}_2 \in \mathcal{D}$, $\bar{\theta} \in (0, 1)$, such that $f(\mathbf{x}_1) \le f(\mathbf{x}_2), f(\mathbf{x}_1) = f(\bar{\mathbf{x}})$. For the restriction $\varphi(\theta) = f(\mathbf{x}_1 + \theta(\mathbf{x}_2 - \mathbf{x}_1))$, $\theta \in (0, 1)$, let us apply (9.A.3) to the points $\bar{\mathbf{x}}$, $\mathbf{x}_2$. This implies that $\varphi'(\bar{\theta}) > 0$ and thus there exists a $\theta_1 \in (0, \bar{\theta})$ such that $\varphi(\theta_1) < \varphi(\bar{\theta})$. But this contradicts the quasiconcavity of $\varphi(\theta)$ on $[0, \bar{\theta}]$. Q.E.D.

We now turn to some important theorems concerning twice differentiable quasiconcave functions. These theorems provide a rationalization for the use of Theorem 9.3. In particular, they address the issue of establishing the quasiconcavity of a function via the investigation of the negative semidefiniteness of a quadratic form on a linear subspace or under a linear constraint. We start with

Theorem 9.A.4

(Necessity) If the function $y = f(\mathbf{x})$: $\mathcal{D} \subset R^n \to R$ is of class $C^{(2)}$, $\mathcal{D}$ open and convex, is quasiconcave on $\mathcal{D}$, then

$$\mathbf{x}_o \in \mathcal{D},\quad \mathbf{u} = \mathbf{x} - \mathbf{x}_o \in R^n, \quad \text{and}$$

$$\nabla f(\mathbf{x}_o)^T\mathbf{u} = 0 \text{ imply } \mathbf{u}^T\nabla^2 f(\mathbf{x}_o)\mathbf{u} \le 0. \tag{9.A.4}$$

Proof: The stipulation that $\mathbf{x}_o \in \mathcal{D}$, $\mathbf{u} \in R^n$, $\nabla f(\mathbf{x}_o)^T\mathbf{u} = 0$, and $\mathbf{u}^T\nabla^2 f(\mathbf{x}_o)\mathbf{u} > 0$ implies that $\mathbf{x}_o$ is a strong local minimum for the restriction of f on a line through $\mathbf{x}_o$ and in the $\mathbf{u}$direction. But this contradicts the stated quasiconcavity of f. Q.E.D.

If we now require the nonexistence of critical points, (9.A.4) is also a sufficient condition, i.e.,

Theorem 9.A.5

(Necessary and Sufficient) Let the function $y = f(\mathbf{x})$: $\mathcal{D} \subset R^n \to R$ of class $C^{(2)}$, $\mathcal{D}$ open and convex, be such that $\nabla f(\mathbf{x}) \ne \mathbf{O}$ for all $\mathbf{x} \subset \mathcal{D}$. Then f is quasiconcave on $\mathcal{D}$ if and only if (9.A.4) holds.

It should be apparent that in (9.A.4) we have $\nabla^2 f(\mathbf{x}_o) = \mathbf{H}_f(\mathbf{x}_o)$ and that this sequence of theorems serves as the foundation of Theorem 9.3.

APPENDIX 9.B **ADDITIONAL THOUGHTS ON DIFFERENTIABLE PSEUDOCONCAVE AND PSEUDOCONVEX FUNCTIONS** (CAMBINI AND MARTEIN [2009]; MANGASARIAN [1969]; BORWEIN AND LEWIS [2000])

A function is pseudoconcave (strictly pseudoconcave) if and only if it is pseudoconcave (strictly pseudoconcave) over each restriction $\varphi(\theta)$, θ contained within an interval $I \subset R$, on a line segment. In this regard, let us consider the pseudoconcavity of a function of a single variable with the goal of obtaining like characterizations for function of more than one argument. To this end we have

Theorem 9.B.1

Let φ be of class $C^{(1)}$ on an interval $I \subset R$. Then $\varphi(\theta)$ is pseudoconcave (strictly pseudoconcave) on I if and only if for every $\theta_o \in I$ such that $\varphi'(\theta_o) = 0$, θ_o is a local minimum (strong local minimum) for φ.

Given this theorem, it follows that

Theorem 9.B.2

Let $y = f(\mathbf{x})$: $\mathcal{D} \subset R^n \to R$, $\mathcal{D}$ open and convex, be of class $C^{(1)}$ on $\mathcal{D}$. Then f is pseudoconcave (strictly pseudoconcave) on $\mathcal{D}$ if and only if for every $\mathbf{x}_o \in \mathcal{D}$ and $\mathbf{u} = (\mathbf{x} - \mathbf{x}_o) \in R^n$ such that $\nabla f(\mathbf{x}_o)^T\mathbf{u} = 0$, the function $\varphi(\theta) = f(\mathbf{x}_o + \theta\mathbf{u})$ attains a local maximum (strong local maximum) at $\theta = 0$.

Let us next consider the behavior of a pseudoconcave function at a critical point. Specifically,

Theorem 9.B.3

Let $y = f(\mathbf{x})$: $\mathcal{D} \subset R^n \to R$, $\mathcal{D}$ open and convex, be of class $C^{(1)}$ on $\mathcal{D}$. Then f is pseudoconcave (strictly pseudoconcave) on $\mathcal{D}$ if and only if:

i. f is quasiconcave on $\mathcal{D}$;

ii. if $\mathbf{x}_o \in \mathcal{D}$ and $\nabla f(\mathbf{x}_o) = 0$, then $\mathbf{x}_o$ is a local maximum (strong local maximum) for f.

Proof: (Necessity) If f is pseudoconcave, then: (i) follows from Theorem 9.13; (ii) follows from Theorem 9.11.

(Sufficiency) Suppose that (i), (ii) hold. By virtue of Theorem 9.B.2 we can demonstrate that if $\mathbf{x}_o \in \mathcal{D}$ and $\mathbf{u} \in R^n$ are such that $\nabla f(\mathbf{x}_o)^T\mathbf{u} = 0$, the restriction $\varphi(\theta) = f(\mathbf{x}_o + \theta\mathbf{u})$ attains a local maximum at $\theta = 0$. Two cases present themselves: (1) if $\nabla f(\mathbf{x}_o) = \mathbf{O}$, then the assertion of the theorem follows from (ii); (2) if $\nabla f(\mathbf{x}_o) \neq \mathbf{O}$, then there exists a neighborhood $\delta(\mathbf{x}_o)$ such that $\nabla f(\mathbf{x}_o) \neq \mathbf{O}$ for all $\mathbf{x} \in \delta(\mathbf{x}_o)$ (since ∇f is continuous near $\mathbf{x}_o$). Theorem 9.13 implies that f is pseudoconcave on $\delta(\mathbf{x}_o)$ while Theorem 9.B.1 implies that $\varphi(\theta)$ attains a local maximum at $\theta = 0$. Q.E.D.

For (strictly) pseudoconcave functions of class $C^{(2)}$, Theorem 9.B.1, 9.B.2 can be expressed in terms of first- and second-order derivatives. To this end we have

Theorem 9.B.4

Let φ be a function of class $C^{(2)}$ defined on an open interval $I \subset R$. Then φ is (strictly) pseudoconcave on I if and only if for every $\theta_o \subset I$ such that $\varphi'(\theta_o) = 0$, either $\varphi''(\theta_o) < 0$ or $\varphi''(\theta_o) = 0$ and θ_o is a (strict) local maximum for φ.

We next have

Theorem 9.B.5

Let $y = f(\mathbf{x})$: $\mathcal{D} \subset R^n \to R$, $\mathcal{D}$ open and convex, be a function of class $C^{(2)}$ on $\mathcal{D}$. Then f is (strictly) pseudoconcave on $\mathcal{D}$ if and only if for every $\mathbf{x}_o \in \mathcal{D}$ and $\mathbf{u} = \mathbf{x} - \mathbf{x}_o \in R^n$ such that $\nabla f(\mathbf{x}_o)^T\mathbf{u} = 0$, either $\mathbf{u}^T\nabla^2 f(\mathbf{x}_o)\mathbf{u} < 0$ or $\mathbf{u}^T\nabla^2 f(\mathbf{x}_o)\mathbf{u} = 0$ and the function $\varphi(\theta) = f(\mathbf{x}_o + \theta\mathbf{u})$ attains a (strong) local maximum at $\theta = 0$.

Given these theorems, let us employ the following depiction for establishing the pseudoconcavity of a function.

Theorem 9.B.6

Let $y = f(\mathbf{x})$: $\mathcal{D} \subset R^n \to R$, $\mathcal{D}$ open and convex, be a function of class $C^{(2)}$ on $\mathcal{D}$. Then f is (strictly) pseudoconcave on $\mathcal{D}$ if and only if the following conditions hold:

$$\text{(i) } \boldsymbol{x} \in \mathcal{D} \text{ and } \boldsymbol{u} \in R^n, \ \ \nabla f(\boldsymbol{x}_o)^T \boldsymbol{u} = 0 \text{ implies } \boldsymbol{u}^T \nabla^2 f(\boldsymbol{x}_o) \boldsymbol{u} \le 0. \tag{9.B.1}$$

ii if $\boldsymbol{x}_o \in \mathcal{D}$ is a critical point for f, then $\boldsymbol{x}_o$ is a (strong) local maximum for f on $\mathcal{D}$.

Proof: (Necessity) If f is (strictly) pseudoconcave, then (i), (ii) follow directly from Theorem 9.B.3, 9.A.4.

(Sufficiency) Given Theorem 9.B.2, we must demonstrate that $\nabla f(\boldsymbol{x}_o)^T \boldsymbol{u} = 0$implies that $\boldsymbol{u}(\theta) = f(\boldsymbol{x}_o + \theta \boldsymbol{u})$ attains a (strong) local maximum $\theta = 0$. If $\nabla f(\boldsymbol{x}_o) = \boldsymbol{O}$, then the assertion of the theorem follows from (ii). If $\nabla f(\boldsymbol{x}_o) \neq \boldsymbol{O}$, we know that $\nabla f(\boldsymbol{x}_o) \neq \boldsymbol{O}$ for all $\boldsymbol{x} \in \delta(\boldsymbol{x}_o)$ (under the continuity of ∇f) so that, from Theorem 9.A.4, (i) implies that f is quasiconcave on $\delta(\boldsymbol{x}_o)$. The assertion follows from Theorems 9.13, 9.B.2. Q.E.D.

It is important to note that condition (9.B.1) is equivalent to determining the negative semi-definiteness of a quadratic form subject to a linear constraint, where $\nabla^2 f(\boldsymbol{x}_o)$ is the Hessian matrix of f at $\boldsymbol{x}_o$.

As a practical (operational) matter, how do we determine if a particular function is pseudoconcave? In what follows we offer a sufficient condition for pseudoconcavity. To this end, suppose $\nabla f \neq \boldsymbol{O}$. Then, since the bordered Hessian of f,

$$\boldsymbol{H} = \begin{bmatrix} 0 & \nabla f(\boldsymbol{x})^T \\ \nabla f(\boldsymbol{x}) & \boldsymbol{H}_f(\boldsymbol{x}) \end{bmatrix},$$

is a real symmetric matrix, it is well known that the conditions:

i. $\nabla f(\boldsymbol{x}_o)^T \boldsymbol{u} = 0, \ \ \boldsymbol{u} \neq \boldsymbol{O}$ implies $\boldsymbol{u}^T \bar{\boldsymbol{H}} \boldsymbol{u} < 0$; and

ii. $(-1)^r |\bar{\boldsymbol{H}}_r| > 0, \ r = 1, \ \ldots, n,$

are equivalent, where $|\bar{\boldsymbol{H}}_r|$ is the rth naturally ordered principal minor of $\bar{\boldsymbol{H}}$. Hence, our sufficient condition for pseudoconcavity is provided by

Theorem 9.B.7

Let $y = f(\boldsymbol{x}): \ \mathcal{D} \subset R^n \to R$, $\mathcal{D}$ open and convex, be a function of class $C^{(2)}$ on $\mathcal{D}$. If

$$(-1)^r |\bar{\boldsymbol{H}}_r| > 0, \ r = 1, \ \ldots, n, \tag{9.B.2}$$

then f is pseudoconcave on $\mathcal{D}$.

Proof: Condition (9.B.2) is equivalent to asserting that for every $\boldsymbol{u} \neq \boldsymbol{O}$ such that $\nabla f(\boldsymbol{x})^T \boldsymbol{u} = 0$ we have $\boldsymbol{u}^T \nabla^2 f(\boldsymbol{x}) \boldsymbol{u} < 0$. Then the assertion of this theorem follows from Theorem 9.B.5. Q.E.D.

(Note that a sufficient condition for the pseudoconvexity of a twice continuously differentiable function f on an open, convex domain involves replacing (9.B.2) by $|\bar{\boldsymbol{H}}_r| < 0, \ r = 1, \ \ldots, n.$)

One final point needs to be reinforced. The requirement (9.B.2) for the pseudoconcavity of f is also sufficient for the quasiconcavity of a function, as evidenced by item no. 6 of Section 9.4 of this chapter.

NOTE

1 Remember that f: $\mathcal{D} \subset R^n \to R$ is lower (upper) semicontinuous on $\mathcal{D}$ if its lower level set $L_f(\alpha)$ (upper level set $U_f(\alpha)$) is closed relative to $\mathcal{D}$ for every α. Equivalently, f is upper semicontinuous on $\mathcal{D}$ if and only if for every $\boldsymbol{x}_o \in \mathcal{D}$ and every sequence $\{\boldsymbol{x}_k\} \subset \mathcal{D}$ converging to $\boldsymbol{x}_o$, the inequality $\limsup_{k\to\infty} f(\boldsymbol{x}_k) \le f\left(\lim_{k\to\infty} \boldsymbol{x}_k\right) = f(\boldsymbol{x}_o)$ holds. For f lower semicontinuous, the preceding inequality must be reversed.

Constrained Extrema: Equality Constraints

10.1 CONSTRAINED EXTREMA: A SINGLE EQUALITY CONSTRAINT IN *N* INDEPENDENT VARIABLES

Throughout chapters six and seven we were interested in maximizing or minimizing a differentiable real-valued function involving n independent variables. In this regard, in attempting to find, say, any (strong) local extrema associated with $y = f(\boldsymbol{x})$: $\mathcal{D} \subset R^n \rightarrow R$, $\mathcal{D}$ open, all variables x_i, $i = 1, \ldots, n$, were *free* to assume any value within $\mathcal{D}$, with the selection of the value of one variable in no way restricting our choice of any other variable. Such problems are classified as *unconstrained optimization problems*. In what follows our objective is to consider the task of optimizing f given that the variables x_i, $i = 1, \ldots, n$, are confined to a specific mode of behavior within $\mathcal{D}$, namely that dictated by a differentiable constraint or side relation involving $\boldsymbol{x}$, $g(\boldsymbol{x}) = b$, $b \in R$. A problem such as this is a *constrained optimization problem*. Before attempting to generate a solution to this type of problem, let us examine the constraint function g itself. As is typically the case, let us rewrite the constraint $g(\boldsymbol{x}) = b$ implicitly as $\hat{g}(\boldsymbol{x}) = b - g(\boldsymbol{x}) = 0$. We consequently state

Definition 10.1: If the function $\hat{g}(\boldsymbol{x}) = 0$, $\boldsymbol{x} \in R^n$, defines x_n as a function of $\hat{\boldsymbol{x}}^T = (x_1, \ldots, x_{n-1}) \in R^{n-1}$, $x_n = \varnothing(\hat{\boldsymbol{x}})$, so that $\hat{g}(\hat{\boldsymbol{x}}, \varnothing(\hat{\boldsymbol{x}})) = 0$, then $x_n = \varnothing(\hat{\boldsymbol{x}})$ is defined **implicitly** as $\hat{g}(\boldsymbol{x}) = 0$.

Under what conditions does $\hat{g}(\boldsymbol{x}) = 0$ define a unique implicit function? That is, can we find a function $x_n = \varnothing(\hat{\boldsymbol{x}})$ such that $\hat{g}(\boldsymbol{x}) = 0$ is satisfied? To answer these questions we look to an existence or Implicit Function Theorem. Specifically,

Theorem 10.1: (**Implicit Function Theorem**) [Taylor and Mann (1972)]. Let $\hat{g}(\boldsymbol{x}) = 0$ be defined over a region $\mathcal{D} \subset R^n$ and let $\boldsymbol{x}_o \in$ int $(\mathcal{D})$. If:

1. $\hat{g}(\boldsymbol{x}_o) = 0$;
2. $\hat{g}_i$, $i = 1, \ldots, n$, are continuous within some spherical δ – neighborhood of $\boldsymbol{x}_o$, $\delta(\boldsymbol{x}_o)$; and
3. $\hat{g}_n(\boldsymbol{x}_o) \neq 0$, then there is some spherical $\bar{\delta}$ – neighborhood of $\hat{\boldsymbol{x}}_o^T = (x_1^o, \ldots, x_{n-1}^o) \in R^{n-1}$, $\bar{\delta}(\hat{\boldsymbol{x}}_o)$, throughout which there exists for $\hat{\boldsymbol{x}} \in \bar{\delta}(\hat{\boldsymbol{x}}_o)$ a unique differentiable function $x_n = \varnothing(\hat{\boldsymbol{x}})$ such that: $x_n^o = \varnothing(\hat{\boldsymbol{x}}_o)$; $\hat{g}(\hat{\boldsymbol{x}}, \varnothing(\hat{\boldsymbol{x}})) = 0$; and which possesses continuous first partial derivatives $\varnothing_i = -\hat{g}_i/\hat{g}_n$, $i = 1, \ldots, n - 1$.

In sum, this theorem states the circumstances under which we can solve $\hat{g}(\boldsymbol{x}) = 0$ for x_n in terms of x_i, $i = 1, \ldots, n - 1$, $x_n = \varnothing(\hat{\boldsymbol{x}})$. It is imperative to note that $\varnothing(\hat{\boldsymbol{x}})$ *only exists locally* over $\bar{\delta}(\hat{\boldsymbol{x}}_o)$. For instance, the contour $\mathcal{G} = \{\boldsymbol{x} | \hat{g}(\boldsymbol{x}) = 0, \boldsymbol{x} \in R^2\} \subset \mathcal{D} \subset R^2$ appears in Figure 10.1 For $\boldsymbol{x}_o \in \mathcal{G}$, there exists a unique differentiable function $x_2 = \varnothing(x_1)$ for all $x_2 \in \bar{\delta}(x_1^o)$ such that the conclusion of the Implicit Function Theorem obtains. Thus: $x_2^o = \varnothing(x_1^o)$; $\hat{g}(x_1, \varnothing(x_1)) = 0$; and

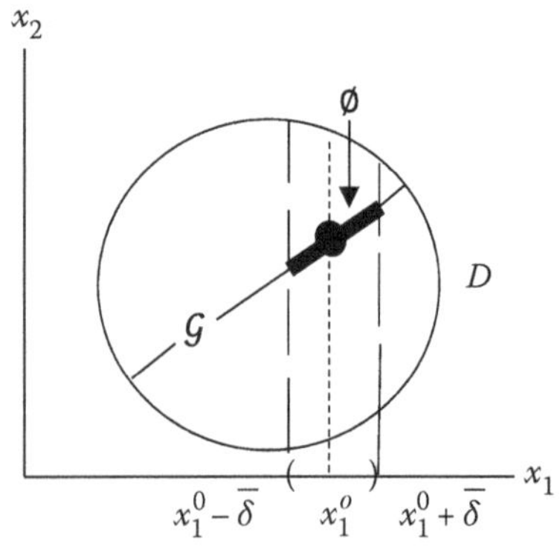

FIGURE 10.1 $\varnothing(x_1)$ exists locally over $\bar{\delta}(x_1^o)$.

$dx_2/dx_1 = -\hat{g}_1/\hat{g}_2$. The graph of $\varnothing$ appears as the shaded portion of $\mathcal{G}$.

How does the introduction of the constraint function $\hat{g} = 0$ impact the optimization of f? The consequence of introducing $\hat{g} = 0$ is twofold: (1) the immediate effect is to reduce the size of the set of admissible solutions, i.e., whereas $\mathcal{D}$ was the set of admissible solutions in the unconstrained problem, the collection of admissible solutions for the constrained problem is only some subset $\mathcal{G} \subset \mathcal{D}$, namely, the set of points of $\mathcal{D}$ which satisfy $\hat{g}(\boldsymbol{x}) = 0$; (2) since the domain of f is restricted, then so is its range.

Let us now look to a classification of the various types of constrained extrema which we shall encounter. As will now be seen, constrained extrema are of the global and local varieties. To this end, let $y = f(\boldsymbol{x})$, $\hat{g}(\boldsymbol{x}) = b - g(\boldsymbol{x}) = 0$, $\boldsymbol{x} \in R^n$, be defined on a closed region $\mathcal{D} \subset R^n$ and let $\boldsymbol{x}_o \in \mathcal{G} \cap \mathcal{D}$, where $\mathcal{G} = \{\boldsymbol{x} | \hat{g}(\boldsymbol{x}) = 0, \boldsymbol{x} \in R^n\}$. Then

Definition 10.2: The real-valued function $y = f(\boldsymbol{x})$ subject to $\hat{g}(\boldsymbol{x}) = 0$ has a **constrained global maximum (minimum)** at $\boldsymbol{x}_o \in \mathcal{G} \cap \mathcal{D}$ if $f(\boldsymbol{x}_o) \geq f(\boldsymbol{x})$ $(\leq f(\boldsymbol{x}))$ for all points $\boldsymbol{x}(\neq \boldsymbol{x}_o) \in \mathcal{G} \cap \mathcal{D}$.

If $f(\boldsymbol{x}_o) \geq f(\boldsymbol{x})(\leq f(\boldsymbol{x}))$ and equality holds between $f(\boldsymbol{x}_o)$ and $f(\boldsymbol{x})$ for at least one point $\boldsymbol{x}(\neq \boldsymbol{x}_o) \in \mathcal{G} \cap \mathcal{D}$, then f **has a constrained weak global maximum (minimum)** at $\boldsymbol{x}_o$. Alternatively, f attains a **constrained strong global maximum (minimum)** at $\boldsymbol{x}_o$ if $f(\boldsymbol{x}_o) > f(\boldsymbol{x})(< f(\boldsymbol{x}))$ for all $\boldsymbol{x}(\neq \boldsymbol{x}_o) \in \mathcal{G} \cap \mathcal{D}$.

Next, for $\mathcal{D} \subset R^n$ open,

Definition 10.3: The real-valued function $y = f(\boldsymbol{x})$ subject to $\hat{g}(\boldsymbol{x}) = 0$ has a **constrained local maximum (minimum)** at $\boldsymbol{x}_o \in \mathcal{G} \cap \mathcal{D}$ if there is some spherical δ – neighborhood about $\boldsymbol{x}_o$, $\delta(\boldsymbol{x}_o)$, such that $f(\boldsymbol{x}_o) \geq f(\boldsymbol{x})(\leq f(\boldsymbol{x}))$ for all points $\boldsymbol{x} \in \delta(\boldsymbol{x}_o) \cap \mathcal{G} \cap \mathcal{D}$.

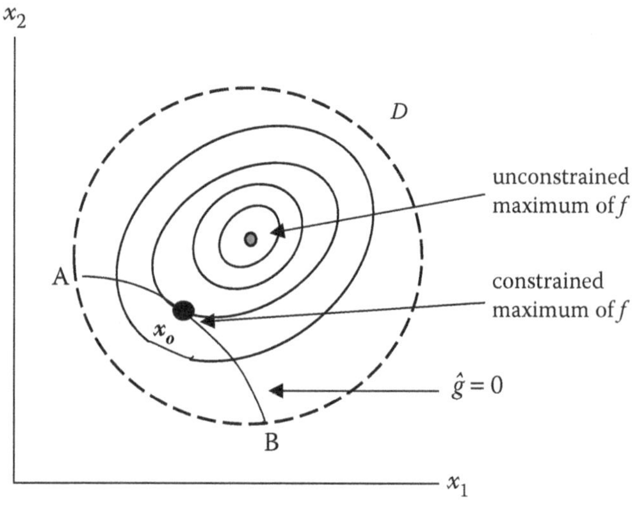

FIGURE 10.2 f has a constrained maximum at $\boldsymbol{x}_o$.

If $f(\boldsymbol{x}_o) \geq f(\boldsymbol{x})(\leq f(\boldsymbol{x}))$ and $f(\boldsymbol{x}_o) = f(\boldsymbol{x})$ for at least one point $\boldsymbol{x}(\neq \boldsymbol{x}_o) \in \delta(\boldsymbol{x}_o) \cap \mathcal{G} \cap \mathcal{D}$, then f has a **constrained weak local maximum (minimum)** at $\boldsymbol{x}_o$, while if $f(\boldsymbol{x}_o) > f(\boldsymbol{x})(< f(\boldsymbol{x}))$ for all $\boldsymbol{x}(\neq \boldsymbol{x}_o) \in \delta(\boldsymbol{x}_o) \cap \mathcal{G} \cap \mathcal{D}$, then f is said to have a **constrained strong local maximum (minimum)** at $\boldsymbol{x}_o$.

What does the attainment of an extreme value of f look like when the x_i, $i = 1, \ldots, n$, are not independent variables but are connected by a side relation of the form $\hat{g}(\boldsymbol{x}) = 0$? Looking to Figure 10.2, suppose that we desire to maximize $y = f(\boldsymbol{x})$, $\boldsymbol{x} \in R^2$ subject to $\hat{g}(\boldsymbol{x}) = 0$. Since the optimizing point $\boldsymbol{x}_o$ must satisfy $\hat{g}(\boldsymbol{x}) = 0$, it follows that we must move along $\hat{g}(\boldsymbol{x}) = 0$ from, say, A to B so as to reach the *highest possible* contour of f. This clearly occurs at point $\boldsymbol{x}_o \in \mathcal{G}$ where a contour of $f(\boldsymbol{x})$ is just tangent to the constraint function $\hat{g}(\boldsymbol{x}) = 0$.

Under what circumstances will a unique point of tangency occur between the contours of f and $\hat{g} = 0$? As we shall now see, f and $\hat{g} = 0$ *cannot be functionally dependent*. More specifically,

Theorem 10.2: Let the differentiable functions $y = f(\boldsymbol{x})$, $w = g(\boldsymbol{x})$, $\boldsymbol{x} \in R^2$, be defined throughout an open region $\mathcal{D} \subset R^2$. Additionally, assume that: (1) $\nabla f, \nabla g \neq \boldsymbol{O}$ on $\mathcal{D}$ so that both f, g have well-defined level curves there; and (2) there exists a function $\varnothing(y, w) = 0$ such that $\varnothing(f(\boldsymbol{x}), g(\boldsymbol{x})) \equiv 0$. Then f, g are **functionally dependent**[1] throughout $\mathcal{D}$ if and only if the Jacobian determinant of f, g identically vanishes at each point of $\mathcal{D}$, i.e.,

$$\left| \frac{\partial(f, g)}{\partial(x_1, x_2)} \right| = \begin{vmatrix} f_1 & f_2 \\ g_1 & g_2 \end{vmatrix} \equiv 0. \qquad (10.1)$$

So, if the Jacobian determinant of f and g is identically zero for all points $x \in \mathcal{D} \subset R^2$, then the level curves of f, g coincide; each level curve of f is a level curve of g and conversely, although their labeling may differ.

Example 10.1: Are the functions $y = f(x) = x_1^2 + x_2^2$ and $w = g(x) = (x_1^2 + x_2^2)^{5/4}$ functionally dependent throughout R^2? Since

$$\begin{vmatrix} 2x_1 & 2x_2 \\ \frac{5}{2}(x_1^2 + x_2^2)^{1/4}x_1 & \frac{5}{2}(x_1^2 + x_2^2)^{1/4}x_2 \end{vmatrix} \equiv 0$$

for all points x, the answer is yes. Obviously, $w = y^{5/4}$ for $x^T \neq O$. ■

Example 10.2: Does $y = f(x) = 2x_1^2 + x_2^2$ possess an extremum over R^2 when x_1, x_2 are connected by the side relation $\hat{g}(x) = 6 - x_1 - x_2 = 0$? From $\hat{g} = 0$ we can obtain $x_2 = \varnothing(x_1) = 6 - x_1$. Forming $F(x_1) = f(x_1, \varnothing(x_1)) = 2x_1^2 + (6 - x_1)^2$ and differentiating yields $F' = 6x_1 - 12$. Setting $F' = 0$ renders the critical value $x_1 = 2$. Then $x_2 = \varnothing(2) = 4$. And with $F'' = 6 > 0$, we see that F has a strong local unconstrained minimum at $x_1^0 = 2$, and thus the constrained minimum of f is at $x_o^T = (2, 4)$.

Let us resolve this problem by operationalizing the discussion underlying Figure 10.2. For $x \in R^2$, if the Jacobian determinant $|\partial(f, \hat{g})/\partial(x_1, x_2)|$ is *not identically zero*, then we may equate it to zero so as to locate a unique point of tangency between a level curve of f and $\hat{g} = 0$. As Figure 10.2 reveals, the coordinates of x_o may be obtained from the simultaneous solution of

$$\left| \frac{\partial(f, \hat{g})}{\partial(x_1, x_2)} \right| = \begin{vmatrix} f_1 & f_2 \\ \hat{g}_1 & \hat{g}_2 \end{vmatrix} = 0, \qquad (10.2)$$

$$\hat{g} = 0.$$

[a necessary condition for f subject to $\hat{g} = 0$ to attain a local extremum]

So given f and $\hat{g} = 0$ from above, (10.2) becomes

$$\begin{vmatrix} f_1 & f_2 \\ \hat{g}_1 & \hat{g}_2 \end{vmatrix} = \begin{vmatrix} 4x_1 & 2x_2 \\ -1 & -1 \end{vmatrix} = 0,$$

$$\hat{g} = 6 - x_1 - x_2 = 0$$

or

$$-4x_1 + 2x_2 = 0$$
$$x_1 + x_2 = 6.$$

The simultaneous solution of this system is $x_o^T = (2, 4)$. ■

Section 10.1 Exercises:

10.1.1 Are the following sets of equations functionally dependent? For $x \in R^2$:

a. $y = g(x) = 4x_1 + x_1 x_2$
   $w = h(x) = x_1 x_2$

b. $y = g(x) = 1 + x_1^{-1}x_2$
   $w = h(x) = x_1 x_2^{-1} + 1$

c. $y = g(x) = x_1^2 - x_2^2$
   $w = h(x) = 2x_1 x_2$

For $x \in R^3$,

d. $y = g(x) = x_1^2 + x_2^2 + x_3^2$
   $w = h(x) = x_1 + x_2 + x_3$
   $z = j(x) = x_1 x_2 + x_2 x_3 + x_1 x_3$

10.1.2 Given $y = g(x)$ and $w = h(x)$, $x \in \mathcal{D} \subset R^2$, let g and h be functions of class $C^{(1)}$ in $\mathcal{D}$. Suppose y and w have a functional relationship between them of the form $F(y, w) = 0$ so that $F[g(x), h(x)] \equiv 0$. Verify that a necessary condition for the existence of a functional relationship between y, w is the vanishing of the Jacobian determinant $|\partial(y, w)/\partial(x_1, x_2)|$ for all $x \in \mathcal{D}$.

10.1.3 Suppose we wish to optimize $y = f(x_1, x_2) = x_1 x_2$ subject to $g(x_1, x_2) = x_1 + x_2 = 10$.

a. Solve this problem by eliminating the variable x_2 (find $x_2 = \varnothing(x_1)$) and optimize $F(x_1) = f(x_1, \varnothing(x_1))$.

b. Obtain the same solution as in (a) by solving system (10.2) for x_1, x_2.

10.2 THE TECHNIQUE OF LAGRANGE

In this section we shall offer an alternative and more flexible procedure for optimizing a real-valued function subject to an equality constraint. We noted above (Figure 10.2 and Equation (10.2)) that if $f(\boldsymbol{x})$, $\boldsymbol{x} \in R^2$, attains a local extremum subject to $\hat{g}(\boldsymbol{x}) = 0$ at a point $\boldsymbol{x}_o \in \mathcal{D}$, with $\mathcal{D} \subset R^2$ an open set, then

$$\frac{f_1^o}{\hat{g}_1^o} = \frac{f_2^o}{\hat{g}_2^o}. \qquad (10.3)$$

What this equality implies is that, at $\boldsymbol{x}_o$, f_1^o is proportional to $\hat{g}_1^0$, and f_2^o is proportional to $\hat{g}_2^o$. In general, for $\boldsymbol{x} \in R^n$, let

$$\frac{f_i^o}{\hat{g}_i^o} = \frac{f_n^o}{\hat{g}_n^o} = \lambda^o, \; i = 1, \ldots, n-1, \qquad (10.4)$$

where the constant of proportionality is λ^o. In this regard we can state

Theorem 10.3: (Necessary Condition). Let the functions $y = f(\boldsymbol{x})$, $\hat{g}(\boldsymbol{x}) = 0$, $\boldsymbol{x} \in R^n$, be defined over an open region $\mathcal{D} \subset R^n$ and of class $C^{(1)}$ throughout some spherical δ – neighborhood of the point $\boldsymbol{x}_o \in \mathcal{D}$. If f has a local extremum subject to $\hat{g} = 0$ at $\boldsymbol{x}_o$, then there exists a scalar λ^o such that

$$\nabla f^o + \lambda^o \nabla \hat{g}^o = \boldsymbol{O} \text{ or } f_i^o + \lambda^o \hat{g}_i^o = 0, \; i = 1, \ldots, n. \qquad (10.5)$$

Proof: Under the conditions of the Implicit Function Theorem we can write $x_n = \varnothing(\hat{\boldsymbol{x}})$, $(\hat{\boldsymbol{x}}, \varnothing(\hat{\boldsymbol{x}})) \in \mathcal{D}$. Then f has a local extremum subject to $\hat{g} = 0$ if and only if $F(\hat{\boldsymbol{x}}) = f(\hat{\boldsymbol{x}}, \varnothing(\hat{\boldsymbol{x}}))$ has an unconstrained local extremum at $\boldsymbol{x}_o$. With

$$F_i = f_i + f_n \varnothing_i, \; i = 1, \ldots, n-1,$$

if F has a local extremum at $\boldsymbol{x}_o$, then

$$F_i(\hat{\boldsymbol{x}}_o) = f_i(\boldsymbol{x}_o) + f_n(\boldsymbol{x}_o)\varnothing_i(\hat{\boldsymbol{x}}_o) = 0, \; i = 1, \ldots, n-1. \qquad (10.6)$$

Since we must restrict our movements throughout $\mathcal{D}$ to those values of $\boldsymbol{x}$ for which $G(\hat{\boldsymbol{x}}) = \hat{g}(\hat{\boldsymbol{x}}, \varnothing(\hat{\boldsymbol{x}})) = 0$, it follows that

$$G_i = \hat{g}_i + \hat{g}_n \varnothing_i = 0, \; i = 1, \ldots, n-1. \qquad (10.7)$$

Then, using (10.6) and (10.7),

$$f_i^o = -f_n^o \varnothing_i^o = -f_n^o(-\hat{g}_i^o/\hat{g}_n^o) = -\hat{g}_i^o(-f_n^o/\hat{g}_n^o),$$
$$\hat{g}_n^o \neq 0, \; i = 1, \ldots, n-1.$$

If we now set $\lambda^o = -f_n^o/\hat{g}_n^o$, then the preceding expression becomes $f_i^o + \lambda^o \hat{g}_i^o$, $i = 1, \ldots, n-1$. Alternatively, for $\lambda^o = -f_i^o/\hat{g}_i^o$, $\hat{g}_i^o \neq 0$, we have $f_n^o + \lambda^o \hat{g}_n^o = 0$. Q. E. D.

Note that in defining λ^o, as either $\lambda^o = -f_n^o/\hat{g}_n^o$ or $\lambda^o = -f_i^o/\hat{g}_i^o$, $i = 1, \ldots, n-1$, it is assumed that at least one of the partial derivatives of $\hat{g} = 0$ is different from zero at $\boldsymbol{x}_o$. This restriction on G will be termed a **constraint qualification** (CQ). In this regard, λ^o*will exist* if at least one of the components of $\nabla \hat{g}$ does not vanish at $\boldsymbol{x}_o$.

As this theorem has revealed, a necessary condition for f subject to $\hat{g} = 0$ to have a local extremum at $\boldsymbol{x}_o$ is

$$\begin{aligned} \nabla f(\boldsymbol{x}_o) + \lambda^o \nabla \hat{g}(\boldsymbol{x}_o) &= \boldsymbol{O}, \\ \hat{g}(\boldsymbol{x}_o) &= 0. \end{aligned} \qquad (10.8)$$

How might an equation system such as this arise? Since the partial derivatives f_i are, respectively, proportional to $\hat{g}_i$, $i = 1, \ldots, n$, let us add to f some multiple λ of $\hat{g} = 0$. That is, we shall form the function

$$L(\boldsymbol{x}, \lambda) = f(\boldsymbol{x}) + \lambda \hat{g}(\boldsymbol{x}), \qquad (10.9)$$

called the **Lagrangian of f**, where L depends upon both $\boldsymbol{x}$ and λ. This is because λ, termed the **Lagrange multiplier**, must be chosen so that (10.5) holds. Then if we differentiate (10.9) with respect to $\boldsymbol{x}$ and λ, we obtain (10.8). The essential feature of the technique of Lagrange is that we may convert a problem involving the optimization of a function f subject to an equality constraint $\hat{g} = 0$ into one without any restriction on the variables x_i, $i = 1, \ldots, n$, by forming the Lagrangian function of f, $L = f + \lambda \hat{g}$. Then optimizing f subject to $\hat{g} = 0$ is equivalent to optimizing the unconstrained Lagrangian as long as $\hat{g} = 0$ is satisfied for the given

λ. And this may be accomplished by deriving system (10.8), the simultaneous solution of which yields x_o, λ^o.

Let us examine the geometry of Equation (10.5). From this expression we can write

$$\nabla f^o = -\lambda^o \nabla \hat{g}^o \text{ or } \nabla f^o = \lambda^o \nabla g^o \qquad (10.5.1)$$

since $\hat{g} = b - g = 0$. So if (10.5.1) holds, then the vectors ∇f^o and ∇g^o, which are *orthogonal* to the tangent lines to the contours of f and g, respectively, are **collinear** or point in the *same direction* (in the case of a constrained maximum) or in *opposite directions* (for a constrained minimum) at $\boldsymbol{x}_o$.

Before we look to the development of criteria with which to distinguish a constrained maximum from a constrained minimum at a point where the above necessary condition holds, let us first develop a bit of terminology and a supporting theorem.

Definition 10.4: Let the functions $y = f(\boldsymbol{x})$, $w = g(\boldsymbol{x})$, $\boldsymbol{x} \in R^n$, be defined over an open region $\mathcal{D} \subset R^n$ The $(n + 1)$ st-order **bordered Hessian matrix** of f is represented, at those points $\boldsymbol{x}$ where g is of class $C^{(1)}$ and f is of class $C^{(2)}$, as

$$\boldsymbol{H}_f^g(\boldsymbol{x}) = \left[\begin{array}{c:c} \boldsymbol{H}_f & \nabla g \\ \hdashline \nabla g^T & 0 \end{array}\right] = \begin{bmatrix} f_{11} & f_{12} & \cdots & f_{1n} & g_1 \\ f_{21} & f_{22} & \cdots & f_{2n} & g_2 \\ \vdots & \vdots & & \vdots & \vdots \\ f_{n1} & f_{n2} & & f_{nn} & g_n \\ g_1 & g_2 & \cdots & g_n & 0 \end{bmatrix}.$$

(Here the Hessian matrix of f, $\boldsymbol{H}_f$, has been bordered by the first partial derivatives of g. In addition, $\boldsymbol{H}_f^g$ is symmetric since $\boldsymbol{H}_f$ is symmetric.)

We next have a necessary and sufficient condition for the sign definiteness of a quadratic from $Q(\boldsymbol{x})$ subject to a linear constraint in $\boldsymbol{x}$ [Debreu (1952)].

Theorem 10.4: (Necessary and Sufficient Condition). The quadratic from $\boldsymbol{x}^T\boldsymbol{A}\boldsymbol{x}$ subject to $\boldsymbol{B}^T\boldsymbol{x} = 0$ is:

a. **positive definite** (> 0) for every $\boldsymbol{x} \neq \boldsymbol{O}$ if and only if

$$\Delta_r = \left|\begin{array}{c:c} \boldsymbol{A}_{rr} & \boldsymbol{B}_r \\ \hdashline \boldsymbol{B}_r^T & 0 \end{array}\right| < 0, r = 2, \ldots, n \ ; \qquad (10.10)$$

b. **negative definite** (< 0) for every $\boldsymbol{x} \neq \boldsymbol{O}$ if and only if

$$\Delta_r = (-1)^r \left|\begin{array}{c:c} \boldsymbol{A}_{rr} & \boldsymbol{B}_r \\ \hdashline \boldsymbol{B}_r^T & 0 \end{array}\right| > 0, r = 2, \ldots, n \ . \qquad (10.11)$$

(Here $\boldsymbol{B}_r$ is obtained from $\boldsymbol{B}$ by retaining only the first r components of $\boldsymbol{B}$, while $\boldsymbol{A}_{rr}$ is obtained from $\boldsymbol{A}$ by keeping only the first r rows and columns of $\boldsymbol{A}$, $r = 2, \ldots, n$.)

Example 10.3: Determine the sign definiteness of $\boldsymbol{x}^T\boldsymbol{A}\boldsymbol{x}$ subject to $\boldsymbol{B}^T\boldsymbol{x} = 3x_1 + 2x_2 + x_3 = 0$ when

$$\boldsymbol{A} = \begin{bmatrix} 1 & 4 & 0 \\ 2 & 1 & 2 \\ 0 & 1 & 0 \end{bmatrix}.$$

Calculating

$$\left|\begin{array}{c:c} \boldsymbol{A}_{22} & \boldsymbol{B}_2 \\ \hdashline \boldsymbol{B}_2^T & 0 \end{array}\right| = \begin{vmatrix} 1 & 4 & 3 \\ 2 & 1 & 2 \\ 3 & 2 & 0 \end{vmatrix} = 23, \left|\begin{array}{c:c} \boldsymbol{A}_{33} & \boldsymbol{B}_3 \\ \hdashline \boldsymbol{B}_3^T & 0 \end{array}\right| = \begin{vmatrix} 1 & 4 & 0 & 3 \\ 2 & 1 & 2 & 2 \\ 1 & 0 & 1 & 1 \\ 3 & 2 & 1 & 0 \end{vmatrix} = -7$$

we see that, from part (b) of Theorem 10.4, $\Delta_2 = 23 > 0$, $\Delta_3 = 7 > 0$ so that $\boldsymbol{x}^T\boldsymbol{A}\boldsymbol{x}$ subject to $\boldsymbol{B}^T\boldsymbol{x} = 0$ is negative definite. ■

Theorem 10.4 is the basis of

Theorem 10.5: (Sufficient Condition). Let $y = f(\boldsymbol{x})$, $\hat{g}(\boldsymbol{x}) = 0$, $\boldsymbol{x} \in R^n$, be of class $C^{(2)}$ throughout a spherical δ−neighborhood of the point $\boldsymbol{x}_o$, $\delta(\boldsymbol{x}_o)$, contained within an open region $\mathcal{D} \subset R^n$. Then the Lagrangian $L(\boldsymbol{x}, \lambda) = f(\boldsymbol{x}) + \lambda\hat{g}(\boldsymbol{x})$ is of class $C^{(2)}$ throughout $\delta(\boldsymbol{x}_o)$ also. If $\nabla f^o + \lambda^o \nabla \hat{g}^o = \boldsymbol{O}$, $L_\lambda^o = 0$, and

$$\left|\begin{array}{c:c} \boldsymbol{H}_L(\boldsymbol{x}_o, \lambda_o)_{rr} & \nabla \hat{g}(\boldsymbol{x}_o)_r \\ \hdashline \nabla \hat{g}(\boldsymbol{x}_o)_r^T & 0 \end{array}\right| \neq 0, r = 2, \ldots, n,$$

then L, and thus f subject to $\hat{g} = 0$, has a strong local extremum at $\boldsymbol{x}_o$. Moreover:

a. if

$$\Delta_r = (-1)^r \left|\begin{array}{c:c} \boldsymbol{H}_L(\boldsymbol{x}_o, \lambda_o)_{rr} & \nabla \hat{g}(\boldsymbol{x}_o)_r \\ \hdashline \nabla \hat{g}(\boldsymbol{x}_o)_r^T & 0 \end{array}\right| > 0, r = 2, \ldots, n, \qquad (10.12)$$

then L, and thus f subject to $\hat{g} = 0$, has a strong local maximum at $\boldsymbol{x}_o$; but

b. if

$$\Delta_r = \left|\begin{array}{c|c} \boldsymbol{H}_L(\boldsymbol{x}_o,\lambda_o)_{rr} & \nabla\hat{g}(\boldsymbol{x}_o)_r \\ \hline \nabla\hat{g}(\boldsymbol{x}_o)_r^T & 0 \end{array}\right| < 0, r = 2,\ldots,n, \quad (10.13)$$

then L, and thus f subject to $\hat{g} = 0$, has a strong local minimum at $\boldsymbol{x}_o$

Proof: To indicate the conditions under which the Lagrangian of f attains a strong local extremum at $\boldsymbol{x}_o$, let us write the second-order Taylor expansion of $L(\boldsymbol{x}, \lambda)$ about $\boldsymbol{x}_o$, with λ^o fixed, as

$$\begin{aligned} L(\boldsymbol{x}_o + \boldsymbol{h}, \lambda^o) &= L(\boldsymbol{x}_o, \lambda^o) + \nabla L(\boldsymbol{x}_o, \lambda^o)^T\boldsymbol{h} \\ &+ \frac{1}{2!}\boldsymbol{h}^T\boldsymbol{H}_L(\boldsymbol{x}_o + \theta\boldsymbol{h}, \lambda^o)\boldsymbol{h},\ 0 < \theta < 1. \end{aligned} \quad (10.14)$$

By hypothesis, $\nabla f^o + \lambda^o\nabla\hat{g}_o = \boldsymbol{O}$ so that (10.14) becomes

$$\begin{aligned} &L(\boldsymbol{x}_o + \boldsymbol{h}, \lambda^o) - L(\boldsymbol{x}_o, \lambda^o) \\ &= \frac{1}{2!}\boldsymbol{h}^T\boldsymbol{H}_L(\boldsymbol{x}_o + \theta\boldsymbol{h}, \lambda^o)\boldsymbol{h},\ 0 < \theta < 1. \end{aligned} \quad (10.14.1)$$

With the second-order partial derivatives of L continuous at $\boldsymbol{x}_o$, the right-hand side of (10.14.1) approaches $\frac{1}{2!}\boldsymbol{h}^T\boldsymbol{H}_L(\boldsymbol{x}_o, \lambda^o)\boldsymbol{h}$ when $\theta \to 0$ and $\|\boldsymbol{h}\| < \delta$. Hence, under this *persistence of sign* property of the second-order partial derivatives of L near $\boldsymbol{x}_o$, the sign of $L(\boldsymbol{x}_o + \boldsymbol{h}, \lambda^o) - L(\boldsymbol{x}_o, \lambda^o)$ depends upon the sign of $\boldsymbol{h}^T\boldsymbol{H}_L(\boldsymbol{x}_o, \lambda^o)\boldsymbol{h}$, provided that the L_{ij}, $i, j = 1, \ldots, n$, are not all zero. However, the arguments of this quadratic form are not independent, i.e., we must determine the sign definiteness of $\boldsymbol{h}^T\boldsymbol{H}_L(\boldsymbol{x}_o, \lambda^o)\boldsymbol{h}$ for those $\boldsymbol{h}$ satisfying $\nabla\hat{g}(\boldsymbol{x}_o)^T\boldsymbol{h} = 0$. In this light, part (b) of Theorem 10.4 informs us that if Equation (10.12) holds, then $L(\boldsymbol{x}_o + \boldsymbol{h}, \lambda^o) - L(\boldsymbol{x}_o, \lambda^o) < 0$ so that L, and thus f subject to $\hat{g} = 0$, has a strong local maximum at $\boldsymbol{x}_o$. Additionally, part (a) of Theorem 10.4 reveals that if Equation (10.13) holds, $L(\boldsymbol{x}_o + \boldsymbol{h}, \lambda^o) - L(\boldsymbol{x}_o, \lambda^o) > 0$ so that L, and thus f subject to $\hat{g} = 0$, has a strong local minimum at $\boldsymbol{x}_o$. Q. E. D.

Example 10.4: Optimize $y = f(\boldsymbol{x}) = 3x_1^2 + x_1x_2 + 4x_2^2$, $\boldsymbol{x} \in R^2$, subject to $3x_1 + x_2 = 6$. From

$$\begin{aligned} L(\boldsymbol{x}, \lambda) &= f(\boldsymbol{x}) + \lambda\hat{g}(\boldsymbol{x}) = 3x_1^2 \\ &+ x_1x_2 + 4x_2^2 + \lambda(6 - 3x_1 - x_2) \end{aligned}$$

we get the linear equation system

$$\begin{aligned} L_1 &= 6x_1 + x_2 - 3\lambda = 0 \\ L_2 &= x_1 + 8x_2 - \lambda = 0 \\ L_\lambda &= 6 - 3x_1 - x_2 = 10, \end{aligned}$$

the simultaneous solution of which is $\boldsymbol{x}_o^T = (1.92, 0.24)$, $\lambda^o = 3.92$. And since

$$\left|\begin{array}{c|c} \boldsymbol{H}_L(\boldsymbol{x}_o,\lambda^o)_{22} & \nabla\hat{g}(\boldsymbol{x}_o)_2 \\ \hline \nabla\hat{g}(\boldsymbol{x}_o)_2^T & 0 \end{array}\right| = \begin{vmatrix} 6 & 1 & -3 \\ 1 & 8 & -1 \\ -3 & -1 & 0 \end{vmatrix} = -72 < 0,$$

we may conclude that f subject to $\hat{g} = 0$ attains a strong local minimum at $\boldsymbol{x}_o$. ■

Example 10.5: Optimize $y = f(\boldsymbol{x}) = 2x_1 + x_2 + 6$ subject to $\hat{g}(\boldsymbol{x}) = 3 - x_1 - 2x_2^2$. Starting with

$$L(\boldsymbol{x}, \lambda) = f(\boldsymbol{x}) + \lambda\hat{g}(\boldsymbol{x}) = 2x_1 + x_2 + 6 + \lambda(3 - x_1 - 2x_2^2)$$

we obtain

$$\begin{aligned} L_1 &= 2 - \lambda = 0 \\ L_2 &= 1 - 4\lambda x_2 = 0 \\ L_\lambda &= 3 - x_1 - 2x_2^2 = 0. \end{aligned}$$

Solving this (nonlinear) system simultaneously yields $\boldsymbol{x}_o^T = (2.97, 0.13)$ with $\lambda^o = 2$. And with

$$\left|\begin{array}{c|c} \boldsymbol{H}_L(\boldsymbol{x}_o,\lambda^o)_{22} & \nabla\hat{g}(\boldsymbol{x}_o)_2 \\ \hline \nabla\hat{g}(\boldsymbol{x}_o)_2^T & 0 \end{array}\right| = \begin{vmatrix} 0 & 0 & -1 \\ 0 & -4\lambda^o & -4x_2^o \\ -1 & -4x_2^o & 0 \end{vmatrix} = 8 > 0,$$

we see that f has a strong local constrained maximum at $\boldsymbol{x}_o$.

Example 10.6: For $\boldsymbol{x} \in R^3$, optimize $y = f(\boldsymbol{x}) = 3x_1 + 4x_3 - x_1^2 - x_2^2$ subject to $x_1 + x_2 + x_3 = 6$. From

$$L(\boldsymbol{x}, \lambda) = f(\boldsymbol{x}) + \lambda \hat{g}(\boldsymbol{x}) = 3x_1 + 4x_3 - x_1^2 - x_2^2 + \lambda(6 - x_1 - x_2 - x_3)$$

we derive the simultaneous linear equation system

$$\begin{aligned} L_1 &= 3 - 2x_1 - \lambda = 0 \\ L_2 &= -2x_2 - \lambda = 0 \\ L_3 &= 4 - \lambda = 0 \\ L_\lambda &= 6 - x_1 - x_2 - x_3 = 0 \end{aligned}$$

with solution $\boldsymbol{x}_o^T = \left(-\frac{1}{2}, -2, \frac{13}{2}\right)$, $\lambda^o = 4$. Additionally,

$$\begin{vmatrix} H_L(\boldsymbol{x}_o, \lambda^o)_{22} & \nabla \hat{g}(\boldsymbol{x}_o)_2 \\ \nabla \hat{g}(\boldsymbol{x}_o)_2^T & 0 \end{vmatrix} = \begin{vmatrix} -2 & 0 & -1 \\ 0 & -2 & -1 \\ -1 & -1 & 0 \end{vmatrix} = 4 > 0,$$

$$\begin{vmatrix} H_L(\boldsymbol{x}_o, \lambda^o)_{33} & \nabla \hat{g}(\boldsymbol{x}_o)_3 \\ \nabla \hat{g}(\boldsymbol{x}_o)_3^T & 0 \end{vmatrix} = \begin{vmatrix} -2 & 0 & 0 & -1 \\ 0 & -2 & 0 & -1 \\ 0 & 0 & 0 & -1 \\ -1 & -1 & -1 & 0 \end{vmatrix} = -4 < 0.$$

Since these bordered principal minors of the bordered Hessian of the Lagrangian alternate in sign, starting positive, we may conclude that f subject to $\hat{g} = 0$ has a strong local maximum at $\boldsymbol{x}_o$.

Section 10.2 Exercises:

10.2.1 Use the technique of Lagrange to

a. Optimize $y = f(\boldsymbol{x}) = 2x_1 - x_2 - x_1 x_2$ subject to $g(\boldsymbol{x}) = x_1 + x_2 = 8$.

b. Optimize $y = f(\boldsymbol{x}) = 2x_1^2 + 2x_1 x_2 + x_2^2$ subject to $g(\boldsymbol{x}) = x_1 - 2x_2 = 10$.

c. Optimize $y = f(\boldsymbol{x}) = x_1 - 2x_2 + 2x_3$ subject to $g(\boldsymbol{x}) = x_1^2 + x_2^2 + x_3^2 = 9$.

10.2.2 Suppose we form the Lagrangian for problem 10.2.1a as $L = f + \lambda(g(x_1, x_2) - 8)$. What is the impact of this respecification of the Lagrangian on the solution to this problem?

10.3 INTERPRETATION OF THE LAGRANGE MULTIPLIER

As the preceding section has indicated, a necessary condition for f to attain an extremum subject to $\hat{g} = 0$ is that

$$f_i + \lambda \hat{g}_i = 0,\ i = 1, \ldots, n, \qquad \hat{g} = 0.$$

And once this equation system is solved simultaneously, we obtain $\boldsymbol{x}_o$, λ^o, and $f(\boldsymbol{x}_o) = f^o$. Let us now consider the following question. What is the effect on the optimal value of f, f^o, of a *small* relaxation in or tightening of the constraint? To answer this question, we must provide an interpretation of λ^o. Since $\hat{g} = b - g = 0$, $db - dg = 0$ or

$$db = dg = \nabla g^T \boldsymbol{h}. \tag{10.15}$$

Evaluating this expression subject to the above necessary condition holding renders, from $f_i + \lambda \hat{g}_i = f_i - \lambda g_i = 0$ or $g_i = f_i / \lambda$, $i = 1, \ldots, n$,

$$db = \sum_{i=1}^{n} (f_i / \lambda)\, h_i = \frac{1}{\lambda} df \tag{10.15.1}$$

and thus, at $\boldsymbol{x}_o$,

$$\lambda^o = df^o / db, \tag{10.16}$$

the total derivative of f^o with respect to b, i.e., λ^o is the marginal or incremental change in f^o with respect to b. In this regard, we may view λ^o as a relative measure of the *force* of the constraint, i.e., it indicates how tightly the constraint is binding. For instance, suppose f^o represents the strong local maximum value of f subject to $\hat{g} = 0$:

a. If $\lambda^o > 0$, a unit increase in b is positively valued since one obtains a higher maximum value of f. In fact, the increase in f^o is exactly equal to λ^o since $df^o = \lambda^o(1) > 0$. Hence, λ^o is the *marginal gain* in f^o due to a one-unit relaxation in the constraint. If b is decreased by one unit, then f must also decrease to a new optimum level with the amount of the decrease being determined by λ^o or $df^o = \lambda^o(-1) < 0$. Thus λ^o may be considered as the *marginal cost*, in terms of f^o, of a tightening of the constraint.

b. if $\lambda^o < 0$, a unit decrease in b is now positively valued i.e., the optimal value of the objective function increases. In this instance the *marginal gain* in f^o due to a tightening of the constraint by one unit is determined by λ^o as $df^o = \lambda^o(-1) > 0$. If b is increased by one unit, the *marginal cost*, in terms of f^o, of a loosening of the constraint is $df^o = \lambda^o(1) < 0$.

In the special case where $\lambda^o \neq 0$ and $db = \pm 1$, λ^o is called the *marginal imputed value of the constraint*, e.g., if f^o denotes the strong local constrained maximum of f, then λ^o, the marginal imputed value of the constraint, is, for $\lambda^o > 0$, the minimum reduction (maximum increase) in f^o precipitated by a one-unit decrease (increase) in b. Alternatively, λ^o may be referred to as a *shadow price* since, for $\lambda^o > 0$, an extra unit of b is worth, at most λ^o units of f. (For $\lambda^o < 0$, one would be willing to pay up to λ^o units of f for the withdrawal of a unit of b.)

Example 10.7: In Example 10.5 we determined that f assumes a strong local constrained maximum at $\boldsymbol{x}_o^T = (2.97, 0.13)$ with $\lambda^o = 2$ and $f^o = 12.07$. Suppose we tighten the constraint by one unit ($db = -1$) so that the new constraint is $\hat{g} = 2 - x_1 - 2x_2^2$. What is the effect of this tightening on the optimal value of f? Since $df^o = \lambda^o db = 2(-1) = -2$, we see that the maximum value of f subject to the modified constraint is $f^o + df^o = 12.07 - 2 = 10.07$.

Let us take another look at the derivation of Equation (10.16). Suppose that the optimization problem at hand is

$$\text{maximize } y = f(\boldsymbol{x}) \text{subject to} g(\boldsymbol{x}) = b.$$

Then the optimal value function for this problem is

$$f^o(b) = \max\{f(\boldsymbol{x}) | g(\boldsymbol{x}) = b\},$$

which is assumed of class $C^{(1)}$. (For the minimization case, $f^o(b) = \min\{f(\boldsymbol{x}) | g(\boldsymbol{x}) = b\}$.) Here $f^o(b)$ is the highest level of f attainable given the restriction $g(\boldsymbol{x}) = b$.

For $b = \bar{b}$, let $\bar{\boldsymbol{x}}$ be the corresponding solution to the problem: maximize f subject to $g = \bar{b}$. Then $f(\bar{\boldsymbol{x}}) = f^o(\bar{b})$ and, for all $\boldsymbol{x}$, $f(\boldsymbol{x}) \leq f^o(\bar{b}) = f^o(g(\boldsymbol{x}))$. Hence, $\eta(\boldsymbol{x}) = f(\boldsymbol{x}) - f^o(g(\boldsymbol{x}))$ has a maximum at $\boldsymbol{x} = \bar{\boldsymbol{x}}$ so that

$$\eta_i(\bar{\boldsymbol{x}}) = f_i(\bar{\boldsymbol{x}}) - \left.\frac{df^o(\bar{b})}{db}\right|_{b=g(\bar{\boldsymbol{x}})} g_i(\bar{\boldsymbol{x}}) = 0,\ i = 1, \ldots, n.$$

If we define $\bar{\lambda} = df^o(\bar{b})/db$, then the preceding equation system reduces to $f_i(\bar{\boldsymbol{x}}) - \bar{\lambda} g_i(\bar{\boldsymbol{x}}) = 0$, $i = 1, \ldots, n$, our necessary condition for a constrained extremum at $\bar{\boldsymbol{x}}$. Hence, we may think of the optimal Lagrange multiplier as the marginal change in the optimal value function with respect to the constraint constant.

Example 10.8: Let us maximize $y = f(x_1, x_2) = x_1 x_2$ subject to $g(x_1, x_2) = 4x_1 + 3x_2 = b$ (= constant). Forming the Lagrangian function we have $L(x_1, x_2, \lambda) = f + \lambda\hat{g} = x_1x_2 + \lambda(b - 4x_1 - 3x_2)$. Then from

$$L_1 = x_2 - 4\lambda = 0$$
$$L_2 = x_1 - 3\lambda = 0$$
$$L_\lambda = b - 4x_1 - 4x_2 = 0$$

we obtain $\lambda^o = b/24$ and thus $x_1^o = 3\lambda^o = b/8$ and $x_2^o = 4\lambda^o = b/6$. And, with the bordered Hessian determinant of L equal to

$$\begin{vmatrix} 0 & 1 & -4 \\ 1 & 0 & -3 \\ -4 & -3 & 0 \end{vmatrix} = 24 > 0,$$

we see that f has a constrained local maximum at (x_1^o, x_2^o).

Using the optimal x_1^o and x_2^o values we can readily form the optimal value function $f^o(b) = x_1^o x_2^o = b^2/48$. Hence, $df^o(b)/db = b/24 = \lambda^o$ as expected.

Section 10.3 Exercises:

10.3.1 Determine the optimal value function for problem 10.2.1a. Let $L = f + \lambda\hat{g} = 2x_1 - x_2 - x_1x_2 + \lambda(b - x_1 - x_2)$.

10.3.2 Suppose we interchange the roles of the constraint function and the original function to be optimized. That is, instead of, say, maximizing $y = f(\boldsymbol{x})$ subject to $\hat{g} = b - g(\boldsymbol{x}) = 0$, let us minimize $g(\boldsymbol{x})$ subject to $\hat{f} = y - f(\boldsymbol{x}) = 0$. This new problem will be termed the **transposed problem** with **Lagrangian transpose** $M(\boldsymbol{x}, \mu) = g(\boldsymbol{x}) + \mu(y - f(\boldsymbol{x}))$, where μ is the Lagrange multiplier. Hence, maximizing (minimizing) $L(\boldsymbol{x}, \lambda)$ is equivalent to minimizing (maximizing) $M(\boldsymbol{x}, \mu)$, where $\mu^o = dg^o/dy = 1/\lambda^o$.

Suppose we want to maximize $y = f(\boldsymbol{x}) = x_1 x_2$ subject to $g(\boldsymbol{x}) = x_1 + x_2 = 10$. Here the optimal solution is $\boldsymbol{x}_o^T = (5, 5)$, $\lambda^o = 5$, $f^o = 25$, the bordered Hessian determinant is equal to 2, and the optimal value function is $f^o(b) = b^2/4$. (So for $b = 10$, $\partial f^o(b)/\partial b = b/2 = 5 = \lambda^o$ as required.) Find the optimal value for the transposed problem and demonstrate that $\mu^o = 1/\lambda^o$.

10.4 CONSTRAINED EXTREMA: *M* EQUALITY CONSTRAINTS IN *N* INDEPENDENT VARIABLES

Our objective in this section is to generalize the technique of Lagrange to the case where a real-valued function in n variables $x_1, \ldots, x_n$ is optimized subject to m equality constraints in those variables. In this regard, we seek to optimize $y = f(\boldsymbol{x})\colon R^n \rightarrow R$ subject to $b_j = g^j(\boldsymbol{x})$ or $\hat{g}^j(\boldsymbol{x}) = b_j - g^j(\boldsymbol{x}) = 0$, $j = 1, \ldots, m$, where we require that $m < n$. (Let us first assume that $m = n$. Since each function $\hat{g}^j$ defines a surface in R^n and the intersection of n surfaces in R^n specifies a single point $\{\boldsymbol{x}^*\} = \bigcap_{j=1}^{n} \{\boldsymbol{x} | \hat{g}^j(\boldsymbol{x}) = 0\}$, it becomes trivial to optimize f subject to n constraints. But if $m < n$, m surfaces in R^n intersect to specify a region or surface (not just a single point) of dimensionality $n - m$, all of whose points are boundary points.)

Before considering a set of optimality conditions for a constrained extremum of f, let us examine the following problem. Given a set of m equations $\hat{g}^j(\boldsymbol{x}) = 0$, $\boldsymbol{x} \in R^n$, $j = 1, \ldots, m$, with $m < n$, under what circumstances can we use these equations to express m of the n variables as functions of the remaining $n - m$ variables? That is, we seek to determine the existence of m functions φ^j, $j = 1, \ldots, m$, such that we can solve for (eliminate) the variables x_j, $j = 1, \ldots, m$, in terms of $x_{m+1}, \ldots, x_n$ as

$$x_j = \varphi^j(x_{m+1}, \ldots, x_n),\ j = 1, \ldots, m.$$

The answer to this question is provided by a necessary and sufficient condition for the existence of the functions φ^j, $j = 1, \ldots, m$, namely the generalized existence or Implicit Function Theorem.

Theorem 10.6: (**Implicit Function Theorem**)[Taylor and Mann (1972)]. Let the functions $\hat{g}^j(\boldsymbol{x}) = 0$, $\boldsymbol{x} \in R^n$, $j = 1, \ldots, m$, $m < n$, be defined over a region $\mathcal{D} \subset R^n$ and let $\boldsymbol{x}_o^T = (x_1^0, \ldots, x_n^o)$ be a point interior to $\mathcal{D}$. If:

1. the functions $g^j(\boldsymbol{x}) = 0$, $j = 1, \ldots, m$, are continuous throughout some spherical δ – neighborhood of $\boldsymbol{x}_o$, $\delta(\boldsymbol{x}_o)$;
2. $g^j(\boldsymbol{x}) = 0$, $j = 1, \ldots, m$; and
3. the rank of the $(n \times m)$ matrix

$$\nabla \mathbf{G}(\boldsymbol{x}_o) = [\nabla \hat{g}^1(\boldsymbol{x}_o), \ldots, \nabla \hat{g}^n(\boldsymbol{x}_o)]$$

equals m (the CQ), where $\mathbf{G}(\boldsymbol{x})$ is an $(m \times 1)$ vector of implicit constraint functions $\hat{g}^j(\boldsymbol{x}) = 0$, $j = 1, \ldots, m$, then there exists a $\bar{\delta}$ – neighborhood of $\hat{\boldsymbol{x}}_o^T = (x_{k_1}^o, \ldots, x_{k_{n-m}}^o) \in R^{n-m}$ (here $k_1, \ldots, k_{n-m}$ is the set of indices from $1, \ldots, n$ on the $n - m$ variables *not associated with* the m linear independent rows of $\nabla \mathbf{G}(\boldsymbol{x}_o)$) such that, for every $\hat{\boldsymbol{x}}^T = (x_{k_1}, \ldots, x_{k_{n-m}}) \in \bar{\delta}(\hat{\boldsymbol{x}}_o)$, there exists a set of functions

$$x_{k_j} = \varphi^j\left(x_{k_1}, \ldots, x_{k_{n-m}}\right),\ j = 1, \ldots, m$$

(k_j, $j = 1, \ldots, m$, is the set of indices from $1, \ldots, n$ on the m variables *associated with* the m linearly independent rows of $\nabla \mathbf{G}(\boldsymbol{x}_o)$) which are single-valued and continuous functions of $\hat{\boldsymbol{x}}$ and which possess the properties:

a. $x_{k_j}^o = \varphi^j(\hat{\boldsymbol{x}}_o)$, $j = 1, \ldots, m$;
b. for any $\hat{\boldsymbol{x}} \in \bar{\delta}(\hat{\boldsymbol{x}}_o)$, the values of x_{k_j}, $j = 1, \ldots, m$, computed from $x_{k_j} = \varphi^j(\hat{\boldsymbol{x}})$, along with $\hat{\boldsymbol{x}}$, yield an $\boldsymbol{x}$ which satisfies $\hat{g}(\boldsymbol{x}) = 0$, $j = 1, \ldots, m$; and
c. $\varphi^j(\hat{\boldsymbol{x}})$ are of class $C^{(1)}$ throughout $\bar{\delta}(\hat{\boldsymbol{x}}_o)$.

The gist of the condition $\rho(\nabla \mathbf{G}(\boldsymbol{x}_o)) = m$ is that the equations $\hat{g}^j(\boldsymbol{x}_o) = 0$, $j = 1, \ldots, m$, are *not functionally dependent*, i.e., none of them can be expressed in terms of any (or all) of the others.

It is instructive to briefly examine the conditions under which functional dependence holds. Specifically, we have

Theorem 10.7: (Necessary and Sufficient Condition). The m functions of n independent variables $m \le n$, are **functionally dependent** in a region $\mathcal{D} \subset R^n$ if every mth-order Jacobian determinant of these m functions with respect to m of the variables vanishes identically in $\mathcal{D}$.

Hence, the requirement that $\rho(\nabla G(x_o)) = m$ implies that the functions $\hat{g}^j(x) = 0, j = 1, \ldots, m$, are functionally independent since, with $\rho(\nabla G(x_o)) = \rho(\nabla G(x_o)^T) = m$, the existence of at least one non-vanishing mth-order Jacobian determinant of $\nabla G(x_o)^T$ is established.

Example 10.9: Given the real-valued functions

$$\hat{g}^1 = 4 - 2x_1x_2 + x_3 = 0$$
$$\hat{g}^2 = 2 - x_1^2x_2 + x_3 = 0,$$

can we solve for any two of the variables x_1, x_2 or x_3 in terms of the third? From $G^T = (\hat{g}^1, \hat{g}^2)$,

$$\nabla G = [\nabla\hat{g}^1, \nabla\hat{g}^2] = \begin{bmatrix} -2x_2 & -2x_1x_2 \\ -2x_1 & -x_1^2 \\ 1 & 1 \end{bmatrix}$$

so that we may find, for example,

$$\begin{vmatrix} -2x_2 & -2x_1x_2 \\ -2x_1 & -x_1^2 \end{vmatrix} = -2x_1^2x_2 0, \not\equiv 0,$$

thus implying that $\rho(\nabla G) = 2$. Hence, both x_1 and x_2 may be eliminated (since they are the variables associated with the two linearly independent rows of ∇G) or

$$x_1 = \varphi^1(x_3) = 2(2 + x_3)(4 + x_3)^{-1}$$
$$x_2 = \varphi^2(x_3) = \tfrac{1}{4}(2 + x_3)^{-1}(4 + x_3)^2$$

as the reader can readily verify. Can any other combination of two variables be expressed in terms of the third?

Once the existence of the function $\varphi^j, j = 1, \ldots, m$, has been determined, we can represent the variables $x_1, \ldots, x_m$ (we may assume, without loss of generality, that the first m variable have been eliminated) as

$$x_j = \varphi^j(\hat{x}), \hat{x} \in \delta(\hat{x}_o), j = 1, \ldots, m.$$

Then the function

$$F(\hat{x}) = f[\varphi^1(\hat{x}), \ldots, \varphi^m(\hat{x})]$$

has an unconstrained local extremum at $\hat{x}_o$, and thus, at x_o. It should be apparent that optimizing $F(\hat{x})$ directly may be an extremely difficult task – it may not be easy to eliminate the m variables if m is large and/or the$\hat{g}^j$ functions have a complicated structure. As an alternative, we can utilize the generalized technique of Lagrange to optimize $f(x)$ subject to $\hat{g}^j(x) = 0, j = 1, \ldots, m$.

Section 10.4 Exercises:

10.4.1 Given the following constraint functions, can x_2 and x_3 be expressed in terms of x_1?

$$\hat{g}^1 = 6 - x_1x_3 - x_2 = 0$$
$$\hat{g}^2 = 4 - x_1^2 - x_2^2 - x_3 = 0.$$

10.4.2 Optimize $y = f(x) = x_1^2 + 2x_1x_2 + x_2x_3$ subject to $g^1(x) = x_1 + 2x_1 = 20$, $g^2(x) = x_1 + x_3 = 10$.

10.4.3 Determine the optimal value function for Exercise 10.4.2. Use it to find λ_1^o, λ_2^o.

10.5 THE GENERALIZED TECHNIQUE OF LAGRANGE

In what follows we shall search exclusively for *local* constrained extrema. To this end, let $y = f(x)$, $\hat{g}^j(x) = 0, j = 1, \ldots, m < n$, be real-valued functions defined on an open region $\mathcal{D} \subset R^n$ with $\mathcal{G} = \{x | \hat{g}^j(x) = 0, j = 1, \ldots, m < n\}$. We then have

Definition 10.5: The real-valued function $y = f(x)$ subject to $\hat{g}^j(x) = 0, x \in R^n, j = 1, \ldots, m < n$, has a **constrained local maximum (minimum)** at $x_o \in \mathcal{G} \cap \mathcal{D}$ if there is some spherical δ – neighborhood $\delta(x_o) \in \mathcal{D}$ about x_o such that $f(x_o) \ge f(x)(\le f(x))$ for all points $x \in \delta(x_o) \cap \mathcal{G} \cap \mathcal{D}$.

If $f(x_o) \ge f(x)(\le f(x))$ and $f(x_o) = f(x)$ for at least one point $x(\ne x_o) \in \delta(x_o) \cap \mathcal{G} \cap \mathcal{D}$, then f has a **constrained weak local maximum (minimum)** at x_o; while if $f(x_o) > f(x)(< f(x))$ for all $x(\ne x_o) \in \delta(x_o) \cap$

$\mathcal{G} \cap \mathcal{D}$, then f is said to have a **constrained strong local maximum (minimum)** at $\boldsymbol{x}_o$.

The **Lagrangian** associated with this constrained optimization problem is the real-valued function (with arguments $\boldsymbol{x}, \boldsymbol{\lambda}$)

$$L(\boldsymbol{x}, \boldsymbol{\lambda}) = f(\boldsymbol{x}) + \sum_{j=1}^{m} \lambda_j \hat{g}^j(\boldsymbol{x}) = f(\boldsymbol{x}) + \boldsymbol{\lambda}^T \boldsymbol{G}(\boldsymbol{x}), \quad (10.17)$$

where $\boldsymbol{x}$ is an $(n \times 1)$ vector of variables $x_1, \dots, x_n$, $\boldsymbol{\lambda}$ is a $(m \times 1)$ vector of Lagrange multipliers λ_j, $j = 1, \dots, m < n$, and $\boldsymbol{G}(\boldsymbol{x})$ is an $(m \times 1)$ vector of implicit constraint functions $\hat{g}^j(\boldsymbol{x}) = 0$.

Our first theorem pertaining to the existence of constrained local extrema is

Theorem 10.8: (Necessary Condition). Let the real-valued functions $y = f(\boldsymbol{x})$, $\hat{g}^j(\boldsymbol{x}) = 0$, $j = 1, \dots, m < n$, be of class $C^{(1)}$ throughout some spherical δ – neighborhood of the point $\boldsymbol{x}_o^T = (x_1^o, \dots, x_n^o)$, with $\delta(\boldsymbol{x}_o)$ contained within an open region $\mathcal{D} \subset R^n$. If f has a local extremum subject to $\hat{g}^j = 0$, $j = 1, \dots, m$, at $\boldsymbol{x}_o$, then $L(\boldsymbol{x}, \boldsymbol{\lambda})$ has an *unconstrained* local extremum at that point and

$$\begin{aligned} \nabla_{\boldsymbol{x}} L^o &= \nabla f(\boldsymbol{x}_o) + \nabla \boldsymbol{G}(\boldsymbol{x}_o)\boldsymbol{\lambda}_o = \boldsymbol{O} \\ \nabla_{\boldsymbol{\lambda}} L^o &= \boldsymbol{G}(\boldsymbol{x}_o) = \boldsymbol{O}, \end{aligned} \quad (10.18)$$

or

$$\begin{aligned} &\nabla f(\boldsymbol{x}_o) + \textstyle\sum_{j=1}^{m} \lambda_j^o \nabla \hat{g}^j(\boldsymbol{x}_o) = \boldsymbol{O} \\ &L_{\lambda_j}^o = \hat{g}^j(\boldsymbol{x}_o) = 0, \; j = 1, \dots, m \end{aligned} \quad (10.18.1)$$

Here $\nabla f(\boldsymbol{x}_o)$ is an $(n \times 1)$ vector of partial derivatives f_i^o, $i = 1, \dots, n$, while the $\nabla \hat{g}^j(\boldsymbol{x}_o)$, $j = 1, \dots, m$, denote $(n \times 1)$ vectors with components $\hat{g}_i^j$. In fact, these latter vectors are themselves the columns of the $(n \times m)$ matrix.

$$\nabla \boldsymbol{G} = [\nabla \hat{g}^1, \dots, \nabla \hat{g}^m].$$

Proof: If f attains a constrained extremum at $\boldsymbol{x}_o$, the differential of f along the $(n - m)$-dimensional constraint surface must vanish there and thus $\nabla f(\boldsymbol{x}_o)^T \boldsymbol{h} = 0$ for all $\boldsymbol{h} \in R^n$ satisfying $\nabla \hat{g}^j(\boldsymbol{x}_o)^T \boldsymbol{h} = 0$, $j = 1, \dots, m$. If we multiply each differential $\nabla \hat{g}^j(\boldsymbol{x}_o)^T \boldsymbol{h} = 0$ by some constant λ_j^o and form the sum

$$\nabla f(\boldsymbol{x}_o)^T \boldsymbol{h} + \sum_{j=1}^{m} \lambda_j^o \nabla \hat{g}^j(\boldsymbol{x}_o)^T \boldsymbol{h} = 0,$$

we obtain

$$\left(\nabla f(\boldsymbol{x}_o) + \sum_{j=1}^{m} \lambda_j^o \nabla \hat{g}^j(\boldsymbol{x}_o) \right)^T \boldsymbol{h} = 0. \quad (10.19)$$

Let us assume, without loss of generality, that the Jacobian determinant associated with the first m columns of $\nabla \boldsymbol{G}(\boldsymbol{x}_o)^T$, $|J_G(x_1^o, \dots, x_m^o)|$, does not vanish identically so that the first m variables $x_1, \dots, x_m$ can be expressed in terms of the last $n - m$ (independent) variables (by the Implicit Function Theorem). Then $\boldsymbol{h}$ may be partitioned as

$$\boldsymbol{h} = \begin{bmatrix} \boldsymbol{h}_1 \\ \boldsymbol{h}_2 \end{bmatrix}, \text{ where } \boldsymbol{h}_1 = \begin{bmatrix} h_1 \\ \vdots \\ h_m \end{bmatrix} \text{ and } \boldsymbol{h}_2 = \begin{bmatrix} h_{m+1} \\ \vdots \\ h_n \end{bmatrix},$$

with the components of $\boldsymbol{h}_1$ being uniquely determined in terms of those in $\boldsymbol{h}_2$, i.e., $h_j = \varphi^j(\boldsymbol{h}_2)$, $j = 1, \dots, m$. Since each component of $\boldsymbol{h}_2$ may be taken to be non-zero (since the variables $x_{m+1}, \dots, x_n$, and their differentials h_k, $k = m + 1, \dots, n$, are all independent), the components of $\boldsymbol{h}_1$, and thus of $\boldsymbol{h}$, are all non-zero. Hence, (10.19) holds only if (10.18.1) does. The solution vector $(\boldsymbol{x}_o, \boldsymbol{\lambda}_o)$ may be found by solving system (10.18.1) simultaneously. Q. E. D.

What this proof implies is that if $\nabla f(\boldsymbol{x}_o)^T \boldsymbol{h} = 0$ (a homogeneous linear equation in the h_is) and $\nabla \hat{g}^j(\boldsymbol{x}_o)^T \boldsymbol{h} = 0$ (a homogeneous system of m linear equations in the h_is), $j = 1, \dots, m$, are to hold simultaneously, then the vector $\nabla f(\boldsymbol{x}_o)$ must be linearly dependent upon the vectors $\nabla \hat{g}^j(\boldsymbol{x}_o)$, $j = 1, \dots, m$, i.e., $\nabla f(\boldsymbol{x}_o)$ can be written as a linear combination of the vectors $\nabla \hat{g}^j(\boldsymbol{x}_o)$. But if this is true, then there must exist scalars λ_j^o, $j = 1, \dots, m$, not all zero such that (10.18.1) holds. That the constants λ_j^o, $j = 1, \dots, m$, are uniquely determined follows from the requirement that $\rho(\nabla \boldsymbol{G}(\boldsymbol{x}_o)) = m$. To see this we need only note that

$$\nabla \boldsymbol{G}(\boldsymbol{x}_o)\boldsymbol{\lambda}_o = -\nabla f(\boldsymbol{x}_o)$$

implies

$$J_G(x_1^o,\ldots,x_m^o)^T\boldsymbol{\lambda}_o = -\nabla f(x_1^o,\ldots,x_m^o)$$

or

$$\boldsymbol{\lambda}_o = [J_G(x_1^o,\ldots,x_m^o)^T]^{-1}(-\nabla f(x_1^o,\ldots,x_m^o))$$

since the indicated Jacobian

$$J_G = [\partial(\hat{g}^1,\ldots,\hat{g}^m)/\partial(x_1,\ldots,x_m)]$$

was assumed to be non-singular.

Before proceeding to specify a sufficient condition for a strong local extremum of f subject to $\hat{g}^j = 0, j = 1, \ldots, m$, let us consider a theorem concerning constrained quadratic forms (Debreu [1952]). In particular, we shall consider the sign definiteness of a quadratic form $Q(\boldsymbol{x})$, $\boldsymbol{x} \in R^n$, subject to $m < n$ linear constraints in $\boldsymbol{x}$. Given that the symmetric matrix $\boldsymbol{A}$ is of order $(n \times n)$, $\boldsymbol{B}$ is of order $(n \times m)$ with $\rho(\boldsymbol{B}) = m$, $\boldsymbol{x}$ is a $(n \times 1)$ vector, and $\boldsymbol{B}^T\boldsymbol{x} = \boldsymbol{O}$ represents a homogeneous system of m linear equations in n unknowns, we have

Theorem 10.9: (Necessary and Sufficient Condition). The quadratic form $Q(\boldsymbol{x}) = \boldsymbol{x}^T\boldsymbol{A}\boldsymbol{x}$ subject to $\boldsymbol{B}^T\boldsymbol{x} = \boldsymbol{O}$:

a. is positive definite (> 0) for every $\boldsymbol{x} \neq \boldsymbol{O}$ if and only if

$$\Delta_r = (-1)^m \begin{vmatrix} \boldsymbol{A}_{rr} & \boldsymbol{B}_{rm} \\ \boldsymbol{B}_{rm}^T & \boldsymbol{O} \end{vmatrix} > 0, r = m+1,\ldots,n. \quad (10.20)$$

b. is negative definite (< 0) for every $\boldsymbol{x} \neq \boldsymbol{O}$ if and only if

$$\Delta_r = (-1)^r \begin{vmatrix} \boldsymbol{A}_{rr} & \boldsymbol{B}_{rm} \\ \boldsymbol{B}_{rm}^T & \boldsymbol{O} \end{vmatrix} > 0, r = m+1,\ldots,n. \quad (10.21)$$

Here the double subscripts on the elements of the partitioned determinants indicate the number of rows and columns which are retained as $r = m + 1, \ldots, n$, e.g., $\boldsymbol{M}_{st}$ is obtained from $\boldsymbol{M}$ by keeping only the elements in the first s rows and first t columns of $\boldsymbol{M}$.

Example 10.10: Determine the sign definiteness of $\boldsymbol{x}^T\boldsymbol{A}\boldsymbol{x}$ subject to $\boldsymbol{B}^T\boldsymbol{x} = \boldsymbol{O}$ when

$$\underset{(3\times3)}{\boldsymbol{A}} = \begin{bmatrix} 2 & 1 & 0 \\ 1 & 3 & 1 \\ 0 & 1 & 2 \end{bmatrix}, \quad \underset{(3\times2)}{\boldsymbol{B}} = \begin{bmatrix} 1 & 0 \\ 2 & 1 \\ 1 & 3 \end{bmatrix}.$$

Here $m + 1 = 3 = n$ so that

$$\begin{vmatrix} \boldsymbol{A}_{33} & \boldsymbol{B}_{32} \\ \boldsymbol{B}_{32}^T & \boldsymbol{O} \end{vmatrix} = \begin{vmatrix} 2 & 1 & 0 & 1 & 0 \\ 1 & 3 & 1 & 2 & 1 \\ 0 & 1 & 2 & 1 & 3 \\ 1 & 2 & 1 & 0 & 0 \\ 0 & 1 & 3 & 0 & 0 \end{vmatrix} = 24 > 0.$$

And since, from (10.20), $\Delta_3 = (-1)^2 24 > 0$, we can conclude that $\boldsymbol{x}^T\boldsymbol{A}\boldsymbol{x}$ subject to $\boldsymbol{B}^T\boldsymbol{x} = \boldsymbol{O}$ is positive definite.

We noted above that $\nabla\hat{g}^j(\boldsymbol{x}_o)^T\boldsymbol{h} = 0, j = 1, \ldots, m$, represents a homogeneous system of m linear equations in n unknowns $h_i, i = 1, \ldots, n$. That is, $\nabla\hat{g}^j(\boldsymbol{x}_o)^T\boldsymbol{h} = 0, j = 1, \ldots, m$, depicts the system $J_G\boldsymbol{h} = \nabla\boldsymbol{G}(\boldsymbol{x}_o)^T\boldsymbol{h} = \boldsymbol{O}$. In the following sufficient condition for a strong local constrained extremum use will be made of the coefficient matrix ∇G^T of this homogeneous linear system and the Hessian of the Lagrangian of f. To this end we have

Theorem 10.10: (Sufficient Condition). Let the real-valued functions $y = f(\boldsymbol{x})$, $\hat{g}^j(\boldsymbol{x}) = 0$, $\boldsymbol{x} \in R^n$, $j = 1, \ldots, m < n$, be of class $C^{(2)}$ throughout a spherical δ-neighborhood of the point $\boldsymbol{x}_o$, $\delta(\boldsymbol{x}_o)$, contained within an open region $\mathcal{D} \subset R^n$. Then the Lagrangian $L(\boldsymbol{x}, \boldsymbol{\lambda}) = f(\boldsymbol{x}) + \boldsymbol{\lambda}^T\boldsymbol{G}(\boldsymbol{x})$ is of class $C^{(2)}$ throughout $\delta(\boldsymbol{x}_o)$ also. If

$$\nabla_x L^o = \nabla f(\boldsymbol{x}_o) + \nabla\boldsymbol{G}(\boldsymbol{x}_o)\boldsymbol{\lambda}_o = \boldsymbol{O}, \quad \nabla_\lambda L^o = \boldsymbol{G}(\boldsymbol{x}_o) = \boldsymbol{O},$$

and

$$\begin{vmatrix} \boldsymbol{H}_L(\boldsymbol{x}_o,\boldsymbol{\lambda}_o)_{rr} & \nabla\boldsymbol{G}(\boldsymbol{x}_o)_{rm} \\ \nabla\boldsymbol{G}(\boldsymbol{x}_o)_{rm}^T & \boldsymbol{O} \end{vmatrix} \neq 0, r = m+1,\ldots,n,$$

then L, and thus f subject to $\hat{g}^j = 0, j = 1, \ldots, m$, has a strong local extremum at $\boldsymbol{x}_o$. Moreover:

a. if

$$\Delta_r = (-1)^r \begin{vmatrix} H_L(x_o,\lambda_o)_{rr} & \nabla G(x_o)_{rm} \\ \nabla G(x_o)_{rm}^T & O \end{vmatrix} > 0, r = m+1,\ldots,n, \quad (10.22)$$

then L, and thus f subject to $\hat{g}^j = 0, j = 1, \ldots, m$, has a strong local maximum at x_o; but

b. if

$$\Delta_r = (-1)^m \begin{vmatrix} H_L(x_o,\lambda_o)_{rr} & \nabla G(x_o)_{rm} \\ \nabla G(x_o)_{rm}^T & O \end{vmatrix} > 0, r = m+1,\ldots,n, \quad (10.23)$$

then L, and thus f subject to $\hat{g}^j = 0, j = 1, \ldots, m$, has a strong local minimum at x_o. Here $H_L(x_o, \lambda_o)_{rr}$ is obtained from $H_L(x_o, \lambda_o)$ by retaining only the elements in its first r rows and columns while $\nabla G(x_o)_{rm}$ is obtained from $\nabla G(x_o)$ by keeping only the elements in its first r rows and first m columns.

Proof: Using a second-order Taylor expansion of L, with λ_o fixed, we have

$$\begin{aligned} &L(x_o + h, \lambda_o) - L(x_o, \lambda_o) \\ &= \frac{1}{2!} h^T H_L(x_o + \theta h, \lambda_o) h, \ 0 < \theta < 1, \quad (10.24) \end{aligned}$$

since, by hypothesis, $\nabla_x L(x_o, \lambda_o) = O$. With the second-order partial derivatives of L continuous at x_o, the right-hand side of (10.24) approaches $Q(h) = \frac{1}{2!} h^T H_L(x_o, \lambda_o) h$ when $\theta \to 0$ and $\|h\| < \delta$ (remember the notion of the *sign persistence* of derivatives). Hence, $L(x_o + h, \lambda_o) - L(x_o, \lambda_o)$ has the same sign as $h^T H_L(x_o, \lambda_o) h$, provided that the partial derivatives L_{ij}^o, $i, j = 1, \ldots, n$, are not all zero. However, the arguments of this quadratic form are not independent, i.e., we must determine the sign definiteness of $Q(h)$ for those h which satisfy $\hat{g}^j(x_o + h) = 0, j = 1, \ldots, m$. But this is equivalent to determining the sign of $Q(h) = h^T H_L(x_o, \lambda_o) h$ subject to $\nabla G(x_o)^T h = 0$ with $\rho(\nabla G(x_o)) = m$. Now, if (10.22) (10.23) holds, then $L(x_o + h, \lambda_o) - L(x_o, \lambda_o) < 0 (>0)$ and thus f subject to $\hat{g}^j = 0, j = 1, \ldots, m$, has a strong local maximum (minimum) at x_o. Q. E. D.

Let us elaborate on the preceding determinantal conditions. Equations (10.22) and (10.23) depict a sequence of signed bordered principal minors of the $(m + n \times n + m)$ bordered Hessian matrix

$$H_L^G(x_o, \lambda_o) = \begin{bmatrix} H_L(x_o,\lambda_o) & \nabla G(x_o) \\ \nabla G(x_o)^T & O \end{bmatrix},$$

where $\nabla G = [\nabla \hat{g}^1, \ldots, \nabla \hat{g}^m]$. For a strong local maximum, the determinants in (10.22) should alternate in sign, the first having the sing of $(-1)^r$. For a strong local minimum, the determinants in (10.23) must each have the same sign as $(-1)^m$. (Remember that m is the number of constraints.)

Example 10.11: Does the real-valued function $y = f(x) = x_1^2 + x_2^2 + x_3^2 + x_4^2$, $x \in R^4$, subject to

$$\hat{g}^1(x) = 1 - x_1 - x_2 - x_3 - x_4 = 0$$
$$\hat{g}^2(x) = 2 - x_1 + 2x_3 = 0$$

attain a local extremum within R^4? Looking to (10.17) we have

$$\begin{aligned} L(x, \lambda) &= f(x) + \lambda^T G(x) \\ &= x_1^2 + x_2^2 + x_3^2 + x_4^2 + \lambda_1(1 - x_1 - x_2 - x_3 - x_4) \\ &\quad + \lambda_2(2 - x_1 + 2x_3). \end{aligned}$$

And from (10.18),

$$\begin{bmatrix} \nabla_x L \\ \nabla_\lambda L \end{bmatrix} = O \text{ or } \begin{array}{ll} L_1 = 2x_1 - \lambda_1 - \lambda_2 & = 0 \\ L_2 = 2x_2 - \lambda_1 & = 0 \\ L_3 = 2x_3 - \lambda_1 + 2\lambda_2 & = 0 \\ L_4 = 2x_4 - \lambda_1 & = 0 \\ L_{\lambda 1} = 1 - x_1 - x_1 - x_3 - x_4 & = 0 \\ L_{\lambda 2} = 2 - x_1 + 2x_3 & = 0 \end{array}$$

the simultaneous solution of which is $x_o^T = \left(\frac{16}{19}, \frac{7}{19}, -\frac{11}{19}, \frac{7}{19}\right)$ and $\lambda_o^T = \left(\frac{14}{19}, \frac{18}{19}\right)$.

Moreover, it can be shown that

$$\begin{vmatrix} H_L(x_o,\lambda_o)_{33} & \nabla G(x_o)_{32} \\ \nabla G(x_o)_{32}^T & O \end{vmatrix} = 28 > 0,$$

$$\begin{vmatrix} H_L(x_o,\lambda_o)_{44} & \nabla G(x_o)_{42} \\ \nabla G(x_o)_{42}^T & O \end{vmatrix} = 76 > 0,$$

thus implying, via (10.23), that f attains a strong local constrained minimum at $\boldsymbol{x}_o$. ■

10.6 INTERPRETATION OF THE LAGRANGE MULTIPLIERS λ_j

We determined in Section 10.3 that $\lambda^o = df^o/db$. A similar interpretation is offered for λ_j^o, i.e., λ_j^o is a measure of the infinitesimal change in $f^o = f(\boldsymbol{x}_o)$ with respect to a small relaxation in or tightening of the jth constraint $\hat{g}^j = b_j - g^j = 0$, given that the remaining constraints are unchanged. Hence, $\lambda_j^o = df^o/db_j$.

To verify this result, let us first consider the effect of a change in b_j upon the optimal value of f, f^o. Thus

$$\frac{\partial f^o}{\partial b_j} = \sum_{i=1}^{n} \frac{\partial f^o(\boldsymbol{x}_o)}{\partial x_i}\frac{dx_i}{db_j}$$

$$= \nabla f(\boldsymbol{x}_o)^T \begin{bmatrix} dx_1/db_j \\ \vdots \\ dx_n/db_j \end{bmatrix} = \nabla f(\boldsymbol{x}_o)^T \left(\frac{d\boldsymbol{x}}{db_j}\right).$$

Since, at optimality, $\nabla f(\boldsymbol{x}_o) = -\boldsymbol{G}(\boldsymbol{x}_o)\boldsymbol{\lambda}_o$, the preceding expression becomes

$$\frac{\partial f^o}{\partial b_j} = -\boldsymbol{\lambda}_o^T \boldsymbol{G}(\boldsymbol{x}_o)^T \left(\frac{d\boldsymbol{x}}{db_j}\right)$$

$$= \sum_{j=1}^{m} \lambda_j^o \nabla g^j(\boldsymbol{x}_o)^T \left(\frac{d\boldsymbol{x}}{db_j}\right). \tag{10.25}$$

From the jth constraint $b_j - \hat{g}^j(\boldsymbol{x}) = 0$, $j = 1, \ldots, m$,

$$\frac{d(b_j - \hat{g}^j(\boldsymbol{x}_o))}{db_k} = \frac{db_j}{db_k} - \sum_{i=1}^{n} \frac{\partial g^j(\boldsymbol{x}_o)}{\partial x_i}\frac{dx_i}{db_k} = 0$$

or

$$\nabla g^j(\boldsymbol{x}_o)^T \left(\frac{d\boldsymbol{x}}{db_k}\right) = \begin{cases} 1, & k = j; \\ 0, & k \neq j. \end{cases}$$

Multiplying this last expression by λ_j^o yields

$$\lambda_j^o \nabla g^j(\boldsymbol{x}_o)^T \left(\frac{d\boldsymbol{x}}{db_k}\right) = \begin{cases} \lambda_j^o, & k = j; \\ 0, & k \neq j. \end{cases} \tag{10.26}$$

For $k = j$, (10.26) corresponds to the jth term (in fact, the only non-zero term) in (10.25) and thus, as expected, $df^o/db_j = \lambda_j^o$.

We can also obtain the result $\partial f^o/\partial b_j = \lambda_j$, $j = 1, \ldots, m$, by working with the **optimal value function** for the problem: maximize $y = f(\boldsymbol{x})$ subject to $\boldsymbol{x} \in \mathcal{G}$. Let us express the said function as

$$f^o(\boldsymbol{x}) = \max\{f(\boldsymbol{x}) | \boldsymbol{x} \in \mathcal{G}\},$$

which is taken to be of class $C^{(1)}$. (For a minimization problem, $f^o(\boldsymbol{b}) = \min\{f(\boldsymbol{x}) | \boldsymbol{x} \in \mathcal{G}\}$.) Clearly $f^o(\boldsymbol{b})$ depicts the largest value of f obtainable given the restrictions $g^j(\boldsymbol{x}) = b_j$, $j = 1, \ldots, m < n$.

For $\boldsymbol{b} = \bar{\boldsymbol{b}}$, let $\bar{\boldsymbol{x}}$ be the solution associated with the problem: maximize $f(\boldsymbol{x})$ subject to $\boldsymbol{g}(\boldsymbol{x}) = \bar{\boldsymbol{b}}$, where the jth component of the $(m \times 1)$ vector-valued function $\boldsymbol{g}$ is $g^j(\boldsymbol{x})$, $j = 1, \ldots, m$. Then $f(\bar{\boldsymbol{x}}) = f^o(\bar{\boldsymbol{b}})$ and, for all $\boldsymbol{x}, f(\boldsymbol{x}) \leq f^o(\bar{\boldsymbol{b}}) = f^o(\boldsymbol{g}(\boldsymbol{x}))$. Then $\eta(\boldsymbol{x}) = f(\boldsymbol{x}) - f^o(\mathrm{g}(\boldsymbol{x}))$ has a maximum at $\boldsymbol{x} = \bar{\boldsymbol{x}}$ so that

$$\eta_i(\bar{\boldsymbol{x}}) = f_i(\bar{\boldsymbol{x}}) - \textstyle\sum_{j=1}^{m} \left.\dfrac{\partial f^o(\bar{\boldsymbol{b}})}{\partial b_j}\right|_{b=g(\bar{x})} g_i^j(\bar{\boldsymbol{x}}) = 0, \ i = 1, \ldots, n.$$

Define $\bar{\lambda}_j = \partial f^o(\bar{\boldsymbol{b}})/\partial b_j$. Then the preceding equation system becomes $\nabla f(\bar{\boldsymbol{x}}) + \nabla \boldsymbol{G}(\bar{\boldsymbol{x}})\bar{\boldsymbol{\lambda}} = \boldsymbol{O}$ or the first equation in (10.18) - - a necessary condition for a constrained extremum at $\bar{\boldsymbol{x}}$. We may thus view the optimal Lagrange multiplier λ_j as the marginal change in the optimal value function with respect to the constraint constant b_j, $j = 1, \ldots, m$.

10.7 ECONOMIC APPLICATIONS

10.7.1 Household Equilibrium

Consider a household's **direct utility function** $u = f(q_1, q_2)$ - - a relationship representing a preference ordering over all admissible combinations of commodities q_1 and q_2. (See Section 3.3 of E-Chapter 3 for further details on preferences and their description via a utility function.) If the level of utility u is held fixed at $u = \bar{u}=$ constant, then the expression $\bar{u} = f(q_1, q_2)$ (specifying an **indifference curve** labeled $\bar{u}$ in the q_1, q_2-plane) *implicitly* defines a function mapping q_1 onto q_2 (Figure 10.3a) so that we may calculate

$$f_1 + f_2\frac{dq_2}{dq_1} = 0$$

or

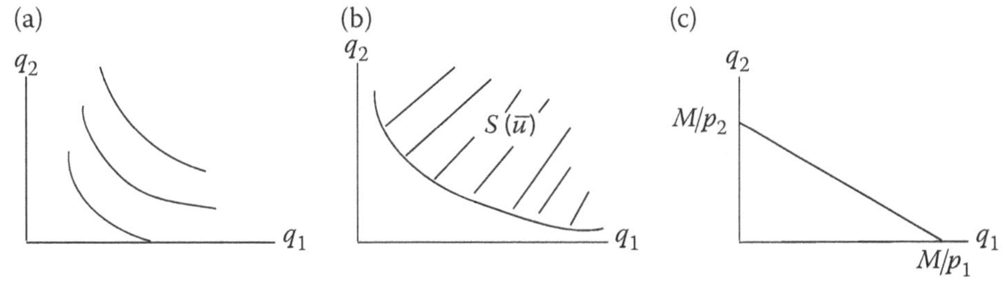

FIGURE 10.3 (a) the $\bar{u}$ indifference curve; (b) upper level set $S(\bar{u})$; (c) the budget line.

$$\frac{dq_2}{dq_1} = -\frac{f_1}{f_2} = -\frac{MU_1}{MU_2}, \tag{10.27}$$

where $MU_1\,(MU_2)$ is the **marginal utility** of commodity one (two) in consumption. The expression $-dq_2/dq_1 = MU_1/MU_2$ is termed the **marginal rate of substitution** (MRS_{q_2,q_1}) between q_1 and q_2 - - the rate at which q_1 must be substituted for q_2 (or q_2 for q_1) in order to keep utility constant at $\bar{u}$.

In what follows we shall assume that f is strictly quasi-concave; $f_1, f_2 > 0$; $f_{11}, f_{22} < 0$ (diminishing marginal utilities); and $f_{12}, f_{21} > 0$. Under these assumptions it follows that the slope of $\bar{u}$ or $d^2q_2/dq_1 < 0$ while

$$\frac{d^2q_2}{dq_1^2} = -\frac{[f_2^2 f_{11} - 2f_1f_2f_{12} + f_1^2 f_{22}]}{f_2^3} > 0, f_2 \neq 0, \tag{10.28}$$

or $-\,d^2q_2/dq_1^2 < 0$ if $[\cdot] < 0$, i.e., under strict quasi-concavity, we have a diminishing MRS_{q_2,q_1} as we move along the indifference curve $\bar{u}$. But this implies that the indifference curve $\bar{u}$ is strictly convex to the origin; the boundary of the upper level set $s(\bar{u}) = \{(q_1, q_2)|f(q_1, q_2) \geq \bar{u}\}$ is strictly convex (Figure 10.3b) (On all this see Example 9.4.)

Given $u = f(q_1, q_2)$, the household is assumed to maximize this utility function subject to a budget constraint

$$p_1q_1 + p_2q_2 = M, \tag{10.29}$$

where p_1 and q_1 are constant commodity prices and M is current (constant) disposable money income for the given planning period (Figure 10.3c). Hence, the household's objective is to

$$\underset{q_1,q_2}{\text{maximize}}\ u = f(q_1, q_2) \text{ subject to } p_1q_1 + p_2q_2 = M. \tag{10.30}$$

From the Lagrangian function

$$L(q_1, q_2, \lambda) = f(q_1, q_2) + \lambda(M - p_1q_1 - p_2q_2)$$

we obtain the first-order condition

$$\begin{aligned} &\text{(a) } L_1 = f_1 - \lambda p_1 = 0 \\ &\text{(b) } L_2 = f_2 - \lambda p_2 = 0 \\ &\text{(c) } L_\lambda = M - p_1q_1 - p_2q_2 = 0. \end{aligned} \tag{10.31}$$

From (10.31.a,b) we obtain

$$\frac{f_1^0}{f_2^0} = \frac{p_1}{p_2} \tag{10.32}$$

or, at point A of Figure 10.4, the slope of the budget line must equal the slope of the indifference curve u^o. Hence, at point A, the optimal commodity combination is (q_1^0, q_2^0). How should we interpret Equation (10.32)?

Specifically, the rate at which the market substitutes q_1 for $q_2\,(p_1/p_2)$ must equal the rate at which the consumer substitutes q_1 for $q_2\,(f_1^0/f_2^0)$ along u^o; or, upon rearranging Equation (10.32), $f_1^0/p_1 = f_2^0/p_2$ - - the marginal utility per dollar's worth of q_1 equals the marginal utility per dollar's worth of q_2. Thus there is no gain in utility by transferring expenditure from q_1 to q_2 (or from q_2 to q_1).

Once system (10.31) is solved simultaneously, we obtain the optimal Lagrange multiplier λ^o. As we have seen earlier $\lambda^o = df^o/dM$ - - λ^o is the **marginal utility of money**. It depicts the increase in total utility (at optimality) resulting from an additional dollar of money income. Looked at in another fashion, we may

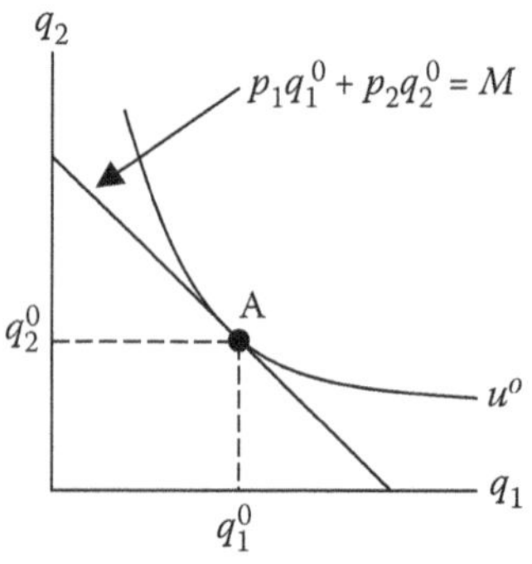

FIGURE 10.4 Household equilibrium.

view λ^o as a *shadow (fictitious) price* - - the consumer would be willing to pay up to $\$\lambda^o$ to have the budget constraint relaxed by one unit (\$1).

What about the second-order condition for a constrained maximum of the utility function $u = f(q_1, q_2)$? Under the assumption that f is strictly quasiconcave, the bordered Hessian determinant of f

$$|\boldsymbol{H}_f^{\hat{g}}| = \begin{vmatrix} f_{11} & f_{12} & -p_1 \\ f_{21} & f_{22} & -p_2 \\ -p_1 & -p_2 & 0 \end{vmatrix} > 0 \quad \text{(from Equation (10.11))} \tag{10.33}$$

as required for a constrained maximum of f, where $\hat{g} = M - p_1q_1 - p_2q_2 = 0$. To see this, let us expand

$$|\boldsymbol{H}_f^{\hat{g}}| = -(f_{11}p_2^2 - 2p_1p_2f_{12} + f_{22}p_1^2) > 0.$$

If we set $p_1 = (f_1/f_2)p_2$ (the first-order condition for a constrained maximum holds) in Equation (10.33), then the preceding inequality becomes

$$|\boldsymbol{H}_f^{\hat{g}}| = -(p_2^2/f_2^2)[f_{11}f_2^2 - 2f_1f_2f_{12} + f_{22}f_1^2] > 0.$$

If the term $[\cdot]$ is strictly negative, then $|H_f^{\hat{g}}| > 0$ as required for a constrained maximum of $u = f(q_1, q_2)$. But $[\cdot] < 0$ is the requirement for strict quasiconcavity of the utility function. In this regard, under strict quasiconcavity, we need only consider the first-order condition for a constrained maximum of utility - - the second-order condition will automatically be satisfied.

10.7.2 Marshallian Demand Functions

If we solve system (10.31) simultaneously for q_1 and q_2, then we obtain the **Marshallian or ordinary (income-constrained) demand functions** for these commodities, which will be denoted as

$$\begin{aligned} q_1 &= g^1(p_1, p_2, M) \\ q_2 &= g^2(p_1, p_2, M). \end{aligned} \tag{10.34}$$

These Marshallian demands possess the properties:

1. g^1, g^2 are *homogeneous of degree zero in prices and money income*, i.e., proportionate changes in p_1, p_2, and M leave the quantities demanded unchanged (we remain at point A in Figure 10.4). For example, $g^1(tp_1, tp_2, tM) = g^1(p_1, p_2, M)$. (More on homogeneity in Chapter 15.)

2. ordinary demand functions *add up* -- the weighted (by prices) sum of g^1 and g^2 equals M, i.e., $p_1g^1 + p_2g^2 = M$.

3. for q_1, q_2 normal goods, the ordinary demand curve is *negatively sloped.*

4. *symmetrical cross effects:*

$$\frac{\partial g^1}{\partial p_2} + q_2\frac{\partial g^1}{\partial M} = \frac{\partial g^2}{\partial p_1} + q_1\frac{\partial g^2}{\partial M}. \qquad (10.35)$$

If the Marshallian demands (10.34) are substituted into the direct utility function $u = f(q_1, q_2)$, then we obtain the **indirect utility function** (the *optimal value function*)

$$u = f(g^1, g^2) = k(p_1, p_2, M), \qquad (10.36)$$

an expression indicating the maximum level of utility attainable for a total expenditure of \$$M$ at given prices. The indirect utility function has the following properties:

1. *homogeneity of degree zero in* p_1, p_2, and M.
2. strictly monotonically increasing in M, monotonically non-increasing in p_1 and p_2, and monotonically decreasing in at least one price.
3. *strictly quasiconcave in* p_1 and p_2 (the upper contour set $\mathcal{J}(\bar{u}) = \{(p_1, p_2) | k(p_1, p_2, M) \geq \bar{u}\}$ is convex or the lower boundary of $\mathcal{J}(\bar{u})$ is convex to the origin).
4. *continuous* in p_1, p_2, and M.
5. if k is of class $C^{(1)}$, then the indirect utility function yields the ordinary or Marshallian demands via **Roy's identity**:

   the ith Marshallian demand is

$$q_i = g^i(p_1, p_2, M) \equiv \frac{\partial k/\partial p_i}{\partial k/\partial M}, \ i = 1, 2. \qquad (10.37)$$

10.7.3 Hicksian Demand Functions

Suppose we now consider the transpose of problem (10.30), i.e., we interchange the objective function and constraint function. Hence, we desire to minimize the cost of attaining a fixed level of utility u or

$$\underset{q_1, q_2}{\text{minimize}}\ p_1q_1 + p_2q_2 \ \text{subject to}\ f(q_1, q_2) = u. \qquad (10.38)$$

Now the Lagrangian is

$$L^* = p_1q_1 + p_2q_2 + \lambda_*(u - f(q_1, q_2)).$$

From the first-order condition for a constrained minimum we get

$$\begin{aligned} &(a)\quad L_1^* = p_1 - \lambda_* f_1 = 0 \\ &(b)\quad L_2^* = p_2 - \lambda_* f_2 = 0 \\ &(c)\quad L_{\lambda_*}^* = u - f(q_1, q_2) = 0. \end{aligned} \qquad (10.39)$$

And from (10.39a,b), we again obtain the equality $f_1^0/f_2^0 = p_1/p_2$ (Figure 10.4). Moreover, it can be demonstrated that $\lambda_* = 1/\lambda^o$ (the inverse of the marginal utility of money). Thus λ_* is the rate at which total expenditure would have to increase if an additional unit of utility were required or $\lambda_* = dM/du$ (the marginal change in M with respect to u).

Looking to the second-order condition for a constrained minimum, we require that the bordered Hessian

$$\begin{vmatrix} -\lambda_* f_{11} & -\lambda_* f_{12} & -f_1 \\ -\lambda_* f_{21} & -\lambda_* f_{22} & -f_2 \\ -f_1 & -f_2 & 0 \end{vmatrix} = \lambda_*[f_{11}f_2^2 - 2f_1f_2f_{12} + f_{22}f_1^2] \qquad (10.40)$$

be negative. Since $\lambda_* > 0$, it follows that Equation (10.39) will be negative if the term $[\cdot] < 0$. But since $u = f(q_1, q_2)$ was deemed strictly quasiconcave, we see that the second-order condition for a constrained minimum holds.

Solving system (10.39) simultaneously for q_1, q_2 we obtain the **compensated (utility-constrained) or Hicksian demand functions**

$$\begin{aligned} q_1 &= h^1(p_1, p_2, u) \\ q_2 &= h^2(p_1, p_2, u). \end{aligned} \qquad (10.41)$$

As far as the properties of these expressions are concerned, they:

1. are *homogeneous of degree zero in commodity prices* (given that M changes proportionately to keep u constant).
2. satisfy the *law of downward sloping demand*, i.e., for $q_i = h^i(p_1, p_2, u)$, $\partial h^i/\partial p_i < 0$ or $\varepsilon_{ii} < 0$, $i = 1, 2$. Stated alternatively, the substitution effect of an own-price increase must be negative.

3. have *symmetrical cross-price effects* - - the cross-price derivatives of the compensated demands are symmetrical or

$$\partial h^1/\partial p_2 = \partial h^2/\partial p_1. \tag{10.42}$$

4. *add up* - - the compensated demands (weighted by commodity prices) exhaust total expenditure or

$$p_1 h^1 + p_2 h^2 = M = E(p_1, p_2, u). \tag{10.43}$$

Here $M = E(p_1, p_2, M)$ is called the **expenditure function** (an *optimal value function*), which indicates the minimum monetary expenditure needed to attain utility level u at the given prices p_1, p_2.

Let us examine the salient features of (10.43). The expenditure function:

1. is *homogeneous of degree one in commodity prices*. Thus, proportionate changes in prices cause total expenditure to increase by the same proportion.

2. is strictly monotonically increasing in u, monotonically non-decreasing in p_1 and p_2, and monotonically increasing in at least one price. Thus, a higher level of utility can be attained by the household only at a higher level of expenditure; and an increase in at least one price will never reduce the cost of attaining u.

3. is *concave in prices* - - as prices increase, total expenditure increases at most linearly, i.e., if all prices save one are held fixed, E lies everywhere on or below each of its possible tangent lines (See Figure 10.5 with $p_2 = \bar{p}_2 =$ constant and $u = \bar{u}$ constant).

4. is *continuous in p_1, p_2 and u* with the first- and second-order price derivatives existing except possibly at a finite number of points.

5. possesses the **derivative property** (**Shephard's lemma**) [Shephard (1970)]) - - where they exist, the partial derivatives of E with respect to p_1 and $p_2(\geq 0)$ yield the equilibrium compensated demands (10.41) or

$$\begin{aligned} &\text{(a)} \quad \frac{\partial E(p_1, p_2, u)}{\partial p_1} \equiv h^1(p_1, p_2, u) = q_1 \\ &\text{(b)} \quad \frac{\partial E(p_1, p_2, u)}{\partial p_2} \equiv h^2(p_1, p_2, u) = q_2. \end{aligned} \tag{10.44}$$

Then under the concavity of E, $\partial h^1/\partial p_1 \leq 0$ and $\partial h^2/\partial p_2 \leq 0$ (for an elementary proof of these results see Appendix 10.7.1.A).

We noted earlier that since E is continuous and strictly monotonically increasing in u, E must have a *unique inverse* which, conveniently, is the indirect utility function $u = k(p_1, p_2, M)$. This inverse relationship is called **utility-expenditure duality**, meaning that at the set of prices for which E yields the minimum expenditure required to achieve u and k yields the maximum utility attainable given M, we must have $E(p_1, p_2, u) = k(p_1, p_2, M)$.

Is there a connection between the Marshallian (ordinary) demand functions and the Hicksian (compensated) demand functions? It can be readily shown that the ordinary demands can be obtained from the compensated demands by a simple substitution, and vice versa. Specifically, given the compensated demand function $q_1 = h^1(p_1, p_2, u)$, a substitution of the indirect utility function $u = k(p_1, p_2, M)$ into $h^1(h^2)$ for u gives us

$$\begin{aligned} q_1 &= h^1(p_1, p_2, u) = h^1(p_1, p_2, k(p_1, p_2, M)) = g^1(p_1, p_2, M), \\ q_2 &= h^2(p_1, p_2, u) = h^2(p_1, p_2, k(p_1, p_2, M)) = g^2(p_1, p_2, M). \end{aligned} \tag{10.45}$$

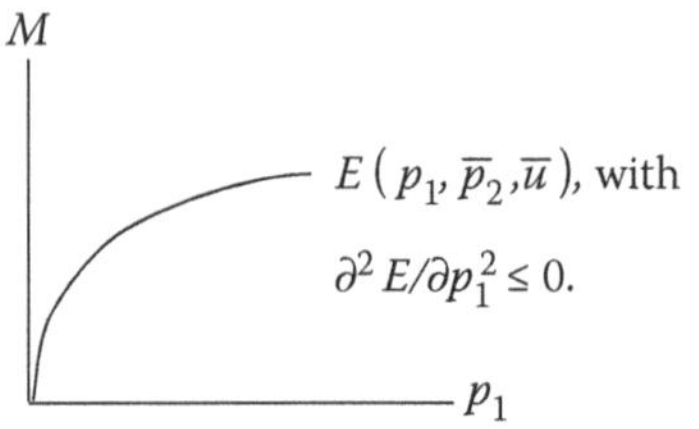

FIGURE 10.5 $E(p_1, \bar{p}_2, \bar{u})$ is concave in prices, with $\partial^2 E/\partial p_1^2 \leq 0$; also, $\partial^2 E/\partial p_2^2 \leq 0$.

Similarly, given the ordinary demand function $q_1 = g^1(p_1, p_2, M)$, a substitution of the expenditure function $M = E(p_1, p_2, u)$ into $g^1 (g^2)$ for M yields

$$\begin{aligned} q_1 &= g^1(p_1, p_2, M) = g^1(p_1, p_2, E(p_1, p_2, u)) = h^1(p_1, p_2, u); \\ q_2 &= g^2(p_1, p_2, M) = g^2(p_1, p_2, E(p_1, p_2, u)) = h^2(p_1, p_2, u). \end{aligned} \tag{10.46}$$

Can we derive the direct utility function from the expenditure function? The answer to this question is based upon the so-called **integrability problem**, which addresses the issue of determining the circumstances under which we can transform a given set of demand functions involving prices into a utility function involving commodity levels.[2] Let us start with the **integrability condition**

$$\left|\frac{\partial(E_1, E_2)}{\partial(p_1, p_2)}\right| = \begin{vmatrix} E_{11} & E_{12} \\ E_{21} & E_{22} \end{vmatrix} \equiv 0 \text{ for all } p_1, p_2 > 0. \tag{10.47}$$

If this expression is satisfied, then the compensated demands $q_1 = \partial E/\partial p_1, q_2 = \partial E/\partial p_2$ (via the derivative property) are *functionally dependent*, i.e., they satisfy $\emptyset(q_1, q_2) = u - f(q_1, q_2) = 0$. So if we can eliminate p_1, p_2 from the system

$$\begin{aligned} \partial E/\partial p_1 &= q_1 \\ \partial E/\partial p_2 &= q_2, \end{aligned} \tag{10.48}$$

then we get an expression involving q_1, q_2, and u - - the direct utility function.

Example 10.12: Given the direct utility function $u = f(q_1, q_2) = q_1 q_2$ and budget constraint $p_1 q_1 + p_2 q_2 = M$:

a. determine the Marshallian demands.

b. determine the indirect utility function.

c. does Roy's identity hold?

d. determine the Hicksian demands.

e. determine the expenditure function (two ways).

f. does Shephard's lemma hold?

g. derive the Hicksian demands from the Marshallian demands and conversely.

h. obtain the direct utility function from the Hicksian demands.

Let Figure 10.6 be your guide.

a. From system (10.31),

$$\begin{aligned} q_1 &= g^1(p_1, p_2, M) = M/2p_1 \\ q_2 &= g^2(p_1, p_2, M) = M/2p_2 \\ \lambda &= M/2p_1 p_2 \\ |\boldsymbol{H}_f^{\hat{g}}| &= p_1 p_2 > 0. \end{aligned}$$

b. $u = f(g^1, g^2) = k(p_1, p_2, M) = \frac{1}{4}\frac{M^2}{p_1 p_2}$.

c. yes.

d. From system (10.39),

$$\begin{aligned} q_1 &= h^1(p_1, p_2, u) = (p_2 p_1^{-1} u)^{1/2} \\ q_2 &= h^2(p_1, p_2, u) = (p_1 p_2^{-1} u)^{1/2} \\ \lambda_* &= 2 p_1 p_2 M^{-1} (=1/\lambda). \end{aligned}$$

For $M = p_1 q_1 + p_2 q_2$ and $\hat{u} = u - f(q_1, q_2)$,

$$|\boldsymbol{H}_M^{\hat{u}}| = -2\lambda_* q_1 q_2 < 0.$$

e. $M = p_1 h^1 + p_2 h^2 = E(p_1, p_2, u) = 2 p_1^{1/2} p_2^{1/2} u^{1/2}$ or $E = k^{-1}$.

f. yes.

g. Equations (10.45), (10.46) give the desired results.

h. $\begin{vmatrix} E_{11} & E_{12} \\ E_{21} & E_{22} \end{vmatrix} \equiv 0$ while $u = f(h^1, h^2) = h^1 h^2$. ■

Note that we have come full circle – we started with the direct utility function and we ended up with the same. All this again reinforces the notion of **utility-expenditure duality**. A salient feature of this duality is a *change of variables*: we have the choice of working with *commodity variables* q_1 and q_2 as arguments in the direct utility function and in the linear expenditure constraint (M and product prices constant) or with *prices* p_1 and p_2 (and u or M) in the expenditure

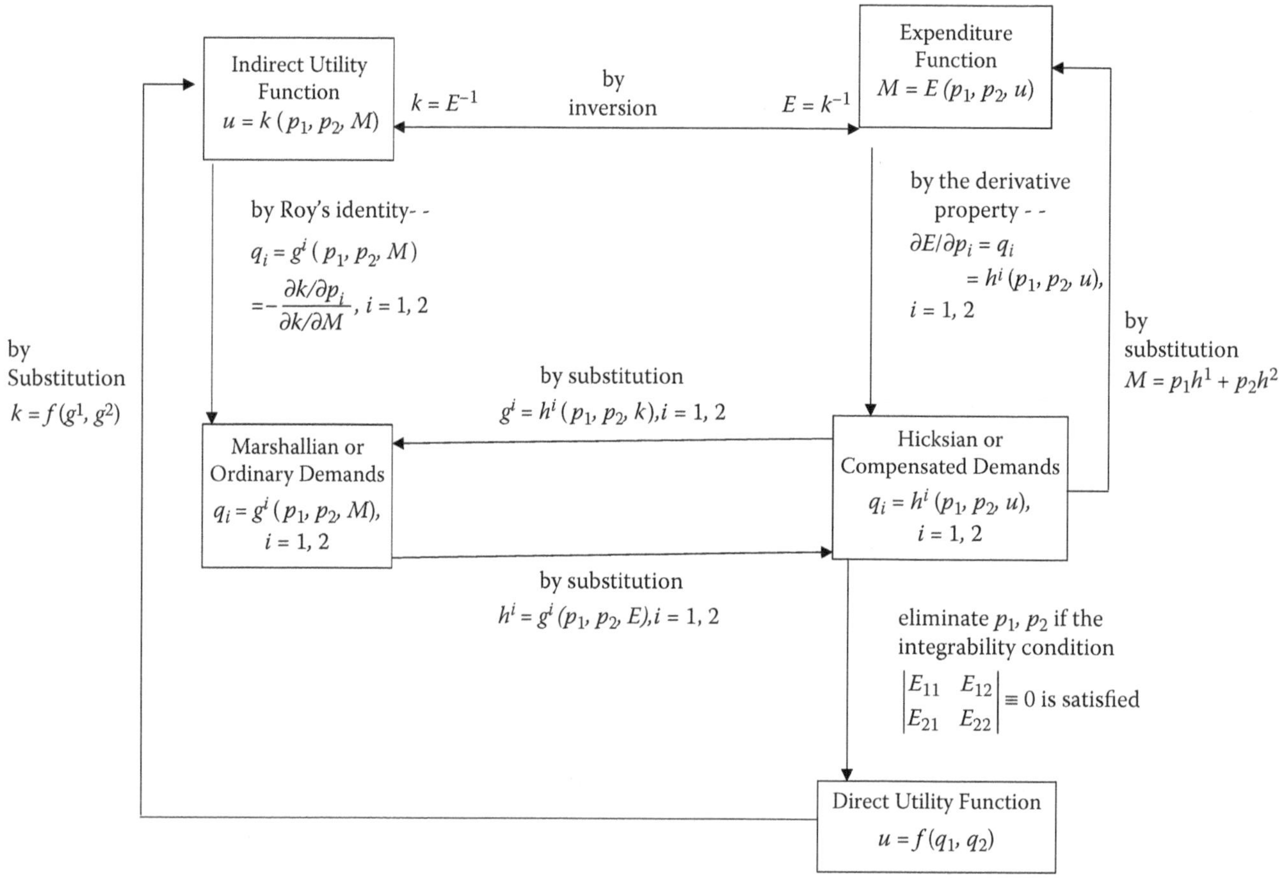

FIGURE 10.6 Utility-expenditure duality.

function or in the indirect utility function, respectively. All of the information contained in any one of these representations is readily obtainable from the other. On all this see Figure 10.6.

10.7.4 Constrained Cost Minimization, Constrained Output Maximization, and Long-Run Profit Maximization

Let us consider first the technology of the firm:

1. **Production function** $q = F(l, k)$ -- a technical relation indicating the maximum quantity of output (q) forthcoming from various combinations of labor (l) and capital (k). It is assumed that F is:
   a. single-valued;
   b. continuous in l, k;
   c. of class $C^{(2)}$; and
   d. strictly quasiconcave.
2. **Production possibility set** $T = \{((l, k), q) | q \le F(l, k)\}$ - - the collection if input bundles (l, k) and output levels q which are technologically feasible.
3. **Producible output set** $T^* = \{q | ((l, k), q) \in T, q \ge 0\}$ - - collection of all non-negative feasible output levels.
4. **Input requirement set** $L(\bar{q}) = \{(l, k) | ((l, k), \bar{q}) \in T; l, k \ge 0\}$ - - contains all technically feasible input combinations which can produce q. $L(q)$ has the properties:
   a. *monotonicity* – if $(l, k) \in L(q)$ and $(\bar{l}, \bar{k}) \ge (l, k)$, then $(\bar{l}, \bar{k}) \in L(q)$, or the input marginal products $F_l, F_k \ge 0$.
   b. *strict convexity* - - since F is taken to be strictly quasiconcave (Figure 10.7a). If $\bar{q} = F(l, k)$, q = constant, we get the **isoquant** or constant product curve $L(\bar{q}) = \{(l, k) | ((l, k), \bar{q}) \in T, \bar{q} = F(l, k); l, k \ge 0\}$

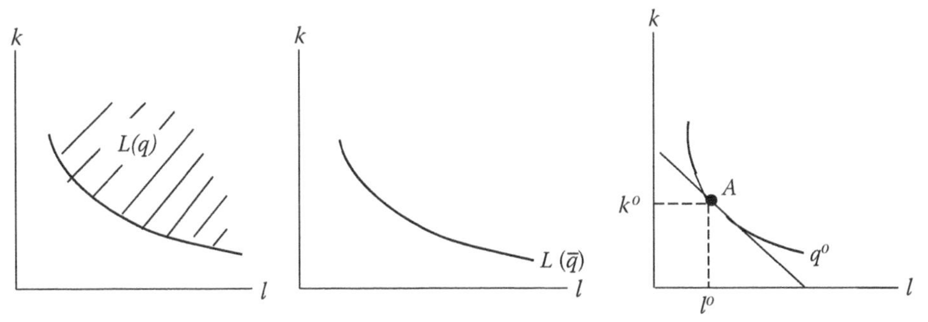

FIGURE 10.7 (a) The input requirement set; (b) the $L(\bar{q})$ isoquant; and (c) constrained cost equilibrium.

- - the locus of input combinations yielding the same level of output $\bar{q}$ (Figure 10.7b).

c. it is **regular** - - $L(q) \neq \varnothing$ and closed (it contains all of its boundary points) for $q \geq 0$.

5. T is **input-regular** if: (a) $T^* \neq \varnothing$; (b) $q > 0$ requires $l, k > 0$; and (c) for each $q \in T^*$, $L(q)$ is regular.

6. T is **input-conventional** if: (a) it is input-regular; and (b) $L(q)$ obeys the monotonicity and strict convexity properties. (Note that if T is a convex set, then so is $L(q)$, but not conversely since T has greater dimensionality than $L(q)$.)

In the discussion which follows it will be assumed that *T is input conventional* so that: $T^* \neq \varnothing$; $q > 0$ requires $l, k > 0$; $L(q)$, when $q \in T^*$, is monotonic, strictly convex (since F is strictly quasiconcave), and regular.

For $q = \bar{q} =$ constant, $dF = F_l\, dl + F_k\, dk = 0$ or $dk/dl = -F_l/F_k$ (the slope of the $\bar{q}$ isoquant).

Then $-\, dk/dl = F_1/F_k$ will be termed the **technical rate of substitution of l for k** ($TRS_{l,k}$) - - the rate at which l must be substituted for k (or k for l) to maintain the same level of output $\bar{q}$. With F strictly quasiconcave, $L(\bar{q})$ is (strictly) concave to the origin, i.e., there exists a diminishing technical rate of substitution of l for k along $L(\bar{q})$ or $-d^2k/dl^2 < 0$.

Given the technological setting just described, let us examine the firm's problem of minimizing the cost of producing a fixed level of output $q = \bar{q}$. It will be assumed that there exists a state of perfect competition in the product and input markets so that product price (p), the money wage rate (w), and the price of a unit of capital services in production (r) are all constant. We can now express the firm's decision problem as:

$$\begin{aligned} &\underset{l,k}{\text{minimize}}\ C = wl + rk \ \text{ subject to} \\ &F(l, k) = q \ \text{ or } \ \hat{g}(l, k) = q - F(l, k) = 0. \end{aligned} \tag{10.49}$$

Forming the Lagrangian function $L = (l, k, \lambda) = wl + rk + \lambda(q - F(l, k))$ we may express the first-order conditions for a constrained minimum as

$$\begin{aligned} &\text{(a)}\quad L_l = w - \lambda F_l = 0 \\ &\text{(b)}\quad L_k = r - \lambda F_k = 0 \\ &\text{(c)}\quad L_\lambda = q - F(l, k) = 0 \end{aligned} \tag{10.50}$$

From equations (10.50a,b) we can easily obtain the firm's optimality or equilibrium condition

$$\frac{F_l^0}{F_k^0} = \frac{w}{r}. \tag{10.51}$$

i.e., to minimize the cost of producing a fixed output level q^o, the firm should operate at the point at which the q^o isoquant is tangent to the (lowest possible) cost equation $C = wl + rk$ (point A in Figure 10.7c). As Equation (10.51) dictates, at point A, the rate at which the firm's technology requires it to substitute l for k in order to keep output constant at q^o (the $TRS_{l,k} = F_l^0/F_k^0$) equals the rate at which market opportunities dictate the substitution of l for k (w/r). Moreover, if system (10.50) is solved simultaneously, then, in addition to (l^o, k^o), we also obtain $\lambda^o = dC/dq$ or **marginal cost** - - the minimum increase in total cost resulting from a one unit increase in q^o.

Looking to the second-order condition for a constrained minimum at (l^o, k^o) we require that

$$|H_L^{\hat{g}}| = \begin{vmatrix} L_{ll} & L_{lk} & \hat{g}_l \\ L_{kl} & L_{kk} & \hat{g}_k \\ \hat{g}_l & \hat{g}_k & 0 \end{vmatrix} = \begin{vmatrix} -\lambda F_{ll} & -\lambda F_{lk} & -F_l \\ -\lambda F_{kl} & -\lambda F_{kk} & -F_k \\ -F_l & -F_k & 0 \end{vmatrix} < 0.$$

Expanding this bordered Hessian determinant yields

$$|H_L^{\hat{g}}| = \lambda\,[F_k^2 F_{ll} - 2F_l F_k F_{lk} + F_l^2 F_{kk}] < 0$$

if $[\cdot] < 0$. However, $[\cdot] < 0$ is the condition for the strict quasiconcavity of F. So under the above assumption that F is strictly quasiconcave, the second-order conditions for a constrained minimum of cost is automatically satisfied.

From the first-order condition

$$\frac{F_l}{F_k} = \frac{w}{r}$$
$$q - F(l, k) = 0$$

or system (10.50) we can obtain the set of conditional (on q) or output constrained input demand functions

$$\begin{aligned} l &= j^l(w, r, q) \\ k &= j^k(w, r, q) \end{aligned} \tag{10.52}$$

and the **marginal cost function** $MC = \lambda = m(w, r, q)$, where m depicts the minimum increment in cost resulting from a unit increase in the minimum output requirement q. It is assumed to be *homogeneous of degree one in w and r* - - proportionate changes in input prices cause total cost to change by the same proportion.

As far as the input demands (10.52) are concerned, it is assumed that they possess the following properties:

a. *homogeneity of degree one in input prices* - - proportionate changes in w, r leave the optimal input quantities unchanged (we remain at point A in Figure 10.7c).

b. the output constrained or conditional input demands *add up*, i.e., $wj^l + rj^k = C$.

c. the conditional input demands are *downward sloping* - - $\partial j^l/\partial w < 0$ and $\partial j^k/\partial r < 0$ - - the substitution effect of an own-price increase must be negative.

d. *symmetrical cross-price derivatives* - - $\partial j^l/\partial r = \partial j^k/\partial w$. Moreover, inputs l and k are *substitutes (complements)* if $\partial j^l/\partial r > 0(<0)$.

e. display *output effects* (changes in input usage attributed to a change in output): if l and k are *normal inputs*, then $\partial j^l/\partial q > 0$, $\partial j^k/\partial q > 0$; and if, say $\partial j^k/\partial q < 0$, then k is an *inferior input* - - its quantity demanded decreases (increases) as output increases (decreases).

Additionally, $\partial j^l/\partial q \equiv \partial MC/\partial w$ and $\partial j^k/\partial q \equiv \partial MC/\partial r$. If, say, k is an inferior input, then $\partial MC/\partial r < 0$.

Next, under the additivity property of the output constrained input demands, we may form the **cost function** as $wj^l + rj^k = C = C(w, r, q)$. This function indicates the minimum cost of producing output q at the given input prices w, r. In fact, $C\,(w, r, q)$ is the **optimal value function** under constrained cost minimization. It exhibits the properties:

a. *homogeneity of degree one in input prices* - - proportionate increases in input prices cause total cost to increase by the same proportion.

b. it is *strictly monotonically increasing in q; monotonically non-decreasing in w and r;* and *monotonically increasing in at least one input price.*

c. *concave in input prices* - - as w and r increase, total cost increases at most linearly. Hence, $\partial C/\partial w \geq 0$, $\partial^2 C/\partial w^2 \leq 0$; $\partial C/\partial r \geq 0$, $\partial^2 C/\partial r^2 \leq 0$.

d. C is continuous in w, r, and q and of class $C^{(2)}$ with respect to w, r.

e. possesses the *derivative property* (**Shephard's lemma**) - - the partial derivatives of C with respect to w and r yield the input constrained (cost-minimizing) input demands

$$\frac{\partial C(w, r, q)}{\partial w} \equiv j^l(w, r, q) = l$$
$$\frac{\partial C(w, r, q)}{\partial r} \equiv j^k(w, r, q) = k.$$

In fact, given the concavity of C and Shephard's lemma, these conditional input demands are negatively sloped, e.g., with $\partial C/\partial w = j^l$, $\partial^2 C/\partial w^2 = \partial j^l/\partial w \leq 0$.

Suppose we now consider the transpose of the preceding constrained cost minimization problem, i.e., we seek to solve the constrained output maximization problem

$$\underset{l,k}{\text{maximize}}\ q = F(l, k) \text{ subject to}$$
$$wl + rk = C \text{ or } \hat{g}(l, k) = C - wl - rk$$

given the same market conditions and under the assumption that F is strictly quasiconcave. Forming the associated Lagrangian function $L^*(l, k, \lambda_*) = F(l, k) + \lambda_*(C - wl - rk)$ we may derive the first-order conditions for a constrained maximum as

$$\begin{aligned} &\text{(a)} \quad L_l^* = F_l - \lambda_* w = 0 \\ &\text{(b)} \quad L_k^* = F_k - \lambda_* r = 0 \\ &\text{(c)} \quad L_{\lambda_*}^* = C - wl - rk = 0. \end{aligned} \tag{10.53}$$

From Equations (10.52a,b) we obtain the optimality or equilibrium condition $F_l/F_k = w/r$ holding at (l^o, k^o) (see Figure 7.10c).

For a cost-constrained maximum of $F(l, k)$ at (l^o, k^o) it is sufficient that the following second-order condition be satisfied:

$$|\boldsymbol{H}_{L^*}^{\hat{g}}| = \begin{vmatrix} L_{ll}^* & L_{lk}^* & \hat{g}_l \\ L_{kl}^* & L_{kk}^* & \hat{g}_k \\ \hat{g}_l & \hat{g}_k & 0 \end{vmatrix} = \begin{vmatrix} F_{ll} & F_{lk} & -w \\ F_{kl} & F_{kk} & -r \\ -w_l & -r & 0 \end{vmatrix} > 0.$$

Since

$$|\boldsymbol{H}_{L^*}^{\hat{g}}| = -(1/\lambda_*^2)[F_k^2 F_{ll} - 2F_l F_k F_{lk} + F_l^2 F_{kk}] > 0$$

if $[\cdot] < 0$ (the condition for the strict quasiconcavity of F), we see that the second-order condition for a cost-constrained maximum is automatically satisfied if F is deemed strictly quasiconcave.

Once system (10.52) is solved simultaneously, we obtain the **conditional (on C) or cost-constrained input demand functions**

$$l = h^l(w, r, C)$$
$$k = h^k(w, r, C)$$

and the inverse of the MC function

$$\begin{aligned} \lambda_*^o = 1/\lambda^o &= u(w, r, C) = u(w, r, C(w, r, q)) \\ &= m_*(w, r, q) \end{aligned} \tag{10.54}$$

which yields the maximum increase in output resulting from a \$1 increase in expenditure C. The properties of these cost constrained input demands are:

1. *homogeneous of degree zero in input prices and expenditure C* (which keeps us at the equilibrium point A in Figure 10.7c).
2. these cost-constrained input demands, weighted by own-input prices, *add up*, i.e., $wh^l + rh^k = C$.
3. *downward sloping* - - $\partial h^l/\partial w < 0$, $\partial h^k/\partial r < 0$.
4. the cost-constrained input demands exhibit *symmetrical cross-price derivatives* - - $\partial h^l/\partial r = \partial h^k/\partial w$.

We next introduce the **indirect production function** (the *optimal value function* under constrained output maximization)

$$\begin{aligned} q = F(l, k) &= F[h^l(w, r, C), h^k(w, r, C)] \\ &= Q(w, r, C), \end{aligned} \tag{10.55}$$

which gives the maximum output attainable from an expenditure level C at fixed input prices w, r. It has the properties:

1. *homogeneous of degree zero in w, r, and C.*
2. *strictly monotonically increasing in C; monotonically non-increasing in w and r; and monotonically decreasing in at least one input price.*
3. *quasiconcave in w and r* - - the upper contour set $\mathcal{G} = \{(w, r) | v(w, r, C) \geq b\}$ is convex for all real b or the contour $v(w, r, C) = b$ is convex to the origin.
4. *continuous in w, r, and C.*
5. if differentiable, yields the cost-constrained input demands via **Roy's identity**:

$$l = h^l(w, r, C) = \frac{-\partial Q/\partial w}{\partial Q/\partial C};$$
$$k = h^k(w, r, C) = \frac{-\partial Q/\partial r}{\partial Q/\partial C}.$$

It is important to note further that the indirect production function $q = Q(w, r, C)$ and the cost function $C = \mathcal{C}(w, r, q)$ are mutual inverses or $Q = \mathcal{C}^{-1}$ and $\mathcal{C} = Q^{-1}$. Hence, the output-constrained input demands may be obtained directly from cost-constrained input demands and vice versa or

$$j^l(w,r,q) = h^l(w,r,\mathcal{C}(w,r,q)),$$
$$j^k(w,r,q) = h^k(w,r,\mathcal{C}(w,r,q));$$

and

$$h^l(w,r,C) = j^l(w,r,Q(w,r,C)),$$
$$h^k(w,r,C) = j^k(w,r,Q(w,r,C)).$$

All of the preceding developments are summarized in Figure 10.8 under the rubric of production-cost duality. Additional details on the same can be found in Appendix 10.7.4A.

Example 10.13: Suppose a firm's objective is to minimize $C = wl + rk$ subject to $q = 3l^{\frac{1}{2}} + 5k^{\frac{1}{2}}$. From the Lagrangian $L = wl + rk + \lambda\left(q - 3l^{\frac{1}{2}} - 5k^{\frac{1}{2}}\right)$ we obtain

$$L_l = w - \frac{3}{2}\lambda l^{-\frac{1}{2}} = 0$$
$$L_k = r - \frac{5}{2}\lambda k^{-\frac{1}{2}} = 0$$
$$L_\lambda = q - 3l^{\frac{1}{2}} - 5k^{\frac{1}{2}} = 0.$$

From the first two equations of this system we obtain

$$l^o = \left(\frac{2}{3}\frac{w}{\lambda}\right)^{-2}$$
$$k^o = \left(\frac{2}{5}\frac{r}{\lambda}\right)^{-2}.$$

Then substituting l^o, k^o into the constraint yields

$$\lambda^o = MC = m(w,r,q) = \left[\frac{2wr}{9r + 25w}\right]q$$

with $dMC/dq > 0$.
The output constrained input demands are

$$l = j^l(w,r,q) = \left[\frac{3rq}{9r + 25w}\right]^2$$
$$k = j^k(w,r,q) = \left[\frac{5wq}{9r + 25w}\right]^2.$$

Under the additivity property of j^l and j^k we may obtain the cost function as

$$wj^l + rj^k = C(w,r,q) = \frac{wrq^2}{9r + 25w},$$

with $\partial C/\partial q = MC = \lambda^o$. ∎

Example 10.14: Assume that the firm's objective is to maximize $Q = 3l^{\frac{1}{2}} + 5k^{\frac{1}{2}}$ subject to $C = wl + rk$. Given the Lagrangian $L^* = 3l^{\frac{1}{2}} + 5k^{\frac{1}{2}} + \lambda_*(C - wl - rk)$ we find that

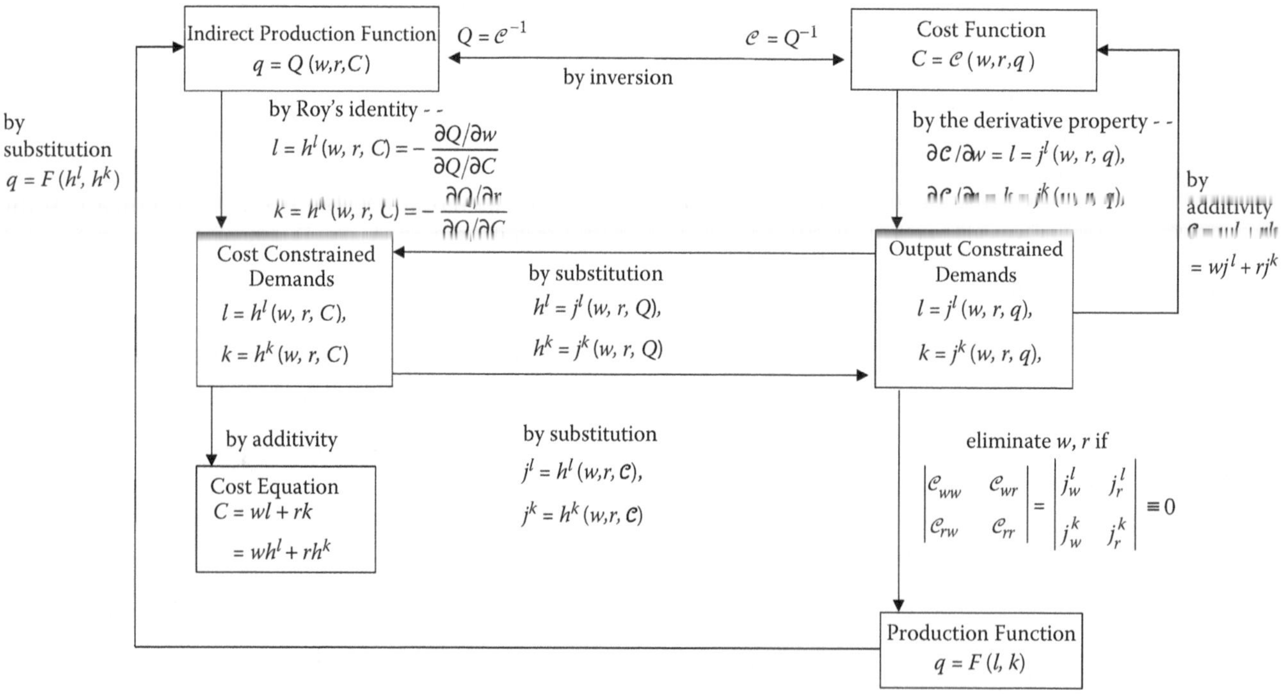

FIGURE 10.8 Production–cost duality.

$$L_l^* = \frac{3}{2}l^{-\frac{1}{2}} - w\lambda_* = 0$$
$$L_k^* = \frac{5}{2}k^{-\frac{1}{2}} - r\lambda_* = 0$$
$$L_{\lambda_*}^* = C - wl - rk = 0.$$

The first two equation of this system render

$$l^o = \left(\frac{2}{3}w\lambda_*\right)^{-2}$$
$$k^o = \left(\frac{2}{5}r\lambda_*\right)^{-2}.$$

A substitution of l^o and k^o into the constraint gives us

$$\lambda_*^o = \frac{1}{2}\left[\frac{9r + 25w}{wr}\right]^{\frac{1}{2}} C^{-\frac{1}{2}}.$$

The cost-constrained input demand functions are

$$l = h^l(w, r, C) = \frac{9w^{-1}rC}{9r + 25w}$$
$$k = h^k(w, r, C) = \frac{25r^{-1}wC}{9r + 25w}.$$

Note that if we employ the cost function $C(w, r, q)$ from Example 10.13, then

$$\lambda_* = \frac{1}{2}\left[\frac{9r + 25w}{wr}\right]^{\frac{1}{2}}\left[\frac{9r + 25w}{wrq^2}\right]^{\frac{1}{2}} = \frac{1}{2}\left[\frac{9r + 25w}{wr}\right]q^{-1}$$
$$= m_*[w, r, q] = \frac{1}{\lambda}.$$

To obtain the indirect product function we simply find

$$q = F(h^1, h^k) = 3(h^l)^{\frac{1}{2}} + 5(h^k)^{\frac{1}{2}}$$
$$= \frac{C^{\frac{1}{2}}(9r + 25w)^{\frac{1}{2}}}{(wr)^{\frac{1}{2}}} = Q(w, r, C).$$ ■

In addition to considering output-constrained cost minimization and cost-constrained output maximization, the firm may also adopt a strategy of long-run profit maximization. Again, assuming constant product and input prices (p, w, and r), the long-run profit function can be expressed as the difference between total revenue and total cost or

$$\Pi(l, k) = TR(l, k) - TC(l, k)$$
$$= pq - wl - rk$$
$$= pF(l, k) - wl - rk,$$

where output q is determined by a *strictly concave* production function $F(l, k)$.

Looking to the first-order condition for profit maximization we have

$$\left.\begin{array}{l}\Pi_l = pF_l - w = 0\\ \Pi_k = pF_k - r = 0\end{array}\right\} \text{ or } \left\{\begin{array}{l}VMP_l = w\\ VMP_k = r.\end{array}\right. \quad (10.56)$$

Thus, for long-run equilibrium, the firm's *optimal decision rule* is: to maximize profit (under the stated assumptions) the firm should hire each input up to the point at which the value of its marginal product equals its unit price.

As to the second-order condition for profit maximization, we require that the naturally ordered principal minors of the Hessian of the profit function,

$$\mathbf{H}_\Pi = \begin{bmatrix} \Pi_{ll} & \Pi_{lk} \\ \Pi_{kl} & \Pi_{kk} \end{bmatrix},$$

alternate in sign, beginning negative, or

$$\Pi_{ll} < 0, \quad \begin{vmatrix} \Pi_{ll} & \Pi_{lk} \\ \Pi_{kl} & \Pi_{kk} \end{vmatrix} > 0. \quad (10.57)$$

But since

$$\mathbf{H}_\Pi = \begin{bmatrix} pF_{ll} & pF_{lk} \\ pF_{kl} & pF_{kk} \end{bmatrix} = p\mathbf{H}_F,$$

We see that the Hessian of the profit function is negative definite if the Hessian of the production function is negative definite ($p > 0$). So if F is strictly concave, then Π is strictly concave and thus the profit function attains a unique global maximum.

If we eliminate p from system (10.56), then we get $F_l/F_k = w/r$ - - the first-order condition for, say, output-constrained cost minimization. This observation leads us to conclude that if the firm maximizes profit, then it also minimizes cost subject to an output constraint. It is important to note the following:

a. under output-constrained cost minimization we need F to be strictly quasiconcave so that the isoquants have the correct shape.

b. under profit maximization we need F to be strictly concave (the isoquants have the correct shape and there exists diminishing input marginal products).

c. Since strict concavity implies strict quasi-concavity, the second-order condition for profit maximization is more restrictive than that for output-constrained cost minimization.

If we solve system (10.56) simultaneously, we obtain the **ordinary input demands**

$$\begin{aligned} l &= y^l(w, r, p) \\ k &= y^k(w, r, p) \end{aligned} \tag{10.58}$$

with properties:

a. *homogeneity of degree zero in product and input prices.*

b. downward sloping – the own-price derivatives are negative or $\partial j^l/\partial w < 0$, $\partial j^k/\partial r < 0$.

c. *symmetric cross-price derivatives* - - $\partial y^l/\partial r = \partial y^k/\partial w$. Inputs l, k are *substitutes (complements)* if $\partial y^l/\partial r > 0(<0)$.

d. *product-price derivatives are usually positive* - - usually $\partial y^l/\partial p > 0$, $\partial y^k/\partial p > 0$. If $l(k)$ is an *inferior input*, then $\partial y^l/\partial p < 0(\partial y^k/\partial p < 0)$.

To find the **indirect production function** we simply make the substitution

$$q = F\left[y^l(w, r, p), y^k(w, r, p)\right] = \Phi(w, r, p). \tag{10.59}$$

This expression may be viewed as a **generalized supply function**, which is homogeneous of degree zero in *all* prices. (Note that if w and r are held constant at $\bar{w}$ and $\bar{r}$, respectively, then

Equation (10.59) can be expressed as $q = \Phi(\bar{w}, \bar{r}, p) = \varnothing(p)$, $dq/dp > 0$. If we write $p = \varnothing^{-1}(q)$, $dp/dq > 0$, then we have the **ordinary supply function** for q.)

Differentiating Equation (10.59) with respect to product price p yields

$$\frac{\partial \Phi}{\partial p} = \frac{\partial F}{\partial l}\frac{\partial y^l}{\partial p} + \frac{\partial F}{\partial k}\frac{\partial y^k}{\partial p} > 0$$

(the generalized supply function must be positively sloped) so either $\partial y^l/\partial p > 0$ or $\partial y^k/\partial p > 0$ or both of these derivatives must be positive, i.e., there must be an increase in the demand for at least one input or all inputs cannot be inferior.

Finally, let us consider the **indirect profit function** (the *optimal value function*)

$$\begin{aligned} \Pi = p\Phi(w, r, p) &- wy^l(w, r, p) \\ &- ry^k(w, r, p) = P(w, r, p) \end{aligned}$$

which represents the maximum profit attainable given the technology and fixed product and factor prices. Its properties are:

a. *non-decreasing in p and non-increasing in w, r.*

b. *homogeneous of degree one in all prices.*

c. *convex in all prices*, i.e.,

$$\begin{aligned} \frac{\partial^2 P}{\partial p^2} &= \frac{\partial \Phi}{\partial p} \geq 0; \\ \frac{\partial^2 P}{\partial w^2} &= -\frac{\partial y^l}{\partial w} \geq 0; \quad \text{and} \\ \frac{\partial^2 P}{\partial r^2} &= -\frac{\partial y^k}{\partial r} \geq 0. \end{aligned}$$

d. *continuous in all prices* for p, w, $r > 0$.

e. possesses the **derivative property (Hotelling's lemma)** - - if P is differentiable with respect to p, w, and r then

$\frac{\partial P}{\partial p} = \Phi(w, r, p)$ (an increase in p increases profit, with the rate of increase being the optimal output level); $\frac{\partial P}{\partial w} = -y^l(w, r, p)$ and $\frac{\partial P}{\partial r} = -y^k(w, r, p)$ (profit decreases with an increase in the price of an input, with the rate of decrease being the optimal input employment level).

Example 10.15: Given the production function $q = F(l, k) = 3l^{\frac{1}{2}} + 5k^{\frac{1}{2}}$, the firm's profit function can be written as

$$\Pi(l, k) = pq - wl - rk = p\left(3l^{\frac{1}{2}} + 5k^{\frac{1}{2}}\right) - wl - rk.$$

The first-order conditions for maximizing Π are

$$\Pi_l = \frac{3}{2}pl^{-\frac{1}{2}} - w = 0$$
$$\Pi_k = \frac{5}{2}pk^{-\frac{1}{2}} - r = 0.$$

Solving this system simultaneously gives the ordinary input demand functions

$$l = \frac{9}{4}\left(\frac{p}{w}\right)^2 = y^l(w, r, p)$$
$$k = \frac{25}{4}\left(\frac{p}{r}\right)^2 = y^k(w, r, p).$$

Going to the second-order condition for profit maximization we have

$$\Pi_{ll} = -\frac{3}{4}pl^{-\frac{3}{2}} < 0, \quad \begin{vmatrix} \Pi_{ll} & \Pi_{lk} \\ \Pi_{kl} & \Pi_{kk} \end{vmatrix} = \begin{vmatrix} -\frac{3}{4}pl^{-\frac{3}{2}} & 0 \\ 0 & -\frac{5}{4}pk^{-\frac{3}{2}} \end{vmatrix} = \frac{15}{16}p^2 l^{-\frac{3}{2}}k^{-\frac{3}{2}}.$$

If we substitute y^l and y^k for l and k, respectively, into these expressions we obtain

$$\Pi_{ll} = -\frac{2}{9}\frac{w}{p^2} < 0, \quad \begin{vmatrix} \Pi_{ll} & \Pi_{lk} \\ \Pi_{kl} & \Pi_{kk} \end{vmatrix} = \frac{4}{225}\frac{(wr^3)}{p^4} > 0.$$

Hence, Π attains a strong global maximum at (l, k). (Remember, however, that the strict concavity of F guarantees that a strong global maximum of Π occurs). If we form the ratio

$$\frac{k}{l} = \frac{y^k}{y^l} = \frac{25}{9}\left(\frac{w}{r}\right)^2,$$

then we may determine the firm's **long-run expansion path**

$$k = e(l) = \frac{25}{9}\left(\frac{w}{r}\right)^2 l,$$

the locus of economically efficient ($F_l/F_k = w/r$ at all points along the expansion path) input combinations corresponding to successively higher levels of total expenditure C.

Consider next the indirect production function

$$q = 3\left(\frac{9}{4}\left(\frac{p}{w}\right)^2\right)^{\frac{1}{2}} + 5\left(\frac{25}{4}\left(\frac{p}{r}\right)^2\right)^{\frac{1}{2}}$$
$$= \frac{p}{2}\left(\frac{9}{w} + \frac{25}{r}\right) = \Phi(w, r, p).$$

Solving for p yields

$$p = \frac{2qwr}{9r + 25w}(= \lambda = MC = m(w, r, q)).$$

And if $\bar{w}$, $\bar{r}$ are held constant at $\bar{w}$ and $\bar{r}$, respectively, then the ordinary supply function for q appears as

$$p = \varnothing^{-1}(q) = \frac{2q\bar{w}\bar{r}}{9\bar{r} + 25\bar{w}}, \quad \frac{dp}{dq} > 0.$$

Finally, we may determine the profit function (optimal value function) as

$$\Pi = p\Phi(w, r, p) - wy^l - ry^k$$
$$= \frac{p^2}{4}(9w^{-1} + 25r^{-1}) = P(w, r, p).$$

APPENDIX 10.7.3.A THE HICKSIAN DEMANDS POSSESS THE DERIVATIVE PROPERTY

Consider equation system (10.44). Let q_1^0, q_2^0 represent the equilibrium or expenditure minimizing combination of commodities which yield u = constant at prices p_1^0 and p_2^0 so that $E(p_1^0, p_1^0, u) = p_1^0 q_1^0 + p_2^0 q_2^0$. Additionally, let p_1 and p_2 be any other set of non-negative prices with corresponding equilibrium commodity levels

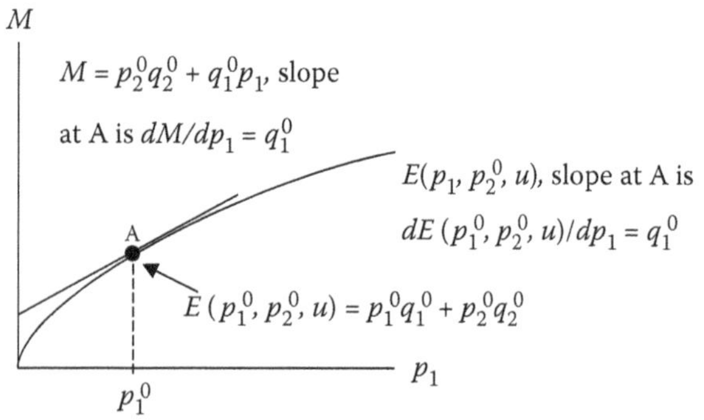

FIGURE 10.A.1 The derivative property illustrated.

q_1 and q_2 which also yields u. Since at these *new prices* $E\,(p_1, p_2, u)$ is the least expensive way to achieve u, we must have

$$E\,(p_1, p_2, u) = p_1 q_1 + p_2 q_2 \le p_1^0 q_1^0 + p_2^0 q_2^0.$$

From this expression we may form the function

$$z\,(p_1, p_2) = E\,(p_1, p_2, u) - p_1^0 q_1^0 - p_2^0 q_2^0 \le 0.$$

Clearly $z\,(p_1^0, p_2^0) = 0$ and, for all other p_1 and p_2 values, $z\,(p_1, p_2) < 0$. Since $p_1 = p_1^0, p_2 = p_2^0$ maximize z, its derivatives must vanish at this point or

$$\frac{\partial z}{\partial p_1} = \frac{\partial E}{\partial p_1} - q_1^0 = 0, \quad \frac{\partial z}{\partial p_2} = \frac{\partial E}{\partial p_2} - q_2^0 = 0,$$

the set of equilibrium Hicksian or compensated demands appearing in system (10.44).

Looked at geometrically, consider Equation (10.44a). For $p_2 = p_2^0$ and u constant, the expenditure function $M = E\,(p_1, p_2^0, u)$ and the linear expenditure constraint $p_1^0 q_1^0 + p_2^0 q_2^0 = M$ appear in Figure 10.A.1, where, as above q_1^0 and q_2^0 depict the expenditure minimizing commodity combination which generates u at prices p_1^0, p_2^0. Since this expenditure constraint is tangent to E for $p_1 = p_1^0$ (point A), their slopes must be equal or Equation (10.44a) obtains.

We noted above that the compensated demands are negatively sloped. From the concavity of the expenditure function and Shephard's lemma we have, from Equation (10.44a), $\partial E/\partial p_1 = q_1 = h^1(p_1, p_2, u)$. With E concave in prices, $\partial^2 E/\partial p_1^2 = \partial h^1/\partial p_1 \le 0$ as required. We also noted from Equation (10.42) that the cross-price derivatives of the compensated demands are symmetric. To see this we have, from Shephard's lemma,

$$\frac{\partial q_1}{\partial p_2} = \frac{\partial^2 E}{\partial p_2 \partial p_1} = \frac{\partial^2 E}{\partial p_1 \partial p_2} = \frac{\partial q_2}{\partial p_1} \tag{10.A.1}$$

(by Young's theorem). (The reader should verify this result using the compensated demands and expenditure function obtained in Example 10.12.)

APPENDIX 10.7.3.B THE INCOME AND SUBSTITUTION EFFECTS

Consider the Marshallian (ordinary) demand function for commodity q_1 or $q_1 = g^1\,(p_1, p_2, M)$. Let us set $p_2 = \bar{p}_2 =$ constant and $M = \bar{M} =$ constant. Then the preceding expression becomes what we have called the own-price demand function $q_1 = g\,(p_1)$.

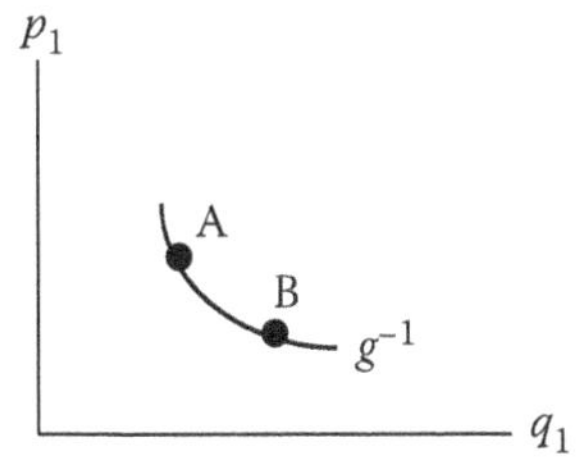

FIGURE 10.B.1 The law of demand in operation.

We previously described the **law of demand** in the following fashion: as the price of a product falls, other things being equal, the corresponding quantity demanded (generally) increases, i.e., generally $dq_1/dp_1 < 0$. With g continuous and single-valued, using $p_1 = g^{-1}(q_1)$, we have $dp_1/dq_1 < 0$. So under the stated *ceteris paribus* price change, we move downward the inverse demand curve from point A to point B (Figure 10.B.1). In the movement from A to B, two forces are operating:

1. **pure income effect** - - at the lower unit price the household can afford more q_1 without giving up the consumption of any other commodities; the price decrease has increased the purchasing power of the households money income (its real income has increased).
2. **pure substitution effect** - - at the lower price the household has an incentive to substitute the cheaper good for similar goods which are relatively more expensive.

So as the own price of a good falls, the substitution and income effects combine to generally induce the household to purchase more of the product at a lower price than at a higher one. Interestingly enough, the income and substitution effects are *additive*:

$\frac{dq_1}{dp_1}$	=	pure substitution effect +	pure income effect.
[Total rate of change in q_1 as its price changes.]		[Rate at which q_1 is substituted for q_2 when p_1 changes. Operates as if real income is constant and only relative prices change.]	[Rate at which purchases of q_1 would change with changes in income. Operates as if prices are constant and real income has change.]

So if, say, p_1 decreases, then the substitution effect *always* calls for an increase in q_1 consumption; however, under the income effect, either more or less of q_1 will be consumed. The total (final) effect of an own-price decrease might be indeterminate. For a *normal or superior good*, the income effect reinforces the substitution effect - - $\partial q_1/\partial p_1 < 0$ and the law of demand holds. For an *inferior good*, the income effect tends to offset somewhat the strength of the substitution effect; expenditure on q_1 declines as real income increases - - on balance $\partial q_1/\partial p_1 < 0$ and the law of demand is still valid. And for a *Giffen good* (the case of extreme inferiority), the income effect dominates the substitution effect; the quantity demanded of q_1 decreases as its own price decreases over some limited range - - $\partial q_1/\partial p_1 > 0$ so the law of demand is violated.

Consider next the (income -) compensated or Hicksian demand for q_1 or $q_1 = h^1(p_1, p_2, u)$.

While the Marshallian demand curve (MDC) is *observed*, the compensated demand function (CDC) is actually *unobserved*. However, we may generate the latter by performing the following *experiment*: eliminate the income effect of a price change by keeping a household's utility or real income constant so that only the substitution effect is operative. To this end, suppose some authority imposes a lump-sum tax or dispenses a subsidy so that the household's *utility is unchanged* after a price decrease or increase, respectively (Figure 10.B.2)

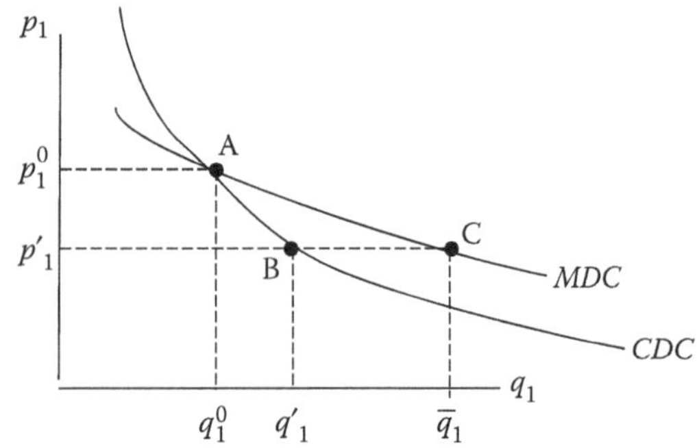

FIGURE 10.B.2 Marshallian (ordinary) vs. Hicksian (compensated) demand functions.

Suppose the household is initially at Point A, where the maximum utility achieved by purchasing q_1^0 on MDC equals the value of u in CDC while the income value M in MDC coincides with the minimum expenditure necessary to achieve u in CDC. Now, let p_1 decrease from p_1^0 to p'_1. Since real income increases, compensation will be negative (a lump sum tax increase causes the increase in real income to disappear). Hence, CDC yields a lower quantity demanded than MDC. Clearly the pure income effect of a price decrease ($q'_1\bar{q}_1$) is eliminated; only the pure substitution effect ($q_1^0 q'_1$) remains. Hence, we move to point B on CDC. To summarize: since MDC is a *constant money income* demand curve, the total effect of a price change will be greater than that for CDC since movements along MDC reflect the operation of both the income and substitution effects; and since CDC is a *constant real income* (utility) demand curve, once compensation takes place, the income effect is eliminated - - only the (negative) substitution effect remains. So, for price decreases (increases), CDC lies below (above) MDC.

Let us now examine the analytical version of this discussion. Again looking to commodity q_1, we know that, at equilibrium, $g^1(p_1, p_2, M) = h^1(p_1, p_2, u)$, i.e., for a given M, at optimality, q_1 is the same regardless of whether it is determined from the Marshallian or Hicksian demand function. Let us hold p_2 constant. Then from the preceding equality,

$$\frac{\partial g^1}{\partial p_1}dp_1 + \frac{\partial g^1}{\partial M}dM = \frac{\partial h^1}{\partial p_1}dp_1 + \frac{\partial h^1}{\partial u}du. \tag{10.B.1}$$

Since real income or utility is held constant, Equation (10.B.1) becomes

$$\frac{\partial g^1}{\partial p_1}dp_1 + \frac{\partial g^1}{\partial M}dM = \frac{\partial h^1}{\partial p_1}dp_1. \tag{10.B.2}$$

Suppose the household is compensated by the amount $dM = q_1 dp_1$. Then Equation (10.B.2) appears as

$$\frac{\partial g^1}{\partial p_1}dp_1 + \frac{\partial g^1}{\partial M}q_1 dp_1 = \frac{\partial h^1}{\partial p_1}dp_1$$

or

$$\frac{dh^1}{dp_1} = \frac{\partial g^1}{\partial p_1} + q_1\frac{\partial g^1}{\partial M} \leq 0, \tag{10.B.3}$$

i.e., the slope of the compensated demand curve is non-positive since the expenditure function is concave in prices.

Since (income) compensated demand is unobserved, let us recast Equation (10.B.3) in terms of the observable Marshallian demand function or

$$\underset{\text{[Total own–price effect]}}{\frac{\partial g^1}{dp_1}} = \underset{\text{[Substitution effect]}}{\frac{\partial h^1}{dp_1}} \underbrace{-q_1\frac{\partial g^1}{dM}}_{\text{[Income effect]}}. \tag{10.B.4}$$

Here Equation (10.B.4) is known as the **Hicks-Slutsky (H-S) decomposition of an own-price effect** or simply the **H-S equation.**

In sum, the total effect of an uncompensated response to, say, an own-price decrease equals the compensated response or substitution effect of the decrease plus the effect of withdrawing the compensation or income effect. Thus the substitution effect eliminates the increase in real income (the household is negatively compensated by an amount sufficient to restore its original level of utility) so that we can isolate the change in the commodity mix attributed solely to a change in real prices while the income effect depicts the impact on the commodity mix due to a change in purchasing power.

Looking to Figure 10.B.2, the total effect of a price decrease is a movement from A to C. The pure substitution effect is the movement along a compensated demand function from A to B; the pure income effect involves the movement from CDC to MDC or B to C. Hence, the total effect is

$$\underset{\text{[Total effect]}}{A \to C} = \underset{\text{[Substitution effect]}}{(A \to B)} + \underset{\text{[Income effect]}}{(B \to C)}$$

or

$$\underset{\text{[Total effect]}}{q_1^0\bar{q}_1} = \underset{\text{[Substitution effect]}}{q_1^0 q'_1} + \underset{\text{[Income effect]}}{q'_1\bar{q}_1}.$$

In terms of elasticities, the H-S decomposition of an own-price change can be written in terms of own-price and income elasticities as

$$\in_{11} = \in_{11}^* - w_1 \in_{1M},$$

where

$$\in_{11}^* = \frac{\partial h^1}{\partial p_1}\frac{p_1}{h^1}$$

is the own-price elasticity of the *compensated* demand function.

APPENDIX 10.7.4.A **PRODUCTION-COST DUALITY: A CLOSER LOOK** (MCFADDEN [19781]; FUSS AND MCFADDEN [1978]; DIEWERT [1973, 1982]; ARRIEL, ET AL. [2010]; AND JORGENSON AND LAU [1974])

At the outset of Section 10.7.4 we introduced concepts such as the production possibility (technology) set (T), the producible output set (T^*), and the input requirement set ($L(q)$). Additionally, the notions of T being input-regular and input-conventional were also offered. To address the cost side of production-cost duality, we must also consider the concept of an **input-conventional cost structure**, consisting of:

a. $T^* \neq \emptyset$;

b. $C = C(w, r, q)$ is defined for $q \in T^*$ and $w, r > 0$ with C non-negative and strictly positive for each $q > 0$ and possessing properties:

i. homogeneous of degree one in w, r;

ii. strictly monotonically increasing in q, monotonically non-decreasing in w and r and monotonically increasing in at least one input price;

iii. concave in w, r;

iv. continuous in w, r, and q and of class $C^{(2)}$ in w, r; and

v. possesses the derivative property (Shephard's lemma).

Given an input-conventional cost structure, we may define, for each $q \in T^*$, the **implicit input requirement set**

$$L^*(q) = \{(l, k) | l, k \geq 0, wl + rk \geq C(w, r, q) \text{ for all } w, r > 0\}.$$

And for $L^*(q) \neq \emptyset$, the associated **implicit production possibility set** is

$$\hat{T} = \{((l, k), q) | q \in T^*, (l, k) \in L^*(q)\}.$$

Given these set-theoretic considerations, we first note that **production-cost duality** essentially states that a hypothesis about the structure of an input-conventional production possibility set is a hypothesis concerning the form of the (indirect) cost function. That is, corresponding to every input-conventional production possibility set T is a cost function C which is "well-behaved" (it is homogeneous, monotonic, concave, continuous, and exhibits the derivative property); and, conversely, for a function C with these five properties, there exists an input-conventional production possibility set such that the given function C is its minimum cost function.

More formally, the preceding discussion may be focused a bit more sharply by the following two duality principles, the first of which establishes the notion that *each* input-conventional cost function determines an input-conventional (implicit) production possibility set. Thus the production function can be derived from a "well-behaved" cost function. In this regard we have

Production-Cost Duality Principle 1

If $C(w, r, q)$ is an input-conventional cost function defined for $q \in T^*$, then the implicit input requirement set $L^*(q) \neq \emptyset$ for each $q \in T^*$, and thus the implicit production possibility set $\hat{T}$ is input-conventional.

To set the stage for our next duality result let us consider the concept of the **cost mapping**, a function

$$C = C(w, r, q) = \min_{(l,k)}\{wl + rk | l, k \in L(q)\} \tag{10.7.4.A.1}$$

from the class of input-conventional production possibility sets into the class of input-conventional cost structures. Here C is, for given factor prices, the function which gives the lowest cost level associated with the input combination (l, k).

Similarly, a **technology mapping** is a function

$$L^*(q) = \{(l, k) | l, k \geq 0, wl + rk \geq C(w, r, q) \text{ for all } w, r > 0\} \tag{10.7.4.A.2}$$

from the class of input-conventional cost structures into the class of input-conventional production possibility sets. Note that the technology mapping is structured so as to eliminate input bundles (l, k)which cost less than $C(w, r, q)$ and thus could not possibly yield q. Thus $L^*(q)$ represents the technology which is compatible with C in that if $(l, k) \in L^*(q)$, then it is at least as costly as any input combination yielding the minimum in the cost mapping. In the light of this discussion the second duality result informs us that there exists a one-to-one relationship between input-conventional production possibility sets and input-conventional cost structures. Specifically,

Production-Cost Duality Principle 2

Applying the cost mapping to an input-conventional production possibility set yields an input-conventional cost structure. Application of the technology mapping to this cost structure yields the initial production possibility set. Conversely, application of the technology mapping to an input-conventional cost structure yields an input-conventional production possibility set. Application of the cost mapping to this production possibility set yields the initial cost structure.

This one-to-one transformation between these two input-conventional classes means that the cost and technology mappings are mutual inverses, i.e., all structural features of the technology are translated via the cost mapping into the specification of the cost function and retrieved from the latter by the technology mapping. To summarize:

a. Applying the cost mapping (10.7.4.A.1) to an input-conventional production possibility set produces a cost function which mirrors the technology;

b. Applying the technology mapping (10.7.4.A.2) to an input-conventional cost structure recovers all salient features of the production possibility set.

Example 10.7.4.A.1:

Given the production function $q = F(l, k) = l^{\frac{1}{5}}k^{\frac{3}{5}}$, the cost mapping produces the cost function $C = C(w, r, q) = 1.7547\ w^{\frac{1}{4}}r^{\frac{3}{4}}q^{\frac{5}{4}}$. To move in the reverse direction and recover the production function, let

$$\begin{vmatrix} C_{ww} & C_{wr} \\ C_{rw} & C_{rr} \end{vmatrix} = C_{ww}C_{rr} - C_{wr}^2 \equiv 0$$

for all $w, r > 0$, i.e., this condition requires that the expression $\partial C/\partial w = l$, $\partial C/\partial r = k$ (Shephard's lemma) are functionally dependent and thus satisfy a relationship of the form $\psi(l, k) = q - F(l, k) = 0$ where ψ may be obtained by eliminating w, r from the system

$$\partial C/\partial w = l,\ \partial C/\partial r = k.$$

In this regard, for

$$\begin{aligned} \partial C/\partial w &= 0.4387\ w^{-\frac{3}{4}}r^{\frac{3}{4}}q^{\frac{5}{4}} = l = j^l(w, r, q) \\ \partial C/\partial r &= 1.3160\ w^{\frac{1}{4}}r^{-\frac{1}{4}}q^{\frac{5}{4}} = k = j^k(w, r, q) \end{aligned}$$

it is easily shown that $C_{ww}C_{rr} - C_{wr}^2 \equiv 0$. Then from this system we may solve for q in terms of l, k to obtain the above production function. ■

Given the preceding discussion, some additional comments pertaining to duality in economic analysis are in order. Specifically, consider the *primal problem* of optimizing an objective function subject to a functional constraint (e.g., maximizing output subject to a cost constraint). When either the objective function or the constraint is *affine* (of the form $\boldsymbol{a}^T\boldsymbol{x} - \alpha = 0$, $\alpha \in R$), it turns out that the maximum value function has as its arguments the parameters of the affine function involved in the primal optimization. Then the *dual function* is the maximum value function whose arguments are these same parameters. Given certain restrictions or regularity conditions, the dual function can be used to recover the nonlinear function that appeared in the primal (Avriel, et al. [2010]).

Section 10.7.1–10.7.3 Exercises:

10.7.1 Given the linear budget constraint $p_1 q_1 + p_2 q_2 = M$, we know that the Marshallian or ordinary demands *add up*, e.g., for $q_1 = g^1(p_1, p_2, M)$ and $q_2 = g^2(p_1, p_2, M)$, $p_1 g^1 + p_2 g^2 = M$. Use this latter expression to derive the **Engel aggregation condition** $w_1 \in_{1M} + w_2 \in_{2M} = 1$ - - the sum of the individual income elasticities, weighted by the expenditure proportions or shares, equals unity. What is the implication of this condition?

10.7.2 The equations

$$w_1 \in_{11} + w_2 \in_{21} + w_1 = 0,$$
$$w_1 \in_{12} + w_2 \in_{22} + w_2 = 0$$

constitute the **Cournot aggregation conditions** - - for commodity i the weighted sum of its own-price elasticity and cross-price elasticity for product $j(i \neq j)$ equals the negative of its expenditure share, with the expenditure shares serving as weights. Derive these conditions and explain why they are important. How are they connected to the Engle aggregation condition?

10.7.3 Suppose $q_1 = Ap_1^{\alpha_1} p_2^{\alpha_2} M^{\alpha_3}$, $q_2 = Bp_1^{\beta_1} p_2^{\beta_2} M^{\beta_3}$. What conditions must be placed upon α_i ***and*** β_i, $i = 1, 2, 3$, in order for the Engel and Cournot aggregation conditions to hold? (Note: $\partial q_i / \partial M \gtrless 0$ if commodity i is a normal (inferior) good; $\partial q_i / \partial p_j \gtrless 0$ if commodity $j\,(\neq i)$ is a substitute (complement).

10.7.4 What restrictions must be imposed on c and d in the direct utility function $u = f(q_1, q_2) = q_1^c q_2^d$ in order for both q_1 and q_2 to exhibit diminishing marginal utility? Are these restrictions sufficient to ensure that u is strictly quasiconcave?

10.7.5 Given the direct utility function $u = f(q_1, q_2) = q_1 q_2^{1/2}$, find:

a. Marshallian demands.

b. The marginal utility of money.

c. Indirect utility function.

d. Verify that Roy's identity holds.

e. Hicksian demands.

f. Expenditure function.

g. Verify that the derivative property (Shephard's lemma) holds.

h. Are q_1, q_2 normal goods; substitutes or complements?

i. Determine the pure substitution effect and the pure income effect. Does the Hicks- Slutsky equation (10.B.4) hold?

10.7.6 Suppose the direct utility function has the form $u = f(q_1, q_2) = q_1 q_2$. We previously determined that

$$g^1(p_1, p_2, M) = \tfrac{1}{2} M\, p_1^{-1},$$
$$h^1(p_1, p_2, u) = p_1^{-\frac{1}{2}} p_2^{\frac{1}{2}} u^{\frac{1}{2}}.$$

Does the Hicks-Slutsky equation (10.B.4) hold?

10.7.7 Given the Hicksian demands

$$h^1(p_1, p_2, u) = 2^{\frac{1}{3}} p_1^{-\frac{1}{3}} p_2^{\frac{1}{3}} u^{\frac{2}{3}}$$
$$h^2(p_1, p_2, u) = 2^{-\frac{2}{3}} p_1^{\frac{2}{3}} p_2^{-\frac{2}{3}} u^{\frac{2}{3}},$$

verify that the cross-price derivatives are symmetric.

10.7.8 Given the expenditure function $E(p_1, p_2, u) = 3\left(2^{-\frac{2}{3}}\right) p_1^{\frac{2}{3}} p_2^{\frac{1}{3}} u^{\frac{2}{3}}$, verify that the cross-price derivatives of the Hicksian demands are symmetric.

10.7.9 Given the Hicksian demands $q_1 = h^1(p_1, p_2, u)$ and $q_2 = h^2(p_1, p_2, u)$, verify that the **substitution matrix** (containing all **substitution terms**, where the own-price substitution terms appear on the main diagonal and cross-price derivatives are off-diagonal)

$$\mathbf{S}(p_1, p_2, u) = \frac{\partial(h^1, h^2)}{\partial(p_1, p_2)} = \begin{bmatrix} \frac{\partial h^1}{\partial p_1} & \frac{\partial h^1}{\partial p_2} \\ \frac{\partial h^2}{\partial p_1} & \frac{\partial h^2}{\partial p_2} \end{bmatrix}$$

is negative semidefinite.

10.7.10 Given $u = f(q_1, q_2)$ and $\hat{g} = M - p_1 q_1 - p_2 q_2 = 0$, verify that $\lambda = df/dM$. (Hint: find df, $d\hat{g} = 0$ under the assumption that Equation (10.32) holds.)

Section 10.7.4 Exercises:

10.7.11 The **Cobb-Douglas (C-D) production function** has the form

$$q = F(l, k) = Al^{\alpha}k^{\beta},$$

where A, α and β are constants and, usually $\alpha + \beta < 1$. For this function:

a. find $MP_l = F_l$ and $MP_k = F_k$.

b. are these input marginal products decreasing for $\alpha + \beta < 1$?

c. verify that the partial elasticity of q with respect to $l\,(k)$ is $\alpha\,(\beta)$.

d. find $TRS_{l,k}$.

e. the **elasticity of substitution of *l* for *k*** (a measure of the similarity of these inputs from a technological viewpoint) is defined as

$$\sigma = \frac{dx}{ds}\frac{s}{x},$$

the elasticity of the input ratio ($x = k/l$) with respect to the technical rate of substitution ($TRS_{l,k} = s = F_l/F_k$). Verify that $\sigma = 1$. (Hint: take the total differential of $\ln s$.)

10.7.12 The **constant elasticity of substitution (*CES*) production function** has the form

$$q = F(l, k) = A\left[\delta k^{-\rho} + (1 - \delta) l^{-\rho}\right]^{-h/\rho}$$

where A is a constant, $\delta\,(0 < \delta < 1)$ is a distribution parameter, h (assumed <1) is the degree of homogeneity of the function, and ρ is the substitution parameter (a transformation of σ or $\rho = \left(1 - \frac{1}{\sigma}\right)$). For this function:

a. find $MP_l = F_l$ and $MP_k = F_k$.

b. find $TRS_{l,k}$.

c. check that σ is the elasticity of substitution for this production function. (Hint: take the total differential of ln s.)

10.7.13 For Example 10.13, verify that $|\boldsymbol{H}_L^{\hat{g}}| < 0$. Is $q = F(l, k)$ strictly concave? Is F strictly quasiconcave? Check for the strict quasiconcavity of F directly. Does F exhibit a diminishing technical rate of substitution between l and k?

10.7.14 Verify that $\partial j^l/\partial q \equiv \partial MC/\partial w$.

10.7.15 For Example 10.13, are the output-constrained input demands downward sloping? Are inputs l, k substitutes? Does the equality $\partial j^l/\partial r = \partial j^k/\partial w$ hold? Are l, k inferior inputs?

10.7.16 For Example 10.13, verify that the derivative property of the cost function holds. Demonstrate that the concavity of the cost function and the derivative property imply that the output-constrained demands are downward sloping.

10.7.17 For Example 10.14, verify that the cost-constrained demands are downward sloping. Are inputs l, k substitutes? Does $\partial h^l/\partial r = \partial h^k/\partial w$?

10.7.18 For Example 10.14, do h^l and h^k add up?

10.7.19 Given the indirect production function $q = Q(w, r, C)$ from Example 10.14, verify that $\partial Q/\partial C > 0$ and $\partial Q/\partial w < 0$, $\partial Q/\partial r < 0$. What can we conclude from the signs of these derivatives?

10.7.20 Verify that for Example 10.14, Roy's identity yields the cost-constrained input demand functions.

10.7.21 Verify that the cost function obtained in Example 10.13 is the inverse of the indirect production function obtained in Example 10.14. Derive the output-constrained demands obtained in Example 10.13 using the cost function and the cost-constrained demands determined in Example 10.14.

10.7.22 For Example 10.15, verify that the ordinary input demand functions are downward sloping. Next, determine if l and k are unrelated inputs. Also verify that l, k are not inferior inputs.

10.7.23 For Example 10.15, use the generalized supply function to determine the effect of an increase in w or r on quantity supplied.

10.7.24 For Example 10.15, is the profit function π convex in all prices? Also demonstrate that Hotelling's lemma holds.

10.7.25 For Example 10.15, demonstrate that since the profit function is convex in input prices, the ordinary demand functions do not slope upwards.

NOTES

1 If y, w are functionally dependent and either $f_1 \neq 0$ or $f_2 \neq 0$, then we can write $f = \alpha\,(g)$. And if $g_1 \neq 0$ or $g_2 \neq 0$, the $g = \beta\,(f)$.

2 This problem is so-named because, by virtue of Shephard's lemma, it is the compensated demands which are the derivatives of the expenditure function and which consequently must integrate back to the same. Hence, the restriction which forces the compensated demands to be consistent with household preferences, as reflected by utility in terms of q_1 and q_2, is known as the integrability condition.

Constrained Extrema: Inequality Constraints

11.1 CONSTRAINED EXTREMA: m INEQUALITY CONSTRAINTS IN n NON-NEGATIVE INDEPENDENT VARIABLES

Let us consider the following problem. We desire to maximize the real-valued function $y = f(x)$: $\mathcal{D} \subset R^n \rightarrow R$ subject to m inequality constraints $\hat{g}^j(\boldsymbol{x}) = b_j - g^j(\boldsymbol{x}) \geq 0$, $j = 1, \ldots, m$, with $\hat{g}^j(\boldsymbol{x})$: $\mathcal{D} \subset R^n \rightarrow R$, and n non-negativity conditions $\boldsymbol{x} \geq \boldsymbol{O}$, $\boldsymbol{x}^T = (x_1, \ldots, x_n) \in R^n$. Here f and the $\hat{g}^j(\boldsymbol{x})$, $j = 1, \ldots, m$, are assumed of class $C^{(1)}$ on the open set $\mathcal{D}$. Forming the $(m \times 1)$ vector-valued function

$$\boldsymbol{G}(\boldsymbol{x}) = \begin{bmatrix} \hat{g}^1(\boldsymbol{x}) \\ \vdots \\ \hat{g}^m(\boldsymbol{x}) \end{bmatrix} \geq \boldsymbol{O}$$

enables us to rewrite this problem as

$$\begin{aligned} &\text{maximize } y = f(\boldsymbol{x}) \\ &\text{subject to } \boldsymbol{G}(\boldsymbol{x}) \geq \boldsymbol{O}, \quad \boldsymbol{x} \geq \boldsymbol{O}, \boldsymbol{x} \in \mathcal{D}. \end{aligned} \quad (11.1)$$

Here $\boldsymbol{G}(\boldsymbol{x}) \geq \boldsymbol{O}$, $\boldsymbol{x} \geq \boldsymbol{O}$, and $\boldsymbol{x} \in \mathcal{D}$ serve to define a region of feasible or admissible solutions in R^n. That is,

Definition 11.1: The **feasible region** $\mathcal{K}$ is the set of points $\boldsymbol{x}$ which simultaneously satisfy all of the inequality constraints and non-negativity conditions or

$$\mathcal{K} = \{\boldsymbol{x} | \boldsymbol{G}(\boldsymbol{x}) \geq \boldsymbol{O}, \boldsymbol{x} \geq \boldsymbol{O}, \boldsymbol{x} \in \mathcal{D}\}.$$

(Note that we do not require that $m < n$ as was the case when we optimized f subject to m equality constraints. Here the inequalities $\hat{g}^j(\boldsymbol{x}) \geq 0$, $j = 1, \ldots, m$, simply define a region the size of which varies inversely with m.) Hence, our problem can be written simply as: maximize $y = f(\boldsymbol{x})$ subject to $\boldsymbol{x} \in \mathcal{K}$. Additionally,

Definition 11.2: A point $\boldsymbol{x}$ is called a **feasible point** if $\boldsymbol{x} \in \mathcal{K}$.

For convenience the restriction $\boldsymbol{x} \geq \boldsymbol{O}$ will be rewritten $x_i = \hat{g}^{m+i}(\boldsymbol{x}) \geq 0$, $i = 1, \ldots, n$. Hence, we ultimately

$$\begin{aligned} \text{maximize } y = f(\boldsymbol{x}) \text{ subject to } \boldsymbol{x} \in \overline{\mathcal{K}} \\ = \{\boldsymbol{x} | \overline{\boldsymbol{G}}(\boldsymbol{x}) \geq \boldsymbol{O}, \boldsymbol{x} \in \mathcal{D}\}, \end{aligned} \quad (11.1.1)$$

where

$$\overline{\boldsymbol{G}}(\boldsymbol{x}) = \begin{bmatrix} \hat{g}^1(\boldsymbol{x}) \\ \vdots \\ \hat{g}^m(\boldsymbol{x}) \\ \hat{g}^{m+1}(\boldsymbol{x}) \\ \vdots \\ \hat{g}^{m+n}(\boldsymbol{x}) \end{bmatrix} \geq \boldsymbol{O}$$

is an $(m + n \times 1)$ vector-valued function whose jth component is $\hat{g}^j(\boldsymbol{x})$, $j = 1, \ldots, m + n$. In this formulation $\boldsymbol{x}$ is unrestricted.

If a point x_o provides a **local solution** to (11.1.1), then $f(x_o) \geq f(x)$ for all $x \in \delta(x_o) \cap \overline{\mathcal{K}}$, where $\delta(x_o)$ is a suitably restricted spherical $\delta-$ neighborhood about x_o. A **global solution** to (11.1.1) emerges when $f(x_o) \geq f(x)$ for all $x \in \overline{\mathcal{K}}$.

It was mentioned in Chapter 3 (Section 3.3) that any vector $x \in R^n$ may be used to specify a direction. In this regard, a feasible direction may be taken to be a vector such that a small movement along it violates no constraint. So if x is feasible, a feasible direction at x is a vector$h$ such that $x + th$ is feasible for t sufficiently small, i.e., if $x + th$ is feasible whenever x is, then h specifies a feasible direction. Hence, we can move a short distance from x in the h-direction and still remain within the feasible region $\overline{\mathcal{K}}$. More formally,

Definition 11.3: Let $\delta(x_o)$ be a suitably restricted spherical $\delta-$ neighborhood about the feasible point $x_o \in R^n$. Then the $(n \times 1)$ vector h is a **feasible direction** at x_o if there exists a scalar t, $0 \leq t < \delta$, such that $x_o + th$ is feasible.

In this regard we have,

Definition 11.4: For $\delta(x_o)$ a suitably restricted spherical $\delta-$ neighborhood about the feasible point $x_o \in R^n$, the **set of feasible directions at** x_o, $\mathcal{D}(x_o)$, is the collection of all $(n \times 1)$ vector $h$such that $x_o + th$ is feasible for t sufficiently small, i.e.,

$$\mathcal{D}(x_o) = \{h \in R^n | \text{if } x_o \in \overline{\mathcal{K}},$$
$$\text{then } x_o + th \in \overline{\mathcal{K}}, \quad 0 \leq t < \delta, h \neq O.\}$$

Additionally, given $y = f(x)$: $\mathcal{D} \subset R^n \to R$, the **cone of ascent directions at** x_o is specified as

$$\mathcal{F} = \{h \in R^n | f(x_o + th) > f(x_o), 0 < t < \delta, h \neq O\}.$$

Here each $h \in \mathcal{F}$ is called an **ascent direction** of f at x_o since a small movement from x_o along some $h \in \mathcal{F}$ leads to solutions of improving values of f.

Geometrically, $\mathcal{D}(x_o)$ is a **tangent support cone** approximating $\overline{\mathcal{K}}$ in the neighborhood of x_o, i.e., it is the finite cone containing all feasible directions at x_o and is generated by the tangent or supporting hyperplanes to $\overline{\mathcal{K}}$ at x_o (Figure 11.1). To see this we need only remember that each supporting hyperplane to $\overline{\mathcal{K}}$ at

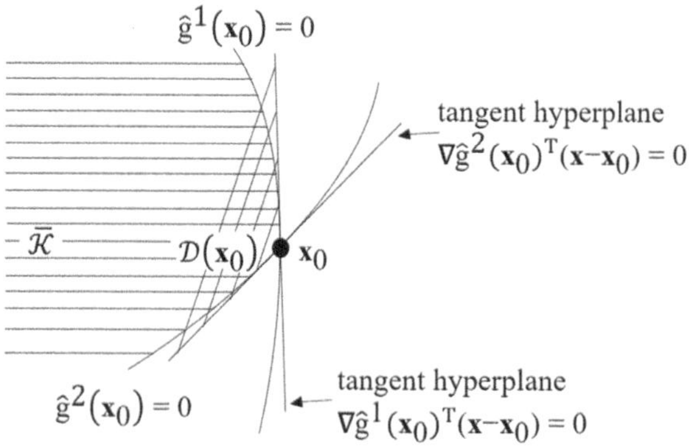

FIGURE 11.1 The feasible region $\overline{\mathcal{K}}$ and the tangent support cone $\mathcal{D}(x_o)$.

x_o, $\nabla\hat{g}^j(x_o)^T(x - x_o) = 0$, specifies a closed half-space $\nabla\hat{g}^j(x_o)^T(x - x_o) \geq 0$. The tangent support cone $\mathcal{D}(x_o)$ represents that portion of the intersection of these closed half-spaces in the immediate vicinity of x_o.

What is the connection between $\mathcal{D}(x_o)$, where x_o is assumed to be a (local) maximal point, and: (a) $f(x_o)$; and (b) $\overline{G}(x_o) \geq O$? To answer part (a) we have

Theorem 11.1: If the real-valued function $y = f(x)$, $x \in \overline{\mathcal{K}}$, is of class $C^{(1)}$ and attains a local maximum at x_o, then any departure from x_o along a feasible direction h cannot increase the value of f or $df(x_o) = \nabla f(x_o)^T h \leq 0$, $h \in \mathcal{D}(x_o)$.

Proof: If $f(x_o)$ is a local maximum of f, $x_o \in \overline{\mathcal{K}}$, then $f(x_o + th) - f(x_o) \leq 0$, $h \in \mathcal{D}(x_o)$. Hence

$$[f(x_o + th) - f(x_o)]/t \leq 0$$

and

$$\lim_{t\to 0}[f(x_o + th) - f(x_o)]/t = \nabla f(x_o)^T h \leq 0. \text{ Q. E D.}$$

Thus the directional derivative of f at x_o, $\nabla f(x_o)^T h$, exhibits a decrease in any feasible direction. Since ∇f indicates (locally) the direction of maximum increase in f, a small movement h in any direction such that $\nabla f(x_o)^T h > 0$ (the angle between $\nabla f(x_o)$ and h is acute $(<\pi/2)$) increases f. So if the local maximum of f is at x_o, $\nabla f(x_o{}^T)h \leq 0$ for all $h \in \mathcal{D}(x_o)$. Hence, *no* feasible direction may form an acute angle between itself and $\nabla f(x_o)$. Let us define $\mathcal{F}_o = \{h \in R^n | \nabla f(x_o)^T h > 0\}$. So if $h \in \mathcal{F}_o$, it follows that there exists a point $x = x_o + th$, sufficiently close to $\boldsymbol{x}_o$, such that $f(\boldsymbol{x}_o) < f(\boldsymbol{x})$. Hence, $\mathcal{F}_o$ characterizes the set of improving directions $\mathcal{F}$ so that $\mathcal{F}_o \subseteq \mathcal{F}$.

Next, for any optimal $\boldsymbol{x} \in \overline{\mathcal{K}}$, it is usually the case that not all of the inequality constraints are **binding** or hold as a strict equality. To incorporate this observation into our analysis, let us divide the components of $\overline{\boldsymbol{G}}(\boldsymbol{x}) \geq \boldsymbol{O}$ into two mutually exclusive classes: those constraints which are binding at $\boldsymbol{x}$, $\hat{g}^j(\boldsymbol{x}) = 0$, and those which are not, $\hat{g}^j(\boldsymbol{x}) > 0$, $j = 1, \ldots, m+n$. If we let $\mathcal{J} = \{j|\hat{g}^j(\boldsymbol{x}) = 0, \boldsymbol{x} \in R^n\}$ depict the *index set of binding constraints*, then $\hat{g}^j(\boldsymbol{x}) > 0$ if $j \notin \mathcal{J}$. In what follows we shall ignore those constraints $\hat{g}^j(\boldsymbol{x}) \geq 0$ for which $j \notin \mathcal{J}$ since, if $\hat{g}^j(\boldsymbol{x}) > 0$, then $\hat{g}^j(\boldsymbol{x} + t\boldsymbol{h}) > 0$ for t sufficiently small and positive and $\boldsymbol{h} \neq \boldsymbol{O}$. So to determine the link between $\mathcal{D}(\boldsymbol{x}_o)$ ($\boldsymbol{x}_o$ optimal) and $\overline{\boldsymbol{G}}(\boldsymbol{x}_o) \geq \boldsymbol{O}$, we need only consider those constraints with indices $j \in \mathcal{J}$ since, if any constraint is not binding at $\boldsymbol{x}_o$, then for all $\boldsymbol{x} = \boldsymbol{x}_o + t\boldsymbol{h} \in \delta(\boldsymbol{x}_o)$, the constraint remains so. To answer part (b) of the above question we have

Theorem 11.2: Suppose the functions $\hat{g}^j(\boldsymbol{x})$, $j \in \mathcal{J}$, are of class $C^{(1)}$ and the $\hat{g}^j(\boldsymbol{x})$, $j \notin \mathcal{J}$, are continuous at $\boldsymbol{x}_o \in \overline{\mathcal{K}}$. If $\boldsymbol{h} \in \mathcal{D}(\boldsymbol{x}_o)$, with $\boldsymbol{x}_o$ optimal, then $\nabla \hat{g}^j(\boldsymbol{x}_o)^T\boldsymbol{h} \geq 0$, $j \in \mathcal{J}$.

Proof: For $\boldsymbol{h} \in \mathcal{D}(\boldsymbol{x}_o)$, $\hat{g}^j(\boldsymbol{x}_o + t\boldsymbol{h}) \geq 0$ for $\boldsymbol{h} \neq \boldsymbol{O}$ and t small, $j \in \mathcal{J}$. Using Taylor's formula,

$$\hat{g}^j(\boldsymbol{x}_o + t\boldsymbol{h}) = \hat{g}^j(\boldsymbol{x}_o) + t\nabla \hat{g}^j(\boldsymbol{x}_o + t\theta\boldsymbol{h})^T\boldsymbol{h},$$
$$0 \leq t < \delta, 0 < \theta < 1, j \in \mathcal{J}.$$

With $\hat{g}^j(\boldsymbol{x}_o) = 0$, $\hat{g}^j(\boldsymbol{x}_o + t\boldsymbol{h}) \geq 0$ for $\boldsymbol{h} \in \mathcal{D}(\boldsymbol{x}_o)$ implies $\nabla \hat{g}^j(\boldsymbol{x}_o + t\theta\boldsymbol{h})^T\boldsymbol{h} \geq 0$, $j \in \mathcal{J}$, and thus, for t sufficiently small, $\nabla \hat{g}^j(\boldsymbol{x}_o)^T\boldsymbol{h} \geq 0$ (or $-\nabla \hat{g}^j(\boldsymbol{x}_o)^T\boldsymbol{h} \leq 0$), $j \in \mathcal{J}$. Q. E. D.

Let us define $\mathcal{G}_o = \{\boldsymbol{h} \in R^n|\nabla \hat{g}^j(\boldsymbol{x}_o)^T\boldsymbol{h} > 0, j \in \mathcal{J}\} \subseteq \mathcal{D}(\boldsymbol{x}_o)$. (Note that $\mathcal{G}_o$ is essentially an (open) **linearizing cone** of $\overline{\mathcal{K}}$ at $\boldsymbol{x}_o$ since it is generated by linearizing the constraint functions at $\boldsymbol{x}_o$.)

Section 11.1 Exercises:

11.1.1 Suppose we have the following set of constraints:

$$3x_1 + x_2 \leq 0$$
$$-x_1 + x_2 - 3 \geq 0$$
$$x_1^2 - x_2 \leq -1$$
$$x_1 \leq 0$$
$$x_2 \geq 0$$

Find $\overline{\boldsymbol{G}}(\boldsymbol{x}) \geq \boldsymbol{O}$ and $\nabla\overline{\boldsymbol{G}}$. What is $\rho(\nabla\overline{\boldsymbol{G}})$ at $\boldsymbol{x}_o^T = (1, 1)$?

11.1.2 For the constraints given in Exercise 11.1.1, is $\overline{\boldsymbol{x}}^T = (-3, 2)$ a feasible point? How about $\boldsymbol{x}_o^T = (-1, 2)$?

11.1.3 Given the constraints appearing in Exercise 11.1.1, determine if $\boldsymbol{h}^T = (h_1, h_2) = (0, 3)$ is a feasible direction at $\boldsymbol{x}_o^T = (-1, 2)$.

11.1.4 Suppose $f(\boldsymbol{x}) = x_1 - \frac{1}{2}x_2$, $\boldsymbol{x} \in R^2$. Let $\boldsymbol{x}_o^T = (1, 1)$. Is $\boldsymbol{h}^T = (2, -1)$ an ascent direction of f at $\boldsymbol{x}_o$? Is it an improving direction for f?

11.1.5 Given the hyperplane $\mathcal{H} = \{\boldsymbol{x} \in R^2|x_1 + 2x_2 = 6\}$, find the vector normal to $\mathcal{H}$. Express $\mathcal{H}$ with reference to the point (2, 2).

11.1.6 Suppose $\hat{g}^1(\boldsymbol{x}) = 3x_1 - x_2 \geq 0$, $\hat{g}^2(\boldsymbol{x}) = -3 - x_1 + x_2 \geq 0$, $\hat{g}^3(\boldsymbol{x}) = x_1 \geq 0$, $\hat{g}^4(\boldsymbol{x}) = x_2 \geq 0$, and $\boldsymbol{x}_o^T = (3/2, 9/2)$, $\boldsymbol{x} \in R^2$. Find the supporting hyperplanes to $\overline{\mathcal{K}}$ at $\boldsymbol{x}_o$.

11.1.7 Suppose the constraints $\hat{g}^1(\boldsymbol{x}) = 6 - x_1 - 3x_2 \geq 0$, $\hat{g}^2(\boldsymbol{x}) = 16 - 4x_1^2 - x_2^2 \geq 0$ are binding at $\boldsymbol{x}_o^T = (1.88, 1.37)$. Is $\boldsymbol{h}^T = (2, -1)$ an element of the linearizing cone $\mathcal{G}_o$ at $\boldsymbol{x}_o$?

11.2 NECESSARY OPTIMALITY CONDITIONS

We begin with

Theorem 11.3: Consider the problem

$$\text{maximize } f(\boldsymbol{x}) \text{ subject to } \boldsymbol{x} \in \mathcal{K} \quad (11.1.1)$$

Let f and $\hat{g}^j$, $j \in \mathcal{J}$, be of class $C^{(1)}$ at a point $\boldsymbol{x}_o \in \overline{\mathcal{K}}$ with the $\hat{g}^j$, $j \notin \mathcal{J}$, continuous at $\boldsymbol{x}_o$. If $\boldsymbol{x}_o$ is a local maximal solution, then $\mathcal{F}_o \cap \mathcal{D}(\boldsymbol{x}_o) = \varnothing$.

Proof: By contradiction, suppose $\boldsymbol{h} \in \mathcal{F}_o \cap \mathcal{D}(\boldsymbol{x}_o)$ so that there is an $\boldsymbol{h}(\neq\boldsymbol{O})$ such that $\nabla f(\boldsymbol{x}_o)^T\boldsymbol{h} > 0$. But this implies (from Definition 11.4) that there exists a $\delta_1 > 0$ such that $f(\boldsymbol{x}_o + t\boldsymbol{h}) > f(\boldsymbol{x}_o)$, $0 < t < \delta_1$, or that $\boldsymbol{h}$ is an ascent direction of f at $\boldsymbol{x}_o$. Also, from Definition 11.4, there exists a $\delta_2 > 0$ such that $\boldsymbol{x}_o + t\boldsymbol{h} \in \overline{\mathcal{K}}$, $0 < t < \delta_2$.

Clearly, these results contradict the assumption that $\mathbf{x}_o$ solves Problem (11.1.1). Q. E. D.

Thus a necessary condition for a local maximum of f subject to $\mathbf{x} \in \overline{\mathcal{K}}$ at $\mathbf{x}_o$ is that $\mathcal{F}_o \cap \mathcal{D}(\mathbf{x}_o) = \emptyset$. But since $\mathcal{G}_o \subseteq \mathcal{D}(\mathbf{x}_o)$, it must also be true that $\mathcal{F}_o \cap \mathcal{G}_o = \emptyset$ is also a necessary condition for a local maximum of f subject to $\mathbf{x} \in \overline{\mathcal{K}}$. That is

Theorem 11.4: Consider Problem (11.1.1) with $\mathbf{x}_o \in \overline{\mathcal{K}}$. Additionally, let $f(\mathbf{x})$ and $\hat{g}^j(\mathbf{x}), j \in \mathcal{J}$, be of class $C^{(1)}$ at $\mathbf{x}_o$ with $\hat{g}^j(\mathbf{x}), j \notin \mathcal{J}$, continuous at $\mathbf{x}_o$. If $\mathbf{x}_o$ is a local optimal solution to Problem (11.1.1), then $\mathcal{F}_o \cap \mathcal{G}_o = \emptyset$.

Proof: If $\mathbf{x}_o$ constitutes a local maximal solution to Problem (11.1.1), then $\mathcal{F}_o \cap \mathcal{D}(\mathbf{x}_o) = \emptyset$ (via the preceding theorem) and thus, since $\mathcal{G}_o \subseteq \mathcal{D}(\mathbf{x}_o)$, $\mathcal{F}_o \cap \mathcal{G}_o = \emptyset$. Q. E. D.

Section 11.2 Exercises:

11.2.1 Suppose our objective is to maximize $y = f(\mathbf{x}) = -2x_1^2 - 3_2^2 + 4x_1x_2 + 4x_1 + 2x_2$ subject to $\hat{g}^1(\mathbf{x}) = 6 - x_1 - 3x_2 \geq 0, \hat{g}^2(\mathbf{x}) = 16 - 4x_1^2 - x_2^2 \geq 0, \mathbf{x} \in R^2$. Suppose both constraints are binding at the maximal point $\mathbf{x}_o^T = (1.88, 1.37)$. Does $\mathcal{F}_o \cap \mathcal{G}_o = \emptyset$?

11.2.2 Let us maximize $y = f(\mathbf{x}) = x_1x_2$ subject to $\hat{g}^1(\mathbf{x}) = 100 - x_1 - x_2 \geq 0, \hat{g}^2(\mathbf{x}) = 50 - x_1 \geq 0$, $\mathbf{x} \in R^2$.

If both constraints are binding at the maximal point $\mathbf{x}_o^T = (50, 50)$, demonstrate that $\mathcal{F}_o \cap \mathcal{G}_o = \emptyset$.

11.3 FRITZ JOHN (FJ) OPTIMALITY CONDITIONS (MANGASARIAN [1969]; MANGASARIAN AND FROMOVITZ [1967]; JOHN [1948])

Our objective in this section is to transform the necessary condition $\mathcal{F}_o \cap \mathcal{G}_o = \emptyset$ into a statement regarding the gradients of the objective function $f(\mathbf{x})$ and of the binding constraints $\hat{g}^j(\mathbf{x}) = 0, j \in \mathcal{J}$. To this end we have

Theorem 11.5: (**FJ Necessary Conditions**). Let $\mathbf{x}_o$ be a feasible solution to Problem (11.1.1) and suppose that $f(\mathbf{x})$ and $\hat{g}^j(\mathbf{x}), j \in \mathcal{J}$, are of class $C^{(1)}$ at $\mathbf{x}_o$ with the $\hat{g}^j(\mathbf{x}), j \notin \mathcal{J}$, continuous at $\mathbf{x}_o$. If $\mathbf{x}_o$ is a local solution to Problem (11.1.1), then there exist scalars λ_o and $\bar{\lambda}_j, j \in \mathcal{J}$, such that:

a. $\lambda_o \nabla f(\mathbf{x}_o) + \sum_{j \in \mathcal{J}} \bar{\lambda}_j \nabla \hat{g}^j(\mathbf{x}_o) = \mathbf{O}$,

b. $\lambda_o, \bar{\lambda}_j \geq 0, j \in \mathcal{J}$, **[FJ Optimality Conditions]** (11.2)

c. $\begin{bmatrix} \lambda_o \\ \bar{\boldsymbol{\lambda}} \end{bmatrix} \neq \begin{bmatrix} 0 \\ \mathbf{O} \end{bmatrix}$,

where $\bar{\boldsymbol{\lambda}}$ is the vector which houses the $\bar{\lambda}_j, j \in \mathcal{J}$. And if the $\hat{g}^j(\mathbf{x}), j \notin \mathcal{J}$, are also of class $C^{(1)}$ at $\mathbf{x}_o$, then (11.2) can be rewritten as:

a. $\lambda_o \nabla f(\mathbf{x}_o) + \sum_{j=1}^{m+n} \bar{\lambda}_j \nabla \hat{g}^j(\mathbf{x}_o) = \mathbf{O}$

b. $\bar{\lambda}_j \hat{g}^j(\mathbf{x}_o) = 0$

c. $\hat{g}^j(\mathbf{x}_o) \geq 0$ $\quad j = 1, \ldots, m + n$ **[FJ Optimality Conditions]** (11.2.1)

d. $\lambda_o, \bar{\lambda}_j \geq 0$

e. $\begin{bmatrix} \lambda_o \\ \bar{\boldsymbol{\lambda}} \end{bmatrix} \neq \begin{bmatrix} 0 \\ \mathbf{O} \end{bmatrix}$,

where now $\bar{\boldsymbol{\lambda}}^T = (\bar{\lambda}_1, \ldots, \bar{\lambda}_{m+n})$.

Proof: (of (11.2)). With $\mathbf{x}_o$ an optimal solution to Problem (11.1.1), Theorem 11.4 informs us that $\mathcal{F}_o \cap \mathcal{G}_o = \emptyset$, i.e., there exists no direction $\mathbf{h}$ such that $\nabla f(\mathbf{x}_o)^T \mathbf{h} > 0$ and $\nabla \hat{g}^j(\mathbf{x}_o)^T \mathbf{h} > 0, j \in \mathcal{J}$. Let the matrix $\mathbf{A}$ have as its rows the vectors $\nabla f(\mathbf{x}_o)^T$ and $\nabla \hat{g}^j(\mathbf{x}_o)^T, j \in \mathcal{J}$. Then the necessary optimality condition $\mathcal{F}_o \cap \mathcal{G}_o = \emptyset$ is equivalent to the requirement that the system $\mathbf{A}\mathbf{h} > \mathbf{O}$ is inconsistent. By Gordan's Theorem of the Alternative[1] (Theorem 11.6), there exists a vector $\mathbf{y} \geq \mathbf{O}\ (\mathbf{y} \neq \mathbf{O})$ such that $\mathbf{A}^T\mathbf{y} = \mathbf{O}$, where $\mathbf{y}^T = (\lambda_o, \bar{\boldsymbol{\lambda}})$ or (11.2a). Q. E. D.

In system (11.2.1) λ_o and the $\bar{\lambda}_j, j = 1, \ldots, m$, are **Lagrange multipliers**, the requirement that $\mathbf{x}_o$ be feasible is called the **primal feasibility condition,** (11.2a) is termed the **dual feasibility condition**, and (11.2.1b) constitutes the so-called **complementary slackness (CS) conditions**. In this regard:

a. if $\hat{g}^j(\mathbf{x}_o) > 0$ (the jth constraint is not active or binding at $\mathbf{x}_o$), then we must have $\bar{\lambda}_j = 0$; and

b. if $\hat{g}^j(\boldsymbol{x}_o) = 0$ (the jth constraint is binding at $\boldsymbol{x}_o$), then $\bar{\lambda}_j$ can be positive.

Example 11.1: Let us maximize the real-valued function $y = f(\boldsymbol{x}) = -2x_1^2 - 3x_2^2 + 4x_1x_2 + 4x_1 + 2x_2$ subject to

$$x_1 + 3x_2 \leq 6$$
$$4x_1^2 + x_2^2 \leq 16$$
$$x_1, x_2 \geq 0.$$

Then

$$\hat{g}^1(\boldsymbol{x}) = 6 - x_1 - 3x_2 \geq 0$$
$$\hat{g}^2(\boldsymbol{x}) = 16 - 4x_1^2 - x_2^2 \geq 0$$
$$\hat{g}^3(\boldsymbol{x}) = x_1 \geq 0$$
$$\hat{g}^4(\boldsymbol{x}) = x_2 \geq 0$$

(see Figure 11.2). Do the FJ conditions (11.2) hold at the optimal solution point A, with $\boldsymbol{x}_o^T = (1.88, 1.37)$? At this point it is evident that $\mathcal{J} = \{1,2\}$. With

$$\nabla f(\boldsymbol{x}_o) = \begin{bmatrix} 1.96 \\ 1.30 \end{bmatrix}, \nabla \hat{g}^1(\boldsymbol{x}_o) = \begin{bmatrix} -1 \\ -3 \end{bmatrix}, \nabla \hat{g}^2(\boldsymbol{x}_o) = \begin{bmatrix} -15.04 \\ -2.74 \end{bmatrix},$$

and $\bar{\lambda}_3 = \bar{\lambda}_4 = 0$, the FJ conditions require the existence of Lagrange multipliers $\lambda_o, \bar{\lambda}_1, \bar{\lambda}_2 \geq 0$ such that

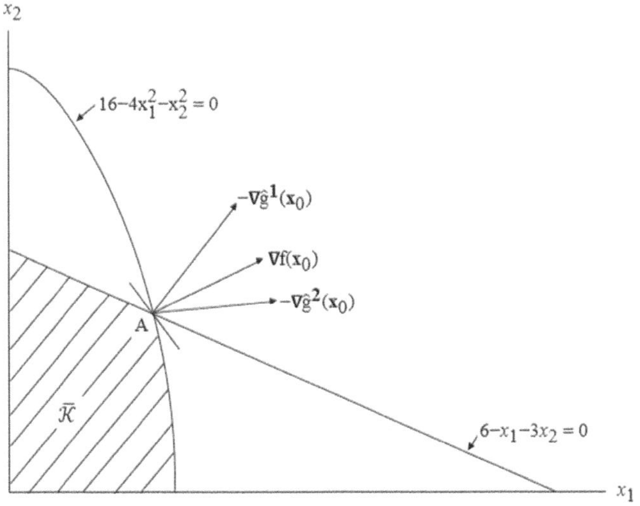

FIGURE 11.2 Point A renders the local constrained maximal solution.

$$\lambda_o \begin{bmatrix} 1.96 \\ 1.30 \end{bmatrix} + \bar{\lambda}_1 \begin{bmatrix} -1 \\ -3 \end{bmatrix} + \bar{\lambda}_2 \begin{bmatrix} -15.04 \\ -2.74 \end{bmatrix} = \begin{bmatrix} 0 \\ 0 \end{bmatrix}$$

or that $\bar{\lambda}_1 = 0.3342\lambda_o > 0$ and $\bar{\lambda}_2 = 0.1081\lambda_o > 0$ for any $\lambda_o > 0$. Hence, the FJ optimality conditions hold. ■

It was determined earlier in this chapter that if the point $\boldsymbol{x}_o \in \bar{\mathcal{K}}$ is a local optimal solution to Problem (11.1.1), then $\mathcal{F}_o \cap \mathcal{G}_o = \varnothing$ and $\boldsymbol{x}_o$ satisfies the FJ necessary optimality conditions, i.e., $\boldsymbol{x}_o$ is a **Fritz-John point**. Since $\mathcal{F}_o \cap \mathcal{G}_o = \varnothing$ is satisfied at any $\boldsymbol{x}_o \in \bar{\mathcal{K}}$ for which $\mathcal{G}_o = \varnothing$, $(\boldsymbol{x}, \lambda_o, \bar{\boldsymbol{\lambda}})$ must also be a FJ point. And this holds true for *any* objective function f so that ∇f becomes irrelevant in the FJ necessary conditions. In this regard, to require that ∇f play an active role in the optimality conditions, we shall introduce the Karush-Kuhn-Tucker (KKT) conditions that admit FJ points for which $\lambda_o > 0$. As will be seen below, the KKT conditions encompass (11.2) with the additional restriction that $\lambda_o > 0$. In particular, and without loss of generality, we can assume that $\lambda_o \equiv 1$. So if $\mathcal{G}_o \neq \varnothing$ at the local maximal point $\boldsymbol{x}_o$, then $(\boldsymbol{x}_o, \boldsymbol{\lambda}_o)$ is a **Karush-Kuhn-Tucker (KKT) point** as well as a FJ point with $\lambda_o > 0$ (more about all this later). The requirement that $\mathcal{G}_o \neq \varnothing$ will be termed **a constraint qualification (CQ)** – a regularity condition on the binding constraints at $\boldsymbol{x}_o$. So if $\mathcal{G}_o \neq \varnothing$, then $\mathcal{F}_o \cap \mathcal{G}_o = \varnothing$ as required at an optimal point.

Following Fiacco and McCormick (FM) [1968], a sufficient condition which guarantees that $\mathcal{F}_o \cap \mathcal{G}_o = \varnothing$ is the linear independence of the gradient vector's of the binding constraints. So if the set of vectors $\{\nabla \hat{g}^j(\boldsymbol{x}_o), j \in \mathcal{J}\}$ is linearly independent, then $\mathcal{G}_o \neq \varnothing$ and thus $\mathcal{F}_o \cap \mathcal{G}_o = \varnothing$.

Keep in mind that the **FM regularity condition** on the constrains is one of many so-called constraint qualifications that have been offered over the years, especially for validating the KKT necessary conditions for an optimum. For an assortment of additional constraint qualifications see, for instance, Bazaraa, *et al.* [2006], Mangasarian [1969], Abadie [2003], and Kuhn and Tucker [1951].

Section 11.3 Exercises:

11.3.1 Suppose our objective is to maximize $y = f(\boldsymbol{x}) = -(x_1 - 6)^2 - (x_2 - 5)^2$ subject to $\hat{g}^1(\boldsymbol{x}) = 3 - x_1 \geq 0$ and $\hat{g}^2(\boldsymbol{x}) = 4 - x_2 \geq 0$, $\boldsymbol{x} \in R^2$. Verify that $\boldsymbol{x}_o^T = (3, 4)$ is the optimal point. Also, determine if $\boldsymbol{x}^T = (3, 0)$ is a FJ point.

11.3.2 Let us maximize $y = f(\boldsymbol{x}) = x_1^2 + x_1x_2$ subject to $\hat{g}^1(\boldsymbol{x}) = 2 - x_1 - x_2^2 \geq 0$, $\hat{g}^2(\boldsymbol{x}) = 1 -$

$x_1 - x_2 \geq 0$, $\hat{g}^3(\boldsymbol{x}) = x_1 \geq 0$, and $\hat{g}^4(\boldsymbol{x}) = x_2 \geq 0$. Is $\boldsymbol{x}_o^T = (1, 1)$ a FJ point?

11.3.3 Suppose we have an optimization problem of the form: for $\boldsymbol{x} \in R^n$, minimize $y = f(\boldsymbol{x})$ subject to $\hat{g}^j(\boldsymbol{x}) \leq 0, j = 1, \ldots, m$, and $\boldsymbol{x} \geq \boldsymbol{O}$, e.g., for $\boldsymbol{x} \in R^2$, minimize $f(\boldsymbol{x}) = (x_1 - 3)^2 + (x_2 - 2)^2$ subject to

$$\hat{g}^1(\boldsymbol{x}) = x_1^2 + x_2^2 - 5 \leq 0$$
$$\hat{g}^2(\boldsymbol{x}) = x_1 + 2x_2 - 4 \leq 0$$
$$\hat{g}^3(\boldsymbol{x}) = -x_1 \leq 0$$
$$\hat{g}^4(\boldsymbol{x}) = -x_2 \leq 0.$$

Determine if $\boldsymbol{x}_o^T = (2, 1)$ is a FJ point.

11.4 KARUSH-KUHN-TUCKER (KKT) OPTIMALITY CONDITIONS (KUHN AND TUCKER [1951]; TUCKER [1956]; ARROW *ET AL.* [1961])

To set the stage for the development of the KKT conditions, let us harken back to Theorem 11.2. How may we interpret this theorem? Since $\nabla\hat{g}^j(\boldsymbol{x}_o)$ is normal to the support of $\hat{g}^j(\boldsymbol{x}) = 0$ at $\boldsymbol{x}_o$ and points in the direction of maximum increase of $\hat{g}^j(\boldsymbol{x}) = 0, j \in \mathcal{J}$, it may be thought of as an inward-pointing or interior normal to the boundary of $\overline{\mathcal{K}}$ at $\boldsymbol{x}_o$. So if $\boldsymbol{h}$ is a feasible direction, it makes a non-obtuse angle ($\leq\pi/2$) with all the interior normals to the boundary of $\overline{\mathcal{K}}$ at $\boldsymbol{x}_o$. These interior normals or gradients of the binding constraints form a finite cone containing all feasible directions making non-obtuse angles with the supports to $\overline{\mathcal{K}}$ at $\mathcal{D}(\boldsymbol{x}_o)$. This cone is polar to the tangent support cone $\mathcal{D}(\boldsymbol{x}_o)$ and will be termed the **polar support cone** $\mathcal{D}(\boldsymbol{x}_o)^+$ (see Figure 11.3a). So given the tangent support cone $\mathcal{D}(\boldsymbol{x}_o)$ of feasible directions $\boldsymbol{h}$, the polar support cone $\mathcal{D}(\boldsymbol{x}_o)^+$ is the cone spanned by the gradients $\nabla\hat{g}^j(\boldsymbol{x}_o)$ such that, for $\boldsymbol{h} \in \mathcal{D}(\boldsymbol{x}_o)$, $\nabla\hat{g}^j(\boldsymbol{x}_o)^T\boldsymbol{h} \geq 0, j \in \mathcal{J}$. Looked at in an alternative fashion $-\nabla\hat{g}^j(\boldsymbol{x}_o)$ indicates the direction of maximum decrease of $\hat{g}^j(\boldsymbol{x}_o) = 0, j \in \mathcal{J}$, and depicts an outward-pointing or exterior normal to the boundary of $\overline{\mathcal{K}}$ at $\boldsymbol{x}_o$. If $\boldsymbol{h}$ is a feasible direction, it must now make a non-acute angle ($\geq\pi/2$) with all the outward-pointing normals to the boundary of $\overline{\mathcal{K}}$ at $\boldsymbol{x}_o$. These exterior normals or negative gradients of the binding constrains form a finite cone containing all feasible directions making non-acute angles with the supports to $\mathcal{K}$ at $\boldsymbol{x}_o$. This cone is the dual of the tangent support cone $\mathcal{D}(\boldsymbol{x}_o)$ and it termed the **dual support cone** $\mathcal{D}(\boldsymbol{x}_o)^*$ (Figure 11.3b). Hence, the dual support cone $\mathcal{D}(\boldsymbol{x}_o)^*$ is the cone spanned by the negative gradients $-\nabla\hat{g}^j(\boldsymbol{x}_o)$ such that, for each $\boldsymbol{h} \in \mathcal{D}(\boldsymbol{x}_o)$, $-\nabla\hat{g}^j(\boldsymbol{x}_o)^T\boldsymbol{h} \leq 0, j \in \mathcal{J}$.

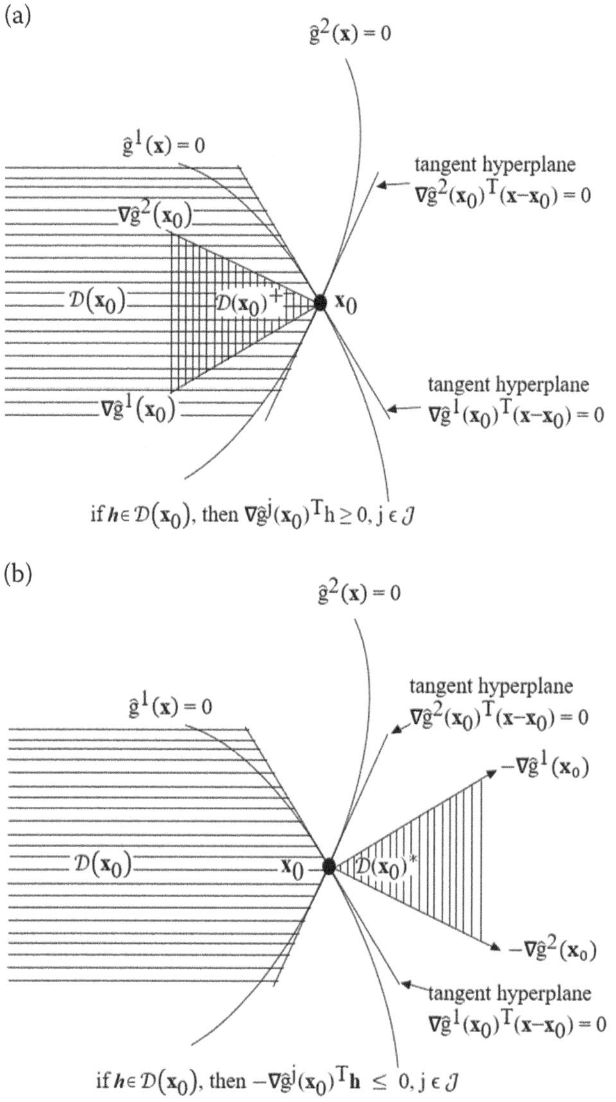

FIGURE 11.3 a. The polar support cone $\mathcal{D}(\boldsymbol{x}_o)^+$; b. the dual support cone $\mathcal{D}(\boldsymbol{x}_o)^*$.

The collection of results from Theorems 11.1, 11.2 may be stated as: if $f(\boldsymbol{x})$ subject to $\nabla\hat{g}^j(\boldsymbol{x}) = 0, j \in \mathcal{J}$, assumes a local maximum at $\boldsymbol{x}_o$, then

$$\nabla f(\boldsymbol{x}_o)^T\boldsymbol{h} \leq 0 \text{ for all } \boldsymbol{h} \text{ satisfying}$$
$$\nabla\hat{g}^j(\boldsymbol{x}_o)^T\boldsymbol{h} \geq 0, j \in \mathcal{J}, \quad \boldsymbol{h} \in \mathcal{D}(\boldsymbol{x}_o). \quad (11.3)$$

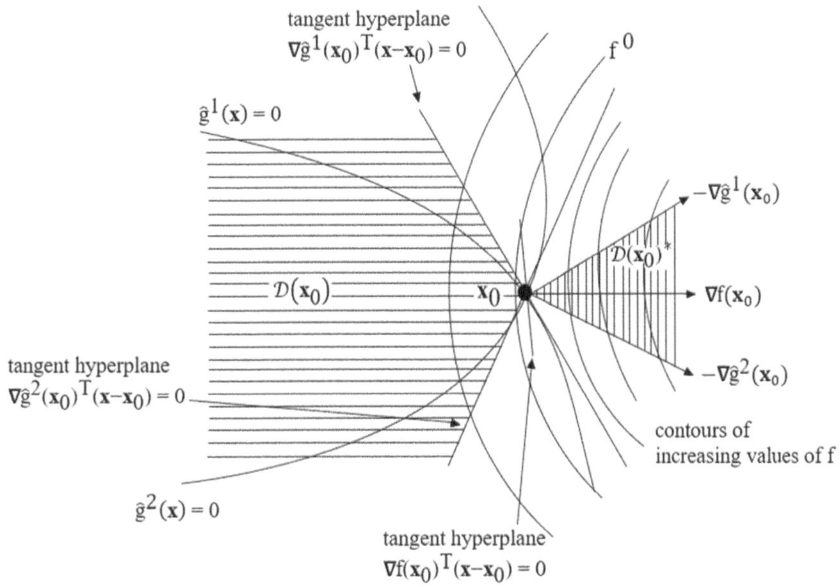

FIGURE 11.4 $\nabla f(\boldsymbol{x}_o) \in \mathcal{D}(\boldsymbol{x}_o)^*$.

How may we interpret this necessary condition? Given any $\boldsymbol{h} \in \mathcal{D}(\boldsymbol{x}_o)$, (11.3) will hold if $\nabla f(\boldsymbol{x}_o)$ lies within the finite cone spanned by the exterior normals $-\nabla\hat{g}^j(\boldsymbol{x}_o), j \in \mathcal{J}$, i.e., $\nabla f(\boldsymbol{x}_o) \in \mathcal{D}(\boldsymbol{x}_o)^*$ (or if $-\nabla f(\boldsymbol{x}_o)$ is contained within the finite cone generated by the interior normals $\nabla\hat{g}^j(\boldsymbol{x}_o), j \in \mathcal{J}$, i.e., $-\nabla f(\boldsymbol{x}_o) \in \mathcal{D}(\boldsymbol{x}_o)^+$). Hence, (11.3) requires that the gradient of f be a non-negative linear combination of the negative gradients of the binding constraints at $\boldsymbol{x}_o$ (Figure 11.4). In this regard, there must exist real numbers $\lambda_j^o \geq 0$ such that

$$\nabla f(\boldsymbol{x}_o) = \sum_j \lambda_j^o(-\nabla\hat{g}^j(\boldsymbol{x}_o)), j \in \mathcal{J}. \tag{11.4}$$

Under what conditions will these Lagrange multipliers $\lambda_j^o \geq 0, j \in \mathcal{J}$, exist? Invoking Farkas' Theorem (see Section 1.8), let $\boldsymbol{b} = \nabla f(\boldsymbol{x}_o)$; the vectors $-\nabla\hat{g}^j(\boldsymbol{x}_o)$, $j \in \mathcal{J}$, constitute the columns of the matrix $\boldsymbol{A}$; and the $\lambda_j^o, j \in \mathcal{J}$, serve as the elements of the vector $\boldsymbol{\lambda} \geq \boldsymbol{O}$. Then, given that the FM regularity condition holds ($\mathcal{G}_o \neq \varnothing$ implies $\mathcal{F}_o \cap \mathcal{G}_o = \varnothing$), a necessary and sufficient condition for $\nabla f(\boldsymbol{x}_o)$ to lie within the finite cone spanned by the vectors $-\nabla\hat{g}^j(\boldsymbol{x}_o), j \in \mathcal{J}$, is that (11.3) holds. Hence, there must exist real numbers $\lambda_j^o \geq 0$ such that (11.4) also holds.

We may incorporate all of the constraints (active or not) $\hat{g}^j(\boldsymbol{x}) \geq 0, j = 1, \ldots, m+n$, into our discussion by defining the scalar λ_j^o as zero whenever $\hat{g}^j(\boldsymbol{x}_o) > 0$ or $j \notin \mathcal{J}$. Then (11.4) is equivalent to the system

$$\nabla f(\boldsymbol{x}_o) = \sum_{j=1}^{m+n} \lambda_j^o(-\nabla\hat{g}^j(\boldsymbol{x}_o))$$

$$\left.\begin{array}{l} \lambda_j^o\,\hat{g}^j(\boldsymbol{x}_o) = 0 \\ \hat{g}^j(\boldsymbol{x}_o) \geq 0 \\ \lambda_j^o \geq 0 \end{array}\right\} j = 1, \ldots, m+n \;\textbf{[KKT Optimality Conditions]}$$

$$\boldsymbol{x}_o \text{ unrestricted.} \tag{11.5}$$

(Note that the CS condition $\lambda_j^o\,\hat{g}^j(\boldsymbol{x}_o) = 0$ holds for all values of j since, if $\hat{g}^j(\boldsymbol{x}_o) > 0$ then $\lambda_j^o = 0$, and if $\hat{g}^j(\boldsymbol{x}_o) = 0$, then $\lambda_j^o \geq \boldsymbol{0}$.)

How might a system such as (11.5) arise? An answer to this question can be obtained via the **classical derivation of the KKT necessary optimality conditions**. We thus have

Theorem 11.6: **(KKT Necessary Conditions)**. Let $y = f(\boldsymbol{x})$ and the vector-valued function $\overline{\boldsymbol{G}}(\boldsymbol{x})$ be of class $C^{(1)}$ throughout some spherical δ-neighborhood of the point $\boldsymbol{x}_o \in \overline{\mathcal{K}} = \{\boldsymbol{x} | \overline{\boldsymbol{G}}(\boldsymbol{x}) \geq \boldsymbol{O}, \boldsymbol{x} \in R^n\}$. If $\boldsymbol{x}_o$ is a local optimal solution to Problem (11.1.1), then, subject to the FM regularity condition, $\mathcal{F}_o \cap \mathcal{G}_o = \varnothing$ and

$$\begin{aligned}&\nabla f(\boldsymbol{x}_o) + \nabla \overline{\boldsymbol{G}}(\boldsymbol{x}_o)\boldsymbol{\lambda}_o = \boldsymbol{O}\\ &\boldsymbol{\lambda}_o^T \overline{\boldsymbol{G}}(\boldsymbol{x}_o) = 0\\ &\overline{\boldsymbol{G}}(\boldsymbol{x}_o) \geq \boldsymbol{O}\\ &\boldsymbol{\lambda}_o \geq \boldsymbol{O}\\ &\boldsymbol{x}_o \text{ unrestricted}\\ &[\textit{KKT Optimality Conditions}]\end{aligned} \quad \text{or} \quad \begin{aligned}&\nabla f(\boldsymbol{x}_o) + \textstyle\sum_{j=1}^{m+n} \lambda_j^o \nabla \hat{g}^j(\boldsymbol{x}_o) = \boldsymbol{O}\\ &\textstyle\sum_{j=1}^{m+n} \lambda_j^o \hat{g}^j(\boldsymbol{x}_o) = 0\\ &\left.\begin{aligned}\hat{g}^j(\boldsymbol{x}_o) \geq 0\\ \lambda_j^0 \geq 0\end{aligned}\right\} j = 1,\ldots,m+n\\ &x_i^o \text{ unrestricted, } i = 1,\ldots,n,\end{aligned} \tag{11.6}$$

where $\nabla \overline{\boldsymbol{G}}(\boldsymbol{x}_o)$ is an $(n \times m+n)$ matrix whose jth column is the vector $\nabla \hat{g}^j(\boldsymbol{x}_o)$, $j = 1, \ldots, m+n$, or

$$\nabla \overline{\boldsymbol{G}}(\boldsymbol{x}_o) = [\nabla \hat{g}^1(\boldsymbol{x}_o), \ldots, \nabla \hat{g}^{m+n}(\boldsymbol{x}_o)].$$

(Note that $\nabla \overline{\boldsymbol{G}}(\boldsymbol{x}_o)$ is the transpose of the Jacobian of $\overline{\boldsymbol{G}}(\boldsymbol{x})$ at $\boldsymbol{x}_o$.)

Proof: The *augmented or extended Lagrangian* associated with Problem (11.1.1) is

$$L(\boldsymbol{x}, \boldsymbol{\lambda}, \boldsymbol{s}) = f(\boldsymbol{x}) + \boldsymbol{\lambda}^T(\overline{\boldsymbol{G}}(\boldsymbol{x}) - \boldsymbol{s}),$$

where the inequality constraint $\overline{\boldsymbol{G}}(\boldsymbol{x}) \geq \boldsymbol{O}$ has been converted to an equality constraint $\overline{\boldsymbol{G}}(\boldsymbol{x}) - \boldsymbol{s} = \boldsymbol{O}$ by subtracting an $(m+n \times 1)$ vector of non-negative slack variables $\boldsymbol{s}$, with $\boldsymbol{s}^T = (s_1^2, \ldots, s_{m+n}^2)$. Then

$$\begin{aligned}&\nabla_{\boldsymbol{x}} L = \nabla f(\boldsymbol{x}) + \nabla \overline{\boldsymbol{G}}(\boldsymbol{x})\boldsymbol{\lambda} = \boldsymbol{O}\\ &\nabla_{\boldsymbol{\lambda}} L = \overline{\boldsymbol{G}}(\boldsymbol{x}) - \boldsymbol{s} = \boldsymbol{O}\\ &\frac{\partial L}{\partial s_j} = -2\lambda_j s_j = 0,\ j = 1,\ldots,m+n.\end{aligned}$$

Given that the FM regularity requirement holds at $\boldsymbol{x}_o$, Farkas' Theorem implies the existence of a $\boldsymbol{\lambda}_o \geq \boldsymbol{O}$ such that $\nabla f(\boldsymbol{x}_o) + \nabla \overline{\boldsymbol{G}}(\boldsymbol{x}_o)\boldsymbol{\lambda}_o = \boldsymbol{O}$ if and only if (11.3) holds or $\mathcal{F}_o \cap \mathcal{G}_o = \varnothing$. Next, from $\nabla_{\boldsymbol{\lambda}} L = \boldsymbol{O}$ we obtain $\boldsymbol{s} = \overline{\boldsymbol{G}}(\boldsymbol{x}_o)$ while from $\partial L/\partial s_j = 0$ we have $\lambda_j s_j = 0$ or $\lambda_j s_j^2 = 0$, $j = 1, \ldots, m+n$. Then $\sum_{j=1}^{m+n} \lambda_j s_j^2 = 0$ or, at $\boldsymbol{x}_o$, $\boldsymbol{\lambda}^T \overline{\boldsymbol{G}}(\boldsymbol{x}_o) = 0$. So if the jth constraint $\hat{g}^j(\boldsymbol{x}_o) \geq 0$, $j = 1, \ldots, m+n$, is not binding at $\boldsymbol{x}_o (j \notin \mathcal{J})$, then $s_j^2 > 0$, $\lambda_j^o = 0$; and if $(j \in \mathcal{J})$, then $s_j^2 = 0$, $\lambda_j^o \geq \boldsymbol{0}$. Hence, at least one element of each pair $(\lambda_j^o, \hat{g}^j(\boldsymbol{x}_o))$ vanishes, thus guaranteeing that $\boldsymbol{\lambda}_o^T \overline{\boldsymbol{G}}(\boldsymbol{x}_o) = 0$. Q. E. D.

If $\overline{\boldsymbol{G}}(\boldsymbol{x}) \geq \boldsymbol{O}$ is replaced by $\boldsymbol{G}(\boldsymbol{x}) \geq \boldsymbol{O}$ so that the n non-negativity conditions $\boldsymbol{x} \geq \boldsymbol{O}$ appear explicitly and are not treated like the constraints $\hat{g}^j(\boldsymbol{x}) \geq 0$, $j = 1, \ldots, m$ (we do not convert $x_i \geq 0$ to $\hat{g}^{m+i}(\boldsymbol{x}) \geq 0$, $i = 1, \ldots, n$), then (11.6) is replaced by (11.7) or

$$\begin{aligned}&\nabla f(\boldsymbol{x}_o) + \nabla \boldsymbol{G}(\boldsymbol{x}_o)\lambda_o \leq \boldsymbol{O}\\ &\boldsymbol{x}_o^T(\nabla f(\boldsymbol{x}_o) + \nabla \boldsymbol{G}(\boldsymbol{x}_o)\lambda_o) = 0\\ &\lambda_o^T \boldsymbol{G}(\boldsymbol{x}_o) = 0\\ &\boldsymbol{G}(x_o) \geq \boldsymbol{O}\\ &\lambda_o \geq \boldsymbol{O}\\ &\boldsymbol{x}_o \geq \boldsymbol{O}\\ &[\textbf{KKT Optimality Conditions}]\end{aligned} \quad \text{or} \quad \begin{aligned}&\nabla f(\boldsymbol{x}_o) + \textstyle\sum_{j=1}^{m} \lambda_j^o \nabla \hat{g}^j(\boldsymbol{x}_o) \leq \boldsymbol{O}\\ &\textstyle\sum_{i=1}^{n} x_i^o (f_i^o + \sum_{j=1}^{m} \lambda_j^o \hat{g}_i^j(\boldsymbol{x}_o)) = 0\\ &\textstyle\sum_{j=1}^{m} \lambda_j^o \nabla \hat{g}^j(\boldsymbol{x}_o) = 0\\ &\left.\begin{aligned}\hat{g}^j(\boldsymbol{x}_o) \geq 0\\ \lambda_j^0 \geq 0\end{aligned}\right\} j = 1,\ldots,m\\ &x_i^o \geq 0,\ i = 1,\ldots,n,\end{aligned} \tag{11.7}$$

where $\nabla \overline{\boldsymbol{G}}(\boldsymbol{x}_o) = [\nabla \hat{g}^1(\boldsymbol{x}_o), \ldots, \nabla \hat{g}^m(\boldsymbol{x}_o)]$ is of order $(n \times m)$ (it is the transpose of the Jacobian of $\boldsymbol{G}(\boldsymbol{x})$ at $\boldsymbol{x}_o$). Looking to the second equation in system (11.7), for $i = 1, \ldots, n$, either $x_i^o = 0$, thus llowing f_i^o to be less than $-\sum_{j=1}^m \lambda_j^o \hat{g}_i^j(\boldsymbol{x}_o)$, or $f_i^o = -\sum_{j=1}^m \lambda_j^o \hat{g}_i^j(\boldsymbol{x}_o)$, in which case x_i^o may be positive. Since at least one of these alternatives must hold, it follows that

$$\sum_{i=1}^n x_i^o \left(f_i^o + \sum_{j=1}^m \lambda_j^o \hat{g}_i^j(\boldsymbol{x}_o) \right) = 0.$$

Additionally, if the constraint $\hat{g}^j(\boldsymbol{x}) \geq 0, j = 1, \ldots, m,$ is not binding at $\boldsymbol{x}_o (s_j^2 > 0)$, we require that $\lambda_j^o = 0$; while if this constraint is active at $\boldsymbol{x}_o (s_j^2 = 0)$, then $\lambda_j^o \geq 0$. Hence, at least one of these alternatives holds so that $\sum_{j=1}^m \lambda_j^o \hat{g}_i^j(\boldsymbol{x}_o) = 0$.

Example 11.2: Using the objective function and constraints appearing in Example 11.1, system (11.6) appears as

$$\nabla f(\boldsymbol{x}) + \nabla \overline{\boldsymbol{G}}(\boldsymbol{x})\boldsymbol{\lambda} = \begin{bmatrix} -4x_1 + 4x_2 + 4 - \lambda_1 - 8\lambda_2 x_1 + \lambda_3 \\ -6x_2 + 4x_1 + 2 - 3\lambda_1 - 2\lambda_2 x_2 + \lambda_4 \end{bmatrix} = \boldsymbol{O}$$

$$\boldsymbol{\lambda}^T \overline{\boldsymbol{G}}(\boldsymbol{x}) = \lambda_1(6 - x_1 - 3x_2) + \lambda_2(16 - 4x_1^2 - x_2^2) + \lambda_3 x_1 + \lambda_4 x_2 = 0$$

$$\overline{\boldsymbol{G}}(\boldsymbol{x}) = \begin{bmatrix} 6 - x_1 - 3x_2 \\ 16 - 4x_1^2 - x_2^2 \\ x_1 \\ x_2 \end{bmatrix} \geq \boldsymbol{O}, \boldsymbol{\lambda}^T = (\lambda_1, \lambda_2, \lambda_3, \lambda_4) \geq \boldsymbol{O}^T.$$

Since (11.6) constitutes only a necessary condition for optimality, we need to test various points of $\overline{\mathcal{K}} = \{\boldsymbol{x} | \overline{\boldsymbol{G}}(\boldsymbol{x}) \geq \boldsymbol{O}, \boldsymbol{x} \in R^2\}$ for optimality. Again selecting point A in Figure 11.2 (here $\mathcal{J} = \{1,2\}$), the preceding system becomes

$$\begin{aligned} &\text{a.} \quad -4x_1 + 4x_2 + 4 - \lambda_1 - 8\lambda_2 x_1 = 0 \\ &\text{b.} \quad -6x_2 + 4x_1 + 2 - 3\lambda_1 - 2\lambda_2 x_2 = 0 \\ &\text{c.} \quad \lambda_1(0) + \lambda_2(0) + 0(x_1) + 0(x_2) = 0 \\ &\text{d.} \quad 6 - x_1 - 3x_2 = 0, \lambda_1 \geq 0 \\ &\text{e.} \quad 16 - 4x_1^2 - x_2^2 = 0, \lambda_2 \geq 0 \\ &\text{f.} \quad x_1 > 0, \lambda_3 = 0 \\ &\text{g.} \quad x_2 > 0, \lambda_4 = 0 \end{aligned} \tag{11.8}$$

From (11.8) we obtain $x_0^T = (1.88, 1.37)$. A substitution of these values into (8.8a,b) yields $\boldsymbol{\lambda}_0^T = (0.31, 0.11)$. Additionally, $f(\boldsymbol{x}_o) = 7.87$.

Does the FM regularity requirement or CQ hold at $\boldsymbol{x}_o$? With $\hat{g}^1(\boldsymbol{x}) = 6 - x_1 - 3x_2 \geq 0$ and $\hat{g}^2(\boldsymbol{x}) = 16 - 4x_1^2 - x_2^2 \geq 0$ we see that

$$\nabla \hat{g}^1(\boldsymbol{x}) = \begin{bmatrix} -1 \\ -3 \end{bmatrix}, \nabla \hat{g}^2(\boldsymbol{x}) = \begin{bmatrix} -8x_1 \\ -2x_2 \end{bmatrix}.$$

Then $\hat{g}^1(\boldsymbol{x}_o)^T = (-1, -3)$, $\hat{g}^2(\boldsymbol{x}_o)^T = (-15.04, -2.74)$, and thus

$$\begin{vmatrix} -1 & -15.04 \\ -3 & -2.74 \end{vmatrix} \neq 0.$$

Hence, the gradient vectors of the binding constraints at $\boldsymbol{x}_o$ are linearly independent so that the FM regularity condition is in effect. ■

How do the KKT necessary optimality conditions compare with the FJ necessary optimality conditions?

First, the **KKT Lagrangian** associated with the **KKT problem**

KKT: Find $\boldsymbol{x} \in \overline{\mathcal{K}}$ and $\boldsymbol{\lambda} \in R^{m+n}$

is

$$L(\boldsymbol{x}, \boldsymbol{\lambda}) = f(\boldsymbol{x}) + \boldsymbol{\lambda}^T \overline{\boldsymbol{G}}(\boldsymbol{x});$$

while the FJ Lagrangian for the FJ problem

FJ: Find $\boldsymbol{x} \in \overline{\mathcal{K}}$ and $\lambda_o \in R$, and $\overline{\boldsymbol{\lambda}} \in R_{m+n}$

is

$$L(\boldsymbol{x}, \lambda_o, \overline{\boldsymbol{\lambda}}) = \lambda_o f(\boldsymbol{x}) + \overline{\boldsymbol{\lambda}} \overline{\boldsymbol{G}}(\boldsymbol{x}).$$

Secondly, the necessary optimality conditions for each technique may be summarized as

KKT Optimality ((11.6))	FJ Optimality ((11.2.1))
$\nabla f(\boldsymbol{x}_o) + \nabla \overline{\boldsymbol{G}}(\boldsymbol{x}_o)\boldsymbol{\lambda}_o = \boldsymbol{O}$	$\lambda_o \nabla f(\boldsymbol{x}_o) + \overline{\boldsymbol{\lambda}}^T \nabla \overline{\boldsymbol{G}}(\boldsymbol{x}_o) = \boldsymbol{O}$
$\boldsymbol{\lambda}_o^T \overline{\boldsymbol{G}}(\boldsymbol{x}_o) = 0$	$\overline{\boldsymbol{G}}(\boldsymbol{x}) \geq \boldsymbol{O}$
$\overline{\boldsymbol{G}}(\boldsymbol{x}) \geq \boldsymbol{O}$	$\begin{bmatrix} \lambda_o \\ \overline{\boldsymbol{\lambda}} \end{bmatrix} \geq \begin{bmatrix} 0 \\ \boldsymbol{O} \end{bmatrix}.$
$\boldsymbol{x}_o$ unrestricted	

So if $(\boldsymbol{x}, \lambda_o, \overline{\boldsymbol{\lambda}})$ is a FJ point with $\lambda_o > 0$, then $(\boldsymbol{x}_o, \overline{\boldsymbol{\lambda}}/\lambda_o)$ is also a KKT point. And if $(\boldsymbol{x}, \boldsymbol{\lambda}_o)$ is a KKT point, then $(\boldsymbol{\lambda}_o, 1, \boldsymbol{\lambda}_o)$ is a FJ point.

For the most part we shall primarily deal with the KKT necessary conditions for an inequality constrained optimization problem. In this regard, let us now examine the circumstance under which these conditions are also sufficient to ensure a constrained local (and global) extremum. The sufficiency argument requires that we introduce some additional assumptions dealing with the characteristics (e.g., functional forms) of $y = f(\mathbf{x})$ and $\hat{g}^j(\mathbf{x}) \geq 0, j = 1, \cdots, m, \mathbf{x} \in R^n$. To aid us in this undertaking, we offer (without proof) a set of theorems which will be useful for determining the KKT sufficient and necessary *and* sufficient conditions for maximizing $f(\mathbf{x})$ subject to $\hat{g}^j(\mathbf{x}) \geq 0, j = 1, \cdots, m, \mathbf{x} \geq \mathbf{O}, \mathbf{x} \in R^n$.

Section 11.4 Exercises:

11.4.1 Maximize $f(\mathbf{x}) = x_1x_2$ subject to $2x_1^2 + x_2 \leq 4$.

11.4.2 Maximize $f(\mathbf{x}) = -x_1^2 - 2x_2^2 - x_1x_2 - x_2x_3$ subject to

$$x_1 + x_2 + x_3 \leq 10$$
$$x_i \geq 0, \ i = 1, 2, 3.$$

Then resolve this problem with the non-negativity conditions absent.

11.4.3 Minimize $f(\mathbf{x}) = x_1^2 + x_2^2 - 2x_1 - 3x_2$ subject to

$$x_1 + 4x_2 \leq 6$$
$$x_1, x_2 \geq 0.$$

11.4.4 Maximize $f(\mathrm{x}) = x_1x_2$ subject to

$$2x_1 + x_2 \leq 12$$
$$x_1 + 2x_2 \leq 10$$
$$x_1, x_2 \geq 0.$$

11.5 KKT SUFFICIENT OPTIMALITY CONDITIONS

We have seen how the KKT first-order or necessary conditions relate ∇f to the $\nabla \hat{g}^j$s (via $\boldsymbol{\lambda} \geq \mathbf{O}$) at an optimal solution point $\mathbf{x}_o$. In fact, we know that, at an optimal solution point to problem (11.1.1), we have $\mathcal{F}_o \cap \mathcal{G} = \varnothing$ so that there does not exist an admissible direction $\mathbf{h}\ (\neq \mathbf{O})$ such that $\nabla f(\mathbf{x}_o)^T\mathbf{h} > 0$. However, it may be the case that $\nabla f(\mathbf{x}_o)^T\mathbf{h} = 0$. In this latter instance, a movement along the direction $\mathbf{h}$ may *increase or decrease* the value of f. To handle this *undecided* outcome, we need to rely upon some second-order information concerning f and the $\hat{g}^j$ constraints. Such information allows us to determine something about the curvature of the Lagrangian at $(\mathbf{x}_o, \boldsymbol{\lambda}_o)$ so that we can distinguish a constrained maximum from a constrained minimum. To this end, in what follows we shall assume that f and $\hat{g}^1, \ldots, \hat{g}^{m+n}$ are of class $C^{(2)}$ on the open region

$$\mathcal{D} \subset R^n.$$

To set the stage for the statement of a key theorem of this section, let us define:

$\mathcal{J}(\mathbf{x}_o) = \{j|\hat{g}^j(\mathbf{x}_o) = 0\}$,

$\hat{\mathcal{J}}(\mathbf{x}_o) = \{j|\hat{g}^j(\mathbf{x}_o) = 0 \text{ and } \nabla_{\mathbf{x}} L(\mathbf{x}_o, \boldsymbol{\lambda}_o) = \mathbf{O}, \lambda_j^o \hat{g}^j(\mathbf{x}_o) = 0,$ $\lambda_j^0 \geq 0$ hold for $\boldsymbol{\lambda}_o > \mathbf{O}\}$, and

$\tilde{\mathcal{J}}(\mathbf{x}_o) = \{\mathbf{h}|\nabla\hat{g}^j(\mathbf{x}_o)^T\mathbf{h} = 0, j \in \hat{\mathcal{J}}(\mathbf{x}_o); \nabla\hat{g}^j(\mathbf{x}_o{}^T)\mathbf{h} \geq 0,$ $j \in \mathcal{J}(\mathbf{x}_o)\}$.

We can now state

Theorem 11.7: (KKT Second-Order Sufficient Condition). Let $\mathbf{x}_o$ be a feasible solution to Problem (11.1.1). If there exists a vector $\boldsymbol{\lambda}_o$ satisfying (11.5) and for every $\mathbf{h}\ (\neq \mathbf{O})$ such that $\mathbf{h} \in \tilde{\mathcal{J}}(\mathbf{x}_o)$ we have

$$\mathbf{h}^T\nabla^2_{\mathbf{xx}} L(\mathbf{x}_o, \boldsymbol{\lambda}_o)\mathbf{h} < 0(>0), \tag{11.9}$$

then $\mathbf{x}_o$ is a strong local maximum (minimum) for Problem (11.1.1), where

$$\nabla^2_{\mathbf{xx}} L(\mathbf{x}_o, \boldsymbol{\lambda}_o) = \nabla^2 f(\mathbf{x}_o) - \sum_{j=1}^{m+n} \lambda_j^o \nabla^2 \hat{g}^j(\mathbf{x}_o)$$

is the Hessian matrix of L evaluated at $(\mathbf{x}_o, \boldsymbol{\lambda}_o)$. So if $\nabla^2_{\mathbf{xx}} L(\mathbf{x}_o, \boldsymbol{\lambda}_o)$ is negative definite (positive definite) then f has a strong local constrained maximum (minimum) at $\mathbf{x}_o$.[2]

Example 11.3: Maximize $f(\mathbf{x}) = -x_1 - x_2$ subject to $\hat{g}(\mathbf{x}) = 2 - x_1^2 - x_2^2 \geq 0, \mathbf{x} \in R^2$. Forming the Lagrangian yields $L = f + \lambda\hat{g} = -x_1 - x_2 + \lambda(2 - x_1^2 - x_2^2)$. Then

$$\nabla_{\mathbf{x}} L = \begin{bmatrix} L_{x_1} \\ L_{x_2} \end{bmatrix} = \begin{bmatrix} -1 - 2\lambda x_1 \\ -1 - 2\lambda x_2 \end{bmatrix} = \mathbf{O},$$
$$L_\lambda = 2 - x_1^2 - x_2^2 = 0.$$

Solving this system simultaneously yields $\mathbf{x}_o^T = (-1, -1)$ and $\lambda_o = 1/2$. (It is easily checked that (11.5) is satisfied.)

Next,

$$\nabla^2_{\mathbf{xx}} L = \begin{bmatrix} L_{x_1x_1} & L_{x_1x_2} \\ L_{x_2x_1} & L_{x_2x_2} \end{bmatrix} = \begin{bmatrix} -2\lambda & 0 \\ 0 & -2\lambda \end{bmatrix}.$$

At $(\mathbf{x}_o, \lambda_o) = \left(-1, -1, \frac{1}{2}\right)$ we see that

$$\nabla_{xx}^2 L(\boldsymbol{x}_o, \lambda_o) = \begin{bmatrix} -1 & 0 \\ 0 & -1 \end{bmatrix},$$

a negative definite matrix. Thus for $\lambda_o = \frac{1}{2}$, f subject to $\hat{g} \geq 0$ has a strong local maximum at $\boldsymbol{x}_o^T = (-1, -1)$.

Is the FM constraint qualification in force at $\boldsymbol{x}_o$? Since $\nabla \hat{g}(\boldsymbol{x})^T = (-2x_1, -2x_2)$ and $\nabla \hat{g}(\boldsymbol{x}_o)^T = (2, 2) \neq \boldsymbol{O}$, the answer is *yes.* Why an answer in the affirmative? Simply because a single vector that is not the null vector is taken to be linearly independent. ■

Next, let us take an alternative approach to developing KKT sufficient conditions for optimality. That is, we shall admit the notion of concavity to our discussion. As will now be seen, the KKT necessary conditions for optimality are sufficient when dealing with a *concave program.* As a prelude to the next optimality theorem (Theorem 11.11), we offer the following set of (three) theorems.

We start with the conditions under which the set of all points satisfying $\hat{g}^j(\boldsymbol{x}) \geq 0$, $j = 1, \ldots, m$, is a convex set.

Theorem 11.8: If the real-valued function $\hat{g}^j(\boldsymbol{x}) \geq 0$, $\boldsymbol{x} \in R^n$, is concave, then the set $\mathcal{K}_j = \{\boldsymbol{x} | \hat{g}^j(\boldsymbol{x}) \geq 0, \boldsymbol{x} \in R^n\}$ is convex, $j = 1, \ldots, m$.

(Note that for $\hat{g}^j = b_j - g^j$, $j = 1, \ldots, m$, to be concave over a convex set, g^j must be convex since, if $-g^j$ is concave, then g^j is convex and conversely.) Next,

Theorem 11.9: Let the real-valued functions $\varnothing^i(\boldsymbol{x})$, $i = 1, \ldots, k$, $\boldsymbol{x} \in R^n$, be concave over the convex region $\mathcal{K} \subset R^n$. Then any non-negative linear combination of the $\varnothing^i$ is concave over $\mathcal{K}$, i.e., if $\varnothing^1, \ldots, \varnothing^k$ are all concave on $\mathcal{K}$, then so is

$$\varnothing(\boldsymbol{x}) = \sum_{i=1}^k \alpha_i \varnothing^i(\boldsymbol{x}), \ \alpha_i \geq 0, \ \alpha_i \in R, \ i = 1, \ldots, k.$$

Also,

Theorem 11.10: The intersection of a finite number of convex sets is convex, i.e., if the sets $\mathcal{S}_i$, $i = 1, \ldots, k$, are all convex, then $\mathcal{S} = \cap_{i=1}^k \mathcal{S}_i$ is convex.

We know that the feasible region for Problem (11.1.1) is $\overline{\mathcal{K}} = \{\boldsymbol{x} | \hat{g}^j(\boldsymbol{x}) \geq 0, j = 1, \ldots, m + n, \boldsymbol{x} \in R^n\}$.

If each g^j is convex, then $\hat{g}^j \geq 0$, $j = 1, \ldots, m$, is concave. Obviously each $\hat{g}^{m+i}(\boldsymbol{x}) \geq 0$, $i = 1, \ldots, n$, is concave.

By Theorem 11.8, $\mathcal{K}_j = \{\boldsymbol{x} | \hat{g}^j(\boldsymbol{x}) \geq 0, \boldsymbol{x} \in R^n\}$, $j = 1, \ldots, m + n$, is convex. With $\overline{\mathcal{K}} = \cap_{j=1}^{m+n} \mathcal{K}_j$, $\overline{\mathcal{K}}$ must also be a convex set via Theorem 11.10. In the light of these remarks, if the real-valued functions $y = f(\boldsymbol{x})$, $\hat{g}^j(\boldsymbol{x}) \geq 0$, $j = 1, \ldots, m$, are strictly concave over the convex region $\overline{\mathcal{K}} = \{\boldsymbol{x} | \overline{\boldsymbol{G}}(\boldsymbol{x}) \geq \boldsymbol{O}, \boldsymbol{x} \in R^n\}$, then system (11.6) becomes a necessary and sufficient condition for a local (and thus global) constrained maximum of f. Specifically,

Theorem 11.11: **(KKT Necessary and Sufficient Conditions)**. Let the real-valued function $y = f(\boldsymbol{x})$ and the vector-valued function $\overline{\boldsymbol{G}}(\boldsymbol{x})$, $\boldsymbol{x} \in R^n$, be strictly concave over the convex region $\overline{\mathcal{K}} = \{\boldsymbol{x} | \overline{\boldsymbol{G}}(\boldsymbol{x}) \geq \boldsymbol{O}, \boldsymbol{x} \in R^n\}$, int $(\overline{\mathcal{K}}) \neq \varnothing$,[3] and of class $C^{(1)}$ at $\boldsymbol{x}_o \in \overline{\mathcal{K}}$ with $\mathcal{F}_o \cap \mathcal{G}_o = \varnothing$. Then f subject to $\overline{\boldsymbol{G}}(\boldsymbol{x}) \geq \boldsymbol{O}$ has a (strong) global maximum at $\boldsymbol{x}_o$ if and only if (11.6) holds.

Proof: (Necessity) If (11.6) is not satisfied, then $\boldsymbol{x}_o$ cannot be a constrained extremum of f. (Sufficiency) For $\boldsymbol{x}_o \in \overline{\mathcal{K}}$, $f(\boldsymbol{x}) \leq f(\boldsymbol{x}) + \boldsymbol{\lambda}_o^T \overline{\boldsymbol{G}}(\boldsymbol{x}) = L(\boldsymbol{x}, \boldsymbol{\lambda}_o)$.

With $L(\boldsymbol{x}, \boldsymbol{\lambda}_o)$ strictly concave in $\boldsymbol{x}$ over $\overline{\mathcal{K}}$,

$$\begin{aligned} L(\boldsymbol{x}, \boldsymbol{\lambda}_o) &< L(\boldsymbol{x}, \boldsymbol{\lambda}_o) + \nabla_{\boldsymbol{x}} L(\boldsymbol{x}, \boldsymbol{\lambda}_o)^T (\boldsymbol{x} - \boldsymbol{x}_o) \\ &= f(\boldsymbol{x}_o) + \boldsymbol{\lambda}_o^T \overline{\boldsymbol{G}}(\boldsymbol{x}_o) + (\nabla f(\boldsymbol{x}_o) + \nabla \overline{\boldsymbol{G}}(\boldsymbol{x}_o)\boldsymbol{\lambda}_o)^T (\boldsymbol{x} - \boldsymbol{x}_o) \\ &= f(\boldsymbol{x}_o). \end{aligned}$$

Hence, $f(\boldsymbol{x}_o) > L(\boldsymbol{x}, \boldsymbol{\lambda}_o) \geq f(\boldsymbol{x})$, $\boldsymbol{x} \in \overline{\mathcal{K}}$, and thus f has a (strong) global constrained maximum at $\boldsymbol{x}_o$. Q. E. D.

Sometimes the preceding theorem is stated under the weaker assumption that f is pseudoconcave over $\overline{\mathcal{K}}$ and the $\hat{g}^j$, $j \in \mathcal{J}$, are differentiable and quasiconcave over $\overline{\mathcal{K}}$ so that global optimality still holds. More on this situation later (Chapter 14) when we discuss *generalized concave programs.*

Section 11.5 Exercises:

11.5.1 For Exercises 11.4.1–11.4.4 determine if the KKT second-order condition (11.9) holds.

11.5.2 For each of the Exercises 11.4.1 – 11.4.4 verify that the FM constraint qualification holds.

11.5.3 Prove Theorem 11.9.

11.5.4 Prove Theorem 11.10.

11.6 THE OPTIMAL VALUE FUNCTION: LAGRANGE MULTIPLIERS REVISITED

It was mentioned earlier that λ_j^o is the optimal Lagrange multiplier associated with the constraint $\hat{g}^j(\boldsymbol{x}) = b_j - g^j(\boldsymbol{x}) \geq 0$, $j = 1, \ldots, m$. As we shall now see, this same result can be obtained by working with the **optimal value function** for Problem (11.1.1),

$$f^o(\boldsymbol{b}) = \boldsymbol{max}\{f(\boldsymbol{x})|\boldsymbol{G}(\boldsymbol{x}) = \boldsymbol{b} - \boldsymbol{g}(\boldsymbol{x}) \geq \boldsymbol{O},$$

$$\boldsymbol{x} \geq \boldsymbol{O}, \boldsymbol{x} \in R^n\},$$

where the jth component of the $(m \times 1)$ vector-valued function $\boldsymbol{g}$ is $g^j(\boldsymbol{x})$, $j = 1, \ldots, m$. Hence, $f^o(\boldsymbol{b})$ is assumed of class $C^{(1)}$ and is non-decreasing in the variables b_j, $j = 1, \ldots, m$. (Clearly, f^o must be non-decreasing in $\boldsymbol{b}$ since, if any component of $\boldsymbol{b}$ is increased in value with the others held fixed, the region of admissible solutions $\mathcal{K} = \{\boldsymbol{x}|\boldsymbol{G}(\boldsymbol{x}) \geq \boldsymbol{O}, \boldsymbol{x} \geq \boldsymbol{O}, \boldsymbol{x} \in R^n\}$ must increase in size.)

For $\boldsymbol{b} = \bar{\boldsymbol{b}}$, let $\bar{\boldsymbol{x}}$ be the solution associated with the problem: maximize $f(\boldsymbol{x})$ subject to $\bar{\boldsymbol{b}} \geq \boldsymbol{g}(\boldsymbol{x})$, $\boldsymbol{x} \geq \boldsymbol{O}$, $\boldsymbol{x} \in R^n$. Then $f(\bar{\boldsymbol{x}}) = f^o(\bar{\boldsymbol{b}})$ and, for all $\boldsymbol{x} \geq \boldsymbol{O}$, $f(\boldsymbol{x}) \leq f^o(\bar{\boldsymbol{b}}) = f^o(\boldsymbol{g}(\boldsymbol{x})) \leq f^o(\boldsymbol{g}(\boldsymbol{x}) + \bar{\boldsymbol{b}} - \boldsymbol{g}(\bar{\boldsymbol{x}}))$ since $\bar{\boldsymbol{b}} \geq \boldsymbol{g}(\bar{\boldsymbol{x}})$ and f^o is non-decreasing. Then the function $\eta(\boldsymbol{x}) = f(\boldsymbol{x}) - f^o(\boldsymbol{g}(\boldsymbol{x}) + \bar{\boldsymbol{b}} - \boldsymbol{g}(\bar{\boldsymbol{x}})) \leq 0$ for all $\boldsymbol{x} \geq \boldsymbol{O}$ and, with $\eta(\bar{\boldsymbol{x}}) = f(\bar{\boldsymbol{x}}) - f^o(\bar{\boldsymbol{b}}) = 0$, we see that $\eta(\boldsymbol{x})$ attains a maximum at $\boldsymbol{x} = \bar{\boldsymbol{x}}$ when

$$\eta_i(\bar{\boldsymbol{x}}) = f_i(\bar{\boldsymbol{x}}) - \sum_{j=1}^{m} \left. \frac{\partial f^o(\bar{\boldsymbol{b}})}{\partial b_j} \right|_{\boldsymbol{b}=\boldsymbol{g}(\bar{\boldsymbol{x}})} g_i^j(\bar{\boldsymbol{x}}) = 0, \; j = 1, \ldots, m.$$

Define $\bar{\lambda}_j = \partial f^o(\bar{\boldsymbol{b}})/\partial b_j$, $j = 1, \ldots, m$. Then the preceding equation system satisfies $\nabla f(\bar{\boldsymbol{x}}) + \nabla \boldsymbol{G}(\bar{\boldsymbol{x}})\bar{\boldsymbol{\lambda}} \leq \boldsymbol{O}$ (see Equation 11.7), where $\boldsymbol{G}(\bar{\boldsymbol{x}}) = \boldsymbol{b} - \boldsymbol{g}(\boldsymbol{x}) \geq \boldsymbol{O}$, $\boldsymbol{x} \geq \boldsymbol{O}$, $\boldsymbol{x} \in R^n$.

11.7 ECONOMIC APPLICATIONS

11.7.1 Optimal Resource Allocation

Suppose a firm faces the following decision problem: determine the product mix $\boldsymbol{x} \geq \boldsymbol{O}$ which maximizes profit subject to a set of resource constraints or

$$\text{maximize } y = f(\boldsymbol{x}) \text{ subject to}$$

$$\hat{g}^j(\boldsymbol{x}) = b_j - g^j(\boldsymbol{x}) \geq 0,$$

$$j = 1, \ldots, m, \; \boldsymbol{x} \geq \boldsymbol{O}, \; \boldsymbol{x} \in R^n, \tag{11.10}$$

where: f is a strictly concave profit (or, generally, a benefit) function; x_i is the ith output or production activity; and $b_j - g^j(\boldsymbol{x}) \geq 0$ is the jth strictly concave resource constraint or restriction on production. For this constraint, b_j is the amount of the jth resource available and $g^j(\boldsymbol{x})$ is the amount of the jth resource used in operating the jth restriction at the activity level $\boldsymbol{x}$.

From (11.7), the KKT necessary and sufficient conditions for optimality at $(\boldsymbol{x}_0, \boldsymbol{\lambda}_o)$ are:

$$\begin{array}{ll}
\text{(a)} & f_i(\boldsymbol{x}_0) - \sum_{j=1}^{m} \lambda_j^o g_i^j(\boldsymbol{x}_0) \leq 0, \; i = 1, \ldots, n, \\
\text{(b)} & \sum_{i=1}^{n} x_i^o (f_i(\boldsymbol{x}_0) - \sum_{j=1}^{m} \lambda_j^o g_i^j(\boldsymbol{x}_0)) = 0 \\
\text{(c)} & \sum_{j=1}^{m} \lambda_j^o (b_j - g^j(\boldsymbol{x}_0)) = 0 \\
\text{(d)} & \hat{g}(\boldsymbol{x}_0) = b_j - g^j(\boldsymbol{x}_0) \geq 0, \; j = 1, \ldots, m \\
\text{(e)} & \lambda_j^0 \geq 0, \; j = 1, \ldots, m \\
\text{(f)} & x_i^0 \geq 0, \; i = 1, \ldots, m
\end{array} \tag{11.11}$$

Let us examine some of the key components of system (11.11):

1. $b_j - g^j(\boldsymbol{x}) = \hat{g}^j(\boldsymbol{x})$ is the *surplus or unused amount of resource j*; with $\hat{g}^j$ a decreasing function of $\boldsymbol{x}$.

2. f_i is the *marginal profitability* of product or activity i.

3. $\hat{g}_i^j(<0)$ is the marginal effect of the ith activity on the jth restriction. Then $-\hat{g}_i^j = g_i^j(>0)$ is the *marginal cost* of the ith output in terms of the jth constraint; it is a *real resource cost*, i.e., the marginal quantity of input j needed to produce an additional unit of product i.

   Note:

   i. if $\hat{g}^j$ is linear, the marginal cost of the jth resource is constant.

   ii. of $\hat{g}^j$ is concave, the marginal cost of the jth resource is either constant or increasing.

   iii. if $\hat{g}^j$ is strictly concave, the marginal cost of the jth resource is strictly increasing.

4. $\lambda_j (=\partial f/\partial b_j)$ is the *marginal imputed value of the jth constraint* (the *shadow* or *accounting price* associated it input j). Thus λ_j determined the maximum amount by which total profit would increase if the jth constraint were relaxed by one unit *or* the minimum reduction in f made necessary by a one-unit reduction of resource j. Hence, it is an *implicit opportunity cost in terms of foregone profit.*

5. $-\lambda_j \hat{g}_i^j = \lambda_j g_i^j$ is the *marginal imputed cost* of the jth input in the performance of activity i (an accounting cost and not a real resource cost). It is the *marginal accounting value* of the amount in input j needed to produce an extra unit of output i.

   Note: it is a composite – the marginal imputed value of the jth constraint times the marginal cost of the ith output in terms of the jth constraint.

6. $-\sum_{j=1}^{m}\lambda_j \hat{g}^j = -\sum_{j=1}^{m}\lambda_j g^j$ is the *marginal imputed cost of activity i* - - the sum of the marginal accounting values of *all* inputs needed to produce an extra unit of product i.

Consider next the KKT conditions (11.11):

(11.11a) $f_i(x_0) - \sum_{j=1e}^{m}\lambda_j^o g_i^j(x_0) \leq 0$ - - the marginal imputed cost of the ith activity is at least as great as its marginal profitability.

(11.11b) $x_i^o(f_i(\boldsymbol{x}_0) - \sum_{j=1}^{m}\lambda_j^o g_i^j(\boldsymbol{x}_0)) = 0$ - - if $f_i(\boldsymbol{x}_0) < \sum_{j=1}^{m}\lambda_j^o g_i^j(\boldsymbol{x}_0)$, then $x_i^o = 0$ (if the marginal imputed cost of activity i exceeds it's marginal profitability, then activity i is not undertaken); if $x_i^o > 0$, then $f_i(\boldsymbol{x}_0) = \sum_{j=1}^{m}\lambda_j^o g_i^j(\boldsymbol{x}_0)$ (activity i is undertaken if its marginal profitability equals its marginal imputed cost, i.e., activity i breaks even in terms of shadow prices).

(11.11c) $\sum_{j=1}^{m}\lambda_j^o \hat{g}_i^j(\boldsymbol{x}_0) = \sum_{j=1}^{m}\lambda_j^o(b_j - g^j(\boldsymbol{x}_0)) = 0$- - considering the jth term:

i. if $b_j > g^j(\boldsymbol{x}_0)$, then $\lambda_j^o = 0$ (if at optimality there is an unused amount of input j, its associated shadow price is zero, e.g., we can ask: "What would the firm be willing to pay for an extra unit of a resource which is already in excess supply?").

ii. if $\lambda_j^o > 0$ then $b_j - g^j(\boldsymbol{x}_0) = 0$ (if all of the jth input is used at optimality, then the firm would be willing to pay up to $\$\lambda_j^o$ to get the jth constraint relaxed by one unit).

This is because, for case (ii) ($\lambda_j^o > 0$), a lack of additional supplies of input j places a limitation on the firm's ability to increase profit. This is in contrast to case (i) ($\lambda_j^o = 0$) in which input j is essentially treated *internally* by the firm as a "free good," even though its *market price* is positive.

11.7.2 Resource Allocation with Generalized Lagrange Multipliers

Consider a decision problem in which a firm seeks to maximize profit subject to a set of resource limitation or seeks to

$$\begin{aligned}&\text{maximize } y = f(\boldsymbol{x}) \text{ subject to}\\ &\boldsymbol{G}(\boldsymbol{x}) \leq \boldsymbol{b},\ \boldsymbol{x} \in S \subseteq R^n,\end{aligned} \tag{11.12}$$

where: f is a profit function; $\boldsymbol{x}$ is an $(n \times 1)$ vector of output or production levels; $\boldsymbol{b}$ is an $(m \times 1)$ vector of resource availabilities; $\boldsymbol{G}$ is an $(m \times 1)$ vector-valued function of resource utilization functions $g^j(\boldsymbol{x})$ used to operate the jth restriction at the activity level $\boldsymbol{x}$, $j = 1, \ldots, m$; and $S\ (\neq\varnothing)$ is an arbitrary set.

Let $\hat{\boldsymbol{\lambda}}^T = (\hat{\lambda}_1, \hat{\lambda}_2, \ldots, \hat{\lambda}_m)$ be a *fixed* vector of non-negative Lagrange multipliers. Then to solve problem (11.12), let us solve the **unconstrained Lagrange problem** (Everett [1963])

$$\text{maximize } L(\boldsymbol{x}, \hat{\boldsymbol{\lambda}}) = f(\boldsymbol{x}) - \hat{\boldsymbol{\lambda}}^T \boldsymbol{G}(\boldsymbol{x}),\ \boldsymbol{x} \in S, \tag{11.13}$$

where $L(\boldsymbol{x}, \hat{\boldsymbol{\lambda}})$ is the **Everett Lagrangian**.

What is the connection between the solution to the unconstrained Lagrange problem (11.13) and the original problem (11.12)? The answer to this question is provided by

Everett's Sufficient Condition: For $\hat{\boldsymbol{\lambda}}$ a vector of non-negative Lagrange multipliers, if $\hat{\boldsymbol{x}} \equiv \boldsymbol{x}(\hat{\boldsymbol{\lambda}})$ is a solution to the unconstrained Lagrange problem, then $\hat{\boldsymbol{x}}$ solves the **surrogate problem**

$$\begin{aligned}&\text{maximize } y = f(\boldsymbol{x}) \text{ subject to}\\ &\boldsymbol{G}(\boldsymbol{x}) \leq \boldsymbol{G}(\hat{\boldsymbol{x}}),\ \boldsymbol{x} \in S,\end{aligned} \tag{11.14}$$

where the constraint constants $\boldsymbol{G}(\hat{\boldsymbol{x}})$ are the actual amounts of the resources expended in achieving the constrained maximum.

So if $\hat{\boldsymbol{x}}$ yields an optimal solution to the unconstrained Lagrange problem using the resource level $\boldsymbol{G}(\hat{\boldsymbol{x}})$, then $f(\hat{\boldsymbol{x}})$ is the highest value of f which can be attained without using any more of the available resources than $\hat{\boldsymbol{x}}$ does.

The exact computational procedure will be as follows:

1. choose $\hat{\boldsymbol{\lambda}} \geq \boldsymbol{O}$;
2. find an $\hat{\boldsymbol{x}} \equiv \boldsymbol{x}(\hat{\boldsymbol{\lambda}}) \in S$ that maximizes $L(\boldsymbol{x}, \hat{\boldsymbol{\lambda}}) = f(\boldsymbol{x}) - \hat{\boldsymbol{\lambda}}^T \boldsymbol{G}(\boldsymbol{x})$;
3. for $\boldsymbol{x} = \hat{\boldsymbol{x}}$, calculate $\boldsymbol{G}(\hat{\boldsymbol{x}})$. If $\boldsymbol{G}(\hat{\boldsymbol{x}}) = \boldsymbol{b}$, an optimal solution to the original problem has been obtained. If $\boldsymbol{G}(\hat{\boldsymbol{x}}) \neq \boldsymbol{b}$, repeat steps (1) - (3) using another $\boldsymbol{\lambda}$ vector.

Note that if all resource utilization levels except the kth are held fixed, then g^k is a monotonic decreasing function of λ_k. So if $g^k > b_k$, increase the value of λ_k; if $g^k < b_k$, decrease the value of λ_k. Clearly, our goal is to choose the λ_j's in a fashion such that we approach the levels of the original set of constraint constants b_j, $j = 1, \ldots, m$. STOPPING CRITERON: the iterations are terminated when $|\boldsymbol{G}(\boldsymbol{x}_i) - \boldsymbol{b}| < \eta$, $i = 1, 2, \ldots$, where η is a pre-determined and arbitrarily small value.

Example 11.4: Consider the program

$$\text{maximize}\, f(x) = 8x - 3x^2 \text{ subject to}$$
$$g(x) = x^2 + 5x \leq 6(=b).$$

For $\lambda = \hat{\lambda}$, the unconstrained Lagrangian is

$$L(x, \hat{\lambda}) = 8x - 3x^2 - \hat{\lambda}(x^2 + 5x).$$

Step 1. Set $\lambda_1 = 1$, $L(x, 1) = 3x - 4x^2$.

$$\frac{dL}{dx} = 3 - 8x = 0, \ \frac{d^2L}{dx^2} = -8 < 0,$$
$$x_1 \equiv x(\lambda_1) = 0.38.$$
$$g(x_1) = 2.04 < b. \ \text{Decrease } \lambda.$$

Step 2. Set $\lambda_2 = 0.5$, $L(x, 0.5) = 5.5x - 3.5x^2$.

$$\frac{dL}{dx} = 5.5 - 7x = 0, \ \frac{d^2L}{dx^2} = -7 < 0,$$
$$x_2 \equiv x(\lambda_2) = 0.79.$$
$$g(x_2) = 4.57 < b. \ \text{Again decrease } \lambda.$$

Step 3. Set $\lambda_3 = 0.25$, $L(x, 0.25) = 6.75x - 3.25x^2$.

$$\frac{dL}{dx} = 6.75 - 6.50x = 0, \ \frac{d^2L}{dx^2} = -6.50 < 0,$$
$$x_3 \equiv x(\lambda_3) = 1.038.$$
$$g(x_3) = 6.27 > b. \ \text{Increase } \lambda.$$

Step 4. Set $\lambda_4 = 0.3$, $L(x, 0.3) = 6.5x - 3.3x^2$.

$$\frac{dL}{dx} = 6.5 - 6.6x = 0, \ \frac{d^2L}{dx^2} = -6.6 < 0,$$
$$x_4 \equiv x(\lambda_4) = 0.98,$$
$$g(x_4) = 5.86 < b.$$

This iterative process converges to $\lambda_o = 0.286$, $x_o \equiv x(\lambda_o) = 1$, and $f(x_o) = 5$. ■

Section 11.7 Exercises:

11.7.1 Solve the Program

$$\text{maximize}\, f(\boldsymbol{x}) = 6x_1 + 3x_2 - x_1^2 - x_2^2 \text{ subject to}$$
$$g^1(\boldsymbol{x}) = 2x_1 - x_2 \leq 4$$
$$g^2(\boldsymbol{x}) = x_1 + 2x_2 \leq 3$$
$$x_1, x_2 \geq 0$$

using the technique of generalized Lagrange multipliers. Start with $\boldsymbol{\lambda}_1^T = (1, 1)$.

NOTES

1 This theorem considers the circumstances under which there exists a semipositive solution $\boldsymbol{y}(\boldsymbol{y} \geq \boldsymbol{O}$ and $\boldsymbol{y} \neq \boldsymbol{O})$ to a homogeneous equation system or Theorem 11.6 (**Gordan's Theorem of the Alternative**). For any $(m \times n)$ matrix $\boldsymbol{A}$, either

I. $Ah > O$ has a solution $\boldsymbol{h} \in R^n$, or

II. $A^t v = O$, $v \geq O$ and $v \neq O$ has a solution $v \in R^m$, but never both. (For additional details see Panik [1993].)

2 If we introduce a constraint qualification (i.e., the FM linear independence requirement for the gradient vectors of the binding constraints) and replace definiteness by semi-definiteness in (11.9) ("< 0" is replaced by "≤ 0" and " > 0" is replaced by "≥ 0"), then we obtain a *KKT second-order necessary condition* for a strong local constrained extremum of *f*.

3 The implication of this requirement is that $\bar{\mathcal{K}}$ has a non-empty interior. This is known as **Slater's CQ**, i.e., $\bar{\boldsymbol{G}}(\boldsymbol{x}) \geq \boldsymbol{O}$ satisfies Slater's CQ on $\bar{\mathcal{K}}$ if there exists an $x^* \in \bar{\mathcal{K}}$ such that $\bar{\boldsymbol{G}}(x^*) > \boldsymbol{O}$ (Slater [1950]).

Constrained Extrema: Mixed Constraints

12.1 PROGRAMS WITH m INEQUALITY AND p EQUALITY SIDE RELATIONS IN n INDEPENDENT VARIABLES

Suppose our optimization problem is structured as: maximize the real-valued function $y = f(\mathbf{x})$ subject to: (1) m inequality constraints $\hat{g}^j(\mathbf{x}) = b_j - g^j(\mathbf{x}) \geq 0$, $j = 1, \ldots, m$; (2) p equality constraints $\hat{h}^r(\mathbf{x}) = c_r - h_r(\mathbf{x}) = 0$, $r = 1, \ldots, p$; and (3) n non-negativity conditions $\mathbf{x} \geq \mathbf{O}$, $\mathbf{x}^T = (x_1, \ldots, x_n) \in R^n$. Additionally, it is assumed that f and the $\hat{g}^j(\mathbf{x})$, $j = 1, \ldots, m$, and the $\hat{h}^r(\mathbf{x})$, $r = 1, \ldots, p$, are all defined and of class $C^{(1)}$ on an open set $\mathcal{D} \subset R^n$. Upon forming the $(m \times 1)$ and $(p \times 1)$ vector-valued functions

$$\mathbf{G}(\mathbf{x}) = \begin{bmatrix} b_1 - g^1(\mathbf{x}) \\ \vdots \\ b_m - g^m(\mathbf{x}) \end{bmatrix} = \begin{bmatrix} \hat{g}^1(\mathbf{x}) \\ \vdots \\ \hat{g}^m(\mathbf{x}) \end{bmatrix} \geq \mathbf{O},$$

$$\mathbf{H}(\mathbf{x}) = \begin{bmatrix} c_1 - h^1(\mathbf{x}) \\ \vdots \\ c_p - h^p(\mathbf{x}) \end{bmatrix} = \begin{bmatrix} \hat{h}^1(\mathbf{x}) \\ \vdots \\ \hat{h}^p(\mathbf{x}) \end{bmatrix} = \mathbf{O},$$

respectively, our problem may be rewritten as:

$$\begin{aligned} &\text{maximize } y = f(\mathbf{x}) \text{ subject to} \\ &\mathbf{x} \in \mathcal{K} = \{\mathbf{x} \mid \mathbf{G}(\mathbf{x}) \geq \mathbf{O}, \mathbf{H}(\mathbf{x}) = \mathbf{O}, \mathbf{x} \in \mathcal{D}\}, \end{aligned} \tag{12.1}$$

where $\mathcal{K}$ is the **feasible region** or set of admissible solutions in R^n. If the ith non-negativity condition is written as $x_i = \hat{g}^{m+i}(\mathbf{x}) \geq 0$, $i = 1, \ldots, n$, then we ultimately:

$$\begin{aligned} &\text{maximize } y = f(\mathbf{x}) \text{ subject to} \\ &\mathbf{x} \in \bar{\mathcal{K}} = \{\mathbf{x} \mid \bar{\mathbf{G}}(\mathbf{x}) \geq \mathbf{O}, \mathbf{H}(\mathbf{x}) = \mathbf{O}, \mathbf{x} \in \mathcal{D}\}, \end{aligned} \tag{12.1.1}$$

where

$$\bar{G}(\mathbf{x}) = \begin{bmatrix} \hat{g}^1(\mathbf{x}) \\ \vdots \\ \hat{g}^m(\mathbf{x}) \\ \hat{g}^{m+1}(\mathbf{x}) \\ \vdots \\ \hat{g}^{m+n}(\mathbf{x}) \end{bmatrix} \geq \mathbf{O}$$

and $\mathbf{x}$ is unrestricted in sign.

If f attains a local maximum at $\mathbf{x}_o \in \bar{\mathcal{K}}$, then, as noted in the previous chapter, any movement from $\mathbf{x}_o$ along a feasible direction $\mathbf{h}$ cannot increase the value of f or $\nabla f(\mathbf{x}_o)^T\mathbf{h} \leq 0$, $\mathbf{h} \in \mathcal{D}(\mathbf{x}_o)$. Moreover, given that $\mathbf{x}_o$ is optimal, if $\mathbf{h} \in \mathcal{D}(\mathbf{x}_o)$, then it is also the case that $\nabla \hat{g}^j(\mathbf{x}_o)^T\mathbf{h} \geq 0$, $j \in \mathcal{J} = \{j \mid \hat{g}^j(\mathbf{x}) = 0, \mathbf{x} \in R^n\}$. If we now admit the equality constraints $\hat{h}^r(\mathbf{x}) = 0$, $r = 1, \ldots, p$, to our discussion, then, for $\mathbf{h}$ a feasible direction, $\nabla \hat{h}^r(\mathbf{x})^T\mathbf{h} = 0$, $r = 1, \ldots, p$. To summarize: if $f(\mathbf{x})$ subject to $\hat{g}^j(\mathbf{x}_o) = 0$, $j \in \mathcal{J}$, $\hat{h}^r(\mathbf{x}) = 0$, $r = 1, \ldots, p$, assumes its local maximum at $\mathbf{x}_o$, then

$$\begin{aligned} &\nabla f(\mathbf{x}_o)^T\mathbf{h} \leq 0 \text{ for all } \mathbf{h} \text{ satisfying} \\ &\nabla \hat{g}^j(\mathbf{x}_o)^T\mathbf{h} \geq 0, j \in \mathcal{J}, \nabla \hat{h}^r(\mathbf{x}_o)^T\mathbf{h} = 0, \\ &r = 1, \ldots, p, \mathbf{h} \in \mathcal{D}(\mathbf{x}_o). \end{aligned} \tag{12.2}$$

Under what circumstances will (12.2) hold? Given any $\mathbf{h} \in \mathcal{D}(\mathbf{x}_o)$, (12.2) will hold if $\nabla f(\mathbf{x}_o)$ lies within the finite cone spanned by the exterior normals (to the boundary of $\bar{\mathcal{K}}$ at $\boldsymbol{x}_o$)

$$-\nabla\hat{g}^j(\boldsymbol{x}_o),\ j\in\mathcal{J},\ -\nabla\hat{h}^r(\boldsymbol{x}_o),\ r=1,\ \dots,p.$$

Hence, the gradient of f is a linear combination of the negative gradients of the binding constraints $\hat{g}^j,\ j\in\mathcal{J}$, and equality constraints $\hat{h}^r,\ r=1,\ \dots,p$, at $\boldsymbol{x}_o$. In this regard, there must exist, respectively, real numbers $\lambda_j^o\geq 0,\ j\in\mathcal{J}$, and μ_r^o (unrestricted in sign), $r=1,\ \dots,p$, such that

$$\nabla f(\boldsymbol{x}_o)=\sum_{j\in\mathcal{J}}\lambda_j^o(-\nabla\hat{g}^j(\boldsymbol{x}_o))+\sum_{r=1}^{p}\mu_r^o(-\nabla\hat{h}^r(\boldsymbol{x}_o)).\quad(12.3)$$

What is the *prerequisite condition* for the numbers $\lambda_j^o\geq 0,\ j\in\mathcal{J}$, and $\mu_r^o,\ r=1,\ \dots,p$, to exist? As in the preceding chapter, let us write

$$\mathcal{F}_o=\{\boldsymbol{h}|\nabla f(\boldsymbol{x}_o)^T\boldsymbol{h}>0\}\text{ and }\mathcal{G}_o=\{\boldsymbol{h}|\nabla\hat{g}^j(\boldsymbol{x}_o)^T\boldsymbol{h}>0,\ j\in\mathcal{J}\}.$$

Additionally, let

$$\mathcal{H}_o=\{\boldsymbol{h}|\nabla\hat{\boldsymbol{h}}^r(\boldsymbol{x}_o)^T\boldsymbol{h}=0,\ r=1,\dots,p\}.$$

Then the prerequisite condition is that $\mathcal{F}_o\bigcap\mathcal{G}_o\bigcap\mathcal{H}_o=\varnothing$, i.e., there do not exist directions $\boldsymbol{h}$ such that $\nabla f(\boldsymbol{x}_o)^T\boldsymbol{h}>0$ for $\nabla\hat{g}^j(\boldsymbol{x}_o)^T\boldsymbol{h}>0,\ j\in\mathcal{J},\ \nabla\hat{h}^r(\boldsymbol{x}_o)^T\boldsymbol{h}=0$, $r=1,\ \dots,p$. By Farkas' Theorem[1], a necessary and sufficient condition for $\nabla f(\boldsymbol{x}_o)$ to lie within the finite cone spanned by the vectors $-\nabla\hat{g}^j(\boldsymbol{x}_o),\ j\in\mathcal{J}$, $-\nabla h^r(\boldsymbol{x}_o),\ r=1,\ \dots,p$, is that $\nabla f(\boldsymbol{x}_o)^T\boldsymbol{h}\leq 0$ for all $\boldsymbol{h}$ satisfying $-\nabla\hat{g}^j(\boldsymbol{x}_o)^T\boldsymbol{h}\leq 0,\ j\in\mathcal{J},\ -\nabla\hat{h}^r(\boldsymbol{x}_o)^T=0$, $r=1,\ \dots,p$. Hence, there exist real numbers $\lambda_j^o\geq 0,\ \mu_r^o$ unrestricted, such that (12.3) holds.

A condition that is sufficient to guarantee that $\mathcal{F}_o\bigcap\mathcal{G}_o\bigcap\mathcal{H}_o=\varnothing$ and the existence of vectors $\boldsymbol{\lambda}_o\geq\boldsymbol{O},\ \boldsymbol{\mu}_o$ unrestricted such that (12.3) obtains is the linear independence of the gradients of all the binding constraints at $\boldsymbol{x}_o$. This observation is incorporated in the **Fiacco-McCormick (FM) regularity condition** [1968]. Specifically, let $\boldsymbol{x}_o\in\bar{\mathcal{K}}=\{\boldsymbol{x}|\bar{\boldsymbol{G}}(\boldsymbol{x})\geq\boldsymbol{O},\ \boldsymbol{H}(\boldsymbol{x})=\boldsymbol{O},\ \boldsymbol{x}\in R^n\}$ with $\bar{\boldsymbol{G}}(\boldsymbol{x}),\ \boldsymbol{H}(\boldsymbol{x})$ of class $C^{(1)}$ throughout a spherical δ-neighborhood of $\boldsymbol{x}_o,\ \delta(\boldsymbol{x}_o)$. If at $\boldsymbol{x}_o$ the vectors $\{\nabla\hat{g}^j(\boldsymbol{x}_o),\ j\in\mathcal{J}\},\ \{\nabla\hat{h}^r(\boldsymbol{x}_o),\ r=1,\dots,p\}$ are linearly independent, then $\mathcal{G}_o\bigcap\mathcal{H}_o\neq\varnothing$ and thus $\mathcal{F}_o\bigcap\mathcal{G}_o\bigcap\mathcal{H}_o=\varnothing$.

Given these considerations, the principal theorem of this chapter is

Theorem 12.2: (**A Necessary Condition**). Let the real-valued function $y=f(\boldsymbol{x})$ and the vector-valued functions $y=f(\boldsymbol{x})$, be of class $C^{(1)}$ throughout some spherical δ-neighborhood of the point $\boldsymbol{x}_o\in\bar{\mathcal{K}}=\{\boldsymbol{x}|\bar{\boldsymbol{G}}(\boldsymbol{x})\geq\boldsymbol{O},\ \boldsymbol{H}(\boldsymbol{x})=\boldsymbol{O},\ \boldsymbol{x}\in R^n\}$. If f subject to $\boldsymbol{x}_o\in\bar{\mathcal{K}}$ has a local maximum at $\boldsymbol{x}_o$, then, provided $\mathcal{F}_o\bigcap\mathcal{G}_o\bigcap\mathcal{H}_o=\varnothing$,

$$\begin{aligned}&\nabla f(\boldsymbol{x}_o)+\nabla\bar{\boldsymbol{G}}(\boldsymbol{x}_o)\boldsymbol{\lambda}_o+\nabla\boldsymbol{H}(\boldsymbol{x}_o)\boldsymbol{\mu}_o=\boldsymbol{O}\\&\boldsymbol{\lambda}_o^T\bar{\boldsymbol{G}}(\boldsymbol{x}_o)=0\\&\bar{\boldsymbol{G}}(\boldsymbol{x}_o)\geq\boldsymbol{O}\quad\textbf{[KKT Optimality}\\&\boldsymbol{H}(\boldsymbol{x}_o)=\boldsymbol{O}\quad\textbf{Conditions]}\\&\boldsymbol{\lambda}_o\geq\boldsymbol{O}\\&\boldsymbol{\mu}_o\ \text{ unrestricted}\\&\boldsymbol{x}_o\ \text{ unrestricted.}\end{aligned}\qquad(12.4)$$

or

$$\begin{aligned}&\nabla f(\boldsymbol{x}_o)+\textstyle\sum_{j=1}^{m+n}\lambda_j^o\nabla\hat{g}^j(\boldsymbol{x}_o)+\sum_{r=1}^{p}\mu_r^o\nabla\hat{h}^r(\boldsymbol{x}_o)=\boldsymbol{O}\\&\textstyle\sum_{j=1}^{m+n}\lambda_j^o\nabla\hat{g}^j(\boldsymbol{x}_o)=0\\&\left.\begin{aligned}&\hat{g}^j(\boldsymbol{x}_o)\geq 0\\&\lambda_j^o\geq 0\end{aligned}\right\}j=1,\dots,m+n\\&\left.\begin{aligned}&\hat{h}^r(\boldsymbol{x}_o)=0\\&\mu_r^o\text{ unrestricted}\end{aligned}\right\}r=1,\dots,p\\&x_i^o\text{ unrestricted},\ i=1,\dots,n.\end{aligned}\qquad(12.4.1)$$

Here $\nabla\bar{\boldsymbol{G}}(\boldsymbol{x}_o)=[\nabla\hat{g}^1(\boldsymbol{x}_o),\dots,\nabla\hat{g}^{m+n}(\boldsymbol{x}_o)]$ is of order $(n\times m+n)$, $\nabla\boldsymbol{H}(\boldsymbol{x}_o)=[\nabla\hat{\boldsymbol{h}}^1(\boldsymbol{x}_o),\dots,\nabla\hat{\boldsymbol{h}}^p(\boldsymbol{x}_o)]$ is of order $(n\times p)$, and $\boldsymbol{\lambda}_o$ and $\boldsymbol{\mu}_o$ are, respectively, $(m+n\times 1)$ and $(p\times 1)$ vectors of Lagrange multipliers.

Proof: The augmented or extended Lagrangian associated with Problem 12.1.1 is

$$L(\boldsymbol{x},\boldsymbol{\lambda},\boldsymbol{\mu},\boldsymbol{s})=f(\boldsymbol{x})+\boldsymbol{\lambda}^T(\bar{\boldsymbol{G}}(\boldsymbol{x})-\boldsymbol{s})+\boldsymbol{\mu}^T\boldsymbol{H}(\boldsymbol{x}),$$

where $\bar{\boldsymbol{G}}(\boldsymbol{x})\geq\boldsymbol{O}$ has been converted to $\bar{\boldsymbol{G}}(x)-\boldsymbol{s}=\boldsymbol{O}$ via the $(m+n\times 1)$ vector of non-negative slack variables

$$\boldsymbol{s}=\begin{bmatrix}s_1^2\\ \vdots\\ s_{m+n}^2\end{bmatrix}\geq\boldsymbol{O}.$$

Then

$$\begin{aligned} &\nabla_{\boldsymbol{x}} L = \nabla f(\boldsymbol{x}) + \nabla \bar{\boldsymbol{G}}(\boldsymbol{x})\boldsymbol{\lambda} + \nabla \boldsymbol{H}(\boldsymbol{x})\boldsymbol{\mu} = \boldsymbol{O} \\ &\nabla_{\lambda} L = \bar{\boldsymbol{G}}(\boldsymbol{x}) - \boldsymbol{s} = \boldsymbol{O} \\ &\nabla_{\mu} L = \boldsymbol{H}(\boldsymbol{x}) = \boldsymbol{O} \\ &\partial L / \partial s_j = -2\lambda_j s_j = 0,\ j = 1,\ldots,m+n. \end{aligned}$$

Suppose the FM regularity condition holds at $\boldsymbol{x}_o$. Then Farkas' Theorem 12.1 implies the existence of vectors $\boldsymbol{\lambda}_o \geq 0$, $\boldsymbol{\mu}_o$ unrestricted such that $\nabla f(\boldsymbol{x}_o) + \nabla \bar{\boldsymbol{G}}(\boldsymbol{x}_o)\boldsymbol{\lambda}_o + \nabla \boldsymbol{H}(\boldsymbol{x}_o)\boldsymbol{\mu}_o = \boldsymbol{O}$ if and only if (12.2) holds or $\mathcal{F}_o \cap \mathcal{G}_o \cap \mathcal{H}_o = \varnothing$. From $\nabla_\lambda L = \boldsymbol{O}$ we have $\boldsymbol{s} = \bar{\boldsymbol{G}}(\boldsymbol{x}_o)$ while from $\partial L/\partial s_j = 0$ we have $\lambda_j s_j = 0$ or $\lambda_j s_j^2 = 0$, $j = 1, \ldots, m+n$. Hence, $\sum_{j=1}^{m+n} \lambda_j s_j^2 = 0$ or, at $\boldsymbol{x}_o$, $\boldsymbol{\lambda}_o^T \bar{\boldsymbol{G}}(\boldsymbol{x}_o) = 0$. So if $\hat{g}^j(\boldsymbol{x}_o) \geq 0$ is not binding at $\boldsymbol{x}_o$ ($j \notin \mathcal{J}$), then $\hat{g}^j(\boldsymbol{x}_o) \geq 0$; and if $j \in \mathcal{J}$, then $s_j^2 = 0$, $\lambda_j^0 \geq 0$.

Thus at least one member of each pair $(\lambda_j^o, \hat{g}^j(\boldsymbol{x}_o))$ must vanish or $\lambda_j^o \hat{g}^j(\boldsymbol{x}_o) = 0$ for all j (the complementary slackness (CS) condition holds), thus guaranteeing that $\boldsymbol{\lambda}_o^T \bar{\boldsymbol{G}}(\boldsymbol{x}_o) = 0$. Q. E. D.

If the n non-negativity conditions are handled explicitly as $\boldsymbol{x} \geq \boldsymbol{O}$, then (12.4) appears as

$$\begin{aligned} &\nabla f(\boldsymbol{x}_o) + \nabla \bar{\boldsymbol{G}}(\boldsymbol{x}_o)\boldsymbol{\lambda}_o + \nabla \boldsymbol{H}(\boldsymbol{x}_o)\boldsymbol{\mu}_o \leq \boldsymbol{O} \\ &\boldsymbol{x}_o^T(\nabla f(\boldsymbol{x}_o) + \nabla \bar{\boldsymbol{G}}(\boldsymbol{x}_o)\boldsymbol{\lambda}_o + \nabla \boldsymbol{H}(\boldsymbol{x}_o)\boldsymbol{\mu}_o) = \boldsymbol{O} \\ &\boldsymbol{\lambda}_o^T \boldsymbol{G}(\boldsymbol{x}_o) = 0 \\ &\boldsymbol{G}(\boldsymbol{x}_o) \geq \boldsymbol{O} \quad \textbf{[KKT Optimality} \\ &\boldsymbol{H}(\boldsymbol{x}_o) = \boldsymbol{O} \quad \textbf{Conditions]} \\ &\boldsymbol{\lambda}_o \geq \boldsymbol{O} \\ &\boldsymbol{\mu}_o \ \text{unrestricted} \\ &\boldsymbol{x}_o \geq \boldsymbol{O} \end{aligned} \tag{12.5}$$

or

$$\begin{aligned} &\nabla f(\boldsymbol{x}_o) + \textstyle\sum_{j=1}^m \lambda_j^o \nabla \hat{g}^j(\boldsymbol{x}_o) + \sum_{r=1}^p \mu_r^o \nabla \hat{h}^r(\boldsymbol{x}_o) \leq \boldsymbol{O} \\ &\textstyle\sum_{i=1}^n x_i^o (f_i^o + \sum_{j=1}^m \lambda_j^o \nabla \hat{g}^j(\boldsymbol{x}_o) + \sum_{r=1}^p \mu_r^o \nabla \hat{h}^r(\boldsymbol{x}_o)) = 0 \\ &\textstyle\sum_{j=1}^m \lambda_j^o \hat{g}^j(\boldsymbol{x}_o) = 0 \\ &\left.\begin{aligned} &\hat{g}^j(\boldsymbol{x}_o) \geq 0 \\ &\lambda_j^o \geq 0 \end{aligned}\right\} j = 1,\ldots,m+n \\ &\left.\begin{aligned} &\hat{h}^r(\boldsymbol{x}_o) = 0 \\ &\mu_r^o \ \text{unrestricted} \end{aligned}\right\} r = 1,\ldots,p \\ &x_i^o \geq 0,\ i = 1,\ldots,n, \end{aligned} \tag{12.5.1}$$

where

$$\boldsymbol{G}(\boldsymbol{x}_o) = \begin{bmatrix} \hat{g}^1(\boldsymbol{x}_o) \\ \vdots \\ \hat{g}^m(\boldsymbol{x}_o) \end{bmatrix} \geq \boldsymbol{O}$$

is of order $(n \times m)$; and $\boldsymbol{\lambda}_o$ is an $(m \times 1)$ vector of Lagrange multipliers. In this regard, for $i = 1, \ldots, n$, either $x_i = 0$, thus allowing $i = 1, \ldots, n$ to be less than $-\sum_{j=1}^m \lambda_j^o \hat{g}_i^j(\boldsymbol{x}_o) - \sum_{r=1}^p \mu_r^o \hat{h}_i^r(\boldsymbol{x}_o)$ or

$$f_i^o = -\textstyle\sum_{j=1}^m \lambda_j^o \hat{g}_i^j(\boldsymbol{x}_o) - \sum_{r=1}^p \mu_r^o \hat{h}_i^r(\boldsymbol{x}_o),$$

in which case x_i^o may be positive. Since at least one of these alternatives must hold, it follows that

$$\textstyle\sum_{i=1}^n x_i^0 \left(f_i^o + \sum_{j=1}^m \lambda_j^o \hat{g}_i^j(\boldsymbol{x}_o) + \sum_{r=1}^p \mu_r^o \hat{h}(\boldsymbol{x}_o) \right) = 0.$$

In addition, if the jth constraint $\hat{g}^j(\boldsymbol{x}) \geq 0$, $j = 1, \ldots, m$, is inactive at $\boldsymbol{x}_o$, then $s_j^2 > 0$, $\lambda_j^o = 0$; while if it is binding there, then $s_j^2 = 0$, $\lambda_j^o \geq 0$. Here too, at least one of these alternatives holds, thus implying that $\sum_{j=1}^m \lambda_j^o \hat{g}^j(\boldsymbol{x}_o) = 0$.

Example 12.1: Minimize the real-valued function $y = f(\boldsymbol{x}) = \frac{1}{8}(x_1 - 5)^2 + (x_2 - 4)^2$ subject to

$$\begin{aligned} &\hat{g}^1(\boldsymbol{x}) = 3 - x_1^2 + x_2 \geq 0 \qquad \hat{g}^3(\boldsymbol{x}) = x_2 \geq 0 \\ &\hat{g}^2(\boldsymbol{x}) = x_1 \geq 0 \qquad \hat{h}^1(\boldsymbol{x}) = 6 - x_1 - x_2 = 0. \end{aligned}$$

Forming

$$\bar{\boldsymbol{G}}(\boldsymbol{x}) = \begin{bmatrix} \hat{g}^1(\boldsymbol{x}) \\ \hat{g}^2(\boldsymbol{x}) \\ \hat{g}^3(\boldsymbol{x}) \end{bmatrix} = \begin{bmatrix} 3 - x_1^2 + x_2 \\ x_1 \\ x_2 \end{bmatrix} \geq \boldsymbol{O}$$

$$\boldsymbol{H}(\boldsymbol{x}) = \begin{bmatrix} \hat{h}^1(\boldsymbol{x}) \end{bmatrix} = [6 - x_1 - x_2] = 0$$

we obtain, from (12.4),

$$\nabla f(\boldsymbol{x}) + \nabla \bar{\mathbf{G}}(\boldsymbol{x})\boldsymbol{\lambda} + \nabla \boldsymbol{H}(\boldsymbol{x})\boldsymbol{\mu} = \begin{bmatrix} \frac{1}{4}(x_1 - 5) - 2\lambda_1 x_1 + \lambda_2 - \mu_1 \\ 2(x_2 - 4) + \lambda_1 + \lambda_3 - \mu_1 \end{bmatrix} = \boldsymbol{O}$$

$$\boldsymbol{\lambda}^T \bar{\mathbf{G}}(\boldsymbol{x}) = \lambda_1(3 - x_1^2 + x_2) + \lambda_2 x_1 + \lambda_3 x_2 = 0$$

$$\bar{\mathbf{G}}(\boldsymbol{x}) = \begin{bmatrix} 3 - x_1^2 + x_2 \\ x_1 \\ x_2 \end{bmatrix} \geq \boldsymbol{O} \qquad (12.6)$$

$$\mathcal{H}(\boldsymbol{x}) = [6 - x_1 - x_2] = 0$$

$$\boldsymbol{\lambda}^T = (\lambda_1, \lambda_2, \lambda_3) \geq \boldsymbol{O}^T$$

$$\boldsymbol{\mu} = [\mu_1] \text{ unrestricted.}$$

1. For $j = 1 \in \mathcal{J}$,

$$\begin{aligned}
&\text{(a)} \quad \tfrac{1}{4}(x_1 - 5) - 2\lambda_1 x_1 - \mu_1 = 0 \\
&\text{(b)} \quad 2(x_2 - 4) + \lambda_1 - \mu_1 = 0 \\
&\qquad \lambda_1(0) + 0(x_1) + 0(x_2) = 0 \\
&\text{(c)} \quad 3 - x_1^2 + x_2 = 0, \lambda_1 \geq 0 \qquad (12.6.1) \\
&\text{(d)} \quad 6 - x_1 - x_2 = 0, \mu_1 \text{ unrestricted} \\
&\qquad x_1 > 0, \lambda_2 = 0 \\
&\qquad x_2 > 0, \lambda_3 = 0.
\end{aligned}$$

From (12.6.1) we obtain $\boldsymbol{x}_o^T = (2.54, 3.46)$. A substitution of these values into (12.6.1) yields $\lambda_1^o = 0.08$, $\mu_1^o = -1.00$. Hence, an admissible solution point occurs at $\boldsymbol{x}_o$ with $f(\boldsymbol{x}_o) = 1.05$.

2. For $j = 2 \in \mathcal{J}$, (12.6) becomes

$$\begin{aligned}
&\text{(a)} \quad -\tfrac{5}{4} + \lambda_2 - \mu_1 = 0 \\
&\text{(b)} \quad 2(x_2 - 4) - \mu_1 = 0 \\
&\qquad 0(3 - x_1^2 + x_2) + \lambda_2(0) + 0(x_2) = 0 \\
&\text{(c)} \quad x_1 = 0, \lambda_2 \geq 0 \qquad (12.6.2) \\
&\text{(d)} \quad 6 - x_1 - x_2 = 0, \mu_1 \text{ unrestricted} \\
&\qquad 3 - x_1^2 + x_2 > 0, \lambda_1 = 0 \\
&\qquad x_2 > 0, \lambda_3 = 0
\end{aligned}$$

In this instance $\boldsymbol{x}_1^T = (0, 6)$ is a second admissible solution point with $\lambda_2^1 = \frac{21}{4}$, $\mu_1^1 = 4$, and $f(\boldsymbol{x}_1) = 7.12$.

3. Setting $j = 3 \in \mathcal{J}$ transforms (12.6) to

$$\begin{aligned}
&\text{(a)} \quad \tfrac{1}{4}(x_1 - 5) - \mu_1 = 0 \\
&\text{(b)} \quad -8 + \lambda_3 - \mu_1 = 0 \\
&\qquad 0(3 - x_1^2 + x_2) + 0(x_1) + \lambda_3(0) = 0 \\
&\text{(c)} \quad x_2 = 0, \lambda_3 \geq 0 \\
&\text{(d)} \quad 6 - x_1 - x_2 = 0, \mu_1 \text{ unrestricted} \\
&\qquad 3 - x_1^2 + x_2 > 0, \lambda_1 = 0 \\
&\qquad x_1 > 0, \lambda_2 = 0
\end{aligned}$$

We now get $\boldsymbol{x}_2^T = (6, 0)$ as a third admissible solution point. Additionally, $\lambda_3^2 = 33/4$, $\mu_1 = 1/4$, and $f(\boldsymbol{x}_2) = 16.1$. For the cases where $j = 1, 2 \in \mathcal{J}$, $j = 1, 3 \in \mathcal{J}$, $j = 2, 3 \in \mathcal{J}$, and $j = 1, 2, 3 \in \mathcal{J}$, the reader can readily verify that infeasibilities/inconsistencies emerge.

To summarize, the three solution candidates for a constrained minimum of f are $f(\boldsymbol{x}_o) = 1.05$, $f(\boldsymbol{x}_1) = 7.12$, and $f(\boldsymbol{x}_2) = 16.1$. Clearly, $\boldsymbol{x}_o$ renders the smallest constrained value of f.

Does the FM constraint qualification hold at $\boldsymbol{x}_o$**?** To answer this question, we need to determine if $\nabla \hat{g}^1(\boldsymbol{x})^T = (-2x_1, 1)$ and $\nabla \hat{h}^1(\boldsymbol{x})^T = (-1, -1)$ are linearly independent at $\boldsymbol{x}_o$. Since

$$\nabla \hat{g}^1(\boldsymbol{x}_o) = \begin{bmatrix} -5.08 \\ 1 \end{bmatrix}, \nabla \hat{h}^1(\boldsymbol{x}_o)\begin{bmatrix} -1 \\ -1 \end{bmatrix}$$

and

$$\begin{vmatrix} -5.08 & -1 \\ 1 & -1 \end{vmatrix} \neq 0,$$

we see that the FM constraint qualification is in effect at $\boldsymbol{x}_o$. (The reader can easily check that the FM CQ is also in force at $\boldsymbol{x}_1$ and $\boldsymbol{x}_2$). ■

Section 12.1 Exercises:

12.1.1 Maximize $y = f(\boldsymbol{x}) = -\frac{1}{2}x_1^2 - (x_2 - 4)^2$ subject to

$$\hat{g}^1(\boldsymbol{x}) = 5 - 2x_1 - x_2^2 \geq 0 \quad \hat{g}^3(\boldsymbol{x}) = x_2 \geq 0$$
$$\hat{g}^2(\boldsymbol{x}) = x_1 \geq 0 \quad \hat{h}^1(\boldsymbol{x}) = x_1 - 2x_2 = 0.$$

12.1.2 Maximize $y = f(\boldsymbol{x}) = x_1$ subject to

$$\hat{g}^1(\boldsymbol{x}) = 1 - x_1^2 - x_2^2 \geq 0$$
$$\hat{h}^1(\boldsymbol{x}) = 1 - x_1 + x_2 = 0.$$

12.1.3 Minimize $y = f(\boldsymbol{x}) = 4x_1 - 3x_2$ subject to

$$\hat{g}^1(\boldsymbol{x}) = 1 + (x_1 - 3)^2 - x_2^2 \geq 0$$
$$\hat{h}^1(\boldsymbol{x}) = 4 - x_1 - x_2 = 0.$$

The solution to this problem can be obtained by solving: maximize $-f(\boldsymbol{x}) = -4x_1 + 3x_2$ subject to the indicated constraints.

12.1.4 Maximize $y = f(\boldsymbol{x}) = -(x_1 + 2)^2 - (x_2 - 1)^2$ subject to

$$\hat{g}^1(\boldsymbol{x}) = 9 - 2x_1^2 - x_2^2 \geq 0 \quad \hat{g}^3(\boldsymbol{x}) = x_2 \geq 0$$
$$\hat{g}^2(\boldsymbol{x}) = x_1 \geq 0 \quad \hat{h}^1(\boldsymbol{x}) = 3 - x_1 - x_2 = 0.$$

12.1.5 Does the FM constraint qualification hold at the optional solution point for Exercise 12.1.1?

12.1.6 Does the FM constraint qualification hold at optionality for Exercise 12.1.2?

12.2 KKT SUFFICIENT AND NECESSARY AND SUFFICIENT OPTIMALITY CONDITIONS

While Theorem 12.2 utilized first-order information to deduce properties of f and the binding constraints at an optimal point $\boldsymbol{x}_o$, let us now examine conditions on these functions that ensure that $\boldsymbol{x}_o$ is a local optimal solution to problem (12.1.1). Here second-order details will be relied upon to describe the curvature of the Lagrangian near $(\boldsymbol{x}_o, \boldsymbol{\lambda}_o, \boldsymbol{\mu}_o)$ so that we can distinguish a constrained maximum from a constrained minimum.

In what follows we shall assume that $f, \hat{g}^1, \ldots, \hat{g}^{m+n}$, and $\hat{h}^1, \ldots, \hat{h}^p$ are of class $C^{(n)}$ on a convex region $\mathcal{D} \subset R^n$. Additionally, let us define:

$$\mathcal{J}(\boldsymbol{x}_o) = \{j|\hat{g}^j(\boldsymbol{x}_o) = 0\},$$
$$\widehat{\mathcal{J}}'(\boldsymbol{x}_o) = \{j|\hat{g}^j(\boldsymbol{x}_o) = 0 \text{ and } \nabla_{\boldsymbol{x}} L(\boldsymbol{x}_o, \boldsymbol{\lambda}_o, \boldsymbol{\mu}_o) = \boldsymbol{O},$$
$$\lambda_j^o \hat{g}^j(\boldsymbol{x}_o) = 0,\ \lambda_j^o \geq 0 \text{ hold for } \boldsymbol{\lambda}_o > \boldsymbol{O}\},$$
$$\tilde{\mathcal{J}}'(\boldsymbol{x}_o) = \{\boldsymbol{h}|\nabla \hat{g}^j(\boldsymbol{x}_o)^T\boldsymbol{h} = 0,\ j \in \widehat{\mathcal{J}}'(\boldsymbol{x}_o);\ \hat{g}^j(\boldsymbol{x}_o) \geq 0,$$
$$j \in \mathcal{J}(\boldsymbol{x}_o),\ \nabla \hat{h}^r(\boldsymbol{x}_o)^T\boldsymbol{h} = 0,\ r = 1, \ldots, p\}.$$

We may now state

Theorem 12.3: (KKT Second-Order Sufficient Condition). Let $\boldsymbol{x}_o$ be a feasible point for problem (12.1.1). If there exist vectors $\boldsymbol{\lambda}_o, \boldsymbol{\mu}_o$ satisfying (12.4) and for every $\boldsymbol{h}(\neq \boldsymbol{O})$ such that $\boldsymbol{h} \in \widetilde{\mathcal{J}}(\boldsymbol{x}_o)$ we have

$$\boldsymbol{h}^T \nabla_{\boldsymbol{xx}}^2 L(\boldsymbol{x}_o, \boldsymbol{\lambda}_o, \boldsymbol{\mu}_o)\boldsymbol{h} < 0(>0), \qquad (12.7)$$

then $\boldsymbol{x}_o$ is a strong local maximum (minimum) for problem (12.1.1), where

$$\nabla_{\boldsymbol{xx}}^2 L(\boldsymbol{x}_o, \boldsymbol{\lambda}_o, \boldsymbol{\mu}_o) = \nabla^2 f(\boldsymbol{x}_o) - \Sigma_{j=1}^{m+n} \lambda_j^o \nabla^2 \hat{g}^j(\boldsymbol{x}_o) - \Sigma_{r=1}^{p} \mu_r^o \nabla^2 \hat{h}^r(\boldsymbol{x}_o)$$

is the Hessian matrix of L evaluated at $(\boldsymbol{x}_o, \boldsymbol{\lambda}_o, \boldsymbol{\mu}_o)$. So if $\nabla_{\boldsymbol{xx}}^2 L(\boldsymbol{x}_o, \boldsymbol{\lambda}_o, \boldsymbol{\mu}_o)$ is negative definite (positive definite), then f has a strong local constrained maximum (minimum) at $\boldsymbol{x}_o$.

Example 12.2: In Example 12.1 we found that the constrained minimum of $f(\boldsymbol{x})$ (based upon comparing the value of f at various candidate points) occurred at $\boldsymbol{x}_o^T = (2.54, 3.46)$ with $\lambda_1^o = 0.08$, $\mu_1^o = -1.00$, and $f(\boldsymbol{x}_o) = 1.05$. Does (12.7) hold at $\boldsymbol{x}_o$? Forming the Lagrangian associated with this problem we have

$$\begin{aligned} L(\boldsymbol{x}, \boldsymbol{\lambda}, \boldsymbol{\mu}) &= f(\boldsymbol{x}) + \Sigma_{j=1}^3 \lambda_j \hat{g}^j(\boldsymbol{x}) + \mu_1 \hat{h}^1(\boldsymbol{x}) \\ &= \tfrac{1}{8}(x_1 - 5)^2 + (x_2 - 4)^2 + \lambda_1(3 - x_1^2 + x_2) \\ &\quad + \lambda_2(x_1) + \lambda_3(x_2) + \mu_1(6 - x_1 - x_2). \end{aligned}$$

Then

$$\nabla_{\boldsymbol{x}} L = \begin{bmatrix} L_{x_1} \\ L_{x_2} \end{bmatrix} = \begin{bmatrix} \frac{1}{4}(x_1 - 5) - 2\lambda_1 x_1 + \lambda_2 - \mu_1 \\ 2(x_2 - 4) + \lambda_1 + \lambda_3 - \mu_1 \end{bmatrix}$$

and thus

$$\nabla_{\boldsymbol{xx}}^2 L = \begin{bmatrix} L_{x_1x_1} & L_{x_1x_2} \\ L_{x_2x_1} & L_{x_2x_2} \end{bmatrix} = \begin{bmatrix} \frac{1}{4} - 2\lambda_1 & 0 \\ 0 & 2 \end{bmatrix}.$$

At $(\boldsymbol{x}_o, \boldsymbol{\lambda}_o, \mu_1^o)^T = (2.54, 3.46, 0.08, 0, 0, -1)$ we have

$$\nabla_{\boldsymbol{xx}}^2 L(\boldsymbol{x}_o, \boldsymbol{\lambda}_o, \mu_1^o) = \begin{bmatrix} 0.09 & 0 \\ 0 & 2 \end{bmatrix},$$

which is positive definite. Hence, f has a strong local constrained minimum at $\boldsymbol{x}_o$. ■

We next turn to the conditions under which the KKT system (12.4) is also sufficient to ensure a constrained local (and global) extremum. To address the sufficiency requirement, we must introduce some supplemental assumptions involving the functional characteristics of the objective function and the constraint functions. In particular, we shall rely upon the assumption of the concavity of the objective function and constraint functions. Then the KKT necessary conditions for optimality are also sufficient for *concave programs*. In this regard, the following theorem represents necessary *and* sufficient conditions for determining a maximum of $f(x)$ subject to $\hat{g}^j(x) \geq 0$, $j = 1, \ldots, m + n$, and $\hat{h}^r(x) = 0$, $r = 1, \ldots, p$. Specifically,

Theorem 12.3: (KKT Necessary and Sufficient Conditions). Let the real-valued function $y = f(x)$ and the vector-valued function $\bar{G}(x)$, $x \in R^n$, be strictly concave, and let the vector-valued function $H(x)$ be *affine linear* (each $\hat{h}^r = c_r - a_j^T x$, $c_r \in R$, $a_j^T = (a_{j1}, \ldots, a_{jn}) \in R^n$, $r = 1, \ldots, p$) over the convex region $\bar{\mathcal{K}} = \{x | \bar{G}(x) \geq O, H(x) = O, x \in R^n\}$, int $(\bar{\mathcal{K}}) \neq \emptyset^2$, and of class $C^{(1)}$ at $x_o \in \bar{\mathcal{K}}$ with $\mathcal{F}_o \cap \mathcal{G}_o \cap \mathcal{H}_o = \emptyset$. Then f subject to $x \in \bar{\mathcal{K}}$ has a (strong) global maximum at x_o if and only if (12.4) holds.

Proof: (Necessity) If (12.4) is not satisfied, then x_o cannot be a constrained extremum of f.
(Sufficiency) Let $x \in \bar{\mathcal{K}}$. Then

$$f(x) \leq f(x) + \lambda_o^T \bar{G}(x) + \mu_o^T H(x) = L(x, \lambda_o, \mu_o).$$

With $L(x, \lambda_o, \mu_o)$ strictly concave in x over $\bar{\mathcal{K}}$,

$$\begin{aligned} L(x, \lambda_o, \mu_o) &< L(x_o, \lambda_o, \mu_o) + \nabla_x L(x_o, \lambda_o, \mu_o)^T (x - x_o) \\ &= f(x_o) + \lambda_o^T \bar{G}(x_o) + \mu_o^T H(x_o) \\ &\quad + (\nabla f(x_o) + \nabla \bar{G}(x_o)\lambda_o + \nabla H(x_o)\mu_o)^T (x - x_o) \\ &= f(x_o). \end{aligned}$$

Thus $f(x_o) > L(x, \lambda_o, \mu_o) \geq f(x)$, $x \in \bar{\mathcal{K}}$, so that f has a (strong) global constrained maximum at x_o. Q. E. D.

We may restate the preceding theorem by employing the weaker assumption that if f is pseudoconcave over $\bar{\mathcal{K}}$, $\hat{g}^j$, $j \in \mathcal{J}$, are of class $C^{(1)}$ and quasiconcave over $\bar{\mathcal{K}}$, and the $\hat{h}^r$, $r = 1, \ldots, p$, are affine, then x_o is a global optimal solution to the problem of maximizing $f(x)$ subject to $x \in \bar{\mathcal{K}}$. If these generalized concavity assumptions are restricted to $\delta(x_o)$, then obviously x_o represents a local maximal solution to this optimization problem (see Chapter 14 for details).

Section 12.2 Exercises:

Does the KKT second-order sufficient condition hold for the following exercises?

12.2.1 Exercise 12.1.1.
12.2.2 Exercise 12.1.2.
12.2.3 Exercise 12.1.3.
12.2.4 Exercise 12.1.4.

12.3 THE OPTIMAL VALUE FUNCTION: LAGRANGE MULTIPLIERS REVISITED

One final point is in order. Let us define the **optimal value function** for problem (12.1) as

$$\begin{aligned} f^o(b, c) = \text{maximize } \{f(x) | G(x) = b - g(x) \geq O, H(x) \\ = c - h(x) = O, x \geq O, x \in R^n\}, \end{aligned}$$

where $g^j(x)$, $j = 1, \ldots, m$. is the jth component of the $(m \times 1)$ vector-valued function $g(x)$ and $h^r(x)$, $r = 1, \ldots, p$ is the rth component of the $(p \times 1)$ vector-valued function $h(x)$. Here $f^o(b, c)$ is assumed of class $C^{(1)}$ and non-decreasing in b, c. (f^o must be non-decreasing since, with $c$fixed and any individual component of b is increased in value, with the remaining components held constant, then the feasible region $\mathcal{K} = \{x | b - g(x) \geq O, c - h(x) = O, x \geq O, x \in R^n\}$ must increase in size. Moreover, for fixed b, f^o is the highest level of f attainable given the requirement that $h(x) = c$.)

For $b = \bar{b}$ and $c = \bar{c}$, let $\bar{x}$ be the solution associated with the problem: maximize $f(x)$ subject to $\bar{b} \geq g(x)$, $\bar{c} - h(x) = O$, $x \geq O$, $x \in R^n$. Then $f(\bar{x}) = f^o(\bar{b}, \bar{c})$ and, for all $x \geq O$, $f(x) \leq f^o(\bar{b}, \bar{c}) = f^o(g(x), h(x)) \leq f^o(g(x) + \bar{b} - g(\bar{x}), h(x) + \bar{c} - h(\bar{x}))$. Since $\bar{b} \geq g(\bar{x})$ and $\bar{c} = h(\bar{x})$, f^o must be non-decreasing. Then the function $\eta(x) = f(x) - f^o(g(x) + \bar{b} - g(\bar{x}), h(x) + \bar{c} - h(\bar{x})) \leq 0$ for all $x \geq O$ and, with $\eta(\bar{x}) = f(\bar{x}) - f^o(\bar{b}, \bar{c}) = 0$, we see that $\eta(x)$ attains a maximum at $x = x$ with

$$\eta_i(\bar{x}) = f_i(\bar{x}) - \sum_{j=1}^m \left.\frac{\partial f^o(\bar{b}, \bar{c})}{\partial b_j}\right|_{b=g(\bar{x})} g_i^j - \sum_{r=1}^p \left.\frac{\partial f^o(\bar{b}, \bar{c})}{\partial c_r}\right|_{c=h(\bar{x})} h_i^r = 0, \; i = 1, \ldots, n.$$

Define $\bar{\lambda}_j = \partial f^o(\bar{b}, \bar{c})/\partial b_j$, $j = 1, \ldots, m$, and $\bar{\mu}_r = \partial f^o(\bar{b}, \bar{c})/\partial c_r$, $r = 1, \ldots, p$. Then this equation system satisfies $\nabla f(\bar{x}) + \nabla G(\bar{x})\bar{\lambda} + \nabla H(\bar{x})\bar{\mu} \leq O$ (see (12.5)), where $G(x) = b - g(x) \geq O$, $H(x) = c - h(x) = O$, $x \geq O$, $x \in R^n$.

NOTES

1 A modification of Farkas' Theorem 1.8.5a is

Theorem 12.1 (**Extended Farkas' Theorem of the Alternative**). Let $\mathcal{C}$ be a finite cone defined by

$$\mathcal{C} = \{\mathbf{g} | \mathbf{g} = \mathbf{A}\mathbf{y} + \mathbf{B}\mathbf{w}, \mathbf{y} \geq \mathbf{O}, \mathbf{g} \in R^n\},$$

where $\mathbf{A}$ and $\mathbf{B}$ are, respectively, matrices of order $(n \times m)$ and $(n \times p)$ and $\mathbf{g}$, $\mathbf{y}$ and $\mathbf{w}$ are, respectively, $(n \times 1)$, $(m \times 1)$ and $(p \times 1)$ vectors. Given a vector $\mathbf{g} \in R^n$, either:

(I) $\mathbf{g}^T\mathbf{x} > 0$, $\mathbf{A}^T\mathbf{x} \leq \mathbf{O}$, $\mathbf{B}^T\mathbf{x} = 0$ has a solution $\mathbf{x} \in R^n$; or

(II) $\mathbf{g} = \mathbf{A}\mathbf{y} + \mathbf{B}\mathbf{w}$, $\mathbf{y} \geq \mathbf{O}$, $\mathbf{w}$ unrestricted, has a solution $\mathbf{y} \in R^m$ and $\mathbf{w} \in R^p$, but never both.

So given $\mathbf{g} \in R^n$, either $\mathbf{g} \in \mathcal{C}$ (i. e. , (II) holds), or there exists an $\mathbf{x} \in R^n$ such that (I) is satisfied, but never both.

To employ this version of Farkas' Theorem, let:, the vectors $-\nabla \hat{g}^j(\mathbf{x}_o)$, $j \in \mathcal{J}$, depict the columns of $\mathbf{A}$, the vectors $-\nabla \hat{h}^r(\mathbf{x}_o)$ constitute the columns of $\mathbf{B}$, and the λ_j^o and μ_r^o, $r = 1, \ldots, p$, represent, respectively, the components of $\mathbf{y}$ and $\mathbf{w}$.

2 That $\bar{\mathcal{K}}$ has a non-empty interior is alternatively known as **Slater's CQ**: $\bar{\mathbf{G}}(\mathbf{x}) \geq 0$, $\mathbf{H}(\mathbf{x}) = 0$ satisfy Slater's CQ if there exists a $\mathbf{x}_* \in \bar{\mathcal{K}}$ such that $\bar{\mathbf{G}}(\mathbf{x}_*) > 0$, $\mathbf{H}(\mathbf{x}_*) = \mathbf{O}$.

CHAPTER 13

Lagrangian Saddle Points and Duality

13.1 INTRODUCTION

In this chapter, emphasis will be placed on solving the saddle-point problem (i.e., finding a saddle point $(\boldsymbol{x}_o, \boldsymbol{\lambda}_o)$ of the Lagrangian function $L(\boldsymbol{x}, \boldsymbol{\lambda}) = f(\boldsymbol{x}) + \boldsymbol{\lambda}^T \bar{\boldsymbol{G}}(\boldsymbol{x})$) and the specification and solution of two closely allied constrained optimization problems, namely the primal problem and its associated dual. In general, if the primal is a constrained maximization (minimization) problem, then its dual involves constrained minimization (maximization). (For our purposes the primal will be specified as a maximization problem and thus the dual will be a minimization problem.) Moreover, this correspondence is such that: under appropriate concavity conditions on $f(\boldsymbol{x})$ and $\bar{\boldsymbol{G}}(\boldsymbol{x})$; the assumed convexity of $\bar{\mathcal{K}}$; and a CQ, the existence of an optimal solution to one of these problems guarantees an optimal solution to the other, with the result that their extreme values are equal. Hence, it is possible to solve the primal indirectly by solving the dual. And it is also the case that, under the aforementioned restrictions, $\boldsymbol{x}_o$ solves the primal problem if and only if there exists a $\boldsymbol{\lambda}_o \geq \boldsymbol{O}$ such that $(\boldsymbol{x}_o, \boldsymbol{\lambda}_o)$ solves the saddle-point problem. Furthermore, if $f(\boldsymbol{x})$ and $\bar{\boldsymbol{G}}(\boldsymbol{x})$are of class $C^{(1)}$, then the KKT conditions (11.6) are necessary and sufficient for a (global) solution to both the primal and dual problems and to the saddle-point problem.

From a computational viewpoint, the choice between solving the primal versus the dual problem is typically made on the basis of computational expediency. Hence, ostensibly, one could solve the primal problem by actually solving its dual if the structure of the latter was simpler. In this regard, as will be demonstrated below, the dual approach to solving the primal focuses on the optimality conditions that are in force at a maximal solution to a primal problem, with the Lagrange multipliers serving as dual variables; and these variables are also optimal in a minimization problem that is dual to the said primal problem. So if $\bar{\mathcal{K}}$ is convex, the primal is concave and obeys a CQ, then the optimal dual solution can be used to find an optimal primal solution.

An additional characteristic of the dual problem is that it provides an upper bound on the primal objective function in a (concave) constrained maximization problem. Moreover, the values of the dual variables reflect the sensitivity of the primal objective function to the constraint constants. And via the CS conditions, we consequently know which constraints are binding at optimality, and which are not.

13.2 LAGRANGIAN SADDLE POINTS (LASDON [1970]; KUHN AND TUCKER [1951]; ARROW *ET AL.* [1958]; UZAWA [1958]; KÜNZI *ET AL.* (1966); AND GEOFFRION [1972])

In this section we shall, for the most part, assume that $\boldsymbol{x} \in \bar{\mathcal{K}} \subset \mathcal{D} \subset R^n$, $\mathcal{D}$ an open set on which the functions, $f, \hat{g}^j, j = 1, \ldots, m$, are defined and of class $C^{(1)}$, and the vector of Lagrange multipliers $\boldsymbol{\lambda} \in R^{n+m}, \boldsymbol{\lambda} \geq \boldsymbol{O}$. (Note that if $\boldsymbol{x} \in \mathcal{K} \subset \mathcal{D} \subset R^n$, then $\boldsymbol{\lambda} \in R^m$.) We start with a couple of definitions.

Definition 13.1: A point $(\boldsymbol{x}_o, \boldsymbol{\lambda}_o) \in R^{2n+m}$ is termed a **local saddle point** of the Lagrangian $L(\boldsymbol{x}, \boldsymbol{\lambda}) = f(\boldsymbol{x}) + \boldsymbol{\lambda}^T \bar{\boldsymbol{G}}(\boldsymbol{x})$

if there exists a spherical δ-neighborhood about $(\boldsymbol{x}_o, \boldsymbol{\lambda}_o)$, $\delta(\boldsymbol{x}_o, \boldsymbol{\lambda}_o)$, such that

$$L(\boldsymbol{x}, \boldsymbol{\lambda}_o) \le L(\boldsymbol{x}_o, \boldsymbol{\lambda}_o) \le L(\boldsymbol{x}_o, \boldsymbol{\lambda}) \qquad (13.1)$$

for all $(\boldsymbol{x}, \boldsymbol{\lambda}) \in \delta(\boldsymbol{x}_o, \boldsymbol{\lambda}_o)$, $\boldsymbol{x} \in \bar{\mathcal{K}}$, $\boldsymbol{\lambda} \ge \boldsymbol{O}$.

Next, we have

Definition 13.2: A point $(\boldsymbol{x}_o, \boldsymbol{\lambda}_o) \in R^{2n+m}$ is termed a **global saddle point** of the Lagrangian $L(\boldsymbol{x}, \boldsymbol{\lambda}) = f(\boldsymbol{x}) + \boldsymbol{\lambda}^T \bar{\boldsymbol{G}}(\boldsymbol{x})$ if

$$L(\boldsymbol{x}, \boldsymbol{\lambda}_o) \le L(\boldsymbol{x}_o, \boldsymbol{\lambda}_o) \le L(\boldsymbol{x}_o, \boldsymbol{\lambda}) \qquad (13.2)$$

for all $(\boldsymbol{x}, \boldsymbol{\lambda}) \in R^{2n+m}$, $\boldsymbol{x} \in \bar{\mathcal{K}}$, $\boldsymbol{\lambda} \ge \boldsymbol{O}$.

These definitions imply that $L(\boldsymbol{x}, \boldsymbol{\lambda})$ simultaneously attains, either locally or globally, a maximum with respect to $\boldsymbol{x}$, and a minimum with respect to $\boldsymbol{\lambda}$. In this regard, (13.1) and (13.2) may be rewritten as

$$L(\boldsymbol{x}_o, \boldsymbol{\lambda}_o) = \max_{\boldsymbol{x} \in \bar{\mathcal{K}}}\{\min_{\boldsymbol{\lambda} \ge \boldsymbol{O}} L(\boldsymbol{x}, \boldsymbol{\lambda})\} = \min_{\boldsymbol{\lambda} \ge \boldsymbol{O}}\{\max_{\boldsymbol{x} \in \bar{\mathcal{K}}} L(\boldsymbol{x}, \boldsymbol{\lambda})\}, \qquad (13.3)$$

and interpreted in either a local or global sense.

Under what conditions will there exist a local (global) saddle point at $(\boldsymbol{x}_o, \boldsymbol{\lambda}_o)$? That is, under what conditions can we obtain a solution to the

> **SADDLE-POINT PROBLEM**. To find a point $(\boldsymbol{x}_o, \boldsymbol{\lambda}_o)$ such that Equation (13.1) holds for all $(\boldsymbol{x}, \boldsymbol{\lambda}) \in \delta(\boldsymbol{x}_o, \boldsymbol{\lambda}_o)$ or Equation (13.2) holds for all $(\boldsymbol{x}, \boldsymbol{\lambda}) \in R^{2n+m}$, $\boldsymbol{x} \in \bar{\mathcal{K}}$, $\boldsymbol{\lambda} \ge \boldsymbol{O}$.

According to (13.3) we shall now see that determining a saddle point of the Lagrangian corresponds to maxi-minimizing or mini-maximizing it. To this end we state the following three theorems. First,

Theorem 13.1: (Necessary Condition). Let $\nabla_{\boldsymbol{x}} L$ and $\nabla_{\boldsymbol{\lambda}} L$ be defined, respectively, for all $\boldsymbol{x} \in \delta(\boldsymbol{x}_o) \subset \mathcal{K}$ and $\boldsymbol{\lambda} \in \varepsilon(\boldsymbol{\lambda}_o)$. If $L(\boldsymbol{x}, \boldsymbol{\lambda})$ has a local saddle point at $(\boldsymbol{x}_o, \boldsymbol{\lambda}_o)$, then:

$$\begin{aligned} &\text{(a)} \quad \nabla_{\boldsymbol{x}} L^o \le \boldsymbol{O},\ (\nabla_{\boldsymbol{x}} L^o)^T \boldsymbol{x}_o = 0,\ \boldsymbol{x}_o \ge \boldsymbol{O}; \\ &\text{(b)} \quad \nabla_{\boldsymbol{\lambda}} L^o \ge \boldsymbol{O},\ (\nabla_{\boldsymbol{\lambda}} L^o)^T \boldsymbol{\lambda}_o = 0,\ \boldsymbol{\lambda}_o \ge \boldsymbol{O}. \end{aligned} \qquad (13.4)$$

Proof: If $L(\boldsymbol{x}, \boldsymbol{\lambda}_o)$ has a local maximum in the $\boldsymbol{x}$-direction at $\boldsymbol{x}_o \ge \boldsymbol{O}$, then (13.4a) holds. That is, either $x_i^o = 0$, in which case $L_{x_i}^o \le 0$, or $x_i^o > 0$, thus implying that $L_{x_i}^o = 0$. If $L(\boldsymbol{x}_o, \boldsymbol{\lambda})$ has a local minimum in the $\boldsymbol{\lambda}$-direction at $\boldsymbol{\lambda}_o \ge \boldsymbol{O}$, then (13.4b) must hold. In this instance $L_{\lambda_j}^o \ge 0$, or $\lambda_j^o > 0$, thus implying that $L_{\lambda_j}^o = 0$. Q. E D.

We next consider

Theorem 13.2: (Sufficient Condition). Let $\nabla_{\boldsymbol{x}} L$ and $\nabla_{\boldsymbol{\lambda}} L$ be defined, respectively, for all $\boldsymbol{x} \in \delta(\boldsymbol{x}_o) \subset \mathcal{K}$ and $\boldsymbol{\lambda} \in \varepsilon(\boldsymbol{\lambda}_o)$ with $\boldsymbol{x} \ge \boldsymbol{O}$, $\boldsymbol{\lambda} \ge \boldsymbol{O}$. If (13.4) holds, then $L(\boldsymbol{x}, \boldsymbol{\lambda})$ has a local saddle point at $(\boldsymbol{x}_o, \boldsymbol{\lambda}_o)$ if:

$$\begin{aligned} &\text{(a)} \quad L(\boldsymbol{x}, \boldsymbol{\lambda}_o) \le L(\boldsymbol{x}_o, \boldsymbol{\lambda}_o) + (\nabla_{\boldsymbol{x}} L^o)^T (\boldsymbol{x} - \boldsymbol{x}_o); \\ &\text{(b)} \quad L(\boldsymbol{x}_o, \boldsymbol{\lambda}) \ge L(\boldsymbol{x}_o, \boldsymbol{\lambda}_o) + (\nabla_{\boldsymbol{\lambda}} L^o)^T (\boldsymbol{\lambda} - \boldsymbol{\lambda}_o). \end{aligned} \qquad (13.5)$$

Proof: Equations (13.5a,b) may be rewritten as

$$\begin{aligned} L(\boldsymbol{x}, \boldsymbol{\lambda}_o) &\le L(\boldsymbol{x}_o, \boldsymbol{\lambda}_o) + (\nabla_{\boldsymbol{x}} L^o)^T \boldsymbol{x} - (\nabla_{\boldsymbol{x}} L^o)^T \boldsymbol{x}_o, \\ L(\boldsymbol{x}_o, \boldsymbol{\lambda}) &\ge L(\boldsymbol{x}_o, \boldsymbol{\lambda}_o) + (\nabla_{\boldsymbol{\lambda}} L^o)^T \boldsymbol{\lambda} - (\nabla_{\boldsymbol{\lambda}} L^o)^T \boldsymbol{\lambda}_o. \end{aligned}$$

From (13.4a)

$$L(\boldsymbol{x}, \boldsymbol{\lambda}_o) \le L(\boldsymbol{x}_o, \boldsymbol{\lambda}_o) + (\nabla_{\boldsymbol{x}} L^o)^T \boldsymbol{x} \le L(\boldsymbol{x}_o, \boldsymbol{\lambda}_o),$$

while from (13.4b),

$$L(\boldsymbol{x}_o, \boldsymbol{\lambda}) \ge L(\boldsymbol{x}_o, \boldsymbol{\lambda}_o) + (\nabla_{\boldsymbol{\lambda}} L^o)^T \boldsymbol{\lambda} \ge L(\boldsymbol{x}_o, \boldsymbol{\lambda}_o).$$

Combining these latter two inequalities yields (13.1). Q. E. D.

Finally, for the third theorem we have

Theorem 13.3: (Necessary and Sufficient Condition). Let $L(\boldsymbol{x}, \boldsymbol{\lambda})$ be defined for all $(\boldsymbol{x}, \boldsymbol{\lambda}) \in \delta(\boldsymbol{x}_o, \boldsymbol{\lambda}_o)$, where $\boldsymbol{x} \in \bar{\mathcal{K}}$ is unrestricted and $\boldsymbol{\lambda} \ge \boldsymbol{O}$. Then $L(\boldsymbol{x}, \boldsymbol{\lambda})$ has a local saddle point at $(\boldsymbol{x}_o, \boldsymbol{\lambda}_o)$ if and only if the **saddle-point optimality conditions** hold:

$$\begin{aligned} &\text{(a)} \quad L(\boldsymbol{x}, \boldsymbol{\lambda}_o) \text{ attains a local maximum at } \boldsymbol{x}_o,\ \boldsymbol{x} \in \delta(\boldsymbol{x}_o); \\ &\text{(b)} \quad \bar{\boldsymbol{G}}(\boldsymbol{x}_o) \ge \boldsymbol{O}; \text{ and} \\ &\text{(c)} \quad \boldsymbol{\lambda}_o^T \bar{\boldsymbol{G}}(\boldsymbol{x}_o) = 0. \end{aligned} \qquad (13.6)$$

Proof: (Necessity) The left-hand inequality of (13.1) is equivalent to (13.6a). The right-hand inequality in (13.1) implies that $f(\boldsymbol{x}_o) + \boldsymbol{\lambda}^T \bar{\boldsymbol{G}}(\boldsymbol{x}_o) \ge f(\boldsymbol{x}_o) + \boldsymbol{\lambda}_o^T \bar{\boldsymbol{G}}(\boldsymbol{x}_o)$ or $(\boldsymbol{\lambda} - \boldsymbol{\lambda}_o)^T \bar{\boldsymbol{G}}(\boldsymbol{x}_o) \ge 0$. If $\bar{\boldsymbol{G}}(\boldsymbol{x}_o) \le 0$, $\boldsymbol{\lambda} \in \delta(\boldsymbol{\lambda}_o)$ may be chosen sufficiently large so that $\boldsymbol{\lambda} - \boldsymbol{\lambda}_o \ge \boldsymbol{O}$ or $(\boldsymbol{\lambda} - \boldsymbol{\lambda}_o)^T \bar{\boldsymbol{G}}(\boldsymbol{x}_o) \le 0$. But then $L(\boldsymbol{x}_o, \boldsymbol{\lambda}) \ge L(\boldsymbol{x}_o, \boldsymbol{\lambda}_o)$ is violated. Hence, (13.6b) must hold. If $\boldsymbol{\lambda} = \boldsymbol{O}$, $(\boldsymbol{\lambda} - \boldsymbol{\lambda}_o)^T \bar{\mathbf{G}}(\boldsymbol{x}_o) = \boldsymbol{\lambda}_o^T \bar{\mathbf{G}}(\boldsymbol{x}_o) \leq 0$. But since $\boldsymbol{\lambda} \geq \boldsymbol{O}$, $\bar{\mathbf{G}}(\boldsymbol{x}_o) \geq \boldsymbol{O}$ together imply $\boldsymbol{\lambda}_o^T \bar{\mathbf{G}}(\boldsymbol{x}_o) \geq 0$, we see that (13.6c) holds as well. So if $L(\boldsymbol{x}, \boldsymbol{\lambda})$ has a local saddle point at $(\boldsymbol{x}_o, \boldsymbol{\lambda}_o)$, (13.6) holds.

(Sufficiency). If $\boldsymbol{x}_o$ maximizes $L(\boldsymbol{x}, \boldsymbol{\lambda}_o), \boldsymbol{x} \in \delta(\boldsymbol{x}_o)$, then $L(\boldsymbol{x}_o, \boldsymbol{\lambda}_o) \geq L(\boldsymbol{x}, \boldsymbol{\lambda}_o)$. If $\boldsymbol{\lambda}_o^T \bar{\mathbf{G}}(\boldsymbol{x}_o) = 0$, then $L(\boldsymbol{x}_o, \boldsymbol{\lambda}_o) = f(\boldsymbol{x}_o)$. Hence, $L(\boldsymbol{x}_o, \boldsymbol{\lambda}) = f(\boldsymbol{x}_o) + \boldsymbol{\lambda}^T \bar{\mathbf{G}}(\boldsymbol{x}_o) = L(\boldsymbol{x}_o, \boldsymbol{\lambda}_o) + \boldsymbol{\lambda}^T \bar{\mathbf{G}}(\boldsymbol{x}_o)$ or $L(\boldsymbol{x}_o, \boldsymbol{\lambda}) \geq L(\boldsymbol{x}_o, \boldsymbol{\lambda}_o)$ since $\boldsymbol{\lambda}^T \bar{\mathbf{G}}(\boldsymbol{x}_o) \geq 0$ if $\bar{\mathbf{G}}(\boldsymbol{x}_o) \geq \boldsymbol{O}$. So if (13.6) holds, $L(\boldsymbol{x}, \boldsymbol{\lambda})$ has a local saddle point at $(\boldsymbol{x}_o, \boldsymbol{\lambda}_o)$. Q. E. D.

Up to this point in our discussion of saddle points we have not made any assumption(s) about the form of either $L(\boldsymbol{x}, \boldsymbol{\lambda}_o)$ or $L(\boldsymbol{x}_o, \boldsymbol{\lambda})$. Suppose that $L(\boldsymbol{x}, \boldsymbol{\lambda}_o)$ is a concave function of $\boldsymbol{x}$ and $L(\boldsymbol{x}_o, \boldsymbol{\lambda})$ is a convex function of $\boldsymbol{\lambda}$. In this regard, for $L(\boldsymbol{x}, \boldsymbol{\lambda}_o)$ concave, $L(\boldsymbol{x}, \boldsymbol{\lambda}_o) \leq L(\boldsymbol{x}_o, \boldsymbol{\lambda}_o) + (\nabla_{\boldsymbol{x}} L^o)^T (\boldsymbol{x} - \boldsymbol{x}_o)$; and if $L(\boldsymbol{x}_o, \boldsymbol{\lambda})$ is convex, $L(\boldsymbol{x}_o, \boldsymbol{\lambda}) \geq L(\boldsymbol{x}_o, \boldsymbol{\lambda}_o) + (\nabla_{\boldsymbol{\lambda}} L^o)^T (\boldsymbol{\lambda} - \boldsymbol{\lambda}_o)$. Here the former (respectively, latter) inequality depicts the equation of a hyperplane in R^{n+m} which is tangent to the surface $L(\boldsymbol{x}, \boldsymbol{\lambda})$ at $(\boldsymbol{x}_o, \boldsymbol{\lambda}_o)$ and lies above (respectively, below) it for all $\boldsymbol{x} \neq \boldsymbol{x}_o$ (respectively, $\boldsymbol{\lambda} \neq \boldsymbol{\lambda}_o$). But these inequalities are just (13.5a, b). Hence, (13.5a, b) always hold at $(\boldsymbol{x}_o, \boldsymbol{\lambda}_o)$ if $L(\boldsymbol{x}, \boldsymbol{\lambda})$ is concave in $\boldsymbol{x}$ (for $\boldsymbol{\lambda} = \boldsymbol{\lambda}_o$) and convex in $\boldsymbol{\lambda}$ (for $\boldsymbol{x} = \boldsymbol{x}_o$).[1] These observations are incorporated in

Theorem 13.4: (Necessary and Sufficient Condition). Let $L(\boldsymbol{x}, \boldsymbol{\lambda}_o)$ be concave in $\boldsymbol{x}$ and $L(\boldsymbol{x}_o, \boldsymbol{\lambda})$ convex in $\boldsymbol{\lambda}$ with $\nabla_{\boldsymbol{x}} L$ and $\nabla_{\boldsymbol{\lambda}} L$ defined, respectively, for all $\boldsymbol{x} \in \delta(\boldsymbol{x}_o)$ and $\boldsymbol{\lambda} \in \varepsilon(\boldsymbol{\lambda}_o)$, with $\boldsymbol{x} \in \mathcal{K}$, $\boldsymbol{x} \geq \boldsymbol{O}, \boldsymbol{\lambda} \geq \boldsymbol{O}$. Then $(\boldsymbol{x}_o, \boldsymbol{\lambda}_o)$ is a local saddle point of $L(\boldsymbol{x}, \boldsymbol{\lambda})$ if and only if (13.4) holds.

Proof. (Necessity) Use Theorem 13.1. (Sufficiency) Employ Theorem 13.2. Q. E. D.

It is important to recognize three essential differences between Theorems 13.3, 13.4:

1. Theorem 13.3 does not require differentiability of the Lagrangian whereas Theorem 13.4 does.
2. Theorem 13.4 hypothesizes that $L(\boldsymbol{x}, \boldsymbol{\lambda}_o)$ is concave in $\boldsymbol{x}$ and $L(\boldsymbol{x}_o, \boldsymbol{\lambda})$ is convex in $\boldsymbol{\lambda}$ (hence, the local saddle point obtained at $(\boldsymbol{x}_o, \boldsymbol{\lambda}_o)$ is also of the global variety under the concavity/convexity assumption); and
3. In Theorem 13.3, $\boldsymbol{x}$ is unrestricted whereas, in Theorem 13.4, we specifically require that $\boldsymbol{x} \geq \boldsymbol{O}$.

In Chapter 11 we determined a set of necessary conditions for the solution to a problem of the form: maximize the real-valued function $y = f(\boldsymbol{x})$ subject to a set of (m) inequality constraints and (n) non-negativity conditions. This problem will now be termed the (local)

> **MAXIMUM PROBLEM.** Find a point $\boldsymbol{x}_o \in \bar{\mathcal{K}} \subset \mathcal{D} \subset R^n$ which maximizes $f(\boldsymbol{x})$ subject to $\bar{\mathbf{G}}(\boldsymbol{x}) \geq \boldsymbol{O}$ with $\boldsymbol{x}$ unrestricted, or a point $\boldsymbol{x}_o \in \mathcal{K} \subset \mathcal{D} \subset R^n$ which maximizes $f(\boldsymbol{x})$ subject to $\mathbf{G}(\boldsymbol{x}) \geq \boldsymbol{O}, \boldsymbol{x} \geq \boldsymbol{O}$.

What is the connection between the solution of the (local) maximum problem and the solution of the (local) saddle-point problem? In what follows we shall consider a battery of important theorems which serve to establish a bond between them. First,

Theorem 13.5: (Sufficient Conditions.) Let $L(\boldsymbol{x}, \boldsymbol{\lambda})$ be defined for all $(\boldsymbol{x}, \boldsymbol{\lambda}) \in \delta(\boldsymbol{x}_o, \boldsymbol{\lambda}_o)$, where $\boldsymbol{x} \in \bar{\mathcal{K}}$ is unrestricted and $\boldsymbol{\lambda} \geq \boldsymbol{O}$. If $(\boldsymbol{x}_o, \boldsymbol{\lambda}_o)$ is a local saddle point for $L(\boldsymbol{x}, \boldsymbol{\lambda})$, then $\boldsymbol{x}_o$ solves the local maximum problem.

Proof: Let $(\boldsymbol{x}_o, \boldsymbol{\lambda}_o)$ be a local saddle point for $L(\boldsymbol{x}, \boldsymbol{\lambda})$. Then (13.6) holds. In this regard, $L(\boldsymbol{x}_o, \boldsymbol{\lambda}_o) \geq L(\boldsymbol{x}, \boldsymbol{\lambda}_o)$ or $f(\boldsymbol{x}_o) + \boldsymbol{\lambda}_o^T \bar{\mathbf{G}}(\boldsymbol{x}_o) = f(\boldsymbol{x}_o) \geq f(\boldsymbol{x}) + \boldsymbol{\lambda}_o^T \bar{\mathbf{G}}(\boldsymbol{x})$. Since $\boldsymbol{\lambda}_o^T \bar{\mathbf{G}}(\boldsymbol{x}) \geq 0, f(\boldsymbol{x}_o) \geq f(\boldsymbol{x})$ for all $\boldsymbol{x} \in \delta(\boldsymbol{x}_o)$. Q. E. D.

This theorem does not require that the Lagrangian be differentiable or that $\boldsymbol{x} \geq$ O. An alternative sufficient condition which does require these characteristics is

Theorem 13.6: (Sufficient Condition.) Let $L(\boldsymbol{x}, \boldsymbol{\lambda})$ be defined for all $(\boldsymbol{x}, \boldsymbol{\lambda}) \in \delta(\boldsymbol{x}_o, \boldsymbol{\lambda}_o)$, with $\boldsymbol{x} \in \mathcal{K}$, $\boldsymbol{x} \geq \boldsymbol{O}, \boldsymbol{\lambda} \geq \boldsymbol{O}$, and with $\nabla_{\boldsymbol{x}} L$ defined for all $\boldsymbol{x} \in \delta(\boldsymbol{x}_o)$. If (13.4) and (13.5a) hold for $\boldsymbol{x}_o$ and some $\boldsymbol{\lambda}_o$, then $\boldsymbol{x}_o$ solves the local maximum problem.

Proof: Since

$$\begin{aligned} L(\boldsymbol{x}, \boldsymbol{\lambda}_o) &= f(\mathrm{x}) + \boldsymbol{\lambda}_o^T \mathbf{G}(\boldsymbol{x}) \leq L(\boldsymbol{x}_o, \boldsymbol{\lambda}_o) + (\nabla_{\boldsymbol{x}} L^o)^T (\boldsymbol{x} - \boldsymbol{x}_o) \\ &= L(\boldsymbol{x}_o, \boldsymbol{\lambda}_o) + (\nabla_{\boldsymbol{x}} L^o)^T \boldsymbol{x} \leq L(\boldsymbol{x}_o, \boldsymbol{\lambda}_o) = f(\boldsymbol{x}_o), \end{aligned}$$

and $\boldsymbol{\lambda}_o^T \mathbf{G}(\boldsymbol{x}) \geq 0$ because $\mathbf{G}(\boldsymbol{x}) \geq \boldsymbol{O}$, it follows that $f(\boldsymbol{x}) + \boldsymbol{\lambda}_o^T \mathbf{G}(\boldsymbol{x}) \leq f(\boldsymbol{x}_o)$ or $f(\boldsymbol{x}_o) \geq f(\boldsymbol{x})$ for all $\boldsymbol{x} \in \delta(\boldsymbol{x}_o)$. Q. E. D.

The importance of the material presented in this section is that it sets the stage for a discussion of the Kuhn-Tucker equivalence theorem. (Actually, two versions of this theorem will be offered; the first version (given without proof) *does not* require differentiability of the Lagrangian, while the second version does.) This theorem establishes the notion that the existence of an optimal solution to the local maximum problem is equivalent (under certain restrictions) to the existence of a local saddle point of the associated Lagrangian of f. Hence, maximization subject to inequality constraints is equivalent to maxi-minimizing (mini-maximizing) the Lagrangian subject to no restraint.

We saw above (Theorem 13.5) that if $(\boldsymbol{x}_o, \boldsymbol{\lambda}_o)$ solves the local saddle-point problem, then $\boldsymbol{x}_o$ solves the local maximum problem. Moreover, Theorem 13.5 holds without any restrictions whatever on the Lagrangian (e.g., concavity) or on the constraint set (i.e., a constraint qualification). What about the converse statement? That is, given that $\boldsymbol{x}_o$ solves the (local) maximum problem, is it possible to find a $\boldsymbol{\lambda}_o$ for which $(\boldsymbol{x}_o, \boldsymbol{\lambda}_o)$ solves the (local) saddle-point problem? In general, the answer is *no*. However, it the Lagrangian is deemed concave *and* a constraint qualification is imposed on the constraint set, the answer is in the affirmative. Moreover, the local constrained maximum of f and the local saddle point of L are necessarily global in character by virtue of the concavity assumption. In this regard we look to

Theorem 13.7: [Kuhn-Tucker Equivalence Theorem I (Necessary and Sufficient Condition)]. Let the real-valued function $y = f(\boldsymbol{x})$ and the vector-valued function $\bar{\boldsymbol{G}}(\boldsymbol{x}) \geq \boldsymbol{O}$, $\boldsymbol{x} \in R^n$, be concave over the convex region $\bar{\mathcal{K}} = \{\boldsymbol{x} | \bar{\boldsymbol{G}}(\boldsymbol{x}) \geq \boldsymbol{O}, \boldsymbol{x} \in R^n\}$ with the property that for some $\boldsymbol{x}_* \in \bar{\mathcal{K}}$, $\bar{\boldsymbol{G}}(\boldsymbol{x}_*) > \boldsymbol{O}$ (Slater's *CQ*). Then x_o solves the (global) maximum problem if and only if there exists a $\boldsymbol{\lambda}_o \geq \boldsymbol{O}$ such that $(\boldsymbol{x}_o, \boldsymbol{\lambda}_o)$ solves the (global) saddle-point problem.[2]

To summarize, $\boldsymbol{x}_o$ solves the global maximum problem if $(\boldsymbol{x}_o, \boldsymbol{\lambda}_o)$ solves the global saddle-point problem, while $\boldsymbol{x}_o$ solves the maximum problem only if $\boldsymbol{x}_o$ and some $\boldsymbol{\lambda}_o$ solve the saddle-point problem, given that $L(\boldsymbol{x}, \boldsymbol{\lambda})$ is concave and a constraint qualification is imposed on the constraint set.

The above version of the Kuhn-Tucker equivalence theorem did not require differentiability of the Lagrangian. If $L(\boldsymbol{x}, \boldsymbol{\lambda}) = f(\boldsymbol{x}) + \boldsymbol{\lambda}^T \bar{\boldsymbol{G}}(\boldsymbol{x})$ is of class $C^{(1)}$ as well as concave, then, subject to the above constraint qualification, the global saddle-point inequalities are equivalent to the KKT conditions (11.6), which are now both necessary and sufficient to ensure that $\boldsymbol{x}_o$ is a solution to the global maximum problem. More formally, we have

Theorem 13.8: [Kuhn-Tucker Equivalence Theorem II (Necessary and Sufficient Condition)]. Let the real-valued function $y = f(\boldsymbol{x})$ and the vector-valued function $\bar{\boldsymbol{G}}(\boldsymbol{x}) \geq \boldsymbol{O}$, $\boldsymbol{x} \in R^n$, be concave and of class $C^{(1)}$ over the convex region $\bar{\mathcal{K}} = \{\boldsymbol{x} | \bar{\boldsymbol{G}}(\boldsymbol{x}) \geq \boldsymbol{O}, \boldsymbol{x} \in R^n\}$ with the property that for some $\boldsymbol{x}_* \in \bar{\mathcal{K}}$, $\bar{\boldsymbol{G}}(\boldsymbol{x}_*) > \boldsymbol{O}$. Then $\boldsymbol{x}_o$ solves the global maximum problem if and only if there exists a $\boldsymbol{\lambda}_o \geq \boldsymbol{O}$ such that $(\boldsymbol{x}_o, \boldsymbol{\lambda}_o)$ solves the saddle-point problem with

$$\nabla f(\boldsymbol{x}_o) + \nabla \bar{\boldsymbol{G}}(\boldsymbol{x}_o)\boldsymbol{\lambda}_o = \boldsymbol{O}, \ \bar{\boldsymbol{G}}(\boldsymbol{x}_o) \geq \boldsymbol{O}, \ \boldsymbol{\lambda}_o^T \bar{\boldsymbol{G}}(\boldsymbol{x}_o) = 0.$$

Proof: By the preceding theorem, $\boldsymbol{x}_o$ solves the maximum problem if and only if there exists a $\boldsymbol{\lambda}_o \geq \boldsymbol{O}$ such that $(\boldsymbol{x}_o, \boldsymbol{\lambda}_o)$ solves the saddle-point problem. With the Lagrangian of class $C^{(1)}$, if $L(\boldsymbol{x}, \boldsymbol{\lambda}_o)$ attains a global maximum at $\boldsymbol{x}_o$, then $L(\boldsymbol{x}, \boldsymbol{\lambda}_o)$ is stationary at $(\boldsymbol{x}_o, \boldsymbol{\lambda}_o)$, and thus $\nabla_{\boldsymbol{x}} L^o = \nabla f(\boldsymbol{x}_o) + \nabla \bar{\boldsymbol{G}}(\boldsymbol{x}_o)\boldsymbol{\lambda}_o = \boldsymbol{O}$ replaces (13.6a). Hence, (13.6) becomes

$$\left.\begin{array}{ll} \text{(a)} & \nabla f(\boldsymbol{x}_o) + \nabla \bar{\boldsymbol{G}}(\boldsymbol{x}_o)\boldsymbol{\lambda}_o = \boldsymbol{O} \\ \text{(b)} & \bar{\boldsymbol{G}}(\boldsymbol{x}_o) \geq \boldsymbol{O} \\ \text{(c)} & \boldsymbol{\lambda}_o^T \bar{\boldsymbol{G}}(\boldsymbol{x}_o) = 0 \end{array}\right\} \text{or (11.6) since } \boldsymbol{\lambda}_o \geq \boldsymbol{O}. \qquad (13.6.1)$$

Thus the KKT conditions are necessary and sufficient for a global maximum at $\boldsymbol{x}_o$ if the Lagrangian is concave and of class $C^{(1)}$ and the constraint set satisfies a constraint qualification. Q. E. D.

13.3 SADDLE POINTS REVISITED: PERTURBATION FUNCTIONS

Suppose the aforementioned maximum problem (denoted (P)) or

$$(P) \quad \text{maximize } y = f(\boldsymbol{x}) \text{ subject to}$$
$$\boldsymbol{x} \in \bar{\mathcal{K}} = \{\boldsymbol{x} | \bar{\boldsymbol{G}}(\boldsymbol{x}) \geq \boldsymbol{O}, \boldsymbol{x} \in R^n\} \subset \mathcal{D}$$

is a subset of a larger family of problems termed perturbed problems. To form such problems, let $\boldsymbol{y}$ be a

$(m + n \times 1)$ **perturbation vector**, where each component of $\boldsymbol{y}$, y_j, is associated with a constraint function $\hat{g}_j(\boldsymbol{x}) \geq 0$, $j = 1, \ldots, n + m$. That is, given (P), the family of **perturbed problems** (P') appears as

$$(P') \quad \text{maximize } y = f(\boldsymbol{x}) \text{ subject to } \boldsymbol{x} \in \bar{\mathcal{K}}_{\boldsymbol{y}} = \{\boldsymbol{x} | \bar{\boldsymbol{G}}(\boldsymbol{x}) \geq \boldsymbol{y}, \boldsymbol{x} \in \bar{\mathcal{K}}\}.$$

For $\boldsymbol{y} \in R^{n+m}$, the optimal value of the perturbed problem (P') is a function $\Phi(\boldsymbol{y})$, called the **perturbation function**, which expresses the optimal value of a program as a function of perturbations of its right-hand side. For $\boldsymbol{y} = \boldsymbol{O}$ we have problem (P) with $\Phi(\boldsymbol{O}) = \max(P)$. The perturbation function is monotonic: if $\boldsymbol{y}' < \boldsymbol{y}$, then $\Phi(\boldsymbol{y}') \leq \Phi(\boldsymbol{y})$ since $\mathcal{K}_{\boldsymbol{y}'} \subseteq \bar{\mathcal{K}}_{\boldsymbol{y}}$. Additionally, if (P) is a concave program, then $\Phi(\boldsymbol{y})$ is a concave function of $\boldsymbol{y}$.

Let us now turn to an important characterization of the vector of Lagrange multipliers for problem (P). Specifically,

Theorem 13.9: (Necessary and Sufficient Condition). Let (P) have an optimal solution $\boldsymbol{x}_o \in \bar{\mathcal{K}}$ with $f(\boldsymbol{x}_o) < +\infty$. Then $\boldsymbol{\lambda}_o$ is a **saddle-point multiplier** if and only if the hyperplane $z = \Phi(\boldsymbol{O}) - \boldsymbol{\lambda}_o^T \boldsymbol{y}$ is a supporting hyperplane at $\boldsymbol{y} = \boldsymbol{O}$ of the graph of $\Phi(\boldsymbol{y})$ (Figure 13.1), i.e., if and only if for $\boldsymbol{y} \in R^{n+m}$,

$$\Phi(\boldsymbol{y}) \leq \Phi(\boldsymbol{O}) - \boldsymbol{\lambda}_o^T \boldsymbol{y}. \tag{13.7}$$

Proof: (Necessity) Suppose $(\boldsymbol{x}_o, \boldsymbol{\lambda}_o)$ is a saddle point. By definition, $f(\boldsymbol{x}) + \boldsymbol{\lambda}_o^T \bar{\boldsymbol{G}}(\boldsymbol{x}) \leq f(\boldsymbol{x}_o) + \boldsymbol{\lambda}_o^T \bar{\boldsymbol{G}}(\boldsymbol{x}_o)$, $\boldsymbol{x} \in \bar{\mathcal{K}} \subset \mathcal{D}$.

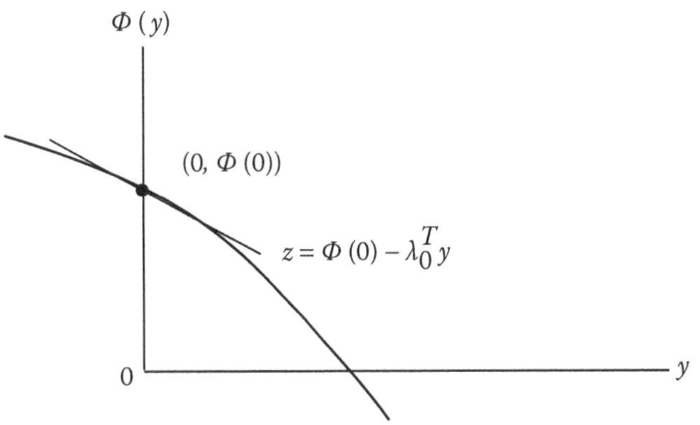

FIGURE 13.1 $\Phi(\boldsymbol{y})$ has a supporting hyperplane $z = \Phi(\boldsymbol{O}) - \boldsymbol{\lambda}_o^T \boldsymbol{y}$ at $\boldsymbol{y} = \boldsymbol{O}$.

With $\boldsymbol{\lambda}_o^T \bar{\boldsymbol{G}}(\boldsymbol{x}_o) = 0$, it follows that $f(\boldsymbol{x}) + \boldsymbol{\lambda}_o^T \bar{\boldsymbol{G}}(\boldsymbol{x}) \leq f(\boldsymbol{x}_o) = \Phi(\boldsymbol{O}) = \max(P)$ and thus $f(\boldsymbol{x}) \leq \Phi(\boldsymbol{O}) - \boldsymbol{\lambda}_o^T \bar{\boldsymbol{G}}(\boldsymbol{x})$, $\boldsymbol{x} \in \bar{\mathcal{K}}$. Specifically, for every $\boldsymbol{y}$ such that (P') has a solution, and for every $\boldsymbol{x}$ which is a solution to (P'),

$$f(\boldsymbol{x}) \leq \Phi(\boldsymbol{O}) - \boldsymbol{\lambda}_o^T \bar{\boldsymbol{G}}(\boldsymbol{x}) \leq \Phi(\boldsymbol{O}) - \boldsymbol{\lambda}_o^T \boldsymbol{y}$$

and thus $\Phi(\boldsymbol{y}) \leq \Phi(\boldsymbol{O}) - \boldsymbol{\lambda}_o^T \boldsymbol{y}$.

(Sufficiency) Let (13.7) hold. Then we must have $\boldsymbol{\lambda}_o \geq \boldsymbol{O}$ since otherwise the monotonicity property of Φ is violated. By the definition of $\Phi(\boldsymbol{y})$, $f(\boldsymbol{x}) \leq \Phi(\bar{\boldsymbol{G}}(\boldsymbol{x}))$ so that, for all $\boldsymbol{x} \in \bar{\mathcal{K}}_{\boldsymbol{y}}$,

$$f(x) \leq \Phi(\bar{\boldsymbol{G}}(\boldsymbol{x})) \leq \Phi(\boldsymbol{O}) - \boldsymbol{\lambda}_o^T \bar{\boldsymbol{G}}(\boldsymbol{x})$$

and thus

$$f(\boldsymbol{x}) + \boldsymbol{\lambda}_o^T \bar{\boldsymbol{G}}(\boldsymbol{x}) \leq f(\boldsymbol{x}_o)(=\Phi(\boldsymbol{O})), \ \boldsymbol{x} \in \bar{\mathcal{K}}_{\boldsymbol{y}}.$$

Then, at $\boldsymbol{x} = \boldsymbol{x}_o$,

$$f(\boldsymbol{x}) + \boldsymbol{\lambda}_o^T \bar{\boldsymbol{G}}(\boldsymbol{x}) \leq f(\boldsymbol{x}_o) = f(\boldsymbol{x}_o) + \boldsymbol{\lambda}_o^T \bar{\boldsymbol{G}}(\boldsymbol{x}_o), \ \boldsymbol{x} \in \bar{\mathcal{K}},$$

implies

$$L(\boldsymbol{x}_o, \boldsymbol{\lambda}_o) = \max_{\boldsymbol{x} \in \bar{\mathcal{K}}} L(\boldsymbol{x}, \boldsymbol{\lambda}_o)$$

so that $(\boldsymbol{x}_o, \boldsymbol{\lambda}_o)$ is a saddle point, with $\boldsymbol{\lambda}_o$ a saddle-point multiplier. Hence, the concavity of Φ is sufficient to ensure that a saddle point of the Lagrangian exists. Q. E. D.

Next, suppose problem (P) is a concave program and that Slater's CQ is satisfied. As we shall now see, in this instance, the Lagrange multipliers are subgradients of the perturbation function $\Phi(\boldsymbol{y})$ at the points of no perturbation (i.e., at $\boldsymbol{y} = \boldsymbol{O}$). To verify this, remember that, under the aforementioned assumptions, if (P) has an optimal solution $\boldsymbol{x}_o$, then there exists a $\boldsymbol{\lambda}_o \geq \boldsymbol{O}$ such that $(\boldsymbol{x}_o, \boldsymbol{\lambda}_o)$ is a saddle point of the Lagrangian function $\boldsymbol{L}(\boldsymbol{x}, \boldsymbol{\lambda})$. Additionally, these assumptions imply that Φ is concave at that $\boldsymbol{y} = \boldsymbol{O} \in \operatorname{int}(\operatorname{dom}(\Phi))$. We also know, from earlier discussions, that $\Phi(\boldsymbol{y})$ has a subgradient $\boldsymbol{\gamma}$ at $\boldsymbol{y} = \boldsymbol{O}$ which satisfies

$$\Phi(\boldsymbol{y}) \leq \Phi(\boldsymbol{O}) + \boldsymbol{\gamma}^T \boldsymbol{y}, \ \boldsymbol{y} \in R^{n+m}.$$

If we set $\boldsymbol{\lambda}_o = -\boldsymbol{y}$, then the existence of a saddle point is confirmed. In sum, if (P) is a concave program, then $\boldsymbol{\lambda}_o$ is a saddle point multiplier if and only if $-\boldsymbol{\lambda}_o$ is a subgradient of the perturbation function Φ at $\boldsymbol{y} = \boldsymbol{O}$,

Example 13.1: Let us maximize the function $f(\boldsymbol{x}) = -x_1^2 - x_2^2$ subject to $g(\boldsymbol{x}) = 2x_1 + x_2 \leq -6$ ***or*** $\hat{g}(\boldsymbol{x}) = -6 - 2x_1 - x_2 \geq 0$, $\boldsymbol{x} \in R^2$. (Note that we have a concave program.) Let us first form the Lagrangian of f or

$$L(\boldsymbol{x}, \lambda) = f(\boldsymbol{x}) + \lambda \hat{g}(\boldsymbol{x}) = -x_1^2 - x_2^2 + \lambda(-6 - 2x_1 - x_2).$$

From the set of first-order conditions for a maximum we have $x_1 = -\lambda$, $x_2 = -\frac{1}{2}\lambda$, and $\lambda = \frac{12}{5}$ so that $x_1 = -\frac{12}{5}$, $x_2 = -\frac{6}{5}$, and the maximum value of f is $-\frac{36}{5}$.

Turning to the perturbed version of this problem (P') we have

$$(P') \quad \Phi(y) = \max\ \{-x_1^2 - x_2^2\} \text{ subject to } -6 - 2x_1 - x_2 \geq y.$$

Again, looking to the set of first-order conditions, we obtain $x_1 = -\lambda$, $x_2 = -\frac{1}{2}\lambda$, and $\lambda = \frac{12+2y}{5}$ so that $x_1 = \frac{-12-2y}{5}$ and $x_2 = \frac{-6-y}{5}$. Then

$$\Phi(y) = \begin{cases} -\left(\frac{-12-2y}{5}\right)^2 - \left(\frac{-6-y}{5}\right)^2 = -\frac{36}{5} - \frac{12}{5}y - \frac{1}{5}y^2, & y > -6; \\ 0, & y \leq -6 \end{cases}$$

with $\Phi(0) = -\frac{36}{5} = \max(P)$. Additionally, we know from (13.7) that $z = \Phi(0) - \lambda y = -\frac{36}{5} - \frac{12}{5}y$ is a supporting hyperplane to $\Phi(y)$ at $y = 0$ (Figure 13.2). ■

One final point merits our attention. We can engage in **post-optimality analysis** of problem (P) by examining the behavior of the perturbation function $\Phi(\boldsymbol{y})$ for small perturbations or changes in the right-hand side of the constraint functions $\hat{g}^j(\boldsymbol{x}) \geq y_j$, $j = 1, \ldots, n+m$, near $\mathrm{y} = \boldsymbol{O}$, i.e., we seek to examine variations in the optimal value of (P) under suitably restricted variations in $\boldsymbol{y}$. To this end, suppose $\Phi(\boldsymbol{y})$ is of class $C^{(1)}$ at $\boldsymbol{y} = \boldsymbol{O}$ and that (P) has a saddle point $(\boldsymbol{x}_o, \boldsymbol{\lambda}_o)$. The supporting hyperplane to the graph of $\Phi(\boldsymbol{y})$ at the point $(\boldsymbol{O}, \Phi(\boldsymbol{O}))$ is obviously the tangent hyperplane.

$$z = \Phi(\boldsymbol{O}) + (\nabla_y \Phi(\boldsymbol{O}))^T(\boldsymbol{y} - \boldsymbol{O}).$$

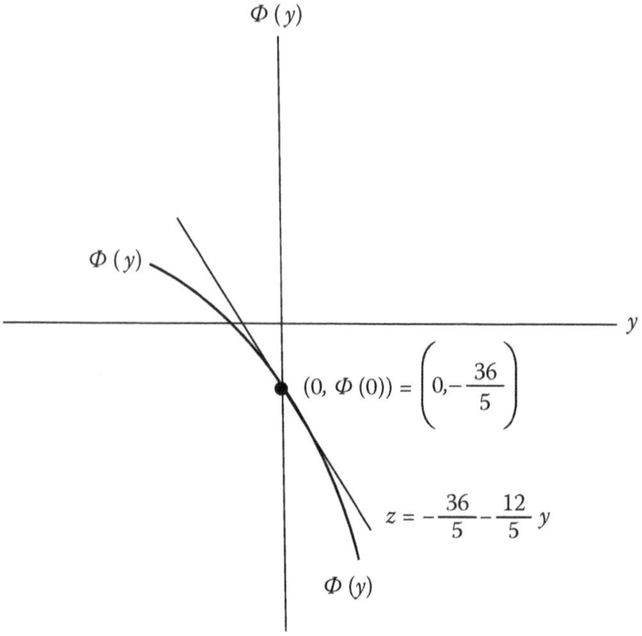

FIGURE 13.2 Supporting hyperplane $z = \Phi(0) - \lambda y = -\frac{36}{5} - \frac{12}{5}y$ to $\Phi(y)$ at $y = 0$.

Then $\boldsymbol{\lambda}_o^T = -(\nabla_y \Phi(\boldsymbol{O}))^T$! or

$$\lambda_j^o = -\frac{\partial \Phi(\boldsymbol{O})}{\partial y_j}, \; j = 1, \ldots, n+m,$$

the saddle-point multiplier associated with the perturbation y_j of the jth constraint. (As evidenced in Example 13.1, $z = \Phi(0) - \lambda y$ and thus $\lambda = -\partial z / \partial y = 12/5$ at $y = 0$.)

Section 13.3 Exercises:

For each primal problem (P) given below, find:

a. the associated perturbed problem (P');

b. the perturbation function $\Phi(\boldsymbol{y})$;

c. the supporting hyperplane to $\Phi(\boldsymbol{y})$ at $\boldsymbol{y} = \boldsymbol{O}$;

d. the saddle point multiplier(s) associated with perturbing the constrains(s).

13.3.1 (P) maximize $y = f(\boldsymbol{x}) = -(x_1 - 3)^2 - (x_2 - 3)^2$ subject to $\hat{g}(\boldsymbol{x}) = 6 - 3x_1 - 2x_2 \geq 0$.

13.3.2 (P) maximize $y = f(\boldsymbol{x}) = x_1 x_2$ subject to

$$\hat{g}^1(x) = 100 - x_1 - x_2 \ge 0$$
$$\hat{g}^2(x) = 50 - x_1 \ge 0$$

13.3.3. (*P*) maximize $y = f(x) = -x_1^2 - x_2^2$ subject to

$$\hat{g}^1(x) = 2x_1 + x_2 \ge 0$$
$$\hat{g}^2(x) = -4 - x_1 - x_2 \ge 0$$

13.4 LAGRANGIAN SADDLE POINTS WITH MIXED CONSTRAINTS

We shall refer to the following optimization problem as the

> **MAXIMUM PROBLEM.** Find a point $\mathbf{x}_o \in \bar{\mathcal{K}} \subset \mathcal{D} \subset R^n$ which maximizes $f(\mathbf{x})$ subject to $\bar{\mathbf{G}}(\mathbf{x}) \ge \mathbf{O}$, $\mathbf{H}(\mathbf{x}) = \mathbf{O}$, with $\mathbf{x}$ unrestricted or a point $\mathbf{x}_o \in \mathcal{K} \subset \mathcal{D} \subset R^n$ which maximizes $f(\mathbf{x})$ subject to $\mathbf{G}(\mathbf{x}) \ge \mathbf{O}$, $\mathbf{H}(\mathbf{x}) = \mathbf{O}$, with $\mathbf{x} \ge \mathbf{O}$.

(A point of reference for this problem is either (12.1) or (12.1.1).) Let us write the Lagrangian associated with this maximum problem as

$$L(\mathbf{x}, \boldsymbol{\lambda}, \boldsymbol{\mu}) = f(\mathbf{x}) + \sum_{j=1}^{m+n} \lambda_j \hat{g}^j(\mathbf{x}) + \sum_{r=1}^{p} \mu_r \hat{h}^r(\mathbf{x}) = f(\mathbf{x}) + \boldsymbol{\lambda}^T \bar{\mathbf{G}}(\mathbf{x}) + \boldsymbol{\mu}^T \mathbf{H}(\mathbf{x}), \tag{13.8}$$

where the vectors of Lagrange multipliers $\boldsymbol{\lambda} \ge \mathbf{O}$ and $\boldsymbol{\mu}$ are, respectively, of order $(n + m \times 1)$ and $(p \times 1)$. For this Lagrangian we have $\mathbf{x} \in \bar{\mathcal{K}} = \{\mathbf{x} | \bar{\mathbf{G}}(\mathbf{x}) \ge \mathbf{O}, \mathbf{H}(\mathbf{x}) = \mathbf{O}, \mathbf{x} \in R^n\} \subset \mathcal{D}$.

We next have

Definition 13.3: A point $(\mathbf{x}_o, \boldsymbol{\lambda}_o, \boldsymbol{\mu}_o) \in R^{2n+m+p}$ is a **local saddle point** of the Lagrangian $L(\mathbf{x}, \boldsymbol{\lambda}, \boldsymbol{\mu}) = f(\mathbf{x}) + \boldsymbol{\lambda}^T \bar{\mathbf{G}}(\mathbf{x}) + \boldsymbol{\mu}^T \mathbf{H}(\mathbf{x})$ if there exists a spherical δ – neighborhood about $(\mathbf{x}_o, \boldsymbol{\lambda}_o, \boldsymbol{\mu}_o)$, $\delta(\mathbf{x}_o, \boldsymbol{\lambda}_o, \boldsymbol{\mu}_o)$, such that

$$L(\mathbf{x}, \boldsymbol{\lambda}_o, \boldsymbol{\mu}_o) \le L(\mathbf{x}_o, \boldsymbol{\lambda}_o, \boldsymbol{\mu}_o) \le L(\mathbf{x}_o, \boldsymbol{\lambda}, \boldsymbol{\mu}) \tag{13.9}$$

for all $(\mathbf{x}, \boldsymbol{\lambda}, \boldsymbol{\mu}) \in \delta(\mathbf{x}_o, \boldsymbol{\lambda}_o, \boldsymbol{\mu}_o)$, $\mathbf{x} \in \bar{\mathcal{K}}$, $\boldsymbol{\lambda} \ge \mathbf{O}$.

Additionally,

Definition 13.4: A point $(\mathbf{x}_o, \boldsymbol{\lambda}_o, \boldsymbol{\mu}_o) \in R^{2n+m+p}$ is a **global saddle point** of the Lagrangian $L(\mathbf{x}, \boldsymbol{\lambda}, \boldsymbol{\mu}) = f(\mathbf{x}) + \boldsymbol{\lambda}^T \bar{\mathbf{G}}(\mathbf{x}) + \boldsymbol{\mu}^T \mathbf{H}(\mathbf{x})$ if

$$L(\mathbf{x}, \boldsymbol{\lambda}_o, \boldsymbol{\mu}_o) \le L(\mathbf{x}_o, \boldsymbol{\lambda}_o, \boldsymbol{\mu}_o) \le L(\mathbf{x}_o, \boldsymbol{\lambda}, \boldsymbol{\mu}) \tag{13.9.1}$$

for all $(\mathbf{x}, \boldsymbol{\lambda}, \boldsymbol{\mu}) \in R^{2n+m+p}$, $\mathbf{x} \in \bar{\mathcal{K}}$, $\boldsymbol{\lambda} \ge \mathbf{O}$.

So whether we define a saddle point in either a local or global sense: $\mathbf{x}_o$ maximizes L over $\bar{\mathcal{K}}$ when $(\boldsymbol{\lambda}, \boldsymbol{\mu})$ is held fixed at $(\boldsymbol{\lambda}_o, \boldsymbol{\mu}_o)$, while $(\boldsymbol{\lambda}_o, \boldsymbol{\mu}_o)$ minimizes L over $\bar{\mathcal{K}}$ when $\mathbf{x}$ is held fixed at $\mathbf{x}_o$.

Given Equations 13.9, 13.9.1, we can now specify the

> **SADDLE-POINT PROBLEM.** To find a point $(\mathbf{x}_o, \boldsymbol{\lambda}_o, \boldsymbol{\mu}_o)$ such that (13.9) holds for all $(\mathbf{x}, \boldsymbol{\lambda}, \boldsymbol{\mu}) \in \delta(\mathbf{x}_o, \boldsymbol{\lambda}_o, \boldsymbol{\mu}_o)$ or (13.9.1) holds for all $(\mathbf{x}, \boldsymbol{\lambda}, \boldsymbol{\mu}) \in R^{2n+m+p}$, $\mathbf{x} \in \bar{\mathcal{K}}$, $\boldsymbol{\lambda} \ge \mathbf{O}$.

What are the characteristics of a solution to the saddle-point problem? The answer is provided by

Theorem 13.10: (Necessary and Sufficient Condition). The triple $(\mathbf{x}_o, \boldsymbol{\lambda}_o, \boldsymbol{\mu}_o)$, with $\mathbf{x} \in \bar{\mathcal{K}}$ and $\boldsymbol{\lambda} \ge \mathbf{O}$, is a saddle point for the Lagrangian $L(\mathbf{x}, \boldsymbol{\lambda}, \boldsymbol{\mu})$ if and only if the **saddle-point optimality conditions** hold:

$$\begin{aligned} &\text{(a)} \quad L(\mathbf{x}, \boldsymbol{\lambda}_o, \boldsymbol{\mu}_o) \text{ attains a maximum at } \mathbf{x}_o \in \bar{\mathcal{K}}; \\ &\text{(b)} \quad \bar{\mathbf{G}}(\mathbf{x}_o) \ge \mathbf{O}, \mathbf{H}(\mathbf{x}_o) = \mathbf{O}; \text{ and} \\ &\text{(c)} \quad \boldsymbol{\lambda}_o^T \bar{\mathbf{G}}(\mathbf{x}_o) = \mathbf{O}. \end{aligned} \tag{13.10}$$

Proof: (Necessity) Let $(\mathbf{x}_o, \boldsymbol{\lambda}_o, \boldsymbol{\mu}_o)$ constitute a saddle point of $L(\mathbf{x}, \boldsymbol{\lambda}, \boldsymbol{\mu})$. Then (13.9.1) holds or

$$\begin{aligned} f(\mathbf{x}) + \boldsymbol{\lambda}_o^T \bar{\mathbf{G}}(\mathbf{x}) + \boldsymbol{\mu}_o^T \mathbf{H}(\mathbf{x}) &\le f(\mathbf{x}_o) + \boldsymbol{\lambda}_o^T \bar{\mathbf{G}}(\mathbf{x}_o) + \boldsymbol{\mu}_o^T \mathbf{H}(\mathbf{x}_o) \\ &\le f(\mathbf{x}_o) + \boldsymbol{\lambda}^T \bar{\mathbf{G}}(\mathbf{x}_o) + \boldsymbol{\mu}^T \mathbf{H}(\mathbf{x}_o). \end{aligned} \tag{13.11}$$

Then the left-hand inequality in (13.11) is equivalent to (13.10a). Rearranging the second inequality in (13.11) yields

$$0 \ge (\boldsymbol{\lambda}_o - \boldsymbol{\lambda})^T \bar{\mathbf{G}}(\mathbf{x}_o) + (\boldsymbol{\mu}_o - \boldsymbol{\mu})^T \mathbf{H}(\mathbf{x}_o) \tag{13.12}$$

for all $\boldsymbol{\lambda} \in R^{n+m}$, $\boldsymbol{\lambda} \ge \mathbf{O}$, and $\boldsymbol{\mu} \in R^p$. This inequality implies that we must have $\bar{\mathbf{G}}(\mathbf{x}_o) \ge \mathbf{O}$ and $\mathbf{H}(\mathbf{x}_o) = \mathbf{O}$ since it can be violated by making a comparent of $\boldsymbol{\lambda}$ or $\boldsymbol{\mu}$ sufficiently large in magnitude. Next, upon taking $\boldsymbol{\lambda} = \mathbf{O}$ in (13.12) we have $0 \ge \boldsymbol{\lambda}_o^T \bar{\mathbf{G}}(\mathbf{x}_o)$.

However, since $\boldsymbol{\lambda} \geq \boldsymbol{O}$ and $\bar{\mathbf{G}}(\boldsymbol{x}_o) \geq \boldsymbol{O}$ imply that $\boldsymbol{\lambda}_o^T \bar{\mathbf{G}}(\boldsymbol{x}_o) \geq 0$, we must have $\boldsymbol{\lambda}_o^T \bar{\mathbf{G}}(\boldsymbol{x}_o) = 0$. Clearly conditions (13.10) hold true.

(Sufficiency.) If $\boldsymbol{x}_o$ maximizes $L(\boldsymbol{x}, \boldsymbol{\lambda}_o, \boldsymbol{\mu}_o)$, then $L(\boldsymbol{x}_o, \boldsymbol{\lambda}_o, \boldsymbol{\mu}_o) \geq L(\boldsymbol{x}, \boldsymbol{\lambda}_o, \boldsymbol{\mu}_o)$. If $\boldsymbol{\lambda}_o^T \bar{\mathbf{G}}(\boldsymbol{x}_o) = 0$ and $\boldsymbol{\mu}_o^T \boldsymbol{H}(\boldsymbol{x}_o) = 0$, then $L(\boldsymbol{x}_o, \boldsymbol{\lambda}_o, \boldsymbol{\mu}_o) = f(\boldsymbol{x}_o)$. Hence, $L(\boldsymbol{x}_o, \boldsymbol{\lambda}, \boldsymbol{\mu}) = f(\boldsymbol{x}_o) + \boldsymbol{\lambda}^T \bar{\mathbf{G}}(\boldsymbol{x}_o) + \boldsymbol{\mu}^T \boldsymbol{H}(\boldsymbol{x}_o) = L(\boldsymbol{x}_o, \boldsymbol{\lambda}_o, \boldsymbol{\mu}_o) + \boldsymbol{\lambda}^T \bar{\mathbf{G}}(\boldsymbol{x}_o) + \boldsymbol{\mu}^T \boldsymbol{H}(\boldsymbol{x}_o)$ or $L(\boldsymbol{x}_o, \boldsymbol{\lambda}, \boldsymbol{\mu}) \geq L(\boldsymbol{x}_o, \boldsymbol{\lambda}_o, \boldsymbol{\mu}_o)$ since $\boldsymbol{\lambda}^T \bar{\mathbf{G}}(\boldsymbol{x}_o) \geq 0$ if $\bar{\mathbf{G}}(\boldsymbol{x}_o) \geq \boldsymbol{O}$ and $\boldsymbol{\mu}_o^T \boldsymbol{H}(\boldsymbol{x}_o) = 0$ if $\boldsymbol{H}(\boldsymbol{x}_o) = \boldsymbol{O}$. So if Equation (13.10) holds, then $L(\boldsymbol{x}, \boldsymbol{\lambda}, \boldsymbol{\mu})$ has a saddle point at $(\boldsymbol{x}_o, \boldsymbol{\lambda}_o, \boldsymbol{\mu}_o)$. Q. E. D.

What is the connection between the solution of the maximum problem and the solution of the saddle-point problem? The answer to this question is provided by the next two theorems. First,

Theorem 13.11: (Sufficient Condition). Let $L(\boldsymbol{x}, \boldsymbol{\lambda}, \boldsymbol{\mu})$ be defined for all $\boldsymbol{x} \in R^n, \boldsymbol{\lambda} \in R^{n+m}, \boldsymbol{\lambda} \geq \boldsymbol{O}$, and $\boldsymbol{\mu} \in R^p$. If $(\boldsymbol{x}_o, \boldsymbol{\lambda}_o, \boldsymbol{\mu}_o)$ is a saddle point for $L(\boldsymbol{x}, \boldsymbol{\lambda}, \boldsymbol{\mu})$, then $\boldsymbol{x}_o$ solves the maximum problem.

Proof: Suppose $(\boldsymbol{x}_o, \boldsymbol{\lambda}_o, \boldsymbol{\mu}_o)$ is a solution to the saddle-point problem. Then (13.9) holds so that (13.11) obtains. By an argument similar to that offered in support of the *necessity* portion of Theorem 13.10 we see that we must have $\bar{\mathbf{G}}(\boldsymbol{x}_o) \geq \boldsymbol{O}$, $\boldsymbol{H}(\boldsymbol{x}_o) = \boldsymbol{O}$, $\boldsymbol{\lambda}_o^T \bar{\mathbf{G}}(\boldsymbol{x}_o) = 0$, and $\boldsymbol{\lambda} \geq \boldsymbol{O}$. Thus $\boldsymbol{x}_o$ must be a feasible solution to the maximum problem. From the second inequality in (13.11) we get, under the preceding discussion,

$$f(\boldsymbol{x}) + \boldsymbol{\lambda}_o^T \bar{\mathbf{G}}(\boldsymbol{x}) + \boldsymbol{\mu}_o^T \boldsymbol{H}(\boldsymbol{x}) \leq f(\boldsymbol{x}_o). \quad (13.13)$$

If $\boldsymbol{x}$ is a feasible solution for the maximum problem, then $f(\boldsymbol{x}_o) \geq f(\boldsymbol{x})$ so that $\boldsymbol{x}_o$ is a maximal solution. Q. E. D.

To argue in the opposite direction we need to assume that $\mathcal{D}$ is a convex set and that f and the $\hat{g}^j, j \in \mathcal{J}$, are concave with the $\hat{h}^r$, $r = 1, \ldots, p$, affine. A CQ will also be needed. To this end we have

Theorem 13.12: (Necessary Condition). Suppose $\mathcal{D}$ is a convex set and $f(\boldsymbol{x})$ and $\bar{\mathbf{G}}(\boldsymbol{x})$ are concave functions with $\boldsymbol{H}(\boldsymbol{x})$ affine. Also, suppose $\boldsymbol{O} \in \text{int}\,(\boldsymbol{H}(\bar{\mathcal{K}}))$, with $\boldsymbol{H}(\bar{\mathcal{K}}) = \{\boldsymbol{H}(\boldsymbol{x}) | \boldsymbol{x} \in \bar{\mathcal{K}}\}$, and there exists an $\tilde{\boldsymbol{x}} \in \bar{\mathcal{K}}$ with $\bar{\mathbf{G}}(\tilde{\boldsymbol{x}}) > \boldsymbol{O}$ and $\boldsymbol{H}(\tilde{\boldsymbol{x}}) = \boldsymbol{O}$ (a generalization of Slater's CQ). If $\boldsymbol{x}_o$ is an optimal solution to the maximum problem, then there exists a vector $(\boldsymbol{\lambda}_o, \boldsymbol{\mu}_o)$ with $\boldsymbol{\lambda}_o \geq \boldsymbol{O}$ such that $(\boldsymbol{x}_o, \boldsymbol{\lambda}_o, \boldsymbol{\mu}_o)$ is a saddle point of $L(\boldsymbol{x}, \boldsymbol{\lambda}, \boldsymbol{\mu})$.

A proof of this theorem will not be given since the next theorem actually demonstrates this result while establishing the relationship between the saddle point characteristics given in Theorem 13.10 and the KKT (first-order) conditions given in Chapter 12.

Theorem 13.13: (Necessary and Sufficient Condition). Let us write the maximum problem as: maximize $y = f(\boldsymbol{x})$ subject to $\boldsymbol{x} \in \bar{\mathcal{K}} \subset \mathcal{D}$. If $\boldsymbol{x}_o \in \bar{\mathcal{K}}$ satisfies the KKT conditions

$$\begin{aligned} \nabla f(\boldsymbol{x}_o) + \nabla \bar{\mathbf{G}}(\boldsymbol{x}_o)\boldsymbol{\lambda}_o + \nabla \boldsymbol{H}(\boldsymbol{x}_o)\,\boldsymbol{\mu}_o &= \boldsymbol{O} \\ \boldsymbol{\lambda}_o^T \bar{\mathbf{G}}(\boldsymbol{x}_o) = 0, \boldsymbol{\lambda}_o &\geq \boldsymbol{O}, \end{aligned} \quad (13.14)$$

$f(\boldsymbol{x})$ and $\hat{g}^j(\boldsymbol{x}), j \in \mathcal{J}$, are concave at $\boldsymbol{x}_o$, and $\hat{h}^r(\boldsymbol{x})$ is affine for $\mu_r^o \neq 0$, then $(\boldsymbol{x}_o, \boldsymbol{\lambda}_o, \boldsymbol{\mu}_o)$ is a saddle point of $L(\boldsymbol{x}, \boldsymbol{\lambda}, \boldsymbol{\mu})$.

Conversely, if $(\boldsymbol{x}_o, \boldsymbol{\lambda}_o, \boldsymbol{\mu}_o)$, with $\boldsymbol{x}_o \in \text{int}\,(\mathcal{D})$ (Slater's CQ) and $\boldsymbol{\lambda}_o \geq \boldsymbol{O}$, solves the saddle point problem, then $\boldsymbol{x}_o$ is feasible for the maximum problem and (13.14) is satisfied.

Proof: (Necessity) Let $(\boldsymbol{x}_o, \boldsymbol{\lambda}_o, \boldsymbol{\mu}_o)$, with $\boldsymbol{x} \in \bar{\mathcal{K}}$ and $\boldsymbol{\lambda}_o \geq \boldsymbol{O}$, satisfy (13.14). With f and $\hat{g}^j, j \in \mathcal{J}$, concave at $\boldsymbol{x}_o$ and with $\hat{h}^r$ affine for $\mu_r^o \neq 0$, we have, for $\boldsymbol{x} \in \bar{\mathcal{K}}$:

$$\begin{aligned} &\text{(a)} \quad f(\boldsymbol{x}) \leq f(\boldsymbol{x}_o) + \nabla f(\boldsymbol{x}_o)^T(\boldsymbol{x} - \boldsymbol{x}_o) \\ &\text{(b)} \quad \hat{g}^j(\boldsymbol{x}) \leq \hat{g}^j(\boldsymbol{x}_o) + \nabla \hat{g}^j(\boldsymbol{x}_o)(\boldsymbol{x} - \boldsymbol{x}_o), j \in \mathcal{J}, \\ &\text{(c)} \quad \hat{h}^r(\boldsymbol{x}) \leq \hat{h}^r(\boldsymbol{x}_o) + \nabla \hat{h}^r(\boldsymbol{x}_o)(\boldsymbol{x} - \boldsymbol{x}_o), \mu_r^o \neq 0, r = 1, \ldots, p. \end{aligned} \quad (13.15)$$

Now, if we add λ_j^o times (13.15b) and μ_r^o times (13.15c) to (13.15a), we obtain

$$\begin{aligned} f(\boldsymbol{x}) + \textstyle\sum_j \lambda_j^o \hat{g}^j(\boldsymbol{x}) + \sum_{r=1}^p \mu_r^o \hat{h}^r(\boldsymbol{x}) &\leq f(\boldsymbol{x}_o) + \textstyle\sum_j \lambda_j^o \hat{g}^j(\boldsymbol{x}_o) + \sum_{r=1}^p \mu_r^o \hat{h}^r(\boldsymbol{x}_o) \\ &+ \left(\nabla f(\boldsymbol{x}_o) + \textstyle\sum_j \lambda_j^o \nabla \hat{g}^j(\boldsymbol{x}_o) + \sum_{r=1}^p \mu_r^o \nabla \hat{h}^r(\boldsymbol{x}_o)\right)^T(\boldsymbol{x} - \boldsymbol{x}_o) \end{aligned}$$

or $L(\boldsymbol{x}, \boldsymbol{\lambda}_o, \boldsymbol{\mu}_o) \leq L(\boldsymbol{x}_o, \boldsymbol{\lambda}_o, \boldsymbol{\mu}_o)$. And with $\bar{\mathbf{G}}(\boldsymbol{x}_o) \geq \boldsymbol{O}$, $\boldsymbol{H}(\boldsymbol{x}_o) = \boldsymbol{O}$, and $\boldsymbol{\lambda}_o^T \bar{\mathbf{G}}(\boldsymbol{x}_o) = 0$, it follows that $L(\boldsymbol{x}_o, \boldsymbol{\lambda}_o, \boldsymbol{\mu}_o) \leq L(\boldsymbol{x}_o, \boldsymbol{\lambda}, \boldsymbol{\mu})$ for all $\boldsymbol{\lambda} \geq \boldsymbol{O}$ and $\boldsymbol{\mu}$. Thus $(\boldsymbol{x}_o, \boldsymbol{\lambda}_o, \boldsymbol{\mu}_o)$ satisfies the saddle-point condition (13.10).

(Sufficiency) Suppose $(\mathbf{x}_o, \boldsymbol{\lambda}_o, \boldsymbol{\mu}_o)$, with $\mathbf{x}_o \in \text{int}\,\bar{\mathcal{K}} \neq \emptyset$ and $\boldsymbol{\lambda}_o \geq \mathbf{O}$, is a saddle point. Then $\mathbf{x}_o$ is feasible for the maximum problem since $L(\mathbf{x}_o, \boldsymbol{\lambda}_o, \boldsymbol{\mu}_o) \leq L(\mathbf{x}_o, \boldsymbol{\lambda}, \boldsymbol{\mu})$ for all $\boldsymbol{\lambda} \geq \mathbf{O}$ and all $\boldsymbol{\mu}$ and, via Theorem 13.10, $\bar{\mathbf{G}}(\mathbf{x}_o) \geq \mathbf{O}$, $\mathbf{H}(\mathbf{x}_o) = \mathbf{O}$, and $\boldsymbol{\lambda}_o^T \bar{\mathbf{G}}(\mathbf{x}_o) = 0$. With $L(\mathbf{x}, \boldsymbol{\lambda}_o, \boldsymbol{\mu}_o) \leq L(\mathbf{x}_o, \boldsymbol{\lambda}_o, \boldsymbol{\mu}_o)$, $\mathbf{x} \in \bar{\mathcal{K}}$, $\mathbf{x}_o$ maximizes $L(\mathbf{x}, \boldsymbol{\lambda}_o, \boldsymbol{\mu}_o)$ subject to $\mathbf{x} \in \bar{\mathcal{K}}$. And with $\mathbf{x} \in \text{int}\,(\bar{\mathcal{K}})$, $\nabla_{\mathbf{x}} L(\mathbf{x}_o, \boldsymbol{\lambda}_o, \boldsymbol{\mu}_o) = \mathbf{O}$ or (13.14) obtains. Q. E. D.

13.5 LAGRANGIAN DUALITY WITH INEQUALITY CONSTRAINTS (GRAVES AND WOLFE [1963]; LASDON [1970]; GEOFFRION [1972]; MANGASARIAN [1962]; WOLFE [1961]; BAZARAA *ET AL.* [2006]; MINOUX [1986]; FIACCO AND MCCORMICK [1968])

Let us write the **primal problem** (P) as (11.1.1) or

$$(P) \quad \text{maximize } y = f(\mathbf{x}) \text{ subject to}$$
$$\mathbf{x} \in \bar{\mathcal{K}} = \{\mathbf{x} | \bar{\mathbf{G}}(\mathbf{x}) \geq \mathbf{O}, \mathbf{x} \in R^n\} \subset \mathcal{D},$$

where f and $\hat{g}^j(\mathbf{x}), j = 1, \ldots, m+n,$ are assumed defined and of class $C^{(1)}$ on an open set $\mathcal{D} \subset R^n$. Given that the **Lagrangian** associated with (P) is

$$L(\mathbf{x}, \boldsymbol{\lambda}) = f(\mathbf{x}) + \boldsymbol{\lambda}^T \bar{\mathbf{G}}(\mathbf{x}), \boldsymbol{\lambda} \geq \mathbf{O}, \boldsymbol{\lambda} \in R^{m+n}, \quad (13.16)$$

let us express the **Lagrangian dual function** as

$$g(\boldsymbol{\lambda}) = \sup_{\mathbf{x}} L(\mathbf{x}, \boldsymbol{\lambda}) = \sup_{\mathbf{x}} [f(\mathbf{x}) + \boldsymbol{\lambda}^T \bar{\mathbf{G}}(\mathbf{x})]. \quad (13.17)$$

Then we may structure the **dual problem** (D) as[3]

$$(D) \quad \text{minimize } g(\boldsymbol{\lambda}) \text{ subject to } \boldsymbol{\lambda} \geq \mathbf{O},$$

where $\boldsymbol{\lambda}$ is an $(m+n \times 1)$ vector of **dual variables**. Hence, the primal-dual pair of problems can be written more succinctly as

$$(P) \max_{\mathbf{x} \in \bar{\mathcal{K}}} f(\mathbf{x}), \quad (D) \min_{\lambda \geq O} g(\lambda).$$

Note that $g(\boldsymbol{\lambda})$ is convex in $\boldsymbol{\lambda}$ alone since it is the pointwise supremum of a collection (indexed by $\mathbf{x}$) of functions linear in $\boldsymbol{\lambda}$, and this holds without any concavity assumptions on the primal problem. Hence, the dual problem is always a convex program.

What is the relationship between the optimal primal objective function value $\left(f^o = \max_{x \in \bar{\mathcal{K}}} f(x)\right)$ and the optimal dual objective function value $\left(g^o = \min_{\lambda \geq 0} g(\lambda)\right)$? The answer to this question is provided by a couple of basic duality theorems. First,

Theorem 13.14: (**Weak Duality**). For any $\boldsymbol{\lambda} \geq \mathbf{O}$,

$$f^o \leq g(\boldsymbol{\lambda}), \quad (13.18)$$

where, as indicated above, f^o is the maximal objective function value for problem (P).

Proof: For some $\bar{\mathbf{x}} \in \bar{\mathcal{K}}$,

$$f(\bar{\mathbf{x}}) \leq f(\bar{\mathbf{x}}) + \sum_j \lambda_j \hat{g}^j(\bar{\mathbf{x}}) = L(\bar{\mathbf{x}}, \boldsymbol{\lambda})$$

since $\lambda_j \geq 0$, $\hat{g}^j(\bar{\mathbf{x}}) \geq 0$. Hence

$$L(\bar{\mathbf{x}}, \boldsymbol{\lambda}) \leq \sup_{\mathbf{x}} [f(\mathbf{x}) + \sum_j \lambda_j \hat{g}(\bar{\mathbf{x}})] = g(\boldsymbol{\lambda})$$

and thus $f(\bar{\mathbf{x}}) \leq L(\bar{\mathbf{x}}, \boldsymbol{\lambda}) \leq g(\boldsymbol{\lambda})$ for all feasible $\bar{\mathbf{x}}$ values so that $f^o \leq g(\boldsymbol{\lambda})$. Q. E. D.

As this theorem indicates, the dual objective function $g(\boldsymbol{\lambda})$, $\boldsymbol{\lambda} \geq \mathbf{O}$, *always* provides an upper bound on the primal objective $f(\mathbf{x})$. In this regard, the weak duality theorem implies that (P) is infeasible if the infimal value of (D) is $-\infty$. While each $g(\boldsymbol{\lambda})$ is an upper bound for f^o, the *best* upper bound for f^o is g^o, the optimal value of the Lagrangian dual program. Hence, $f^o \leq g^o$, the **weak duality inequality**. The difference $g^o - f^o$ will be termed the **optimal duality gap**.

The notion of **strong duality** emerges if the optimal duality gap is zero or $f^o = g^o$. It is imperative to note that strong duality *does not hold* for general non-linear programs. However, if we assume that $f(\mathbf{x})$, $\hat{g}^j(\mathbf{x}), j = 1, \ldots, m+n,$ are all concave functions defined on a convex set $\mathcal{D}$, and we also invoke a CQ (Slater's CQ, that int $(\bar{\mathcal{K}}) \neq \emptyset$ or that there exists a feasible $\bar{\mathbf{x}}$ such that $\bar{\mathbf{G}}(\bar{\mathbf{x}}) > \mathbf{O}$, does nicely), then strong duality holds. These observations ae incorporated in

Theorem 13.15: (**Strong Duality**). For problem (P):

a. let $f(\boldsymbol{x}), \hat{g}^j(\boldsymbol{x}), j = 1, \ldots, m+n,$ be concave functions on a convex set $\mathcal{D}$;

b. let Slater's CQ hold so that $f^o < +\infty$;
Then

$$\min_{\boldsymbol{\lambda} \geq \boldsymbol{O}} g(\boldsymbol{\lambda}) = \sup_{\boldsymbol{x} \in \bar{\mathcal{K}}} [f(\boldsymbol{x}) + \boldsymbol{\lambda}^T \bar{\boldsymbol{G}}(\boldsymbol{x})].$$

c. And if f^o is bounded (from above), then the optimal dual objective value g^o is attained at $\boldsymbol{\lambda}^o \geq \boldsymbol{O}$.

Proof: Let f^o be the optimal value of f in (P). From Farkas Theorem 1.8.5 we know that there exists a vector $\bar{\boldsymbol{\lambda}} \geq \boldsymbol{O}$ such that

$$L(\boldsymbol{x}, \bar{\boldsymbol{\lambda}}) = f(\boldsymbol{x}) + \sum_j \bar{\lambda}_j \hat{g}^j(\boldsymbol{x}) \leq f^o, \ \boldsymbol{x} \in \bar{\mathcal{K}},$$

or

$$\sup_{\boldsymbol{x} \in \bar{\mathcal{K}}} L(\boldsymbol{x}, \bar{\boldsymbol{\lambda}}) = g(\bar{\boldsymbol{\lambda}}) \leq f^o.$$

So from the weak duality theorem (Theorem 13.14), $g(\bar{\boldsymbol{\lambda}}) \geq g^o \geq f^o$ and thus $f^o = g(\bar{\boldsymbol{\lambda}})$ so that $\bar{\boldsymbol{\lambda}}$ is a minimizer for the dual problem (D). Q. E. D.

The implications of the strong duality theorem are the following. Under the given set of assumptions: (1) (D) has an optimal solution; (2) $g^o = f^o$ (no duality gap); (3) the optimal solution vector $\boldsymbol{\lambda}_o$ for (D) represents the optimal multiplier vector for (P); and (4) any optimal solution for (D) enables us to recover all optimal solutions for (P) as the maximizers over $\bar{\mathcal{K}}$ of the associated Lagrangian function for which the CS conditions hold, it follows that these results enable (D) to represent an appropriate substitute or surrogate problem for (P).

The final theorem of this section deals with the connection between Lagrangian saddle points and the optimal objective function values of the primal and dual problems. Specifically, we have

Theorem 13.16: [**Duality Theorem** (Necessary and Sufficient Condition)] The primal problem (P) has a saddle point $(\boldsymbol{x}_o, \boldsymbol{\lambda}_o)$ if and only if the optimal objective function value for (P) equals the optimal objective function value for the dual problem (D) or

$$g^o = g(\boldsymbol{\lambda}_o) = f(\boldsymbol{x}_o) = f^o.$$

Proof: (Necessity) With $(\boldsymbol{x}_o, \boldsymbol{\lambda}_o)$ a saddle point,

$$\begin{aligned} L(\boldsymbol{x}_o, \boldsymbol{\lambda}_o) &= f(\boldsymbol{x}_o) + \boldsymbol{\lambda}_o^T \bar{G}(\boldsymbol{x}_o) = f(\boldsymbol{x}_o) \\ &= \sup_{\boldsymbol{x}} L(\boldsymbol{x}, \boldsymbol{\lambda}_o) = g(\boldsymbol{\lambda}_o). \end{aligned}$$

By the weak duality property, we have for all $\boldsymbol{\lambda} \geq \boldsymbol{O}, f(\boldsymbol{x}_o) \leq g(\boldsymbol{\lambda})$. Hence, $f(\boldsymbol{x}_o) = g(\boldsymbol{\lambda}_o) = \min_{\boldsymbol{\lambda} \geq \boldsymbol{O}} g(\boldsymbol{\lambda})$.

(Sufficiency) Suppose there exists a solution $\boldsymbol{x}_o$ of (P) and a $\boldsymbol{\lambda}_o \geq \boldsymbol{O}$ such that $g(\boldsymbol{\lambda}_o) = f(\boldsymbol{x}_o)$. By the definition of $g(\boldsymbol{\lambda}_o)$, $g(\boldsymbol{\lambda}_o) = f(\boldsymbol{x}_o) \geq f(\boldsymbol{x}_o) + \boldsymbol{\lambda}_o^T \bar{\boldsymbol{G}}(\boldsymbol{x}_o)$ so that $\boldsymbol{\lambda}_o^T \bar{\boldsymbol{G}}(\boldsymbol{x}_o) \leq 0$. But since $\boldsymbol{\lambda}_o \geq \boldsymbol{O}$ and $\bar{\boldsymbol{G}}(\boldsymbol{x}_o) \geq \boldsymbol{O}$, it also follows that $\boldsymbol{\lambda}_o^T \bar{\boldsymbol{G}}(\boldsymbol{x}_o) \geq 0$ and thus $\boldsymbol{\lambda}_o^T \bar{\boldsymbol{G}}(\boldsymbol{x}_o) = 0$. And since $\lambda_j^o \hat{g}^j(\boldsymbol{x}_o) = 0$ for all $j = 1, \ldots, m+n$, Theorem 13.3 enables us to conclude that $(\boldsymbol{x}_o, \boldsymbol{\lambda}_o)$ is a saddle point. Q. E. D.

Example 13.2: Let us maximize $y = f(\boldsymbol{x}) = -\frac{1}{2}x_1^2 - \frac{1}{2}x_2^2$ subject to $\hat{g}(\boldsymbol{x}) = x_1 + x_2 - 1 \leq 0$, $\boldsymbol{x} \in R^2$. Forming the Lagrangian we have

$$L(\boldsymbol{x}, \lambda) = f(\boldsymbol{x}) + \lambda \hat{g}(\boldsymbol{x}) = -\frac{1}{2}x_1^2 - \frac{1}{2}x_2^2 + \lambda(x_1 + x_2 - 1).$$

Holding λ fixed (in which case L becomes a concave function of $\boldsymbol{x}$), the maximum of L with respect to $\boldsymbol{x}$ is achieved when $\nabla_x L = \boldsymbol{O}$ or when $x_1 = x_2 = \lambda$. Substituting these x_1 and x_2 values into $L(\boldsymbol{x}, \lambda)$ renders the Lagrangian dual function $g(\lambda) = \lambda^2 - \lambda$ and thus the dual problem is

$$\min_{\lambda} g(\lambda) = \min_{\lambda} \{\lambda^2 - \lambda\} \ \text{subject to} \ \lambda \geq \boldsymbol{0}.$$

Its solution is $2\lambda - 1 = 0$ or $\lambda^o = \frac{1}{2}$ with $g(\lambda^o) = -\frac{1}{4} = g^o$. And with $x_1^o = x_2^o = \frac{1}{2}$, the optimal f value is $f^o = -\frac{1}{4}$ so that, as required at an optimal primal-dual solution, $f^o = g^o$ (no duality gap). ■

Example 13.3: Let us reexamine Example 13.1 in the light of our duality considerations. In this example we specified:

$$f(\boldsymbol{x}) = -x_1^2 - x_2^2,$$
$$\hat{g}(\boldsymbol{x}) = -6 - 2x_1 - x_2 \geq 0, \quad \text{and}$$
$$L(\boldsymbol{x}, \lambda) = f(\boldsymbol{x}) + \lambda \hat{g}(\boldsymbol{x}) = -x_1^2 - x_2^2 + \lambda(-6 - 2x_1 - x_2).$$

Upon optimizing L with respect to$\boldsymbol{x}$ we obtained $x_1 = -\lambda$, $x_2 = -\frac{1}{2}\lambda$. If we substitute x_1 and x_2 into $L(\boldsymbol{x}, \boldsymbol{\lambda})$ we obtain the Lagrangian dual function $g(\lambda) = \frac{5}{4}\lambda^2 - 6\lambda$. Then the dual problem is

$$\min_{\lambda} g(\lambda) = \min_{\lambda}\left\{\frac{5}{4}\lambda^2 - 6\lambda\right\} \text{subject to } \lambda \geq \boldsymbol{0}.$$

The solution to this dual problem is $\frac{5}{2}\lambda - 6 = 0$ or $\lambda^o = \frac{12}{5}$ with $g(\lambda^o) = -\frac{36}{5} = g^o$. Hence, $x_1^o = -\frac{12}{5}$ and $x_2^o = -\frac{6}{5}$ with $f^o = -\frac{36}{5}$. Thus, as expected, $f^o = g^o$ (no duality gap). ■

A summary of some of the key features of Lagrangian saddle points and duality are provided in Table 13.1.

Section 13.5 Exercises:

13.5.1 Find the Lagrange dual problem associated with the primal problem

TABLE 13.1 Lagrangian Saddle Points and Duality

Maximum Problem	Minimum Problem
$(P) \max f(\boldsymbol{x})$ subject to $\bar{G}(\boldsymbol{x}) \geq \boldsymbol{O}$, $\boldsymbol{x} \in R^n$, or subject to $\boldsymbol{x} \in \bar{\mathcal{K}}=$ $\{\boldsymbol{x} \mid \bar{G}(\boldsymbol{x}) \geq \boldsymbol{O}, \boldsymbol{x} \in R^n\}$.	$(P) \min f(\boldsymbol{x})$subject to $\bar{G}(\boldsymbol{x}) \leq \boldsymbol{O}$, $\boldsymbol{x} \in R^n$, or subject to $\boldsymbol{x} \in \bar{\mathcal{K}}=$ $\{\boldsymbol{x} \mid \bar{G}(\boldsymbol{x}) \leq \boldsymbol{O}, \boldsymbol{x} \in R^n\}$.
Lagrangian Function $L(\boldsymbol{x}, \boldsymbol{\lambda}) = f(\boldsymbol{x}) + \boldsymbol{\lambda}^T \bar{G}(\boldsymbol{x})$, $\boldsymbol{\lambda} \in R^{m+n}$.	**Lagrangian Function** $L(\boldsymbol{x}, \boldsymbol{\lambda}) = f(\boldsymbol{x}) + \boldsymbol{\lambda}^T \bar{G}(\boldsymbol{x})$, $\boldsymbol{\lambda} \in R^{m+n}$.
Lagrange Dual Function $g(\boldsymbol{\lambda}) = \max_{\boldsymbol{x} \in \bar{\mathcal{K}}} L(\boldsymbol{x}, \boldsymbol{\lambda})$	**Lagrange Dual Function** $g(\boldsymbol{\lambda}) = \min_{\boldsymbol{x} \in \bar{\mathcal{K}}} L(\boldsymbol{x}, \boldsymbol{\lambda})$
Dual Problem $(D) \quad \min_{\lambda} g(\boldsymbol{\lambda})$ subject to $\boldsymbol{\lambda} \geq \boldsymbol{O}$.	**Dual Problem** $(D) \quad \max_{\lambda} g(\boldsymbol{\lambda})$ subject to $\boldsymbol{\lambda} \geq \boldsymbol{O}$.
Weak Duality For any $\boldsymbol{\lambda} \geq \boldsymbol{O}$ $f^o = \max_{\boldsymbol{x} \in \bar{\mathcal{K}}} f(\boldsymbol{x}) \leq g(\boldsymbol{\lambda})$.	**Weak Duality** For any $\boldsymbol{\lambda} \geq \boldsymbol{O}$ $f^o = \min_{\boldsymbol{x} \in \bar{\mathcal{K}}} f(\boldsymbol{x}) \geq g(\boldsymbol{\lambda})$.
Strong Duality Under restrictions on $\bar{\mathcal{K}}, f, \hat{g}^j, j \in \mathcal{J}$, $\min_{\boldsymbol{\lambda} \geq \boldsymbol{O}} g(\boldsymbol{\lambda}) = \sup_{\boldsymbol{x} \in \bar{\mathcal{K}}} L(\boldsymbol{x}, \boldsymbol{\lambda})$.	**Strong Duality** Under restrictions on $\bar{\mathcal{K}}, f, \hat{g}^j, j \in \mathcal{J}$, $\max_{\boldsymbol{\lambda} \geq \boldsymbol{O}} g(\boldsymbol{\lambda}) = \inf_{\boldsymbol{x} \in \bar{\mathcal{K}}} L(\boldsymbol{x}, \boldsymbol{\lambda})$.
(Local) Saddle-Point Problem Find a point $(\boldsymbol{x}_o, \boldsymbol{\lambda}_o)$ such that $L(\boldsymbol{x}, \boldsymbol{\lambda}_o) \leq L(\boldsymbol{x}_o, \boldsymbol{\lambda}_o) \leq L(\boldsymbol{x}_o, \boldsymbol{\lambda})$, $(\boldsymbol{x}, \boldsymbol{\lambda}) \in \delta(\boldsymbol{x}_o, \boldsymbol{\lambda}_o)$, $\boldsymbol{x} \in \bar{\mathcal{K}}$, $\boldsymbol{\lambda} \geq \boldsymbol{O}$.	**(Local) Saddle Point Problem** Find a point $(\boldsymbol{x}_o, \boldsymbol{\lambda}_o)$ such that $L(\boldsymbol{x}, \boldsymbol{\lambda}_o) \geq L(\boldsymbol{x}_o, \boldsymbol{\lambda}_o) \geq L(\boldsymbol{x}_o, \boldsymbol{\lambda})$, $(\boldsymbol{x}, \boldsymbol{\lambda}) \in \delta(\boldsymbol{x}_o, \boldsymbol{\lambda}_o)$, $\boldsymbol{x} \in \bar{\mathcal{K}}$, ,$\boldsymbol{\lambda} \geq \boldsymbol{O}$.
(Local) Saddle-Point Optimality $L(\boldsymbol{x}, \boldsymbol{\lambda})$ has a saddle point at $(\boldsymbol{x}_o, \boldsymbol{\lambda}_o)$, $\boldsymbol{\lambda}_o \geq \boldsymbol{O}$, if and only if: a. $L(\boldsymbol{x}, \boldsymbol{\lambda}_o)$ has a local maximum at $\boldsymbol{x}_o$, $\boldsymbol{x} \in \delta(\boldsymbol{x}_o)$; b. $\bar{G}(\boldsymbol{x}_o) \geq \boldsymbol{O}$; and c. $\boldsymbol{\lambda}_o^T \bar{G}(\boldsymbol{x}_o) = 0$.	**(Local) Saddle-Point Optimality** $L(\boldsymbol{x}, \boldsymbol{\lambda})$ has a saddle point at $(\boldsymbol{x}_o, \boldsymbol{\lambda}_o)$, $\boldsymbol{\lambda}_o \geq \boldsymbol{O}$, if and only if: a. (a) $L(\boldsymbol{x}, \boldsymbol{\lambda}_o)$ has a local minimum at $\boldsymbol{x}_o$, $\boldsymbol{x} \in \delta(\boldsymbol{x}_o)$; b. $\bar{G}(\boldsymbol{x}_o) \leq \boldsymbol{O}$; and c. $\boldsymbol{\lambda}_o^T \bar{G}(\boldsymbol{x}_o) = 0$.

Duality Theorem

Primal problem (P) has a saddle point $(\boldsymbol{x}_o, \boldsymbol{\lambda}_o)$ if and only if
$g^o = g(\boldsymbol{\lambda}_o) = f(x_o) = f^o$.

KKT Equivalence Theorem

Under restrictions on $\bar{\mathcal{K}}, f, \hat{g}^j, j \in \mathcal{J}$, and a CQ, $\boldsymbol{x}_o$ solves (P)
if and only if there exists a $\boldsymbol{\lambda}_o \geq \boldsymbol{O}$ such that $(\boldsymbol{x}_o, \boldsymbol{\lambda}_o)$ solves the saddle-point problem with
$\nabla f(\boldsymbol{x}_o) + \nabla \bar{G}(\boldsymbol{x}_o)\boldsymbol{\lambda}_o = \boldsymbol{O}$, $\bar{G}(\boldsymbol{x}_o) \geq \boldsymbol{O}$, $\boldsymbol{\lambda}_o^T \bar{G}(\boldsymbol{x}_o) = 0$.

$$\text{maximize } f(\boldsymbol{x}) = x_1x_2 \text{ subject to}$$
$$(P) \quad \hat{g}^1(\boldsymbol{x}) = 100 - x_1 - x_2 \geq 0$$
$$\hat{g}^2(\boldsymbol{x}) = 50 - x_1 \geq 0.$$

What is the optimal solution to the dual problem?

13.5.2 Determine the Lagrange dual problem associated with the primal problem

$$\text{maximize } f(\boldsymbol{x}) = -(x_1 - 6)^2 - (x_2 - 5)^2 \text{ subject to}$$
$$(P) \quad \hat{g}^1(\boldsymbol{x}) = 3 - x_1 \geq 0$$
$$\hat{g}^2(\boldsymbol{x}) = 4 - x_2 \geq 0.$$

What is the optimal solution to the dual problem?

13.3.3 Find the Lagrange dual problem given that the primal problem appears as

$$\text{maximize } f(\boldsymbol{x}) = x_1x_2 \text{ subject to}$$
$$(P) \quad \hat{g}^1(\boldsymbol{x}) = 12 - 2x_1 - x_2 \geq 0$$
$$\hat{g}^2(\boldsymbol{x}) = 15 - x_1 - 2x_2 \geq 0.$$

Determine the optimal solution to the dual problem.

13.5.4 Determine the Lagrange dual problem for the primal linear program

$$(P) \text{ maximize } f(\boldsymbol{x}) = \boldsymbol{C}^T\boldsymbol{x} \text{ subject to } \boldsymbol{Ax} \leq \boldsymbol{b}.$$

13.5.5 Determine the Lagrange dual problem for the primal quadratic program

$$(P) \text{ maximize } f(\boldsymbol{x}) = \frac{1}{2}\boldsymbol{x}^T \boldsymbol{Qx} \text{ subject to } \boldsymbol{Ax} \leq \boldsymbol{b},$$

where $\boldsymbol{Q}$ is an $(n \times n)$ symmetric negative definite matrix.

13.6 LAGRANGIAN DUALITY REVISITED

Our objective in his section is to examine some aspects of the geometry of duality theory and to relate these results to the notion of the conjugate of a particular function. To this end, let us express the primal problem as the *constrained* optimization problem

$$(P) \text{ maximize } y = f(\boldsymbol{x}) \text{ subject to } \bar{G}(\boldsymbol{x}) \geq \boldsymbol{O}, \boldsymbol{x} \in V,$$

where V is a finite dimensional normed vector space and f is assumed to be a proper closed concave function. As utilized in earlier chapters, a convenient device for eliminating the constraints is to introduce a "correction term" when the constraints are violated. This renders a new primal problem of the form

$$(P') \text{ maximize } L(\boldsymbol{x}, \boldsymbol{\lambda}) = f(\boldsymbol{x}) + \boldsymbol{\lambda}^T\bar{\boldsymbol{G}}(\boldsymbol{x}), \boldsymbol{\lambda} \geq \boldsymbol{O}.$$

For a given $\boldsymbol{\lambda} \geq \boldsymbol{O}$, the maximum value of this *unconstrained* problem is superoptimal relative to the maximum value of the constrained problem (P).

This construct introduces a new problem which will be called the dual problem: find a $\boldsymbol{\lambda} \geq \boldsymbol{O}$ so that the maximum value of the unconstrained problem (P') is as small as possible. Let us again define the **Lagrangian dual function** as

$$g(\boldsymbol{\lambda}) = \sup_{x} L(\boldsymbol{x}, \boldsymbol{\lambda}).$$

Then the dual problem is

$$(D) \text{ minimize } g(\boldsymbol{\lambda}) \text{ subject to } \boldsymbol{\lambda} \geq \boldsymbol{O}.$$

We know from our previous discussion of convexity that, for a concave function f, the hypograph of f (hyp (f)) is a convex set. Moreover, from the Representation Theorem (see Example 1.8.5), a closed convex set S can be represented as the intersection of all the closed half spaces containing it. Hence, S can be viewed in terms of its supporting hyperplanes, with the set of all supporting hyperplanes to S taken as its *dual representation*. But this means that hyp (f) can be depicted in terms of affine functions $(\boldsymbol{a}^*)^Tx - \alpha$ which are minorized by f, where the slope $\boldsymbol{a}^* \in V^*$ and V^* is taken to be the **dual space** of V. (Here V^* is the space of all continuous linear functionals on that space.)

For a given $\boldsymbol{a}^* \in V^*$, what is the "best" choice for α? From

$$f(\boldsymbol{x}) \leq (\boldsymbol{a}^*)^T\boldsymbol{x} - \alpha, \boldsymbol{x} \in V,$$

we get

$$\alpha \leq (\boldsymbol{a}^*)^T\boldsymbol{x} - f(\boldsymbol{x}), \boldsymbol{x} \in V,$$
$$\alpha \leq \boldsymbol{inf}_{x \in V} \{(\boldsymbol{a}^*)^T\boldsymbol{x} - f(\boldsymbol{x})\}$$

and the "best" choice of is

$$f^*(a^*) = \inf_{x \in V} \{(a^*)^T x - f(x)\},$$

where f^* is the (closed concave) conjugate function of f. If this infimum is finite, then $f(x) \le (a^*)^T x - f^*(a^*)$ is the "best" affine majorant of f with slope $(a^*) \in V^*$. And if f is a proper concave function, then f can be obtained from f^* as $f = f^{**}$.

Looked at in another fashion, given (P), we can obtain the Lagrangian dual function $g(\lambda)$ by perturbing (P) (see Section 13.3). That is, first set

$$g(\lambda) = \sup_x L(x, \lambda) = \sup_x \{f(x) + \lambda^T \bar{G}(x)\}.$$

Then, for $\bar{G}(x) \ge y$,

$$\begin{aligned} g(\lambda) &= \sup_y \left[\sup_x \{f(x) + \lambda^T y\} \right] = \sup_y \{\Phi(y) + \lambda^T y\} \\ &= -\inf_y \{(-\lambda)^T y - \Phi(y)\} = -\Phi^*(-\lambda), \end{aligned}$$

where $\Phi(y)$, $y \in R^{n+m}$, is the perturbation function with $\Phi(O) = \max(P)$. Thus the Lagrangian dual function can be expressed as the negative of the conjugate of the perturbations function Φ evaluated at $-\lambda$. Hence, a restatement of the Lagrangian dual problem is

$$(D')\, \underset{\lambda}{\text{minimize}} - \Phi^*(-\lambda)$$

or

$$(D'')\, \underset{\lambda}{\text{maximize}}\, \Phi^*(-\lambda).$$

Then an appropriate perturbation of the dual problem enables us to recover the primal problem (provided $\Phi^{**} = \Phi$).

We previously found that λ_o is a saddle-point multiplier if and only if $z = \Phi(O) - \lambda_o^T y$ is a supporting hyperplane at $y = O$ of the graph of $\Phi(y)$, i.e., if and only if $\Phi(y) \le \Phi(O) - \lambda_o^T y$ (since Φ is concave if (P) is). How should $\Phi(O)$ be chosen? Rewriting this inequality as

$$\begin{aligned} -\Phi(O) &\le -\lambda_o^T y - \Phi(y), \\ -\Phi(O) &\le \inf_y \{(-\lambda_o)^T y - \Phi(y)\} \end{aligned}$$

we see that the "best" choice of $-\Phi(O)$ is

$$\Phi^*(-\lambda_o) = \inf_y \{(-\lambda_o)^T y - \Phi(y)\}.$$

Thus the "best" affine majorant of Φ is $\Phi(y) \le \Phi^*(-\lambda_o) - \lambda_o^T y$.

Section 13.6 Exercises:

13.6.1 Find the Lagrangian dual problem associated with the primal problem

$$\begin{aligned} &\text{maximize } f(x) = x_1 \text{ subject to} \\ (P)\quad &\hat{g}^1(x) = 1 - x_1^2 - x_2^2 \ge 0 \\ &\hat{h}^1(x) = 1 - x_1 + x_2 = 0. \end{aligned}$$

What is the optimal solution to the dual problem?

13.6.2 Find the Lagrangian dual problem associated with the primal problem

$$(P)\, \text{maximize } f(x) = C^T x \text{ subject to } Ax = b,\, x \ge O.$$

13.6.3 Determine the Lagrangian dual problem for the primal problem

$$(P)\, \text{maximize } f(x) = x^T x \text{ subject to } Ax = b,\, x \ge O.$$

13.7 LAGRANGIAN DUALITY WITH MIXED CONSTRAINTS

Suppose $\mathcal{D} \subset R^n$ is an open set on which the functions $f, \hat{g}^j, j = 1, \ldots, m+n, \hat{h}^r, r = 1, \ldots, p,$ are defined and differentiable and that the **primal problem** (P) appears as (12.1.1) or

$$(P)\, \text{maximize } y = f(x) \text{ subject to}$$
$$x \in \bar{\mathcal{K}} = \{x | \bar{G}(x) \ge O,\, H(x) = O,\, x \in R^n\} \subset \mathcal{D}.$$

If we write the **Lagrangian** for (P) as

$$L(x, \lambda, \mu) = f(x) + \lambda^T \bar{G}(x) + \mu^T H(x),\ \lambda \ge O,\ \mu \text{ unrestricted}, \tag{13.19}$$

where $\lambda \in R^{m+n}$ and $\mu \in R^p$, then the **Lagrangian dual function** can be expressed as

$$g(\lambda, \mu) = \sup_x L(x, \lambda, \mu) = \sup_x [f(x) + \lambda^T \bar{G}(x) + \mu^T H(x)] \tag{13.20}$$

and thus we may write the **dual problem** (D) as

$$(D)\text{minimize } g(\boldsymbol{\lambda}, \boldsymbol{\mu}) \text{ subject to } \boldsymbol{\lambda} \geq \boldsymbol{O}, \quad \boldsymbol{\mu} \text{ unrestricted,}$$

where it is assumed that the minimum exists and the **dual variables** are the components of the vectors $\boldsymbol{\lambda}$ and $\boldsymbol{\mu}$. Here the primal-dual pair of problems appear as

$$(P) \max_{\boldsymbol{x} \in \bar{\mathcal{K}}} f(\boldsymbol{x}) \quad (D) \min_{\boldsymbol{\lambda} \geq \boldsymbol{O}, \boldsymbol{\mu}} g(\boldsymbol{\lambda}, \boldsymbol{\mu})$$

Let us express the optimal primal objective function value as $f^o = \max_{\boldsymbol{x} \in \bar{\mathcal{K}}} f(\boldsymbol{x})$ while the optimal dual objective function value is denoted as $g^o = \min_{\boldsymbol{\lambda} \geq \boldsymbol{O}, \boldsymbol{\mu}} g(\boldsymbol{\lambda}, \boldsymbol{\mu})$. To establish the connection between f^o and g^o, as well as that between problems (P) and (D) in general, we can state:

1. **Weak Duality**. For any $\boldsymbol{\lambda} \geq \boldsymbol{O}$ and $\boldsymbol{\mu}, f^o \leq g(\boldsymbol{\lambda}, \boldsymbol{\mu})$, i.e., the dual objective function value provides a upper bound for f^o. Moreover, the *best* upper bound for f^o is g^o. Hence, $f^o \leq g^o$ serves as the **weak duality inequality**, with the difference $g^o - f^o$ termed the **optimal duality gap**. Additionally, the weak duality notion implies that (P) is infeasible if the infimal value of (D) is $-\infty$.

2. **Strong Duality**. (a) Let $f(\boldsymbol{x}), \hat{g}^j(\boldsymbol{x}), j = 1, \ldots, m+n$, be concave functions with $\boldsymbol{H}(\boldsymbol{x})$ affine on a convex set $\mathcal{D}$; (b) let the generalized Slater's CQ hold (there is an $\bar{\boldsymbol{x}} \in \bar{\mathcal{K}}$ such that $\bar{\boldsymbol{G}}(\bar{\boldsymbol{x}}) \geq \boldsymbol{O}, \boldsymbol{H}(\bar{\boldsymbol{x}}) = \boldsymbol{O}$, and $\boldsymbol{O} \in \text{int}(\boldsymbol{H}(\bar{\mathcal{K}}))$ so that $f(\boldsymbol{x}) < +\infty$ and thus $\min_{\boldsymbol{\lambda} \geq \boldsymbol{O}, \boldsymbol{\mu}} g(\boldsymbol{\lambda}, \boldsymbol{\mu}) = \sup_{\boldsymbol{x}} [f(\boldsymbol{x}) + \boldsymbol{\lambda}^T \bar{\boldsymbol{G}}(\boldsymbol{x}) + \boldsymbol{\mu}^T \boldsymbol{H}(\boldsymbol{x})]$; and (c) if f^o is bounded from above, then the optimal dual objective function value g^o is attained at $\boldsymbol{\lambda}_o \geq \boldsymbol{O}$ and $\boldsymbol{\mu}_o$.

   So under strong duality: (D) has an optimal solution with $g^o = f^o$ (no duality gap); and $\boldsymbol{\lambda}_o \geq \boldsymbol{O}$, $\boldsymbol{\mu}_o$ represent the optimal multiplier vectors for (P) so that $\boldsymbol{\lambda}_o^T \bar{\boldsymbol{G}}(\boldsymbol{x}) = 0$. Hence, it is possible to solve problem (P) indirectly by solving (D).

3. **Duality Theorem**. The Lagrangian of the primal problem (P) has a saddle point $(\boldsymbol{x}_o, \boldsymbol{\lambda}_o, \boldsymbol{\mu}_o)$ if and only if $g^o = g(\boldsymbol{\lambda}_o, \boldsymbol{\mu}_o) = f(\boldsymbol{x}_o) = f^o$.

4. Given the preceding strong duality conclusions along with the notion that if, under certain conditions, $\boldsymbol{x}_o$ is a KKT solution to (P), then the Lagrange multipliers for this solution are **saddle-point multipliers**, and conversely. Hence, the optimal dual variables for (D) are the Lagrange multipliers for the optimal KKT solution to (P) as well as the multipliers for the saddle-point optimality conditions.

Given the primal problem (P) stated at the beginning of this section, let us embed (P) in a larger class of problems called perturbed problems. To accomplish this, let us first introduce, respectively, $(m + n \times 1)$ and $(p \times 1)$ **perturbation vectors** $\boldsymbol{y}_1^T = (y_1^1, \ldots, y_{m+n}^1)$ and $\boldsymbol{y}_2^T = (y_1^2, \ldots, y_p^2)$, with $\boldsymbol{y}^T = (\boldsymbol{y}_1^T, \boldsymbol{y}_2^T)$. Here y_j^1 is associated with the constraint function $\hat{g}^j(\boldsymbol{x}) \geq 0$, $j = 1, \ldots, m+n$, and y_r^2 is associated with the constraint function $\hat{h}^r(\boldsymbol{x}) = 0$, $r = 1, \ldots, p$. So given (P), the family of **perturbed problems** (P') appears as

$$(P') \text{ maximize } y = f(\boldsymbol{x}) \text{ subject to}$$
$$\boldsymbol{x} \in \bar{\mathcal{K}}_y = \{\boldsymbol{x} | \bar{\boldsymbol{G}}(\boldsymbol{x}) \geq \boldsymbol{y}_1, \boldsymbol{H}(\boldsymbol{x}) = \boldsymbol{y}_2, \boldsymbol{x} \in \bar{\mathcal{K}}\}.$$

For $\boldsymbol{y} \in R^{m+n+p}$, the optimal value of this perturbed problem (P') is a function $\Phi(\mathrm{y})$, termed the **perturbation function**; it expresses the optimal value of a program as a function of perturbations of its right-hand side. For $\boldsymbol{y} = \boldsymbol{O}$, we obtain problem (P) with $\Phi(\boldsymbol{O}) = \max(P)$.

Suppose that $\boldsymbol{x}_o$ is an optimal solution to (P) with $f(\boldsymbol{x}) < +\infty$. Then it can be shown that $(\boldsymbol{x}_o, \boldsymbol{\lambda}_o, \boldsymbol{\mu}_o)$ is a saddle point of $L(\boldsymbol{x}, \boldsymbol{\lambda}, \boldsymbol{\mu})$ ((13.19)), with $\boldsymbol{\lambda}_o$ and $\boldsymbol{\mu}_o$ serving as saddle point multipliers, if and only if the hyperplane $z = \Phi(\boldsymbol{O}) - (\boldsymbol{\lambda}_o^T, \boldsymbol{\mu}_o^T)\boldsymbol{y}, \boldsymbol{y} \in R^{m+n+p}$, is a supporting hyperplane at $\boldsymbol{y} = \boldsymbol{O}$ of the graph of $\Phi(\boldsymbol{y})$, i.e., if and only if $\Phi(\boldsymbol{y}) \leq \Phi(\boldsymbol{O}) - (\boldsymbol{\lambda}_o^T, \boldsymbol{\mu}_o^T)\boldsymbol{y}$. So if (P) has an optimal solution, the existence of a saddle-point solution and thus the absence of a duality gap, is equivalent to the existence of a supporting hyperplane for the graph of $\Phi(\boldsymbol{y})$ at the optimal point $(\boldsymbol{O}, \Phi(\boldsymbol{O}))$.

Next, suppose (P) is a concave program (i.e., $\mathcal{D}$ is a convex set, f and $\hat{g}^j, j = 1, \ldots, m+n$, are concave functions and $\hat{h}^r, r = 1, \ldots, p$, is affine). Then Φ is concave and thus the Lagrange multipliers are subgradients of the perturbation function $\Phi(\boldsymbol{y})$ at the point of no perturbation. (Remember that $\Phi(\boldsymbol{y})$ has a subgradient $\gamma^T = (\gamma_1^T, \gamma_2^T)$ at $y = O$ which satisfies $\Phi(y) \le \Phi(O) + \gamma^T y$, $y \in R^{m+n+p}$.) If we set $\lambda_o = -\gamma_1$ and $\mu_o = -\gamma_2$. then the existence of a saddle point is confirmed.

One final point is in order – a **post-optimal analysis** of problem (P). Let us consider the behavior of the perturbation function $\Phi(y)$ for small perturbations in the right-hand side of the constraint functions $\hat{g}^j(x) \ge y_j^1$, $j = 1, \ldots, m+n$, and $\hat{h}^r(x) = y_r^2$, $r = 1, \ldots, p$, near $y = O$. In this regard, suppose $\Phi(y)$ is of class $C^{(1)}$ at $y = O$ and (P) has a saddle point (x_o, λ_o, μ_o). The supporting hyperplane to the graph of $\Phi(y)$ at the point $(O, \Phi(O))$ is the tangent hyperplane

$$z = \Phi(O) + (\nabla_y \Phi(O))^T (y - O).$$

Then $(\lambda_o^T, \mu_o^T) = -(\nabla_y \Phi(O))^T$ or

$$\lambda_j^o = -\frac{\partial \Phi(O)}{\partial y_j^1}, \; j = 1, \ldots, m+n;$$

$$\mu_r^o = -\frac{\partial \Phi(O)}{\partial y_r^2}, \; r = 1, \ldots, p.$$

Hence, the optimal saddle-point multipliers are the negatives of the **marginal rates of change** in the optimal objective function value for (P) with respect to perturbations in the right-hand sides of the constraints.

13.8 CONSTRAINED OUTPUT MAXIMIZATION: A LAGRANGIAN DUAL APPROACH

Suppose we frame our **constrained optimization problem** as

$$\underset{l,k}{\text{maximize}}\; q = F(l, k) = l^{1/5}k^{3/5} \text{ subject to}$$
$$(l, k) \in \mathcal{K} = \{(l, k) | C - wl - rk \ge 0, (l, k) \in R^2\}.$$

Then the *Lagrangian function* is

$$L(l, k, \lambda) = l^{1/5}k^{3/5} + \lambda(C - wl - rk).$$

To form the Lagrangian dual function let us maximize L with respect to l, k, and λ. (Note that since the objective or production function exponents on l and k sum to 4/5 < 1, the Lagrangian function is concave in l, k.)

From the first-order conditions for a maximum we get

$$\begin{aligned} &\text{(a)} \quad L_l = \tfrac{1}{5}l^{-4/5}k^{3/5} - w\lambda = 0 \\ &\text{(b)} \quad L_k = \tfrac{3}{5}l^{1/5}k^{-2/5} - r\lambda = 0 \\ &\text{(c)} \quad L_\lambda = C - wl - rk = 0 \end{aligned} \qquad (13.21)$$

(under the assumption that l and k are positive and the cost constraint is binding). Solving Equations (13.21a, b) for l and k we obtain, respectively

$$\begin{aligned} l &= \left(\tfrac{5}{3}r\lambda\right)^{-3}(5w\lambda)^{-2} \\ k &= \left(\tfrac{5}{3}r\lambda\right)^{-4}(5w\lambda)^{-1}. \end{aligned} \qquad (13.22)$$

Given l and k from system (13.22), we may form the *Lagrangian dual function* as

$$\begin{aligned} g(\lambda) &= \max_{(l,k)\in\mathcal{K}} L(l, k, \lambda) \\ &= \tfrac{1}{5}\left(\tfrac{5}{3}r\right)^{-3}(5w)^{-1}\lambda^{-4} + C\lambda. \end{aligned}$$

Then the dual problem appears as

$$\underset{\lambda}{\text{minimize}}\; g(\lambda) \text{ subject to } \lambda \ge 0.$$

Upon setting $g'(\lambda = 0)$ and solving for λ (assuming $\lambda > 0$) we get

$$\lambda^o = \left(\frac{4}{5}\frac{1}{C}\right)^{1/5}\left(\frac{5}{3}r\right)^{-3/5}(5w)^{-1/5}. \qquad (13.23)$$

If we now substitute λ^o into system (13.22) we obtain

$$\begin{aligned} l^o &= \tfrac{1}{4}Cw^{-1} \\ k^o &= \tfrac{3}{4}Cr^{-1}. \end{aligned} \qquad (13.24)$$

Finally, as expected,

$$\begin{aligned} F^o &= F(l^o, k^o) = g^o = g(\lambda^o) \\ &= 0.6376w^{-1/5}r^{-3/5}C^{4/5} \end{aligned}$$

so that there is no duality gap.

NOTES

1 To see this let f, G be concave. (Note: $G = b - g$ will be concave if g is convex.) Then, for $x \in \mathcal{K}$, $x \ge O$,

$$f(\boldsymbol{x}) \le f(\boldsymbol{x}_o) + \nabla f(\boldsymbol{x}_o)^T(\boldsymbol{x} - \boldsymbol{x}_o)$$
$$\boldsymbol{G}(\boldsymbol{x}) \le \boldsymbol{G}(\boldsymbol{x}_o) + \nabla \boldsymbol{G}(\boldsymbol{x}_o)^T(\boldsymbol{x} - \boldsymbol{x}_o).$$

For $\lambda_o \ge \boldsymbol{O}$, $\boldsymbol{\lambda}_o^T\boldsymbol{G}(\boldsymbol{x}) \le \boldsymbol{\lambda}_o^T\boldsymbol{G}(\boldsymbol{x}_o) + \boldsymbol{\lambda}_o^T\nabla\boldsymbol{G}(\boldsymbol{x}_o)^T(\boldsymbol{x} - \boldsymbol{x}_o)$ and thus

$$\begin{aligned} L(\boldsymbol{x}, \boldsymbol{\lambda}_o) &= f(\boldsymbol{x}) + \boldsymbol{\lambda}_o^T\boldsymbol{G}(\boldsymbol{x}) \le \nabla f(\boldsymbol{x}_o) + \boldsymbol{\lambda}_o^T\boldsymbol{G}(\boldsymbol{x}_o) + \nabla f(\boldsymbol{x}_o)^T(\boldsymbol{x} - \boldsymbol{x}_o) \\ &\quad + \boldsymbol{\lambda}_o^T\nabla\boldsymbol{G}(\boldsymbol{x}_o)^T(\boldsymbol{x} - \boldsymbol{x}_o) \\ &= L(\boldsymbol{x}_o, \boldsymbol{\lambda}_o) + (\nabla f(\boldsymbol{x}_o) + \nabla\boldsymbol{G}(\boldsymbol{x}_o)\boldsymbol{\lambda}_o)^T(\boldsymbol{x} - \boldsymbol{x}_o) \\ &= L(\boldsymbol{x}_o, \boldsymbol{\lambda}_o) + (\nabla_x L^o)^T(\boldsymbol{x} - \boldsymbol{x}_o). \end{aligned}$$

Hence, (13.5a). Since $L = f + \boldsymbol{\lambda}^T\boldsymbol{G}$ is linear in $\boldsymbol{\lambda}$, $L(\boldsymbol{x}_o, \boldsymbol{\lambda}) = L(\boldsymbol{x}_o, \boldsymbol{\lambda}_o) + (\nabla_\lambda L^o)^T(\boldsymbol{\lambda} - \boldsymbol{\lambda}_o)$ and thus (13.5b) holds as well.

2 For a proof see Uzawa (1958, pp. 34-45) or Künzi et al (1966, pp. 63–66).

3 We are implicitly assuming that the minimum exists.

CHAPTER 14

Generalized Concave Optimization

14.1 INTRODUCTION

Chapter 9 introduced two classes of functions that, among others, serve as a generalization of concave (convex) functions, namely those characterized as quasiconcave (quasiconvex) and pseudoconcave (pseudoconvex). Our objective in this chapter is to utilize, say, the aforementioned generalized concave functions to develop optimality criteria for the following nonlinear programming problem. That is, given the class of $C^{(1)}$ functions $y = f(\boldsymbol{x})$: $\mathcal{D} \subset R^n \rightarrow R$, $\hat{g}^j(\boldsymbol{x})$: $\mathcal{D} \subset R^n \rightarrow R$, $j = 1, \ldots, m$, with $\mathcal{D}$ an open set, let us

$$\text{maximize } y = f(\boldsymbol{x}) \text{ subject to}$$
$$\boldsymbol{x} \in \bar{K} = \{\boldsymbol{x} | \bar{\boldsymbol{G}}(\boldsymbol{x}) \geq \boldsymbol{O}, \boldsymbol{x} \in \mathcal{D}\},$$

where

$$\bar{\boldsymbol{G}}(\boldsymbol{x}) = \begin{bmatrix} \hat{g}^1(\boldsymbol{x}) \\ \vdots \\ \hat{g}^m(\boldsymbol{x}) \\ \hat{g}^{m+1}(\boldsymbol{x}) \\ \vdots \\ \hat{g}^{m+n}(\boldsymbol{x}) \end{bmatrix} \tag{14.1}$$

is an $(m + n \times 1)$ vector-valued function, $\hat{g}^j(\boldsymbol{x}) = b_j - g^j(\boldsymbol{x}) \geq 0$, $j = 1, \ldots, m$, and $\hat{g}^{m+i}(\boldsymbol{x}) = x_i \geq 0$, $i = 1, \ldots, n$.

To put our current discussion in perspective, let us briefly review some of our previous KKT optimality conditions. As established earlier, these criteria are either necessary or sufficient or both necessary and sufficient for the attainment of a constrained maximum. In particular, for the sufficient optimality conditions to hold, we need the assumptions of *differentability and concavity.* For our necessary optimality conditions, we need *differentiability and a constraint qualification* (CQ)[1] (concavity is not needed to specify necessary optimality criteria).

We start with a necessary condition.

Theorem 14.1: (KKT Necessary Optimality Condition). Let $y = f(\boldsymbol{x})$ and the vector-valued function $\bar{\boldsymbol{G}}(\boldsymbol{x})$, $\boldsymbol{x} \in R^n$, be of class $C^{(1)}$ throughout some spherical δ – neighborhood of the point $\boldsymbol{x}_o \in (\bar{\mathcal{K}})$. If $\boldsymbol{x}_o$ is a local optimal solution to Problem (14.1), then, subject to the FM constraint qualification, there exist scalars $\lambda_1^o, \ldots, \lambda_{m+n}^o$ such that

$$\begin{aligned} &\nabla f(\boldsymbol{x}_o) + \nabla \bar{\boldsymbol{G}}(\boldsymbol{x}_o)\boldsymbol{\lambda}_o = \boldsymbol{O} \\ &\boldsymbol{\lambda}_o^T \bar{\boldsymbol{G}}(\boldsymbol{x}_o) = 0 \\ &\bar{\boldsymbol{G}}(\boldsymbol{x}_o) \geq \boldsymbol{O} \\ &\boldsymbol{\lambda}_o \geq \boldsymbol{O} \\ &\boldsymbol{x}_o \text{ unrestricted,} \end{aligned} \tag{14.2}$$

where

$$\nabla \bar{\boldsymbol{G}}(\boldsymbol{x}_o) = [\nabla \hat{g}^1(\boldsymbol{x}_o), \ldots, \nabla \hat{g}^{m+n}(\boldsymbol{x}_o)].$$

We next have a sufficient condition.

Theorem 14.2: (KKT Sufficient Optimality Condition). Let $y = f(\boldsymbol{x})$ and the vector-valued function $\bar{\boldsymbol{G}}(\boldsymbol{x})$, $\boldsymbol{x} \in R^n$, be of class $C^{(1)}$ throughout a spherical

δ−neighborhood of the point $\boldsymbol{x}_o \in \bar{\mathcal{K}}$ and concave at $\boldsymbol{x}_o$. If (14.2) holds, then Problem (14.1) has a local optimal solution at $\boldsymbol{x}_o$.

Finally, a necessary and sufficient condition is offered. That is, if a CQ is in effect and f and $\bar{\boldsymbol{G}}$ are of class $C^{(1)}$ and concave, then (14.2) depicts a necessary and sufficient condition for a local optimal solution. Specifically,

Theorem 14.3: (KKT Necessary and Sufficient Optimality Condition). Let $y = f(\boldsymbol{x})$ and the vector-valued function $\bar{\boldsymbol{G}}(\boldsymbol{x})$, $\boldsymbol{x} \in R^n$, be of class $C^{(1)}$ and concave over a convex region $\bar{\mathcal{K}}$, with int $(\bar{\mathcal{K}}) \neq \emptyset$. Then Problem (14.1) admits a (strong) local, and thus global, optimal solution if and only if (14.2) holds.

These theorems now provide us with a convenient point of reference when we next look to their subsequent generalization.

Section 14.1 Exercise:

14.1 For a quasiconcave program, can any of the following expressions serve as the objective function? (Hint: suppose $f(\boldsymbol{x})$: $\mathcal{D} \subset R^n_+ \to R$, $\mathcal{D}$ open and convex, is twice differentiable on $\mathcal{D}$. Then f is quasiconcave on $\mathcal{D}$ if $(-1)|\bar{\boldsymbol{H}}_r| > 0$, $r = 1, \ldots, n$, where

$$|\bar{\boldsymbol{H}}_r| = \begin{vmatrix} 0 & f_1 & \cdots & f_r \\ f_1 & f_{11} & \cdots & f_{1r} \\ \vdots & \vdots & & \vdots \\ f_r & f_{r1} & \cdots & f_{rr} \end{vmatrix}.)$$

a. $f(\boldsymbol{x}) = -(x_1 + 2)^2 - (x_2 + 2)^2$, $\boldsymbol{x} \in R^2_+$.

b. $f(\boldsymbol{x}) = x_2(x_1^2 + 3)$, $\boldsymbol{x} \in R^2_+$.

c. $f(\boldsymbol{x}) = 6x_1^{-2}x_2$, $\boldsymbol{x} \in R^2_+$.

d. $f(\boldsymbol{x}) = x_1 x_2$, $\boldsymbol{x} \in R^2_+$.

14.2 QUASICONCAVE PROGRAMMING

The pioneering work involving a generalization of concave programming was undertaken by Arrow and Enthoven (AE) [1961]. Their study relaxed the concavity assumption by requiring the objective function $f(\boldsymbol{x})$ and the constraint functions $\hat{g}^j(\boldsymbol{x}) \geq 0$, $j = 1, \ldots, m$, to be quasiconcave. For the sake of completeness, let us reproduce their theorem.

AE start with the problem

$$\begin{aligned} &\text{maximize } f(\boldsymbol{x}) \text{ subject to} \\ \boldsymbol{x} \in \mathcal{K} = &\{\boldsymbol{x} | \boldsymbol{G}(\boldsymbol{x}) \geq \boldsymbol{O}, \boldsymbol{x} \geq \boldsymbol{O}\}, \\ &\text{where } \boldsymbol{G}(\boldsymbol{x})^T = [\hat{g}^1(\boldsymbol{x}), \ldots, \hat{g}^m(\boldsymbol{x})]. \end{aligned} \tag{P}$$

In addition, let us write the KKT conditions for an optimal solution to (P) at $\boldsymbol{x}_o$, $\boldsymbol{\lambda}_o$ as

$$\begin{aligned} &\nabla f(\boldsymbol{x}_o) + \nabla \boldsymbol{G}(\boldsymbol{x}_o)\boldsymbol{\lambda}_o \leq \boldsymbol{O} \\ &\boldsymbol{x}_o^T(\nabla f(\boldsymbol{x}_o) + \nabla \boldsymbol{G}(\boldsymbol{x}_o)\boldsymbol{\lambda}_o) = 0 \\ &\boldsymbol{\lambda}_o^T \boldsymbol{G}(\boldsymbol{x}_o) = 0, \ \boldsymbol{\lambda}_o \geq \boldsymbol{O}, \ \boldsymbol{x}_o \geq \boldsymbol{O}. \end{aligned} \tag{KKT}$$

We then can state

Theorem 14.4: (AE Sufficient Optimality Condition). Let $f(\boldsymbol{x})$ be a $C^{(1)}$ quasiconcave function of the n-dimensional vector $\boldsymbol{x}$, and let $\boldsymbol{G}(\boldsymbol{x})$ be an m-dimensional vector-valued $C^{(1)}$ quasiconcave function, both defined for $\boldsymbol{x} \geq \boldsymbol{O}$. Let $\boldsymbol{x}_o$, $\boldsymbol{\lambda}_o$ satisfy (KKT), and let one of the following conditions be satisfied:

a. $\partial(\boldsymbol{x}_o)/\partial x_i < 0$ for some $i \in \{1, \ldots, n\}$;

b. $\partial(\boldsymbol{x}_o)/\partial x_i > 0$ for some **relevant variable**x_i that can take on a positive value without violating the constraints;

c. $\nabla f(\boldsymbol{x}_o) \neq \boldsymbol{O}$ and f is of class $C^{(2)}$ near $\boldsymbol{x}_o$; and

d. $f(\boldsymbol{x})$ is concave.

Then $\boldsymbol{x}_o$ solves Problem (P).

(Note that from (a) and (b) it follows that $\nabla f(\boldsymbol{x}_o) \neq \boldsymbol{O}$ is sufficient if all variables x_i, $i = 1, \ldots, n$, are relevant.) Clearly this maximization problem under quasiconcavity is a bit less restrictive than the one considered above (Theorem 14.2) under concavity.

A mild modification of the preceding theorem is also valid, i.e., if f and $\hat{g}^j$, $j = 1, \ldots, m$, are of class $C^{(1)}$ and quasiconcave for $\boldsymbol{x} \geq \boldsymbol{O}$ and (a), (b), and (d) hold, then (KKT) is sufficient for a maximal solution to Problem (P).

AE also offer a second theorem which provides a CQ test in the presence of quasiconcave constraints. Specifically,

Theorem 14.5: (AE Generalized CQ). Let $\boldsymbol{G}(\boldsymbol{x})$ be an m-dimensional $C^{(1)}$and quasiconcave vector function. Let $\boldsymbol{G}(\bar{\boldsymbol{x}}) > \boldsymbol{O}$ for some $\bar{\boldsymbol{x}} \geq \boldsymbol{O}$ (Slater's CQ), and, for each j, let either:

i. $\hat{g}^j(\boldsymbol{x})$ is concave; or

ii. for each $\boldsymbol{x} \in \mathcal{K}$, the partial derivatives of $\hat{g}^j(\boldsymbol{x}), j = 1, \ldots, m$, are not all zero hold.

Note that if the $\hat{g}^j$ constraints are all linear, then this CQ is automatically satisfied.

Now, if the hypotheses of Theorems 14.4, 14.5 are satisfied, then (KKT) is both necessary and sufficient for a constrained optimal solution to Problem (P).

While the proof of the AE Theorem 14.4 is somewhat involved, let us offer a slight variation of their result.

Theorem 14.6: (Sufficient Optimality Condition). Suppose $f(\boldsymbol{x})$ and $\bar{\boldsymbol{G}}(\boldsymbol{x})$ are $C^{(1)}$ quasiconcave functions. If there exist vectors $\boldsymbol{x}_o, \boldsymbol{\lambda}_o$ such that $\boldsymbol{x}_o$ is feasible and (14.2) holds with $\nabla f(\boldsymbol{x}_o) \neq \boldsymbol{O}$, then $\boldsymbol{x}_o, \boldsymbol{\lambda}_o$ are optimal for Problem (14.1).

Proof: Suppose that $f(\boldsymbol{x}) > f(\boldsymbol{x}_o)$ for all $\boldsymbol{x} \in \bar{\mathcal{K}}$. Then it must be true that $\nabla f(\boldsymbol{x}_o)^T(\boldsymbol{x} - \boldsymbol{x}_o) > 0$, i.e., the directional derivative of f at $\boldsymbol{x}_o$ must increase locally in the $\boldsymbol{x} - \boldsymbol{x}_o$ direction. To see this, for a sufficiently small movement (θ) from $\boldsymbol{x}$ in the direction of maximum decrease in f at $\boldsymbol{x}_o$, $-\nabla f(\boldsymbol{x}_o)$, we have $f(\boldsymbol{x} - \theta \nabla f(\boldsymbol{x}_o)) \geq f(\boldsymbol{x}_o)$ and thus $\nabla f(\boldsymbol{x}_o)^T(\boldsymbol{x} - \theta \nabla f(\boldsymbol{x}_o) - \boldsymbol{x}_o) \geq 0$ or

$$\nabla f(\boldsymbol{x}_o)^T(\boldsymbol{x} - \boldsymbol{x}_o) - \nabla f(\boldsymbol{x}_o)^T \theta \nabla f(\boldsymbol{x}_o) \geq 0$$

and thus

$$\nabla f(\boldsymbol{x}_o)^T(\boldsymbol{x} - \boldsymbol{x}_o) \geq \theta \nabla f(\boldsymbol{x}_o)^T \nabla f(\boldsymbol{x}_o) > 0$$

since $\nabla f(\boldsymbol{x}_o) \neq \boldsymbol{O}$.

Let $\boldsymbol{x} \in \bar{\mathcal{K}}$. If $j \in \mathcal{J}$, then $\lambda_j^o \hat{g}^j(\boldsymbol{x}) \geq \lambda_j^o \hat{g}^j(\boldsymbol{x}_o)$ or $-\lambda_j^o \hat{g}^j(\boldsymbol{x}) \leq -\lambda_j^o \hat{g}^j(\boldsymbol{x}_o)$. If $j \notin \mathcal{J}$, this last inequality also holds for $\lambda_j^o = 0$. Since each $-\lambda_j^o \hat{g}^j(\boldsymbol{x})$ is quasiconvex, we have $-\boldsymbol{\lambda}_o^T \nabla \bar{\boldsymbol{G}}(\boldsymbol{x}_o)^T(\boldsymbol{x} - \boldsymbol{x}_o) \leq 0$ or, from (14.2), $\nabla f(\boldsymbol{x}_o)^T(\boldsymbol{x} - \boldsymbol{x}_o) \leq 0$. But by virtue of our discussion in the first paragraph of this proof, it follows that $f(\boldsymbol{x}) - f(\boldsymbol{x}_o) \leq 0$ or f has a constrained maximum at $\boldsymbol{x}_o$. Q. E. D.

Note that if f is strictly quasiconcave and the $\hat{g}^j$ are all quasiconcave, then $(\boldsymbol{x}_o, \boldsymbol{\lambda}_o)$ is the unique optimal solution to Problem 14.1.

Section 14.2 Exercise:

14.2 Demonstrate that for the program maximize $f(\boldsymbol{x}) = 10 - (x_1 - 2)^2 - (x_2 - 4)^2$ subject to $\hat{g}^1(\boldsymbol{x}) = 2 - x_1^2 - x_2^2, \boldsymbol{x} \geq \boldsymbol{O}$, Theorems 14.6, 14.7 are applicable. Find the optimal solution. (Hint: suppose $f(\boldsymbol{x})$: $D \subset R^n \rightarrow R$, D open and convex, is twice differentiable on D. Then f is strictly concave (and thus concave) on D if $(-1)|H_{f,r}| > 0$, $r = 1, \ldots, n$, where

$$|\bar{\boldsymbol{H}}_{f,r}| = \begin{vmatrix} f_{11} & \cdots & f_{1r} \\ f_{21} & \cdots & f_{2r} \\ \vdots & & \vdots \\ f_{r1} & \cdots & f_{rr} \end{vmatrix}.)$$

14.3 EXTENSIONS OF QUASICONCAVE PROGRAMMING

We noted earlier (Chapter 11) that the KKT conditions (Equations 11.6, 11.7) are sufficient for a global maximum if $f(\boldsymbol{x})$ and the constraint function $\hat{g}^j(\boldsymbol{x}) \geq 0, j = 1, \ldots, m + n$, are all concave over a convex region $\bar{\mathcal{K}}$. A slight generalization of this result is provided by

Theorem 14.7: (Sufficient Condition). Let $f(\boldsymbol{x})$ be concave and let $\bar{\boldsymbol{G}}(\boldsymbol{x})$ be quasiconcave over $\bar{\mathcal{K}}$. If $\boldsymbol{x}_o \in \bar{\mathcal{K}}$ and $\boldsymbol{x}_o, \boldsymbol{\lambda}_o$ satisfy the KKT conditions (14.2), then $\boldsymbol{x}_o$ is an optimal solution to problem (14.1).

Proof: With f concave,

$$\begin{aligned} f(\boldsymbol{x}) - f(\boldsymbol{x}_o) &\leq \nabla f(\boldsymbol{x}_o)^T(\boldsymbol{x} - \boldsymbol{x}_o) \\ &= -\boldsymbol{\lambda}_o^T \nabla \bar{\boldsymbol{G}}(\boldsymbol{x}_o)^T(\boldsymbol{x} - \boldsymbol{x}_o) \end{aligned}$$

for all feasible $\boldsymbol{x}$. We must demonstrate that for all such $\boldsymbol{x}$, if $-\boldsymbol{\lambda}_o^T \bar{\boldsymbol{G}}(\boldsymbol{x}_o)^T(\boldsymbol{x} - \boldsymbol{x}_o) \leq 0$, then $f(\boldsymbol{x}) - f(\boldsymbol{x}_o) \leq 0$. This inequality holds for all $j \notin \mathcal{J}$ since then $\lambda_j^o = 0$. And for those $j \in \mathcal{J}$, we have, for feasible $\boldsymbol{x}$, $\hat{g}^j(\boldsymbol{x}) \geq \hat{g}^j(\boldsymbol{x}_o)$. Hence, $-\lambda_j^o \hat{g}^j(\boldsymbol{x}) \leq -\lambda_j^o \hat{g}^j(\boldsymbol{x}_o)$. Since $-\lambda_j^o \hat{g}^j(\boldsymbol{x})$ is quasiconvex, it follows that $-\boldsymbol{\lambda}_o^T \nabla \bar{\boldsymbol{G}}(\boldsymbol{x}_o)^T(\boldsymbol{x} - \boldsymbol{x}_o) \leq 0$ and thus

$\nabla f(\boldsymbol{x}_o)^T(\boldsymbol{x}-\boldsymbol{x}_o) \leq 0$ so that $f(\boldsymbol{x}) - f(\boldsymbol{x}_o) \leq 0$, i.e., f has a (global) constrained maximum at $\boldsymbol{x}_o$. Q. E. D.

Another generalization of the KKT sufficient optimality conditions (which is also more comprehensive than the sufficiency results of AE) can be offered by replacing the concavity assumptions concerning f and $\bar{\boldsymbol{G}}$ by the less restrictive assumptions of pseudoconcavity and quasiconcavity, respectively. To this end we have

Theorem 14.8: (Sufficient Optimality Condition). Let $f(\boldsymbol{x})$ and the vector-valued function $\bar{\boldsymbol{G}}(\boldsymbol{x})$, $\boldsymbol{x} \in R^n$, be of class $C^{(1)}$on the convex set $\bar{\mathcal{K}}$. Suppose $f(\boldsymbol{x})$ is pseudoconcave and $\bar{\boldsymbol{G}}(\boldsymbol{x})$ is quasiconcave on $\bar{\mathcal{K}}$. If there exist vectors $\boldsymbol{x}_o \in R^n$ and $\boldsymbol{\lambda}_o \in R^{m+n}$ such that $\boldsymbol{x}_o$, $\boldsymbol{\lambda}_o$ satisfy (14.2), then $\boldsymbol{x}_o$ is an optimal solution to Problem (14.1).

Proof: With each $\hat{g}^j(\boldsymbol{x})$ quasiconcave, the level sets $U(\hat{g}^j) = \{\boldsymbol{x}|\hat{g}^j(\boldsymbol{x}) \geq 0, j = 1,\ldots,m+n, \boldsymbol{x} \in D\}$ are convex sets. And since $\cap_{j=1}^{m+n} U(\hat{g}^j)$ is convex, it follows that $\bar{\mathcal{K}}$ is a convex set. Let $\mathcal{J} = \{j|\hat{g}^j(\boldsymbol{x}) = 0\}$ be the index set of active constraints at $\boldsymbol{x}_o$. For $\boldsymbol{x}(\neq \boldsymbol{x}_o) \in \bar{\mathcal{K}}$, the convex combination $\theta\boldsymbol{x} + (1-\theta)\boldsymbol{x}_o = \boldsymbol{x}_o + \theta(\boldsymbol{x}-\boldsymbol{x}_o) \in \bar{\mathcal{K}}$, $\theta \in (0, 1)$. For $j \in \mathcal{J}$,

$$\hat{g}^j(\boldsymbol{x}_o + \theta(\boldsymbol{x}-\boldsymbol{x}_o)) \geq 0 = \hat{g}^j(\boldsymbol{x}_o),\ \theta \in (0, 1).$$

Since the value of $\hat{g}^j(\bullet)$ does not decrease by moving a bit in the feasible direction $\boldsymbol{x}-\boldsymbol{x}_o$, we must have $\nabla\hat{g}^j(\boldsymbol{x}_o)^T(\boldsymbol{x}-\boldsymbol{x}_o) \geq 0$, $j \in \mathcal{J}$, or $\nabla\bar{\boldsymbol{G}}(\boldsymbol{x}_o)^T(\boldsymbol{x}-\boldsymbol{x}_o) \geq \boldsymbol{O}$ for all j. Thus, from (14.2),

$$\nabla f(\boldsymbol{x}_o)^T(\boldsymbol{x}-\boldsymbol{x}_o) = -[\nabla\bar{\boldsymbol{G}}(\boldsymbol{x}_o)\boldsymbol{\lambda}_o]^T(\boldsymbol{x}-\boldsymbol{x}_o) \leq 0.$$

With f pseudoconcave, the preceding inequality implies $f(\boldsymbol{x}) \leq f(\boldsymbol{x}_o)$, $\boldsymbol{x} \in \bar{\mathcal{K}}$. Thus, f attains a constrained (global) maximum at $\boldsymbol{x}_o$. Q. E. D.

It is instructive to note that we can state a generalized necessary optimality theorem by *softening* Slater's CQ ($\bar{\boldsymbol{G}}(\boldsymbol{x})$ is convex and int $(\bar{\mathcal{K}}) \neq \varnothing$) by replacing it by

> **Slater's Weak CQ:** $\bar{\boldsymbol{G}}(\boldsymbol{x})$ satisfies Slater's weak CQ at $\bar{\boldsymbol{x}} \in \bar{\mathcal{K}}$ if $\bar{\boldsymbol{G}}(\boldsymbol{x})$is of class $C^{(1)}$ at $\bar{\boldsymbol{x}}$, $\bar{\boldsymbol{G}}(\boldsymbol{x})^T = [\hat{g}^1(\boldsymbol{x}), \ldots, \hat{g}^{m+n}(\boldsymbol{x})]$ is pseudoconcave at $\bar{\boldsymbol{x}}$ for $j \in \mathcal{J}$, and there exists an $\hat{\boldsymbol{x}} \in \bar{\mathcal{K}}$ such that $\bar{\boldsymbol{G}}(\hat{\boldsymbol{x}}) > \boldsymbol{O}$ for $j \in \mathcal{J}$.

We can now state

Theorem 14.9: (Generalized Necessary Optimality Condition). Let $f(\boldsymbol{x})$ and $\bar{\boldsymbol{G}}(\boldsymbol{x})$ be defined on $\bar{\mathcal{K}}$ with $f(\boldsymbol{x})$, $\bar{\boldsymbol{G}}(\boldsymbol{x})$ of class $C^{(1)}$at $\boldsymbol{x}_o \in \bar{\mathcal{K}}$. If $\boldsymbol{x}_o$ solves Problem (14.1) and $\bar{\boldsymbol{G}}(\boldsymbol{x})$ satisfies Slater's weak CQ at $\boldsymbol{x}_o$, then there exists a vector $\boldsymbol{\lambda}_o \in R^{m+n}$ such that $\boldsymbol{x}_o$, $\boldsymbol{\lambda}_o$ satisfy (14.2).

Section 14.3 Exercise:

14.3 Is the following problem a quasiconcave program? If so, determine its solution.

$$\text{maximize } f(\boldsymbol{x}) = x_1x_2 \text{ subject to}$$
$$\hat{g}^1(\boldsymbol{x}) = 1 - \tfrac{1}{3}x_1^3 - x_2 \geq 0,\ \boldsymbol{x} \geq \boldsymbol{O}.$$

Does the AE generalized CQ hold for this program?

14.4 EXTENSIONS OF QUASICONCAVE PROGRAMMING TO MIXED CONSTRAINTS

Our objective in this section is to employ generalized concave functions to develop optimization criteria for the following program. Given the $C^{(1)}$ functions $y = f(\boldsymbol{x})$: $\mathcal{D} \subset R^n \to R$, $\hat{g}^j(\boldsymbol{x})$: $\mathcal{D} \subset R^n \to R$, $j = 1, \ldots, m$, $\hat{h}^r(\boldsymbol{x}) = c_r - h^r(\mathrm{x})$: $\mathcal{D} \subset R^n \to R$, $r = 1, \ldots, p$, and the n non-negativity conditions $\boldsymbol{x} \geq \boldsymbol{O}$, with $\mathcal{D}$ an open set, let us

$$\text{maximize } y = f(\boldsymbol{x}) \text{ subject to}$$
$$\boldsymbol{x} \in \bar{\mathcal{K}} = \{\boldsymbol{x}|\bar{\boldsymbol{G}}(\boldsymbol{x}) \geq \boldsymbol{O}, \boldsymbol{H}(\boldsymbol{x}) = \boldsymbol{O}, \boldsymbol{x} \in \mathcal{D}\},$$

where

$$\bar{\boldsymbol{G}}(\boldsymbol{x}) = \begin{bmatrix} \hat{g}^1(\boldsymbol{x}) \\ \vdots \\ \hat{g}^{m+n}(\boldsymbol{x}) \end{bmatrix}, \boldsymbol{H}(\boldsymbol{x}) = \begin{bmatrix} \hat{h}^1(\boldsymbol{x}) \\ \vdots \\ \hat{h}^p(\boldsymbol{x}) \end{bmatrix}, \quad (14.1.1)$$

and the $\hat{h}^r(\boldsymbol{x})$, $r = 1, \ldots, p$, are taken to be affine linear constraints.

Given this optimization problem, remember that the KKT conditions developed in Chapter 12 appear as

$$
\begin{aligned}
&\nabla f(x_o) + \nabla \bar{G}(x_o)\lambda_o + \nabla H(x_o)\mu_o = O \\
&\lambda_o^T \bar{G}(x_o) = 0 \\
&\bar{G}(x_o) \geq O \\
&H(x_o) = O \\
&\lambda_o \geq O \\
&\mu_o \text{ unrestricted} \\
&x_o \text{ unrestricted.}
\end{aligned}
\tag{14.3}
$$

Consider the following generalized optimization theorem.

Theorem 14.10: (Generalized Sufficient Optimality Condition). Let $y = f(x)$ and the vector-valued functions $\bar{G}(x)$ and $H(x)$, $x \in R^n$, be of class $C^{(1)}$ on the convex set $\bar{\mathcal{K}}$. Suppose $f(x)$ is pseudoconcave, $\bar{G}(x)$ is quasiconcave, and $H(x)$ is affine linear on $\bar{\mathcal{K}}$. If there exists a vector $x_o \in R^n$ and multipliers $\lambda_o \in R^{m+n}$, $\mu_o \in R^p$ such that (x_o, λ_o, μ_o) satisfies (14.3), then x_o is a (global) optimal solution to Problem (14.1.1).

Proof: With each $\hat{g}^j(x) \geq 0$, the level sets $U(\hat{g}^j) = \{x | \hat{g}^j(x) \geq 0, j = 1,\ldots,m+n, x \in \mathcal{D}\}$ are convex sets. In addition, with $\hat{h}^r(x) = 0$, $r = 1, \ldots, p$, affine linear, the sets $S(\hat{h}^r) = \{x | \hat{h}^r(x) = 0, r = 1,\ldots,p, x \in \mathcal{D}\}$ are convex. And since $(\cap_{j=1}^{m+n} U(\hat{g}^j)) \cap (\cap_{r=1}^{p} S(\hat{h}^r))$ is also convex, it follows that $\bar{\mathcal{K}}$ is a convex set.

Let $J = \{j | \hat{g}^j(x) = 0\}$ represent the index set of binding constraints at x_o. For $x (\neq x_o) \in \bar{\mathcal{K}}$, the convex combination $\theta x + (1-\theta)x_o = x_o + \theta(x - x_o) \in \bar{\mathcal{K}}$, $\theta \in (0, 1)$. For $j \in \mathcal{J}$,

$$\hat{g}^j(x_o + \theta(x - x_o)) \geq 0 = \hat{g}^j(x_o), \ \theta \in (0, 1).$$

Since the value of $\hat{g}^j(x_o + \theta(x - x_o))$ does not decrease by moving slightly in the feasible direction $x - x_o$, we must have

$$\nabla \hat{g}^j(x_o)^T(x - x_o) \geq 0, \ j \in \mathcal{J},$$

or $\nabla \bar{G}(x_o)^T(x - x_o) \geq O$ for all j. In like fashion,

$$\hat{h}^r(x_o + \theta(x - x_o)) = 0, \ r = 1, \ldots, p,$$

and thus $\nabla \hat{h}^r(x_o)^T(x - x_o) = 0$ or $\nabla H(x_o)^T(x - x_o) = O$. Then, from (14.3)

$$\nabla f(x_o)^T(x - x_o) = -[\nabla \bar{G}(x_o)\lambda_o + \nabla H(x_o)\mu_o]^T(x - x_o) \leq 0.$$

With f pseudoconcave, the preceding inequality implies $f(x) \leq f(x_o)$, $x \in \bar{\mathcal{K}}$. Thus, f attains a constrained (global) maximum at x_o. Q. E. D.

Section 14.4 Exercise:

14.4 Is the following problem a quasiconcave program? If so, determine its solution.

$$
\begin{aligned}
\text{maximize } & f(x) = x_1 x_2 + 4x_1 + x_2 \text{ subject to} \\
& \hat{g}^1(x) = 4 - x_2^2 - x_2^2 \geq 0 \\
& \hat{h}^1(x) = 2 - x_1 - x_2 = 0 \\
& x \geq O.
\end{aligned}
$$

NOTE

1 Remember that our necessary condition requires the satisfaction of a CQ. For our purposes we have been using either of two CQs: (a) the Fiacco-McCormick CQ – the gradient vectors of the binding constraints at the optimum point must be linearly independent, and (b) Slater's CQ – the feasible region $\bar{\mathcal{K}}$ has a nonempty interior. However, other regularity conditions may be utilized, such as: (c) the $\hat{g}^j s$ are linear; or (d) the $\hat{g}^j s$ are concave, among others.

CHAPTER 15

Homogeneous, Homothetic, and Almost Homogeneous Functions

15.1 HOMOGENEITY DEFINED

In Chapter 1 we specified a (convex) cone as a set $\mathcal{D} \subset R^n$ such that, if $\boldsymbol{x} \in \mathcal{D}$, then $t\boldsymbol{x} \in \mathcal{D}$, $t > 0$. In what follows the functions under discussion are all defined on a cone in R^n. In this regard, a function $y = f(\boldsymbol{x})$, $\boldsymbol{x} \in R^n$, defined on a cone $\mathcal{D}$ is homogeneous of degree $\lambda \geq 0$ if, when each of its arguments is multiplied by a scalar $t > 0$, the value of the function itself is multiplied by t^λ. More formally,

Definition 15.1: Let the function $y = f(\boldsymbol{x})$: $\mathcal{D} \subset R^n \to R$ be defined on a convex cone $\mathcal{D}$. Then f is **homogeneous of degree λ** if $f(t\boldsymbol{x}) \equiv t^\lambda f(\boldsymbol{x})$ or

$$f(tx_1, \ldots, tx_n) \equiv t^\lambda f(x_1, \ldots, x_n),\ t > 0. \tag{15.1}$$

("$\equiv$" means that t changes the value of the function by an amount t^λ *identically*. Thus (15.1) holds for *all* points $\boldsymbol{x}$ and all $t > 0$ – we do not have an *equation* that holds for a limited number of points.)

If $\lambda = 1$, then f is said to be **linearly homogeneous** (or homogeneous of degree one), i.e., if all variables are increased by a factor t, f increases by the same factor or $f(t\boldsymbol{x}) = t\ f(\boldsymbol{x})$; and if $\lambda = 0$ (homogeneity of degree zero), the function value is unchanged under a proportionate change in all variables or $f(t\boldsymbol{x}) = f(\boldsymbol{x})$. For instance, let $q = F(l, k) = Al^\alpha k^\beta$, $A > 0$ and $A \in R$, for $\mathcal{D} = \{(l, k) | l, k > 0\}$. Here F describes the **Cobb-Douglas production function**, where q represents output, A depicts neutral technical change, and l, k are, respectively, labor and capital inputs. Then

$$F(tl, tk) = A(tl)^\alpha (tk)^\beta = t^{\alpha+\beta}(Al^\alpha k^\beta)$$

and thus F is homogeneous of degree $\lambda = \alpha + \beta$ in l and k.

Section 15.1 Exercises:

15.1.1 Demonstrate that the Cobb-Douglas function $y = f(\boldsymbol{x}) = 10x_1^{\frac{1}{4}}x_2^{\frac{1}{2}}$ is homogeneous of degree $\lambda = 3/4$.

15.1.2 Demonstrate that the constant elasticity of substitution (CES) function

$$y = f(x) = A(\delta_1 x_1^{-\rho} + \delta_2 x_2^{-\rho})^{v/\rho}$$

is homogeneous of degree $v > 0$, where: A is constant; the substitution parameter $\rho \neq 0$ and $\rho > -1$; and the distribution parameters $\delta_i > 0$, $i = 1, 2$, and $\delta_1 + \delta_2 = 1$.

15.1.3 Determine the degree of homogeneity of the following functions.

a. $f = 2x^2$

b. $f = 2x_1^3 x_2$

c. $f = 4x_1^2 x_2^3 x_3^4$

d. $f = x_1^3 + x_1^2 x_2 - x_1 x_2^2$

e. $f = x_1^2 x_2 + 4x_1^2 x_2^3$

f. $f = 4x_1 + 6x_2 - 5x_3^2$

g. $f = x_1^{-2}x_2^4 + x_1^{3/2}x_2^{1/2}$

h. $f = \frac{x_1^4 - x_1x_2^3}{x_1^2 + x_1x_2 - x_1^7x_2^{-5}}$

15.2 PROPERTIES OF HOMOGENEOUS FUNCTIONS

We now look to the following sequence of theorems.

Theorem 15.1: Suppose $y = f(\boldsymbol{x}): \mathcal{D} \subset R^n \to R$, $\mathcal{D}$ open, is of class $C^{(1)}$. If f is homogeneous of degree λ, then the first partial derivatives of f, f_i, $i = 1, \ldots, n$, are homogeneous of degree $\lambda - 1$.
Proof: Differentiating both sides of (15.1) with respect to x_i yields

$$t\, f_i(tx_1, \ldots, tx_n) = t^{\lambda} f_i(x_1, \ldots, x_n)$$

or

$$f_i(tx_1, \ldots, tx_n) = t^{\lambda-1} f_i(x_1, \ldots, x_n). \quad \text{Q. E. D.} \quad (15.2)$$

So if $q = F(l, k) = Al^{\alpha}k^{\beta}$, we know that $F(tl, tk) = t^{\alpha+\beta}Al^{\alpha}k^{\beta}$ and thus $F_l = \alpha A l^{\alpha-1}k^{\beta}$. Then

$$\begin{aligned} F_l(tl, tk) &= \alpha A(tl)^{\alpha-1}(tk)^{\beta} = \alpha t^{\alpha+\beta-1}Al^{\alpha-1}k^{\beta} \\ &= t^{\alpha+\beta-1}F_l(l, k). \end{aligned}$$

Hence, the *marginal product of labor*, F_l, is homogeneous of degree $\lambda - 1 = \alpha + \beta - 1$.

Theorem 15.2: Any function $y = f(\boldsymbol{x}): \mathcal{D} \subset R^n \to R$ homogeneous of degree zero is expressible as a function of the ratio between the independent variables and any one such variable.
Proof: For f homogeneous of degree zero we have $f(tx_1, tx_2, \ldots, tx_n) = f(x_1, x_2, \ldots, x_n)$.

Set, say, $t = 1/x_1$. Then

$$f\left(1, \frac{x_2}{x_1}, \ldots, \frac{x_n}{x_1}\right) = f(x_1, x_2, \ldots, x_n). \text{ Q. E. D.}$$

Theorem 15.3: (Euler's Theorem). Suppose $y = f(\boldsymbol{x}): \mathcal{D} \subset R^n \to R$, $\mathcal{D}$ open, is of class $C^{(1)}$. If f is homogeneous of degree λ, then

$$\textstyle\sum_{i=1}^{n} f_i x_i = \nabla f^T \boldsymbol{x} \equiv \lambda f(x_1, \ldots, x_n). \qquad (15.3)$$

Proof: Differentiate both sides of (15.1) with respect to t. Then

$$f_1 x_1 + f_2 x_2 + \cdots + f_n x_n = \lambda t^{\lambda-1} f(x_1, x_2, \ldots, x_n).$$

If we set $t = 1$ in the preceding expression, then (15.3) obtains. Q. E. D.

Example 15.1: For the Cobb-Douglas production function $q = F(l, k) = Al^{\alpha}k^{\beta}$, with $\alpha + \beta = 1$ (homogeneity of degree 1), the factor marginal products are, respectively,

$$\begin{aligned} F_l &= \alpha A l^{\alpha-1}k^{\beta} \\ F_k &= \beta A l^{\alpha}k^{\beta-1}. \end{aligned}$$

Then, via Euler's Theorem,

$$F_l l + F_k k = \alpha A l^{\alpha}k^{\beta} + \beta A l^{\alpha}k^{\beta} = (\alpha + \beta)Al^{\alpha}k^{\beta} = q$$

and thus

$$\frac{F_l l}{q} + \frac{F_k k}{q} = S_l + S_k = 1,$$

where $S_l = F_l l/q$ is the share of labor in total product and $S_k = F_k k/q$ is the share of capital in total product. So, under linear homogeneity of the production function, Euler's Theorem informs us that if the factors of production l and k are paid their marginal products, then *total product is exhausted* by the factors of production.

More generally, under homogeneity of degree 1, $F(tl, tk) = tF(l, k)$ so that, for $t = 1/l$,

$$F\left(1, \frac{k}{l}\right) = \frac{1}{l}F(l, k)$$

or

$$F(l, k) = lF(1, x) = lf(x),$$

where $x = k/l$ can be thought of as *capital intensity*. Then

$$\begin{aligned} F_l &= f(x) + lf'(x)(dx/dl) \\ &= f(x) - xf'(x); \\ F_k &= lf'(x)(dx/dk) \\ &= f'(x). \end{aligned}$$

So under homogeneity of degree 1, the factor marginal products F_l, F_k depend upon capital intensity $x = k/l$. What is their degree of homogeneity? Also, do the factor average products $AP_l = q/l, AP_k = q/k$ depend exclusively on x?

Theorem 15.4: Let the real-valued function f be multiplicatively separable, i.e., $f(\mathbf{x}, \mathbf{y}) = g(\mathbf{x})\cdot h(\mathbf{y})$. Then f is homogeneous of degree λ in $(\mathbf{x}, \mathbf{y})$ if and only if $g(\mathbf{x})$ is homogeneous of degree p and $h(\mathbf{y})$ is homogeneous of degree $\lambda - p$.

Proof: Let

$$g(t\mathbf{x}) = t^p g(\mathbf{x}),$$
$$h(t\mathbf{y}) = t^{\lambda-p} h(\mathbf{y}).$$

Then

$$\begin{aligned} f(t\mathbf{x}, t\mathbf{y}) &= g(t\mathbf{x})\cdot h(t\mathbf{y}) \\ &= t^p g(\mathbf{x})\cdot t^{\lambda-p} h(\mathbf{y}) \\ &= t^\lambda g(\mathbf{x})\cdot h(\mathbf{y}) = t^\lambda f(\mathbf{x}, \mathbf{y}). \quad \text{Q.E.D} \end{aligned}$$

Theorem 15.5: Let $y = f(\mathbf{x}): \mathcal{D} \subset R^n \to R$ and let $G(G' \neq 0)$ be a monotonic transformation of f yielding a new function $z = G(y) = G[f(\mathbf{x})] = h(\mathbf{x})$. If f is homogeneous of degree λ and h is homogeneous of degree s, then G must be of the form

$$G(y) = cy^{s/\lambda},\ c = \text{constant}.$$

(This is the only functional form which will transform one homogeneous function into another.)

For instance, let $y = f(\mathbf{x}) = 2x_1^{1/3}x_2^{1/3}$. Then

$$f(t\mathbf{x}) = 2(tx_1)^{1/3}(tx_2)^{1/3} = 2t^{\frac{2}{3}}x_1^{\frac{1}{3}}x_2^{\frac{1}{3}} = t^{\frac{2}{3}}f(\mathbf{x})$$

so that f is homogeneous of degree $\lambda = 2/3$. Let $G(y) = 4x_1^{2/3}x_2^{2/3} = h(\mathbf{x})$. Then

$$h(t\mathbf{x}) = 4(tx_1)^{2/3}(tx_2)^{2/3} = 4t^{\frac{4}{3}}x_1^{\frac{2}{3}}x_2^{\frac{2}{3}} = t^{\frac{4}{3}}h(\mathbf{x}).$$

Clearly $h(\mathbf{x})$ is homogeneous of degree $s = 4/3$ and thus

$$G(y) = y^{s/\lambda} = y^{(4/3)/(2/3)} = y^2.$$

We close this section with one final theorem. A function $f(x): R \to R$ is homogeneous of degree λ if and only if it is of the form $f(x) = Ax^\lambda$. What is the degree of homogeneity of the function $f(x_1, x_2, x_3) = Ax_1^{\lambda_1}x_2^{\lambda_2}x_3^{\lambda_3}$? The answer $(\lambda = \lambda_1 + \lambda_2 + \lambda_3)$ is provided by

Theorem 15.6: Suppose $y = f^j(\mathbf{x}): \mathcal{D} \subset R^n \to R$, $\mathcal{D}$ a convex cone, are, respectively, homogeneous functions of degree λ_j, $j = 1, \ldots, m$. Then $g(\mathbf{x}) = \Pi_{j=1}^m f^j(\mathbf{x})$ is homogeneous of degree $\lambda = \Sigma_{j=1}^m \lambda_j$.If $\lambda_j = \lambda$ for all $j = 1, \ldots, m$, then $h(\mathbf{x}) = (\Sigma_{j=1}^m f^j(\mathbf{x}))^\alpha$ is homogeneous of degree $\lambda\alpha$.

Section 15.2 Exercises:

15.2.1

a. For the function given in Exercise 15.1.1, demonstrate that its first partial derivative $\partial f/\partial x_1$ is homogeneous of degree $\lambda - 1$.

b. What is the degree of homogeneity of $\partial^2 f/\partial x_1^2$?

c. What is the degree of homogeneity of $\partial^n f/\partial x_1^n$?

15.2.2 The following theorem demonstrates how an arbitrary function f can be taken to be the *restriction of a homogeneous function F* that is defined on a higher dimensional space.

Theorem. Let the function $y = f(\mathbf{x}): \mathcal{D} \subset R^n \to R$ be defined on a convex cone $\mathcal{D}$. Let F be function of the $n + 1$ arguments $\mathbf{x} \in R^n$ and $z \in R_+$. Then

$$F(\mathbf{x}, z) = z^\lambda f\left(\frac{\mathbf{x}}{z}\right), \qquad (*)$$

where F is homogeneous of degree $\lambda > 0$ on the convex cone $\mathcal{D} \times R_+ \subset R^{n+1}$. Hence, $f(\mathbf{x}) = F(\mathbf{x}, 1)$, $\mathbf{x} \in \mathcal{D}$, is the restriction of F to the subset $\mathcal{D}$ of R^{n+1}.

a. Verify (*).

b. Verify that the converse of the preceding theorem is also true, i.e., if $F(\mathbf{x}, z)$ is homogeneous of degree $\lambda > 0$ on the set $\mathcal{D} \times R_+ \subset R^{n+1}$ with $f(\mathbf{x}) = F(\mathbf{x}, 1)$ for all $\mathbf{x} \in \mathcal{D}$, then $F(\mathbf{x}, z) = z^\lambda f\left(\frac{\mathbf{x}}{z}\right)$ for all $(\mathbf{x}, z) \in \mathcal{D} \times R_+$.

c. Suppose $f(x) = 2x^{1/3}$, $x \in R_+$. Find its homogenization $F(x, z)$ of degree $\lambda = 1$.

d. Suppose $f(x) = x + 3x^2$. Find its homogenization $F(x, z)$ of degree $\lambda = 1$.

e. Let $f(x) = 4x^2 - 6x$. Find its homogenization $F(x, z)$ of degree $\lambda = 2$.

f. Let $f(x) = 3x_1^2 + x_1x_2^3$. Find its homogenization $F(x, z)$ of degree $\lambda = 1$.

g. Suppose $y = F(x_1, x_2, x_3)$ is homogeneous of degree 1. Find the restriction f of F to the subset $\mathcal{D}$ (a convex cone) of R^2.

15.2.3 Given $f(x_1, x_2, x_3) = Ax_1^{\theta_1}x_2^{\theta_2}x_3^{\theta_3}$, determine its degree of homogeneity and verify that Euler's theorem holds.

15.2.4 Let $g(x) = 4x^{\theta_1}$ and $h(y) = 2y^{\theta_2}$. If the function $f(x, y) = g{\cdot}h$ is to be homogeneous of degree $\lambda = \frac{1}{4}$, what must θ_2 be equal to?

15.2.5 Let $y = f(x_1, x_2) = x_1^{2/3}x_2$ with $h(x_1, x_2) = x_1^{4/3}x_2^2$. Use Theorem 15.5 to determine $G(y)$.

15.3 HOMOTHETIC FUNCTIONS

We start with

Definition 15.2: A function $h(\boldsymbol{x})$: $R_+^n \to R$ is **homothetic** in $\boldsymbol{x}$ if it can be expressed as a monotonic increasing transformation of a homogeneous function, i.e., if there is a monotonic transformation $G(z)$: $R_+ \to R$, $G' > 0$, and a homogeneous function $f(\boldsymbol{x})$: $R_+^n \to R$ of any degree λ in $\boldsymbol{x}$ such that $h(\boldsymbol{x}) = G(f(\boldsymbol{x}))$.

(Without loss of generality, we can always restrict f to being homogeneous of degree 1, i.e., if f is homogeneous of degree λ in $\boldsymbol{x}$, then $g(\boldsymbol{x}) = f^{1/\lambda}$ is homogeneous of degree 1 and thus we may define H as a transformation of a linearly homogeneous function g by $G[f(\boldsymbol{x})] \equiv H[g(\boldsymbol{x})] \equiv H[f(\boldsymbol{x})^{1/\lambda}]$.)

It should be evident that homotheticity is a more general condition than homogeneity, i.e., every homogeneous function is homothetic, but not conversely. Homotheticity is a weaker condition.

It is important to note that *not all* monotonic transformations preserve the homogeneity property of a function. For instance, suppose $y = f(\boldsymbol{x}) = x_1^{\alpha}x_2^{1-\alpha}$. While $g = \ln y$ is homothetic, g is not homogeneous. However, any monotonic transformation of a homothetic function is homothetic.

We next consider a few important theorems concerning homothetic functions.

Theorem 15.7 Let $y = h(\boldsymbol{x})$: $R^n \to R$ be a homothetic function. Then $h(\boldsymbol{x}_1) = h(\boldsymbol{x}_2)$ implies that $h(t\boldsymbol{x}_1) = h(t\boldsymbol{x}_2)$, $t > 0$.

Proof: If h is homothetic, then $h = G[f(\boldsymbol{x})]$, where $G' > 0$ and f is linearly homogeneous. Thus $h(\boldsymbol{x}_1) = h(\boldsymbol{x}_2)$ if and only if $f(\boldsymbol{x}_1) = f(\boldsymbol{x}_2)$. So if $h(\boldsymbol{x}_1) = h(\boldsymbol{x}_2)$, then, via the linear homogeneity of f, we have $f(t\boldsymbol{x}_1) = tf(\boldsymbol{x}_1) = tf(\boldsymbol{x}_2) = f(t\boldsymbol{x}_2)$ so that $h(t\boldsymbol{x}_1) = G[f(t\boldsymbol{x}_1)] = G[f(t\boldsymbol{x}_2)] = h(t\boldsymbol{x}_2)$. Q. E. D.

We next use Theorem 15.1 to demonstrate that the gradient of a homothetic function varies proportionately along a ray from the origin. Specifically,

Theorem 15.8: Let $h(\boldsymbol{x})$: $R^n \to R$ be homothetic and of class $C^{(1)}$. Then for $\boldsymbol{x} \in R^n$ and $t > 0$, there is a scalar $k > 0$ such that

$$\nabla h(t\boldsymbol{x}) = k\nabla h(\boldsymbol{x}). \tag{15.4}$$

Proof: If h is homothetic, then $h(\boldsymbol{x}) = G[f(\boldsymbol{x})]$, $G' > 0$. For $\boldsymbol{x} \in R^n$ and $\lambda > 0$,

$$\frac{dh(\boldsymbol{x})}{d\boldsymbol{x}} = G'[f(\boldsymbol{x})]\frac{df(\boldsymbol{x})}{d\boldsymbol{x}} \text{ and } \frac{dh(t\boldsymbol{x})}{d\boldsymbol{x}} = G'[f(t\boldsymbol{x})]\frac{df(t\boldsymbol{x})}{d\boldsymbol{x}}.$$

If f is homogeneous of degree 1, then each f_i is homogeneous of degree 0 so that $df(t\boldsymbol{x})/d\boldsymbol{x} = df(\boldsymbol{x})/d\boldsymbol{x}$ and

$$\frac{dh(t\boldsymbol{x})}{d\boldsymbol{x}} = G'[f(t\boldsymbol{x})]\frac{df(t\boldsymbol{x})}{d\boldsymbol{x}} = G'[f(t\boldsymbol{x})]\frac{df(\boldsymbol{x})}{d\boldsymbol{x}}$$
$$= \left(\frac{G'[f(t\boldsymbol{x})]}{G'[f(\boldsymbol{x})]}\right)G'[f(\boldsymbol{x})]\frac{df(\boldsymbol{x})}{d\boldsymbol{x}} = \left(\frac{G'[f(t\boldsymbol{x})]}{G'[f(\boldsymbol{x})]}\right)\frac{dh(\boldsymbol{x})}{d\boldsymbol{x}}$$

or

$$\nabla h(t\boldsymbol{x}) = k\nabla h(\boldsymbol{x}),$$

where obviously $k = (G'[f(t\boldsymbol{x})]/G'[f(\boldsymbol{x})])$. Q.E.D.

Let us consider again the production function $q = F(l, k,)$ where F_l, F_k are, respectively, the marginal

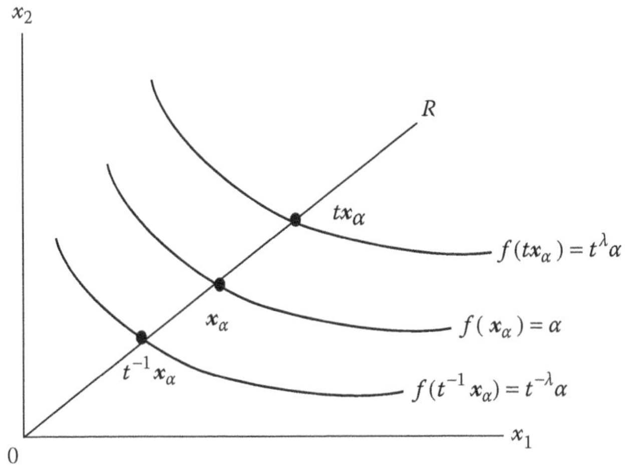

FIGURE 15.1 Level curves of a function homogeneous of degree λ.

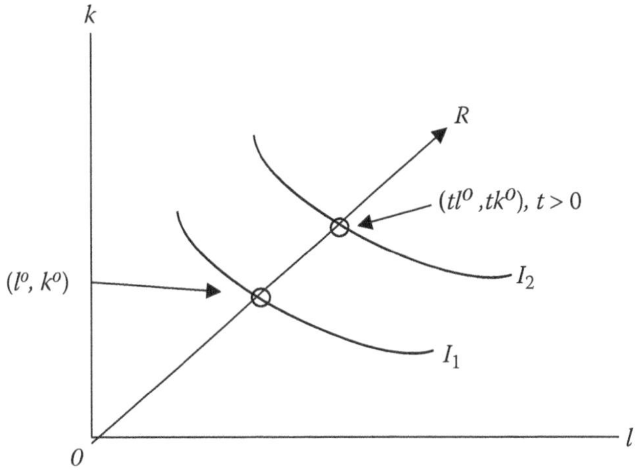

FIGURE 15.2 Isoquant I_2 is a radial expansion of isoquant I_1 and (tl^o, tk^o) is a radial translation of (l^o, k^o).

products of labor and capital. These factor marginal products are homogeneous of degree zero if F is homogeneous of degree 1, i.e., increasing each input by a factor t (we move along a ray OR from the origin) leaves factor marginal products unchanged at every point along OR. Moreover, homogeneity of any degree implies that:

a. the level curves (e.g., isoquants or indifference curves) are radial expansions and contractions of each other (Figure 15.1). To see this, remember that the *upper level set* of a function $f(\boldsymbol{x})$, $\boldsymbol{x} \in R^2$, is defined as $U_f(\alpha) = \{\boldsymbol{x} | f(\boldsymbol{x}) \geq \alpha, \boldsymbol{x} \in \mathcal{D}, \alpha \in R\}$ while the α *level curve* is $I(\alpha) = \{\boldsymbol{x} | f(\boldsymbol{x}) = \alpha, \boldsymbol{x} \in \mathcal{D}, \alpha \in R\}$ - - it is the locus of all points $\boldsymbol{x}$ such that $f(\boldsymbol{x}) = \alpha$. Now, suppose f is homogeneous of degree λ, with $\boldsymbol{x}_\alpha$ a point on level curve $I(\alpha) = f(\boldsymbol{x}_\alpha) = \alpha$, with ray OR passing through $\boldsymbol{x}_\alpha$. From $\boldsymbol{x}_\alpha$, let us move to the point $t\boldsymbol{x}_\alpha$, $t > 0$, along OR to the higher level $f(t\boldsymbol{x}_\alpha) = t^\lambda f(\boldsymbol{x}_\alpha) = t^\lambda \alpha$. Under a contraction from $\boldsymbol{x}_\alpha$ along OR, we move to the point $t^{-1}\boldsymbol{x}_\alpha$ on the lower level curve $f(t^{-1}\boldsymbol{x}_\alpha) = t^{-\lambda} f(\boldsymbol{x}_\alpha) = t^{-\lambda}\alpha$

b. the slopes of the level curves (e.g., isoquants)[1] are unchanged along a ray OR through the origin, i.e., the ratios of marginal products are unchanged along a ray from the origin or, for $\boldsymbol{x}^T = (l, k)$ and $F(\boldsymbol{x})$ homogeneous of degree λ,

$$\frac{F_l(t\boldsymbol{x})}{F_k(t\boldsymbol{x})} = \frac{t^{\lambda-1}F_l(\boldsymbol{x})}{t^{\lambda-1}F_k(\boldsymbol{x})} = \frac{F_l(\boldsymbol{x})}{F_k(\boldsymbol{x})}.$$

Thus the slope of an isoquant at (l^o, k^o) is the same as at (tl^o, tk^o) (Figure 15.2).

c. Suppose $h(\boldsymbol{x}) = G[F(\boldsymbol{x})]$, $\boldsymbol{x}^T = (l, k)$, is homothetic, with $F(\boldsymbol{x})$ homogeneous of degree 1. Then

$$h(t\boldsymbol{x}) = G[F(t\boldsymbol{x})] = G[tF(\boldsymbol{x})].$$

Upon differentiating this expression with respect to t we obtain

$$h_l l + h_k k \equiv G'F(l, k),$$

[**Euler's Theorem for homothetic production functions**]

If we now assume that $F(l, k)$ is homothetic, then the level curves or isoquants of F are also radial projections of each other and, as was the case for a homogeneous production function, the tangent lines to the isoquants are parallel along a ray from the origin (the isoquant slopes along a radial projection remain unchanged). In fact, this is the most general class of production functions which exhibit this property. To see this, suppose that $h(\boldsymbol{x})$: $R^n \to R$ is a differentiable homothetic production function, where $\boldsymbol{x}^T = (l, k)$. Let us write the **technical rate of substitution between l and k** as

$$TRS_{l,k} = -\left.\frac{dk}{dl}\right|_{h(\boldsymbol{x})=c} = \frac{h_l}{h_k}, \quad c = \text{constant}.$$

Then for $\boldsymbol{x} \in R^n$ and $t > 0$, $TRS_{l,k}(l, k) = TRS_{l,k}(tl, tk)$. This equality follows directly from (15.4) since

$$TRS_{l,k}(\boldsymbol{tx}) = \frac{h_l(\boldsymbol{tx})}{h_k(\boldsymbol{tx})} = \frac{k\, h_l(\boldsymbol{x})}{k\, h_k(\boldsymbol{x})} = \frac{h_l(\boldsymbol{x})}{h_k(\boldsymbol{x})}$$
$$= TRS_{l,k}(\boldsymbol{x}).$$

In sum, if $h(\boldsymbol{x}) = G[f(\boldsymbol{x})]$, $\boldsymbol{x} \in R^n$, where $G' > 0$ and f is a homogeneous function, the family of level curves of homothetic h is the same as that of f - - homothetic functions display the geometric properties of homogeneous functions.

Section 15.3 Exercises:

15.3.1 For $x \in R_{++}$, are the following expressions monotonic transformations?

a. $x^3 + x$

b. $x^2 - x$

c. $\sqrt{x} + 1$

d. $(x^4 + 2)^{1/4}$

15.3.2 Suppose $f(x_1, x_2) = x_1^2 x_2^2$. Transform f to a function $g(x_1, x_2)$ which is homogeneous of degree 1.

15.3.3 Suppose $f(x_1, x_2) = x_1 x_2$. Find a homothetic function $h(x_1, x_2) = G(f(x_1, x_2))$ when $G(z)$ has the following forms:

a. $G(z) = z^4 + z$

b. $G(z) = z + 6$

c. $G(z) = z^{1/2}$

What monotonic transformation $G(z)$ yielded $h(x_1, x_2) = x_1^4 x_2^2 + 2x_1^5 x_2 + x_1^6 + x_1^2 x_2 + x_1^3$ when $f(x_1, x_2) = x_1^2 x_2 + x_1^3$?

15.3.4 A sufficient condition for the homotheticity of a given function is the following:

Suppose $f(\boldsymbol{x})$: $R_+^n \to R$ is a differentiable function. If for all i, j and all $t > 0$ we have

$$\frac{\partial f(\boldsymbol{tx})}{\partial x_i} / \frac{\partial f(\boldsymbol{tx})}{\partial x_j} = \frac{\partial f(\boldsymbol{x})}{\partial x_i} / \frac{\partial f(\boldsymbol{x})}{\partial x_j}, \qquad (**)$$

then f is homothetic. Given (**), for $\boldsymbol{x} \in R_+^2$, determine if the following functions are homothetic:

a. $f = \exp\left(4x_1^2 x_2 + 2x_1 x_2^2\right)$

b. $f = \ln x_1 + \ln x_2$

c. $f = x_1 x_2 + x_1 x_2^2$

d. $f = x_1^2 x_2^4 + 6x_1 x_2^2 - 10$

15.4 ALMOST HOMOGENEOUS FUNCTIONS

We determine above that a function $y = f(\boldsymbol{x})$ is said to be homogeneous of degree λ if, when each of its arguments is multiplied by a scalar $t > 0$, the value of the function is multiplied by t^λ, i.e., in a production function context, as each input is increased proportionately, total product increases by t^λ. However, under almost homogeneity, we are looking at the output response to a disproportionate (percentage) increase in factor inputs. Specifically,

Definition 15.3: The production function $q = F(l, k)$ is **almost homogeneous** with respect to positive scalars θ_l, θ_k, and λ if and only if

$$F\left(t^{\theta_l} l, t^{\theta_k} k\right) \equiv t^\lambda F(l, k),\ t > 0. \qquad (15.5)$$

So if l is increased by a power θ_l of t and k is increased by a power θ_k of t, then F will increase by another power λ of t. (For $\theta_l = \theta_k = \lambda = 1$, F is homogeneous of degree 1 in l, k.)

For almost homogeneous functions, the following modification of Euler's Theorem holds:

$$\theta_l F_l l + \theta_k F_k k \equiv \lambda q. \qquad (15.6)$$

[Euler's Theorem for Almost Homogeneous Production Functions]

Additionally, if we set $t = l^{-1/\theta_l}$ so that $t^{\theta_k} = l^{-\theta_k/\theta_l}$, then

$$F\left(1, k l^{-\theta_k/\theta_l}\right) = l^{-\lambda/\theta_l} F(l, k)$$

and thus this expression can be written, for $y = k l^{-\theta_k/\theta_l}$, as

$$q = l^{\lambda/\theta_l} f(y).$$

From this equation it is readily demonstrated that factor marginal products are

$$q_l = l^{\frac{\lambda}{\theta_l}-1}\left(\frac{\lambda}{\theta_l}f - \frac{\theta_k}{\theta_l}yf'\right),$$
$$q_k = l^{\frac{\lambda}{\theta_l}-\frac{\theta_k}{\theta_l}}f'$$

and thus, the technical rate of substitution between l, k is

$$TRS_{l,k} = \frac{q_l}{q_k} = l^{\frac{\theta_k}{\theta_l}-1}\left(\frac{\lambda}{\theta_l}\frac{f}{f'} - \frac{\theta_k}{\theta_l}y\right). \tag{15.7}$$

15.5 HOMOGENEITY AND CONCAVITY (CONVEXITY)

We begin with

Definition 15.4: A function $y = f(\boldsymbol{x})$: $\mathcal{D} \subset R^n \to R$ is **subadditive** if

$$f(\boldsymbol{x}_1 + \boldsymbol{x}_2) \leq f(\boldsymbol{x}_1) + f(\boldsymbol{x}_2) \text{ for all } \boldsymbol{x}_1, \boldsymbol{x}_2 \in \mathcal{D}. \tag{15.8}$$

It is **superadditive** if

$$f(\boldsymbol{x}_1 + \boldsymbol{x}_2) \geq f(\boldsymbol{x}_1) + f(\boldsymbol{x}_2) \text{ for all } \boldsymbol{x}_1, \boldsymbol{x}_2 \in \mathcal{D}. \tag{15.9}$$

Note that if $\boldsymbol{O} \in \mathcal{D}$ and f is subadditive (superadditive), then $f(\boldsymbol{O}) > 0(f(\boldsymbol{O}) < 0)$.

If we now admit homogeneity to our discussion, we then have

Theorem 15.9: Let the function $y = f(\boldsymbol{x})$: $\mathcal{D} \subset R^n \to R$, $\mathcal{D}$ a convex cone, be linearly homogeneous. Then f is concave if and only if for every $\boldsymbol{x}_1, \boldsymbol{x}_2 \in \mathcal{D}$, f is superadditive or $f(\boldsymbol{x}_1 + \boldsymbol{x}_2) \geq f(\boldsymbol{x}_1) + f(\boldsymbol{x}_2)$.

Proof: Linear homogeneity and concavity imply superadditivity since

$$f(\boldsymbol{x}_1 + \boldsymbol{x}_2) = f\left(2\left(\tfrac{\boldsymbol{x}_1 + \boldsymbol{x}_2}{2}\right)\right) = 2f\left(\tfrac{\boldsymbol{x}_1 + \boldsymbol{x}_2}{2}\right)$$
$$\geq 2\left(\tfrac{1}{2}f(\boldsymbol{x}_1) + \tfrac{1}{2}f(\boldsymbol{x}_2)\right).$$

Also, linear homogeneity and superadditivity imply concavity since

$$f(\lambda\boldsymbol{x}_1 + (1-\lambda)\boldsymbol{x}_2) \geq f(\lambda\boldsymbol{x}_1) + f((1-\lambda)\boldsymbol{x}_2)$$
$$= \lambda f(\boldsymbol{x}_1) + (1-\lambda)f(\boldsymbol{x}_2). \quad \text{Q.E.D.}$$

Clearly this Theorem can easily be modified to read that if f is a linear homogeneous function, then f is convex if and only if for all $\boldsymbol{x}_1, \boldsymbol{x}_2 \in \mathcal{D}$, f is subadditive or $f(\boldsymbol{x}_1 + \boldsymbol{x}_2) \leq f(\boldsymbol{x}_1) + f(\boldsymbol{x}_2)$.

Let us now consider the instance in which linear homogeneity and quasiconcavity imply concavity.

Theorem 15.10: Let the function $y = f(\boldsymbol{x})$: $\mathcal{D} \subset R^n \to R$, $\mathcal{D}$ a convex set, be linearly homogeneous. If $f(\boldsymbol{x}) > 0(\boldsymbol{x} \neq \boldsymbol{O})$ for all $\boldsymbol{x} \in \mathcal{D}$, then f is quasiconcave if and only if it is concave.

Proof: Let us first consider the notion that linear homogeneity and quasiconcavity imply superadditivity or, for every $\boldsymbol{x}_1, \boldsymbol{x}_2 \in \mathcal{D}$, $f(\boldsymbol{x}_1 + \boldsymbol{x}_2) \geq f(\boldsymbol{x}_1) + f(\boldsymbol{x}_2)$. Let $y_i = f(\boldsymbol{x}_i) > 0$, $i = 1, 2$. With f homogeneous of degree 1, $f(\boldsymbol{x}_1/y_1) = f(\boldsymbol{x}_2/y_1) = 1$ so that the quasiconcavity of f implies that

$$f\left(\lambda\left(\frac{\boldsymbol{x}_1}{y_1}\right) + (1-\lambda)\left(\frac{\boldsymbol{x}_2}{y_2}\right)\right) \geq 1 \text{ for all } \lambda \in (0, 1).$$

Set $\lambda = y_2/(y_1 + y_2)$. We then have $1 - \lambda = y_1/(y_1 + y_2)$ and thus

$$f\left(\frac{\boldsymbol{x}_1}{y_1 + y_2} + \frac{\boldsymbol{x}_2}{y_1 + y_2}\right) = \frac{1}{y_1 + y_2}f(\boldsymbol{x}_1 + \boldsymbol{x}_2) \geq 1$$

or

$$f(\boldsymbol{x}_1 + \boldsymbol{x}_2) \geq y_1 + y_2 = f(\boldsymbol{x}_1) + f(\boldsymbol{x}_2).$$

Next, we demonstrate that linear homogeneity and superadditivity imply concavity (since if f is concave on $\mathcal{D}$, then it is quasiconcave on $\mathcal{D}$). Thus

$$f(\lambda\boldsymbol{x}_1 + (1-\lambda)\boldsymbol{x}_2) \geq f(\lambda\boldsymbol{x}_1) + f((1-\lambda)\boldsymbol{x}_2)$$
$$= \lambda f(\boldsymbol{x}_1) + (1-\lambda)f(\boldsymbol{x}_2). \quad \text{Q.E.D.}$$

A similar result is also true in that if f is linear homogeneous and $f(\boldsymbol{x}) > 0(\boldsymbol{x} \neq \boldsymbol{O})$ for all $\boldsymbol{x} \in \mathcal{D}$, then f is quasiconvex if and only if it is convex.

An extension of Theorem 15.10 is

Theorem 15.11: Let $y = f(\boldsymbol{x})$: $\mathcal{D} \subset R^n \to R$, $\mathcal{D}$ a convex set, be homogeneous of degree λ, $0 < \lambda \leq 1$. If

$f(\mathbf{x}) > 0(\mathbf{x} \neq \mathbf{O})$ for all $\mathbf{x} \in \mathcal{D}$, then f is quasiconcave if and only if it is concave.

Proof: Given the proof of Theorem 15.10, we need only consider the instance where $\lambda < 1$. Define a new function $g(\mathbf{x}) = (f(\mathbf{x}))^{1/\lambda}$ for all $\mathbf{x} \in \mathcal{D}$. Then g is quasiconcave and linearly homogeneous with $g(\mathbf{x}) > 0$ for $\mathbf{x} \neq \mathbf{O}$ so that, as argued above, g is concave and $f = g^{\lambda}$ is also concave. Q.E.D.

Following the line of reasoning provided by this theorem we can also state that if f is homogeneous of degree $\lambda \geq 1$ with $f(\mathbf{x}) > 0$ for all $\mathbf{x} \in \mathcal{D}$, then f is quasiconvex if and only if it is convex.

Example 15.2: Let $f(x_1, x_2) = (x_1^2 + x_2^2)^{\alpha}$, $\mathbf{x} \in R^2_{++}$, $\alpha > 0$. For what value of α is f convex? To answer this question, let us employ the argument offered after the proof of Theorem 11.15. Since $x_1^2 + x_2^2$ is convex (the Hessian matrix associated with this expression is positive definite), it follows that f is quasiconvex (given that $(x_1^2 + x_2^2)^{\alpha}$ is non-decreasing). Since f is homogeneous of degree $\lambda = 2\alpha$, it follows that f is convex for $2\alpha \geq 1$ or $\alpha \geq 1/2$. ∎

Example 15.3: Suppose $f = Ax_1^{\alpha_1}x_2^{\alpha_2}$, with $\mathbf{x} \in R^2_{++}$, $A > 0$, $\alpha_i > 0$, $i = 1, 2$. We know that f is homogeneous of degree $\alpha = \alpha_1 + \alpha_2$. Let us verify that f is quasiconcave and that it is concave if and only if $\alpha \leq 1$. Since $\ln f = \ln A + \alpha_1 \ln x_1 + \alpha_2 \ln x_2$ is a positive linear combination of concave functions, it is concave. Then $f(\mathbf{x}) = \exp f(\mathbf{x})$ is quasiconcave and Theorem 11.15 applies. ∎

15.6 HOMOGENEOUS PROGRAMMING (LASSERRE AND HIRIART-URRUTY [2002]; ZHAO AND LI [2012])

Given the functions $y = f(\mathbf{x}): \mathcal{D} \subset R^n \to R$ and $g^j(\mathbf{x}): \mathcal{D} \subset R^n \to R$, $j = 1, \ldots, m$, let us form the optimization problem

$$\begin{aligned} &\text{minimize } f(\mathbf{x}) \text{ subject to} \\ &g^j(\mathbf{x}) \leq b_j, \ j = 1, \ldots, m. \end{aligned} \tag{P}$$

If f and the g^j, $j = 1, \ldots, m$, are homogeneous on $\mathcal{D}$, then (P) will be regarded as **a homogeneous constrained optimization problem**. (Remember that a function $h(\mathbf{x})$ is homogeneous of degree p on $\mathcal{D}$ if $h(t\mathbf{x}) = t^p h(\mathbf{x})$ for all $t > 0$ and $\mathbf{x} \in \mathcal{D}$.)

Consider the following variant of problem (P):

$$\begin{aligned} &\text{minimize } f(\mathbf{x}) \text{ subject to} \\ &g^j(\mathbf{x}) = b_j, \ j = 1, \ldots, m, \end{aligned} \tag{P.1}$$

where f and the g^j are all continuous and homogeneous of the same degree $p > 0$ on $\mathcal{D}$.

For a scalar $\alpha \in (0, 1]$, consider the **auxiliary problem**

$$\begin{aligned} &\text{minimize } F(\mathbf{x}, u; \alpha) = (u + \alpha) f(\mathbf{x}) + \tfrac{1}{2}(u + \alpha - 1)^2 \\ &\text{subject to} \\ &(u + \alpha) g^j(\mathbf{x}) = b_j, \ j = 1, \ldots, m, \ u \geq 0. \end{aligned} \tag{P.2}$$

Then an optimal solution to (P.1) is in one-to-one correspondence with an optimal solution to (P.2). To verify this assertion, we have

Theorem 15.12: (Zhao and Li [2012]). Let $f(\mathbf{x})$ and $g^j(\mathbf{x})$, $j = 1, \cdots, m$, be of class $C^{(n)}$ and homogeneous of degree $p > 0$ on $\mathcal{D} \subset R^n$.

i. if $\mathbf{x}_o$ is an optimal solution to (P.1) (i.e., a KKT solution) with associated Lagrange multiplier $\boldsymbol{\lambda}_o \in R^m$, then $(\hat{\mathbf{x}}, \hat{u}) = (\mathbf{x}_o, 1 - \alpha)$ is an optimal solution to (P.2) with Lagrange multiplier $(\hat{\boldsymbol{\lambda}}, \hat{\mu}) = (\boldsymbol{\lambda}_o, 0) \in R^m \times R_+$.

ii. Conversely, if $(\hat{\mathbf{x}}, \hat{u})$ is an optimal solution to (P.2) with associated Lagrange multiplier $(\hat{\boldsymbol{\lambda}}, \hat{\mu}) \in R^m \times R_+$, then we must have $\hat{u} = 1 - \alpha$ and $\hat{\mu} = 0$ and thus $\hat{\mathbf{x}}$ is an optimal solution to (P.1), with $\hat{\boldsymbol{\lambda}}$ the Lagrange multiplier corresponding to $\hat{\mathbf{x}}$.

Proof: We know that $\mathbf{x}_0$ is a KKT solution to (P.1) with Lagrange multiplier $\boldsymbol{\lambda}_o$ if and only if the following conditions hold:

$$\nabla f(\mathbf{x}_o) + \textstyle\sum_{j=1}^m \lambda_j^o \nabla g^j(\mathbf{x}_o) = \mathbf{O} \tag{15.10}$$

$$g^j(\mathbf{x}_o) = b_j, \ j = 1, \ldots, m, \tag{15.11}$$

and that $(\hat{\mathbf{x}}, \hat{u})$ is a KKT solution to (P.2) with Lagrange multiplier $(\hat{\boldsymbol{\lambda}}, \hat{\mu})$ if and only if[2]

$$(\hat{u} + \alpha)\left[\nabla f(\hat{\mathbf{x}}) + \textstyle\sum_{j=1}^m \hat{\lambda}_j \nabla g^j(\hat{\mathbf{x}})\right] = \mathbf{O} \tag{15.12}$$

$$f(\hat{x}) + (\hat{u} + \alpha - 1) + \sum_{j=1}^{m} \hat{\lambda}_j g^j(\hat{x}) + \hat{\mu} = 0 \quad (15.13)$$

$$(\hat{u} + \alpha) g^j(\hat{x}) = b_j, \; j = 1, \ldots, m \quad (15.14)$$

$$\hat{\mu} \geq 0, \; \hat{u} \geq 0, \; \hat{u}\hat{\mu} = 0, \quad (15.15)$$

where $\hat{u}\hat{\mu} = 0$ is a *complementary slackness condition.*

i. Let $\boldsymbol{x}_o$ solve (P.1) with Lagrange multiplier $\boldsymbol{\lambda}_o$. Then $(\hat{\boldsymbol{x}}, \hat{u}, \hat{\boldsymbol{\lambda}}, \hat{\mu}) = (\boldsymbol{x}_o, 1 - \alpha, \boldsymbol{\lambda}_o, 0)$ satisfies (15.12), (15.14) and (15.1). Equation (15.13) is also satisfied since, if we multiply both sides of (15.10) by $\boldsymbol{x}_o$, we obtain

$$x_o^T \nabla f(\boldsymbol{x}_o) + \sum_{j=1}^{m} \lambda_j^o \boldsymbol{x}_o^T \nabla g^j(\boldsymbol{x}_o) = \boldsymbol{O}.$$

By Euler's theorem, the preceding expression becomes

$$p\left(f(\boldsymbol{x}_o) + \sum_{j=1}^{m} \lambda_j^o g^j(x_o) \right) = 0,$$

which implies that (15.13) holds with $(\hat{\boldsymbol{x}}, \hat{u}, \hat{\boldsymbol{\lambda}}, \hat{\mu}) = (\boldsymbol{x}_o, 1 - \alpha, \boldsymbol{\lambda}_o, 0)$.

ii. Suppose $(\hat{\boldsymbol{x}}, \hat{u}, \hat{\boldsymbol{\lambda}}, \hat{\mu})$ satisfies (15.12) - (15.15). Then we must have $(\boldsymbol{x}_o, \boldsymbol{\lambda}_o) = (\hat{\boldsymbol{x}}, \hat{\boldsymbol{\lambda}})$ satisfying (15.10) and (15.11), i.e., since $\hat{u} + \alpha > 0$, (15.12) becomes

$$\nabla f(\hat{\boldsymbol{x}}) + \sum_{j=1}^{m} \hat{\lambda}_j \nabla g^j(\hat{\boldsymbol{x}}) = \boldsymbol{O}$$

or (15.10) holds. Additionally, by again invoking Euler's theorem, the preceding equality implies that

$$f(\hat{\boldsymbol{x}}) + \sum_{j=1}^{m} \hat{\lambda}_j g^j(\hat{\boldsymbol{x}}) = 0$$

and thus, from (15.13), $\hat{\mu} = \hat{u} + \alpha - 1$. Since $\hat{\mu} \geq 0$, it follows that $\hat{u} \geq 1 - \alpha \geq 0$. If we multiply both sides of $\hat{\mu} = \hat{u} + \alpha - 1$ by $\hat{u}$ (given that $\hat{u}\hat{\mu} = 0$) we obtain $\hat{u}(\hat{u} + \alpha - 1) = 0$. If we couple this result with $\hat{u} \geq 1 - \alpha \geq 0$, we see that $\hat{\mu} = 1 - \alpha$ implies $\hat{\mu} = 0$. Hence, (15.13) reduces to (15.11). Q. E. D.

Two additional points are of importance:

1. If $\alpha = 1$ in the preceding theorem, then:
   i. if $\boldsymbol{x}_o$ solves (P.1) with associated Lagrange multiplier $\boldsymbol{\lambda}_o \in R^m$, then $(\hat{\boldsymbol{x}}, \hat{u}) = (\boldsymbol{x}_o, 0)$ solves (P.2).
   ii. conversely, if $(\hat{\boldsymbol{x}}, \hat{u})$solves (P.2) with associated Lagrange multiplier $(\hat{\boldsymbol{\lambda}}, \hat{\mu}) \in R^m \times R_{+}$, then $\hat{u} = \hat{\mu} = 0$ and hence $\hat{\boldsymbol{x}}$ solves (P.1) with Lagrange multiplier $\hat{\boldsymbol{\lambda}}$.
2. Given the assumptions of Theorem 15.12, $\boldsymbol{x}_o$ is a global optimal solution to problem (P.1) if and only if $(\boldsymbol{x}_o, u_o)$, with $u_o = 1 - \alpha$, is a global optimal solution to (P.2). Both problems attain the same global optimal point values.

Let us now consider the instance where Theorem 15.12 can be extended to the case involving m inequality constraints with the objective function and constraint functions having different degrees of homogeneity. That is, consider problem (P) above wherein f is homogeneous of degree p and g^j is homogeneous of degree q_j, $j = 1, \ldots, m$

Let the scalar $\alpha \in [0, 1)$. Upon introducing the variable u into (P) we obtain the auxiliary problem

$$\begin{aligned} &\text{minimize } F(\boldsymbol{x}, u; \alpha) = (u + \alpha) f(\boldsymbol{x}) + \tfrac{1}{2}(u + \alpha - 1)^2 \\ &\text{subject to} \\ &(u + \alpha)\left[g^j(\boldsymbol{x}) - b_j\left(1 - q_j/p\right) \right] \leq b_j q_j / p, \\ &j = 1, \ldots, m, \; u \geq 0. \end{aligned} \quad \text{(P.3)}$$

To see how the optimal or KKT solutions to (P) and (P.3) are related, we look to

Theorem 15.13 (Zhao and Li [2012]). Let $f(\boldsymbol{x})$ and $g^j(\boldsymbol{x})$, $j = 1, \ldots, m$, be of class $C^{(1)}$ and homogeneous on $\mathcal{D} \subset R^n$, with f homogeneous of degree p and g^j homogeneous of degree q_j.

i. if $\boldsymbol{x}_o$ is an optimal or KKT solution to (P) with associated Lagrange multiplier $\boldsymbol{\lambda}_o \in R_+^m$, then $(\boldsymbol{x}_o, u_o)$, with $u_o = 1 - \alpha$, is an optimal solution to (P.3) with associated Lagrange multiplier $(\boldsymbol{\lambda}_o, u_o = 0)$.

ii. if $(\bar{\boldsymbol{x}}, \bar{u})$ is an optimal solution of (P.3) with associated Lagrange multiplier $(\bar{\boldsymbol{\lambda}}, \bar{\mu}) \in R_+^{m+1}$, and

$$\sum_{j=1}^{m} \bar{\lambda}_j b_j \left(1 - q_j/p\right)\left(q_j/p\right) < \bar{u} + \alpha, \quad (15.16)$$

then $\bar{u} = 1 - \alpha$, $\bar{\mu} = 0$, and thus $\bar{\boldsymbol{x}}$ is an optimal solution of (P) and $\bar{\boldsymbol{\lambda}}$ is the associated Lagrange multiplier.

Note that (15.16) holds trivially if one of the following is true:

$$\begin{aligned}
&\text{(a)}\quad \text{the } g^j \text{ have the same degree of homogeneity or } p = q_j,\ j = 1,\ldots,m;\\
&\text{(b)}\quad \boldsymbol{b}^T = (b_1,\ldots,b_m) = \boldsymbol{O}^T;\\
&\text{(c)}\quad b_j\left(p - q_j\right) \le 0,\ j = 1,\ldots,m;\\
&\text{(d)}\quad \boldsymbol{b} \ge \boldsymbol{O} \text{ and } p \le \min_{1\le j\le m} q_j.
\end{aligned} \quad (15.17)$$

So for cases (a)–(d), the optimal solution of (P) is in one-to-one correspondence to the optimal solution to (P.3), where the latter are provided by the KKT system

$$(\bar{u} + \alpha)[\nabla f(\bar{\boldsymbol{x}}) + \textstyle\sum_{-1}^{m} \bar{\lambda}_j \nabla g^j(\bar{\boldsymbol{x}})] = \boldsymbol{O} \quad (15.18)$$

$$f(\bar{\boldsymbol{x}}) + (\bar{u} + \alpha - 1) + \sum_{j=1}^{m} \bar{\lambda}_j \left[g^j(\bar{\boldsymbol{x}}) - b_j\left(1 - q_j/p\right)\right] + \bar{\mu} = 0 \quad (15.19)$$

$$\bar{\lambda}_j \left\{\left[g^j(\bar{\boldsymbol{x}}) - b_j\left(1 - q_j/p\right)\right](\bar{u} + \alpha) - \left(q_j/p\right)b_j\right\} = 0,\ j = 1,\ \ldots, m \quad (15.20)$$

$$\bar{\mu} \ge 0,\ \bar{u} \ge 0,\ \bar{\mu}\bar{u} = 0. \quad (15.21)$$

Example 15.4: Suppose our problem is to

$$\begin{aligned}
&\text{minimize } f(\boldsymbol{x}) = x_1 + x_2 \text{ subject to}\\
&g^1(\boldsymbol{x}) = x_1 x_2 \le 2 = b_1\\
&g^2(\boldsymbol{x}) = x_1^2 \le 4 = b_2.
\end{aligned}$$

Here f is homogeneous of degree $p = 1$ while g^1, g^2 are each homogeneous of degree $q_1 = q_2 = 2$. The auxiliary problem (P.3) associated with this program is

$$\begin{aligned}
&\text{minimize } F(\boldsymbol{x}, u; \alpha) = (u + \alpha)(x_1 + x_2) + \tfrac{1}{2}(u + \alpha - 1)^2\\
&\quad \text{subject to}\\
&(u + \alpha)(x_1 x_2 + 2) \le 4\\
&(u + \alpha)(x_1^2 + 4) \le 8\\
&\quad u \ge 0.
\end{aligned}$$

We know that a point $(\bar{\boldsymbol{x}}, \bar{u}) \in R^2 \times R_+$ is a KKT solution to (P.3) with Lagrange multiplier $(\bar{\boldsymbol{\lambda}}, \bar{\mu}) \in R_+^{m+1}$ if equations (15.18)–(15.21) hold or

$$\begin{aligned}
&(\bar{u} + \alpha)\left(\begin{bmatrix}1\\1\end{bmatrix} + \bar{\lambda}_1\begin{bmatrix}\bar{x}_2\\ \bar{x}_1\end{bmatrix} + \bar{\lambda}_2\begin{bmatrix}2\bar{x}_1\\0\end{bmatrix}\right) = \boldsymbol{O}\\
&(\bar{x}_1 + \bar{x}_2) + (\bar{u} + \alpha - 1) + \bar{\lambda}_1(\bar{x}_1\bar{x}_2 + 2) + \bar{\lambda}_2(\bar{x}_1^2 + 4) + \bar{\mu} = 0\\
&\bar{\lambda}_1[(\bar{x}_1\bar{x}_2 + 2)(\bar{u} + \alpha) - 4] = 0\\
&\bar{\lambda}_2[(\bar{x}_1^2 + 4)(\bar{u} + \alpha) - 8] = 0\\
&\bar{\mu} \ge 0,\ \bar{u} \ge 0,\ \bar{\mu}\bar{u} = 0.
\end{aligned}$$

With both constraints binding ($\bar{\lambda}_1 > 0$, $\bar{\lambda}_2 > 0$), $\bar{\mu} = 0$, and $\bar{u} + \alpha = 1$, one can readily check that the optimal solution to the original problem is $\bar{x}_1 = 2$, $\bar{x}_2 = 1$, $\bar{\lambda}_1 = \frac{1}{2}$, $\bar{\lambda}_2 = \frac{1}{8}$, and $f(\bar{\boldsymbol{x}}) = 3$. ∎

Section 15.6 Exercises:

15.6.1 Suppose we want to minimize $f(\boldsymbol{x}) = x_1^2$ subject to $2x_1^2 + x_2^2 = 10$. Solve this problem via the homogeneous auxiliary program (P.2).

15.6.2 (Laseure and Hiriart-Urruty [2002].) Minimize $f(\boldsymbol{x}) = x_1$ subject to $x_1^2 - x_2^2 \le 1$, $x_1^2 + x_2^2 \le 7$.

Solve this problem via the homogeneous auxiliary program (P.3).

15.7 ECONOMIC APPLICATIONS

15.7.1 The Long-Run Expansion Path

Let us assume that the firm (a price taker in the product and factor markets) maximizes output $q = F(l, k)$ subject to a cost constraint $C = wl + rk$ and that both l and k are freely variable. If the firm desires to expand its output, what combination of factors should it hire at each stage of the expansion? The answer is that, for fixed input prices w and r and increasing levels of expenditure C, the firm moves along its **long-run expansion path** – the locus of *economically efficient* (in the sense that at each point along the expansion path

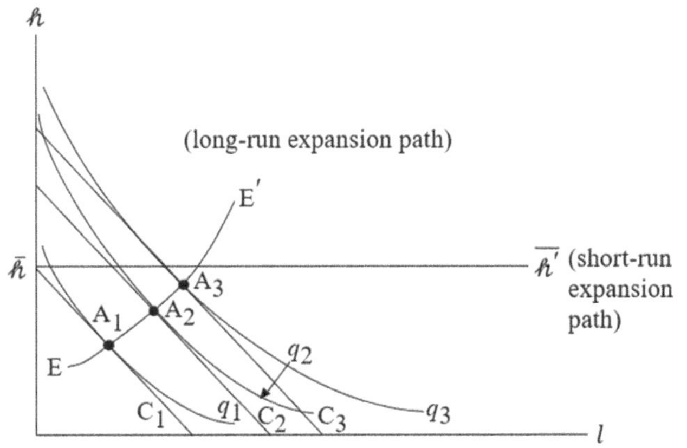

FIGURE 15.3 Short- and long-run expansion paths.

the equality $F_l/F_k = w/r$ holds) input combinations A_1, A_2 and A_3 corresponding to successively higher levels of total expenditure C_1, C_2, and C_3. Thus, the expansion path is the curve EE' in Figure 15.3. (Note that in the short run capital is held fixed at the $k = \bar{k}$ level so that the **short-run expansion path** is $\bar{k}\,\bar{k}'$.)

Given that $F_l/F_k = w/r$ at each point along the expansion path, this equality may be solved to yield

$$k = h(l) \text{ or } l = j(k) \qquad (15.22)$$
[long – run expansion path in *explicit form*]

or

$$g(l, k) = 0. \qquad (15.23)$$
[long – run expansion path in *implicit form*]

Additionally, the expansion path can be written in terms of total expenditure, i.e., both l and k can be expressed as a function of C. To see this, let us start with the cost equation $wl + rk = C$. If for l in this expression we substitute $j(k)$ from equation (15.22), then $wj(k) + rk = C$ or $k = h^k(w, r, C)$. And if we replace k in the cost equation by $h(l)$ from Equation (15.22), then $wl + rk(l) = C$ or $l = h^l(w, r, C)$. But h^l and h^k are the cost-constrained factor demands derived earlier (Section 10.7.2) and thus represent the parametric form (in terms of C) of the long-run expansion path.

We mentioned at the outset of this section that the firm is interested in efficiently expanding its output over the long run and thus, by virtue of the requirement that $F_l/F_k = w/r$, restricts its movement to the expansion path EE' along which both l and k increase. Let us assume that *l and k increase proportionately.* Under this restriction we can discuss the notion of **returns to scale** – the output response to a proportionate increase in *all* inputs. In this regard:

a. if output increases by the same proportion, returns to scale are *constant*;

b. if output increases by a greater proportion, returns to scale are *increasing*; and

c. if output increases by a smaller proportion, returns to scale are *decreasing*.

The type of returns to scale in production may be easily determined by examining the distances between the isoquants as measured along a ray OR emanating from the origin (Figures 15.1, 15.2). (Note that along ray OR input or factor proportions are held constant.) In fact, a given production function can exhibit all three types of returns to scale. For constant returns to scale the isoquants are equally spaced; for increasing returns to scale, the distance between successive isoquants decreases for equal increments in output; and for decreasing returns to scale, the distance between successive isoquants becomes greater as output increases by equal amounts.

To characterize returns to scale mathematically, we need only rely upon the concept of a homogeneous production function. Given that F is homogeneous of degree λ in l and k, it follows that $F(tl, tk) = t^{\lambda}F(l, k)$. So if:

a. $\lambda = 1$, returns to scale are *constant*;

b. $\lambda > 1$, returns to scale are *increasing*;

c. $\lambda < 1$, returns to scale are *decreasing*.

What is the connection between the notion of homogeneity of F and the long-run expansion path? Suppose F is homogeneous of *any* degree. We know that if (l^o, k^o) is the optimal input combination, then, under homogeneity, the point (tl^o, tk^o), $t > 0$, must lie on a ray OR passing through (l^o, k^o) so that the technical rate of substitution between l and k at (l^o, k^o) is the same as at (tl^o, tk^o). But this means that if the production function is homogeneous of *any* degree, then the long-run expansion path is linear (although a linear expansion path does not imply that the underlying production function is homogeneous).

Example 15.5: The production function $q = F(l, k) = 10\, l^{1/4}k^{1/2}$ exhibits decreasing returns to scale since $\lambda = 3/4 < 1$. If the firm maximizes output subject to a cost constraint, then it operates where

$$\frac{F_l}{F_k} = \frac{1}{4}\frac{k}{l} = \frac{w}{r}.$$

Hence, the long-run expansion path in explicit form is

$$k = h(l) = (4w/r)l \text{ or } l = j(k) = (r/4w)k.$$

In implicit form the long-run expansion path appears as

$$g(l, k) = k - (4w/r)l = 0.$$

So, for F homogeneous, the expansion path is clearly linear.

Using $wl + rk = w(r/4w)k + rk = C$ we obtain $k = (4/5r)C$; and from $wl + r(4w/r)l = C$ we have $l = (1/5w)C$. Hence, the parametric version of the long-run expansion path is

$$l = (1/5w)C,$$
$$k = (4/5r)C.$$

■

15.7.2 The Short-Run Cost Functions

In earlier chapters (7 and 10) the concept of short-run cost was explored and a variety of examples concerning the derivation and form of a short-run cost function were offered. The procedure which was implicitly employed, and which we shall now formally generalize, was to start with the following equation system:

(a) the production function;
(b) the cost equation; and (15.24)
(c) the expansion path.

In the short-run, with capital held fixed at $k = \bar{k}$, the preceding system becomes:

$$\begin{aligned}&(a)\quad q = F(l, k);\\ &(b)\quad C = wl + rk; \text{ and}\\ &(c)\quad k = \bar{k}\,(\text{short-run expansion path}).\end{aligned} \qquad (15.24.1)$$

Then (15.24.1) becomes $q = F(l, \bar{k})$ and, with F continuous and single-values, $l = F^{-1}(q)$. Inserting this inverse function into (15.24.1b) gives us the **short-run cost function (with a single variable input)**

$$C = wF^{-1}(q) + r\bar{k} = \gamma(q) + b, \qquad (15.25)$$

where $b = r\bar{k}$ = constant. This expression yields the *minimum cost* of producing a given level of output since we are on the short-run expansion path.

If the firm employs two variable inputs (l and k) in the presence of additional fixed factors or inputs whose aggregate cost will simply be written as b = constant, then (15.24) becomes;

$$\begin{aligned}&(a)\quad q = F(l, k);\\ &(b)\quad C = wl + rk + b; \text{ and}\\ &(c)\quad k = h(l) \text{ or } l = j(k)\,(\text{long-run expansion path in explicit form}).\end{aligned} \qquad (15.24.2)$$

Proceeding along the lines used to derive equation (15.25) we have either

1. $q = F(l, h(l)) = \bar{F}(l)$, $l = \bar{F}^{-1}(q)$ ($\bar{F}$ is continuous and single valued) and thus

$$\begin{aligned}C &= w\bar{F}^{-1}(q) + rh(l) + b\\ &= w\bar{F}^{-1}(q) + rh(\hat{F}^{-1}(q)) + b = \bar{\gamma}(q) + b;\end{aligned} \qquad (15.26a)$$

or

2. $q = F(j(k), k) = \hat{F}(k)$, $k = \hat{F}^{-1}(q)$ ($\hat{F}$ is continuous and single-valued), and thus

$$\begin{aligned}C &= wj(k) + r\hat{F}^{-1}(q) + b\\ &= wj(\hat{F}^{-1}(q)) + r\hat{F}^{-1}(q) + b = \bar{\gamma}(q) + b.\end{aligned} \qquad (15.26b)$$

Here (15.26a, b) each depict the **short-run cost function (with two variable inputs)** and represent the minimum cost of producing an output level q since we are restricting our movements to points along the expansion path EE' (Figure 15.3).

Example 15.6: For the production function $q = F(l, k) = l^{0.2}k^{0.6}$ it is readily shown that (15.24.2) appears as

(a) $q = F(l, k) = l^{0.2}k^{0.6}$;
(b) $C = wl + rk + b$; and (15.27)
(c) $k = h(l) = (3w/r)l$.

Then from (15.27 a, c)

$$q = l^{0.2}h(l)^{0.6} = (3w/r)^{0.6}l^{0.8}$$

and thus

$$l = \bar{F}^{-1}(q) = (3w/r)^{-0.75}q^{1.25}.$$

Substituting this latter expression into (15.27 b, c) yields (15.26a) or

$$\begin{aligned} C &= wl + rk + b \\ &= 1.7548w^{0.25}r^{075}q^{1.25} + b = \bar{y}(q) + b. \end{aligned}$$ ■

15.7.3 The Elasticity of Substitution Between Labor and Capital: Another Look

In Exercise 10.7.2.1 we defined the elasticity of substitution of labor for capital as

$$\sigma = \frac{dx}{ds}\frac{s}{x},$$

the elasticity of the factor input ratio ($x = k/l$) with respect to the technical rate of substitution ($TRS_{l,k} = s = F_l/F_k$). Clearly this expression is independent of the form of the production function F. What sort of expression for σ do we get if F is taken to be almost homogeneous? As equation (15.5) revealed, the production function $q = F(l, k)$ is almost homogeneous with respect to positive scalars θ_l, θ_k, and λ if and only if

$$F\left(t^{\theta_l}l, t^{\theta_k}k\right) \equiv t^{\lambda}F(l, k), \; t > 0.$$

Let us set $t = l^{-1/\theta_l}$ so that $t^{\theta_k} = l^{-\theta_k/\theta_l}$ and thus

$$F\left(1, kl^{-\theta_k/\theta_l}\right) = l^{-\lambda/\theta_l}F(l, k).$$

Then this expression can be rewritten, for $y = kl^{-\theta_k/\theta_l}$, as

$$q = l^{\lambda/\theta_l}f(y). \tag{15.28}$$

From equation (15.28) it is readily demonstrated that factor marginal products are

$$q_l = l^{\frac{\lambda}{\theta_l}-1}\left(\frac{\lambda}{\theta_l}f - \frac{\theta_k}{\theta_l}f'\right),$$

$$q_k = l^{\frac{\lambda}{\theta_l}-\frac{\theta_k}{\theta_l}}f',$$

and the rate of technical substitution between l and k is, from equation (15.7),

$$s = \frac{q_l}{q_k} = l^{\frac{\theta_k}{\theta_l}-1}\left(\frac{\lambda}{\theta_l}\frac{f}{f'} - \frac{\theta_k}{\theta_l}y\right).$$

In addition, it can be shown that

$$dy = \left[l^{\frac{-\theta_k}{\theta_l}}(-s) - k\frac{\theta_k}{\theta_l}l^{\frac{-\theta_k}{\theta_l}-1}\right]dl, \tag{15.29}$$

$$ds = \left[\frac{\partial s}{\partial l} + \frac{\partial s}{\partial k}(-s)\right]dl. \tag{15.30}$$

Given that $\sigma = (dx/ds)(s/x)$, from equations (15.7, (15.29), and (15.30) we have

$$\sigma = \frac{s}{k}\frac{-s - k(\theta_k/\theta_l)l^{-1}}{(\partial s/\partial l) + (\partial s/\partial k)(-s)}.$$

Some additional calculations render

$$\sigma = \frac{q_lq_k}{lk}\frac{lq_l + k(\theta_k/\theta_l)q_k}{T}, \tag{15.31}$$

Where $T = -(q_k^2q_{ll} - 2q_lq_kq_{kl} + q_l^2q_{kk})$. Using (15.6) allows us to write equation (15.31) as

$$\sigma = \frac{q_lq_k}{lk}\frac{(\lambda/\theta_l)q}{T}. \tag{15.31.1}$$

Finally, another set of algebraic manipulations allow us to transform equation (15.31.1) into

$$\sigma = \frac{q_lq_k}{q_lq_k\left[\lambda^2q - \theta_l\theta_k(kq_k + lq_l)\right] - q_{kl}(\lambda q)^2}, \tag{15.32}$$

Here (15.32) depicts the elasticity of input substitution corresponding to the almost homogeneous production function (15.5).

For $\theta_l = \theta_k = 1$ (F is homogeneous of degree λ in l and k), equation (15.32) becomes

$$\sigma = \frac{q_l q_k}{(1-\lambda) q_l q_k + \lambda q q_{kl}}; \quad (15.32.1)$$

and for $\theta_l = \theta_k = \lambda = 1$ (F is homogeneous of degree 1 in l and k), equation (15.32) appears as

$$\sigma = \frac{q_l q_k}{q q_{kl}}, \quad (15.32.2)$$

the expression derived by Allen [1960].

Section 15.7 Exercises:

15.7.1 Given the production function $q = F(l, k) = l^{0.2}k^{0.6}$, determine the long-run expansion path in explicit form.

15.7.2 Rework Example 15.6 for a long-run expansion path of the form $l = j(k)$.

15.7.3 Determine the long-run expansion path $k = h(l)$ for the CES production function

$$q = [\delta l^{-\rho} + (1-\delta)k^{-\rho}]^{-h/\rho}.$$

NOTES

1 For the level curve $F(l, k) =$ constant, we can determine its slope as follows. From

$$dF = F_l\, dl + F_k\, dk = 0$$

we readily obtain the slope as $\frac{dk}{dl} = -\frac{F_l}{F_k}$ or $-\frac{dk}{dl} = \frac{F_l}{F_k}$.

2 Here the Lagrangian of $F(\mathbf{x}, u; \alpha)$ is

$$L(\mathbf{x}, u, \boldsymbol{\lambda}, \mu; \alpha) = (u+\alpha) f(\mathbf{x}) + \frac{1}{2}(u+\alpha-1)^2 - \Sigma_{j=1}^{m} \lambda_j (b_j - (u+\alpha) g^j(\mathbf{x})) + \mu u.$$

Then equations (15.12), (15.13) result, respectively, by setting $\partial L/\partial \mathbf{x} = \mathbf{O}$, $\partial L/\partial u = 0$.

CHAPTER 16

Envelope Theorems

16.1 INTRODUCTION

In this chapter we consider the optimization of a function (with or without constraints) which depends upon a set of arguments (the independent variables) as well as a set of parameters and/or exogenous variables. We may view the value of a **parameter** as given (a constant) under any particular discussion or state of affairs. But as the discussion changes, the value of the parameter changes as well. For instance, α is a parameter in the expression $f(x) = \alpha x^2$. Clearly, α can assume any admissible value in a particularization of f, e.g., $f = 4x^2$.

The value of an **exogenous variable** is determined by factors external to the function or model. Thus, exogenous variables have values that are taken as datum.

For instance, let us frame the basic problem as

$$\text{maximize } y = f(\boldsymbol{x};\ \boldsymbol{\alpha}) \text{ subject to } \boldsymbol{x} \in \mathcal{K}(\boldsymbol{\alpha}), \tag{16.1}$$

where $\boldsymbol{x} \in X \subseteq R^n$, the vector of parameters $\boldsymbol{\alpha}^T = (\alpha_1, \ldots, \alpha_p) \in \Omega \subset R^p$ (here Ω is termed **parameter space**), $f\colon X \times \Omega \subseteq R^{n+p} \rightarrow R$, and the **constraint correspondence** is $\mathcal{K}\colon \Omega \rightarrow X$.[1] So given an $\boldsymbol{\alpha}$, we need to determine an $\boldsymbol{x}$ within the **feasible constraint set** $\mathcal{K}(\boldsymbol{\alpha})$ which houses the feasible values of $\boldsymbol{x}$ for admissible values of $\boldsymbol{\alpha}$. Additionally, let us denote the **solution correspondence** as $S\colon \Omega \rightarrow X$, with $S(\boldsymbol{\alpha})$ depicting the set of **optimal solutions** - - the set whose elements $\boldsymbol{x}_o$ are, for each $\boldsymbol{\alpha}$, the optimal solutions to (16.1). Hence, the solution correspondence can be written as

$$S(\boldsymbol{\alpha}) = \arg\max\{f(\boldsymbol{x};\ \boldsymbol{\alpha})\,|\,\boldsymbol{x}\in\mathcal{K}(\boldsymbol{\alpha})\}$$

(where arg max $\{\bullet\}$ represents the set of all vectors $\boldsymbol{x} \in \mathcal{K}(\boldsymbol{\alpha})$ that maximize $f(\boldsymbol{x};\ \boldsymbol{\alpha})$) and if the solution to (16.1) is unique for each $\boldsymbol{\alpha}$, then the correspondence $S(\boldsymbol{\alpha})$ *becomes a function* and we can simply write $\boldsymbol{x}_o = \boldsymbol{x}_o(\boldsymbol{\alpha})$.

If we now substitute $\boldsymbol{x}_o = \boldsymbol{x}_o(\boldsymbol{\alpha})$ into the objective function f, then we obtain the optimal value function $f^o\colon \Omega \rightarrow R$ or

$$f^o(\boldsymbol{\alpha}) = f(\boldsymbol{x}_o;\ \boldsymbol{\alpha}) = f(\boldsymbol{x}_o(\boldsymbol{\alpha});\ \boldsymbol{\alpha}),\ \boldsymbol{x}_o \in S(\boldsymbol{\alpha}).$$

Clearly, $S(\boldsymbol{\alpha}) = \{\boldsymbol{x} \in \mathcal{K}(\boldsymbol{\alpha})\,|\,f(\boldsymbol{x};\ \boldsymbol{\alpha}) = f^o(\boldsymbol{\alpha})\}$.

Hence, the optimal value function has as its arguments the optimal values $\boldsymbol{x}_o$ of the independent variables $\boldsymbol{x}$. And once the optimal values of the x_is have been substituted into f, the latter "indirectly" becomes a function of $\boldsymbol{\alpha}$. In this regard, the optimal value function is alternatively termed the **indirect objective function**. And since this function displays the path taken by the maximum values of f as $\boldsymbol{\alpha}$ varies, the optimal value function is the **envelope** of this set of optimal objective function values determined by varying the parameters α_r, $r = 1, \ldots, p$.

If we want to obtain meaningful envelope results, then the feasible set $\mathcal{K}(\boldsymbol{\alpha})$ must change continuously as $\boldsymbol{\alpha}$ changes. But how does a set such as $\mathcal{K}(\boldsymbol{\alpha})$ change continuously? As we shall now see, a correspondence is continuous if it is both upper and lower hemicontinuous.

16.2 CONTINUOUS CORRESPONDENCES

How do we describe the notion of a continuous correspondence? And what impact does a continuous correspondence have on our ability to garner envelope results? To answer these questions, we need the concepts of upper and lower hemicontinuity. Two views of upper as well as lower hemicontinuity will be offered. Moreover, both Ω and X will be taken to be topological spaces.

We first address the upper hemicontinuity concept in terms of open sets. Specifically,

Definition 16.2: (Upper hemicontinuity in terms of open sets). The correspondence $\mathcal{K}: \Omega \rightarrow X$ is **upper hemicontinuous** (uhc) at $\boldsymbol{\alpha}_o \in \Omega$ if for each open set $A \subset X$ containing the image set $\mathcal{K}(\boldsymbol{\alpha}_o)(\neq\varnothing)$, there exists a neighborhood $\delta(\boldsymbol{\alpha}_o)$ of $\boldsymbol{\alpha}_o$ such that $\boldsymbol{\alpha} \in \delta(\boldsymbol{\alpha}_o)$ implies $\mathcal{K}(\boldsymbol{\alpha}) \subset A$ (Figure 16.1).

So, for every open subset A of X containing $\mathcal{K}(\boldsymbol{\alpha}_o)$, $\mathcal{K}(\boldsymbol{\alpha})$ is also contained in A. Thus, for any $\boldsymbol{\alpha}$ near $\boldsymbol{\alpha}_o$, $\mathcal{K}(\boldsymbol{\alpha})$ is in A as well.

If $\mathcal{K}: \Omega \rightarrow X$ is **compact-valued** (the image set $\mathcal{K}(\boldsymbol{\alpha})$ is a compact subset of X for every $\boldsymbol{\alpha} \in \Omega$), then upper hemicontinuity in terms of neighborhoods has an equivalent specification in terms of sequences. A necessary and sufficient condition for $\mathcal{K}$ to be uhc at $\boldsymbol{\alpha}_o \in \Omega$ is provided by

Theorem 16.1: (Upper hemicontinuity in terms of convergent sequences). The compact-valued point-to-set function $\mathcal{K}: \Omega \rightarrow X$ is uhc at $\boldsymbol{\alpha}_o \in \Omega$ if and only if for every $\{\boldsymbol{\alpha}_k\} \rightarrow \boldsymbol{\alpha}_o$ and every $\{\boldsymbol{x}_k\}$ with $\boldsymbol{x}_k \in \mathcal{K}(\boldsymbol{\alpha}_k)$ there is some $\{\boldsymbol{x}_{k_j}\} \rightarrow \boldsymbol{x}_o \in \mathcal{K}(\boldsymbol{\alpha}_o)$, $j = 1, 2, \ldots$.

That is, as $\{\boldsymbol{\alpha}_k\}$ converges to $\boldsymbol{\alpha}_o \in \Omega$, the sequence $\{\boldsymbol{x}_k\}$ with $\boldsymbol{x}_k \in \mathcal{K}(\boldsymbol{\alpha}_k)$ admits a convergent subsequence whose limit $\boldsymbol{x}_o \in \mathcal{K}(\boldsymbol{\alpha}_o)$. Looked at in an alternative fashion, $\mathcal{K}$ is uhc at $\boldsymbol{\alpha}_o$ if the image set $\mathcal{K}(\boldsymbol{\alpha}_o)$ is "large enough" to contain all limit points $\boldsymbol{x}_o$ of $\{\boldsymbol{x}_k\}$. The upper hemicontinuity of $\mathcal{K}$ at $\boldsymbol{\alpha}_o$ requires that $\mathcal{K}(\boldsymbol{\alpha}_o)$" does not suddenly become much larger" for some small perturbations in $\boldsymbol{\alpha}_o$.

The correspondence $\mathcal{K}: \Omega \rightarrow X$ is **uhc** if it is uhc at each point of Ω.

We next examine the lower hemicontinuity of a correspondence in terms of open sets. That is,

Definition 16.3: (Lower hemicontinuity in terms of open sets). The point-to-set function $\mathcal{K}: \Omega \rightarrow X$ is termed **lower hemicontinuous** (lhc) at $\boldsymbol{\alpha}_o$ if for each open set $A \subset X$ with $\mathcal{K}(\boldsymbol{\alpha}_o) \cap A \neq \varnothing$, there exists a neighborhood $\delta(\boldsymbol{\alpha}_o)$ of $\boldsymbol{\alpha}_o$ such that $\boldsymbol{\alpha} \in \delta(\boldsymbol{\alpha}_o)$ implies $\mathcal{K}(\boldsymbol{\alpha}) \cap A \neq \varnothing$ (Figure 16.2).

So for every open subset A of X with $\mathcal{K}(\boldsymbol{\alpha}_o) \cap A \neq \varnothing$, the image set $\mathcal{K}(\boldsymbol{\alpha})$ also has a non-null intersection with A when $\boldsymbol{\alpha}$ is near $\boldsymbol{\alpha}_o$.

For $\mathcal{K}: \Omega \rightarrow X$ a set-valued function of Ω into X, lower hemicontinuity in terms of open sets has a parallel representation in terms of sequences. (Unlike the uhc case, the assumption that $\mathcal{K}$ is compact-valued is not needed.)

Theorem 16.2: (Lower hemicontinuity in terms of convergent sequences). The point-to-set function $\mathcal{K}: \Omega \rightarrow X$ is lhc at $\boldsymbol{\alpha}_o \in \Omega$ if and only if $\{\boldsymbol{\alpha}_k\} \rightarrow \boldsymbol{\alpha}_o$, $\{\boldsymbol{x}_k\} \rightarrow \boldsymbol{x}_o$, and $\boldsymbol{x}_k \in \mathcal{K}(\boldsymbol{\alpha}_k)$ for all k imply that $\boldsymbol{x}_o \in \mathcal{K}(\boldsymbol{\alpha}_o)$.

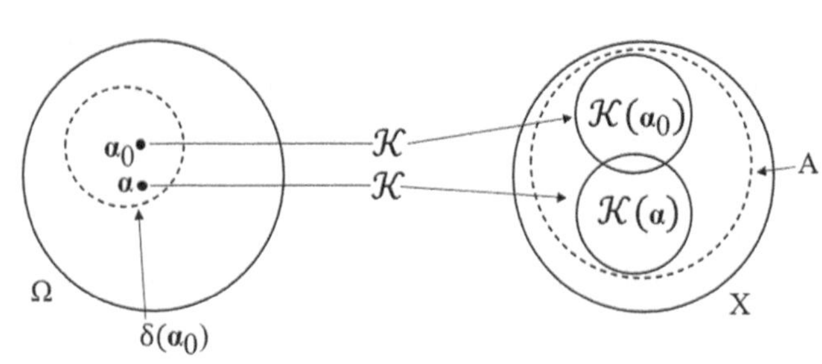

FIGURE 16.1 The correspondence $\mathcal{K}: \Omega \rightarrow X$ is upper hemicontinuous.

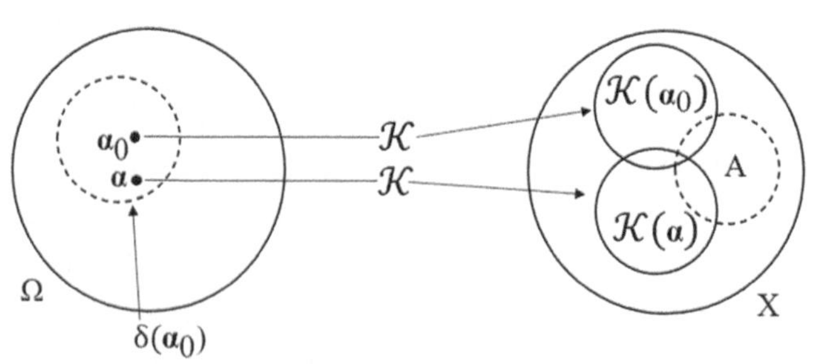

FIGURE 16.2 The correspondence $\mathcal{K}: \Omega \rightarrow X$ is lower hemicontinuous.

Here the lower hemicontinuity of $\mathcal{K}$ at $\boldsymbol{\alpha}_o$ requires that $\mathcal{K}(\boldsymbol{\alpha}_o)$ "does not suddenly become much smaller" for small perturbations in $\boldsymbol{\alpha}_o$.

As one might have anticipated, the correspondence $\mathcal{K}$: $\Omega \to X$ is **lhc** if it is lhc at each point of Ω.

A point-to-set function $\mathcal{K}$: $\Omega \to X$ is **continuous at** $\boldsymbol{\alpha}_o$ if it is both uhc and lhc at $\boldsymbol{\alpha}_o$. For a set-valued function $\mathcal{K}$: $\Omega \to X$, if $\mathcal{K}(\boldsymbol{\alpha})$ always consists of a single element $\{\boldsymbol{x}\} \in X$, then we have a point-to-point function f. In this instance the concepts of upper and lower semicontinuity coincide and are equivalent to the continuity of a point-to-point function f.

A concept closely related to hemicontinuity is that of the "closedness" of a set-valued function $\mathcal{K}$. Let $\mathcal{K}$: $\Omega \to X$ be a point-to-set mapping of Ω into X. Then $\mathcal{K}$ is said to be **closed at** $\boldsymbol{\alpha}_o \in \Omega$ if whenever $\{\boldsymbol{\alpha}_k\} \to \boldsymbol{\alpha}_o$, $\{\boldsymbol{x}_k\} \to \boldsymbol{x}_o$, and $\boldsymbol{x}_k \in \mathcal{K}(\boldsymbol{\alpha}_k)$ for all k, then $\boldsymbol{x}_o \in \mathcal{K}(\boldsymbol{\alpha}_o)$. $\mathcal{K}$ is said to be **closed** if it is closed at every $\boldsymbol{\alpha}$ within the domain of $\mathcal{K}$. Looked at from another perspective, the correspondence $\mathcal{K}$: $\Omega \to X$ is closed (or has the **closed graph property**) if its graph $G_{\mathcal{K}} = \{(\boldsymbol{\alpha}, \boldsymbol{x}) \in \Omega \times X | \boldsymbol{x} \in \mathcal{K}(\boldsymbol{\alpha})\}$ is a closed subset of $\Omega \times X$. A closed point-to-set function is **closed-valued** if $\mathcal{K}(\boldsymbol{\alpha})$ is a closed subset of X for every $\boldsymbol{\alpha} \in \Omega$. In general, the point-to set mapping $\mathcal{K}$: $\Omega \to X$ may be closed but not uhc and conversely. However, if $\mathcal{K}$ is a closed-valued mapping of Ω into a compact set X, then the notions of a closed-valued function and a uhc set-valued function coincide. If $\mathcal{K}$: $\Omega \to X$ is closed-valued and uhc, then $\mathcal{K}$ is closed.

Example 16.1: Suppose the constraint set $\mathcal{K}(\boldsymbol{\alpha}) = \{\boldsymbol{x} \in R^2 | \alpha_1 x_1 + \alpha_2 x_2 \le \alpha_3, \boldsymbol{x} \ge \boldsymbol{0}\}$. Clearly, $\boldsymbol{\alpha}^I = (\alpha_1, \alpha_2, \alpha_3) \to \mathcal{K}(\boldsymbol{\alpha})$ defines a correspondence from $R_{++}^{n+1=3}$ into $R_{+}^{n=2}$ (Figure 16.3). ∎

Section 16.2 Exercises:

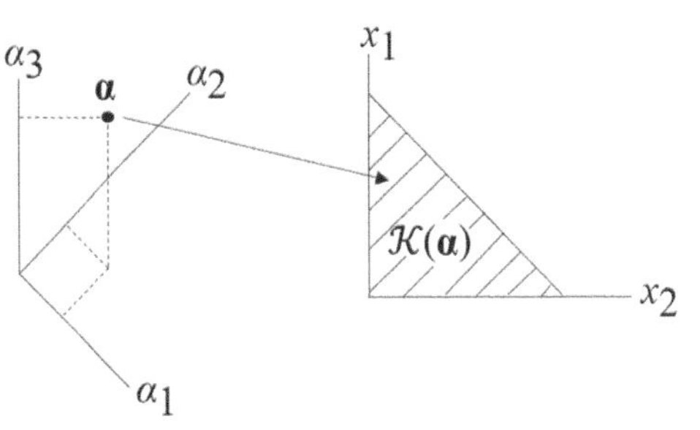

FIGURE 16.3 The correspondence of Ω into X.

Before attempting to complete these exercises, the reader should be reminded that:

a. The correspondence F: $X \to Y$ is uhc at a point $\boldsymbol{x}_o \in X$ if there exists an open set $A \subseteq Y$ containing $F(\boldsymbol{x}_o)$ which also contains $F(\boldsymbol{x}), \boldsymbol{x} \in \delta(\boldsymbol{x}_o)$.

b. The correspondence F: $X \to Y$ is lhc at a point $\boldsymbol{x}_o \in X$ if there exists an open set $A \subseteq Y$ having points in common with $F(\boldsymbol{x}_o)$ and with $F(\boldsymbol{x}), \boldsymbol{x} \in \delta(\boldsymbol{x}_o)$.

c. A correspondence F: $X \to Y$ is compact-valued if $F(\boldsymbol{x})$ is a compact subset of Y.

d. A correspondence F: $X \to Y$ is closed or has the closed-graph property if G_F is a closed subset of $X \times Y$.

16.2.1 For $X = Y = [0, 1]$, let F: $X \to Y$ be defined as

$$F(x) = \left\{y \in Y | \frac{1}{2}x \le y \le x, x \in X\right\}.$$

a. Is F uhc at $x = 0.8$?

b. Is F lhc at $x = 0.8$?

c. Is G_F closed?

d. Is F compact-valued?

16.2.2 Suppose the correspondence F: $X \to Y$ is defined as

$$F(x) = \begin{cases} [0, 1], 0 \le x < 1; \\ \left[0, \frac{1}{3}\right], x = 1. \end{cases}$$

a. Is F uhc at $x = 1$?

b. Is F lhc at $x = 1$?

c. Is F compact valued?

d. Does F have the closed-graph property?

16.2.3 Let the correspondence F: $X \to Y$ be specified as

$$F(x) = \begin{cases} [1, 4], x < 2; \\ \{2\}, x \ge 2. \end{cases}$$

a. Is F uhc at $x = 2$?

b. Is F lhc at $x = 2$?

c. Is F compact valued?

d. Does F exhibit the closed-graph property?

16.3 THE MAXIMUM THEOREM (BERGE [1963])

We now turn to the justification for a parametric optimization to have continuous solutions with respect to the parameters α_r, $r = 1, \ldots, p$, i.e., that problem (16.1) is assured to have solutions that *vary smoothly* with $\boldsymbol{\alpha}$. We thus state

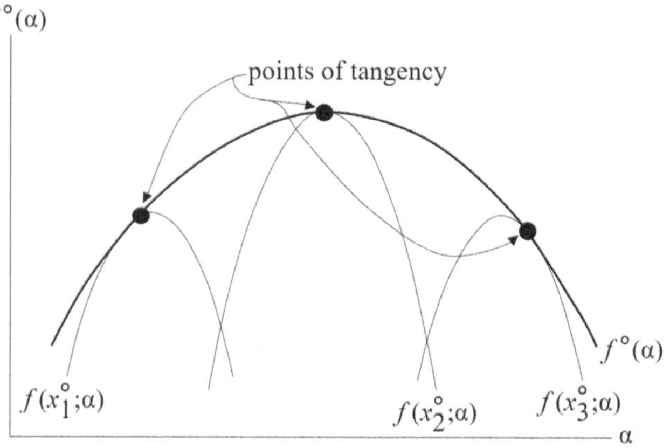

FIGURE 16.4 Upper envelope curve.

Theorem 16.3: **(Berge's Maximum Theorem)**. For the set $X \subseteq R^n$ and parameter set $\Omega \in R^p$, let $f(\boldsymbol{x}; \boldsymbol{\alpha})$: $X \times \Omega \to R$ be a continuous function and let the constraint correspondence $\mathcal{K}$: $\Omega \to X$, with feasible constraints set $\mathcal{K}(\boldsymbol{\alpha})$, be compact-valued and continuous. Given the parametric optimization problem (16.1), the optimal value function $f^o(\boldsymbol{\alpha})$ is continuous and the solution correspondence S: $\Omega \to X$, with optimal solution set $S(\boldsymbol{\alpha})$, is non-empty, compact-valued, and upper hemicontinuous.

Thus the maximum theorem gives conditions (on problem (16.1)) under which the optimal value function $f^o(\boldsymbol{\alpha})$ is continuous and the set of optimal solutions $S(\boldsymbol{\alpha})$ is upper hemicontinuous with compact values and whose elements $x_i^o = x_i^o(\boldsymbol{\alpha})$, $i = 1, \ldots, n$, are single-valued and thus continuous as well. These conditions are that f is continuous as a function of $\boldsymbol{x}$ and $\boldsymbol{\alpha}$ and the feasible constraint set $\mathcal{K}(\boldsymbol{\alpha})$ is compact and changes continuously (since $\mathcal{K}$ is a correspondence that is both upper and lower hemicontinuous) as $\boldsymbol{\alpha}$ does.

For a proof of this theorem see Appendix 16.A.

16.4 THE OPTIMAL VALUE OR ENVELOPE FUNCTION

We noted above that for problem (16.1), the optimal value function f^o: $\Omega \to R$ gives the maximum value of f for each $\boldsymbol{\alpha}$ or

$$
\begin{aligned}
f^o(\boldsymbol{\alpha}) &= \max_{\boldsymbol{x}}\{f(\boldsymbol{x}; \boldsymbol{\alpha}) \,|\, \boldsymbol{x} \in \mathcal{K}(\boldsymbol{\alpha})\} \\
&= f(\boldsymbol{x}_o(\boldsymbol{\alpha}); \boldsymbol{\alpha}),\ \boldsymbol{x}_o \in S(\boldsymbol{\alpha}).
\end{aligned}
$$

So, for each admissible $\boldsymbol{x}$, if we plot f as a function of $\boldsymbol{\alpha}$, then we obtain a family of curves. Then the optimal value function is the **envelope curve** of this family (it is tangent to each member of the family and intersects none of them), i.e., in the case of an unconstrained maximum problem, the envelope curve is the *upper envelope* of the family of curves given in Figure 16.4, where $x, \alpha \in R$.

An important property of the optimal value function is that, under certain conditions, it is a concave function of $\boldsymbol{\alpha}$. To see this we need to make problem (17.1) a bit more specific. That is, let us

$$
\begin{aligned}
&\text{maximize } y = f(\boldsymbol{x}; \boldsymbol{\alpha}) \text{ subject to} \\
&\qquad \mathbf{G}(\boldsymbol{x}; \boldsymbol{\alpha}) \geq \boldsymbol{O},
\end{aligned}
\tag{16.1.1}
$$

where $\mathbf{G}$ is an $(m \times 1)$ vector-valued function with jth component $g^j(\boldsymbol{x}; \boldsymbol{\alpha})$, $j = 1, \ldots, m$.

We can now state

Theorem 16.4: **(Concavity of the Optimal Value Function)**. Let the optimal value function associated with problem (16.1.1) be $f^o(\boldsymbol{\alpha}) = \max\{f(\boldsymbol{x}; \boldsymbol{\alpha}) \,|\, \mathbf{G}(\boldsymbol{x}; \boldsymbol{\alpha}) \geq \boldsymbol{O}\}$ and let f be concave in $\boldsymbol{x}$ and $\boldsymbol{\alpha}$ with all the constraint functions $g^j(\boldsymbol{x}; \boldsymbol{\alpha})$, $j = 1, \ldots, m$, quasiconcave. Then $f^o(\boldsymbol{\alpha})$ is concave in $\boldsymbol{\alpha}$.

Proof: Given $\boldsymbol{x}_1, \boldsymbol{x}_2 \in X$ and $\boldsymbol{\alpha}_1, \boldsymbol{\alpha}_2 \in \Omega$, let $(\boldsymbol{x}_1; \boldsymbol{\alpha}_1)$ and $(\boldsymbol{x}_2; \boldsymbol{\alpha}_2)$ be feasible points in that $g^j(\boldsymbol{x}_1; \boldsymbol{\alpha}_1) \geq 0$ and $g^j(\boldsymbol{x}_2; \boldsymbol{\alpha}_2) \geq \boldsymbol{0}$ for all j. Suppose also that $\boldsymbol{x}_1 = \boldsymbol{x}_1(\boldsymbol{\alpha}_1)$, $\boldsymbol{x}_2 = \boldsymbol{x}_2(\boldsymbol{\alpha}_2)$ represent optimal solutions to Problem (16.1.1).

Additionally, let the point $(\boldsymbol{x}_\theta, \boldsymbol{\alpha}_\theta) \in R^{n+p}$, where $\boldsymbol{x}_\theta = \theta\boldsymbol{x}_1 + (1 - \theta)\boldsymbol{x}_2$ and $\boldsymbol{\alpha}_\theta = \theta\boldsymbol{\alpha}_1 + (1 - \theta)\boldsymbol{\alpha}_2$, $\theta \in [0, 1]$. That $\boldsymbol{x}_\theta$ is feasible for $\boldsymbol{\alpha}_\theta$ (but it may not be optimal) follows from the quasiconcavity of $g^j(\boldsymbol{x}; \boldsymbol{\alpha})$ since

$$g^j(\boldsymbol{x}_o; \boldsymbol{\alpha}_\theta) \geq \min\{g^j(\boldsymbol{x}_1; \boldsymbol{\alpha}_1), g^j(\boldsymbol{x}_2; \boldsymbol{\alpha}_2)\} \geq 0$$

for all j. With

$$\begin{aligned} f^o(\boldsymbol{\alpha}_\theta) &= f(\boldsymbol{x}(\boldsymbol{\alpha}_\theta); \boldsymbol{\alpha}(\theta)) \geq f(\boldsymbol{x}_\theta; \boldsymbol{\alpha}_\theta) \\ &\geq \theta f(\boldsymbol{x}_1(\boldsymbol{\alpha}_1); \boldsymbol{\alpha}_1) + (1-\theta) f(\boldsymbol{x}_2(\boldsymbol{\alpha}_2); \boldsymbol{\alpha}_2) \\ &= \theta f^o(\boldsymbol{\alpha}_1) + (1-\theta) f^o(\boldsymbol{\alpha}_2), \end{aligned}$$

since f was assumed to be concave, it follows that f^o is also concave. Q. E. D.

16.5 ENVELOPE THEOREMS

The following set of theorems addresses the issue of how the optimal value function changes when one of the function parameters changes.

16.5.1 $y = f(x; \alpha)$, $x \in X \subseteq R$, $\alpha \in \Omega \subseteq R$.
Suppose our basic problem is to find an unconstrained local maximum of $f(x; \alpha)$. Two approaches can be taken:

1. Setting $f' = 0$ we can solve for $x_o = x_o(\alpha)$, the optimal level of x (provided, of course, that $f''(x_o) < 0$). Thus, the optimal value function is $f^o(\alpha) = f(x_o(\alpha); \alpha)$. We can then find $df^o(\alpha)/d\alpha$, the total effect of a change in α on f as x adjusts to a new optimal level.

2. Hold x constant at its optimal value $x_o = x_o(\alpha)$, calculate $\partial f/\partial\alpha$, and then evaluate it at $x_o(\alpha)$, i.e., find $\left.\frac{\partial f}{\partial \alpha}\right|_{x_o(\alpha)}$, the partial effect of a change in α on f with x held fixed at its optimal level. This second approach is the basis of the following theorem.

Theorem 16.5: (**Envelope Theorem I**). Let $y = f(x; \alpha)$ be of class $C^{(2)}$ with $x_o = x_o(\alpha)$ yielding a strong local maximum of f. Then the optimal value function $f^o(\alpha) = \max\limits_x f(x; \alpha) = f(x_o(\alpha); \alpha)$ is differentiable with respect to α with

$$\frac{df^o(\alpha)}{d\alpha} = \left.\frac{\partial f}{\partial \alpha}\right|_{x_o(\alpha)}. \tag{16.2}$$

Proof: Given the assumptions of the theorem, x_o is a well-defined differentiable function of α, $x_o = x_o(\alpha)$. If we differentiate the optimal value function $f^o(\alpha) = f(x_o(\alpha); \alpha)$ with respect to α we obtain

$$\frac{df^o(\alpha)}{d\alpha} = \frac{\partial f}{\partial x}\frac{dx_o}{d\alpha} + \frac{\partial f}{\partial \alpha}.$$

With x fixed at its optimal level, $\partial f/\partial x = 0$ and thus

$$\frac{df^o(\alpha)}{d\alpha} = \left.\frac{\partial f}{\partial \alpha}\right|_{x_o(\alpha)}. \text{ Q. E. D.}$$

As this theorem reveals, a marginal change in α affects f^o in two distinct ways: (i) f^o changes *directly* since f^o is a function of α; and (ii) f^o changes *indirectly* through the change in $x_o(\alpha)$. But since $\partial f/\partial x = 0$ at optimality, the indirect change in f^o will be zero for infinitesimal departures in x from x_o. Thus, only the direct effect is pertinent.

Example 16.2: For $= f(x; \alpha) = x^2 - \alpha x$, use the two approaches stated above to determine how the optimal value of f changes when α changes. (Remember that the first involves finding $df^o/d\alpha$ while the second applies the envelope theorem.)

1. Setting $f' = 0$ we obtain $x_o = x_o(\alpha) = \alpha/2$. (With $f'' = 2 > 0$, f has a strong local minimum at x_o.) Then $f^o(\alpha) = f(x_o(\alpha); \alpha) = (\alpha/2)^2 - \alpha(\alpha/2) = -\alpha^2/4$ and thus $df^o(\alpha)/d\alpha = -\frac{1}{2}\alpha$.

2. According to the envelope theorem, we are to hold x constant at its optimal level, calculate $\partial f/\partial\alpha$, and then evaluate this partial derivative at $x_o(\alpha)$. To this end we have $\partial f/\partial\alpha = -x$. With $x_o(\alpha) = \alpha/2$, we obtain, from Equation (16.2),

$$\left.\frac{\partial f}{\partial \alpha}\right|_{x_o(\alpha)} = -x_o(\alpha) = -\frac{1}{2}\alpha.$$

Thus $df^o(\alpha)/d\alpha$ (the *total effect* of a change in α on f as x adjusts to a new optimal level) equals $\left.\frac{\partial f}{\partial \alpha}\right|_{x_o(\alpha)}$ (the partial effect of a change in α on f with x fixed at its optimal value).

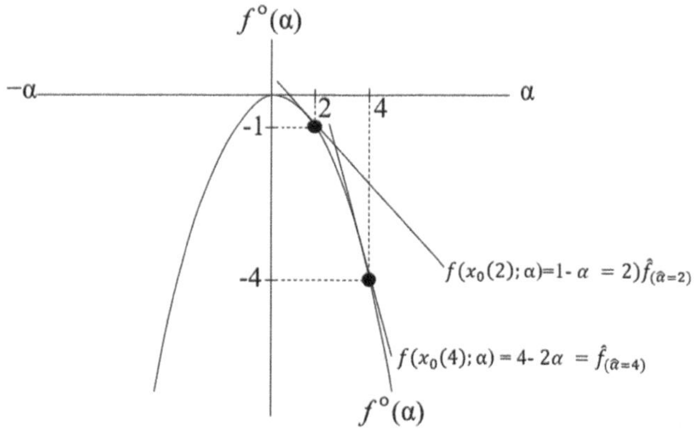

FIGURE 16.5 The envelope curve $f^o(\alpha) = -\alpha^2/4$.

Let's now see why Theorem 16.5 is called an envelope theorem. In what follows we shall work with the optimal value function $f^o(\alpha)$.

Set $\alpha = \hat{\alpha} = 2$. Then $x_o(\hat{\alpha}) = 1, f^o(\hat{\alpha}) = -1$, and $\left.\frac{df}{d\alpha}\right|_{\hat{\alpha}=2} = -1$. Also, $f(x_o(\hat{\alpha}); \alpha) = 1 - \alpha = \hat{f}_{(\hat{\alpha}=2)}$ (Figure 16.5) with $\frac{d\hat{f}}{d\alpha} = -1 = \left.\frac{df^o}{d\alpha}\right|_{\hat{\alpha}=2}$.

Next, set $\alpha = \hat{\alpha} = 4$. Now $x_o(\hat{\alpha}) = 2, f^o(\hat{\alpha}) = -4$, and $\left.\frac{df^o}{d\alpha}\right|_{\hat{\alpha}=4} = -2$. Also, $f(x_o(\hat{\alpha}); \alpha) = 4 - 2\alpha = \hat{f}_{(\hat{\alpha}=4)}$ (Figure 16.5) with $\frac{d\hat{f}}{d\alpha} = -2 = \left.\frac{df^o}{d\alpha}\right|_{\hat{\alpha}=4}$.

Further inspection of Figure 16.5 reveals that $\hat{f} = f(x_o(\hat{\alpha}); \alpha) \geq f^o(\alpha)$ with $\hat{f} = f(x_o(\hat{\alpha}); \hat{\alpha}) = f^o(\hat{\alpha})$ and

$$\left.\frac{df^o}{d\alpha}\right|_{\hat{\alpha}} = \left.\frac{d\hat{f}}{d\alpha}\right|_{\hat{\alpha}}.$$

Thus $f^o(\alpha)$, the optima value function, is the envelope curve of the $\hat{f}$ functions, where $\hat{f}$ holds x constant at its optimal level.

Clearly, $f^o(\alpha)$ is tangent to each $\hat{f}$ function and intersects none of them. ■

17.5.2 $y = f(\boldsymbol{x}; \boldsymbol{\alpha}), \boldsymbol{x} \in X \subseteq R^n, \boldsymbol{\alpha} \in \Omega \subseteq R^p$

Here our basic problem is to find an unconstrained maximum of $f(\boldsymbol{x}; \boldsymbol{\alpha})$. Let's go directly to

Theorem 16.6: (Envelope Theorem II). Let $y = f(\boldsymbol{x}; \boldsymbol{\alpha})$ be of class $C^{(2)}$ with $\boldsymbol{x}_o = \boldsymbol{x}_o(\boldsymbol{\alpha})$ yielding a strong local maximum of f. Then the optimal value function $f^o(\boldsymbol{\alpha}) = \max_{\boldsymbol{x}} f(\boldsymbol{x}; \boldsymbol{\alpha}) = f^o(\boldsymbol{x}_o(\boldsymbol{\alpha}); \boldsymbol{\alpha})$ is differentiable with respect to $\boldsymbol{\alpha}$ and

$$\nabla_{\boldsymbol{\alpha}} f^o(\boldsymbol{\alpha}) = \nabla_{\boldsymbol{\alpha}} f(\boldsymbol{x}_o(\boldsymbol{\alpha}); \boldsymbol{\alpha}). \tag{16.3}$$

Proof: Under the conditions of the theorem, $\boldsymbol{x}_o$ is a well-defined and differentiable function of $\boldsymbol{\alpha}$, $\boldsymbol{x}_o = \boldsymbol{x}_o(\boldsymbol{\alpha})$. If we differentiae the optimal value function $f^o(\boldsymbol{\alpha}) = f(\boldsymbol{x}_o(\boldsymbol{\alpha}); \boldsymbol{\alpha})$ with respect to $\boldsymbol{\alpha}$, we obtain

$$\nabla_{\boldsymbol{\alpha}} f^o(\boldsymbol{\alpha}) = \nabla_{\boldsymbol{x}} f^o(\boldsymbol{x}_o(\boldsymbol{\alpha}); \boldsymbol{\alpha}) \frac{d\boldsymbol{x}_o(\boldsymbol{\alpha})}{d\boldsymbol{\alpha}} + \nabla_{\boldsymbol{\alpha}} f(\boldsymbol{x}_o(\boldsymbol{\alpha}); \boldsymbol{\alpha}).$$

Holding $\boldsymbol{x}$ fixed at its optimal level, $\nabla_{\boldsymbol{x}} f^o(\boldsymbol{x}_o(\boldsymbol{\alpha}); \boldsymbol{\alpha}) = \boldsymbol{O}$ (the *indirect* effect of a change in f^o through the change in $\boldsymbol{x}_o(\boldsymbol{\alpha})$ vanishes) and thus

$$\nabla_{\boldsymbol{\alpha}} f^o(\boldsymbol{\alpha}) = \nabla_{\boldsymbol{\alpha}} f(\boldsymbol{x}_o(\boldsymbol{\alpha}); \boldsymbol{\alpha}). \quad Q.\ E.\ D.$$

As this theorem reveals, we do not need to actually find $f^o(\boldsymbol{\alpha})$; only the *direct* effect of a marginal change in $\boldsymbol{\alpha}$ on f^o survives in Equation (16.3). That is, looking again to Equation (16.3), $\nabla_{\boldsymbol{\alpha}} f^o(\boldsymbol{\alpha})$ represents the total effect of a change in $\boldsymbol{\alpha}$ on f as the x_i, $i = 1, \ldots, n$, adjust to new optimal levels while $\nabla_{\boldsymbol{\alpha}} f(\boldsymbol{x}_o(\boldsymbol{\alpha}); \boldsymbol{\alpha})$ is the partial effect of a change in $\boldsymbol{\alpha}$ on f with the x_i, $i = 1, \ldots, n$, fixed at their optimal levels.

Example 16.3: For $y = f(x_1, x_2; \alpha) = x_1^2 - \alpha x_1 x_2 + \frac{1}{2}x_2^2 - x_1$, use the envelope theorem to determine how the optimal value of f changes as α changes. That is, we need to find $df^o/d\alpha$. According to Theorem 16.6, $\boldsymbol{x}$ is held constant at its optimal level, $df/d\alpha$ is determined, and then this partial derivative is evaluated at $\boldsymbol{x}_o(\boldsymbol{\alpha})$. In this regard, from

$$f_1 = 2x_1 - \alpha x_2 - 1 = 0$$
$$f_2 = -\alpha x_1 + x_2 = 0$$

we obtain $x_1^o = (2 - \alpha^2)^{-1}$ and $x_2^o = \alpha(2 - \alpha^2)^{-1}$. (Note that the Hessian determinant

$$\begin{vmatrix} f_{11} & f_{12} \\ f_{21} & f_{22} \end{vmatrix} = \begin{vmatrix} 2 & -\alpha \\ -\alpha & 1 \end{vmatrix} = 2 - \alpha^2 \not\equiv 0$$

And with $f_{11} > 0$, we see that f attains a strong local minimum at $\boldsymbol{x}_o$ if $2 - \alpha^2 > 0$.) We next find $\partial f/\partial\alpha = -x_1x_2$ and thus

$$\left.\frac{\partial f}{\partial \alpha}\right|_{\boldsymbol{x}_o(\alpha)} = -\alpha(2-\alpha^2)^{-2} = \frac{df^o(\alpha)}{d\alpha}. \quad ■$$

Example 16.4: For $y = f(x_1, x_2; \alpha_1, \alpha_2) = x_1^2 - \alpha_1x_1x_2 + \frac{1}{2}x_2^2 - \alpha_2x_1$, use Envelope Theorem II to determine how the minimal value of f changes as $\boldsymbol{\alpha}^T = (\alpha_1, \alpha_2)$ changes. From

$$f_1 = 2x_1 - \alpha_1x_2 - \alpha_2 = 0$$
$$f_2 = -\alpha_1x_1 + x_2 = 0$$

we obtain

$$\boldsymbol{x}_o(\boldsymbol{\alpha}) = \begin{bmatrix} \alpha_2/(2-\alpha_1^2) \\ \alpha_1\alpha_2/(2-\alpha_1^2) \end{bmatrix}.$$

Next,

$$\nabla_{\boldsymbol{\alpha}} f = \begin{bmatrix} \partial f/\partial\alpha_1 \\ \partial f/\partial\alpha_2 \end{bmatrix} = \begin{bmatrix} -x_1x_2 \\ -x_1 \end{bmatrix}$$

and thus

$$\nabla_{\boldsymbol{\alpha}} f(\boldsymbol{x}_o(\boldsymbol{\alpha}); \boldsymbol{\alpha}) = \begin{bmatrix} -\alpha_1\alpha_2^2/(2-\alpha_1^2)^2 \\ -\alpha_2/(2-\alpha_1^2) \end{bmatrix} = \nabla_{\boldsymbol{\alpha}} f^o(\boldsymbol{\alpha}). \quad ■$$

16.5.3 $y = f(\boldsymbol{x}; \boldsymbol{\alpha}),\ \boldsymbol{G}(\boldsymbol{x}; \boldsymbol{\alpha}) = \boldsymbol{O},\ \boldsymbol{x} \in X \subseteq R^n,\ \boldsymbol{\alpha} \in \Omega \subseteq R^p$

Suppose our basic problem is to maximize $f(\boldsymbol{x}; \boldsymbol{\alpha})$ subject to the set of parameterized equality constraints $\boldsymbol{G}(\boldsymbol{x}; \boldsymbol{\alpha}) = \boldsymbol{O}$, where $\boldsymbol{G}$ is an $(m \times 1)$ vector-valued function with jth component $g^j(\boldsymbol{x}; \boldsymbol{\alpha}),\ j = 1, \ldots, m$. The Lagrangian associated with this problem has the form $L(\boldsymbol{x}, \boldsymbol{\lambda}; \boldsymbol{\alpha}) = f(\boldsymbol{x}; \boldsymbol{\alpha}) + \boldsymbol{\lambda}^T\boldsymbol{G}(\boldsymbol{x}; \boldsymbol{\alpha})$, where $\boldsymbol{\lambda}$ is an $(m \times 1)$ vector of undetermined Lagrange multipliers. We now turn to

Theorem 16.7: (**Envelope Theorem III**). Let f, $\boldsymbol{G}$ be of class $C^{(2)}$. If $\boldsymbol{x}_o(\boldsymbol{\alpha})$ is a strong local constrained maximum of f, then the optimal value function $f^o(\boldsymbol{\alpha}) = \max_{\boldsymbol{x}} \{f(\boldsymbol{x}; \boldsymbol{\alpha}) | \boldsymbol{G}(\boldsymbol{x}; \boldsymbol{\alpha}) = \boldsymbol{O}\}$ is differentiable with respect to $\boldsymbol{\alpha}$ and

$$\begin{aligned}\nabla_{\boldsymbol{\alpha}} f^o(\boldsymbol{\alpha}) &= \nabla_{\boldsymbol{\alpha}} L(\boldsymbol{x}_o(\boldsymbol{\alpha}), \boldsymbol{\lambda}_o(\boldsymbol{\alpha}); \boldsymbol{\alpha}) \\ &= \nabla_{\boldsymbol{\alpha}} f(\boldsymbol{x}_o(\boldsymbol{\alpha}); \boldsymbol{\alpha}) + \nabla_{\boldsymbol{\alpha}} \mathrm{G}(\boldsymbol{x}_o(\boldsymbol{\alpha}); \boldsymbol{\alpha})^T\boldsymbol{\lambda}_o(\boldsymbol{\alpha})\end{aligned}$$

or

$$\nabla_{\boldsymbol{\alpha}} f^o(\boldsymbol{\alpha}) = \nabla_{\boldsymbol{\alpha}} L^o(\boldsymbol{\alpha}) = \nabla_{\boldsymbol{\alpha}} L|_{\boldsymbol{x}_o(\boldsymbol{\alpha}), \boldsymbol{\lambda}_o(\boldsymbol{\alpha})}. \quad (16.4)$$

Proof: Under the assumptions of the theorem, $\boldsymbol{x}_o$ and $\boldsymbol{\lambda}_o$ are well-defined differentiable functions of $\boldsymbol{\alpha}$ with $\boldsymbol{x}_o = \boldsymbol{x}_o(\boldsymbol{\alpha})$, $\boldsymbol{\lambda}_o = \boldsymbol{\lambda}_o(\boldsymbol{\alpha})$. Hence, we can write the optimal value function as $f^o(\boldsymbol{\alpha}) = f(\boldsymbol{x}_o(\boldsymbol{\alpha}); \boldsymbol{\alpha})$ with associated Lagrangian

$$L^o(\boldsymbol{\alpha}) = f(\boldsymbol{x}_o(\boldsymbol{\alpha}); \boldsymbol{\alpha}) + \boldsymbol{\lambda}_o(\boldsymbol{\alpha})^T\boldsymbol{G}(\boldsymbol{x}_o(\boldsymbol{\alpha}); \boldsymbol{\alpha}) = f^o(\boldsymbol{\alpha}).$$

Then

$$\begin{aligned}\nabla_{\boldsymbol{\alpha}} L^o(\boldsymbol{\alpha}) = {} & [\nabla_{\boldsymbol{x}} f(\boldsymbol{x}_o(\boldsymbol{\alpha}); \boldsymbol{\alpha}) + \boldsymbol{\lambda}_o(\boldsymbol{\alpha})^T\nabla_{\boldsymbol{x}}\boldsymbol{G}(\boldsymbol{x}_o(\boldsymbol{\alpha}); \boldsymbol{\alpha})]\frac{d\boldsymbol{x}_o(\boldsymbol{\alpha})}{d\boldsymbol{\alpha}} \\ & + \boldsymbol{G}(\boldsymbol{x}_o(\boldsymbol{\alpha}); \boldsymbol{\alpha})^T\frac{d\boldsymbol{\lambda}_o(\boldsymbol{\alpha})}{d\boldsymbol{\alpha}} \\ & + \nabla_{\boldsymbol{\alpha}} f(\boldsymbol{x}_o(\boldsymbol{\alpha}); \boldsymbol{\alpha}) + \boldsymbol{\lambda}_o(\boldsymbol{\alpha})^T\nabla_{\boldsymbol{\alpha}}\boldsymbol{G}(\boldsymbol{x}_o(\boldsymbol{\alpha}); \boldsymbol{\alpha}).\end{aligned}$$

If $\boldsymbol{x}$ and $\boldsymbol{\lambda}$ are fixed at their optimal levels, then the preceding expression simplifies to

$$\begin{aligned}\nabla_{\boldsymbol{\alpha}} L^o(\boldsymbol{\alpha}) &= \nabla_{\boldsymbol{\alpha}} f(\boldsymbol{x}_o(\boldsymbol{\alpha}); \boldsymbol{\alpha}) + \boldsymbol{\lambda}_o(\boldsymbol{\alpha})^T\nabla_{\boldsymbol{\alpha}}\boldsymbol{G}(\boldsymbol{x}_o(\boldsymbol{\alpha}); \boldsymbol{\alpha}) \\ &= \nabla_{\boldsymbol{\alpha}} L|_{\boldsymbol{x}_o(\boldsymbol{\alpha}), \boldsymbol{\lambda}_o(\boldsymbol{\alpha})}.\end{aligned}$$

Also, with $f^o(\boldsymbol{\alpha}) = f(\boldsymbol{x}_o(\boldsymbol{\alpha}); \boldsymbol{\alpha})$,

$$\begin{aligned}\frac{df^o(\boldsymbol{\alpha})}{d\boldsymbol{\alpha}} &= \nabla_{\boldsymbol{x}} f(\boldsymbol{x}_o(\boldsymbol{\alpha}); \boldsymbol{\alpha})\frac{d\boldsymbol{x}_o(\boldsymbol{\alpha})}{d\boldsymbol{\alpha}} + \nabla_{\boldsymbol{\alpha}} f(\boldsymbol{x}_o(\boldsymbol{\alpha}); \boldsymbol{\alpha}) \\ &= (-\boldsymbol{\lambda}_o(\boldsymbol{\alpha})^T\nabla_{\boldsymbol{x}}\boldsymbol{G}(\boldsymbol{x}_o(\boldsymbol{\alpha}); \boldsymbol{\alpha}))\frac{d\boldsymbol{x}_o(\boldsymbol{\alpha})}{d\boldsymbol{\alpha}} + \nabla_{\boldsymbol{\alpha}} f(\boldsymbol{x}_o(\boldsymbol{\alpha}); \boldsymbol{\alpha}) \\ &= \nabla_{\boldsymbol{\alpha}} f(\boldsymbol{x}_o(\boldsymbol{\alpha}); \boldsymbol{\alpha}) - \boldsymbol{\lambda}_o(\boldsymbol{\alpha})^T\nabla_{\boldsymbol{\alpha}}\boldsymbol{G}(\boldsymbol{x}_o(\boldsymbol{\alpha}); \boldsymbol{\alpha}) \\ &= \nabla_{\boldsymbol{\alpha}} L|_{\boldsymbol{x}_o(\boldsymbol{\alpha}), \boldsymbol{\lambda}_o(\boldsymbol{\alpha})}.^{2}\end{aligned}$$

Then

$$\nabla_{\boldsymbol{\alpha}} f^o(\boldsymbol{\alpha}) = \nabla_{\boldsymbol{\alpha}} L^o(\boldsymbol{\alpha}) = \nabla_{\boldsymbol{\alpha}} L|_{\boldsymbol{x}_o(\boldsymbol{\alpha}), \boldsymbol{\lambda}_o(\boldsymbol{\alpha})}. \text{ Q. E. D.}$$

How are we to interpret Equation (16.4)? Here $\nabla_{\boldsymbol{\alpha}} f^o(\boldsymbol{\alpha})$ is the total effect of a change in $\boldsymbol{\alpha}$ on f as the components of $\boldsymbol{x}$ adjust to new optimal levels; $\nabla_{\boldsymbol{\alpha}} L^o(\boldsymbol{\alpha})$ is the total effect of a change in L as the components of $\boldsymbol{x}$ and $\boldsymbol{\lambda}$ adjust to new optimal levels; and $\nabla_{\boldsymbol{\alpha}} L|_{\boldsymbol{x}_o(\boldsymbol{\alpha}),\boldsymbol{\lambda}_o(\boldsymbol{\alpha})}$ is the partial effect of a change in $\boldsymbol{\alpha}$ on L with the components of $\boldsymbol{x}$ and $\boldsymbol{\lambda}$ held fixed at their optimal levels.

Let us examine the geometry of Theorem 16.7. Given that we have maximized f subject to $\boldsymbol{G} = \boldsymbol{O}$, we may view the optimal value function $f^o(\boldsymbol{\alpha})$ as an *envelope function*. To see this, let $\alpha \in R$ with $m = 1$ or

$$
\begin{aligned}
L(\boldsymbol{x}, \lambda; \alpha) &= f(\boldsymbol{x}; \alpha) + \lambda g(\boldsymbol{x}; \alpha),\\
L^o(\alpha) &= L(\boldsymbol{x}_o(\alpha), \lambda_o(\alpha); \alpha) = f(\boldsymbol{x}_o(\alpha); \alpha)\\
&\quad + \lambda_o(\alpha) g(\boldsymbol{x}_o(\alpha); \alpha)\\
&= f(\boldsymbol{x}_o(\alpha); \alpha) = f^o(\alpha).
\end{aligned}
$$

Since $\boldsymbol{x}_o(\alpha)$ maximizes L,

$$L(\boldsymbol{x}_o(\alpha), \lambda_o(\alpha); \alpha) \geq f(\boldsymbol{x}; \alpha) + \lambda_o(\alpha) g(\boldsymbol{x}; \alpha)$$

or

$$f^o(\alpha) \geq f(\boldsymbol{x}; \alpha) + \lambda_o(\alpha) g(\boldsymbol{x}; \alpha).$$

Suppose we hold $\boldsymbol{x}$ constant in the preceding inequality. Then:

1. for $\alpha = \alpha_1$, $\boldsymbol{x}_o = \boldsymbol{x}_o(\alpha_1)$ maximizes $f(\boldsymbol{x}; \alpha_1)$ subject to $g(\boldsymbol{x}; \alpha_1) = 0$ so that

$$f^o(\alpha) \geq f(\boldsymbol{x}_o(\alpha_1); \alpha) + \lambda_o(\alpha) g(\boldsymbol{x}_o(\alpha_1); \alpha) = l(\alpha_1, \alpha)$$

   and, at α_1, $f^o(\alpha_1) = l(\alpha_1, \alpha_1)$ (Figure 16.6).

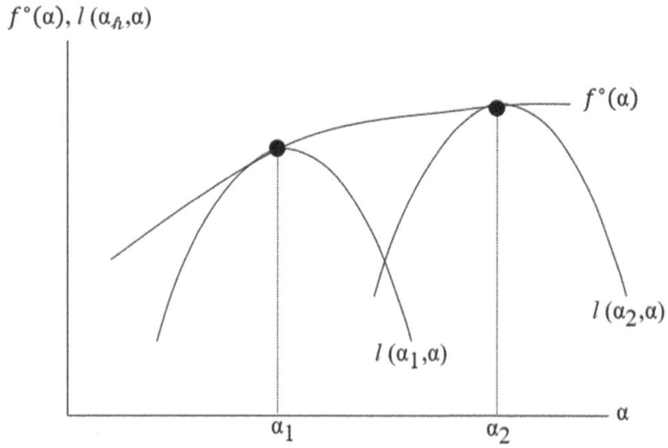

FIGURE 16.6 Upper envelope curve.

2. for $\alpha = \alpha_2$, $\boldsymbol{x}_o = \boldsymbol{x}_o(\alpha_2)$ maximizes $f(\boldsymbol{x}; \alpha_2)$ subject to $g(\boldsymbol{x}; \alpha_2) = 0$ so that

$$f^o(\alpha) \geq f(\boldsymbol{x}_o(\alpha_2); \alpha) + \lambda_o(\alpha) g(\boldsymbol{x}_o(\alpha_2); \alpha) = l(\alpha_2, \alpha)$$

   and, at α_2, $f^o(\alpha_2) = l(\alpha_2, \alpha_2)$ (Figure 16.6). Thus, $f^o(\alpha)$ is the outer or upper envelope of all the $l(\alpha_k, \alpha)$ curves, $k = 1, \ldots, s$.

The preceding envelope theorem affords us an interesting characterization of a Lagrange multiplier. Suppose our problem is to maximize $f(\boldsymbol{x})$ subject to $\hat{g}(\boldsymbol{x}) = b - g(\boldsymbol{x}) = 0$, where $b \in R$ will be treated as a parameter. Given the Lagrangian $L(\boldsymbol{x}, \lambda; b) = f(\boldsymbol{x}) + \lambda \hat{g}(\boldsymbol{x}) = f(\boldsymbol{x}) + \lambda(b - g(\boldsymbol{x}))$, suppose we obtain the optimal solution $\boldsymbol{x}_o = \boldsymbol{x}_o(b)$ and $\lambda_o = \lambda_o(b)$. Then $f^o(b) = f(\boldsymbol{x}_o(b))$ and, from the envelope theorem,

$$\frac{df^o(b)}{db} = \left.\frac{\partial L}{\partial b}\right|_{\boldsymbol{x}_o(b),\lambda_o(b)} = \lambda_o,$$

the marginal change in the optimal value of f with respect to a change in the constraint constant b.

Example 16.5: For the problem maximize $f(\boldsymbol{x}; \alpha) = \alpha x_1 x_2$ subject to

$$\hat{g}(\boldsymbol{x}; \alpha) = b - g(\boldsymbol{x}; \alpha) = 4 - \alpha^2 x_1 - x_2 = 0,$$

use Envelope Theorem III to determine how the optimal value of f changes as $\boldsymbol{\alpha}$ changes. Forming the Lagrangian

$$L = f + \lambda \hat{g} = \alpha x_1 x_2 + \lambda(4 - \alpha^2 x_1 - x_2)$$

we can obtain, upon setting $L_1 = L_2 = L_\lambda = 0$, $x_1^o = 2/\alpha^2$, $x_2^o = 2$, and $\lambda_o = 2/\alpha$. (The reader should verify that the bordered Hessian determinant is positive.) Then $\partial L/\partial \alpha = x_1 x_2 - 2\alpha\lambda x_1$ and

$$\left.\frac{\partial L}{\partial \alpha}\right|_{\boldsymbol{x}_o(\alpha),\lambda_o(\alpha)} = -4/\alpha^2 = \frac{df^o(\alpha)}{d\alpha}.$$

(Note that we did not actually have to find $f^o(\alpha)$.

However, as a check, $f^o(\alpha) = 4/\alpha$ and, as expected, $df^o/d\alpha = -4/\alpha^2$.) ■

Example 16.6: Let $\boldsymbol{x} \in X \subseteq R^3$ and $\boldsymbol{\alpha} \in \Omega \subseteq R^2$. For $y = f(\boldsymbol{x}; \boldsymbol{\alpha}) = -\alpha_1 x_1^2 + x_1 x_2 - x_2^2 - x_3^2$, $\hat{g}^1(\boldsymbol{x}; \boldsymbol{\alpha}) = 20 - \alpha_1 x_2 = 0$, $\hat{g}^2(\boldsymbol{x}; \boldsymbol{\alpha}) = 10 - \alpha_2 x_3 = 0$, use Envelope Theorem III to determine $\nabla_{\boldsymbol{\alpha}} L|_{\boldsymbol{x}_o(\boldsymbol{\alpha}),\boldsymbol{\lambda}_o(\boldsymbol{\alpha})} = \nabla_{\boldsymbol{\alpha}} f^o(\boldsymbol{\alpha})$. From

$$L(\boldsymbol{x}, \boldsymbol{\lambda}; \boldsymbol{\alpha}) = -\alpha_1 x_1^2 + x_1 x_2 - x_2^2 - x_3^2 + \lambda_1(20 - \alpha_1 x_2) + \lambda_2(10 - \alpha_2 x_3)$$

we can obtain, via the first-order conditions for a constrained maximum (or from $\nabla_{\boldsymbol{x}} L = \boldsymbol{O}$, $\nabla_{\boldsymbol{\lambda}} L = \boldsymbol{O}$)

$$\begin{aligned} x_1^o &= 10\alpha_1^{-2} \\ x_2^o &= 20\alpha_1^{-1} \\ x_3^o &= 10\alpha_2^{-1} \end{aligned} \qquad \begin{aligned} \lambda_1^o &= 10\alpha_1^{-2}(\alpha_1^{-1} - 4) \\ \lambda_2^o &= -20\alpha_2^{-2}. \end{aligned}$$

Since

$$\nabla_{\boldsymbol{\alpha}} L = \begin{bmatrix} \partial L/\partial \alpha_1 \\ \partial L/\partial \alpha_2 \end{bmatrix} = \begin{bmatrix} -x_1^2 - \lambda_1 x_2 \\ -\lambda_2 x_3 \end{bmatrix},$$

we can determine

$$\nabla_{\boldsymbol{\alpha}} L|_{\boldsymbol{x}_o(\boldsymbol{\alpha}),\boldsymbol{\lambda}_o(\boldsymbol{\alpha})} = \begin{bmatrix} 800\alpha_1^{-3} - 300\alpha_1^{-4} \\ 200\alpha_2^{-3} \end{bmatrix} = \nabla_{\boldsymbol{\alpha}} f^o(\boldsymbol{\alpha}).$$ ■

Section 16.5 Exercises:

For each of the following problems, use the appropriate envelope theorem to determine the effect on the optimal value of f as the parameter vector $\boldsymbol{\alpha}$ changes:

16.5.1 $f = -x_1^2 - \alpha_1 x_2^2 + \alpha_2 x_1 + 2x_2$.

16.5.2 $f = -2x_1^2 + \alpha_1 x_1 x_2 - x_2^2 - \alpha_2 x_3^2 - 4x_1 - \alpha_3 x_2 + 6x_3$.

16.5.3 minimize $f = \alpha x_1^2 + 4x_1 x_2 + 3x_2^2 - x_1$ subject to $g = x_1 - \alpha x_2 = 0$.

16.5.4 maximize $f = x_1^2 + \alpha_1 x_3$ subject to

$$\begin{aligned} \hat{g}^1 &= 2 - \alpha_2 x_2 - x_3 = 0 \\ \hat{g}^2 &= 8 - \alpha_1 x_1 - x_2 = 0 \end{aligned}$$

16.6 ECONOMIC APPLICATIONS

16.6.1 Long-Run Total Cost: Envelope Results

Before we attempt to derive the firm's long-run total cost curve, let us consider some general envelope results. This exercise will more sharply focus the details of the procedure that will be employed to derive our cost equations.

Let $f(x, y, \theta) = 0$ represent an implicit function of the variables, x, y, and a parameter θ. For θ fixed, f corresponds to a *given* curve in the x, y – plane, with a different curve resulting for each possible value of θ. In this regard, as we vary θ, we generate a *family of curves* in the x, y – plane. Once this family of curves is specified, we can find its **envelope** -- a curve tangent to each member of the family and which intersects none of them (Figure 16.7).

How do we obtain the equation of the envelope? As will be demonstrated shortly, we need only eliminate θ from the system

$$\begin{aligned} f(x, y, \theta) &= 0 \\ \partial f/\partial \theta = f_\theta &= 0 \end{aligned} \qquad (16.5)$$

provided that the Jacobian determinant

$$\left| \frac{\partial(f, f_\theta)}{\partial(x, y)} \right| = \begin{vmatrix} f_x & f_y \\ f_{\theta x} & f_{\theta y} \end{vmatrix} = f_x f_{\theta y} - f_y f_{\theta x} \not\equiv 0, \qquad (16.6)$$

i.e., $f = 0$, $f_\theta = 0$ are not functionally dependent. Actually, the envelope of $f(x, y, \theta) = 0$ is usually represented two ways: (1) in *parametric form*

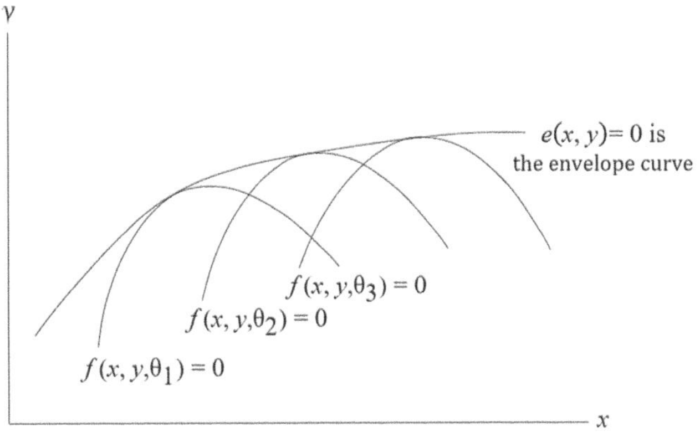

FIGURE 16.7 The envelope curve.

$$x = \alpha(\theta)$$
$$y = \beta(\theta), \tag{16.7}$$

or (2) in implicit form

$$e(x, y) = 0. \tag{16.8}$$

Example 16.7: Find the envelope to the family of curves represented by $f(x, y, \theta) = \theta x + \theta^{-1}y - 4 = 0$. From (16.5),

$$\begin{aligned} f(x, y, \theta) &= \theta x + \theta^{-1}y - 4 = 0 \\ f_\theta &= x - \theta^{-2}y = 0. \end{aligned} \tag{*}$$

With

$$f_x f_{\theta y} - f_y f_{\theta x} = -2\theta^{-1} \not\equiv 0,$$

system (*) may be rewritten as

$$x = \alpha(\theta) = 2\theta^{-1}$$
$$y = \beta(\theta) = 2\theta,$$

or as

$$e(x, y) = xy - 4 = 0.$$

Let us now apply this procedure to the specification of the firm's long-run total cost function. We know that, in the presence of fixed factors of production, the firm's short-run problem centers upon the optimal utilization of a plant of a given size or capacity while, in the long-run, given that all factors are freely variable, the firm attempts to select a plant of optimal size.

If plant size θ is deemed continuously variable, then fixed cost may be expressed as a strictly increasing function of this parameter or

$$\text{fixed cost} = G(\theta),\ G'(\theta) > 0.$$

In view of this specification, we may represent (15.24) as

a. $q = F(l, k, \theta)$;

b. $C = wl + rk + G(\theta)$; and

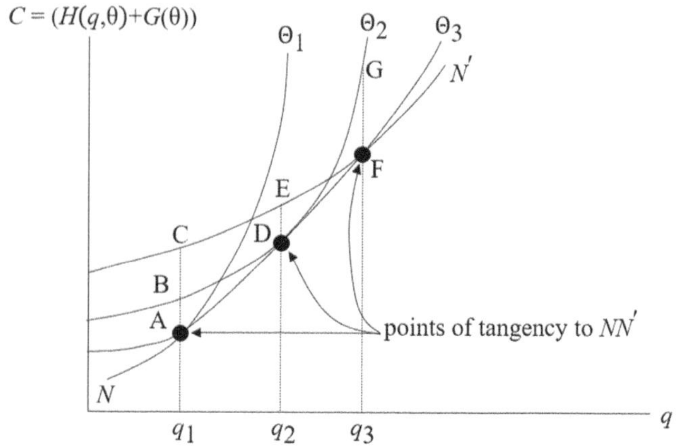

FIGURE 16.8 Long-run total cost curve NN'.

c. $g(l, k, \theta) = 0$ (long-run expansion path in implicit form).

If we eliminate l and k from this system so as to obtain total cost C as a function of q and θ, then the resulting cost function may be written as

$$C = H(q, \theta) + G(\theta), \tag{16.10}$$

a family of total cost curves generated by assigning different values to θ (Figure 16.8) (For $\theta = \bar{\theta} =$ constant, $C = H(q, \bar{\theta}) + G(\bar{\theta}) = \bar{y}(q) + b$.) Here NN' (the envelope of the short-run total cost curves) is the long-run total cost function; it gives the minimum cost of producing each output level if the firm is free to vary its plant size. Clearly, NN' touches each short-run total cost curve and intersects none of them.

Hence, the firm's long-run problem is to choose a plant of optimum size from, say, the set of alternatives depicted in Figure 16.8. The firm's choice criterion is minimum total cost for a given q. That is, if expected output is q_1, then the firm selects scale θ_1 since, at this output level, plants of scale θ_2 and θ_3 have higher costs of production ($Cq_1 > Bq_1 > Aq_1$). Similarly, if a production level of q_2 is deemed appropriate, scale θ_2 will be chosen since it corresponds to a lower total cost of production than a plant of scale θ_3 ($Eq_2 > Dq_2$). And if output level q_3 is forecast, the firm chooses scale θ_3 since ($Gq_3 > Fq_3$). So in the short run the firm's operations are restricted to a plant size of θ_1 or θ_2 or θ_3, etc. But in the long run, with plant size θ variable, the firm chooses a θ value which enables it to produce a given output at minimum cost. Hence, the firm restricts its movements in the long run to its *long-run total cost (envelope) curve NN′*.

To find the envelope of the family of short-run total cost functions, let us write Equation (16.10) in implicit form as $C - H(q, \theta) - G(\theta) = J(C, q, \theta) = 0$. Then the analogue of system (16.5) is

$$\begin{aligned} J(C, q, \theta) &= 0 \\ \partial J/\partial\theta = J_\theta &= 0 \end{aligned} \qquad (16.11)$$

given that

$$J_q J_{\theta C} - J_C J_{\theta q} \not\equiv 0. \qquad (16.12)$$

Upon eliminating θ from system (16.11) and solving for C in terms of q, we obtain the equation of the long-run total cost (envelope) function

$$C = \Phi(q). \qquad (16.13)$$

As indicated above, this function yields the cost of producing a given output in a plant of optimum size.

Example 16.8: Suppose (16.9) appears as

$$\begin{aligned} &\text{(a)} \quad q = l^{0.2}(\theta k)^{0.6}; \\ &\text{(b)} \quad C = wl + rk + \theta^{1.4}; \text{ and} \\ &\text{(c)} \quad k - (3w/r)l = 0. \end{aligned} \qquad (16.14)$$

Let us determine, for this system, the form of Equation (16.13). Substituting Equation (16.14c) into Equation (16.14a) yields

$$q = \bar{F}(l) = l^{0.8}\theta^{0.6}(3w/r)^{0.6}.$$

Then

$$l = \bar{F}^{-1} = (3w/r)^{-0.75}\theta^{-0.75}q^{1.25}$$

and thus, upon inserting this latter expression into Equation (16.14b), we obtain

$$\begin{aligned} C &= H(q, \theta) + G(\theta) \\ &= 4(3w/r)^{-0.75}\theta^{-0.75}q^{1.25} + \theta^{1.4}. \end{aligned}$$

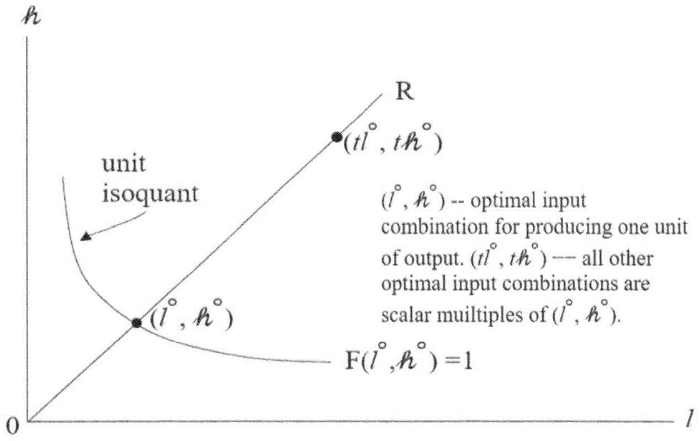

FIGURE 16.9 Long-run expansion path *OR*.

From this parametric short-run cost function we get (16.11) or

$$\begin{aligned} J(C, q, \theta) &= C - 4(3w/r)^{-0.75}\theta^{-0.75}q^{1.25} - \theta^{1.4} = 0 \\ J_\theta &= 3(3w/r)^{-0.75}\theta^{-1.75}q^{1.25} - 1.4\theta^{0.4} = 0. \end{aligned}$$

If we eliminate θ from this system we can get C in terms of q as

$$C = \Phi(q) = 2.7536\ w^{-0.4883}\ r^{0.4883}\ q^{0.814}.$$ ■

If in our discussion of long-run total cost we assume that the production function is homogeneous for any degree λ, then, with all inputs freely variable along a linear expansion path, Equation (16.13) takes on a particularly simple form. In Figure 16.9 let (l^o, k^o) represent the optimal input combination for producing one unit of output so that $F(l^o, k^o) = 1$. Additionally, let the cost of producing this single unit of product be denoted as $C^o = wl^o + rk^o$. Since the long-run expansion path

OR is linear and any point on the expansion path can be written as (tl^o, tk^o), $q = F(tl, tk) = t^\lambda F(l, k)$ evaluated at (l^o, k^o) yields

$$q = F(tl^o, tk^o) = t^\lambda F(l^o, k^o) = t^\lambda,$$

and

$$C = w(tl^o) + r(tk^o) = tC^o,$$

eliminating t from the last two expressions renders

$$C = C^o q^{1/\lambda}, \tag{16.15}$$

the **long-run total cost (envelope) function when F is homogeneous of degree λ.**

Example 16.9: Given that $q = F(l, k) = l^{0.2}k^{0.6}$ is homogeneous of degree $\lambda = 0.8$ in l and k, let us employ the system

(a) $1 = l^{0.2}k^{0.6}$;
(b) $C^o = wl + rk$;
(c) $k = h(l) = (3w/r)l$ (linear long – run expansion path)

(16.16)

to obtain C^o, the cost of a single unit of output. Substituting Equation (16.16c) into Equation (16.16a) gives $l = (3w/r)^{-0.75}$. Inserting this expression along with Equation (16.16c) into Equation (16.16b) yields

$$C^o = wl + rk$$
$$= 1.7548\ w^{0.25}r^{0.75}.$$

Then from Equation (16.15),

$$C = 1.7548\ w^{0.25}r^{0.75}q^{1.25}.$$ ■

Given the expression for long-run total cost provided by Equation (16.13), we can now easily determine **long-run average cost** $C/q = \Phi(q)/q$ and **long-run marginal cost** $dC/dq = d\Phi/dq$. How are these two cost concepts interpreted?

Suppose the firm deems the output level q_1 in Figure 16.10 most profitable. It then produces q_1 in a plant of scale θ_1 at point A on SAC_1 (the short-run average cost curve for scale θ_1), where the unit cost of producing q_1 is lowest. If an expansion of output to q_2 is warranted, unit cost could be reduced by moving down SAC_1 to point B, the minimum of SAC_1. However, unit cost could be reduced even further if, in anticipation of an increase in output to the q_2 level, the firm had built a plant of scale θ_2 as depicted by SAC_2. Now the firm will operate at point C on SAC_2, where the cost per unit of producing q_2 is lowest. If output is expected to increase beyond q_2, the firm's planners select the scale of plant θ_i (as depicted by the short-run average cost curve SAC_i) which yields the least cost of producing the target volume of output. In this regard, the firm's long-run average cost curve is the locus of points (such as A, C, D, and E in Figure 16.10) yielding the least unit cost of producing various levels of output. It is the envelope curve LL' of a family of short-run average cost curves in that it is tangent to each short-run curve (and intersects none of them) at the output level for which the short-run curve represents the optimum or unit cost minimizing plant size.

Next, long-run marginal cost represents the increment in long-run total cost resulting from the production of an additional unit of output given that plant size has been adjusted in order to produce that output level at minimum unit cost. We may obtain the long-run marginal cost curve by again noting that long-run average cost is tangent to each short-run average cost curve at the output level for which the short-run curve represents optimal plant size. Hence, at these product levels long- and short-run average, total, and thus marginal costs are equal. In this regard, long-run marginal cost is the locus of those points on the family of short-run marginal cost curves corresponding to the optimal plant size for each output level (points, F, G, D, and I in Figure 16.10). Note also that long-run marginal cost intersects long-run average cost at the minimum of the latter (output level q_r); at a short-run plant size of θ_r, the minimum of the short-run average cost curve SAC_r coincides with the minimum of long-run average cost. Since at point D the long-and short-run average, as well as total cost curves, are tangent, long-run marginal cost equals short-run marginal cost, with long-run marginal cost-cutting long-run average cost at its minimum.

It is instructive to consider the analytical connection between long-run total cost, long-run average cost, and long-run marginal cost from an additional perspective.

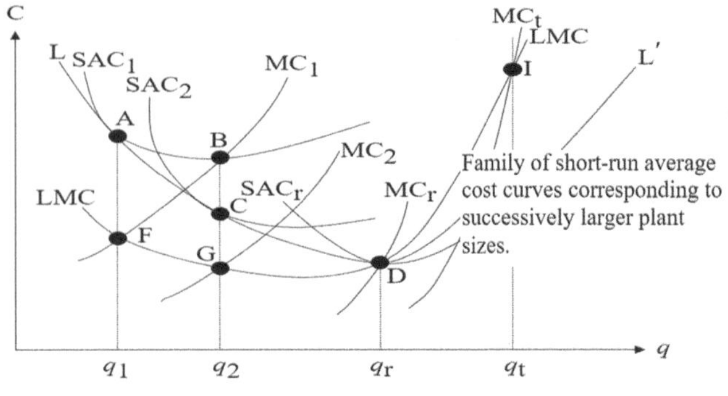

FIGURE 16.10 Long-run average cost and long-run marginal cost.

If we represent the family of short-run average cost curves parametrically as

$$S = C/q = H(q, \theta)/q + G(\theta)/q, \qquad (16.17)$$

"then the envelope of the same can be derived by eliminating θ from the system

$$\begin{aligned} T(S, q, \theta) &= S - H(q, \theta)/q - G(\theta)/q = 0 \\ \partial T/\partial\theta &= T_\theta = 0 \end{aligned} \qquad (16.18)$$

(provided $T_q T_{\theta S} - T_S T_{\theta q} \not\equiv 0$) and solving for S in terms of q as

$$S = \Omega(q). \qquad (16.19)$$

Then long-run total cost can be obtained from Equation (16.19) as $C = q\Omega(q)$. Long-run marginal cost is then $dC/dq = \Omega(q) + q\Omega'(q)$.

Section 16.6 Exercises:

16.6.1 Relative to Example 16.7, use the following table to generate a few members of the family of curves depicted by $\theta x + \theta^{-1}y = 4$ by varying the parameter θ. Plot the resulting curves and, using $e(x, y) = 0$, plot the set of points where e is tangent to each of the said curves. Then connect the set of tangency points to determine the envelope curve.

$\theta\theta x + \theta^{-1}y = 4$
$\frac{1}{2}\frac{1}{2}x + 2y - 4$
1?
2?
3?

16.6.2 Determine the envelope to $x_1 - \theta x_1 x_2 + \theta^2 x_2 = 2$, θ a parameter.

16.6.3 Determine the envelope curve to $\alpha^2 x_1 + \alpha x_1 x_2 + x_2 = 10$, α a parameter.

16.6.4 Given $q = F(l, k) = l^{1/2}k^{1/2}$, determine the long-run total cost function. Use: (a) the homogeneity argument; and (b) system

$$\begin{aligned} q &= l^{1/2}(\theta k)^{1/2} \\ C &= wl + rk + 10\theta \\ k &= h(l). \end{aligned}$$

16.6.5 Given

$$C = H(q, \theta) + G(\theta) = 4(3w/r)^{-0.75}\theta^{-0.75}q^{1.25} + \theta^{1.4}$$

from Example 16.8, find the long-run average cost function using Equations (16.17)–(16.19).

16.6.6 For a production function $q = F(l, k)$, the output elasticities of l and k can be represented, respectively, as

$$\beta_l = \frac{\partial F}{\partial l}\frac{l}{F} = MP_l/AP_l,\ \beta_k = \frac{\partial F}{\partial k}\frac{k}{F} = MP_k/AP_k.$$

Here β_l depicts the proportionate change in output resulting from a given proportionate increase in l with k held constant. Similarly for β_k. However, if we let *all* inputs be increased simultaneously in the *same* proportion dt/t, then we can define the **elasticity of production** or **function coefficient** as

$$\varepsilon = \frac{dF}{dt}\frac{t}{F},$$

the proportionate change in output resulting when all inputs are increased in the same proportion. What is the connection between the numerical value of ε and the concept of returns to scale?

16.6.7 (a) For $q = F(l, k)$, express dq/q in terms of β_l and β_k.
(b) If $dl/l = dk/k = dt/t$, verify that $\varepsilon = \beta_l + \beta_k$.
(c) If F is homogeneous of degree one in l and k, with l, k, and q all changing by the same proportion dt/t, verify that $1 = \beta_l + \beta_k$.

16.6.8 Suppose the production function $q = F(l, k)$ is homogeneous of *any* degree in l and k. Since $\varepsilon = \beta_l + \beta_k$, let us write ε as

$$\begin{aligned} \varepsilon &= F_l\left(\frac{lw}{qw}\right) + F_k\left(\frac{kr}{qr}\right) = \lambda^{-1}\left(\frac{wl}{q} + \frac{rk}{q}\right) \\ &= \lambda^{-1}(C/q) = \frac{\text{long-run average cost } (LRAC)}{\text{long-run marginal cost } (LRMC)} \end{aligned}$$

given that

$$\frac{Fl}{w} = \frac{Fk}{r} = \frac{1}{\lambda} = \frac{1}{MC}$$

holds for movement along the long-run expansion path. (Note that we are implicitly assuming that the firm minimizes cost subject to an output constraint. In this instance the optimal Lagrange multiplier $\lambda = dC/dq$ = marginal cost.) How does the relationship between *LRAC* and *LRMC* dictate the behavior of returns to scale?

16.6.9 The (long-run) **elasticity of total cost** (the proportionate change in long-run total cost $C = \Phi(q)$ relative to a given proportionate change in output) is specified as

$$\kappa = \frac{dC}{dq}\frac{q}{C} = \frac{LRMC}{LRAC} = \frac{1}{\varepsilon}.$$

Hence, $\kappa \lesseqgtr 1$ for $\varepsilon \gtreqless 1$. Additionally, the (long-run) **elasticity of average cost** (the proportionate change in long-run average cost $C/q = \Phi(q)/q$ relative to a given proportionate change in output) is

$$\rho = \frac{d(C/q)}{dq}\frac{q^2}{C} = \kappa - 1 = \frac{1}{\varepsilon} - 1.$$

Thus, $\rho \lesseqgtr 0$ when $\varepsilon \gtreqless 1$. Verify that if F is homogeneous of degree one in l and k, then $\varepsilon = 1$, $\kappa = 1$, and thus $\rho = 0$.

APPENDIX 16.A A PROOF OF BERGE'S MAXIMUM THEOREM

The proof offered below consists of the following sequence of steps:

1. Demonstrate that the set of optimal solutions $S(\alpha) = \arg\max \{f(\boldsymbol{x}; \boldsymbol{\alpha}) | \boldsymbol{x} \in \mathcal{K}(\boldsymbol{\alpha})\}$ or $\boldsymbol{x}_o = \boldsymbol{x}_o(\boldsymbol{\alpha}) \neq \emptyset$.
2. Verify that $S(\boldsymbol{\alpha})$ is compact valued.
3. Show that $S(\boldsymbol{\alpha})$ is uhc.
4. Demonstrate that the optimal value function $f^o(\boldsymbol{\alpha})$ is continuous.

Step 1. For a given $\boldsymbol{\alpha} \in \Omega$, $f(\boldsymbol{x}; \boldsymbol{\alpha})$ is continuous in $\boldsymbol{x}$ and $\mathcal{K}(\boldsymbol{\alpha}) \subseteq X$ is non-empty and compact. Hence, via Weierstrass's theorem, $f^o(\boldsymbol{\alpha})$ is well defined and thus $S(\boldsymbol{\alpha}) \neq \emptyset$.

Step 2. Since $S(\boldsymbol{\alpha}) \subseteq \mathcal{K}(\boldsymbol{\alpha})$ and $\mathcal{K}(\boldsymbol{\alpha})$ is compact, $S(\boldsymbol{\alpha})$ is, by definition, bounded. Let $\{\boldsymbol{x}_k\}$ be a sequence with $\boldsymbol{x}_k \in S(\boldsymbol{\alpha})$ and $\boldsymbol{x}_k \to \boldsymbol{x}$ for all k. With $\mathcal{K}(\boldsymbol{\alpha})$ closed, it follows that $\boldsymbol{x} \in \mathcal{K}(\boldsymbol{\alpha})$. And with $f^o(\boldsymbol{\alpha}) = f(\boldsymbol{x}_k; \boldsymbol{\alpha})$ for all k, $f(\boldsymbol{x}; \boldsymbol{\alpha}) = f^o(\boldsymbol{\alpha})$. Thus $\boldsymbol{x} \in S(\boldsymbol{\alpha})$ so that $S(\boldsymbol{\alpha})$ is closed. So with $S(\boldsymbol{\alpha})$ closed and bounded, it follows that $S(\boldsymbol{\alpha})$ (or $\boldsymbol{x}_o(\boldsymbol{\alpha})$) must be compact valued.

Step 3. For a fixed $\boldsymbol{\alpha} \in \Omega$, let $\{\boldsymbol{\alpha}_k\}$ be a sequence with $\boldsymbol{\alpha}_k \to \boldsymbol{\alpha}$ and let $\{\boldsymbol{x}_k\}$ be a sequence with $\boldsymbol{x}_k \in S(\boldsymbol{\alpha}_k)$ for all k. Since $\mathcal{K}$ is uhc, there exists (via Theorem 16.1) a subsequence $\{\boldsymbol{x}_{k_j}\}$ converging to $\boldsymbol{x} \in \mathcal{K}(\boldsymbol{\alpha})$. Let the point $\hat{\boldsymbol{x}}$ also be an element of $\mathcal{K}(\boldsymbol{\alpha})$. Since correspondence $\mathcal{K}$ is also lhc, there exists (via Theorem 16.2) a sequence $\{\hat{\boldsymbol{x}}_{k_j}\}$ with $\hat{\boldsymbol{x}}_{k_j} \in \mathcal{K}(\boldsymbol{\alpha}_{k_j})$ and $\hat{\boldsymbol{x}}_{k_j} \to \hat{\boldsymbol{x}}$. With (maximal) $f(\boldsymbol{x}_{k_j}; \boldsymbol{\alpha}_{k_j}) \geq f(\hat{\boldsymbol{x}}_{k_j}; \boldsymbol{\alpha}_{k_j})$ for all j, it follows that $f(\boldsymbol{x}; \boldsymbol{\alpha}) \geq f(\hat{\boldsymbol{x}}; \boldsymbol{\alpha})$. Since $\hat{\boldsymbol{x}} \in \mathcal{K}(\boldsymbol{\alpha})$ is arbitrary, we conclude that $\boldsymbol{x} \in S(\boldsymbol{\alpha})$ so that $S(\boldsymbol{\alpha})$ (or $\boldsymbol{x}_o(\boldsymbol{\alpha})$) is uhc.

Step 4. Given $\boldsymbol{\alpha} \in \Omega$, consider the sequence $\{\boldsymbol{\alpha}_k\} \to \boldsymbol{\alpha}$. Let $\boldsymbol{x}_k \in S(\boldsymbol{\alpha}_k)$ for all k with

$$\bar{v} = \overline{\lim_k}\ f^o(\boldsymbol{\alpha}_k),\ \underline{v} = \underline{\lim_k}\ f^o(\boldsymbol{\alpha}_k)$$

so that there exists a subsequence $\{\boldsymbol{x}_{k_j}\}$ such that $\bar{v} = \lim_k f(\boldsymbol{x}_{k_j}; \boldsymbol{\alpha}_{k_j})$. With $S(\boldsymbol{\alpha})$ uhc, there exists a subsequence of $\{\boldsymbol{x}_{k_j}\}$ converging to $\boldsymbol{x} \in S(\boldsymbol{\alpha})$ and thus $\bar{v} = f(\boldsymbol{x}; \boldsymbol{\alpha}) = f^o(\boldsymbol{\alpha})$. In like fashion there exists a subsequence $\{x_{k_j}\}$ such that $\underline{v} = \lim_{k} f(x_{k_j}; \alpha_{k_j})$, and with $S(\alpha)$ uhc, there exists a subsequence of $\{x_{k_j}\}$ converging to $x \in S(\alpha)$. Hence, $\underline{v} = f(x; \alpha) = f^o(\alpha)$ so that $f^o(a_j) \to f^o(\alpha)$ is continuous. Q. E. D.

NOTES

1 In general, a correspondence is a point-to-set or set-valued function. That is,

Definition 16.1

Given a set $X \subseteq R^n$, a **correspondence** $F: X \to Y$ is a rule which associates with a vector $x \in X$ a non-empty subset $F(x)$ such that $F(x)$ contains more than one element $y \in Y$.

Thus a correspondence is a mapping from a set X into a set of non-empty subsets of Y. For $F: X \to Y$, the **domain of F** is the set $D_F = \{x | \varnothing \neq F(x) \subset Y\}$ and the **range space of F** is Y. Here $F(x)$ is the **image set** of $x \in X$ while the set $R_F = \bigcup_{x \in X} F(x)$ is the **range of F.** F is termed **strict** if all image sets $F(x) \neq \varnothing$. And the **graph of F**, G_F, is the subset of $X \times Y$ associated with F or $G_F = \{(x, y) \in X \times Y | y \in F(x)\}$.

2 Since $G(x_o(\alpha); \alpha) = 0$, $\nabla_x G(x_o(\alpha); \alpha) \frac{dx_o(\alpha)}{d\alpha} + \nabla_\alpha G(x_o(\alpha); \alpha) = 0$ and thus $\nabla_\alpha G(x_o(\alpha); \alpha) = -\nabla_x G(x_o(\alpha); \alpha) \frac{dx_o(\alpha)}{d\alpha}$.

CHAPTER 17

The Fixed Point Theorems of Brouwer and Kakutani

17.1 INTRODUCTION

In this chapter, an assortment of fixed point theorems will be presented in a variety of forms that incorporate different sets of assumptions. We may view the notion of a fixed point quite generally as: let $S \subset R^n$ be a set and let f be a mapping such that $f\colon S \to S$. A point $\bar{x} \in S$ is a **fixed point** in S if $\bar{x} = f(\bar{x})$. Many distinctive fixed point theorems exist and differ from each other (or are equivalent) depending upon the assumptions that are made concerning f and the characteristics or properties of S.

The first such theorem offered will be that of Brouwer (1912) (a sufficient condition for the existence of a fixed point); and this fixed-point theorem will then be generalized to that of Kakutani (1941). As will be seen below, in Brouwer's theorem f is a continuous point to point mapping and S is first taken to be a k-simplex $\Delta^k \subset R^n$. And since a nonempty, compact (closed and bounded) convex set $S \subset R^n$ is homeomorphic[1] to Δ^k, it follows that Brouwer's theorem can alternatively be stated as: a continuous mapping of a nonempty, compact convex set S into itself necessarily has a fixed point.

Next comes the fixed-poin t theorem of Kakutani. Here, too, f has a fixed point since it is taken to be an upper hemicontinuous point-to-set (or set-valued) mapping of a set S into the collection of all compact, convex subsets of S. Extensions of both the Brouwer and Kakutani theorems will also be mentioned.

An important application of Brouwer's theorem is that it can be used to establish the existence of a solution to a nonlinear system of equations. We know that a fixed point of the mapping $f\colon S \to S$ is an element $x \in S$ such that $f(x) = x$, where $f^T = (f^1(x), \ldots, f^n(x))$. If the system of equations is written as $g(x) = O$, then we can express $g(x) = f(x) - x$. Then a fixed point $\bar{x}$ of f is a solution to $g(x) = O$ or $g(\bar{x}) = O = f(\bar{x}) - \bar{x}$.

17.2 SIMPLEXES

A set $S \subset R^n$ is termed a **convex polytope** if it is expressible as the convex hull of the finite collection of points $\{x_1, \ldots, x_k\}$ or $S = co\{x_1, \ldots, x_k\}$ (Figure 17.1a).

Additionally, a set of $k + 1$ vectors $\{x_o, x_1, \ldots, x_k\} \subset R^n$ is **affinely independent** if and only if the *translated set* $\{x_1 - x_o, \ldots, x_k - x_o\}$ is linearly independent (Figure 17.1b).

Upon combining these concepts, we can now state that a convex polytope $S \subset R^n$ is a **k-dimensional simplex** (or **k-simplex** for short) if and only if it is the convex hull of a set of $k + 1$ affinely independent vectors $\{x_o, x_1, \ldots, x_k\}$, $k \leq n + 1$. That is, a k-simplex spanned by the points $\{x_o, x_1, \ldots, x_k\} \subset R^n$ will be denoted as

$$\Delta^k = co\{x_o, x_1, \ldots, x_k\} = \{x \mid x = \textstyle\sum_{i=0}^k \lambda_i x_i, \ \sum_{i=0}^k \lambda_i = 1, \ 0 \leq \lambda_i \in R \text{ for all } i\}. \tag{17.1}$$

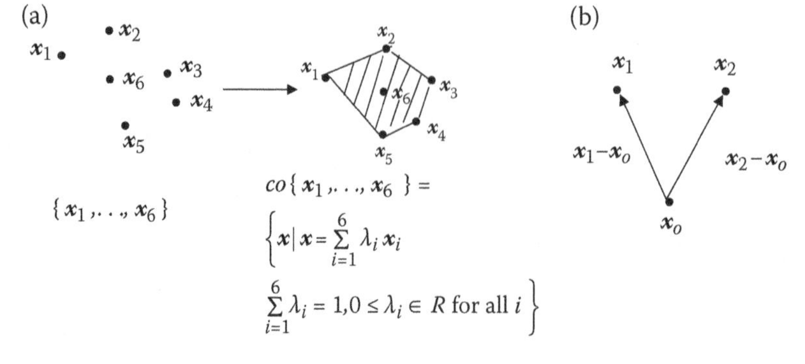

FIGURE 17.1 a. The convex hull of $\{x_1, \ldots, x_6\}$; b. The vectors $\{x_o, x_1, x_2\}$ are affinely independent if and only if the translated set $\{x_1 - x_o, x_2 - x_o\}$ is linearly independent.

In general, dim (Δ^k) = number of vertices less $1 = k$ since the vertices of Δ^k specify an affine subspace of dimension k. Additionally, a subset of Δ^k is termed a **subsimplex** of Δ^k if it is itself a simplex (see Figures 17.2a, b for $k = 2, 3$).

Given the preceding discussion, we can now state that if the set of vectors $\{x_1, \ldots, x_n\} \subset R^n$ is linearly independent, then $co\{O, x_1, \ldots, x_n\}$ is an n-simplex. In particular, the **standard n-simplex** in R^n is defined as $co\{O, e_1, \ldots, e_n\}$, where e_i, $i = 1, \ldots, n$, is the ith unit column vector; and an n-dimensional **unit simplex** in R^{n+1} is formed as $co\{e_1, \ldots, e_n, e_{n+1}\}$ or

$$\Delta^n = \{x \mid \textstyle\sum_{i=1}^{n+1} x_i = 1, O \leq x_i \in R^n \text{ for all } i\}. \quad (17.2)$$

The boundary of a k-dimensional simplex contains simplexes of lower dimension called **faces** and the union of all faces of Δ^k is its **closure**. In fact, the number of faces of dimension l of a k-simplex is given by

$$\binom{k+1}{l+1} = \frac{(k+1)!}{(l+1)!(k-l)!},$$

i.e., for $k = 2$, Δ^2 has $\binom{2+1}{0+1} = 3$ zero-dimensional faces (the vertices); $\binom{2+1}{1+1} = 3$ one-dimensional faces (the sides of the triangle); and $\binom{2+1}{2+1} = 1$ two-dimensional face (which is Δ^2 itself) (Figure 17.2a).

More explicitly, an l-simplex Δ^l is an l-face of a k-simplex Δ^k if all vertices of Δ^l are also vertices of Δ^k. Thus the vertices of Δ^l form a subset of the vertices of Δ^k and the convex hull of this subset is the resulting face. If $l = 0$, $\Delta^l = \Delta^o$ is a vertex of Δ^k; and if $l = k - 1$, Δ^l is a specialized face which is termed a **facet** of Δ^k. In this latter case there is exactly one vertex in Δ^k which is not in Δ^l. Let's denote this vertex as x_i, $0 \leq i \leq k$. Then Δ^l is called the **face (facet) of Δ^k opposite the vertex x_i** and referred to as a $(k$-$1)$-simplex. For $x = \sum_{i=0}^{k} \lambda_i x_i \in co\{x_o, x_1, \ldots, x_k\}$, the index set $\mathcal{J}(x) = \{i \mid \lambda_i > 0, 0 \leq i \leq k\} = \{i_o, \ldots, i_l\}$. Then $x \in \Delta^l$, where the face of Δ^l is termed the **carrier face** of x and is formed as $co\{x_{i_o}, \ldots, x_{i_l}\}$. Note that x belongs to exactly one face of Δ^k and amounts to the face of lower dimension of Δ^k containing x. In Figure 17.2b, the carrier face of x_4 is the face spanned by x_2 and x_3

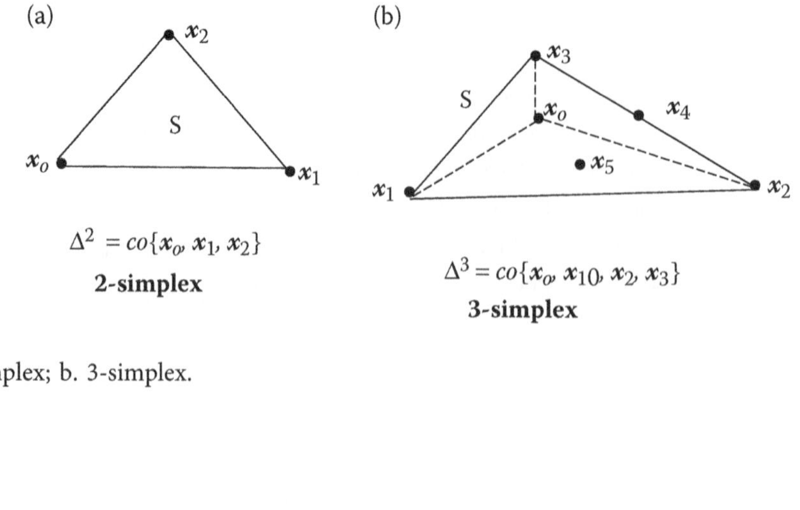

FIGURE 17.2 a. 2-simplex; b. 3-simplex.

(i.e., $x_4 \in co\{x_2, x_3\}$) while the carrier face of x_5 is the face generated by x_o, x_1 and x_2 (i.e., $x_5 \in co\{x_o, x_1, x_2\}$). The carrier face of any interior point of Δ^3 is Δ^3 itself.

Section 17.2 Exercises:

17.2.1 Let

$$x_o = \begin{bmatrix} 1 \\ 0 \\ 1 \end{bmatrix}, x_1 = \begin{bmatrix} -1 \\ 3 \\ 1 \end{bmatrix}, x_2 = \begin{bmatrix} 0 \\ 0 \\ 1 \end{bmatrix}, x_3 = \begin{bmatrix} 1 \\ -1 \\ 5 \end{bmatrix}.$$

Is this set of vectors affinely independent?

17.2.2 Given the vectors

$$x_o = \begin{bmatrix} 2 \\ 3 \\ 0 \end{bmatrix}, x_1 = \begin{bmatrix} 1 \\ 1 \\ 0 \end{bmatrix}, x_2 = \begin{bmatrix} 4 \\ 2 \\ 0 \end{bmatrix}, x_3 = \begin{bmatrix} 2 \\ 2 \\ 3 \end{bmatrix},$$

does $x_4^T = (3, 2, 1)$ lie within $\Delta^3 = co\{x_o, x_1, x_2, x_3\}$?

17.2.3 How many vertices exist for a 3-simplex?

17.3 SIMPLICIAL DECOMPOSITION AND SUBDIVISION

We now turn to the notion that simplexes can be viewed as the fundamental components of more complicated constructs such as complexes and convex polytopes. First, let $S = \{\Delta^{k_1}, \ldots, \Delta^{k_r}\}$ be a finite set of simplexes in R^n. Then S is termed a **finite simplicial complex** $\mathcal{G}$:

a. if simplex $\Delta^{k_j} \in \mathcal{G}$, then each of its faces is also a member of $\mathcal{G}$; and

b. if simplexes $\Delta^{k_j}, \Delta^{k_l} \in \mathcal{G}$, then $\Delta^{k_j} \cap \Delta^{k_l} = \varnothing$.

Thus a complex in R^n is a collection of mutually disjoint simplexes. So if $\Delta^2 = co\{x_o, x_1, x_2\}$ is a 2-simplex, then its associated 2-dimensional complex is

$$\begin{aligned} \mathcal{G} &= \Delta^{-1} \cup \{\Delta_i^o\}_{i=1}^3 \cup \{\Delta_i^1\}_{i=1}^3 \cup \Delta^2 \\ &= \{\Delta^{-1}, x_o, x_1, x_2, co\{x_o, x_1\}, co\{x_o, x_2\}, \\ &\quad co\{x_1, x_2\}, \Delta^2\}, \end{aligned}$$

i.e., $\mathcal{G}$ contains the empty simplex, all vertices and edges of Δ^2, and the 2-simplex itself. In this regard, a complex in R^n will admit 2^{k+1} mutually disjoint simplexes.

Given the preceding discussion, for $S = \{\Delta^{k_1}, \ldots, \Delta^{k_r}\}$ again taken to be a finite set of simplexes in R^n, their union $\mathcal{P} = \cup_{i=1}^r \Delta^{k_i}$ is a **convex polytope**. Moreover, if $\mathcal{P}$ is the union of all simplexes within a simplicial complex $\mathcal{G} \subset R^n$, then $\mathcal{G}$ is said to be a **simplicial decomposition** (or **triangulation**) of $\mathcal{P}$ into disjoint simplexes. So, for the 2-dimensional complex $\mathcal{G}$ specified above, it is evident that $\mathcal{G}$ represents the decomposition of Δ^2. Additionally, the union of all simplexes in $\mathcal{G}$ is the polytope

$$\begin{aligned} \mathcal{P} = \Delta^{-1} &\cup x_o \cup x_1 \cup x_2 \cup co\{x_o, x_1\} \cup co\{x_o, x_2\} \\ &\cup co\{x_1, x_2\} \cup \Delta^2. \end{aligned}$$

If $\mathcal{P}$ has the simplicial decomposition $\mathcal{G}$ and $x \in \mathcal{P}$, then the simplex of smallest dimension of $\mathcal{G}$ which contains x is termed the **carrier simplex** of x.

A **simplicial subdivision**$\mathcal{D}$ of a k-simplex in R^n is a decomposition or finite collection of simplexes $\{\Delta^{k_j} | j \in \mathcal{J}\}$ such that:

a. $\cup_{j \in \mathcal{J}} \Delta^{k_j} = \Delta^k$, i.e., the subsimplexes cover Δ^k;

b. for any $l, m \in \mathcal{J}$, either $\Delta^{k_l} \cap \Delta^{k_m} = \varnothing$ or the intersection is a face common to both simplexes; and

c. the faces of any Δ^{k_j} are members of the subdivision $\mathcal{D}$.

Any facet or $(k-1)$-face of a subsimplex of the subdivision of Δ^k is either interior to Δ^k (it represents the face of exactly two simplexes of the subdivision) or is on the boundary of Δ^k (and a face of exactly one subsimplex). The **mesh** of the subdivision is the diameter of the largest subsimplex; and two different subsimplexes $\Delta^{k_l}, \Delta^{k_m}$ of a given subdivision $\mathcal{D}$ are **adjacent** if they have a common facet or $(k-1)$-dimension face.

If Δ^n is a unit simplex, a **restricted simplicial subdivision** of Δ^n is a subdivision which contains no vertices on the faces of Δ^n other than the unit column vectors $e_1, \ldots, e_{n+1}$. In general, any (sub)simplex obtained as the result of the subdivision of a simplex Δ^k will be called a **derived (sub)simplex**.

An important special type of subdivision of a k-simplex in R^n is a **barycentric subdivision**. To obtain this subdivision, let us first note that for any simplex $\Delta^k \subset R^n$, the

barycenter of Δ^k is the point y_k having the **barycentric weights** $\lambda_o = \lambda_1 = \cdots = \lambda_k = \frac{1}{k+1}$. Then from (17.1),

$$y_k = \frac{1}{k+1} = \sum_{i=0}^{k} x_i. \tag{17.3}$$

If we determine the barycenters of all the faces of dimensions 0, 1, 2, ... of Δ^k, then we obtain a family $\mathcal{D}^{(1)}$ of k-simplexes called the **barycentric subdivision of Δ^k of order 1** having properties:

a. each k-simplex of $\mathcal{D}^{(1)}$ contains the barycenter of a face of dimension 0, the barycenter of a face of dimension 1, and so on.

b. if a k-simplex of $\mathcal{D}^{(1)}$ contains the barycenters of $\Delta^{k_p} = co\{x_{i_o},\ldots,x_{i_p}\}$ and $\Delta^{k_q} = co\{x_{j_o},\ldots,x_{j_q}\}$ with $p < q$, then $\{i_o,\ldots,i_p\} \subset \{j_o,\ldots,j_q\}$.

If we next divide each k-simplex of $\mathcal{D}^{(1)}$ in a similar fashion, we generate a new family of k-simplexes called the **barycentric subdivision of Δ^k of order 2** (denoted $\mathcal{D}^{(2)}$). In general, we may define recursively the vth order barycentric subdivision of Δ^k as: the **barycentric subdivision of Δ^k of order 2, $\mathcal{D}^{(v)}$**, is the first-order barycentric subdivision of $\mathcal{D}^{(v-1)}$, where the latter family of simplexes is the (v-1)th order subdivision of Δ^k. And for increasing v, we obtain a collection of barycentric subdivisions of arbitrarily small mesh.

For instance, the barycentric subdivision of a zero-dimensional simplex $\Delta^o = \{x_o\}$ is itself (trivially) a subdivided simplex. So, from (17.3), $y_o = x_o$. The barycentric subdivision of a one-dimensional simplex $\Delta^1 = co(x_o, x_1)$ consists of two subsimplexes of the same dimension, $co\{x_o, y_1\}$ and $co\{y_1, x_1\}$, where $y_1 = \frac{1}{2}(x_o + x_1)$ is the barycentric center of Δ^1 (see Figure 17.3a). In general, a simplex of dimension k is barycentrically subdivided into $(k+1)!$ subsimplexes of the same dimension, e.g., in Figure 17.3b, $\Delta^2 = co\{x_o, x_1, x_2\}$ is subdivided into $3! = 6$ subsimplexes of dimension two. And if we focus on the shaded area $co\{y_o, y_1, y_2\}$ (here y_o is the barycenter of a face of dimension zero, y_1 is the barycenter of a face of dimension one, and y_2 is the barycenter of a face of dimension two), then a moments reflection reveals that, under barycentric subdivision, this derived subsimplex has associated with it a unique hierarchical sequence $\mathcal{V}_o \subset \mathcal{V}_1 \subset \mathcal{V}_2$ of proper subsets of $\{x_o, x_1, x_2\}$. To generalize, any S-dimensional derived subsimplex Δ^s is in one-to-one correspondence with a unique hierarchical sequence $\mathcal{V}_o(\neq\emptyset) \subset \mathcal{V}_1 \subset \ldots \subset \mathcal{V}_r \subset \ldots \subset \mathcal{V}_s$ of proper subsets of $\{x_o, x_1,\ldots,x_k\}$ and such that Δ^s is uniquely expressible as $co\{y_o, y_1,\ldots,y_r,\ldots,y_s\}$, where y_r is the barycenter of the subsimplex spanned by $\mathcal{V}_r$. [2]

On the basis of this discussion we may note that:

1. A $(k-1)$- dimensional derived simplex Δ^{k-1} in the barycentric subdivision of Δ^k is either:

   a. a face of exactly one k-dimensional derived subsimplex if $\Delta^{(k-1,v)}$ completely lies on the boundary of Δ^k; or

   b. a face common to exactly two k-dimensional derived subsimplexes.

2. Let $\mathcal{D}^{(v)}$ represent a vth order barycentric subdivision of Δ^k. Then a $(k-1)$-dimensional derived subsimplex $\Delta^{(k-1,v)}$ of order v is either:

   a. a face of exactly one k-dimensional derived subsimplex of order v if $\Delta^{(k-1,v)}$ completely lies on the boundary of Δ^k; or
   b. a face common to exactly two k-dimensional derived subsimplexes of order v.

Section 17.3 Exercises:

17.3.1 Suppose $\Delta^3 = co\{x_o, x_1, x_2, x_3\}$ is a 3-simplex in R^n and $\mathcal{G}$ is its associated complex. How many mutually disjoint simplexes will make up $\mathcal{G}$?

17.3.2 Given Δ^3 from Exercise 17.3.1, find $\mathcal{G}$.

17.3.3 Let

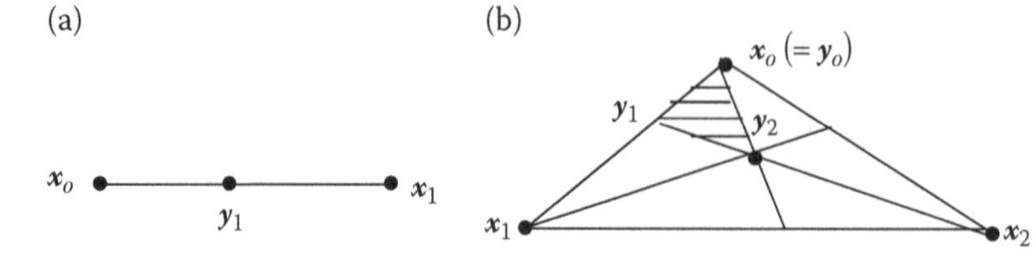

FIGURE 17.3 a. Barycentric subdivision of Δ^1; b. barycentric subdivision of Δ^2.

$$x_o = \begin{bmatrix} 2 \\ 3 \\ 0 \end{bmatrix}, x_1 = \begin{bmatrix} 1 \\ 1 \\ 0 \end{bmatrix}, x_2 = \begin{bmatrix} 4 \\ 2 \\ 0 \end{bmatrix}, x_3 = \begin{bmatrix} 2 \\ 3 \\ 3 \end{bmatrix}.$$

Additionally, let $\Delta^3 = co\{x_o, x_1, x_2, x_3\}$. Find the barycenter of Δ^3.

17.3.4 Given the first-order barycentric subdivision of Δ^2 given in Example 17.A.1 (See Appendix 17.A), select the permutation "1, 2, 0" and form $y = \lambda_1 x_1 + \lambda_2 x_2 + \lambda_o x_o$ with $\lambda_1 \geq \lambda_2 \geq \lambda_0 \geq 0$.

17.4 SIMPLICIAL MAPPINGS AND LABELING

In this section we define a mapping from one simplex (or convex polytope) to another given that a simplicial decomposition has been defined on each of them, i.e., simplexes Δ^{k_1} and Δ^{k_2} can be viewed as the union of all disjoint subsimplexes within their respective simplicial complexes $\mathcal{G}_1$, $\mathcal{G}_2$ in R^n. Specifically $f: \Delta^{k_1} \rightarrow \Delta^{k_2}$ is a single-valued **simplicial mapping** from the complex $\mathcal{G}_1$ to the complex $\mathcal{G}_2$ if:

a. the vertices of subsimplex $\Delta_i^{k_1}$ of Δ^{k_1} are always mapped by f into the vertices of subsimplex $\Delta_j^{k_2}$ of Δ^{k_2}, i.e., f assigns to each vertex x_r, $r = 0, 1, \ldots, k_1$, of $\Delta_i^{k_1}$ a vertex $y_t = f(x_r)$, $t = 0, 1, \ldots, k_2$, of $\Delta_j^{k_2}$; and

b. interior points of subsimplex $\Delta_i^{k_1}$ of Δ^{k_1} are always mapped by f into interior points of subsimplex $\Delta_j^{k_2}$ of Δ^{k_2}, i.e., f assigns to each point

$$\begin{aligned} x \in \Delta_i^{k_1} &= co(x_o, x_1, \ldots, x_{k_1}) \\ &= \{x | x = \Sigma_p \lambda_p x_p, \Sigma_p \lambda_p = 1, \\ &\quad 0 < \lambda_p \in R, p = 0, 1, \ldots, k_1\} \end{aligned}$$

a point $y = f(x) = \Sigma_p \lambda_p f(x_p) \in \Delta^{k_2} = co\{y_o, y_1, \ldots, y_{k_2}\}$.

On the basis of this definition we have

Theorem 17.1: For $\mathcal{G}'$ a subdivision of a simplicial complex $\mathcal{G} \subset R^n$, let the mapping $f: \mathcal{G}' \rightarrow \mathcal{G}$ assign to each vertex $x_i \in \mathcal{G}'$ an arbitrary vertex $y_j = f(x_i)$ of the carrier face of $x_i \in \mathcal{G}$. Then f is a simplicial mapping from $\mathcal{G}'$ to $\mathcal{G}$.

What this theorem implies is that for x_i a vertex of an arbitrary derived k-simplex $\Delta^{k'}$ of $\mathcal{G}'$, there exists a k-simplex Δ^k of $\mathcal{G}$ such that $\Delta^{k'} \subset \Delta^k$ and thus $x_i \in \Delta^k$. Since a carrier simplex of vertex x_i is a face simplex (of lowest dimension) of Δ^k, the image $y_j = f(x_i)$ must be a vertex of Δ^k and thus f is a simplicial mapping.

In order to see how the concept of a simplicial mapping may be utilized, we need to explore the notion of an *integer label*. Let the k-simplex $\Delta^k = co\{x_o, x_1, \ldots, x_k\}$ be simplically subdivided and let $\mathcal{V}$ represent the collection of all vertices of all the derived subsimplexes. A **labeling function** l is a mapping from the set of vertices of $\mathcal{V}$ into the set of integer labels $\mathcal{J} = \{0, 1, \ldots, k\}$. Here, $l(x_i)$ is termed the **label** of $x_i \in \mathcal{V}$ and the label assignment must be such that $x_i \in co\{x_{i_o}, x_{i_1}, \ldots, x_{i_m}\}$ implies $l(x_i) \in \{i_o, i_1, \ldots, i_m\}$. In particular, if $x_{i_1}, \ldots, x_{i_p}$ are the vertices of some simplex Δ^{k_p} in $co\{x_o, x_1, \ldots, x_k\}$, then Δ^{k_p} is characterized as "labeled" by the set $\{l(x_{i_1}), \ldots, l(x_{i_p})\}$. If for all $x_i \in \mathcal{V}$ we have $l(x) \in \mathcal{J}$, then the mapping l is called a **proper labeling** of the subdivision. Moreover, a subsimplex is termed **completely labeled** if $l(x)$ assumes all of the values in $\mathcal{J}$ on its set of vertices.

A derived subsimplex $co\{y_o, y_1, \ldots, y_k\}$ is said to be **regular** if $l(y_i) \neq l(y_j)$, $i \neq j$, i.e., the different vertices in $\mathcal{V}$ have different labels under l. Hence, any regular derived subsimplex must be completely labeled. In addition, a $(k-1)$-dimensional derived subsimplex $co\{\mathcal{V}_o, \mathcal{V}_1, \ldots, \mathcal{V}_{k-1}\}$ is regular if $l(\mathcal{V}_i) \neq l(\mathcal{V}_j)$, $i \neq j$, and $l(\mathcal{V}_i) \neq k$ for $i = 0, 1, \ldots, k-1$.

The next three theorems will be utilized to address the question of the existence of a fixed point for a continuous point-to-point mapping. That is, Sperner's lemma ensures the existence of at least one completely labeled derived subsimplex in a simplicial subdivision. The Knaster-Kuratowski - Mazurkiewicz (KKM) theorem offers a set of assumptions which, via Sperner's lemma, guarantee that the intersection of a finite collection of sets on a simplex is not vacuous, and, by also invoking Sperner's lemma, Brouwer's theorem demonstrates that a continuous mapping of a simplex into itself admits a fixed point.

We start with

Theorem 17.2: (**Sperner's lemma**) [Sperner (1928); Nikaido (1968); Kuhn (1968); and Stoer and Witzgall (1970)]. Let $\mathcal{D}$ be a simplicial subdivision of the k-simplex $\Delta^k = co\{x_o, x_1, \ldots, x_k\}$. If Δ^k is properly labeled by the labeling function l, then there exists an odd number of completely labeled subsimplexes in $\mathcal{D}$.

What Sperner's lemma implies is that if a k-simplex Δ^k is (barycentrically) subdivided and a labeling function l assigns to a vertex y_i in the subdivision a label $l(y_i)$ of a vertex $x_{l(y_i)}$ of a carrier of y_i, then there is some k-dimensional derived simplex $\Delta^{(k)} = co\{y_o, y_1, \ldots, y_k\}$ for which $l(y_i) \neq l(y_j)$, $i \neq j$, i.e., $\Delta^{(k)}$ is regular and thus completely labeled. Moreover, the number of such simplexes is odd.

The next theorem provides conditions which guarantee that the intersection of $k + 1$ closed sets $\mathcal{A}_o$, $\mathcal{A}_1, \ldots, \mathcal{A}_k$ defined on a simplex Δ^k is not empty. Specifically,

Theorem 17.3: **(KKM Theorem)** [KKM (1929); Berge (1963); and Border (1985)]. Let $\Delta^k = co\{x_o, x_1, \ldots, x_k\}$ be a k-simplex in R^n and let $\{\mathcal{A}_o, \mathcal{A}_1, \ldots, \mathcal{A}_k\}$ be a family of closed subsets of Δ^k. If for each index set $\{i_o, i_1, \ldots, i_m\} \subset \{0, 1, \ldots, k\}$ we have $co\{x_{i_o}, x_{i_1}, \ldots, x_{i_m}\} \subset \cup_{j=0}^{m} \mathcal{A}_{i_j}$, then $\cap_{i=0}^{k} \mathcal{A}_i \neq \emptyset$ is compact.

The gist of the KKM theorem is that, if each face simplex of Δ^k is a proper subset of a subcollection of closed sets $\mathcal{A}_o, \mathcal{A}_1, \ldots, \mathcal{A}_k$ in R^n, then the intersection of all $k + 1$ of these sets is nonempty.

17.5 THE EXISTENCE OF FIXED POINTS

We now turn to a demonstration that a continuous point-to-point mapping of a simplex Δ^k into itself has at least one fixed point. The accompanying proof illustrates the application of Sperner's lemma.

Theorem 17.4: **(Brouwer's Fixed-Point Theorem I)** [Brouwer (1912); Stoer and Witzgall (1970); KKM (1929); Border (1985); and Scarf (1973)]. Let f: $\Delta^k \to \Delta^k$ be a continuous point-to-point mapping. Then f has a fixed point $\bar{x} = f(\bar{x})$.

Proof: Let $\Delta^k = co\{x_o, x_1, \ldots, x_k\}$. Then for each $x \in \Delta^k$,

$$x = \sum_{i=0}^{k} \lambda_i x_i, \ \sum_{i=0}^{k} \lambda_i = 1, \ 0 \leq \lambda_i \in R, \ i = 0, 1, \ldots, k.$$

Correspondingly, the image of x under f, $f(x)$ (also within Δ^k), is expressible as

$$f(x) = \sum_{i=0}^{k} \hat{\lambda}_i x_i, \ \sum_{i=0}^{k} \hat{\lambda}_i = 1, \ 0 \leq \hat{\lambda}_i \in R, \ i = 0, 1, \ldots, k.$$

Since we may view the weights λ_i, $\hat{\lambda}_i$ in these convex combinations as continuous functions $\lambda_i(x)$ and $\hat{\lambda}_i(x)$, respectively, on Δ^k, it follows that the sets

$$\mathcal{F}_i = \{x | \hat{\lambda}_i(f(x)) \leq \lambda_i(x), \ x \in \Delta^k\}, \ i = 0, 1, \ldots, k,$$

are closed. Moreover, for every subset $\mathcal{J}(x) = \{i_o, i_1, \ldots, i_l\} \subset \{0, 1, \ldots, k\}$, they satisfy

$$\Delta^l = co\{x_{i_o}, x_{i_1}, \ldots, x_{i_l}\} \subseteq \mathcal{F}_{i_o} \cup \mathcal{F}_{i_1} \cup \cdots \cup \mathcal{F}_{i_l}. \quad (17.4)$$

(If a point $x \in \Delta^k$ is not in $\cup_{j=0}^{l} \mathcal{F}_{i_j}$, then $\hat{\lambda}_{i_h}(f(x)) > \lambda_{i_h}$ for $0 \leq h \leq k$ and thus

$$1 \geq \hat{\lambda}_{i_o}(f(x)) + \cdots + \hat{\lambda}_{i_l}(f(x)) > \lambda_{i_o}(x) + \cdots + \lambda_{i_l}(x)$$

so that $x \notin \Delta^l = co\{x_{i_o}, x_{i_1}, \ldots, x_{i_l}\}$.) Let $\mathcal{D}$ be a simplicial subdivision of Δ^k with u an arbitrary vertex in $\mathcal{D}$. Then there exists a smallest face $co\{x_{i_o}, x_{i_1}, \ldots, x_{i_l}\}$ containing u and, by (17.4), there exists an integer h such that $u \in \mathcal{F}_{i_h}$. For $l(u) = i_h$ a labeling of $\mathcal{D}$, Sperner's lemma yields the existence of a k-simplex $\Delta_m^k \in \mathcal{D}$ with label set $\{0, 1, \cdots, k - 1\}$. Under this labeling $u \in \mathcal{F}_{l(u)}$ and thus Δ_m^k meets each of the closed sets $\mathcal{F}_i$, $i = 0, 1, \ldots, k$.

Consider a sequence of simplicial subdivisions $\{\mathcal{D}^{(\nu)}\}_{\nu=1,2,\ldots}$ of Δ^k such that the mesh of $\mathcal{D}^{(\nu)}$ tends to zero as $\nu \to \infty$. In each $\mathcal{D}^{(\nu)}$ there exists a k-simplex $\Delta_m^{(k,\nu)}$ with vertices $x_o^{(\nu)}, x_1^{(\nu)}, \ldots, x_k^{(\nu)}$ such that $x_i^{(\nu)} \in \mathcal{F}_i$, $i = 0, 1, \ldots, k$. Since Δ^k is compact, $\{\mathcal{D}^{(\nu)}\}_{\nu=1,2,\ldots}$ has a subsequence which admits a convergent sequence of vertices $x_i^{(\nu)} \to x_i^{(\infty)}$, $i = 0, 1, \ldots, k$. With $\mathcal{F}_i$ closed, $x_i^{(\infty)} \in \mathcal{F}_i$, $i = 0, 1, \ldots, k$. Let $\Delta_m^{(k,\nu)} = co\{x_o^{(\nu)}, x_1^{(\nu)}, \ldots, x_k^{(\nu)}\}$. Since $d(\sigma_m^{(k,\nu)}) \to 0$ as $\nu \to \infty$, all vertices converge to a single point $\bar{x} \in \mathcal{F}_i$. And since $\bar{x}$ satisfies $\hat{\lambda}_i(f(\bar{x})) \leq \lambda_i(\bar{x})$ for all i and $\sum_{i=0}^{k} \hat{\lambda}_i(f(\bar{x})) = \sum_{i=0}^{k} \lambda_i(\bar{x}) = 1$, it follows that $\hat{\lambda}_i(f(\bar{x})) = \lambda_i(\bar{x})$ for all i so that $\bar{x} = f(\bar{x})$ is a fixed point of Δ^k. Q. E. D.

To further illustrate some of the salient elements of this theorem we state the following:

1. When we posit that f is a continuous mapping of Δ^k into itself we say that there exists a rule or law of correspondence which associates with each $x \in \Delta^k$ an image $f(x)$ also in Δ^k which depends in a continuous way on x. Note that there is no requirement that every $x \in \Delta^k$ be the image of

some point of Δ^k, i.e., f may only map Δ^k into some small subset of itself.

2. f is completely arbitrary; all that is required is the image of each $\boldsymbol{x}$ be an element of Δ^k. So if $\boldsymbol{x} \in \Delta^k$, then $\boldsymbol{y} = f(\boldsymbol{x})$ or $y_i = f_i(\boldsymbol{x}) \geq 0$ for all i and $\Sigma_i f_i(\boldsymbol{x}) = 1$.

3. It should be evident that the motivation underlying the proof of Brouwer's fixed-point theorem was to first employ Sperner's lemma to establish the existence of a completely labeled subsimplex of the (arbitrary) decomposition $\mathcal{D}$. By taking progressively finer decomposition we obtain a sequence of arbitrarily small completely labeled subsimplexes $\Delta^{(k,v)} = co\{\boldsymbol{x}_1^{(v)}, \boldsymbol{x}_2^{(v)}, \ldots, \boldsymbol{x}_k^{(v)}\}$. From the compactness of Δ^k itself the $\Delta^{(k,v)}$ get smaller and smaller and consequently converge (on a subsequence) in the limit to a point $\bar{\boldsymbol{x}}$, i.e., all vertices of $\Delta^{(k,v)}$ converge to $\bar{\boldsymbol{x}}$ so that $\boldsymbol{x}_i^{(v)} \rightarrow \bar{\boldsymbol{x}}$ as $v \rightarrow \infty$, $i = 0, 1, \ldots, k$.

To extend Brouwer's theorem to general convex sets, let us observe first that two convex sets S_1, S_2 are **homeomorphic** if there exists a one-to-one and onto correspondence f which is continuous in both directions between them, i.e., both the point-to-point mapping f: $S_1 \rightarrow S_2$ and its inverse f^{-1} are continuous. Here, f is called a **homeomorphism** between S_1 and S_2. Additionally, two convex sets $S_1, S_2 \subset R^n$ are homeomorphic if dim (S_1) = dim $(S_2) = k$ while a compact convex set in R^n is homeomorphic to the unit ball $\boldsymbol{B} = \{\boldsymbol{x} \,|\, \|\boldsymbol{x}\| \leq 1\}$ of the same dimension and thus to a simplex of the same dimension. By virtue of this observation we see that Brouwer's theorem is applicable to $\boldsymbol{B}$ and thus, in general, to any set homeomorphic to Δ^k. In the light of this discussion we can state, as corollary to Theorem 17.4,

Theorem 17.5: **(Brouwer's Fixed-Point Theorem II)**. Let set $S \subset R^n$ be homeomorphic to Δ^k with f: $S \rightarrow S$ a continuous point-to-point mapping. Then f has a fixed point $\bar{\boldsymbol{x}} = f(\bar{\boldsymbol{x}})$.

An even stronger version of this corollary is

Theorem 17.6: **(Brouwer's Fixed-Point Theorem III)**. Let set $S \subset R^n$ be a nonempty, compact convex set with f: $S \rightarrow S$ a continuous point-to-point mapping. Then f has a fixed point $\bar{\boldsymbol{x}} = f(\bar{\boldsymbol{x}})$.

Proof: Since S is bounded, it is contained within some simplex $\Delta^k \subset R^n$ and, hence, it is a retraction[3] of Δ^k. Let g represent the retraction of Δ^k onto S. Since f: $S \rightarrow S$ is continuous, f may be extended to a continuous mapping $h = f \circ g$: $\Delta^k \rightarrow \Delta^k$. Hence, any fixed point $\bar{\boldsymbol{x}}$ of h must be a fixed point of f. Q. E. D.

It is interesting to note that Sperner's lemma, the KKM theorem, and Brouwer's fixed-point theorem are all mathematically equivalent. That is, Sperner's lemma can be used to verify the KKM and Brouwer results; the KKM theorem implies Brouwer's theorem; and Brouwer's theorem implies both Sperner's lemma and the KKM result.

Section 17.5 Exercises:

17.5.1 Let $f(x) = \frac{1}{2}(x + 1)$, $x \in [-1, 1] \subset R$. Does f have a fixed point over its domain?

17.5.2 Suppose $y = f(x)$ is a continuous mapping of $[-1, 1]$ into itself. Verify that there must exist a point $\bar{x} \in [-1, 1]$ such that $\bar{x} = f(\bar{x})$.

17.5.3 Let $S = [a, b] \subset R$ with continuous f: $[a, b] \rightarrow [a, b]$. Suppose we remove a point x^1 from $S = [a, b]$ and form $S^1 = [a, b] \backslash \{x^1\}$. Does Brouwer's theorem still hold if S^1 replaces S?

17.5.4 Let $f(x) = x/2$ be a mapping of $S = (0, 1] \subset R$ into itself. Does f have a fixed point?

17.5.5 Let

$$f(x) = \begin{cases} \frac{1}{4} + x, \; x \in \left[0, \frac{1}{4}\right]; \\ \frac{1}{2}, \; x \in \left(\frac{1}{4}, 1\right] \end{cases}$$

be a mapping of $S = [0, 1] \subset R$ onto itself. Does f have a fixed point?

17.6 FIXED POINTS OF COMPACT POINT-TO-POINT FUNCTIONS

A generalization of Brouwer's fixed-point Theorem 17.6 has been developed by Schauder (1930). Here, a continuous point-to-point mapping is replaced by one that is either completely continuous or compact. To define these concepts, let $S (\neq \emptyset) \subset R^n$ with f a continuous point-to-point mapping of f into itself. Then f: $S \rightarrow S$ is **completely continuous** if the image of each bounded subset $\mathcal{A}$ of S is contained within a compact set; and $f: S \to S$ is said to be **compact** if $f(S)$ is compact.

We can now state

Theorem 17.8: **(Schauder's Fixed-Point Theorem I)** [Schauder (1930); and Cronin (1964)]. Let S be a closed convex set in R^n with $f: S \to R$ a compact point-to-point mapping such that $f(S) \subset S$. Then f has a fixed point, i.e., there exists an $\bar{x} \in S$ such that $\bar{x} = f(\bar{x})$.

Schauder's theorem has the following equivalent representations:

Theorem 17.9: **(Schauder's Fixed-Point Theorem II)**. For $S \subset R^n$, let $f: S \to S$ be a completely continuous point-to-point mapping with $f(S)$ bounded. Then f has a fixed point $\bar{x} = f(\bar{x})$.

Theorem 17.10: **(Schauder's Fixed-Point Theorem III)**. If S is a compact convex set in R^n, then every continuous point-to-point mapping $f: S \to S$ has a fixed point $\bar{x} = f(\bar{x})$.

And as a corollary to Schauder's Theorem 17.8 we have

Theorem 17.11: [Cronin (1964)]. Let set S be homeomorphic to a closed convex set in R^n with $f: S \to R$ a compact point-to-point mapping such that $f(S) \subset S$. Then f has a fixed point $\bar{x} = f(\bar{x})$.

(Further generalizations of Schauder's Theorem 17.8 can be found in Tychonoff (1935); Klee (1960); Brouwer (1959 and 1965 a, b) and Fan (1969).)

17.7 FIXED POINTS OF POINT-TO-SET FUNCTIONS

Brouwer's theorem established the existence of a fixed point for a continuous point-to-point mapping f of a compact convex set S into itself. If f is replaced by an upper hemicontinuous point-to-set mapping of S into the set of all compact convex subsets of S, then a generalization of Brouwer's theorem is provided by

Theorem 17.12: **(Kakutani's Fixed Point Theorem I)** [Kakutani (1941)Nikaido (1968); Berge (1963); Klein (1973); and Stoer and Witzgall (1979)]. Let S be a nonempty compact and convex set in R^n and let $f: S \to S$ be an upper hemicontinuous set-valued function with nonempty, compact convex image sets $F(\boldsymbol{x})$ for each $\boldsymbol{x} \in S$. Then F has a fixed point $\bar{\boldsymbol{x}} = F(\bar{\boldsymbol{x}})$.

Additional variants of Kakutani's theorem abound. For instance, one may specify an alternative form of this theorem by substituting the notion of a closed-graph point-to-set mapping for an upper hemicontinuous one,[4] or one can directly employ the concept of lower hemicontinuity of the set-valued mapping or by using the inverse image of the same. Kakutani's fixed-point theorem has been generalized by Berge (1950) and Eileaberg and Montgomery (1946). Further extensions of fixed-point theorems of set-valued functions are offered by Fan (1969).

Section 17.7 Exercises:

17.7.1 Let $F(x)$ be a set-valued function which maps $x \in [0, 1] \subset R$ into $[1 - \frac{1}{2}x, 1 - \frac{1}{3}x]$. Does f have a fixed point?

17.7.2 What is the relationship between Brouwer's theorem and Kakutani's theorem?

17.7.3 Let $F(x) = [1, 12] \to [1, 12] \subset R$ be a set-valued function defined as

$$F(x) = \begin{cases} \left[3 + \frac{1}{2}x, 6 + \frac{1}{2}x\right], & x \in [0, 6); \\ \left[1 + \frac{1}{2}x, 6 + \frac{1}{2}x\right], & x \in (6, 12]. \end{cases}$$

Does f have a fixed point?

17.8 ECONOMIC APPLICATIONS: EXISTENCE OF A COMPETITIVE EQUILIBRIUM (DEBREU [2007]; TAKAYAMA [1987]; MCKENZIE [1959, 1961], NIKAIDO [1960], ARROW AND DEBREU [1954], ARROW AND HAHN [1971]; AND MAS-COLLEL ET AL. [2007])

17.8.1 A Pure Exchange Economy

We know that in a competitive market system all transactions between individuals are facilitated by the market, with each consumer's behavior dictated solely by self-interest. An individual, being a minor player in each market, is thus considered to be a price taker (whether acting as a buyer or seller). Equilibrium in each market taken separately is attained when, at the prevailing market price, buyers' decisions are consonant with sellers' decisions. By extension, equilibrium in the entire market system emerges when, at the set of prevailing market prices, buyers' demand decisions are in harmony with sellers' supply decisions. In fact, under this competitive scheme of decentralized decision making, *only market price information matters.* In what follows then it is assumed that there exists a set of prices at which demands are always met and supplies are always purchased, i.e., a set of prices that simultaneously clears all interdependent markets. The type of competitive market system to which we now turn is termed a **pure exchange economy**.

We need a set of assumptions/characteristics which describe the economic environment associated with a pure exchange economy:

1. no production; each of m consumers is endowed with a given amount of each of n distinct commodities.

2. Let e_i^j = consumer j's initial endowment of commodity i, $i = 1, \ldots, n$; $j = 1, \ldots, m$.

   $(e^j)^T = (e_1^j, \ldots, e_n^j) \geq O^T$ is a vector of consumer j's initial endowment of all n commodities, $j = 1, \ldots, m$.

   $e_i = \sum_{j=1}^m e_i^j$ is the total endowment of commodity i to all consumers, $i = 1, \ldots, n$.

   $\boldsymbol{e}^T = (e_1, \ldots, e_n)$ is a vector of total endowment of all n commodities.

3. Let x_i^j= consumer j's final demand for commodity i (see item (4)), $i = 1, \ldots, n$; $j = 1, \ldots, m$.

   $(\boldsymbol{x}^j)^T = (x_1^j, \ldots, x_n^j)$ is a vector of consumer j's final demands for all n commodities, $j = 1, \ldots, m$.

   $x_i = \sum_{j=1}^m x_i^j$ is the total demand for commodity i by all consumers, $i = 1, \ldots, n$.

   $\boldsymbol{x}^T = (x_1, \ldots, x_n)$ is a vector of total demands of all n commodities.

4. Each consumer has a preference relation (on a consumption set $X_j \in R_+^n$) that is represented by a utility function $u^j = f^j(\boldsymbol{x}^j)$ which is taken to be continuous, strictly increasing, and strictly quasiconcave. Hence, each consumer solves a problem of the form

$$\begin{aligned} &\underset{\boldsymbol{x}^j \in R_+^n}{\text{maximize}}\ f^j(\boldsymbol{x}^j) \text{ subject to} \\ &\boldsymbol{p}^T\boldsymbol{x}^j \leq \boldsymbol{p}^T\boldsymbol{e}^j. \end{aligned} \tag{17.5}$$

where: $\boldsymbol{p} > \boldsymbol{O}$ is an $(n \times 1)$ vector of commodity prices; $\boldsymbol{p}^T\boldsymbol{e}^j$ represents consumer j's income; and the budget constraint simply depicts the notion that total expenditure does not exceed income.

Once (17.5) is solved, we obtain for consumer j a set of continuous (in $\boldsymbol{p}$) ordinary or Marshallian final demand functions on R_+^n

$$x_i^j = x_i^j(\boldsymbol{p}, \boldsymbol{p}^T\boldsymbol{e}^j),\ i = 1, \ldots, n. \tag{17.6}$$

For convenience let us rewrite these final demands as simply $x_i^j = x_i^j(\boldsymbol{p})$, $i = 1, \ldots, n$; $j = 1, \ldots, m$.

While there are n separate markets incorporating a system of demand and supply functions, convention dictates that each individual market be described by a single excess demand function. That is, let us express the **aggregate excess demand function for commodity *i*** as

$$z_i(\boldsymbol{p}) \equiv x_i(\boldsymbol{p}) - e_i = \sum_{j=1}^m x_i^j(\boldsymbol{p}) - \sum_{j=1}^m e_i^j. \tag{17.7}$$

As this expression reveals, total excess demand for commodity i equals the total demand for commodity i by all consumers less the total endowment of commodity i taken over all consumers and, if $z_i(\boldsymbol{p}) > 0(<0)$, then there exists excess demand (excess supply) for commodity i. Additionally, **aggregate excess demand** is thus the vector $\boldsymbol{z}(\boldsymbol{p}^T) \equiv (z_1(\boldsymbol{p}), \ldots, z_n(\boldsymbol{p}))$.

If each consumers utility function f^j is continuous, strictly increasing, and strictly quasiconcave, then for all price vectors $\boldsymbol{p} > \boldsymbol{O}$, it can be shown that $\boldsymbol{z}(\boldsymbol{p})$ is continuous in $\boldsymbol{p}$ as well as homogeneous of degree zero in commodity prices. Moreover, with the budget constraint in (17.5) binding for all consumers, it follows that

$$\boldsymbol{p}^T\boldsymbol{x}^j = \boldsymbol{p}^T\boldsymbol{e}^j \text{ or } \sum_{i=1}^n p_i x_i^j(\boldsymbol{p}) = \sum_{i=1}^n p_i e_i^j.$$

Since this equality holds for all consumers, upon summing this expression over all j values renders

$$\begin{aligned} &\textstyle\sum_{i=1}^n p_i \sum_{j=1}^m x_i^j(\boldsymbol{p}) = \sum_{i=1}^n p_i \sum_{j=1}^m e_i^j, \\ &\textstyle\sum_{i=1}^n p_i x_i(\boldsymbol{p}) = \sum_{i=1}^n p_i e_i, \\ &\textstyle\sum_{i=1}^n p_i (x_i(\boldsymbol{p}) - e_i) = \sum_{i=1}^n p_i z_i(\boldsymbol{p}) \\ &= \boldsymbol{p}^T\boldsymbol{z}(\boldsymbol{p}) \equiv 0. \quad \text{[Walras' law]} \end{aligned} \tag{17.8}$$

Here (17.8) is termed **Walras' Law**: the value of aggregate excess demand will always be zero at *any* set of positive prices. In fact, prices that equate aggregate demand and supply in every market constitute a **Walrasian (general) equilibrium**, i.e., a price vector $\boldsymbol{p}^* > \boldsymbol{O}$ constitutes a Walrasian equilibrium if $\boldsymbol{z}(\boldsymbol{p}^*) = \boldsymbol{O}$. Additionally, from (17.7), for $\boldsymbol{p}^*$ an equilibrium price vector, $z_i(\boldsymbol{p}^*) = x_i(\boldsymbol{p}^*) - e_i \le 0,\ i = 1, \ldots, n$. However, for some commodity k, can we have $p_k^*(x_k(\boldsymbol{p}^*) - e_k) < 0$? The answer is *no* since, by Walras' Law, with $\boldsymbol{p}^*$ an equilibrium price vector, $x_k(\boldsymbol{p}^*) - e_k < 0$ implies that $p_k^* = 0$ for any k. So if any commodity is in excess supply in equilibrium, its equilibrium price must be zero. And since at the prevailing set of prices any excess supply in the system of markets must be matched exactly by an excess demand of equal value somewhere in the system, it follows that if $n - 1$ markets are in equilibrium, then Walras' Law dictates that the nth market must also be in equilibrium.

We noted earlier that the excess demands $\boldsymbol{z}(\boldsymbol{p})$ are homogeneous of degree zero, meaning that, if $\boldsymbol{z}(\boldsymbol{p}) \le \boldsymbol{O}$, then $\boldsymbol{z}(t\boldsymbol{p}) \le \boldsymbol{O},\ t > 0$, and thus, if the combination $(\boldsymbol{x}^*, \boldsymbol{p}^*)$ is a competitive equilibrium, then so is $(\boldsymbol{x}^*, t\boldsymbol{p}^*)$. Hence, the implication of homogeneity is that we can normalize prices so that their sum is always unity, i.e., simply divide $\boldsymbol{p}$ by $\boldsymbol{1}^T\boldsymbol{p}$ ($\boldsymbol{1}$ is the **sum vector** containing all 1's as components) so that $\sum_{i=1}^n (p_i/1^T\boldsymbol{p}) = 1$. But this enables us to restrict our analysis to normalized price vectors defined on the unit simplex $\Delta^{n-1} = \{\boldsymbol{p} | \boldsymbol{p} \in R_+^n,\ 1^T\boldsymbol{p} = 1\} \subset R_+^n$.

With the excess demands $z_i(\boldsymbol{p})$ continuous on Δ^{n-1}, the problem of verifying the existence of a competitive (Walrasian) equilibrium can be stated as:

> Determine a price vector $\boldsymbol{p}^* \in \Delta^{n-1}$ such that $z_i(\boldsymbol{p}^*) \le 0,\ i = 1, \ldots, n$, given that the $z_i(\boldsymbol{p})$ are continuous on Δ^{n-1} with $\sum_{i=1}^n p_i z_i(\boldsymbol{p}) = 0,\ \ \boldsymbol{p} \in \Delta^{n-1}$.

As we shall now see, the proof can be executed by invoking Brouwer's fixed point theorem[5] to verify the existence of $\boldsymbol{p}^*$. As required by Brouwer's theorem, we need to construct a continuous mapping of Δ^{n-1} into itself for which a fixed point corresponds to an equilibrium price vector.

Consider the mapping $\boldsymbol{g}: \Delta^{n-1} \to \Delta^{n-1}$ defined by

$$g_i(\boldsymbol{p}) = \frac{[p_i + \max\{0, z_i(\boldsymbol{p})\}]}{[1 + \sum_{k=1}^n \{0, z_k(\boldsymbol{p})\}]},\ i = 1, \ldots, n, \quad (17.9)$$

where $1 + \sum_{k=1}^n \{0, z_k(\boldsymbol{p})\} \ge 1$. The $g_i(\boldsymbol{p})$ are continuous (since the $z_i(\boldsymbol{p})$ are) with $g_i(\boldsymbol{p}) \in \Delta^{n-1}$ when $\boldsymbol{p} \in \Delta^{n-1}$ since $\sum_{i=1}^n g_i(\boldsymbol{p}) = 1$. With $\boldsymbol{g}$ a continuous mapping from Δ^{n-1} to Δ^{n-1}, by Brouwer's theorem, there exists a $\boldsymbol{p}^*$ such that $\boldsymbol{p}^* = \boldsymbol{g}(\boldsymbol{p}^*)$ or

$$p_i^* = \frac{[p_i^* + \max\{0, z_i(\boldsymbol{p}^*)\}]}{[1 + \sum_{k=1}^n \{0, z_k(\boldsymbol{p}^*)\}]},\ i = 1, \ldots, n. \quad (17.9.1)$$

Let us employ (17.9.1) so as to obtain

$$p_i^*[1 + \textstyle\sum_{k=1}^n \{0, z_k(\boldsymbol{p}^*)\}] = p_i^* + \max\{0, z_i(\boldsymbol{p}^*)\},$$

$$p_i^* \textstyle\sum_{k=1}^n \{0, z_k(\boldsymbol{p}^*)\} = \max\{0, z_i(\boldsymbol{p}^*)\},$$

$$p_i^* z_i(\boldsymbol{p}^*) \textstyle\sum_{k=1}^n \{0, z_k(\boldsymbol{p}^*)\} = z_i(\boldsymbol{p}^*) \max\{0, z_i(\boldsymbol{p}^*)\},$$

$$\textstyle\sum_{i=1}^n p_i^* z_i(\boldsymbol{p}^*) \sum_{k=1}^n \{0, z_k(\boldsymbol{p}^*)\} = \sum_{i=1}^n z_i(\boldsymbol{p}^*) \max\{0, z_i(\boldsymbol{p}^*)\}.$$

By Walras' Law, the preceding expression becomes

$$0 = \textstyle\sum_{i=1}^n z_i(\boldsymbol{p}^*) \max\{0, z_i(\boldsymbol{p}^*)\}. \quad (17.10)$$

Moreover, since each term in Equation (17.10) is nonnegative, the right-hand side sums to zero only if $z_i(\boldsymbol{p}^*) \le 0,\ i = 1, \ldots, n$. With $z_i(\boldsymbol{p}^*) \le 0,\ \boldsymbol{p}^* > \boldsymbol{O}$, and Walras' Law in effect, we thus conclude that each $z_i(\boldsymbol{p}^*) = 0,\ i = 1, \ldots, n$, as required for $\boldsymbol{p}^*$ to be an equilibrium price vector.

17.8.2 A Private Ownership (Production) Economy

Suppose there are n commodities, m consumers, and q firms. Additionally,

1. each consumer has a preference relation defined on a consumption set $X_j,\ j = 1, \ldots, m$.
2. $\boldsymbol{e}^T = (e_1, \ldots, e_n)$ is an $(n \times 1)$ vector of total endowments of all n goods, where $e_i = \sum_{j=1}^m e_i^j$, $i = 1, \ldots, n$, represents the total endowment of commodity i to all consumers.
3. each firm has a production set $Y_k,\ k = 1, \ldots, q$, that is nonempty, closed, strictly convex, and bounded from above. Let y_i^k = amount of the ith commodity produced by firm $k,\ i = 1, \ldots, n;\ k = 1, \ldots, q$.

$(\mathbf{y}^k)^T = (y_1^k, \ldots, y_n^k) \geq \mathbf{O}^T$ is a vector of firm k's production amounts of the various commodities, $k = 1, \ldots, q$.

$y_i = \Sigma_{k=1}^q y_i^k$ is the amount of commodity i produced by all firms, $i = 1, \ldots, n$.

$\mathbf{y}^T = (y_1, \ldots, y_n)$ is a vector of total amounts of all n commodities produced.

Under these considerations, an **allocation** $(\mathbf{x}^T, \mathbf{y}^T) = (\mathbf{x}^1, \ldots, \mathbf{x}^m, \mathbf{y}^1, \ldots, \mathbf{y}^q)$ is a specification of a consumption vector $\mathbf{x}^j$ for each consumer and a production vector $\mathbf{y}^k$ for each firm. The allocation is **feasible** if

$$x_i = e_i + y_i \text{ or } \sum_{j=1}^m x_i^j = \sum_{j=1}^m e_i^j + \sum_{k=1}^q y_i^k,\ i = 1, \ldots, n,$$

or $\mathbf{x} = \mathbf{e} + \mathbf{y}$.

4. each commodity is traded in a market at a known price, with both consumers and firms behaving as price takers. Consumers trade in the marketplace to maximize their utility while firms produce and trade to maximize profits. The wealth of consumers stems from their individual endowments of commodities and from private ownership shares to the profits of the firms, with each firm owned by consumers. Let us write the ownership claim of consumer j to the profits of firm k, π_k, as $\theta_k^j \in [0, 1]$, $\Sigma_{k=1}^q \theta_k^j = 1$. Hence, the total wealth of the jth consumer is

$$w_j = \mathbf{p}^T \mathbf{e}^j + \sum_{k=1}^q \theta_k^j \pi_k.$$

Given a private ownership economy, an allocation $(\mathbf{x}^*, \mathbf{y}^*)$ and a price vector $\mathbf{p}$ constitute a Walrasian (competitive) equilibrium if:

a. for firm k, $(\mathbf{y}^k)^* \in Y_k$ maximizes profit or $\mathbf{p}^T \mathbf{y}^k \leq \mathbf{p}^T (\mathbf{y}^k)^*$ for all $\mathbf{y}^k \in Y_k$.

b. the commodity bundle $(\mathbf{x}^j)^* \in X_j$ maximizes consumer j's utility subject to the budget constraint

$$\mathbf{p}^T \mathbf{x}^j \leq \mathbf{p}^T \mathbf{e}^j + \sum_{k=1}^q \theta_k^j \mathbf{p}^T (\mathbf{y}^k)^* \qquad (17.11)$$

for all $\mathbf{x}^j \in X_j$.

c. for every commodity the excess demand over supply is zero or

$$\bar{z}(\mathbf{p})^* \equiv \sum_{j=1}^m \mathbf{x}^j(\mathbf{p})^* - \sum_{k=1}^q \mathbf{y}^k(\mathbf{p})^* - \sum_{j=1}^m \mathbf{e}^j = \mathbf{O},$$

the **production inclusive aggregate excess demand function**.

To address the issue of the existence of a competitive equilibrium, we shall focus on the **excess demand correspondence** $\varphi(\mathbf{p})$ from the price simplex Δ^{n-1} to commodity space R^n. In what follows (which rests heavily on the analysis in Debreu [2007]) it is assumed that there is a price vector $\mathbf{p}^* \geq \mathbf{O}$ that yields a **free disposal equilibrium** (all commodities can be freely disposed of) if and only if there is a vector $\bar{z}(\mathbf{p}^T)$ in $\varphi(\mathbf{p}^*)$ such that $\bar{z}(\mathbf{p}^*) \leq \mathbf{O}$ and $(\mathbf{p}^*)^T \bar{z}(\mathbf{p}^*) = \mathbf{O}$.

More specifically, let $\xi_j(\mathbf{p})$ denote the set of utility maximizing consumption bundles under the budget constraint (see (17.11b) and $\eta_k(\mathbf{p})$ the set of profit-maximizing production quantities. Thus the jth consumer seeks an element $\mathbf{x}^j \in \xi_j(\mathbf{p})$ and the kth producer seeks an element $\mathbf{y}^k \in \eta_k(\mathbf{p})$. Hence, the associated excess demand $\bar{z} = \Sigma_{j=1}^m \mathbf{x}^j - \Sigma_{k=1}^q \mathbf{y}^k - \Sigma_{j=1}^m \mathbf{e}^j$ is an element of the set $\varphi(\mathbf{p}) = \Sigma_{j=1}^m \xi_j(\mathbf{p}) - \Sigma_{k=1}^q \eta_k(\mathbf{p}) - \Sigma_{j=1}^m \mathbf{e}^j$, where φ is the excess demand correspondence from Δ^{n-1} to commodity space R^n. As indicated earlier, a price vector $\mathbf{p}^*$ yields a free disposal equilibrium if and only if there is a vector $\bar{z}(\mathbf{p}^*) \in \varphi(\mathbf{p}^*)$ such that $\bar{z}(\mathbf{p}^*) \leq \mathbf{O}$ and $(\mathbf{p}^*)^T \bar{z}(\mathbf{p}^*) = \mathbf{O}$. Since under free disposal $\mathbf{z}(\mathbf{p}^*) \in \varphi(\mathbf{p}^*)$, we have, for every j, a vector $(\mathbf{x}^j)^* \in \xi_j(\mathbf{p}^*)$, and, for every k, a vector $(\mathbf{y}^k)^* \in \eta_k(\mathbf{p}^*)$ such that

$$\bar{z}(\mathbf{p}^*) = \sum_{j=1}^m (\mathbf{x}^j)^* - \sum_{k=1}^q (\mathbf{y}^k)^* - \sum_{j=1}^m \mathbf{e}^j.$$

With $\bar{z}(\mathbf{p}^*) \leq \mathbf{O}$, the allocation $((\mathbf{x}^j)^*, (\mathbf{y}^k)^*)$ for $\mathbf{p}^*$ is attainable. With $(\mathbf{x}^j)^*$ attainable, and with consumption non-satiated in the attainable consumption set, it follows that, for every j,

$$(\mathbf{p}^*)^T (\mathbf{x}^j)^* = (\mathbf{p}^*)^T \mathbf{e}^j + \sum_{k=1}^q \theta_k^j (\mathbf{p}^*)^T (\mathbf{y}^k)^*.$$

Then, upon summing over all j, $(\mathbf{p}^*)^T \bar{z}(\mathbf{p}^*) = 0$.

Let us now examine the properties of the excess demand correspondence φ. Given any $p \in \Delta^{n-1}$: for every j, the set $\xi_j(p)$ is nonempty, convex, and compact (since it constitutes the set of maximizers of a strictly quasiconcave utility function on the convex and compact budget set); and for every k, $\eta_k(p)$ is nonempty, convex, and compact (being the set of maximizers of the profit function on the convex and compact production set). Hence, $\varphi(p)$ is itself convex and compact as the sum of convex and compact sets. Debreu [2007] further demonstrates that, for every j, ξ_j is upper hemicontinuous; and, for every k, $\eta_k$is also upper hemicontinuous. Thus φ is upper hemicontinuous since it is the sum of upper hemicontinuous correspondences. One additional observation is in order. For every $p \in \Delta^{n-1}$, for every j and every $x^j \in \xi_j(p)$, for every k and every $y^k \in \eta_k(p)$, we have

$$p^T x^j \leq p^T e^j + \sum_{k=1}^{q} \theta_k^j p^T y^k.$$

Upon summing over j, for every $p \in \Delta^{n-1}$ and every $\bar{z} \in \varphi(p)$, $p^T \bar{z} \leq 0$ or $p^T \varphi(p) \leq 0$. To summarize:

> the correspondence φ is upper hemicontinuous and for every $p \in \Delta^{n-1}$, the set $\varphi(p)$ is compact, convex, and satisfies $p^T \varphi(p) \leq 0$.

Given that φ is upper hemicontinuous and convex-valued, Debreu [1956, 2007] (see also Gale [1955] and Nikaido [1956]) then commences to establish the existence of a price vector $p^* \in \Delta^{n-1}$ yielding an excess demand vector $\bar{z}(p^*) \leq O$ via Kakutani's fixed point theorem.[6] Hence, φ has a fixed point $(p^*, \bar{z}(p^*))$ and thus there exists a competitive equilibrium. In fact, for every p within the domain of φ, Walras' Law holds or $p^T \varphi(p) = 0$ since under our preceding assumptions on the utility function and consumption set, excess demands are zero.

Section 17.8 Exercises:

17.8.1 Given that the constraint in (17.5) holds with strict equality, verify that Walras' Law holds by first summing over commodities and then summing over consumers.

17.8.2 We noted earlier that the purpose of Equation (17.9) was to develop a continuous mapping of Δ^{n-1} into itself in order to produce a fixed point or equilibrium price vector. Will the following price mechanism serve as a legitimate device for achieving this same objective? Let $p \rightarrow \hat{p}$, where

$$\hat{p}_i = p_i + \varnothing_i(p),\ i = 1,\ \ldots, n,$$

is the new or adjusted price.

17.8.3 Suppose consumer j's utility function is continuous, strictly increasing, and strictly quasiconcave on R_+^n. Then the excess demand vector $z(P)$ (see Equation (17.7)) will be homogeneous of degree zero in prices. What is the significance of this property of excess demands?

APPENDIX 17.A **THE BARYCENTRIC SUBDIVISION OF A *K*-SIMPLEX** (SHAPLEY [1973]; AND SCARF [1973])

Given the structure of Δ^k provided by Equation (17.1), the first-order barycentric subdivision of Δ^k is a simplicial subdivision consisting of $(k+1)!$ subsimplexes, each one associated with a particular permutation $i_o, \ldots, i_k$ of the integers $0, \ldots, k$. For each such permutation the derived k-dimensional subsimplex $\Delta_{i_o,\ldots,i_k}^{(k,1)}$ if defined as consisting of all vertices with $\lambda_{i_o} \geq \lambda_{i_1} \geq \cdots \geq \lambda_{i_k} \geq 0$. In general, the vertices of the derived simplexes $\Delta_{i_o,\ldots,i_k}$ of a first-order barycentric subdivision of Δ^k are:

$$\begin{aligned} y_{i_o} &= x_{i_o} \\ y_{i_1} &= \frac{1}{2}(x_{i_o} + x_{i_1}) \\ &\vdots \\ y_{i_k} &= \frac{1}{k+1}(x_{i_o} + x_{i_1} + \cdots + x_{i_k}). \end{aligned} \qquad (17.A.1)$$

Any higher order barycentric subdivisions of Δ^k may be obtained recursively from the immediately preceding one.

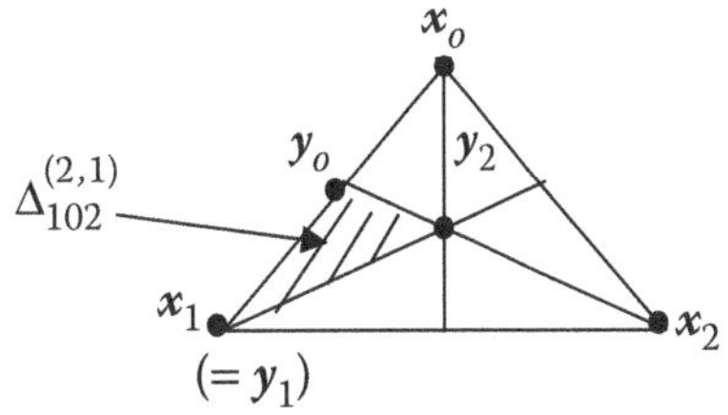

FIGURE 17.A.1 Barycentric subdivision of Δ^2.

Example 17.A.1

Let $\Delta^2 = co\{x_o, x_1, x_3\}$. There are $(k + 1)! = 6$ derived subsimplexes constituting the barycentric subdivision of Δ^2 (Figure 17.A.1). Clearly, the six permutations of the subscripts "0, 1, 2" are

0,1, 2; 1, 0, 2; 2, 1, 0; 0, 2, 1; 2, 0, 1; 1, 2,0.

Consider the permutation "1, 0, 2." We thus must form

$$\lambda_1 x_1 + \lambda_0 x_0 + \lambda_2 x_2 \text{ with } \lambda_1 \geq \lambda_0 \geq \lambda_2 \geq 0.$$

Using (17.A.1), the vertices of $\Delta_{102}^{(2,1)} = co\{y_1, y_0, y_2\}$ are

$$\begin{aligned} y_1 &= x_1 \\ y_o &= \frac{1}{2}(x_1 + x_o) \\ y_2 &= \frac{1}{3}(x_1 + x_o + x_2), \end{aligned}$$

i.e., $\Delta_{102}^{(2,1)}$ consists of all vectors of the form

$$y = \theta_1 y_1 + \theta_o y_o + \theta_2 y_2 \text{ with } \theta_1, \theta_0, \theta_2 \geq 0.$$

Substituting from above we have

$$\begin{aligned} y &= \theta_1 x_1 + \theta_o\left(\frac{1}{2}(x_1 + x_o)\right) + \theta_2\left(\frac{1}{3}(x_1 + x_o + x_2)\right) \\ &= \left(\theta_1 + \frac{1}{2}\theta_o + \frac{1}{3}\theta_2\right) + \left(\frac{1}{2}\theta_o + \frac{1}{3}\theta_2\right)x_o + \frac{1}{3}\theta_2 x_2 \\ &= \lambda_1 x_1 + \lambda_o x_o + \lambda_2 x_2. \end{aligned}$$

And as required,

$$\lambda_1 = \theta_1 + \frac{1}{2}\theta_o + \frac{1}{3}\theta_2 > \lambda_o = \frac{1}{2}\theta_o + \frac{1}{3}\theta_2 > \lambda_2 = \frac{1}{3}\theta_2.$$

■

As just demonstrated, any vector of the form $\theta_1 y + \theta_o y_o + \theta_2 y_2$ with $\theta_i \geq 0$, $i = 0, 1, 2$, can be expressed as $\lambda_1 x_1 + \lambda_o x_o + \lambda_2 x_2$ with $\lambda_1 \geq \lambda_o \geq \lambda_2 \geq 0$ and conversely. The remaining five derived subsimplexes of the first-order barycentric subdivision of Δ^2 are similarly determined. Additionally, the second-order barycentric subdivision of Δ^2 is obtained by subjecting each of the six derived subsimplexes to its own first-order barycentric subdivision. There will be $((k + 1)!)^2$ derived subsimplexes in the barycentric subdivision of Δ^2 of order two.

NOTES

1 A **homeomorphism** between points in two geometric figures or spaces is a one-to-one correspondence that is continuous in both directions.

2 A systematic computational procedure which yields a concise description of the barycentric subdivision of a k-simplex is provided in Appendix 17.A.

3 Let $\mathcal{K}$ be a subset of R^n. A point-to-point mapping g: $R^n \rightarrow \mathcal{K}$ is a retraction mapping on R^n if $g(\mathbf{x}) = \mathbf{x}$ for all $\mathbf{x} \in \mathcal{K}$. Hence, $\mathcal{K}$ is called a **retraction** of R^n. If $\mathcal{K}$ is contained within an arbitrary subset S of R^n, then g: $S \rightarrow \mathcal{K}$ is a retraction of S onto $\mathcal{K}$ if $g(\mathbf{x}) = \mathbf{x}$ for all $\mathbf{x} \in \mathcal{K}$. Under what conditions will there exist a retraction of S onto a set? To answer this question we have Theorem 17.7 [Stoer and Witzgall (1970)]. Let $\mathcal{K}$ be a closed convex subset of an arbitrary set $S \subset R^n$. Then there exists a continuous point-to-point mapping g: $S \rightarrow \mathcal{K}$ which is a retraction of S onto $\mathcal{K}$.

4 It is instructive to state Kakutani's fixed-point theorem in terms of a closed graph. To this end we first offer the following bit of terminology. Specifically, a set function Q has a **closed graph** if the set $\{(\mathbf{x}, \mathbf{y}) | \mathbf{y} \in Q(\mathbf{x})\}$ is a closed subset of $X \times Y$, i.e., for all sequences $\{\mathbf{x}_n\}_{n \in N}$ and $\{\mathbf{y}_n\}_{n \in N}$ such that $\mathbf{x}_n \rightarrow \mathbf{x}$, $\mathbf{y}_n \rightarrow \mathbf{y}$ and $\mathbf{y}_n \in Q(\mathbf{x})$ for all n we have $\mathbf{y} \in Q(\mathbf{x})$. We can now stateTheorem 17.13 (**Kakutani's Fixed-Point Theorem II**). Let S be a nonempty, compact subset of R^n and let Q be a set-valued function on S with a closed graph having the properties that $Q(\mathbf{x})$ is nonempty and convex for all $\mathbf{x} \in S$. Then Q has a fixed point $\bar{\mathbf{x}} = Q(\bar{\mathbf{x}})$.

5 To review: let $S \subset R^n$ be a nonempty, compact, convex set with f a single-valued continuous function from S into itself. Then there exists a point $\hat{\mathbf{x}} \in S$ for which $\hat{\mathbf{x}} = f(\hat{\mathbf{x}})$.

6 To review: Let $S \subset R^n$ be a nonempty, compact, convex set with f an upper hemicontinuous convex-valued correspondence from S to S. Then f has a fixed point.

CHAPTER 18

Dynamic Optimization: Optimal Control Modeling

18.1 INTRODUCTION AND BASIC PROBLEM

Up to this point in our discussion of optimization (constrained or otherwise), we have dealt exclusively with **static optimization** procedures, i.e., we have abstracted completely from the passage of time in that all functions were relevant only at a particular point in time. In this chapter, the explicit inclusion of time (t) into our models dictates that *intertemporal* or *multiperiod* optimization techniques will be introduced. In this regard, **dynamic optimization modeling**, in which the values of the apposite variables (and thus the objective and constraint function(s) are time dependent, will be developed. Hence, decision making is conducted over a **planning period** having an **initial time point** $t = 0$ and a **terminal timepoint** $t = T$ (where an infinite time horizon with $T \to +\infty$ is permissible).

Under static optimization, the optimal levels of the variables are determined at a given point in time. But under dynamic optimization, our goal is to choose, optimally, the *time paths* of the variables; and this will be accomplished via optimal control modeling.

Let us examine the basic structure of an optimal control problem. We consider first the relevant variables:

i. The choice or decision variable will be called the **control variable** and denoted $u(t)$. This *flow variable* is to be determined optimally at each moment of time, with its optimal solution value, denoted $\hat{u}(t)$, depicting its time path over the planning horizon.

ii. The control variable affects another variable, denoted $x(t)$, called the **state variable** (a *stock variable*). The rate of change of $x(t)$, $dx(t)/dt = \dot{x}(t)$, is a first-order differential equation which depends on t, $x(t)$, and $u(t)$ and is expressed as

$$\dot{x}(t) = g(t, x(t), u(t)), \; x(0) = x_o \qquad (18.1)$$

(with x_o a given constant), and serves as a constraint which essentially connects $u(t)$ and $x(t)$, i.e., the selection of $u(t)$ controls the evolution of $x(t)$ via the constraint (18.1). Hence, (18.1) can be described as a **law of motion for *x*** or the **state trajectory**. That is, given the optimal time path of $u(t)$, denoted $\hat{u}(t)$, we can evaluate (18.1) at $\hat{u}(t)$ to obtain

$$\dot{x}(t) = g(t, x(t), \hat{u}(t)), \; x(0) = x_o. \qquad (18.1.1)$$

The solution to (18.1.1), depicted as $\hat{x}(t)$, traces out the optimal time path of $x(t)$ over $[0, T]$. Clearly, the time path of $x(t)$ is determined by the time path of $u(t)$.

iii. Any pair $(x(t), u(t))$ satisfying equation (18.1) is termed an **admissible pair**; the **optimal pair**, determined from the set of admissible pairs, will be represented as $(\hat{x}(t), \hat{u}(t))$.

Given this information, we can express the *basic structure of the continuous-time optimal control problem* as:

$$\begin{aligned}
&\text{(a) } \max_{u(t)} \int_0^T f(t, x(t), u(t))\, dt \text{ subject to} \\
&\text{(b) } \dot{x}(t) = g(t, x(t), u(t)), \\
&\text{(c) } x(0) = x_o = \text{constant}, \\
&\text{(d) } x(T) \text{ a terminal condition}, \\
&\text{(e) } u(t) \in \mathcal{U} \subseteq R \ (\mathcal{U} \text{ an open set}), \quad t \in [0, T],
\end{aligned} \tag{18.2}$$

where:

$$x, u: [0, T] \rightarrow R,$$
$$f, g: R^2 \times [0, T] \rightarrow R$$

and f, g are of class $C^{(n)}$.

Note that the objective function (18.2a) is actually a definite integral evaluated from time $t = 0$ to time $t = T$, with the integrand f specifying how both x and u at time t jointly determine the expression to be maximized; the differential equation (18.2b) illustrates how the control u influences or constrains the state x at time t; the initial condition on x (18.2c) is assumed to be known; the terminal condition on $x(T)$ (18.2d) *typically* appears as:

$$\begin{aligned}
&\text{(a) } x(T) \text{ free}, \\
&\text{(b) } x(T) = x_T, \\
&\text{(c) } x(T) \geq x_T,
\end{aligned} \tag{18.3}$$

with x_T a parameter; and $u(t) \in \mathcal{U}$, $t \in [0, T]$ ((18.2e) restricts the control u to a **region of admissible values** $\mathcal{U}$ over the planning period $[0, T]$).

To summarize: solving the optimal control problem (18.2) requires that we find the optimal control which maximizes the objective function (18.2a) subject to a differential equation (18.2b) connecting the control and state variables along with initial and terminal conditions for the latter (18.2c, d). So among all pairs $(x(t), u(t))$ that satisfy (18.2b)–(18.2e), we need to determine the one that maximizes the objective function. The resulting solution is a function $\hat{u}(t)$ describing the optimal time path of the control. From this function, we can determine the optimal time path $\hat{x}(t)$ for the state variable by solving (18.2b) given the initial and terminal conditions (18.2c, d).

18.2 THE LAGRANGIAN AND HAMILTONIAN FUNCTIONS

We now turn to the functional details of solving the optimal control problem (18.2). In fact, we already have the requisite machinery for solving a constrained optimization problem (see the technique of Lagrange provided in Chapter 10). Obviously, we are interested in maximizing the definite integral in (18.2a) subject to the constraint provided by (18.2b). Since time t is continuous and the optimization must be executed over the time horizon $[0, T]$, it follows that, effectively, we have an infinite number of equality constraints (and thus Lagrangians), one for each $t \in [0, T]$. In this regard, we must have, effectively, a separate Lagrange multiplier λ for each $t \in [0, T]$. To accomplish this we need only express λ as a function of t, $\lambda(t)$ (termed the **multiplier function**), and conjoin it to the constraint (18.2b). In the context of optimal control theory, for each $t \in [0, T]$,$\lambda(t)$is termed a **costate or adjoint variable**, and the term $\lambda(t)(\dot{x}(t) - g(t, x(t), u(t)))$ is used to form the **Lagrangian function**

$$\begin{aligned}
L(t, x(t), u(t), \lambda(t)) = \int_0^T [f(t, x(t), u(t)) - \\
\lambda(t)(\dot{x}(t) - g(t, x(t), u(t)))]\, dt,
\end{aligned} \tag{18.3}$$

where we have, for now, conveniently ignored the initial time requirement on x, namely $x(0) = x_o$. This expression serves as the foundation for the Lagrangian approach to solving the optimal control problem (18.2). In fact, as will be explained in the next section, the maximum principle provides us with a set of first-order (necessary) conditions for obtaining a solution to (18.2).

Since it will be difficult to maximize L in the presence of the term $\dot{x}(t)$, we can eliminate the same by first rewriting (18.3) as

$$\begin{aligned}
L(t, x(t), u(t), \lambda(t)) &= \int_0^T f(t, x(t), u(t))\, dt \\
&- \int_0^T \lambda(t)\, \dot{x}(t)\, dt + \int_0^T \lambda(t) g(t, x(t), u(t))\, dt \\
&= \int_0^T [f(t, x(t), u(t)) + \lambda(t) g(t, x(t), u(t))]\, dt \\
&- \int_0^T \lambda(t)\, \dot{x}(t)\, dt.
\end{aligned} \tag{18.4}$$

Performing integration by parts on the second definite integral on the right-hand side of (18.4)[1] enables us to rewrite the same as

$$L = \int_0^T [f(t, x, u) + \lambda g(t, x, u) + \dot{\lambda} x]\, dt - [\lambda(T)x(T) - \lambda(0)x(0)], \tag{18.5}$$

where, for convenience, the argument t in this expression has been suppressed.

Let us now focus on a key term within (18.5), namely, the **Hamiltonian**

$$H(t, x, u, \lambda) = f(t, x, u) + \lambda g(t, x, u). \quad (18.6)$$

Then (18.5) can be written as

$$L = \int_0^T (H + \dot{\lambda} x)\, dt - [\lambda(T)x(T) - \lambda(0)x(0)]. \quad (18.5.1)$$

18.3 THE MAXIMUM PRINCIPLE (PONTRYAGIN *ET AL.* [1962])

The expression (18.5.1) is maximized over all admissible values of $x(t)$ and $u(t)$ by maximizing $H(t, x, u, \lambda) + \dot{\lambda} x$, $t \in [0, T]$. The legitimacy of this Lagrangian approach is provided by

Theorem 18.1 (Maximum Principle). If $\hat{u}(t)$ and $\hat{x}(t)$ provide an unconstrained maximum for $H + \dot{\lambda} x$ for $\hat{u}(t) \in \mathcal{U}$ and all $t \in [0, T]$ and satisfy the laws of motion for x and λ, then $(\hat{x}(t), \hat{u}(t))$ represents an optimal pair for problem (18.2).

Note that the Maximum Principle in essence enables us to solve a problem involving the maximization of a constrained integral by finding an unconstrained maximum of $H + \dot{\lambda} x$. More specifically, if $\hat{u}(t) \in \mathcal{U}$ solves the optimization problem (18.2) for all $t \in [0, T]$, then there exists a $\lambda(t)$ such that $\hat{u}(t)$ and $\hat{x}(t)$ maximize $H + \dot{\lambda} x$.

Armed with the Maximum Principle, we now turn to the first-order **Hamiltonian (necessary) conditions** for maximizing $H + \dot{\lambda} x$:

$$\begin{aligned}
&\text{(a)} \quad \frac{\partial (H + \dot{\lambda} x)}{\partial u} = \frac{\partial H}{\partial u} = 0; \\
&\text{(b)} \quad \frac{\partial (H + \dot{\lambda} x)}{\partial x} = \frac{\partial H}{\partial x} + \dot{\lambda} = 0 \text{ or } \dot{\lambda} = -\frac{\partial H}{\partial x}; \\
&\text{(c)} \quad \frac{\partial (H + \dot{\lambda} x)}{\partial \lambda} = \frac{\partial H}{\partial \lambda} = g(t, x, u) = \dot{x}.
\end{aligned} \tag{18.7}$$

Clearly, (18.7c) (the **state equation**) requires that the constraint in (18.2) is satisfied. Additionally, (18.7b) (the **costate or adjoint equation**) provides us with the **law of motion for λ**. System (18.7), accompanied by the end-point restrictions $x(0) = x_o$ and $x(T)$ a terminal condition, constitute the **Hamiltonian conditions** for problem (18.2).[2]

What is the circumstance under which the Hamiltonian necessary conditions (18.7) are also sufficient for an optimal solution to (18.2)? The answer is provided by

Theorem 18.2: (Mangasarian [1966]). If $f(t, x(t), u(t))$ is *concave* in u and x for each $t \in [0, T]$ and if $g(t, x(t), u(t))$ is *linear* in u and x, then the Hamiltonian necessary conditions (18.7) are necessary and sufficient for an optimal solution to (18.2) to obtain. Stated alternatively, if the admissible pair $(\hat{x}(t), \hat{u}(t))$ satisfies (18.7) and the Hamiltonian $H(t, x, u, \lambda)$ is *concave* in x and u for every $t \in [0, T]$, then $(\hat{x}, \hat{u})$ is an optimal solution to the problem (18.2). Moreover, if $H(t, x, u, \lambda)$ is *strictly concave* in x and u for every $t \in [0, T]$, then $(\hat{x}, \hat{u})$ represents a unique optimal solution to (18.2).

A less restrictive or weaker version of this theorem is provided by Arrow and Kurz [1970] in the event that the Hamiltonian H is not concave. Specifically, let us denote the **maximized (with respect to *u*) Hamiltonian** as $\hat{H}(t, x, \lambda) = \max_{u \in \mathcal{U}} H(t, x, u, \lambda)$. If $(\hat{x}, \hat{u})$ is an admissible pair which satisfies (18.7) and if $\hat{H}$ is concave in x for all $t \in [0, T]$, then $(\hat{x}, \hat{u})$is an optimal solution to (18.2). (Note that if $\hat{H}$ is not concave in x, then Theorem 18.2 will not hold.)

18.4 END POINT AND TRANSVERSALITY CONDITIONS

A glance back at equations (18.7b, c) reveals that we have two first-order indefinite differential equations whose solutions will include two arbitrary constants of integration. Hence, we will need two boundary conditions to obtain particular solutions to these differential equations. In this regard, if both $x(0)$ and $x(T)$ are provided, then they can be utilized to particularize the two arbitrary constants. However, if the terminal state or boundary condition is not also fixed, then we must rely upon a so-called **transversality condition** that takes the place of the terminal condition. To see how all this works, let us first consider the typical terminal conditions provided by (18.3). From (18.3a), if $x(T)$ is free, then the relevant **transversality conditions**

are $\boldsymbol{x(T) = 0}$ ***and*** $\boldsymbol{\lambda(T) = 0}$. The following example houses the details of this case.

Example 18.1: Let us

$$\underset{u(t)}{\text{maximize}} \int_0^2 \left(3x - \tfrac{1}{2}u^2\right) dt \text{ subject to}$$
$$\dot{x} = u + 1,\ x(0) = 1,\ x(T) = x(2) \text{ free.}$$

For this problem

$$H + \dot{\lambda}x = 3x - \frac{1}{2}u^2 + \lambda(u + 1) + \dot{\lambda}x,$$

where H is the Hamiltonian $H = f + \lambda g$. Looking to the Hamiltonian (necessary) conditions (18.7) we have

$$\text{(a)}\quad \frac{\partial H}{\partial u} = -u + \lambda = 0$$
$$\text{(b)}\quad \frac{\partial H}{\partial x} + \dot{\lambda} = 3 + \dot{\lambda} = 0$$
$$\text{(c)}\quad \frac{\partial H}{\partial \lambda} = u + 1 = \dot{x}.$$

(Note that the Mangasarian sufficient condition for a maximum holds.)

From (a) we obtain $u = \lambda$ while from (b), (c) we have the system of first-order differential equations

$$\text{(d)}\quad \dot{\lambda} = -3$$
$$\text{(e)}\quad \dot{x} = u + 1 = \lambda + 1.$$

The solution to (d) is

$$d\lambda = -3dt,$$
$$\lambda(t) = -3t + C_1,$$

where C_1 is an arbitrary constant. To determine C_1, let us employ the transversality condition $\lambda(T) = \lambda(2) = 0$. Then $\lambda(2) = 0 = -6 + C_1$ or $C_1 = 6$. Hence the optimal time path for $\lambda(t)$ is

$$\text{(f)}\quad \hat{\lambda}(t) = -3t + 6.$$

Next, the solution to (e) is obtained as follows. From (f)

$$\dot{x} = \hat{\lambda} + 1 = -3t + 6 + 1 = -3t + 7.$$

Then

$$dx = -3tdt + 7dt,$$
$$x(t) = -\tfrac{3}{2}t^2 + 7t + C_2,$$

where C_2 is an arbitrary constant. To find C_2, we use the transversality condition $x(T) = x(2) = 0$. Then $x(2) = 0 = -6 + 14 + C_2$ or $C_2 = -8$. Hence the optimal time path for x is

$$\text{(g)}\quad \hat{x}(t) = -\frac{3}{2}t^2 + 7t - 8.$$

And to obtain the optimal time path for u we have

$$\text{(h)}\quad \hat{u}(t) = \hat{\lambda}(t) = -3t + 6.$$

At $t = 0$, $\hat{u}(0) = 6$. Hence u begins at $\hat{u} = 6$ and decreases steadily to $\hat{u}(2) = 0$. ■

Next, suppose $x(0) = x_o$ is given but (18.3b) or $x(T) = x_T$ now holds. In this instance the transversality condition $\lambda(T) = 0$ is not required. Simply set $x(T) = x_T$ within (18.2). To see this we look to

Example 18.2 Let us

$$\underset{u(t)}{\text{maximize}} \int_0^2 \left(3x - \tfrac{1}{2}u^2\right) dt \text{ subject to}$$
$$\dot{x} = u + 1,\ x(0) = 1,\ x(T) = x(2) = 3.$$

(Note that this optimal control problem is the same as the one offered in the preceding example save for the new terminal condition on x.) From the same set of Hamiltonian conditions we again have $u = \lambda$, $\dot{\lambda} = -3$, and $\dot{x} = \lambda + 1$. We found in Example 18.1 that the solution to $\dot{\lambda} = -3$ was

$$\lambda(t) = -3t + C_1.$$

Substituting this expression into $\dot{x} = \lambda + 1$ yields $\dot{x} = -3t + C_1 + 1$ so that

$$dx = -3t\ dt + (C_1 + 1)\ dt.$$

Then

$$x(t) = -\frac{3}{2}t^2 + (C_1 + 1)t + C_2.$$

To find C_2 we simply set $x(0) = 1 = C_2$.
Moreover, setting $x(2) = 3 = -6 + 2C_1 + 3$ gives

$C_1 = 3$. Hence, the optimal time paths for $\lambda(t)$ and $x(t)$ are, respectively,

$$\hat{\lambda}(t) = -3t + 3,$$
$$\hat{x}(t) = -\tfrac{3}{2}t^2 + 4t + 1.$$

Additionally, the optimal time path for $u(t)$ is

$$\hat{u}(t) = -3t + 3.$$

■

The final terminal point condition to consider is when (18.3c) or $x(T) \geq x_T$ holds. To solve an optimal control problem under this restriction:

1. first set $\lambda(T) = 0$ as the transversality condition ($x(T)$ is now treated as free) and determine if $\hat{x}(T) \geq x_T$. If so, the terminal inequality $x(T) \geq x_T$ is satisfied and the problem is solved. If $\hat{x}(T) < x_T$, then
2. treat the problem as a fixed-terminal point problem with $x(T) = x_T$. In this instance the transversality condition $\lambda(T) = 0$ is not needed.

These two cases may be combined and written as the **CS transversality condition**

$$(\hat{x}(T) - x_T)\lambda(T) = 0, \ \lambda(T) \geq 0, \qquad (18.8)$$

i.e., if $\lambda(T) = 0$, then $\hat{x}(T) = x_T$; and if $\hat{x}(T) - x_T = 0$, then $\lambda(T) \geq 0$.

The usual endpoint restrictions (18.3) and the set of requisite transversality conditions are summarized in Table 18.1.

Example 18.3 Suppose we want to

$$\underset{u(t)}{\text{maximize}} \int_0^2 \left(3x - \tfrac{1}{2}u^2\right) dt \ \text{ subject to}$$
$$\dot{x} = u + 1, \ x(0) = 1, \ x(T) = x(2) \geq 4.$$

TABLE 18.1 Endpoint and Corresponding Transversality Conditions

Endpoint Restrictions (x_T a Parameter)	**Transversality Conditions**
$x(T)$ free	$\lambda(T) = 0$
$x(T) = x_T$	$\lambda(T)$ unrestricted
$x(T) \geq x_T$	$(\hat{x}(T) - x_T)\lambda(T) = 0, \ \lambda(T) \geq 0$

To solve this problem, first set $\lambda(T) = 0$. Then we can obtain the Example 18.1 optimal solution for $x(t)$, namely

$$\hat{x}(t) = -\frac{3}{2}t^2 + 7t - 8.$$

Let us check to see if $\hat{x}(T) = x_T$ or $\hat{x}(2) \geq 4$. From the preceding expression, we get $\hat{x}(2) = 0$. Clearly, the inequality $x(T) \geq x_T$ is violated.

Next, set $x(T) = x_T = 4$. We know from Example 18.2 that

$$x(t) = -\frac{3}{2}t^2 + (C_1 + 1)t + C_2.$$

From $x(0) = 1$ we get $C_2 = 1$. And from $x(2) = 4 = -3 + 2C_1$ we get $C_1 = 7/2$. Thus

$$\hat{x}(t) = -\frac{3}{2}t^2 + \frac{9}{2}t + 1.$$

The optimal time paths for λ and u are the same as in Example 18.2. ■

Section 18.4 Exercises:

18.4.1 Determine the optimal control path for the problem

$$\underset{u(t)}{\text{maximize}} \int_0^2 (x - u^2) \, dt \ \text{ subject to}$$
$$\dot{x} = u, \ x(0) = 3, \ x(2) \ \text{free}.$$

18.4.2 Find the optimal control path for

$$\underset{u(t)}{\text{maximize}} \int_0^2 (x - u^2) \, dt \ \text{ subject to}$$
$$\dot{x} = u, \ x(0) = 3, \ x(2) = 5.$$

18.4.3 Determine the optimal control path for

$$\underset{u(t)}{\text{maximize}} \int_0^2 (x - u^2) \, dt \ \text{ subject to}$$
$$\dot{x} = u, \ x(0) = 3, \ x(2) \geq 5.$$

18.4.4 Find the optimal control path which solves

$$\underset{u(t)}{\text{maximize}} \int_0^3 (-2x - 3u^2) \, dt \ \text{ subject to}$$
$$\dot{x} = x + u, \ x(0) = 1, \ x(3) \ \text{free}.$$

18.4.5 Determine the optimal solution to the problem

$$\underset{u(t)}{\text{maximize}} \int_0^2 (x - u^2)\, dt \text{ subject to}$$
$$\dot{x} = u + 1,\ x(0) = 3,\ x(2) = \theta = \text{constant}.$$

18.5 SENSITIVITY ANALYSIS: COSTATE VARIABLES AS SHADOW PRICES

Let us harken back to the Lagrangian expression (18.5.1) or

$$L = \int_o^T (H + \dot{\lambda}x)\, dt - [\lambda(T)x(T) - \lambda(0)x(0)]. \quad (18.5.1)$$

And let us also consider the terminal point condition (18.3c) or $x(T) \geq x_T$. If we rewrite this inequality as $-x_T + x(T) \geq 0$, then we may introduce this constraint into (18.5.1) via an additional Lagrange multiplier μ or (18.5.1) becomes

$$L = \int_o^T (H + \dot{\lambda}x)\, dt - [\lambda(T)x(T) - \lambda(0)x(0)] + \mu(-x_T + x(T)). \quad (18.5.2)$$

Remember that we maximized the Lagrangian given by (18.5.1) over x, u values by maximizing the expression $H + \dot{\lambda}x$ for each $t \in [0, T]$. This enabled us to obtain the Hamiltonian conditions (18.7). But given the new Lagrangian provided by (18.5.2), maximizing this expression over x, u now necessitates the calculation of

$$\frac{\partial L}{\partial x(T)} = -\lambda(T) + \mu = 0$$

or $\mu = \lambda(T)$. If we substitute $\mu = \lambda(T)$ into (18.5.2), the latter simplifies to

$$L = \int_o^T (H + \dot{\lambda}x)\, dt - \lambda(T)x_T + \lambda(0)x(0). \quad (18.5.3)$$

How should the costate or adjoint variable $\lambda(T)$ be interpreted? For problem (18.2), suppose an optimal solution for the same consists of $\hat{x}(t)$, $\hat{u}(t)$, and the adjoint function $\lambda(T)$. Since the optimal time paths and $\hat{x}(t)$ and $\hat{u}(t)$ satisfy the constraints appearing in (18.2), we may express the **optimal value function** as

$$V(x_o, x_T, T) = \int_o^T f(t, \hat{x}(t), \hat{u}(t))\, dt, \quad (18.9)$$

where x_o, x_T, and T are parameters. Moreover, this optimal value function and the Lagrangian (18.5.3) evaluated at $\hat{x}(t)$ and $\hat{u}(t)$ are actually equal or[3]

$$V(x_o, x_T, T) = \int_o^T [f(t, \hat{x}, \hat{u}) + \lambda g(t, \hat{x}, \hat{u}) + \dot{\lambda}x]\, dt - \lambda(T)x_T + \hat{\lambda}(0)\hat{x}(0).$$

As we shall now see, the derivatives of the optimal value function (18.9) with respect to x_o and x_T yield some useful **dynamic envelope results**.

It should be intuitively clear that $\hat{x}(t)$ and $\hat{u}(t)$ will change for $t \in [0, T]$. For V of class $C^{(1)}$, using (18.5.3) enables us to determine that

$$\frac{\partial V(x_o, x_T, T)}{\partial x_o} = \lambda(0), \quad (18.11)$$

i.e., $\lambda(0)$ represents the sensitivity of the optimal value function to a change in the initial state $x(0)$ - - it is the marginal change in V precipitated by a small increase in x_o.

Next, again looking to (18.5.3), for V of class $C^{(1)}$,

$$\frac{\partial V(x_o, x_T, T)}{\partial x_T} = -\lambda(T), \quad (18.12)$$

i.e., $\lambda(T)$ depicts the sensitivity of the optimal value function to a change in the terminal point x_T - - it is the marginal change in V associated with a small reduction in x_T. When the value of x_T is decreased, the constraint $x(T) \geq x_T$ is relaxed so that the optimal value of V cannot be reduced. Looking to the CS transversality condition in Table 18.1, if $x_T \geq x_T$ is not binding, then obviously $\partial V/\partial x_T = -\lambda(T) = 0$; and if this constraint is binding, then $\partial V/\partial x_T = -\lambda(T) < 0$.

Note that if $x(T)$ is free, then no improvement in V is possible since, in this instance, $\lambda(T) = 0$ (see Table 18.1). In fact, for this case, x_T will not even be an argument of V.

We know from our discussion of optimization subject to an equality constraint (Chapter 10) that the sign of the optimal Lagrange multiplier is unrestricted. So if the terminal point condition $x(T) = x_T$ holds, then the sign of $\partial V/\partial x_T$ cannot be specified in advance.

When we consider constrained optimization in earlier chapters, we found that Lagrange multipliers had a shadow price interpretation. Consistent with this notion we can view, say, (18.12) in a similar light. That is, we would be willing to pay, in terms of V, at most $\lambda(T)$ to have the constraint $x(T) \geq x_T$ relaxed by one unit.

Example 18.4: For the problem presented in Example 18.1 we shall demonstrate that, for the optimal value

$$V(x_o, T) = \int_0^2 \left(3\hat{x} - \frac{1}{2}\hat{u}^2\right) dt,$$

Equation (18.11) holds.[4] We found that $\hat{x} = -\frac{3}{2}t^2 + 7t - 8$ with $\hat{u} = \hat{\lambda} = -3t + 6$. Rewriting $\hat{x}$ as $\hat{x} = -\frac{3}{2}t^2 + 7t + x_o$ and substituting into the expression for V yields

$$V = \int_0^2 \left[\left(-\frac{9}{2}t^2 + 21t + 3x_o\right) - \frac{9}{2}t^2 + 18t - 18\right] dt$$
$$= \int_0^2 (-9t^2 + 39t + 3x_o - 18)\, dt.$$

Then via (**),

$$\frac{\partial V}{\partial x_o} = \int_0^2 3\, dt = 3t]_0^2 = 6.$$

But since $\hat{\lambda}(0) = 6$, we see that (18.11) holds. ■

Section 18.5 Exercises:

18.5.1 For problem 18.4.1, determine the optimal value function V (see Equation (18.9)). Also: (1) verify that $V = L$, where L is the Lagrangian expression (18.10); and (2) demonstrate that $\partial V / \partial x_o = \hat{\lambda}(0)$.

18.5.2 Given the problem

$$\underset{u(t)}{\text{maximize}} \int_0^1 (x - u^2)\, dt \text{ subject to}$$
$$\dot{x} = u,\ x(0) = 1,\ x(1) \text{ free}$$

verify that $\partial V / \partial x_o = \hat{\lambda}(0)$.

18.6 AUTONOMOUS OPTIMAL CONTROL: THE CURRENT VALUE HAMILTONIAN

An optimal control problem in which time t is not an explicit argument within f and g is termed an **autonomous problem**, i.e., the *date doesn't matter*. A glance back to Section 18.2 reveals that the Hamiltonian associated with the Lagrangian (18.5) is (18.6) or

$$H(t, x, u, \lambda) = f(t, x, u) + \lambda g(t, x, u).$$

Clearly t can explicitly appear in this expression so that this Hamiltonian might not be associated with an autonomous problem. In what specific context might a non-autonomous problem arise?

A common type of optimal control problem encountered in economic modeling is when time enters into the objective function only through a **discount factor** of the form e^{-rt}, where $r > 0$ is the **discount rate**. The problem being alluded to is of the form

$$\begin{aligned} &\underset{u(t)}{\text{maximize}} \int_0^T F(x, u)\, e^{-rt} dt \text{ subject to} \\ &\dot{x} = G(x, u),\ x(0) = x_o \\ &x(T) \text{ a terminal condition} \\ &u \in \mathcal{U} \in R\ (\mathcal{U} \text{ an open set}),\ t \in [0, T]. \end{aligned} \tag{18.13}$$

Here the Hamiltonian appears as

$$H(t, x, u, \lambda) = F(x, u)\, e^{-rt} + \lambda G(x, u). \tag{18.14}$$

If we now multiply both sides of (18.14) by e^{rt}, then we obtain the current value Hamiltonian H_{cv} or

$$\begin{aligned} e^{rt}\, H(t, x, u, \lambda) &= F(x, u) + \lambda e^{rt} G(x, u) \\ &= F(x, u) + vG(x, u) \\ &= H_{cv}(x, u, v), \end{aligned} \tag{18.15}$$

where the new Lagrange multiplier $v = \lambda e^{rt}$ is termed the **current value Lagrange multiplier** or **current value costate variable**. Thus the current value Hamiltonian is simply $H_{cv} = e^{rt}H$. (Note that since $H = e^{-rt}H_{cv}$, we may view H as the **present value Hamiltonian**.) Obviously, we have transformed (18.13) into an autonomous problem by forming H_{cv}.

Under the (18.15) transformation from H to H_{cv}, we are thus led to offer a revised Maximum Principle or

Theorem 18.3: **(Current Value Maximum Principle).** If $\hat{u}(t)$ and $\hat{x}(t)$ provide an unconstrained maximum for $H_{cv} + \dot{\lambda}x$ for $u(t) \in \mathcal{U}$ and all $t \in [0, T]$ and satisfy the laws of motion for x and λ, then $(\hat{u}(t), \hat{x}(t))$ represents an optimal solution for problem (18.13).

So if $\hat{u}(t) \in \mathcal{U}$ solves the optimal control problem (18.13) for all $t \in [0, T]$, then there exists a continuous function $v(t) = \lambda e^{rt}$ such that $\lambda(t) \neq 0$ and $\hat{u}(t)$ and $\hat{x}(t)$ maximize $H_{cv} + \dot{\lambda}x$.

Given Theorem 18.3, we can now turn to the first-order current value Hamiltonian (necessary) condition for maximizing $H_{cv} + \dot{\lambda}x$. Specifically, given that $H = H_{cv}e^{-rt}$, $\lambda = ve^{-rt}$, and $\dot{\lambda} = \dot{v}e^{-rt} - rve^{-rt}$, it follows that

$$H + \dot{\lambda} x = H_{cv} e^{-rt} + x(\dot{v} e^{-rt} - rv e^{-rt})$$
$$= e^{-rt}[H_{cv} + x(\dot{v} - rv)]. \quad (18.16)$$

Hence, mirroring (18.7) by differentiating (18.16) with respect to u, x, and λ, respectively:

(a) $e^{-rt} \frac{\partial H_{cv}}{\partial u} = 0$ or $\frac{\partial H_{cv}}{\partial u} = 0$;

(b) $e^{-rt}\left[\frac{\partial H_{cv}}{\partial x} + (\dot{v} - rv)\right] = 0$ or $\dot{v} - rv = -\frac{\partial H_{cv}}{\partial x}$;   (18.17)

(c) $e^{-rt} \frac{\partial H_{cv}}{\partial \lambda} = e^{-rt} e^{rt} G(x, u) = G(x, u) = \dot{x}$,

where the transversality conditions (listed to mirror (18.3)) are:

(a)   if $x(T)$ free, then $\lambda(T) = e^{-rt} v(T) = 0$;

(b)   if $x(T) = x_T$, then $\lambda(T) = e^{-rt} v(T)$ unrestricted;   (18.18)

(c)   if $x(T) \geq x_T$, then $\lambda(T) = e^{-rt} v(T) \geq 0$ since $(\hat{x}(T) - x_T)\lambda(T) = 0$ under CS requirements.

System (18.17), along with the restrictions $x(0) = x_o$ and (18.18), constitute **current value Hamiltonian conditions** for problem (18.13).

In Section 18.3 we considered a set of requirements under which the Hamiltonian necessary conditions (18.7) were also sufficient for an optimal solution to the problem (18.2). These conditions were offered by Mangasarian [1966] and Arrow and Kurz [1970]. Let us now extend these sufficient conditions to the system (18.17). Specifically, system (18.17) is sufficient for an admissible solution $(\hat{x}, \hat{u})$ to be optimal for problem (18.13) if:

a. $H_{cv}(x, u, v)$ is *concave* in admissible (x, u) (the **Mangasarian sufficiency criterion**); or

b. $\hat{H}_{cv}(x, v) = \max_{u \in U} H_{cv}(x, u, v)$ is *concave* in x (the **Arrow-Kurz sufficiency criterion**).

We now turn to some sensitivity analysis issues regarding current and present value shadow prices. For optimal control problem (18.13), what will the optimal value function look like? Remember that for this problem the objective function is $\int_o^T F(x, u) e^{-rt} dt$. Suppose the optimal time paths for the variables in (18.13) have been determined using the current value Hamiltonian and appear as $\hat{x}(t)$ and $\hat{u}(t)$. Then the **current value optimal value function** is

$$V_{cv} = e^{rt} V_{pv} = e^{rt} \int_o^T F(\hat{x}, \hat{u}) e^{-rt} dt, \quad (18.19)$$

where V_{pv} is the **present value optimal value function** which has its value discounted back to time $t = 0$.

For instance, we know from our previous sensitivity analysis that $\partial V_{pv} / \partial x_o = \lambda(0)$ - - the **present value shadow price** which gives the marginal value of a unit of x discounted back to time $t = 0$. The **current value shadow price** is $\partial V_{cv} / \partial x_o = v(0)$ (remember that $v(t) = e^{rt} \lambda(t)$) – it gives the marginal value of a unit of x in terms of its value at time $t = 0$.

Example 18.5 To solve the problem

$$\text{maximize} \int_0^3 e^{-rt}(-4x - u^2) \, dt \text{ subject to}$$
$$\dot{x} = 2u, \, x(0) = x_o = 1, \, x(T) = x(3) \text{ free,}$$

we start with the current value Hamiltonian $H_{cv} = e^{rt} H = -4x - u^2 + v(2u)$. Then from (18.17):

(a) $\frac{\partial H_{cv}}{\partial u} = -2u + 2v = 0$ or $u = v$;

(b) $\dot{v} - rv = -\frac{\partial H_{cv}}{\partial x}$ or $\dot{v} = rv + 4$;   and   (18.20)

(c) $\dot{x} = 2u = 2v$ via (18.20a).

(One can easily verify that the Mangasarian sufficient condition for an optimum holds.) From (20.20b) $dv/dt = rv + 4$. Solving this first-order differential equation gives

$$v = C_1 e^{rt} - \frac{4}{r}.$$

Since $x(T)$ is free, we use the transversality condition $\lambda(T) = 0$ or $v(T) = v(3) = 0$. Then

$$v(3) = C_1 e^{r3} - \frac{4}{r} = 0 \text{ or } C_1 = \frac{4}{r} e^{-r3}$$

and thus

$$\hat{v}(t) = \frac{4}{r}(e^{r(t-3)} - 1) = \hat{u}(t). \quad (18.21)$$

From (18.20c),

$$\frac{dx}{dt} = \frac{8}{r}(e^{r(t-3)} - 1).$$

The solution to this first-order differential equation is

$$x = \left(\frac{8}{r}\right)t + \left(\frac{8}{r^2}\right)e^{r(t-3)} + C_2.$$

To obtain C_2, we employ the transversality condition $x(T) = x(3) = 0$ so that

$$x(3) = \frac{24}{r} + \frac{8}{r^2} + C_2 = 0 \text{ or } C_2 = -\frac{24}{r} - \frac{8}{r^2}$$

and thus

$$\hat{x}(t) = \frac{8}{r}[(t-3) - \frac{1}{r}(1 - e^{r(t-3)})]. \quad (18.22)$$

Keep in mind that (18.21) and (18.22) provide us with current value time paths for the control, state, and costate variables. ■

Example 18.6 Determine the solution to the problem

$$\underset{u(t)}{\text{maximize}} \int_0^T e^{-rt}(-x^2 - u^2)\, dt \text{ subject to}$$
$$\dot{x} = u,\ x(0) = x_o,\ x(T) \text{ free}, \quad v(T) = 0.$$

From the current-value Hamiltonian

$$H_{cv} = x^2 - u^2 + v(u)$$

and (18.19) we have

$$\begin{aligned} &\text{(a)} \quad \frac{\partial H_{cv}}{\partial u} = -2u + v = 0 \text{ or } u = \tfrac{1}{2}v; \\ &\text{(b)} \quad \dot{v} - rv = -\frac{\partial H_{cv}}{\partial x} \text{ or } \dot{v} = rv + 2x; \quad \text{and } (18.23) \\ &\text{(c)} \quad \dot{x} = u = \tfrac{1}{2}v. \end{aligned}$$

Hence our system of homogeneous first-order linear differential equations is

$$\begin{aligned} \frac{dv}{dt} &= rv + 2x \\ \frac{dx}{dt} &= \frac{1}{2}v. \end{aligned} \quad (18.24)$$

For this system

$$A = \begin{bmatrix} r & 2 \\ 1/2 & 0 \end{bmatrix}, |A| = -1, \text{ and } tr(A) = r.$$

From $|A - \lambda I_2| = 0$ we obtain $\lambda^2 - r\lambda - 1 = 0$ or

$$\lambda = \frac{1}{2}[r \pm (r^2 + 4)^{1/2}]$$

with

$$\lambda_1 = \tfrac{1}{2}[r + (r^2 + 4)^{1/2}],$$
$$\lambda_2 = \tfrac{1}{2}[r - (r^2 + 4)^{1/2}].$$

Then the solution to (18.24) is, via Theorem 20. B.1,

$$\begin{bmatrix} v(t) \\ x(t) \end{bmatrix} = C_1 h_1 e^{\lambda_1 t} + C_2 h_2 e^{\lambda_2 t}, \quad (18.25)$$

where h_1 and h_2 are the eigenvectors corresponding to the eigenvalues λ_1 and λ_2, respectively. ■

Section 18.6 Exercises:

18.6.1 Determine the solution to

$$\underset{u(t)}{\text{maximize}} \int_0^2 (x - u^2)\, e^{-rt}\, dt \text{ subject to}$$
$$\dot{x} = u,\ x(0) = 2,\ x(2) \text{ free}.$$

18.6.2 Find the solution to

$$\underset{u(t)}{\text{maximize}} \int_0^3 (-x^2 - 2u^2)\, dt \text{ subject to}$$
$$\dot{x} = 3u + 1,\ x(0) = 1,\ x(3) \text{ free}.$$

18.7 INFINITE TIME HORIZON

A common type of optimal control problem frequently encountered in economic modeling is one in which the planning period is of infinite duration. That is, the problem to be solved is structured as

$$\underset{u(t)}{\text{maximize}} \int_0^{+\infty} f(x, u)\, e^{-rt} dt \text{ subject to}$$
$$\begin{aligned} &\text{(a)} \quad \dot{x} = g(x, u); \\ &\text{(b)} \quad x(0) = x_o = \text{constant}, \quad (18.26) \\ &\text{(c)} \quad u(t) \in \mathcal{U} \subseteq R(\mathcal{U} \text{ an open set}), \quad t \in [0, +\infty); \\ &\text{(d)} \quad \lim_{t \to +\infty} x(t) \geq x_*(\text{a fixed value of } x). \end{aligned}$$

Note that this problem is **autonomous** (time enters only through the discount factor e^{-rt}) and (18.26d) replaces "$x(T)$= a terminal condition" appearing in the system (18.2). It requires that the indicated limit exists and must equal or exceed x_*; this requirement precludes $x(t)$ from increasing without bound or exhibiting cyclical behavior. If (18.26d) replaces "$x(T)$ free", then this inequality can possibly be modified to read

$$\lim_{t \to +\infty} x(t) = \bar{x},$$

where $\bar{x}$ is the steady-state value of x. However, in many models, no restriction is placed on $x(t)$ as $t \to +\infty$. We can simply write $x(\infty)$ free.

Clearly the objective function in (18.26) is an **improper integral** since the upper limit of integration is $+\infty$. This said, under what conditions will this integral converge? Suppose $(x(t), u(t))$ is an **admissible solution pair** in that it satisfies (18.26a),..., (18.26d). Then the said integral will converge for all admissible solution pairs if:

i. e^{-rt} with $r > 0$ is continuous and $\lim_{t \to +\infty} e^{-rt} = 0$; and
ii. $f(x, u)$ is continuous and $\int_0^{+\infty} f(x, u)\, dt$ is bounded or $|f(x, u)| \leq M$ for a real scalar M.

Section 18.7 Exercises:

18.7.1 Solve the problem

$$\underset{u(t)}{\text{maximize}} \int_0^{+\infty} (x - u^2)\, e^{-rt}\, dt \text{ subject to}$$
$$\dot{x} = u + 1,\ x(0) = 1,\ x(+\infty) \text{ free.}$$

18.7.2 Determine the solution to

$$\underset{u(t)}{\text{maximize}} \int_0^{+\infty} (x - u^2)\, e^{-t}\, dt \text{ subject to}$$
$$\dot{x} = ue^{-t},\quad x(0) = 1,\quad x(+\infty) \text{ free.}$$

18.8 SUFFICIENT OPTIMALITY CONDITIONS FOR THE INFINITE HORIZON CASE (CAPUTO [2005]; MANGASARIAN [1966])

All of the current-value Hamiltonian (necessary) conditions for the finite horizon problem (18.13) will also hold for the infinite horizon problem (18.26) save for the transversality conditions. Before we consider this exception in greater detail, let us examine

Theorem 18.4 (Sufficient Condition, Infinite Horizon (Caputo [2005]; Mangasarian [1966]). Let $(\hat{x}(t), \hat{u}(t))$ be an admissible pair for problem (18.26) which satisfies:

a. $\frac{\partial H_{cv}}{\partial u} = 0$;

b. $\dot{v} - rv = -\frac{\partial H_{cv}}{\partial x}$;

c. $\dot{x} = g(x, u)$;

d. H_{cv} is concave in $(x(t), u(t))$, $t \in [0, +\infty)$, over an open set containing admissible points $(x(t), u(t))$; and

e. the **transversality condition** $\lim_{t \to +\infty} v(t)e^{-rt}[x(t) - \hat{x}(t)] \geq 0$ holds for all admissible $(x(t), u(t))$.

Then $(\hat{x}(t), \hat{u}(t))$ represents the global maximum for (18.26). (Note that if H_{cv} is strictly concave in $(x(t), u(t))$, then$(\hat{x}(t), \hat{u}(t))$ is the unique global maximum.)

Example 18.7 Let us determine the solution to the infinite horizon problem

$$\underset{u(t)}{\text{maximize}} \int_0^{T} e^{-rt}(-x - u^2)\, dt \text{ subject to}$$
$$\dot{x} = u,\ x(0) = x_o,\ x(\infty) \text{ free.}$$

From the current-value Hamiltonian $H_{cv} = -x - u^2 + vu$ we can obtain

$$\frac{\partial H_{cv}}{\partial u} = -2u + v = 0,\ u = \frac{1}{2}v.$$

Additionally,

$$\dot{v} - rv = -\frac{\partial H_{cv}}{\partial x} \text{ or } \dot{v} = rv + 1.$$

Hence our system of linear first-order autonomous differential equations is

$$\begin{aligned} &\text{(a)} \quad \dot{v} = rv + 1 \\ &\text{(b)} \quad \dot{x} = \tfrac{1}{2}v. \end{aligned} \tag{18.27}$$

Since (18.27a) does not involve x, we can obtain its solution directly: the homogeneous solution is

$$v = C_1\, e^{rt};$$

the steady state solution is, from $\dot{v} = 0$, $\bar{v} = -\frac{1}{r}$; and the complete solution is

$$v = C_1 \; e^{rt} - \frac{1}{r}.$$

Suppose $v(0) = 0$. Then $C_1 = 1/r$ and thus the complete solution satisfying the initial condition is

$$\hat{v}(t) = \frac{1}{r}(e^{rt} - 1). \tag{18.28}$$

From (18.27b) and (18.28) we have

$$\dot{x} = \frac{1}{2r}(e^{rt} - 1)$$

with solution

$$x = \frac{1}{2r^2}e^{rt} - \frac{t}{2r} + C_2.$$

From the initial condition $x(0) = 0$ we get $C_2 = -\frac{1}{2r^2}$ and thus

$$\hat{x}(t) = \frac{1}{2r^2}(e^{rt} - 1) - \frac{t}{2r}.$$

It is easily demonstrated that this solution diverges since $\lim_{t \to +\infty} \hat{x}(t) = +\infty$.

Example 18.8 Let us verify that the problem

$$\underset{u(t)}{\text{maximize}} \int_0^{+\infty} e^{-rt}(x - x^2 - u^2)\, dt \text{ subject to}$$
$$\dot{x} = u - x, \quad x(0) = x_o, \quad x(\infty) \text{ free}$$

displays a saddle-point equilibrium.

From

$$H_{cv} = x - x^2 - u^2 + v(u - x)$$

we have

$$\frac{\partial H_{cv}}{\partial u} = -2u + v = 0 \text{ or } u = \frac{1}{2}v.$$

With

$$\dot{v} - rv = -\frac{\partial H_{cv}}{\partial x} = -(1 - 2x - v)$$

we obtain

$$\dot{v} = (r + 1)v + 2x - 1$$

and thus the autonomous linear first-order system of differential equations is

$$\begin{aligned} &\text{(a)} \quad \dot{v} = (r + 1)v + 2x - 1 \\ &\text{(b)} \quad \dot{x} = \tfrac{1}{2}v - x. \end{aligned} \tag{18.29}$$

For this system

$$\boldsymbol{A} = \begin{bmatrix} r + 1 & 2 \\ \frac{1}{2} & -1 \end{bmatrix} \text{ with } |\boldsymbol{A}| = -(r + 2) < 0.$$

Hence the characteristic roots are of opposite sign and thus the equilibrium or steady-state solution $(\bar{v}, \bar{x})$ is a saddle point, where

$$\begin{bmatrix} \bar{v} \\ \bar{x} \end{bmatrix} = -\boldsymbol{A}^{-1}\boldsymbol{b} = -\begin{bmatrix} \frac{1}{r+2} & \frac{2}{r+2} \\ \frac{1}{2(r+2)} & -\frac{r+1}{r+2} \end{bmatrix} \begin{bmatrix} -1 \\ 0 \end{bmatrix} = \begin{bmatrix} \frac{1}{r+2} \\ \frac{1}{2(r+2)} \end{bmatrix}.$$

As these last two example problems have revealed, with autonomous infinite-horizon, free-endpoint problems, the first-order system

$$\begin{aligned} \dot{v} &= rv - \frac{\partial H_{cv}}{\partial x} \\ \dot{x} &= g(u, x) \end{aligned}$$

either renders a saddle-point, steady-state solution or leads to a divergent (unstable) solution. In the former case $x(t)$ converges to its limiting steady-state value of $\bar{x}$ (or possibly is constant).

18.9 CONSTRAINTS ON THE CONTROL VARIABLE

Probably the most common types of constraints on the control variable u are: (a) a non-negativity constraint $u \geq 0$; (b) a generalized upper bound constraint of the form $h(t, x, u) \geq u$;or (c) $u \in [a, b]$, where a and b are real scalars. If we consider case (b), then our optimal control problem appears as

$$\begin{aligned}&\underset{u(t)}{\text{maximize}} \int_0^T f(t, x(t), u(t))\, dt \text{ subject to}\\ &\dot{x} = g(t, x(t), u(t)),\ x(0) = x_o,\ x(T) \text{ free}\\ &\hat{h}(t, x(t), u(t)) = h(t, x(t), u(t)) - u(t) \geq 0.\end{aligned} \tag{18.30}$$

As a first step in solving this problem we need to form the Lagrangian

$$L = H + \dot{\lambda}x + \theta\hat{h}(t, x, u), \tag{18.31}$$

where the Hamiltonian $H = f + \lambda g$ and θ is an undetermined Lagrange multiplier. Then the set of *expanded* first-order Hamiltonian (necessary) conditions is:

$$\begin{aligned}&\text{(a)}\quad \frac{\partial H}{\partial u} + \theta\frac{\partial \hat{h}}{\partial u} = 0\\ &\text{(b)}\quad \dot{\lambda} = -\frac{\partial H}{\partial x} - \theta\frac{\partial \hat{h}}{\partial x}\\ &\text{(c)}\quad \frac{\partial H}{\partial \lambda} = \dot{x} = g(t, x, u)\\ &\text{(d)}\quad \theta \geq 0\\ &\text{(e)}\quad \hat{h}(t, x, u) \geq 0\\ &\text{(f)}\quad \theta\hat{h}(t, x, u) = 0.\ \text{(CS)}\end{aligned} \tag{18.32}$$

From the complementary slackness (CS) condition (20.32f), if $\theta > 0$, then the upper bound constraint $h(t, x, u) = u$; and if $\theta = 0$, then possibly $h(t, x, u) > u$.

Next, suppose the constraint on the control variable u is of the form $a \leq u \leq b$, with a, b real scalars. Then the relevant Lagrangian is

$$L = H + \dot{\lambda}x + \theta_1(b - u) + \theta_2(u - a), \tag{18.33}$$

where θ_1 and θ_2 are undetermined Lagrange multipliers. Now the *extended* first-order Hamiltonian (necessary) conditions are:

$$\begin{aligned}&\text{(a)}\quad \frac{\partial H}{\partial u} - \theta_1 + \theta_2 = 0\\ &\text{(b)}\quad \dot{\lambda} = -\frac{\partial H}{\partial x}\\ &\text{(c)}\quad \frac{\partial H}{\partial \lambda} = \dot{x} = g(t, x, u)\\ &\text{(d)}\quad \theta_1 \geq 0,\ b - u \geq 0,\ \theta_1(b - u) = 0\ \text{(CS)}\\ &\text{(e)}\quad \theta_2 \geq 0,\ u - a \geq 0,\ \theta_2(u - a) = 0.\ \text{(CS)}\end{aligned} \tag{18.34}$$

From (18.34d), if $u = b$, then $\theta_1 > 0$, $u - a > 0$, and thus $\theta_2 = 0$ so that $\partial H/\partial u + \theta_1 = 0$. And from (18.34e), if $u = a$, then $\theta_2 > 0$, $b - u > 0$, and thus $\theta_1 = 0$. If $a < u < b$, then $\theta_1 = \theta_2 = 0$ so that $\partial H/\partial u = 0$.

Example 18.9 Let us determine the solution to the problem

$$\begin{aligned}&\underset{u(t)}{\text{maximize}} \int_0^3 (x - u)\, dt \text{ subject to}\\ &\dot{x} = x + u,\ x(0) = 5,\ x(3) \text{ free}\\ &u \in [0, 1].\end{aligned}$$

Forming the Lagrangian function we have $L = H + \dot{\lambda}x + \theta_1(1 - u) + \theta_2(u)$, where $H + \dot{\lambda}x = x - u + \lambda(x + u) + \dot{\lambda}x$. Then the extended first-order Hamiltonian conditions are, from (18.33),

$$\begin{aligned}&\text{(a)}\quad \frac{\partial H}{\partial u} - \theta_1 + \theta_2 = 0\\ &\text{(b)}\quad \dot{\lambda} = -\frac{\partial H}{\partial x}\\ &\text{(c)}\quad \frac{\partial H}{\partial \lambda} = \dot{x} = x + u\\ &\text{(d)}\quad \theta_1 \geq 0,\ 1 - u \geq 0,\ \theta_1(1 - u) = 0;\\ &\text{(e)}\quad \theta_2 \geq 0,\ u \geq 0,\ \theta_2(u) = 0\end{aligned} \tag{18.35}$$

or

$$\begin{aligned}&\text{(a)}\quad -1 + \lambda - \theta_1 + \theta_2 = 0\\ &\text{(b)}\quad \dot{\lambda} = -(1 + \lambda)\\ &\text{(c)}\quad \dot{x} = x + u\\ &\text{(d)}\quad \theta_1 \geq 0,\ 1 - u \geq 0,\ \theta_1(1 - u) = 0;\\ &\text{(e)}\quad \theta_2 \geq 0,\ u \geq 0,\ \theta_2(u) = 0.\end{aligned} \tag{18.35.1}$$

Given this system, the reader can readily verify that the case which solves the original problem is when $\theta_1 > 0$, $u = 1$, and $\theta_2 = 0$ (since $u > 0$). Then (18.35.1a-c) render

$$\begin{aligned}&\text{(a)}\quad \dot{\lambda} = -\lambda - 1\\ &\text{(b)}\quad \dot{x} = x + 1.\end{aligned} \tag{18.36}$$

Looking to (18.36a), the homogeneous solution is

$$\lambda(t) = C_1e^{-t}$$

while the particular (steady-state) solution is $\bar{\lambda} = -1$. Hence the complete solution is

$$\lambda(t) = C_1 e^{-t} - 1.$$

For $\lambda(T) = \lambda(3) = 0$, we get $C_1 = e^3$ and thus

$$\lambda(t) = e^{3-t} - 1.$$

From (18.36b), a similar solution procedure yields

$$x(t) = e^{t-3} - 1.$$ ■

Section 18.9 Exercises:

18.9.1 Determine the solution to

$$\begin{aligned} &\underset{u(t)}{\text{maximize}} \int_0^3 x \, dt \text{ subject to} \\ &\dot{x} = x + u, \ x(0) = 0, \ x(3) \text{ free}, \\ &u \in [-1, 1]. \end{aligned}$$

18.9.2 Find the solution to

$$\begin{aligned} &\underset{u(t)}{\text{maximize}} \int_0^2 (-x - u^2) \, dt \text{ subject to} \\ &\dot{x} = u, \ x(0) = 1, \ x(2) = 3, \ u \leq 2. \end{aligned}$$

18.10 ECONOMIC APPLICATIONS

18.10.1 The Neoclassical Optimal Growth Model (Solow [1956]; Swan [1956]; Cass [1965]; Koopmans [1965]; and Takayama [1985, 1996])

Given a linearly homogeneous production function of the form $Y = F(L, K)$, where Y is output, L is the labor input variable (we do not make a distinction between labor and population) and K is the capital input variable. It is assumed that

$$F_L > 0, \ F_K > 0; \quad F_{LL} < 0, \ F_{KK} < 0.$$

Under the linear homogeneity of F, we may express the production function as $y = f(k)$, where $y = Y/L$ and $k = K/L$, with $f' > 0$, $f'' < 0$ for $k > 0$ and

$$\lim_{k \to 0} f' = +\infty, \ \lim_{k \to +\infty} f' = 0.$$

It is further assumed that output is allocated to either consumption C or gross investment I_{gr}, i.e., $Y = C + I_{gr}$. Moreover, net investment (or changes to the capital stock) is simply $I = \dot{K} = I_{gr} - \delta K$, $0 < \delta < 1$, where δ is the rate of depreciation of the capital stock. Given these relationships, we can now write

$$\dot{K} = Y - C - \delta K. \tag{18.37}$$

Upon dividing both sides of this expression by L we get

$$\frac{\dot{K}}{L} = \frac{Y}{L} - \frac{C}{L} - \frac{\delta K}{L} = y - c - \delta k, \tag{18.38}$$

where $c = C/L$. Given (18.38), it can be shown (Exercise 18.9.2) that this equation can be rewritten as

$$\dot{k} = f(k) - c - (n + \delta)k, \tag{18.38.1}$$

the **fundamental equation of neoclassical growth theory**, where n depicts the population or labor force growth rate.

Given this background, the **optimal growth problem** to which we now turn can be stated as: What is the requisite amount of consumption at each point in time that will maximize a given target functional subject to the conditions provided by Equation (18.38.1) and $k(0) = k_o$? The target functional is taken to be an integral of the form

$$\begin{aligned} G &\equiv \int_0^{+\infty} u(c) \ L(t) e^{-\rho t} \, dt = \int_0^{+\infty} u(c) \ L_o e^{nt} e^{-\rho t} dt \\ &= \int_0^{+\infty} u(c) \ e^{-rt} dt, \end{aligned} \tag{18.39}$$

where:

i. $u(c)$ is an instantaneous social utility or welfare function of class $C^{(2)}$ on $(0, +\infty)$ and expressed in terms of per capita consumption that is weighted by the population size at time t, with the initial population level L_o normalized to unity;

ii. $\rho > 0$ is a social discount rate;

iii. $r = \rho - n > 0$ over the entire planning period; and

iv. $u(c)$ is bounded above (ostensibly due to satiation) so that (18.39) converges.

$u(c)$ is taken to reflect the preferences of a representative member of society and exhibits the properties: $u' > 0$, $u'' < 0$ for all $c > 0$ with

$$\lim_{c \to 0} u' = +\infty, \quad \lim_{c \to +\infty} u' = 0.$$

Our optimal growth problem can now be fully expressed as

$$\begin{aligned} &\underset{c(t)}{\text{maximize}} \int_0^{+\infty} u(c)\, e^{-rt}\, dt \text{ subject to} \\ &\dot{k} = f(k) - c - (n + \delta)k \\ &k(0) = k_o,\, c(t) \geq 0,\, t \in [0, +\infty), \end{aligned} \tag{18.40}$$

where $c(t)$ is the control variable and $k(t)$ is the state variable. Since t does not appear explicitly as a separate argument in u or in the state equation ((18.38.1)), we have an autonomous optimal control problem that requires solution by employing the current-value Hamiltonian

$$H_{cv} = e^{rt}H = u(c) + v(f(k) - c - (n + \delta)k),$$

where the current-value Lagrange multiplier is $v = \lambda e^{rt}$. (Note that the constraint $c(t) \geq 0$ has not been incorporated into H_{cv} since we must have $c(t) > 0$ for an optimal consumption plan.) From (18.17):

$$\begin{aligned} &\text{(a)} \quad \tfrac{\partial H_{cv}}{\partial c} = u'(c) - v = 0 \text{ or } v = u'(c); \\ &\text{(b)} \quad \dot{v} - rv = -\tfrac{\partial H_{cv}}{\partial k} \text{ or } \dot{v} = -[f'(k) - (r + n + \delta)]v; \\ &\text{(c)} \quad \dot{k} = f(k) - c - (n + \delta)k, \end{aligned} \tag{18.41}$$

with the transversality condition

$$\lim_{t \to +\infty} u'(c)\, e^{-rt}k(t) = \lim_{t \to +\infty} \lambda(t)k(t) = 0.$$

(The necessary conditions for maximizing H_{cv} ((20.41)) are also sufficient since $u(c)$ and $f(k) - c - (n + \delta)k$ are concave in c and k. See also Exercise 18.9.3.)

As Equation (18.41a) reveals, at optimality, $\lambda = e^{-rt}u'(c)$ or the present value of the marginal utility of consumption equals the shadow price (i.e., the opportunity cost or loss of a unit of consumption due to investing in capital) of capital accumulation. Obviously, $v = e^{rt}\lambda$ is the current opportunity cost of capital accumulation.

A glance back at the equations of motion (18.41b, c) reveals that since u and f are unspecified in form, it is impossible to develop analytical expressions for the time paths of v and k. And since c also appears in this system, we need to eliminate one of these three variables in order to construct a phase diagram in two dimensions. With Equation (18.41a) involving v as a function of c, it seems that possibly v can be eliminated and we can analyze a phase diagram in k, c – space.

To this end, from (18.41a) or $v = u'(c)$,

$$\dot{v} = u''(c)\dot{c}. \tag{18.42}$$

Given (18.41a) and (18.42), we can transform (18.41b) to

$$\dot{c} = -\frac{u'(c)}{u''(c)}[f'(k) - (r + n + \delta)].$$

Hence our revised equations of motion system appears as

$$\begin{aligned} &\text{(a)} \quad \dot{k} = f(k) - c - (n + \delta)k \\ &\text{(b)} \quad \dot{c} = -\tfrac{u'(c)}{u''(c)}[f'(k) - (r + n + \delta)]. \end{aligned} \tag{18.43}$$

We know from Section 18.C.2 that to obtain a phase diagram in k, c – space, we must set $\dot{k} = \dot{c} = 0$ in system (18.43) so as to obtain the equations

$$\begin{aligned} &\text{(a)} \quad c = f(k) - (n + \delta)k \quad (\dot{k} = 0 \text{ isocline}) \\ &\text{(b)} \quad f'(k) = r + n + \delta \quad (\dot{c} = 0 \text{ isocline}) \end{aligned} \tag{18.44}$$

Looking to the illustration of the phase diagram associated with system (18.43), it can be demonstrated that the $\dot{k} = 0$ and $\dot{c} = 0$ isoclines (respectively, Equations (18.44a, b)) plot as indicated in Figure 18.1. At the intersection point A we have the steady-state values of k and c, or $\bar{k}$ and $\bar{c}$, respectively.

Additionally, from (18.43b), $\partial\dot{c}/\partial k = -[u'(c)/u''(c)]\, f''(k) < 0$ (since both $u''(c) < 0, f''(k) < 0$). Hence all points to the left (right) of the $\dot{c} = 0$ line have $\dot{c} > 0(<0)$. And from (18.43a), $\partial\dot{k}/\partial c = -1 < 0$. Clearly, the curves $\dot{c} = 0$ and $\dot{k} = 0$ divide the phase plane into four distinct regions, with each of then exhibiting various pairings of the signs of $\dot{c}$ and $\dot{k}$.

It should be obvious from an examination of Figure 18.1 that we have a saddle point equilibrium at point A or at $(\bar{k}, \bar{c})$. To formally verify this observation, the nonlinear system (18.43) must be linearized near $(\bar{k}, \bar{c})$ and then Theorem 18.C.2 applied (see Exercise 18.9.5).

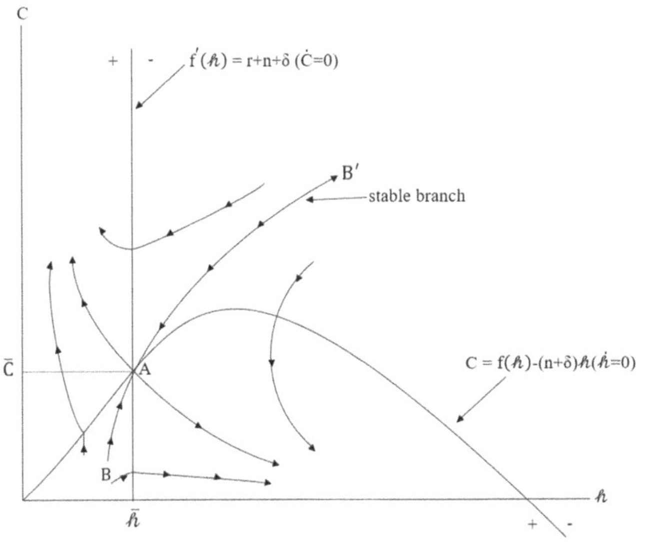

FIGURE 18.1 Phase diagram for system (18.43). Stable branch is BB' with point A serving as a steady-state equilibrium.

18.10.2 A Neoclassical Investment Model With Adjustment Costs (Jorgenson [1967, 1971,1972]; Caputo [2005]; and Takayama [1985, 1996])

As specified in the preceding section, suppose a firm has a linearly homogeneous production function so that $Y = F(L, K)$ (with Y depicting output and with L and K denoting labor and capital inputs, respectively) can be rewritten as $y = f(k)$, where $y = Y/L$ and $k = K/L$. Additionally, let $f' > 0$ and $f'' < 0$ for all $k > 0$ with

$$\lim_{k\to 0} f' = +\infty, \quad \lim_{t\to\infty} f' = 0.$$

Let I depict the flow of investment of the firm with δ the rate of depreciation of the capital stock or $\dot{K} = I - \delta K$, $0 < \delta < 1$. Additionally, let the firm be a price taker in the product and input markets so that p and w are, respectively, the constant price of a unit of output and labor services. Since capital equipment must be *installed*, there are obvious so-called *frictional* or *adjustment costs* associated with capital accumulation. In this regard, suppose the unit adjustment cost of investment C increases with the rate of investment I or $C = C(I)$, where $C(0) = 0$, $C' > 0$ for all I and $C'' > 0$ for all $I > 0$.

If i represents a constant discount rate, then we may write the present value of the sum of future net cash flows as

$$D \equiv \int_0^\infty R(t)\ e^{-it}dt,$$

where $R = pY - wL - C(I)$ depicts net cash flow. Hence the firm's decision problem is to

$$\begin{aligned}&\underset{L(t),I(t)}{\text{maximize}} \int_0^\infty [pF(L, K) - wL - C(I)]\ e^{-it}\ dt \text{ subject to}\\ &\dot{K} = I - \delta K,\ \ L \geq 0,\ \ K \geq 0 \text{ for all } t,\ \ K(0) = k_o.\end{aligned} \tag{18.45}$$

Here L, I are the control variables and K is the state variable.

Let us write the current-value Hamiltonian as

$$H_{cv} = e^{it}H = pF(L, K) - wL - C(I) + v(I - \delta K), \tag{18.46}$$

where the current-value Lagrange multiplier is $v = \lambda e^{it}$. Under the assumptions that: both L and K are positive and bounded for all t; and with H nonlinear in I (i.e., it is convex since $C'' > 0$), it follows that I must also be bounded at optimality. Then, from (18.17):

$$\begin{aligned}&\text{(a)}\quad \frac{\partial H_{cv}}{\partial L} = pF_L - w = 0 \text{ or } F_L = w/p;\\ &\text{(b)}\quad \frac{\partial H_{cv}}{\partial I} = -C' + v = 0 \text{ or } v = C';\\ &\text{(c)}\quad \dot{v} - iv = -\frac{\partial H_{cv}}{\partial K} \text{ or } \dot{v} = (i + \delta)v - pF_K;\\ &\text{(d)}\quad \dot{K} = I - \delta K; \text{ and}\end{aligned} \tag{18.47}$$

the transversality condition appears as

$$\lim_{t\to\infty} v(t)e^{-it}K(t) = \lim_{t\to\infty} \lambda(t)K(t) = 0.$$

With w/p constant and $\bar{L}$, $\bar{K}$ representing the long-run steady-state values of L and K, respectively, we see that $\bar{k} = \bar{K}/\bar{L}$ must also be constant. If we let $\tau \equiv p\bar{F}_K = pf'(\bar{K})$ then (18.47c) can be rewritten as

$$\dot{v} = (i + \delta)v - \tau. \tag{18.48}$$

The steady-state solution to Equation (18.48) is $v = \tau/(i + \delta) = \bar{v} > 0$ (which satisfies the transversality condition) and is a constant for all t. From Equation (18.47b), $C'(I) = \tau/(i + \delta)$ so that $I = \bar{I} > 0$ is also a constant for all t.

From $v = C'$ we can get $\dot{v} = C''(I)\dot{I}$ and therefore the preceding expression becomes

$$C''(I)\dot{I} = (i + \delta)C'(I) - \tau$$

or

$$\dot{I} = \frac{C'(I)}{C''(I)}(i + \delta) - \frac{\tau}{C''(I)}.$$

Hence, we may write our *equations of motion system* as

$$\begin{aligned} &\text{(a)}\quad \dot{K} = I - \delta K \\ &\text{(b)}\quad \dot{I} = \frac{C'(I)}{C''(I)}(i + \delta) - \frac{\tau}{C''(I)}. \end{aligned} \qquad (18.49)$$

Since the form of C is not explicitly specified, we can analyze the phase diagram of system (18.49) in K, I-space. To do so we must set $\dot{K} = \dot{I} = 0$ so as to obtain the equations

$$\begin{aligned} &\text{(a)}\quad I - \delta K = 0 \quad (\dot{K} = 0 \text{ isocline}) \\ &\text{(b)}\quad C'(I)(i + \delta) - \tau = 0 \quad (\dot{I} = 0 \text{ isocline}) \end{aligned} \qquad (18.50)$$

The $\dot{K} = 0$ and $\dot{I} = 0$ isoclines (Equations (18.50 a, b), respectively) are illustrated in Figure 18.2.

At the intersection of these isoclines (point A) we have the steady-state values of K and I or $\bar{K}$ and $\bar{I}$, respectively. For points above (below) the $\dot{K} = 0$ isocline, $\dot{I} > 0(<0)$. Clearly $\bar{K} = \bar{I}/\delta$. An inspection of Figure 18.2 reveals that we have a saddle-point equilibrium at $(\bar{K}, \bar{I})$ (see also Exercise 18.9.6).

Section 18.10 Exercises:

18.10.1 Verify that for homogeneity of degree one, $Y = (L, K)$ can be expressed as $y = f(k)$, with $y = Y/L$, $k = K/L$.

18.10.2 Derive Equation (18.38.1) from Equation (18.38). (Hint: set $K = kL$ and use the product rule of differentiation with respect to t.)

18.10.3 Directly check that the sufficient condition for a maximum of H_{cv} holds by showing that the bordered Hessian determinant of the Lagrangian $L = u(c)e^{-rt} + \lambda(f(k) - c - (n + \delta)k)$ is positive.

18.10.4 Determine the $\dot{c} = 0$ and $\dot{k} = 0$ isoclines for system (18.44). (Hint: Equation (18.44a) is the difference between the functions $f(k)$ and $(n + \delta)k$.)

18.10.5 Given the nonlinear differential equation system

$$\begin{aligned} \dot{x} &= f^1(x, y) \\ \dot{y} &= f^2(x, y), \end{aligned} \qquad (*)$$

with f^1 and f^2 of class $C^{(n)}$, let us assume that this system has a steady-state equilibrium point at $(\bar{x}, \bar{y})$. We may linearize this nonlinear system by employing a Taylor expansion of f^1 and f^2, respectively, near $(\bar{x}, \bar{y})$ or

$$\begin{aligned} \dot{x} &= f^1(\bar{x}, \bar{y}) + \left.\frac{\partial f^1}{\partial x}\right|_{(\bar{x},\bar{y})}(x - \bar{x}) + \left.\frac{\partial f^1}{\partial y}\right|_{(\bar{x},\bar{y})}(y - \bar{y}) + \text{higher order terms} \\ \dot{y} &= f^2(\bar{x}, \bar{y}) + \left.\frac{\partial f^2}{\partial x}\right|_{(\bar{x},\bar{y})}(x - \bar{x}) + \left.\frac{\partial f^2}{\partial y}\right|_{(\bar{x},\bar{y})}(y - \bar{y}) + \text{higher order terms.} \end{aligned} \qquad (**)$$

In matrix terms, and ignoring the higher terms in (**), we have

$$\begin{aligned} \begin{bmatrix} \dot{x} \\ \dot{y} \end{bmatrix} &= \begin{bmatrix} f^1(\bar{x}, \bar{y}) \\ f^2(\bar{x}, \bar{y}) \end{bmatrix} + \begin{bmatrix} \partial f^1/\partial x & \partial f^1/\partial y \\ \partial f^2/\partial x & \partial f^2/\partial y \end{bmatrix}_{(\bar{x},\bar{y})} \begin{bmatrix} x - \bar{x} \\ y - \bar{y} \end{bmatrix} \\ &= \begin{bmatrix} f^1(\bar{x}, \bar{y}) \\ f^2(\bar{x}, \bar{y}) \end{bmatrix} + \left.\frac{\partial(f^1, f^2)}{\partial(x, y)}\right|_{(\bar{x},\bar{y})} \begin{bmatrix} x - \bar{x} \\ y - \bar{y} \end{bmatrix}, \end{aligned} \qquad (***)$$

where $\partial(f^1, f^2)/\partial(x, y)_{(\bar{x},\bar{y})}$ is the Jacobian matrix of f^1, f^2 with respect to x, y.

Given the *linear* differential equation system appearing in (***), let us now invoke Theorem 18.C.2. Specifically, if

$$\left.\frac{\partial f^1}{\partial x}\right|_{(\bar{x},\bar{y})} + \left.\frac{\partial f^2}{\partial y}\right|_{(\bar{x},\bar{y})} < 0 \text{ and } \left.\frac{\partial(f^1, f^2)}{\partial(x, y)}\right|_{(\bar{x},\bar{y})} > 0,$$

then system (***), and thus system (*), is asymptotically stable. However, if the Jacobian determinant $|\partial(f^1, f^2)/\partial(x, y)|_{(\bar{x},\bar{y})} < 0$, then the steady-state point $(\bar{x}, \bar{y})$ is a saddle point. Use this line of argumentation to verify that system (18.43) has a saddle point at $(\bar{k}, \bar{c})$.

18.10.6 Use the argument developed in the preceding exercise to verify that point $(\bar{K}, \bar{I})$ in Figure 18.2 is a saddle point equilibrium.

18.10.7 Given that $\bar{I}$ is constant for all t in the neoclassical investment model, determine the time path of the optimal capital stock K^* given $\dot{K}_* = \bar{I} - \delta K_*$, $K(0) = K_o$. Demonstrate that $K_* \to \bar{K} = \bar{I}/\delta$ as $t \to \infty$.

18.10.8 Determine the signs of the expressions: $\partial\bar{I}/\partial i$, $\partial\bar{I}/\partial\delta$, and $\partial\bar{I}/\partial p$. (Hint: use $C'(I) = \tau/(i + \delta)$ as well as $\tau = pf'(\bar{k})$.)

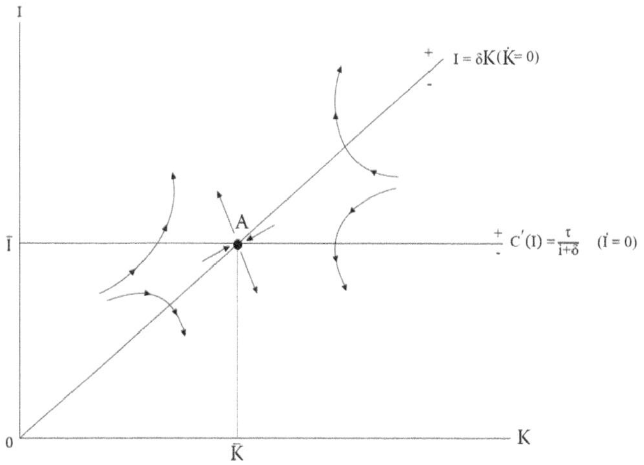

FIGURE 18.2 Phase diagram for System (18.49).

APPENDIX 18. A ORDINARY DIFFERENTIAL EQUATIONS

18.A.1 INTRODUCTION

A fundamental problem encountered in the integral calculus is the determination of a function $y = f(t)$ when its derivative is a given function of t, $g(t)$. In this regard, given

$$\frac{dy}{dt} = g(t) \text{ or } dy = g(t)\ dt, \tag{18.A.1}$$

the function y can be expressed as

$$y = \int g(t)\ dt + C, \quad C \text{ an arbitrary constant.}$$

Equation (18.A.1) is a simple example of an **ordinary differential equation** (ODE).

18.A.2 FIRST-ORDER DIFFERENTIAL EQUATIONS

Consider the ODE of the form

$$F\left(t, y, \frac{dy}{dt}\right) = 0, \tag{18.A.2}$$

where F is a given function of the arguments t, y, and dy/dt. This expression represents a **first-order differential equation** since it contains only the first derivative dy/dt. Hence the **order** of a differential equation is determined by the highest order of derivative in the equation. Additionally, the **degree** of a differential equation is the highest power on dy/dt.

A solution to (18.A.2) will typically be a relation of the form

$$G(t, y, C) = 0, \quad C \text{ an arbitrary constant.} \tag{18.A.3}$$

Given (18.A.3), we may obtain an ODE that is solved by (18.A.3) by differentiating this equation with respect to t,

$$\frac{\partial G}{\partial t} + \frac{\partial G}{\partial y}\frac{dy}{dt} = 0, \tag{18.A.4}$$

and eliminating C from (18.A.3) and (18.A.4).

For instance, suppose

$$G(t, y, C) = y - Ct = 0 \tag{18.A.5}$$

so that

$$\frac{dy}{dt} - C = 0. \tag{18.A.6}$$

If we eliminate C from the preceding two equations, we obtain the ODE

$$y - t\frac{dy}{dt} = 0. \tag{18.A.7}$$

Does (18.A.5) represent *all* solutions to (18.A.7)? The answer to this question is *no.*

A general theorem pertaining to the question of all solution of the ODE $dy/dt = f(t, y)$ is offered by Theorem 18.A.1

Theorem 18.A.1:
(Existence and Uniqueness). Let the functions f, $\partial f/\partial y$ be continuous in some rectangular region $R = \{(t, y)|t \in (a, b), y \in (c, d)\}$ of the xy-plane containing the point (x_o, y_0). Then within some subinterval $t \in (x_o - h, x_o + h)$ of (a, b) there is a unique solution $y = y(t)$ of the ODE $dy/dt = f(t, y)$ with the initial condition $y_0 = y(t_o)$.

The conditions in this theorem are *sufficient* to guarantee the existence of a unique solution of $dy/dt = f(t, y)$, $y_0 = y(t_o)$. (note that if f does not satisfy the hypotheses of this theorem, a unique solution may still exist.) The expression $G(t, y, C) = 0$ that contains all solutions of $F\left(t, y, \frac{dy}{dt}\right) = 0$ in R will be termed the **general solution** of this ODE in R. A solution obtained from the general solution of an ODE by specifying a particular value of C will be called a **particular solution.**

18.A.3 SEPARATION OF VARIABLES

The simplest class of ODEs $dy/dt = f(t, y)$ that can be solved in a straightforward fashion are the equations in which the variables are separable in a way such that

$$N(y)\frac{dy}{dt} + M(t) = 0$$

or

$$M(t)\,dt + N(y)\,dy = 0. \tag{18.A.8}$$

Then the general solution of (18.A.8) is

$$\int M(t)\,dt + \int N(y)\,dy = C, \quad C \text{ an arbitrary constant.}$$

Example 18.A.1:
Suppose

$$\frac{dy}{dt} + te^{-y} = 0.$$

Then (see (18.A.8),

$$\int e^{y}dy + \int t\,dt = C$$

or

$$e^{y} + \frac{1}{2}t^{2} = C.$$

Then solving for y yields

$$y = \ln\left(C - \frac{1}{2}t^{2}\right).$$

(Any restriction on C?). ■

A class of first-order ODEs that can be rendered separable by a change of variable is

$$\frac{dy}{dt} = g\left(\frac{y}{t}\right). \tag{18.A.9}$$

Here g is said to be **homogeneous** in that it depends only on the ratio y/t (or t/y) and not on t and y separately. If we set $v = y/t$ so that $dy/dt = g(v)$, where t is the independent variable and v is taken to be the new dependent variable, then, if we can find v as a function of t, $y = vt$ will ultimately be a function of t and a solution to (18.A.9) follows. To find v, use $y = v(t)$ to obtain

$$\frac{dy}{dt} = v + t\frac{dv}{dt}.$$

Then (18.A.9) becomes

$$v + t\frac{dv}{dt} = g(v),$$

and thus the variables v and t are separable as

$$\frac{dv}{v - g(v)} + \frac{dt}{t} = 0.$$

(See (18.A.8).)

Example 18.A.2:
Let us solve

$$t\frac{dy}{dt} - te^{-y/t} - y = 0.$$

(Note that this expression is homogeneous once both sides are divided by t.) Setting $y = vt$ we obtain

$$t\left(v + t\frac{dv}{dt}\right) - te^{-v} - vt = 0$$

or

$$t\frac{dv}{dt} - e^{-v} = 0$$

and thus

$$e^{v}dv - dt/t = 0. \quad \text{(separability holds)}$$

Upon integrating we have

$$e^{v} - \ln t = \ln C,$$
$$e^{v} = \ln Ct$$
$$v = \ln(\ln Ct)$$

or

$$y = vt = t\ln(\ln Ct),\ Ct > 1.$$

■

18.A.4 LINEAR FIRST-ORDER DIFFERENTIAL EQUATIONS

A differential equation is linear if all "y-related terms" (i.e., y, dy/dt, d^2/dt^2, etc.) are raised only to the first power. Hence a linear first-order differential equation has the general form

$$\frac{dv}{dt} + p(t)y = q(t). \tag{18.A.10}$$

Obviously this equation can appear in a variety of versions, depending on the assumptions made about $p(t)$ and $q(t)$. In what follows we shall consider the progression of these assumptions on $p(t)$ and $q(t)$ from the most basic to the most general. To this end, two versions of (18.A.10) will be considered: **autonomous** (the variable t does not enter (18.A.10) explicitly); and **non-autonomous** (the variable t is an explicit argument of either p or q or both p and q). More specifically, we have:

a. the **linear autonomous first-order differential equation**

$$\frac{dy}{dt} + ay = b, \tag{18.A.10a}$$

where a, b are given constants.

b. the **linear non-autonomous first-order differential equation**

$$\frac{dy}{dt} + p(t)y = q(t), \tag{18.A10}$$

where p, q are continuous functions of t.

18.A.4A AUTONOMOUS EQUATIONS

We start with some important terminology. The **general solution**y to the complete equation (18.A.10a) is the sum of the **general homogeneous solution**y_h (obtained when $b = 0$) and a **particular solution**y_p (typically b/a). Then $y = y_h + y_p$. To obtain the general solution of the homogeneous form of (18.A.10a) we start with

$$\frac{dy}{dt} + ay = 0, \quad a \neq 0. \tag{18.A.10b}$$

Then

$$\frac{dy}{y} = -a\ dt,$$
$$\ln\ y = -at + \ln\ C,\quad C(>0)\ \text{an arbitrary constant},$$
$$y = Ce^{-at}.$$

Hence the general homogeneous solution to (18.A.10b) is

$$y_h(t) = Ce^{-at}. \tag{18.A.10c}$$

To obtain a particular solution to the complete differential equation (18.A.10a), we can utilize its **steady-state value** $\bar{y}$ -- the value of y at which $dy/dt = 0$ or the value of y at which y is stationary. Quite simply, to find $\bar{y}$, set $dy/dt = 0$ in (18.A.10a) so as to obtain $ay = b$ or $\bar{y} = b/a$, $a \neq 0$. Then the general solution to the complete differential equation (18.A.10a) is

$$\begin{aligned} y(t) &= y_h(t) + y_p \\ &= Ce^{-at} + \frac{b}{a}. \end{aligned} \tag{18.A.10d}$$

Suppose the **initial value** of y at time $t = t_o$ is denoted $y(t_o) = y_0$. Also, suppose that we are faced with solving an **initial value problem** – a problem for which (18.A.10d) must satisfy the **initial condition** $y(t_o) = y_0$ for $t = t_o$. From (18.A.10d) it is readily determined that

$$y_0 = Ce^{-at_o} + \frac{b}{a}$$

so that

$$C = \left(y_0 - \frac{b}{a}\right)e^{at_o}.$$

Substituting this expression into (18.A.10d) yields

$$y(t) = \left(y_o - \frac{b}{a}\right)e^{-a(t-t_o)} + \frac{b}{a}. \tag{18.A.10e}$$

For a (frequently chosen) initial value of $t_o = 0$, the preceding expression simplifies to

$$y(t) = \left(y_o - \frac{b}{a}\right)e^{-at} + \frac{b}{a}. \tag{18.A.10f}$$

Example 18.A.3:
Determine the solution to $dy/dt + 2y = 1$ given the initial condition $y(0) = 1$ at $t_o = 0$. Going directly to (18.A.10e) we have

$$\begin{aligned} y(t) &= \left(1 - \tfrac{1}{2}\right)e^{-2t} + \tfrac{1}{2} \\ &= \tfrac{1}{2}(1 - e^{-2t}). \end{aligned}$$

Note that $y(t)$ converges to the steady state value $\bar{y} = b/a = 1/2$ as $t \to \infty$ or

$$\lim_{t\to\infty} y(t) = \lim_{t\to\infty} \frac{1}{2}(1 - e^{-2t}) = \frac{1}{2}.$$

Note also that the limiting behavior of $y(t)$ depends upon the behavior of the term e^{-at} and thus upon the sign of a:

i. if $a > 0$, then $\lim_{t\to\infty} y(t) = \bar{y}$;
ii. if $a < 0$, then $\lim_{t\to\infty} y(t) = +\infty$ provided that $y_o - \bar{y} > 0$; $\lim_{t\to\infty} y(t) = -\infty$ when $(y - \bar{y}) < 0$. ■

18.A.4B NON-AUTONOMOUS EQUATIONS

Remember that we previously described a linear non-autonomous first-order differential equation as one in which time t appears explicitly as an argument in $p(t)$ or $q(t)$ or in both of these functions. Hence, via (18.A.10),

$$\frac{dy}{dt} + p(t)y = q(t). \tag{18.A.10}$$

As will be demonstrated later on (Section 18.A.6), the **general solution** to this differential equation has the form

$$y(t) = e^{-\int p(t)dt}\left[\int e^{\int p(t)dt} q(t)dt + C\right], \quad C \text{ an arbitrary constant.} \tag{18.A.10g}$$

Note that if $p(t) = a$ = constant and $q(t) = b$ = constant, then this expression reduces to (18.A.10d), i.e.,

$$y(t) = e^{-at}\left[\frac{b}{a}e^{at} + C\right] = \frac{b}{a} + Ce^{-at}.$$

Example 18.A.4:
Solve the differential equation $dy/dt + ty = t$ using (18.A.10g). Here $p(t) = q(t) = t$ with

$$e^{\int p(t)dt} = e^{\int tdt} = e^{\frac{1}{2}t^2}.$$

Then, from (18.A.10g),

$$\begin{aligned} y(t) &= e^{-\frac{1}{2}t^2}\left[\int te^{\frac{1}{2}t^2}dt + C\right] \\ &= e^{-\frac{1}{2}t^2}\left[e^{\frac{1}{2}t^2} + C\right] = 1 + Ce^{-\frac{1}{2}t^2}. \end{aligned} \qquad (*)$$

Suppose $y(t)$ must satisfy the initial condition $y(t_o) = y_o$ at $t = t_o$. Then from the preceding equation we have

$$y_o = 1 + Ce^{-\frac{1}{2}t_0^2}$$

and thus

$$C = (y_o - 1)e^{\frac{1}{2}t_0^2}$$

so that (*) becomes

$$y(t) = 1 + (y_o - 1)e^{-\frac{1}{2}(t^2 - t_0^2)}.$$

■

18.A.5 EXACT ORDINARY DIFFERENTIAL EQUATIONS

Given the function

$$g(t, y) = C = \text{constant}, \qquad (18.A.11)$$

the associated ordinary differential equation is

$$dg = \frac{\partial g}{\partial t}dt + \frac{\partial g}{\partial y}dy = 0.$$

Now, suppose we have an ordinary differential equation of the form

$$M(t, y)dt + N(t, y)dy = 0. \qquad (18.A.12)$$

If M and N are the partial derivatives of some function $g(t, y)$, with

$$M(t, y) = \partial g/\partial t, N(t, y) = \partial g/\partial y, \qquad (18.A.13)$$

then we can argue that if (18.A.11) and (18.A.12) hold it must be the case that (18.A.13) also holds. That is, if there exists a function $g(t, y)$ for which (18.A.13) is true, then the solution of (18.A.12) is provided by (18.A.11). In this instance the expression $M(t, y)dt + N(t, y)dy$ is an exact differential, dg, and (18.A.12) is called an **exact ordinary differential equation**.

A necessary and sufficient condition on the functions M and N which ensures that (18.A.12) will be an *exact* ordinary differential is that

$$\frac{\partial M(t,y)}{\partial y} = \frac{\partial N(t,y)}{\partial t}. \tag{18.A.14}$$

When this equality holds, can we always find a function g for which (18.A.13) also holds? To answer this, let's start with determining a g such that

$$\frac{\partial g(t,y)}{\partial t} = M. \tag{18.A.15}$$

Given g such that (18.A.15) holds, can we also find $N = \partial g/\partial y$? From (18.A.15),

$$g(t,y) = \int^t M(u,y)\,du + h(y) \tag{18.A.16}$$

(we integrate partially with respect to t), where $h(y)$ is an arbitrary *function of integration*. Then from (18.A.16),

$$\begin{aligned}\frac{\partial g}{\partial y} &= \frac{\partial}{\partial y}\int^t M(u,y)\,du + h'(y)\\ &= \int^t \frac{\partial M(u,y)}{\partial y}du + h'(y).\end{aligned} \tag{18.A.17}$$

Substituting $\partial g/\partial y = N(t,y)$ into (18.A.17) yields

$$h'(y) = N(t,y) - \int^t \frac{\partial M(u,y)}{\partial y}du. \tag{18.A.17.1}$$

The next step is to determine $h(y)$. To verify that the preceding expression is a function of y alone, let us differentiate the right-hand side with respect to t so as to obtain

$$N_t(t,y) - M_y(t,y) = 0$$

via (18.A.14). Since the right-hand side of (18.A.17.1) is independent of t, a single integration with respect to y gives $h(y)$ or $h(y)$can be obtained from

$$h'(y) = N - \frac{\partial}{\partial y}\int M\ dt,$$

Finally, substituting for $h(y)$ in (18.A.16) gives

$$g(t,y) = \int^t M(u,y)\,du + \int^y \left[N(t,s) - \int^t M_s(u,s)\,du\right]ds. \tag{18.A.18}$$

Example 18.A.5:
Determine if the differential equation

$$(2ty^2 + 2y)\,dt + (2t^2y + 2t)\,dy = 0$$

is exact. If so, obtain its solution.

Let $M(t, y) = 2ty^2 + 2y$ and $N(t, y) = 2t^2y + 2t$.

Since

$$\frac{\partial M}{\partial y} = 4ty + 2 = \frac{\partial N}{\partial t},$$

we see that we have an exact ordinary differential equation. To obtain a solution, we start with

$$\begin{aligned}\frac{\partial g}{\partial t} &= M = 2ty^2 + 2y, \\ g &= \int (2ty^2 + 2y)\,dt + h(y) \\ &= t^2y^2 + 2ty + h(y)\end{aligned}$$

and thus

$$\frac{\partial g}{\partial y} = 2t^2y + 2t + h'(y).$$

Since also

$$\frac{\partial g}{\partial y} = N = 2t^2y + 2t,$$

it follows that

$$\begin{aligned}h'(y) &= 2t^2y + 2t - \frac{\partial}{\partial y}\int M dt \\ &= 2t^2y + 2t - \frac{\partial}{\partial y}(t^2y^2 + 2ty) \\ &= 2t^2y + 2t - (2t^2y + 2t) = 0\end{aligned}$$

or $h(y) = C_o$ = constant. The solution to the original differential equation is thus $g = C$ or

$$g = t^2y^2 + 2ty = C,$$

with C_o being absorbed into the constant C. ■

18.A.6 INTEGRATING FACTORS

Suppose a differential equation is not exact. In this instance our objective is to transform any such differential equation to an exact one by multiplying the non-exact differential equation by a suitable function of t and y. That is, given

$$M(t, y)dt + N(t, y)dy = 0, \tag{18.A.19}$$

with $\partial M/\partial y \neq \partial N/\partial t$, we want to make (18.A.19) exact by forming

$$\mu M(t, y)dt + \mu N(t, y)dy = 0, \tag{18.A.20}$$

where $\mu = \mu(t, y)$. If any such μ exists, then it is called an **integrating factor** of (18.A.19). The *requirement for exactness* of (18.A.20) is

$$\frac{\partial(\mu M)}{\partial y} = \frac{\partial(\mu N)}{\partial t},$$
$$\frac{\partial \mu}{\partial y}M + \frac{\partial M}{\partial y}\mu = \frac{\partial \mu}{\partial t}N + \frac{\partial N}{\partial y}\mu$$

or

$$\frac{\partial \mu}{\partial y}M - \frac{\partial \mu}{\partial t}N = \left(\frac{\partial N}{\partial t} - \frac{\partial M}{\partial y}\right)\mu. \tag{18.A.21}$$

So, if a function μ satisfying (18.A.21) can be found, then (18.A.20) will be exact. The solution of (18.A.20) can then be obtained by the method of the preceding section and will appear implicitly as $g(t, y) = C$. And since μ can be eliminated from all terms of (18.A.20), $g = C$ also defines the solution to (18.A.19).

As an alternative to the procedure of finding a function μ, let us consider the conditions on M and N that will enable us to obtain an integrating factor of a particular form. Specifically, we shall restrict the equation for μ to one that depends on t alone. We thus set $\partial\mu/\partial y = 0$ and rewrite (18.A.21) as

$$\frac{1}{\mu}\frac{d\mu}{dt} = \frac{\partial M/\partial y - \partial N/\partial t}{N}. \tag{18.A.21.1}$$

Since the left-hand side of this equation has t as its argument, we seek an integrating factor $\mu = \mu(t)$ such that the right-hand side of (18.A.21.1) also has t as its sole argument. If we set the right-hand side of (18.A.21.1) equal to $r(t)$, then the integrating factor appears as

$$\frac{1}{\mu}\frac{d\mu}{dt} = r(t)$$

or

$$\mu = e^{\int r(t)dt}. \tag{18.A.22}$$

Example 18.A.6:
Let us solve the differential equation

$$(3ty + y^2)dt + (t^2 + ty)dy = 0,$$

where $M(t, y) = 3ty + y^2$, $N(t, y) = t^2 + ty$, and

$$\frac{\partial M}{\partial y} = 3t + 2y \neq \frac{\partial N}{\partial t} = 2t + y.$$

Since this differential equation is not exact, let us determine the integrating factor from (18.A.21.1) as

$$\frac{1}{\mu}\frac{d\mu}{dt} = \frac{3t + 2y - (2t + y)}{t^2 + ty} = \frac{1}{t}$$

or $d\mu/\mu = dt/t$ and thus $\ln\ \mu = \ln\ t$ or $\mu(t) = t$. Then from (18.A.20)

$$(3t^2y + ty^2)dt = (t^3 + t^2y)dy,$$

where now $M = 3t^2y + ty^2$, $N = t^3 + t^2y$, and $\partial M/\partial y = \partial N/\partial t$. Clearly this revised or new differential equation is exact and may be solved via the technique of the preceding section. Its solution is

$$g(t, y) = t^3y + \frac{1}{2}t^2y^2 = C. \quad ■$$

In Section 18.A.4b we presented an expression (Equation (18.A.10g) for the general solution of the linear non-autonomous differential equation

$$\frac{dy}{dt} + p(t)y = q(t) \tag{18.A.10}$$

or

$$[p(t)y - q(t)]dt + dy = 0, \tag{18.A.23}$$

where $M(t, y) = [\cdot]$ and $N(t, y) = 1$. As we shall now see, for this differential equation, we can find an integrating factor μ which depends only on t. To this end we start with

$$\frac{\partial M/\partial y - \partial N/\partial t}{N} = p(t)$$

so that the **integrating factor** has the form

$$\mu = e^{\int p(t)dt}. \tag{18.A.24}$$

If we multiply both sides of (18.A.10) by μ, we obtain

$$e^{\int p(t)dt}\left[\frac{dy}{dt} + p(t)y\right] = e^{\int p(t)dt}q(t)$$

or

$$\frac{d}{dt}\left(e^{\int p(t)dt}y\right) = e^{\int p(t)dt}q(t).$$

Then from this latter expression we get the solution to (18.A.10) or

$$e^{\int p(t)dt}y = \int e^{\int p(t)dt}q(t)dt + C$$

or

$$y(t) = e^{-\int p(t)dt}\left[\int e^{\int p(t)dt}q(t)dt + C\right]. \tag{18.A.10g}$$

18.A.7 VARIATION OF PARAMETERS

Let us express the homogeneous version of (18.A.10) as

$$\frac{dy_h}{dt} + p(t)y_h = 0$$

or

$$\frac{dy_h}{y_h} + p(t)dt = 0. \tag{18.A.25}$$

The solution to this homogeneous linear differential equation is

$$\begin{aligned} y_h &= \theta \; e^{-\int p(t)dt} \\ &= \theta \; u(t), \quad \theta \text{ a parameter,} \end{aligned} \tag{18.A.26}$$

where

$$u(t) = e^{-\int p(t)dt}. \tag{18.A.27}$$

Suppose we speculate that the full solution to (18.A.10) is of the form

$$y = v(t)u(t). \tag{18.A.28}$$

Here we are varying the parameter θ by substituting in its stead a term $v(t)$ which depends on t. Substituting (18.A.27) into (18.A.10) yields a differential equation for $v(t)$:

$$\begin{aligned} &\frac{dv(t)}{dt}u(t) + v(t)\frac{du(t)}{dt} + p(t)v(t)u(t) = \frac{dv(t)}{dt}u(t) = q(t) \left(\text{since}\frac{dy_h}{dt} + p(t)y_h = 0\right), \frac{dv(t)}{dt} = \frac{q(t)}{u(t)}dt, v(t) \\ &= \int \frac{q(t)}{u(t)}dt + C. \end{aligned} \tag{18.A.29}$$

Inserting (18.A.27) and (18.A.29) into (18.A.28) yields

$$\begin{aligned} y(t) &= u(t)v(t) \\ &= e^{-\int p(t)dt}\left[\int q(t)e^{\int p(t)dt}dt + C\right]. \end{aligned} \tag{18.A.30}$$

Example 18.A.7:
Given the differential equation

$$\frac{dy}{dt} - y = e^t,$$

where $p(t) = -1$ and $q(t) = e^t$, we have, from (18.A.27),

$$u(t) = e^{-\int p(t)dt} = e^{\int dt} = e^t.$$

Then (18.A.30) becomes

$$y(t) = e^t\left[\int e^t e^{-\int dt}dt + C\right] = e^t\left[\int e^t e^{-t}dt + C\right] = e^t(t + C).$$

■

18.A.8 NONLINEAR DIFFERENTIAL EQUATIONS OF THE FIRST ORDER AND FIRST DEGREE

For a linear differential equation, the terms dy/dt and y are restricted to the first degree. Moreover, the product of these terms or $y(dy/dt)$ is completely excluded from any such differential equation. What makes a differential equation *nonlinear* is that y appears to a power higher than one.

Suppose we express a first-order, first-degree nonlinear differential equation as

$$f(y, t)dy + g(y, t)dt = 0 \text{ or } \frac{dy}{dt} = h(y, t), \tag{18.A.31}$$

where dy/dt is a first-order derivative appearing to the first power and y and t may be raised to powers greater than one. What options are there for obtaining a solution to (18.A.31)? Three cases will be considered.

1. The differential equation is *exact.* Consider y appearing to a power greater than one in a nonlinear differential equation, e.g., $3y^2t\ dy + y^3dt = 0$. Section 18.5.A serves as a guide for generating a solution to differential equations such as this one.
2. The differential equation allows *separation of variables.*

If in (18.A.31)f depends on y alone and g depends only on t, then this differential equation has the specific form

$$f(y)dy + g(t)dt = 0$$

and, consequently, is separable in the variables y, t. Section 18.A.3 exhibits how this type of differential equation is solved.

3. The differential equation *reduces to linear form.*

Suppose (18.A.31) or $dy/dt = h(y, t)$ assumes the nonlinear form

$$\frac{dy}{dt} + p(t)y = q(t)y^m, \quad m \neq 0, 1. \tag{18.A.32}$$

This is the (Jacob) **Bernoulli equation** that can always be reduced to a linear differential equation and solved accordingly. The *reduction procedure* (due to Leibnitz) consists of the following steps:

1. divide both sides of (18.A.32) by y^m. This yields

$$y^{-m}\frac{dy}{dt} + p(t)y^{1-m} = q(t). \tag{18.A.33}$$

2. Let $z = y^{1-m}$ with

$$\frac{dz}{dt}\left(=\frac{dz}{dy}\frac{dy}{dt}\right) = (1-m)y^{-m}\frac{dy}{dt}$$

so that

$$\frac{1}{1-m}\frac{dz}{dt} = y^{-m}\frac{dy}{dt}. \tag{18.A.34}$$

3. A substitution of (18.A.34) into (18.A.33) yields

$$\frac{1}{1-m}\frac{dz}{dt} + p(t)z = q(t) \tag{18.A.35}$$

4. Multiplying (18.A.35) by $(1 - m)dt$ renders

$$dz + [(1-m)p(t)z - (1-m)q(t)]dt = 0. \tag{18.A.36}$$

Clearly this is a linear first-order differential equation of the form $dz/dt + u(t)z = w(t)$ (mirroring (18.A.10)) in which z has replaced y. Its solution is provided by (18.A.10g) and, having the solution in terms of $z(t)$, we can transform back to one stated in terms of $y(t)$ since$y = z^{1/(1-m)}$.

APPENDIX 18.B SIMULTANEOUS SYSTEMS OF LINEAR DIFFERENTIAL EQUATIONS

18.B.1 INTRODUCTION

In Appendix 18.A we focused on techniques useful for solving a single differential equation. Our goal now is to extend the single differential equation solution methodologies to systems of (autonomous) differential equations.

A **linear system of (two) autonomous differential equations** can be represented as

$$\begin{aligned}&\text{(a)}\quad \frac{dy_1}{dt} = a_{11}y_1 + a_{12}y_2 + b_1\\ &\text{(b)}\quad \frac{dy_2}{dt} = a_{21}y_1 + a_{22}y_2 + b_2\end{aligned} \tag{18.B.1}$$

As noted in Appendix 18.A, system (18.B.1) is linear because y_1 and y_2 appear only to the first power while it is autonomous since the coefficients a_{ij} and the right-hand side values b_i, $i, j = 1, 2$, are all constants so that time t does not appear explicitly as an argument in either (18.B.1a) or (18.B.1b). Clearly these differential equations must be solved simultaneously since the solution to one of them depends upon the solution to the other.

Also noted in Appendix 18.A was the fact that the *complete solution* to a linear differential equation was formed as the sum of a *homogeneous solution* and a *particular solution*. Here too the complete solution to system 18.B.1 will be expressed as the sum of homogeneous and particular solutions or

$$\begin{aligned}&\text{(a)}\quad y_1(t) = y_1^h + y_1^p\\ &\text{(b)}\quad y_2(t) = y_2^h + y_2^p\end{aligned} \tag{18.B.2}$$

where, as before, $y_i(t)$ is the complete solution while y_i^h and y_i^p denote, respectively, homogeneous and particular solutions, $i = 1, 2$.

While system (18.B.1) is useful for expressing the most basic set of simultaneous linear differential equations, it obviously has an important shortcoming - - it is not efficient when the number of differential equations is greater than two. As one might have guessed, a more manageable way to express large systems of linear differential equations is through the medium of matrix notation. Hence, for sizeable systems, (18.B.1) can be rewritten as

$$\frac{d\boldsymbol{y}}{dt} = \boldsymbol{A}\boldsymbol{y} + \boldsymbol{b}, \tag{18.B.3}$$

where $\boldsymbol{A}$ is an $(n \times n)$ coefficient matrix of constant coefficients a_{ij}, with $i, j = 1, \ldots, n$; $\boldsymbol{b}$ is an $(n \times 1)$ vector of constants b_i, $i = 1, \ldots, n$; $\boldsymbol{y}$ is an $(n \times 1)$ vector of variables y_i, $i = 1, \ldots, n$; and $d\boldsymbol{y}/dt$ is an $(n \times 1)$ vector of derivatives dy_i/dt, $i = 1, \ldots n$. Furthermore, it is assumed that the columns of $\boldsymbol{A}$ are linearly independent or $|\boldsymbol{A}| \neq 0$.

18.B.2 AUTONOMOUS SYSTEMS

Given (18.B.3), let us begin with obtaining a homogeneous solution to this complete system. That is, the **homogeneous form** of the complete system appears as

$$\frac{d\boldsymbol{y}}{dt} = \boldsymbol{A}\boldsymbol{y} \tag{18.B.4}$$

with $\boldsymbol{A} \neq \boldsymbol{O}$, $\boldsymbol{b} = \boldsymbol{O}$. Following the lead of Equation (18.A.10c), we may surmise that the **homogeneous solution** is of the form

$$y_h(t) = \boldsymbol{h}e^{\lambda t}, \tag{18.B.5}$$

where $\boldsymbol{h}$ is an $(n \times 1)$ vector of constants $h_1, \ldots, h_n$ and λ is a scalar. To see if this *trial solution* is correct, let us evaluate (18.B.4) at $\boldsymbol{y} = \boldsymbol{h}e^{\lambda t}$ or, given $d\boldsymbol{y}/dt = \lambda \boldsymbol{h}e^{\lambda t}$,

$$\lambda \boldsymbol{h}e^{\lambda t} = \boldsymbol{A}\boldsymbol{h}e^{\lambda t}$$

or

$$(\boldsymbol{A} - \lambda \boldsymbol{I}_n)\boldsymbol{h} = \boldsymbol{O}. \tag{18.B.6}$$

Here we have an nth order system of *linear homogeneous equations* that will possess a non-trivial solution ($\boldsymbol{h} \neq \boldsymbol{O}$) if and only if $\rho(\boldsymbol{A} - \lambda \boldsymbol{I}_n) = n$ or if

$$|\boldsymbol{A} - \lambda \boldsymbol{I}_n| = 0. \tag{18.B.7}$$

As specified in E - Chapter 2 (Section 2.1.1), for (18.B.6) to have a nontrivial solution, $\boldsymbol{A} - \lambda \boldsymbol{I}_n$ must be singular or (18.B.7) must hold, where this equation, an nth degree polynomial $f(\lambda)$, is called the characteristic equation for matrix $\boldsymbol{A}$. And as also seen in E - Chapter 2 (Section 2.1.1), the solutions to (18.B.7) are the characteristic roots or eigenvalues of $\boldsymbol{A}$ and any vector $\boldsymbol{h}(\neq \boldsymbol{O})$ that satisfies (18.B.6) is an eigenvector or characteristic vector of $\boldsymbol{A}$.

For instance, if $n = 2$, (18.B.7) becomes

$$f(\lambda) = |\boldsymbol{A} - \lambda \boldsymbol{I}_n| = \begin{vmatrix} a_{11} - \lambda & a_{12} \\ a_{21} & a_{22} - \lambda \end{vmatrix}$$
$$= \lambda^2 - tr(\boldsymbol{A})\lambda + |\boldsymbol{A}| = 0$$

and thus the two roots of this expression are, from the quadratic formula,

$$\lambda = \frac{1}{2}[tr(A) \pm (tr(A)^2 - 4|\boldsymbol{A}|)^{1/2}] \tag{18.B.8}$$

(see Example 2.1 in E - Chapter 2). To determine the eigenvectors $\boldsymbol{h}_1$ and $\boldsymbol{h}_2$ corresponding to the eigenvalues λ_1 and λ_2, respectively, we need only solve the homogeneous equation systems

$$(\boldsymbol{A} - \lambda_1 \boldsymbol{I}_2)\boldsymbol{h} = \boldsymbol{O}$$
$$(\boldsymbol{A} - \lambda_2 \boldsymbol{I}_2)\boldsymbol{h} = \boldsymbol{O}.$$

However, if there are n equations and n characteristic roots, then we can expect n solutions to (18.B.4).
More specifically, we have

Theorem 18.B.1:
(**Principle of Superposition**). If the vector functions

$$\boldsymbol{y}_i(t) = \boldsymbol{h}_i e^{\lambda_i t},\ i = 1, \ldots, n,$$

constitute n linearly independent solutions to the homogeneous system (18.B.4), then the general solution to the same is the **superposition** or linear combination

$$\boldsymbol{y}(t) = \Sigma_{i=1}^{n} C_i \boldsymbol{y}_i(t)$$
$$= \Sigma_{i=1}^{n} C_i \boldsymbol{h}_i e^{\lambda_i t}, \tag{18.B.9}$$

where the C_i are uniquely determined constants.

Example 18.B.1:
Suppose we have a linear system of autonomous differential equations of the form

$$\frac{dy_1}{dt} = 3y_1 - y_2 + 6$$
$$\frac{dy_2}{dt} = -4y_1 + 3y_2 + 10.$$

Determine the solution to the associated system of *homogeneous* linear differential equations

$$\frac{dy_1}{dt} = 3y_1 - y_2$$
$$\frac{dy_2}{dt} = -4y_1 + 3y_2.$$

Here

$$\boldsymbol{A} = \begin{bmatrix} 3 & -1 \\ -4 & 3 \end{bmatrix}$$

and

$$|A - \lambda I_2| = \begin{vmatrix} 3-\lambda & -1 \\ -4 & 3-\lambda \end{vmatrix} = 0$$

or $(3-\lambda)^2 - 4 = \lambda^2 - 6\lambda + 5 = 0$. The solution to this quadratic equation is, from (18.B.8), $\lambda = \frac{1}{2}[6 \pm (36 - 4(5))^{1/2}]$ or $\lambda_1 = 5$, $\lambda_2 = 1$. Hence we need to solve the homogeneous equation systems

$$(A - \lambda_1 I_2)h = O \tag{*}$$

$$(A - \lambda_2 I_2)h = O. \tag{**}$$

For $\lambda_1 = 5$ we have $(A - 5I_2)h = O$ or

$$\begin{aligned} -2h_1 - h_2 &= 0 \\ -4h_1 - 2h_2 &= 0 \end{aligned}$$

and thus $h_1^T = \left(-\frac{1}{2}h_2, h_2\right)$. For the sake of specificity, it is often the case that we arbitrarily set $h_1 = 1$ within h_1. Then $h_1^T = (1, -2)$ and thus

$$y_1(t) = h_1 e^{5t} = \begin{bmatrix} 1 \\ -2 \end{bmatrix} e^{5t}.$$

For $\lambda_2 = 1$ we obtain $(A - I_2)h = O$ or

$$\begin{aligned} 2h_1 - h_2 &= 0 \\ -4h_1 + 2h_2 &= 0 \end{aligned}$$

and thus $h_2^T = \left(\frac{1}{2}h_2, h_2\right)$. If we set $h_1 = 1$ in h_2 we get $h_2^T = (1, 2)$ so that

$$y_2(t) = h_2 e^{t} = \begin{bmatrix} 1 \\ 2 \end{bmatrix} e^{t}.$$

With the solutions $y_1(t)$ and $y_2(t)$ linearly independent (since $|A| \neq 0$), the general solution to the linearly homogeneous system is, from (18.B.9),

$$y_h(t) = C_1 \begin{bmatrix} 1 \\ -2 \end{bmatrix} e^{5t} + C_2 \begin{bmatrix} 1 \\ 2 \end{bmatrix} e^{t}.$$ ■

18.B.3 PARTICULAR SOLUTION

It was determined in Section 18.A.4a that a convenient particular solution to an autonomous first-order linear differential equation was the **steady-state solution** $\bar{y}$ obtained by setting $dy/dt = 0$ in the complete differential equation. Mirroring this procedure, let us set $dy/dt = O$ in the complete system of differential equations so as to obtain $A\bar{y} + b = O$ or

$$\bar{y} = -A^{-1}b, \tag{18.B.10}$$

provided A is nonsingular or $|A| \neq 0$.

Example 18.B.2:
Use Equation (18.B.10) to find a particular solution to the system of differential equations given in Example 18.B.1. In the said example we found that

$$A = \begin{bmatrix} 3 & -1 \\ -4 & 3 \end{bmatrix}$$

with $|A| = 5 \neq 0$ so that A is nonsingular and thus possesses an inverse or

$$A^{-1} = \begin{bmatrix} 3/5 & 1/5 \\ 4/5 & 3/5 \end{bmatrix}.$$

Then from (18.B.10),

$$\bar{y} = A^{-1}b = -\begin{bmatrix} 3/5 & 1/5 \\ 4/5 & 3/5 \end{bmatrix}\begin{bmatrix} 6 \\ 10 \end{bmatrix} = \begin{bmatrix} -28/5 \\ -54/5 \end{bmatrix}.$$

The complete solution to the differential equation system given in the preceding example is thus

$$y(t) = y_h(t) + y_p(t) = y_h(t) + \bar{y}$$

or

$$\begin{bmatrix} y_1(t) \\ y_2(t) \end{bmatrix} = C_1\begin{bmatrix} 1 \\ -2 \end{bmatrix}e^{5t} + C_2\begin{bmatrix} 1 \\ 2 \end{bmatrix}e^{t} - \begin{bmatrix} 28/5 \\ 54/5 \end{bmatrix}.$$

Example 18.B.3:
Determine values for C_1, C_2 such that the solutions to Example 18.B.1 satisfy the initial conditions $y_1(0) = 1$, $y_2(0) = 3$. From the complete solutions

$$y_1(t) = C_1e^{5t} + C_2e^{t} - \frac{28}{5}$$
$$y_2(t) = -2C_1e^{5t} + 2C_2e^{t} - \frac{54}{5}$$

let us evaluate these expressions at $t = 0$. This gives

$$y_1(0) = 1 = C_1 + C_2 - \tfrac{28}{5}$$
$$y_2(0) = 3 = -2C_1 + 2C_2 - \tfrac{54}{5}$$

or

$$C_1 - C_2 = \tfrac{33}{5}$$
$$-2C_1 + 2C_2 = \tfrac{69}{5}.$$

The simultaneous solution to this system is $C_1 = -\frac{3}{20}$, $C_2 = -\frac{27}{4}$. Hence the complete solution satisfying the initial conditions is

$$y_1(t) = -\frac{3}{20}e^{5t} + \frac{27}{4}e^{t} - \frac{28}{5}$$

$$y_2(t) = \frac{3}{10}e^{5t} + \frac{27}{2}e^{t} - \frac{54}{5}.$$

■

APPENDIX 18.C QUALITATIVE ANALYSIS OF DIFFERENTIAL EQUATIONS: PHASE DIAGRAMS AND STABILITY

18.C.1 EQUILIBRIUM POINTS AND STABILITY FOR A SINGLE AUTONOMOUS DIFFERENTIAL EQUATION

We previously determined the time path for a variable y by finding a *quantitative solution* to its associated autonomous differential equation $dy/dt = \dot{y} = F(y)$, where $F(y)$ may be a linear or nonlinear expression. Quite often, though, one needs to resort to an examination of the *qualitative properties* of the time path of y, i.e., those properties which indicate if y converges asymptotically to a stable equilibrium point. This can be accomplished via an analysis of a **phase diagram** - - a plot of $\dot{y}$ vs y (Figure 18.C.1a), where the actual graph of $F(y)$ is termed the **phase line.**

What sort of information does the phase line convey? Specifically:

a. any point $(y, \dot{y})$ on the phase line above the y-axis means that $\dot{y} > 0$ so that y must be increasing over time as we move from *left to right* (Figure 18.C.1b). Similarly, a point $(y, \dot{y})$ on the phase line and below the y-axis indicates that $\dot{y} < 0$ and thus y decreases over time as we move from *right to left.* (These *directional propensities* are illustrated by the so-called *arrowheads of motion.*)

Does the differential equation $\dot{y} = F(y)$ have an equilibrium or stationary state? Moreover, if any such state exists, will it be stable? To answer these questions, we look to

b. An **equilibrium level of y,** if it exists, occurs on the y-axis where $\dot{y} = 0$ or where the phase line intersects the y-axis (e.g., points y_A, y_B in Figure 18.C.1). At an equilibrium point, since $\dot{y} = 0$, y must be stationary (neither increasing nor decreasing) there. (It must be remembered that not all points where a phase line intersects the y-axis are equilibrium points.)

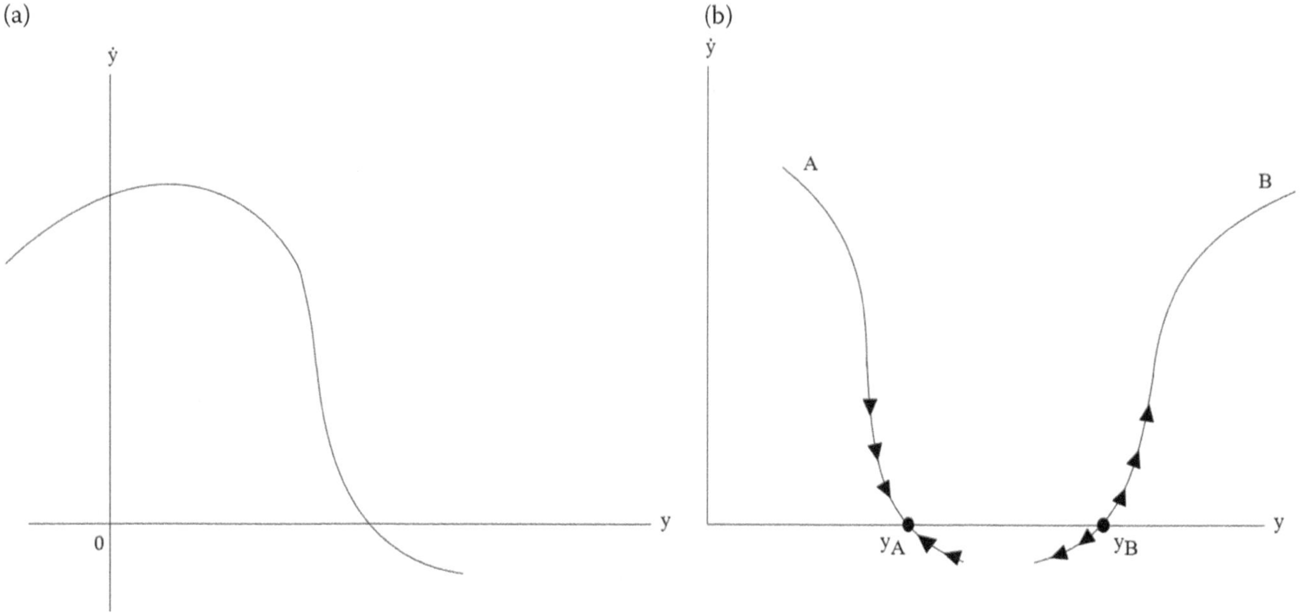

FIGURE 18.C.1 (a) a generic phase diagram; (b) asymptotically stable and unstable equilibria at y_A and y_B, respectively.

Example 18.C.1:
Suppose we have an autonomous differential equation of the form $\dot{y} = F(y) = y - y^2$. Without obtaining a quantitative solution to this differential equation, let us explore the time path of y by considering its qualitative behavior.

From $\dot{y} = 0$ we obtain the equilibrium values $y = 0$ and $y = 1$. And with $d\dot{y}/dy = 1 - 2y = 0$ or $y = \frac{1}{2}$, we see that F has a strong local maximum at $y = \frac{1}{2}$ since $d^2\dot{y}/dy^2 = -2 < 0$ there (Figure 18.C.2). A glance at this figure reveals that: (i) for $y = 0$ or 1, y is stationary since, with $\dot{y} = 0$, y remains at these values; (ii) for $0 < y < \frac{1}{2}$, $\dot{y} > 0$ so that y increases over time; for $y < 0$, $\dot{y} < 0$ and thus y forever decreases over time; (iii) for $y > \frac{1}{2}$, y increases towards 1. Clearly y will converge asymptotically to 1 if it starts at a positive level; and it will diverge to $-\infty$ if it starts at a negative level. ■

Example 18.C.2:
What qualitative information can be obtained from the differential equation

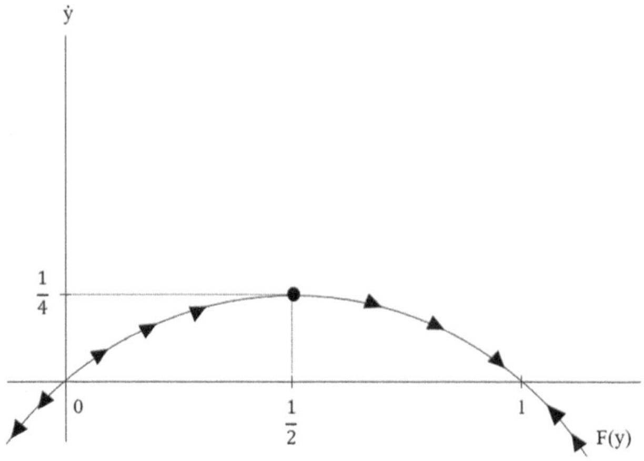

FIGURE 18.C.2 Phase line for $\dot{y} = F(y)$.

$$\frac{dy}{dt} + ay = b \text{ or } \dot{y} = -ay + b$$

without examining its phase diagram? Since $\dot{y} = F(y)$ has a constant slope equal to $-a$, a moment's reflection reveals that:

i. if $a > 0$, y *converges asymptotically* to equilibrium;

ii. if $a < 0$, y *diverges* from equilibrium.

Is this conclusion consistent with the quantitative solution to this differential equation? If $\dot{y} = -ay + b$ is solved, given its initial condition $y(0)$ and its steady-state value of b/a, we have

$$y(t) = (y(0) - b/a)e^{-at} + b/a.$$

If we focus on the term e^{-at}, we see that, starting from a non-equilibrium level of y, we have asymptotic convergence to the equilibrium point b/a since $\lim_{t\to\infty} e^{-at} = 0$ when $a > 0$; if $a < 0$ then $\lim_{t\to\infty} e^{-at} = +\infty$ and thus y diverges away from any equilibrium point. ■

These two examples have revealed that it is the slope of the phase line of a differential equation at its intersection with the y-axis which dictates whether or not the time path of y converges asymptotically to a *stable* equilibrium point. A finite negative (positive) slope indicates dynamic or asymptotic stability (instability).

These observations may be summarized as

Theorem 18.C.1:
(**Asymptotic Stability Condition**). An equilibrium point (occurring at a steady-state value $\bar{y}$ where $\dot{y} = 0$) of a linear or nonlinear first-order autonomous differential equation $\dot{y} = F(y)$ is asymptotically stable (unstable) if $F'(\bar{y}) < 0(>0)$. And if $F(\bar{y}) = F'(\bar{y}) = 0$, the test is inconclusive.

For instance, if $\dot{y} = F(y) = 2y^2 - y$, the equilibrium or steady-state points are, from $\dot{y} = 0$ or $y(2y - 1) = 0$, $y = 0$ and $y = \frac{1}{2}$. Since

$$\frac{d\dot{y}}{dy} = 4y - 1 = \begin{cases} -1, & y = 0; \\ 1, & y = \frac{1}{2}, \end{cases}$$

we see that, by virtue of the preceding theorem, $y = 0$ is an asymptotically stable equilibrium point while $y = \frac{1}{2}$ is an unstable equilibrium point.

18.C.2 EQUILIBRIUM POINTS AND STABILITY FOR AUTONOMOUS LINEAR SYSTEMS

We previously represented an autonomous linear system of two differential equations as (18.B.1) or, with a slight change in notation,

$$\begin{aligned} &\text{(a)} \quad \dot{y}_1 = a_{11}y_1 + a_{12}y_2 + b_1 \\ &\text{(b)} \quad \dot{y}_2 = a_{21}y_1 + a_{22}y_2 + b_2. \end{aligned} \tag{18.C.1}$$

In matrix form, (20.C.1) becomes

$$\begin{bmatrix} \dot{y}_1 \\ \dot{y}_2 \end{bmatrix} = \begin{bmatrix} a_{11} & a_{12} \\ a_{21} & a_{22} \end{bmatrix} \begin{bmatrix} y_1 \\ y_2 \end{bmatrix} + \begin{bmatrix} b_1 \\ b_2 \end{bmatrix} \text{ or } \dot{\boldsymbol{y}} = \boldsymbol{A}\boldsymbol{y} + \boldsymbol{b} \tag{18.C.2}$$

(see (18.B.3) for details).

To determine the equilibrium points for (18.C.1), set $\dot{y}_1 = \dot{y}_2 = 0$ so as to obtain

$$\boldsymbol{A}\boldsymbol{y} = -\boldsymbol{b}. \tag{18.C.3}$$

If $|\boldsymbol{A}| \neq 0$, then (18.C.3) has the unique solution

$$\bar{\boldsymbol{y}} = \begin{bmatrix} \bar{y}_1 \\ \bar{y}_2 \end{bmatrix} = -\boldsymbol{A}^{-1}\boldsymbol{b} \tag{18.C.4}$$

(see also (18.B.10)). To characterize the asymptotic stability of a steady-state or equilibrium combination $(\bar{y}_1, \bar{y}_2)$ we have

Theorem 18.C.2:
(**Asymptotic Stability Condition for Autonomous Linear Systems**). Given (18.C.3) with $|\boldsymbol{A}| \neq 0$, an equilibrium point $(\bar{y}_1, \bar{y}_2)$ for this system is **asymptotically stable** if and only if $tr(\boldsymbol{A}) = a_{11} + a_{22} < 0$ and $|\boldsymbol{A}| > 0$. Equivalently, the said equilibrium point is asymptotically stable if *both* characteristic values or eigenvalues of $\boldsymbol{A}$ have distinct real parts.

To rationalize this result let us note that the solutions to a system of two autonomous linear differential equations contain terms such as $e^{\lambda_1 t}$, $e^{\lambda_2 t}$, $\bar{y}_1$, and $\bar{y}_2$. If λ_1 and λ_2 are real and negative, then $y_1(t)$ converges to $\bar{y}_1$ while $y_2(t)$ converges to $\bar{y}_2$since

$$\lim_{t\to\infty} e^{\lambda_1 t} = 0, \quad \lim_{t\to\infty} e^{\lambda_2 t} = 0.$$

Clearly $y_1(t)$ and $y_2(t)$ converge asymptotically to their steady-state solutions $\bar{y}_1$ and $\bar{y}_2$, respectively, no matter what the initial conditions are. Hence the equilibrium or steady-state solution is *asymptotically stable if both*λ_1, λ_2*are real and negative.*

But this is not the whole story - - other possibilities abound. For instance, if both λ_1 and λ_2 are real and positive, then

$$\lim_{t\to\infty} e^{\lambda_1 t} = +\infty, \quad \lim_{t\to\infty} e^{\lambda_2 t} = +\infty.$$

So provided that C_1 and C_2 are not zero, both solutions $y_1(t)$ and $y_2(t)$ diverge from equilibrium or from the steady state. Thus the equilibrium solutions are *unstable when both* λ_1 *and* λ_2 *are real and positive.*

Next, suppose one eigenvalue (say, λ_1) is positive and the other (λ_2) is negative. Then

$$\lim_{t\to\infty} e^{\lambda_1 t} = +\infty, \quad \lim_{t\to\infty} e^{\lambda_2 t} = 0.$$

So with the characteristic roots real and of opposite signs (as evidenced by $|\boldsymbol{A}| < 0$), there is a unique equilibrium or steady-state point which is a **saddle point**. Thus both solutions diverge from equilibrium save for the case where the constant of integration associated with $e^{\lambda_1 t}$ is zero; the steady state is unstable when only one of the eigenvalues is positive.

Our final case (we are ignoring the instance where the characteristic equation associated with $\boldsymbol{A}$ has imaginary roots) merits our attention - - the case of *real and equal eigenvalues.* Since equal eigenvalues are either both positive or both negative, in the former situation, with $\lambda_1 = \lambda_2 = \lambda > 0$,

$$\lim_{t\to\infty} e^{\lambda t} = +\infty;$$

in the latter instance, with $\lambda < 0$,

$$\lim_{t\to\infty} e^{\lambda t} = 0.$$

Clearly either both solutions converge to a stable equilibrium position or both diverge from an unstable equilibrium point.

Example 18.C.3:
Find an equilibrium point for the linear autonomous system

$$\begin{aligned}\dot{y}_1 &= -2y_1 + y_2 + 1\\ \dot{y}_2 &= -2y_2 + 6.\end{aligned}$$

If an equilibrium solution exists, determine if it is stable. Here

$$\boldsymbol{A} = \begin{bmatrix} -2 & 1 \\ 0 & -2 \end{bmatrix}, \quad |\boldsymbol{A}| = 4 > 0, \quad tr(\boldsymbol{A}) = -4, \quad \text{and}$$

$$\bar{y} = -\boldsymbol{A}^{-1}\boldsymbol{b} = -\begin{bmatrix} -1/2 & -1/4 \\ 0 & -1/2 \end{bmatrix}\begin{bmatrix} 1 \\ 6 \end{bmatrix} = \begin{bmatrix} 2 \\ 3 \end{bmatrix}.$$

Hence $\bar{\boldsymbol{y}}^T = (2, 3)$ is an equilibrium point and, by virtue of Theorem 18.C.2, the said point is asymptotically stable in that all solutions converge to the equilibrium point (2, 3) as $t \to +\infty$. ■

As was the case for single autonomous linear differential equations, we can garner significant *qualitative* information about the time paths of $y_1(t)$ and $y_2(t)$ by examining a **phase diagram** in a **phase plane.** Let us rewrite the autonomous system of differential equations (linear or otherwise) as

$$\begin{aligned}\dot{y}_1 &= f(y_1, y_2)\\ \dot{y}_2 &= g(y_1, y_2),\end{aligned} \tag{18.C.5}$$

where f and g are of class $C^{(1)}$. A solution $\boldsymbol{y}^T = (y_1(t), y_2(t))$ of (18.C.5) represents a **trajectory path** in the y_1, y_2-plane consisting of all points $P = \{(y_1(t), y_2(t)) | t_1 \le t \le t_2\}$, with $(\dot{y}_1, \dot{y}_2)$ uniquely determined at each point $(y_1, y_2) \in P$. If we set $\dot{y}_1 = \dot{y}_2 = 0$, then a point $(\bar{y}_1, \bar{y}_2)$ where $f(\bar{y}_1, \bar{y}_2) = g(\bar{y}_1, \bar{y}_2) = 0$ is termed an equilibrium or steady-state point for (18.C.5). Clearly (y_1, y_2) must be a point of intersection between the curves $f(y_1, y_2) = 0$, $g(y_1, y_2) = 0$. In what follows these curves will be termed **isoclines** (or **nullclines**).

Suppose we have the simultaneous system

$$\begin{aligned}&\text{(a)}\quad \dot{y}_1 = y_2\\ &\text{(b)}\quad \dot{y}_2 = 2y_1 + y_2.\end{aligned} \tag{18.C.6}$$

Our objective is to draw a phase diagram for this system. Setting $\dot{y}_1 = \dot{y}_2 = 0$ we have $f = y_1 = 0$ and $g = 2y_1 + y_2 = 0$ (Figure 18.C.3). It is readily shown that the intersection point between $f = 0$ and $g = 0$ is the origin (0, 0) and represents an equilibrium point. At each point on $f(y_1, y_2) = 0$ we have $\dot{y}_1 = 0$ and thus the **velocity vector** $\dot{\boldsymbol{y}}^T = (0, \dot{y}_2)$ is vertical (its direction is up (down) if $\dot{y}_2 > 0(<0)$). Additionally, at each point on $g(y_1, y_2) = 0$ we have $\dot{y}_2 = 0$ and consequently the velocity vector $\dot{\boldsymbol{y}}^T = (\dot{y}_1, 0)$ is horizontal (its direction is left (right) if $\dot{y}_1 < 0(>0)$).

The isocline functions $f(y_1, y_2) = 0$ and $g(y_1, y_2) = 0$ divide the phase plane into four **regions of motion**. That is, for equation (18.C.6a):

$$\begin{aligned}
&\text{(I)} \quad y_2 > 0 \text{ so that } \dot{y}_1 > 0;\\
&\text{(II)} \quad y_2 > 0 \text{ so that } \dot{y}_1 > 0;\\
&\text{(III)} \quad y_2 < 0 \text{ so that } \dot{y}_1 < 0;\\
&\text{(IV)} \quad y_2 < 0 \text{ so that } \dot{y}_1 < 0.
\end{aligned}$$

And for (18.C.6b)

$$\begin{aligned}
&\text{(I)} \quad 2y_1 + y_2 > 0 \text{ so that } \dot{y}_2 > 0;\\
&\text{(II)} \quad 2y_1 + y_2 < 0 \text{ so that } \dot{y}_2 < 0;\\
&\text{(III)} \quad 2y_1 + y_2 < 0 \text{ so that } \dot{y}_2 < 0;\\
&\text{(IV)} \quad 2y_1 + y_2 > 0 \text{ so that } \dot{y}_2 > 0.
\end{aligned}$$

Given this discussion, the implied *directions of motion* at points on time paths within the phase plane are also provided in Figure 18.C.3.

Example 18.C.4:
Let us examine the phase diagram for the autonomous linear system

$$\begin{aligned}
&\text{(a)} \quad \dot{y}_1 = y_2 - 2\\
&\text{(b)} \quad \dot{y}_2 = \tfrac{1}{2}y_1 - 1.
\end{aligned} \tag{18.C.7}$$

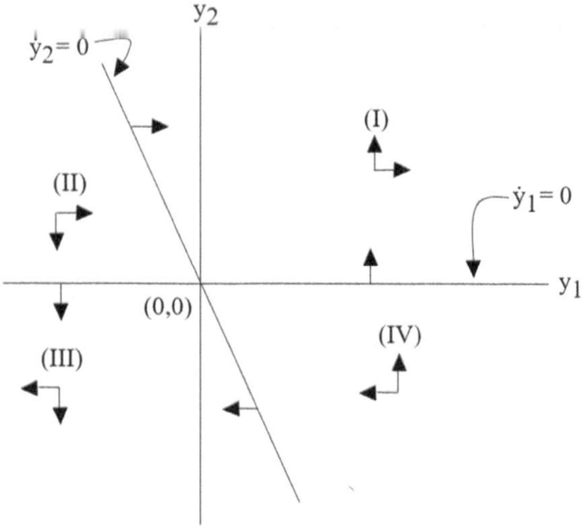

FIGURE 18.C.3 Isoclines and regions of motion.

Here

$$|A| = \begin{vmatrix} 0 & 1 \\ \frac{1}{2} & 0 \end{vmatrix} = -\frac{1}{2} < 0,$$

which indicates the existence of an equilibrium point at (2, 2) which is a saddle point. Alternatively, from the characteristic equation

$$|A - \lambda I_2| = \begin{vmatrix} 0 - \lambda & 1 \\ \frac{1}{2} & 0 - \lambda \end{vmatrix} = \lambda^2 - \frac{1}{2} = 0$$

we can solve for the characteristic roots or eigenvalues which are $\lambda_1 = -1/\sqrt{2}$, $\lambda_2 = 1/\sqrt{2}$. We thus reach the same conclusion as before – since the eigenvalues are of opposite signs, the steady-state equilibrium point is a saddle point.

Next, let us examine the phase diagram for (18.C.7). Setting $\dot{y}_1 = \dot{y}_2 = 0$ we have the isolines $f = y_2 - 2 = 0$ and $g = \frac{1}{2}y_1 - 1 = 0$ (Figure 18.C.4). So for the isocline $f = y_2 - 2$:

(I) $y_2 > 2$ so that $\dot{y}_1 > 0$;
(II) $y_2 > 2$ so that $\dot{y}_1 > 0$;
(III) $y_2 < 2$ so that $\dot{y}_1 < 0$;
(IV) $y_2 < 2$ so that $\dot{y}_1 < 0$.

And for the isocline $g = \frac{1}{2}y_1 - 1 = 0$:

(I) $y_1 > 2$ so that $\dot{y}_2 > 0$;
(II) $y_1 < 2$ so that $\dot{y}_2 < 0$;
(III) $y_1 < 2$ so that $\dot{y}_2 < 0$;
(IV) $y_1 > 2$ so that $\dot{y}_2 > 0$.

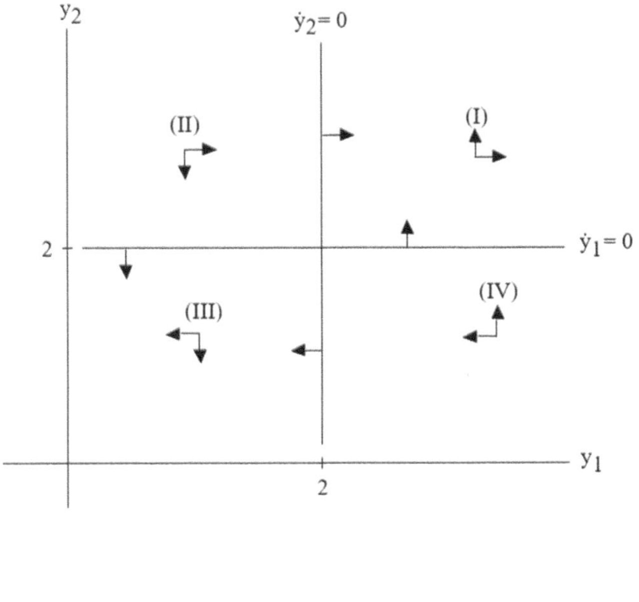

FIGURE 18.C.4 Isoclines and regions of motion.

A glance at Figure 18.C.4 reveals that the arrows or directions of motion in regions (I) and (III) indicate that the y_1 and y_2 time paths move away from the steady-state equilibrium; while the directions of motion in regions (II) and (IV) demonstrate a movement towards said equilibrium.

As just indicated, the arrows of motion provide an indication of the direction of the phase plane paths or trajectories. To plot some exact trajectories in the phase plane requires that we actually solve system (18.C.7). However, in the absence of the solutions we can offer an adequate (albeit "rough") plot of a typical trajectory in the presence of a saddle-point equilibrium (Figure 18.C.5). These trajectories reveal, by virtue of the directions of the arrows of motion, that there is a divergence from the steady-state saddle-point equilibrium. In fact, *most* trajectories for a saddle-point equilibrium will diverge. However, some may converge to equilibrium if the initial conditions $y_1(0)$, $y_2(0)$ satisfy the equation

$$y_2 = \left(\frac{\lambda_1 - a_{11}}{a_{12}}\right)(y_1 - \bar{y}_1) + \bar{y}_2, \tag{18.C.8}$$

where $\lambda_1 < 0$ and $\lambda_2 > 0$. The locus of points (y_1, y_2) specified by this equation is termed a **saddle path**. For this problem the saddle path appears as $y_2 = 3.414 - \frac{1}{\sqrt{2}}y_1$ (see Figure 18.C.5). If the initial conditions $y_1(0)$, $y_2(0)$ lie on the saddle path, then $y_1(t)$, $y_2(t)$ converge to the saddle point. ■

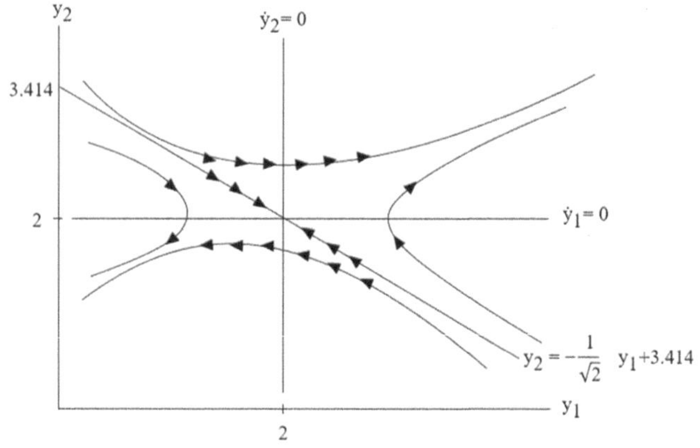

FIGURE 18.C.5 Saddle point trajectories and the saddle path.

NOTES

1 For a definite integral, **integration by parts** renders

$$\int_{t=\alpha}^{t=\beta} y(t)\dot{z}(t)\,dt = \int_{\alpha}^{\beta} y(t)\,dz(t)$$

$$= y(t)z(t)]_{\alpha}^{\beta} - \int_{\alpha}^{\beta} z(t)\dot{y}(t)\,dt. \qquad (*)$$

Hence, following (*),

$$-\int_{o}^{T} \lambda(t)\dot{x}(t)\,dt = -\int_{o}^{T} \lambda(t)\,dx$$

$$= -\left[\int_{o}^{T} \lambda(t)x(t) - \int_{o}^{T} x(t)\dot{\lambda}(t)\,dt\right]$$

$$= -\left[\lambda(t)x(t) - \lambda(0)x(0) - \int_{o}^{T} x(t)\dot{\lambda}(t)\,dt\right].$$

2 If the region of admissible u values $\mathcal{U}$ is a closed set, then (18.7a) still holds if $\hat{u}(t) \in \text{int}(\mathcal{U})$; but if $\hat{u}(t)$ occurs on the boundary of $\mathcal{U}$, then boundary conditions may be imposed and (18.7a) must be replaced by $H(t, x, \hat{u}, \lambda) \geq H(t, x, u, \lambda)$ for all $t \in [0, T]$.

3 It was determined in Example 18.1 that $\hat{x}(t) = -\frac{3}{2}t^2 + 7t - 8$ and $\hat{u}(t) = \hat{\lambda}(t) = -3t + 6$. If these optimal time paths for x and u are substituted into (18.9) we have

$$V = \int_{o}^{2} \left(3\hat{x} - \frac{1}{2}\hat{u}^2\right) dt.$$

Evaluating this expression yields $V = -30$. And if $\hat{x}(t)$ and $\hat{u}(t) = \hat{\lambda}(t)$ are substituted into the Lagrangian (18.5.1) we have

$$L = \int_{o}^{T} \left[\left(3\hat{x} - \frac{1}{2}\hat{u}^2 + \hat{\lambda}(\hat{u} + 1) - 3\hat{x}\right)\right] dt + \hat{\lambda}(0)\hat{x}(0),$$

with $\lambda(T) = 0$ since $x(T)$ was free. Simplifying this equation gives $L = -30(=V)$.

4 To evaluate $\partial V/\partial x_o$ we need to consider a rule for differentiating V with respect to x_o under the integral sign. In general, suppose a function $F(x)$ is defined by means of the integral

$$F(x) = \int_{\alpha(x)}^{\beta(x)} f(x, t)\,dt,$$

where $f(x, t)$ and f_x are defined and continuous for $x \in [a, b]$ and $t \in [c, d]$, and $\alpha(x)$ and $\beta(x)$ are of class $C^{(n)}$ for $x \in (a, b)$. Then for $x \in (a, b)$,

$$F'(x) = f(x, \beta(x))\beta'(x) - f(\alpha, a(x))\alpha'(x) + \int_{\alpha(x)}^{\beta(x)} f_x\,dt. \qquad (*)$$

[Leibnitz'Rule]

So if x changes, then $\alpha(x)$ and $\beta(x)$ concomitantly change along with $f(x, t)$ for each t. Then (*) provides us with the total change in F from changing x.If $\alpha(x) = \alpha =$ constant and $\beta(x) = \beta =$ constant, and $f(x, t)$ and f_x are defined and continuous for $x \in [a, b]$, $t \in [c, d]$, and F is defined by

$$F(x) = \int_{\alpha}^{\beta} f(x, t)\,dt,$$

then $F'(x)$ exists on (a, b) and is given by

$$F'(x) = \int_{\alpha}^{\beta} f_x(x, t)\,dt. \qquad (**)$$

Comparative Statics Revisited

19.1 INTRODUCTION

We previously considered the optimization of a function (unconstrained or otherwise) when that function included certain *parameters* in addition to its arguments. For example, these parameters may be the commodity prices and income level faced by a household or the prices of a firm's inputs or product. In many applications, it is important to determine what happens to the optimal levels of the arguments of a function when an individual parameter changes. This type of analysis is known as **comparative statics**. In fact, the reader may remember that we actually engaged in a comparative statics analysis earlier when we considered problems such as utility maximization, profit maximization, and cost minimization. This was accomplished via the application of a variety of *envelope theorems* (e.g., by presupposing the existence of a utility or production function along with an appropriate behavioral assumption such as utility maximization or cost minimization, notions such as Shephard's lemma, Roy's identity, and Hotelling's lemma, to name but a few, emerged as envelope results).

Traditional comparative statistics results are concerned with the effect of an infinitesimal shift in a parameter. While such results may be a bit more complicated than those obtained by applying envelope theorems, they are no less important.

19.2 THE FUNDAMENTAL EQUATION OF COMPARATIVE STATICS

Let f and $\hat{g}^j$, $j = 1, \ldots, m$, be real-valued functions of class $C^{(n)}$ on R^n. The problem is to choose an $x \in R^n$ to

$$\begin{aligned} &\text{maximize } f(\boldsymbol{x}; \hat{\boldsymbol{\alpha}}) \text{ subject to} \\ &\hat{g}^j(\boldsymbol{x}; \hat{\boldsymbol{\alpha}}) \geq 0, \ j = 1, \ldots, m, \ \boldsymbol{x} \geq \boldsymbol{O}, \end{aligned} \tag{19.1}$$

where $\hat{\boldsymbol{\alpha}} = (\hat{\alpha}_1, \ldots, \hat{\alpha}_l)^T$ is an $(l \times 1)$ vector of real parameters. The Lagrangian associated with this problem is

$$L(\boldsymbol{x}, \boldsymbol{\lambda}; \hat{\boldsymbol{\alpha}}) = f(\boldsymbol{x}; \hat{\boldsymbol{\alpha}}) + \boldsymbol{\lambda}^T \boldsymbol{G}(\boldsymbol{x}; \hat{\boldsymbol{\alpha}}), \tag{19.2}$$

where $\boldsymbol{\lambda} = (\lambda_1, \ldots, \lambda_m)^T$ is an $(m \times 1)$ vector of Lagrange multipliers and

$$\boldsymbol{G}(\boldsymbol{x}; \hat{\boldsymbol{\alpha}}) = \begin{bmatrix} \hat{g}^1(\boldsymbol{x}; \hat{\boldsymbol{\alpha}}) \\ \vdots \\ \hat{g}^m(\boldsymbol{x}; \hat{\boldsymbol{\alpha}}) \end{bmatrix}$$

is an $(m \times 1)$ vector of constraint functions. In what follows we shall assume, for the sake of convenience, that, at optimality, $x_i > 0$, $i = 1, \ldots, n$ (the optimal solution occurs at an interior point of the region of feasible solutions $\mathcal{K} = \{\boldsymbol{x} | \boldsymbol{G}(\boldsymbol{x}; \hat{\boldsymbol{\alpha}}) \geq \boldsymbol{O}, \boldsymbol{x} \geq \boldsymbol{O}\} \subset \mathcal{D}$, where $\mathcal{D}$ is an open set), and $\lambda_j > 0$, $j = 1, \ldots, m$ (all $\hat{g}^j$, $j = 1, \ldots, m$, constraints are binding at optimality).

If we invoke the generalized implicit function theorem, then there exist functions $\boldsymbol{x} = \boldsymbol{x}(\boldsymbol{\alpha})$ and $\boldsymbol{\lambda} = \boldsymbol{\lambda}(\boldsymbol{\alpha})$, each of class $C^{(n)}$, and thus the first-order conditions become

$$\begin{aligned} &\text{(a)} \quad \nabla_x L(\boldsymbol{x}(\boldsymbol{\alpha}), \boldsymbol{\lambda}(\boldsymbol{\alpha}); \boldsymbol{\alpha}) = \boldsymbol{O} \\ &\text{(b)} \quad \nabla_\lambda L(\boldsymbol{x}(\boldsymbol{\alpha}), \boldsymbol{\lambda}(\boldsymbol{\alpha}); \boldsymbol{\alpha}) = \boldsymbol{G}(\boldsymbol{x}(\boldsymbol{\alpha}); \boldsymbol{\alpha}) = \boldsymbol{O} \end{aligned} \tag{19.3}$$

for all $\boldsymbol{\alpha}$ contained within a suitably restricted spherical

δ−neighborhood of $\hat{\boldsymbol{\alpha}}$, $\delta(\hat{\boldsymbol{\alpha}})$. To perform a comparative statics analysis, we must determine the effect of changes in the α_k's, $k = 1, \ldots, l$, on the optimal values of $\boldsymbol{x}(\boldsymbol{\alpha})$ and $\boldsymbol{\lambda}(\boldsymbol{\alpha})$ which satisfy system (19.3). To this end, let us differentiate equations (19.3a, b) with respect to α_k so as to obtain

$$\begin{aligned} \nabla^2_{\boldsymbol{xx}} L \frac{\partial \boldsymbol{x}(\alpha)}{\partial \alpha_k} + \nabla^2_{\boldsymbol{x}\lambda} L \frac{\partial \lambda(\alpha)}{\partial \alpha_k} + \nabla_{\boldsymbol{x}\alpha_k} L &= \boldsymbol{O} \\ \nabla^2_{\lambda \boldsymbol{x}} L \frac{\partial \boldsymbol{x}(\alpha)}{\partial \alpha_k} + \nabla_{\lambda \alpha_k} L &= \boldsymbol{O} \end{aligned} \tag{19.4}$$

or, in matrix terms,

$$\begin{bmatrix} \nabla^2_{xx} L & \nabla^2_{\boldsymbol{x}\lambda} L \\ \nabla^2_{\lambda x} L & \boldsymbol{O} \end{bmatrix} \begin{bmatrix} \frac{\partial \boldsymbol{x}(\alpha)}{\partial \alpha_k} \\ \frac{\partial \lambda(\alpha)}{\partial \alpha_k} \end{bmatrix} + \begin{bmatrix} \nabla_{\boldsymbol{x}\alpha_k} L \\ \nabla_{\lambda \alpha_k} L \end{bmatrix} = \boldsymbol{O}, \tag{19.4.1}$$

where $\nabla^2_{\boldsymbol{x}\lambda} L = \nabla \boldsymbol{G}$ (the transpose of the Jacobian of $\boldsymbol{G}$ or $\partial(\hat{g}^1, \ldots, \hat{g}^m)/\partial(x_1, \ldots, x_{1n})^T$) with $\rho(\nabla \boldsymbol{G}) = m$ and $\nabla^2_{\lambda x} L = \nabla \boldsymbol{G}^T$. It should be evident that $\nabla^2_{\boldsymbol{xx}} L$ is the Hessian matrix ($\boldsymbol{H}_L$) of the Lagrangian function (with all second partial derivatives evaluated at $\boldsymbol{\alpha}$). Then, the coefficient matrix of system (19.4.1) may be written as the bordered Hessian matrix

$$\underset{(n+m)\times(n+m)}{\boldsymbol{H}_L^G} = \begin{bmatrix} \underset{(n\times n)}{\boldsymbol{H}_L} & \underset{(n\times m)}{\nabla \boldsymbol{G}} \\ \underset{(m\times n)}{\nabla \boldsymbol{G}^T} & \underset{(m\times m)}{\boldsymbol{O}} \end{bmatrix}. \tag{19.5}$$

Suppose $|\boldsymbol{H}_L^G| \neq 0$. Then, $[\boldsymbol{H}_L^G]^{-1}$ exists in a δ−neighborhood of $\boldsymbol{\alpha}$ and thus (19.4.1) can be transformed to

$$\underset{(n+m\times 1)}{\begin{bmatrix} \frac{\partial \boldsymbol{x}(\alpha)}{\partial \alpha_k} \\ \frac{\partial \lambda(\alpha)}{\partial \alpha_k} \end{bmatrix}} = -\begin{bmatrix} \boldsymbol{H}_L & \nabla \boldsymbol{G} \\ \nabla \boldsymbol{G}^T & \boldsymbol{O} \end{bmatrix}^{-1} \underset{(n+m\times 1)}{\begin{bmatrix} \nabla_{\boldsymbol{x}\alpha_k} L \\ \nabla_{\lambda \alpha_k} L \end{bmatrix}}, \tag{19.6}$$

[Fundamental Equation of Comparative Statics]

where the ith component of $\nabla_{\boldsymbol{x}\alpha_k} L$ is $\partial^2 L(\boldsymbol{x}(\boldsymbol{\alpha}), \boldsymbol{\lambda}(\boldsymbol{\alpha}); \boldsymbol{\alpha})/\partial x_i \partial \alpha_k$; and the jth component of $\nabla_{\lambda \alpha_k} L$ is $\partial \hat{g}^j(\boldsymbol{x}(\boldsymbol{\alpha}); \boldsymbol{\alpha})/\partial \alpha_k$. With $\boldsymbol{H}_L^G$ symmetric, it follows that $[\boldsymbol{H}_L^G]^{-1}$ is also symmetric. Moreover, this inverse matrix will have negative elements on its main diagonal (see Exercise 19.2.2) if $\boldsymbol{D}_1$ is negative semidefinite and has negative main diagonal components. A formal summary of this discussion is provided by the following stylized theorem.

Non-Singularity Theorem (Carathéodory [1935, 1967]; Samuelson [1983]; Takayama [1985]). Suppose $\boldsymbol{H}_L^G$ is an $(n + m) \times (n + m)$ real symmetric matrix that is partitioned as in (19.5) with $\rho(\nabla \boldsymbol{G}) = m$ and $\boldsymbol{H}_L$ is negative definite subject to $\nabla \boldsymbol{G}^T \boldsymbol{x} = \boldsymbol{O}$. Then, $[\boldsymbol{H}_L^G]^{-1}$ exists and $\boldsymbol{D}_1$ (see Exercise 19.2.2) of this inverse is symmetric and negative semidefinite.

Section 19.2 Exercises:

19.2.1 Suppose $\boldsymbol{H}_L^G$ in (19.5) is rewritten as

$$\boldsymbol{H}_L^G = \begin{bmatrix} \boldsymbol{O} & \nabla \boldsymbol{G} \\ \nabla \boldsymbol{G}^T & \boldsymbol{H}_L \end{bmatrix}.$$

How must (19.6) be rewritten?

19.2.2 Use partitioned matrix inversion to verify that the inverse of (19.5) is

$$\begin{bmatrix} \boldsymbol{H}_L & \nabla \boldsymbol{G} \\ \nabla \boldsymbol{G}^T & \boldsymbol{O} \end{bmatrix}^{-1} = \begin{bmatrix} \underset{(n\times n)}{\boldsymbol{D}_1} & \underset{(n\times m)}{\boldsymbol{D}_2} \\ \underset{(m\times n)}{\boldsymbol{D}_3} & \underset{(m\times m)}{\boldsymbol{D}_4} \end{bmatrix},$$

where:

$$\boldsymbol{D}_1 = \boldsymbol{H}_L^{-1} + \boldsymbol{H}_L^{-1}\ \nabla \boldsymbol{G} \boldsymbol{A}^{-1} \nabla \boldsymbol{G}^T\ \boldsymbol{H}_L^{-1};$$

$$\boldsymbol{D}_2 = -\boldsymbol{H}_L^{-1}\ \nabla \boldsymbol{G}\ \boldsymbol{A}^{-1};$$

$$\boldsymbol{D}_3 = -\boldsymbol{A}^{-1} \nabla \boldsymbol{G}^T \boldsymbol{H}_L^{-1};$$

$$\boldsymbol{D}_4 = \boldsymbol{A}^{-1};$$

and

$$\boldsymbol{A} = -\nabla \boldsymbol{G}^T \boldsymbol{H}_L^{-1}\ \nabla \boldsymbol{G}.$$

19.3 ECONOMIC APPLICATIONS

19.3.1 Constrained Utility Maximization (McKenzie [1957]; Takayama [1977, 1985]; and Silberberg and Suen [2001])

Our problem is to choose a commodity bundle $\boldsymbol{q} \in R^n$ so as to maximize a strictly quasiconcave utility function subject to a budget constraint or

$$\begin{aligned}&\text{maximize } u = f(\boldsymbol{q}) \text{ subject to}\\ &\hat{g}(\boldsymbol{q}) = \hat{M} - \hat{\boldsymbol{p}}^T\boldsymbol{q} = 0,\end{aligned} \tag{19.7}$$

where f and $\hat{g}$ are functions of class $C^{(n)}$ and $\hat{\boldsymbol{p}} > \boldsymbol{O}$ and $\hat{M} > 0$ are parameters. Let $\nabla\hat{g}^T = (-p_1,\dots,-p_n) \neq \boldsymbol{O}$ and $\rho(\nabla\hat{g}) = 1$ for all $\boldsymbol{q}$. The Lagrangian associated with (19.7) is

$$L(\boldsymbol{q}, \lambda; \hat{\boldsymbol{p}}, \hat{M}) = f(\boldsymbol{q}) + \lambda(\hat{M} - \hat{\boldsymbol{p}}^T\boldsymbol{q}), \tag{19.8}$$

where λ is the undetermined Lagrange multiplier.

Given (19.8), the first-order conditions at the optimal point $(\boldsymbol{q}_o, \lambda_o)$ are:

$$\left.\begin{aligned}\nabla_{\boldsymbol{q}} L = \boldsymbol{O}\\ \nabla_{\lambda} L = O\end{aligned}\right\} \text{ or } \left\{\begin{aligned}&f_i^o - \lambda_o\hat{p}_i = 0,\ i = 1,\dots,n;\\ &\hat{M} - \hat{\boldsymbol{p}}^T\boldsymbol{q}_o = 0.\end{aligned}\right. \tag{19.9}$$

Additionally, the bordered Hessian matrix for this problem is

$$\boldsymbol{H}_f^{\hat{g}} = \begin{bmatrix}\boldsymbol{H}_f & \nabla\hat{g}\\ \nabla\hat{g}^T & 0\end{bmatrix} = \begin{bmatrix}\boldsymbol{H}_f & -\boldsymbol{p}\\ -\boldsymbol{p}^T & 0\end{bmatrix} = \begin{bmatrix}f_{11} & f_{12} & \cdots & f_{1n} & -p_1\\ f_{21} & f_{22} & \cdots & f_{2n} & -p_2\\ \vdots & \vdots & & \vdots & \vdots\\ f_{n1} & f_{n2} & \cdots & f_{nn} & -p_n\\ -p_1 & -p_2 & \cdots & -p_n & 0\end{bmatrix}, \tag{19.10}$$

where $\boldsymbol{H}_f$ is the Hessian matrix of f and all second-order partial derivatives are evaluated at $\boldsymbol{q}_o$. It is also assumed that $|\boldsymbol{H}_f^{\hat{g}}| \neq 0$.

Applying the implicit function theorem to system (19.9) renders the $C^{(n)}$ functions $q(\boldsymbol{p}, M)$ and $\lambda(\boldsymbol{p}, M)$ with $\boldsymbol{q}_o = \boldsymbol{q}(\hat{\boldsymbol{p}}, \hat{M})$, $\lambda_o = \lambda(\hat{\boldsymbol{p}}, \hat{M})$, and

$$\begin{aligned}\nabla_{\boldsymbol{q}} f(\boldsymbol{q}(\boldsymbol{p}, M)) - \lambda(\boldsymbol{p}, M) = \boldsymbol{O}\\ M - \boldsymbol{p}^T\boldsymbol{q}(\boldsymbol{p}, M) = 0\end{aligned} \tag{19.9.1}$$

for all $(\boldsymbol{p}, M)$ within a δ−neighborhood of $(\hat{\boldsymbol{p}}, \hat{M})$, $\delta(\hat{\boldsymbol{p}}, \hat{M})$. Partially differentiating (19.9.1) with respect to p_j gives

$$\begin{aligned}\nabla_{\boldsymbol{qq}}^2 f\frac{\partial\boldsymbol{q}}{\partial p_j} - \frac{\partial\lambda}{\partial p_j}\boldsymbol{p} - \lambda\boldsymbol{e}_j = \boldsymbol{O}\\ -\boldsymbol{p}^T\frac{\partial\boldsymbol{q}}{\partial p_j} - q_j = 0\end{aligned} \tag{19.11}$$

or, in matrix terms,

$$\begin{bmatrix}\nabla_{\boldsymbol{qq}}^2 f & -\boldsymbol{p}\\ -\boldsymbol{p}^T & 0\end{bmatrix}\begin{bmatrix}\dfrac{\partial\boldsymbol{q}}{\partial p_j}\\ \dfrac{\partial\lambda}{\partial p_j}\end{bmatrix} = \lambda\begin{bmatrix}\boldsymbol{e}_j\\ 0\end{bmatrix} + q_j\boldsymbol{e}_{j+1}. \tag{19.12}$$

Next, partially differentiating (19.9.1) with respect to M yields

$$\begin{aligned}\nabla_{\boldsymbol{qq}}^2 f\frac{\partial\boldsymbol{q}}{\partial M} - \frac{\partial\lambda}{\partial M}\boldsymbol{p} = \boldsymbol{O}\\ 1 - \boldsymbol{p}^T\frac{\partial\boldsymbol{q}}{\partial M} = 0\end{aligned} \tag{19.13}$$

or

$$\begin{bmatrix}\nabla_{\boldsymbol{qq}}^2 f & -\boldsymbol{p}\\ -\boldsymbol{p}^T & 0\end{bmatrix}\begin{bmatrix}\dfrac{\partial\boldsymbol{q}}{\partial M}\\ \dfrac{\partial\lambda}{\partial M}\end{bmatrix} = -\boldsymbol{e}_{j+1}. \tag{19.14}$$

Given that $[\boldsymbol{H}_f^{\hat{g}}]^{-1}$ exists, we can transform (19.12) and (19.14) to

$$\begin{bmatrix}\dfrac{\partial\boldsymbol{q}}{\partial p_j}\\ \dfrac{\partial\lambda}{\partial p_j}\end{bmatrix} = \lambda\begin{bmatrix}\boldsymbol{H}_f^{\hat{g}}\end{bmatrix}^{-1}\begin{bmatrix}\boldsymbol{e}_j\\ o\end{bmatrix} + q_j\begin{bmatrix}\boldsymbol{H}_f^{\hat{g}}\end{bmatrix}^{-1}\boldsymbol{e}_{j+1} \tag{19.15}$$

$$\begin{bmatrix}\dfrac{\partial\boldsymbol{q}}{\partial M}\\ \dfrac{\partial\lambda}{\partial M}\end{bmatrix} = -\begin{bmatrix}\boldsymbol{H}_f^{\hat{g}}\end{bmatrix}^{-1}\boldsymbol{e}_{j+1}, \tag{19.16}$$

respectively, for all $(\boldsymbol{p}, M)$ within $\delta(\hat{\boldsymbol{p}}, \hat{M})$. Upon substituting (19.16) into (19.15) we obtain

$$\begin{bmatrix}\dfrac{\partial\boldsymbol{q}}{\partial p_j}\\ \dfrac{\partial\lambda}{\partial p_j}\end{bmatrix} = \lambda\begin{bmatrix}\boldsymbol{H}_f^{\hat{g}}\end{bmatrix}^{-1}\begin{bmatrix}\boldsymbol{e}_j\\ o\end{bmatrix} - q_j\begin{bmatrix}\dfrac{\partial\boldsymbol{q}}{\partial M}\\ \dfrac{\partial\lambda}{\partial M}\end{bmatrix}. \tag{19.17}$$

Considering the first n components of (19.17) we get

$$\frac{\partial q_i}{\partial p_j} = \lambda \left[H_f^{\hat{g}}\right]^{-1} e_j - q_j \frac{\partial q_i}{\partial M}, \; i, j = 1, \ldots, n. \quad (19.18)$$

Let us write

$$\left[H_f^{\hat{g}}\right]^{-1} = \frac{[H_f^{\hat{g}}]^{+}}{|H_f^{\hat{g}}|} = \frac{1}{|H_f^{\hat{g}}|}\begin{bmatrix} C_{11} & C_{21} & \cdots & C_{n1} \\ C_{12} & C_{22} & \cdots & C_{n2} \\ \vdots & \vdots & & \vdots \\ C_{1n} & C_{2n} & \cdots & C_{nn} \end{bmatrix},$$

where $[H_f^{\hat{g}}]^{+}$ is the adjoint of $[H_f^{\hat{g}}]$ and C_{ji} is the cofactor of the i-jth element of $[H_f^{\hat{g}}]$. If we define

$$s_{ij} = \lambda \frac{C_{ji}}{|H_f^{\hat{g}}|}, \; i, j = 1, \ldots, n, \quad (19.19)$$

then we may rewrite Equation (19.18) as

$$\frac{\partial q_i}{\partial p_j} = s_{ij} - q_j \frac{\partial q_i}{\partial M}, \; i, j = 1, \ldots, n. \quad (19.20)$$

[Hicks-Slutsky Equation With Cross Effects]

Here, (19.20) is the Hicks-Slutsky equation with cross effects, where s_{ij} is the **(net) substitution term** representing the substitution effect when q_i is adjusted as a result of a change in p_j. As evidenced by this equation, an uncompensated price response is decomposable into a *pure substitution effect* s_{ij} (a compensated price response) and a *pure income effect.*

Note that s_{ij} is the i-jth element of the nth-order symmetric **Slutsky substitution matrix** of compensated price responses $\mathbf{S} = [s_{ij}]$, $i, j = 1, \ldots, n$, where:

a. although $s_{ij} = s_{ji}$, $i, j = 1, \ldots, n$, which corresponds to D_1 in $[H_f^{\hat{g}}]^{-1}$ (see Exercise 19.2.2), the sign of s_{ij} is generally unknown;

b. $s_{ii} < 0$, $i = 1, \ldots, n$, and

c. $\mathbf{S}$ is negative semidefinite.

If $i = j$, then (19.20) becomes

$$\frac{\partial q_i}{\partial p_i} = s_{ii} - q_i \frac{\partial q_i}{\partial M}. \quad (19.21)$$

Now the compensated own-price response s_{ii} is unambiguously negative since the compensated demand curve is downward sloping.

19.3.2 Constrained Cost Minimization

Suppose a firm is a price-taker in the market for its inputs x_i, $i = 1, \ldots, n$, and seeks to determine the input combination $\boldsymbol{x}^T = (x_1, \ldots, x_n)$, $\boldsymbol{x} \in R^n$, which minimizes its total cost subject to an output constraint $y = f(\boldsymbol{x})$, where y represents output and $f(\boldsymbol{x})$ is the firm's strictly quasiconcave production function. Additionally, let $\boldsymbol{w}^T = (w_1, \ldots, w_n)$, $\boldsymbol{w} \in R_+^n$, depict a vector of constant input prices. If we write the production constraint in implicit form, then the firm's objective is to

$$\begin{aligned} &\text{minimize } c = \hat{\boldsymbol{w}}^T \boldsymbol{x} \text{ subject to} \\ &\hat{f}(\boldsymbol{x}) = \hat{y} - f(\boldsymbol{x}) = 0, \end{aligned} \quad (19.22)$$

where both f and $\hat{f}$ are functions of class $C^{(n)}$ and $\hat{\boldsymbol{w}} > \boldsymbol{O}$ and $\hat{y} > 0$ are parameters.

The Lagrangian associated with this problem is

$$L(\boldsymbol{x}, \lambda; \hat{\boldsymbol{w}}, \hat{y}) = \hat{\boldsymbol{w}}^T \boldsymbol{x} + \lambda(\hat{y} - f(\boldsymbol{x})), \quad (19.23)$$

where λ is the undetermined Lagrange multiplier. Looking to the first-order conditions for a constrained minimum we have, at the optimal point $(\boldsymbol{x}_o, \lambda_o)$:

$$\left.\begin{aligned} \nabla_{\boldsymbol{x}} L &= \boldsymbol{O} \\ \nabla_{\lambda} L &= 0 \end{aligned}\right\} \text{ or } \left\{\begin{aligned} &\hat{w}_i - \lambda_o f_i^o = 0, =1, \ldots, n; \\ &\hat{y} - f(\boldsymbol{x}_o) = 0. \end{aligned}\right. \quad (19.24)$$

By virtue of the implicit theorem, we obtain, from system (19.24), the $C^{(n)}$ functions $\boldsymbol{x} = \boldsymbol{x}(\boldsymbol{w}, y)$ and $\lambda = \lambda(\boldsymbol{w}, y)$, with $\boldsymbol{x}_o = \boldsymbol{x}(\hat{\boldsymbol{w}}, \hat{y})$ and $\lambda_o = \lambda(\hat{\boldsymbol{w}}, \hat{y})$, each of class $C^{(n)}$, and the first-order conditions

$$\begin{aligned} &\boldsymbol{w} - \lambda(\boldsymbol{w}, y) \nabla_{\boldsymbol{x}} f(\boldsymbol{x}(\boldsymbol{w}, y)) = \boldsymbol{O} \\ &y - f(\boldsymbol{x}(\boldsymbol{w}, y)) = 0 \end{aligned} \quad (19.24.1)$$

for all $(\boldsymbol{w}, y)$ within a δ–neighborhood of $(\hat{\boldsymbol{w}}, \hat{y})$, $\delta(\hat{\boldsymbol{w}}, \hat{y})$.

Partially differentiating (19.24.1) with respect to w_j yields

$$\begin{aligned} -\lambda\nabla_{xx}^2 f\frac{\partial x}{\partial w_j} - \nabla_x f\frac{\partial \lambda}{\partial w_j} &= -e_j \\ -(\nabla_x f)^T\frac{\partial x}{\partial w_j} \qquad &= 0 \end{aligned} \tag{19.25}$$

or, in matrix terms,

$$\begin{bmatrix} -\nabla_{xx}^2 f & -\nabla_x f \\ -(\nabla_x f)^T & 0 \end{bmatrix}\begin{bmatrix} \frac{\partial x}{\partial w_j} \\ \frac{\partial \lambda}{\partial w_j} \end{bmatrix} = \begin{bmatrix} -e_j \\ 0 \end{bmatrix}. \tag{19.25.1}$$

Also, partially differentiating (19.24.1) with respect to y gives

$$\begin{aligned} -\lambda\nabla_{xx}^2 f\frac{\partial x}{\partial y} - \nabla_x f\frac{\partial \lambda}{\partial y} &= O \\ -(\nabla f_x)^T\frac{\partial x}{\partial y} \qquad &= -1 \end{aligned} \tag{19.26}$$

or

$$\begin{bmatrix} -\nabla_{xx}^2 f & -\nabla_x f \\ -(\nabla_x f)^T & 0 \end{bmatrix}\begin{bmatrix} \frac{\partial x}{\partial y} \\ \frac{\partial \lambda}{\partial y} \end{bmatrix} = -e_{j+1}. \tag{19.26.1}$$

If we denote the bordered coefficient matrix of systems (19.25.1) and (19.26.1) by H, then, with $|H| \neq 0$, H^{-1} exists and thus (19.25.1) and (19.26.1) can be transformed to

$$\begin{bmatrix} \frac{\partial x}{\partial w_j} \\ \frac{\partial \lambda}{\partial w_j} \end{bmatrix} = -H^{-1}\begin{bmatrix} e_j \\ 0 \end{bmatrix}, \tag{19.27}$$

$$\begin{bmatrix} \frac{\partial x}{\partial y} \\ \frac{\partial \lambda}{\partial y} \end{bmatrix} = -H^{-1}e_{j+1}, \tag{19.28}$$

respectively, for all $(w, y) \in \delta(\hat{w}, \hat{y})$.

Focusing on the first n components of (19.27) we obtain

$$\frac{\partial x_i}{\partial w_j} = -\frac{C_{ji}}{|H|}, \; i, j = 1, \ldots, n, \tag{19.29}$$

where C_{ji} is the cofactor of the i-jth element within H.

Since $|H| < 0$, the sign of C_{ji} determines the sign of $\partial x_i/\partial w_j$, $i, j = 1, \ldots, n$. In general, the sign of C_{ji}, $i \neq j$, is indeterminate and thus the sign of $\partial x_i/\partial w_j$, $i \neq j$ is also indeterminate. However, if $i = j$, $C_{ii} < 0$ and thus $\partial x_i/\partial w_i < 0$ (i.e., the conditional input demand function is downward sloping).[1] Moreover, $\partial x_i/\partial w_j = \partial x_j/\partial w_i$ for all $i, j = 1, \ldots, n$. If we let $x_{ij} \equiv \partial x_i/\partial w_j$, then $S = [x_{ij}]$ represents the symmetric negative semidefinite **substitution matrix** with $x_{ii} < 0$ for all · For the two-input case, $\partial x_1/\partial w_2 = \partial x_2/\partial w_1 > 0$.

From (19.28)

$$\frac{\partial x_i}{\partial y} = -\frac{C_{j+1,\,i}}{|H|}, \; i = 1, \ldots, n. \tag{19.30}$$

Also, $\partial x_i/\partial y = \partial \lambda/\partial w_i$ for all i, j.

19.3.3 Long-Run Profit Maximization

Suppose a firm is a price taker in the product and input markets and seeks to maximize total profit $\pi(x; \hat{p}, \hat{w}) = \hat{p}f(x) - \hat{w}^T x$, where: $f(x)$ is a strictly concave production function of class $C^{(n)}$, $x \in R^n$ is an $(n \times 1)$ vector of inputs, and the parameters $\hat{p} \in R_+$ and $\hat{w} \in R_+^n$ are, respectively, the constant product price and $(n \times 1)$ vector of constant input prices. Hence the firm's problem is to choose an x which maximizes the profit function or

$$\underset{x}{\text{maximize}} \; \pi(x; \hat{p}, \hat{w}) = \hat{p}f(x) - \hat{w}^T x. \tag{19.31}$$

Note that π represents long-run profit since there are no fixed inputs.

The first-order condition at the optimal input combination x_o is

$$\nabla_x \pi = O \text{ or } \hat{p}f(x_o) - \hat{w} = O. \tag{19.32}$$

By the implicit function theorem there is a $C^{(n)}$ function $x(p, w)$ with $x_o = x(\hat{p}, \hat{w})$ and

$$p\nabla_x f(x(p, w)) - w = O \tag{19.32.1}$$

for all $x \in \delta(x_o)$. (With f strictly concave near x_o, the second-order conditions for a maximum must hold).

Partially differentiating (19.32.1) with respect to w_j gives

$$p\nabla^2_{xx} f \frac{\partial \boldsymbol{x}}{\partial w_j} - \boldsymbol{e}_j = \boldsymbol{O} \text{ or } \boldsymbol{H}\frac{\partial \boldsymbol{x}}{\partial w_j} = \boldsymbol{e}_j, \quad (19.33)$$

where $\boldsymbol{H} = p\nabla^2_{xx} f$. Given that $\boldsymbol{H}^{-1}$ exists, (23.33) becomes

$$\frac{\partial \boldsymbol{x}}{\partial w_j} = \boldsymbol{H}^{-1}\boldsymbol{e}_j. \quad (19.34)$$

Next, partially differentiating (19.32.1) with respect to p yields

$$\nabla_x f + p\nabla^2_{xx} f \frac{\partial \boldsymbol{x}}{\partial p} = \boldsymbol{O}$$

or

$$\boldsymbol{H}\frac{\partial \boldsymbol{x}}{\partial p} = -\nabla_x f$$

and, thus

$$\frac{\partial \boldsymbol{x}}{\partial p} = -\boldsymbol{H}^{-1}\nabla_x f. \quad (19.35)$$

Looking to (19.34) we can write

$$\frac{\partial x_i}{\partial w_j} = \frac{C_{ji}}{|\boldsymbol{H}|}, \; i, j = 1, \ldots, n, \quad (19.34.1)$$

where C_{ji} is the cofactor of the i-jth element of $\boldsymbol{H}$ and $|\boldsymbol{H}|$ has the sign of $(-1)^n$ (as dictated by the second-order condition for a maximum of π). For $i \neq j$, the sign of C_{ji} is indeterminate. However, since $\boldsymbol{H}$ is a symmetrical matrix, it is easily seen that

$$\frac{\partial x_i}{\partial w_j} = \frac{C_{ji}}{|\boldsymbol{H}|} = \frac{C_{ij}}{|\boldsymbol{H}|} = \frac{\partial x_j}{\partial w_i}, \; i, j = 1, \ldots, n. \quad (19.36)$$

For $i = j$,

$$\frac{\partial x_i}{\partial w_i} = \frac{C_{ii}}{|\boldsymbol{H}|} < 0, \; i = 1, \ldots, n. \quad (19.37)$$

The negativity of this expression results because C_{ii} is a *principal cofactor* and thus has the sign of $(-1)^{n-1}$. But since $(-1)^n$ and $(-1)^{n-1}$ must be of opposite signs, clearly the response of an input's own price change must be in the direction that is opposite to the change.

From (19.35) we can write

$$\frac{\partial x_i}{\partial p} = -\frac{1}{|\boldsymbol{H}|}\sum_{j=1}^{n} C_{ji} f_j, \; i, j = 1, \ldots, n. \quad (19.38)$$

Obviously the sign of this derivative is indeterminate.

Section 19.3 Exercises:

19.3.1 We know that $q_i(\boldsymbol{p}, M)$ is homogeneous of degree zero in $\boldsymbol{p}$ and M. Using Euler's theorem and Equation (19.20), verify that $\sum_{i=1}^{n} s_{ij} p_j = 0$ for all $(\boldsymbol{p}, M) \in \delta(\hat{\boldsymbol{p}}, \hat{M})$.

19.3.2 We noted that for the two-input constrained cost minimization model, $\partial x_1/\partial w_2 = \partial x_2/\partial w_1 > 0$. Verify this result and rationalize why this is true given that $\partial x_1/\partial w_1 < 0$, $\partial x_2/\partial w_2 < 0$.

19.3.4 The Le Châtelier Effect (Samuelson [1960, 1983]; Currier [2000]; Silberberg [1971]; Silberberg and Suen [2001])

Let $\boldsymbol{x} = \boldsymbol{x}^l(p, \boldsymbol{w})$ represent the *long-run input demand vector* determined from the maximization of long-run profit (19.31). Additionally, let us re-express (19.37) as

$$\frac{\partial x_i^l}{\partial w_i}, \; i = 1, \ldots, n,$$

where the "l" superscript denotes the "long-run" situation where there are *no fixed inputs*. Suppose the firm is currently operating at $\boldsymbol{x}^l$ but input x_n^l is *now held fixed* or $x_n = x_n^l$ is treated as a parameter. Obviously the firm's *short-run problem* (with relevant particulars displaying an "s" superscript) appears as

$$\underset{\boldsymbol{x}^s}{\text{maximize}} \; \pi^s(\boldsymbol{x}^s; \hat{p}, \hat{\boldsymbol{w}}^s, \hat{x}_n^l) = \hat{p} f(\boldsymbol{x}^s; \hat{x}_n^l) - (\hat{\boldsymbol{w}}^s)^T \boldsymbol{x}^s, \quad (19.39)$$

where $(\boldsymbol{x}^s)^T = (x_1, \ldots, x_{n-1})$ and $(\boldsymbol{w}^s)^T = (w_1, \ldots, w_{n-1})$.

From the first-order condition which holds at a local optimal point $\boldsymbol{x}_o^s$ we have

$$\nabla_{\boldsymbol{x}^s} \pi = \boldsymbol{O} \text{ or } \hat{p}\nabla_{\boldsymbol{x}^s} f(\boldsymbol{x}_o^s; \hat{x}_n^l) - \hat{\boldsymbol{w}} = \boldsymbol{O}. \quad (19.40)$$

Again relying on the implicit function theorem we see that there is a class $C^{(n)}$ function $\boldsymbol{x}(p, \boldsymbol{w}^s, x_n^l)$ with $\boldsymbol{x}_o^s = \boldsymbol{x}(\hat{p}, \hat{\boldsymbol{w}}^s, \hat{x}_n^l)$ and

$$p\nabla_{\boldsymbol{x}^s} f(\boldsymbol{x}^s(p, \boldsymbol{w}^s, x_n^l); x_n^l) - \boldsymbol{w}^s = \boldsymbol{O} \quad (19.40.1)$$

for all $\boldsymbol{x}^s \in \delta(\boldsymbol{x}_o^s)$.

Differentiating (19.40.1) with respect to w_i^s yields

$$p\nabla^2_{\boldsymbol{x}^s\boldsymbol{x}^s} f \frac{\partial \boldsymbol{x}^s}{\partial w_i^s} - \boldsymbol{e}_i = \boldsymbol{O}$$

or

$$\boldsymbol{H}^s \frac{\partial \boldsymbol{x}^s}{\partial w_i^s} = \boldsymbol{e}_i, \; i = 1, \ldots, n-1, \quad (19.41)$$

where

$$\boldsymbol{H}^s = \begin{bmatrix} pf_{11} & \cdots & pf_{1,n-1} \\ pf_{21} & \cdots & pf_{2,n-1} \\ \vdots & & \vdots \\ pf_{n-1,1} & \cdots & pf_{n-1,n-1} \end{bmatrix}. \quad (19.42)$$

Given that $|\boldsymbol{H}^s| \neq 0$, $[\boldsymbol{H}^s]^{-1}$ exists and thus (19.41) is transformed to

$$\frac{\partial \boldsymbol{x}^s}{\partial w_i^s} = [\boldsymbol{H}^s]^{-1}\boldsymbol{e}_i, \; i = 1, \ldots, n-1, \quad (19.43)$$

for all $\boldsymbol{x}^s \in \delta(\boldsymbol{x}_o^s)$ or

$$\frac{\partial x_i^s}{\partial w_i^s} = \frac{C_{ii}^s}{|\boldsymbol{H}^s|} < 0, \; i = 1, \ldots, n-1, \quad (19.43.1)$$

since $|\boldsymbol{H}^s|$ has the sign of $(-1)^{n-1}$ and C_{ii}^s has the sign of $(-1)^{n-2}$. Additionally, we know that

$$\frac{\partial x_i^l}{\partial w_i} = \frac{C_{ii}}{|\boldsymbol{H}|} < 0, \; i = 1, \ldots, n-1, \quad (19.44)$$

since $|\boldsymbol{H}|$ has the sign of $(-1)^n$ and C_{ii} has the sign of $(-1)^{n-1}$.

Let us next consider the relative magnitudes of the own-price derivatives appearing in (19.43.1) and (19.44). To this end, let us write

$$\boldsymbol{H} = \begin{bmatrix} H^s & pf_{1n} \\ & \vdots \\ pf_{n1} & pf_{nn} \end{bmatrix}. \quad 19.45$$

Given this matrix, we know, by a specialization of a theorem by Jacobi,[2] that $|\boldsymbol{H}| = |\boldsymbol{H}^s|(|\boldsymbol{H}|/C_{nn})$. Then,

$$\begin{aligned} \frac{\partial x_i^l}{\partial w_i} - \frac{\partial x_i^s}{\partial w_i^s} &= \frac{C_{ii}}{|\boldsymbol{H}|} - \frac{C_{ii}^s}{|\boldsymbol{H}^s|} = \frac{C_{ii}C_{nn}}{|\boldsymbol{H}^s||\boldsymbol{H}|} - \frac{C_{ii}^s}{|\boldsymbol{H}^s|} \\ &= \frac{C_{ii}C_{nn} - |\boldsymbol{H}|C_{ii}^s}{|\boldsymbol{H}^s||\boldsymbol{H}|}. \end{aligned} \quad (19.46)$$

If we invoke the **Desnanot-Jacobi (D-J) identity,**[3] then Equation (19.46) becomes

$$|\boldsymbol{H}_i^n||\boldsymbol{H}_n^i|/|\boldsymbol{H}^s||\boldsymbol{H}| = |\boldsymbol{H}_i^n|^2/|\boldsymbol{H}^s||\boldsymbol{H}| < 0$$

or

$$\frac{\partial x_i^l}{\partial w_i} < \frac{\partial x_i^s}{\partial w_i^s}, \; i = 1, \ldots, n-1. \quad (19.47)$$

This result has been termed the **Le Châtelier effect**: the long-run own-price derivative of an input is more negative than its short-run counterpart, i.e., locally, the response in x_i due to a change in its price is greater (in absolute value) when x_n is variable (in the long run) than when x_n is held fixed (in the short run).

Section 19.4 Exercises:

19.3.4.1 Let $y = f(x_1, x_2)$ be a strictly concave production function with w_1 and w_2 serving as constant input prices and p representing the constant unit price of a profit-maximizing firm's output.

a. Determine the first-order and second-order conditions for long-run profit maximization.

b. Let the long-run profit-maximizing input demands be represented as

$$x_1 = x_1^l(p, w_1, w_2),$$
$$x_2 = x_2^l(p, w_1, w_2).$$

Find $\frac{\partial x_1^l}{\partial w_1}$. Show that this derivative is negative.

c. Let us fix x_2 at the level x_2^l. Reoptimize with respect to x_1 (x_2^l is a parameter) and specify the second-order condition for a maximum.

d. Verify that

$$\frac{\partial x_1^l}{\partial w_1} \leq \frac{\partial x_1^s}{\partial w_1^s}.$$

19.3.4.2 Given the matrix

$$A = \begin{bmatrix} 1 & 2 & 3 \\ 2 & 2 & 3 \\ 1 & 0 & 6 \end{bmatrix} \text{ with } B = \begin{bmatrix} 1 & 2 \\ 2 & 2 \end{bmatrix},$$

Verify that $|B| = |A||Z|$ (footnote 2).

19.3.4.3 Given the matrix

$$A = \begin{bmatrix} 1 & 2 & 3 \\ 2 & 2 & 5 \\ 1 & 0 & 6 \end{bmatrix}$$

Verify that the D-J identity holds (footnote 3), where $|A_{i,n}^{i,n}| = |A_{2,3}^{2,3}|$, etc.

NOTES

1 $C_{11}, C_{22}, \ldots, C_{nn}$ are **border-preserving principal cofactors** of H so that they are all negative. The $C_{ij} = C_{ji}$, $i \neq j$, are not so - - their signs are typically indeterminate, e.g., for $i = 1, 2$, and

$$H = \begin{bmatrix} -\lambda f_{11} & -\lambda f_{12} & -f_1 \\ -\lambda f_{21} & -\lambda f_{22} & -f_2 \\ -f_1 & -f_2 & 0 \end{bmatrix}, \; C_{11} = (1)^2 \begin{vmatrix} -\lambda f_{22} & -f_2 \\ -f_2 & 0 \end{vmatrix}$$

is border-preserving whereas

$$C_{12} = (1)^3 \begin{vmatrix} -\lambda f_{21} & -f_2 \\ -f_1 & 0 \end{vmatrix}$$

is not.

2 For a nonsingular $(n \times n)$ matrix A, where

$$A = \begin{bmatrix} B & C \\ D & E \end{bmatrix}, \; A^{-1} = \begin{bmatrix} W & X \\ Y & Z \end{bmatrix},$$

and B and W are each of order $(k \times k)$, $k < n$, it follows that $|B| = |A||Z|$.

3 For an $(n \times n)$ matrix A, let A_i^j be the $(n - 1) \times (n - 1)$ submatrix obtained by deleting row i and column j from A. Then the D-J identity is

$$|A||A_{i,n}^{i,n}| \equiv |A_i^i||A_n^n| - |A_i^n||A_n^i|.$$

References

Abadie, J. On the Kuhn-Tucker Theorem. In *Nonlinear Programming*, J. Abadie, ed. Amsterdam: North-Holland (1967).

Arrow, K., A. Enthoven. Quasi-Concave Programming. *Econometrica* 29: 779–800 (1961).

Arrow, K., F. Hahn. *General Competitive Analysis*. San Francisco: Holden Day (1971).

Arrow, K., G. Debreu. Existence of an Equilibrium for a Competitive Economy. *Econometrica* 22: 265–290 (1954).

Arrow, K., L. Hurwicz, H. Uzawa. Constraint Qualifications in Maximization Programs. *Naval Research Logistics Quarterly* 8: 175–191 (1961).

Arrow, K., L. Hurwicz (eds.) Studies in Linear and Nonlinear Programming. Stanford University Press (1958).

Arrow, K., M. Kurz. *Public Investment, the Rate of Return, and Optimal Fiscal Policy*. Baltimore: Johns Hopkins University Press (1970).

Avriel, M. *Nonlinear Programming: Methods and Analysis*. New York: Dover Publications (2003).

Avriel, M., E.W. Diewert, S. Schaible, I. Zang. *Generalized Concavity*. Philadelphia: SIAM (2010).

Bagirov, A., N. Karmitsa, M. Mäkelä. *Introduction to Nonsmooth Optimization*. New York: Springer (2014).

Barr, D. Hand Computation of Generalized Inverses. *Mathematics Magazine* 58: 102–107 (1985).

Bazaraa, M., H. Sherali, C. Shetty. *Nonlinear Programming*, 3rd ed. Hoboken, New Jersey: John Wiley and Sons, Inc. (2006).

Beckmann, M., K. Kapau. Conjugate Duality: Some Applications to Economic Theory. *Journal of Economic Theory* 6: 292–302 (1972).

Berge, C. *Topological Spaces*. New York: The Macmillan Co. (1963).

Blume, L.E. Duality. *The New Palgrave Dictionary of Economics*, 2nd ed., S. Durlauf, L. Blume, eds. New York: Palgrave, Macmillan (2008).

Border, K. *Fixed Point Theorems With Applications To Economics and Game Theory*. New York: Cambridge University Press (1985).

Borwein, J.M., A. Lewis. *Convex Analysis and Nonlinear Optimization*. New York: Springer-Verlag Ing. (2000).

Boyd, S., L. Vanderberghe. *Convex Optimization*. Cambridge: Cambridge University Press (2008).

Brouwer, L.E.J. Über Abbildung von Mannigfaltigkeiten. *Mathematischen Annalen* 71: 97–115 (1912).

Browder, F.E. Another Generalization of the Schauder Fixed Point Theorem. *Duke Mathematical Journal* 32: 399–406 (1965).

Browder, F.E. On a Generalization of the Schauder Fixed Point Theorem. *Duke Mathematical Journal* 26: 291–303 (1959).

Cambini, A., L. Martein. *Generalized Convexity and Optimization*. Berlin: Springer-Verlag (2009).

Caputo, M. *Foundations of Dynamic Economic Analysis*. New York: Cambridge University Press (2005).

Carathéodory, C. *Calculus of Variations and Partial Differential Equations of the First Order*. Trans. R. Dean. San Francisco: Holden Day (1967).

Cass, D. Optimal Growth in an Aggregate Model of Capital Accumulation. Review of Economic Studies 32: 233–240 (1965).

Clarke, F.H. Generalized Gradients of Lipschitz Functionals. *Advances in Mathematics* 40: 52–67 (1981).

Clarke, F.H. Generalized Gradients and Applications. *Transactions of the America Mathematical Society* 205: 247–262 (1975).

Clarke, F. *Optimization and Nonsmooth Analysis*. Philadelphia: SIAM (1990).

Cronin, J. *Fixed Points and Topological Degree in Nonlinear Analysis*. Providence, R. I.: American Math. Society (1964).

Currier, K.M. *Comparative Statics Analysis in Economics*. New Jersey: World Scientific (2000).

Deaton, A. The Distance Function and Consumer Behavior With Applications to Index Numbers and Optimal Taxation. *Review of Economic Studies* 46: 391–405 (1979).

Deaton, A., J. Muellbauer. *Economics and Consumer Behavior*. Cambridge: Cambridge University Press (2009).

Debreu, G.A. Social Equilibrium Existence Theorem. *Proceedings of the National Academy of Sciences* 38: 886–893 (1952).

Debreu, G. Market Equilibrium. *Proceedings of the National Academy of Sciences of the U. S. A.* 42: 876–878 (1956).

Debreu, G. New Concepts and Techniques for Equilibrium Analysis. *International Economic Review* 3: 257–273 (1962).

Debreu, G. Definite and Semi-Definite Quadratic Forms. *Econometrica* 20: 285–300 (1952).

Debreu, G. Existence of Competitive Equilibrium. In Handbook of Mathematical Economics, vol. 2, eds.Arrow, K., M. Intriligator. Amsterdam: North-Holland (2007).

Diewert, W. *Duality in Economics.* Amsterdam: North Holland (1982).

Diewert, W.E. Applications of Duality Theory. In *Frontiers of Quantitative Economics*, M. Intriligator, D.A. Kendrick, eds. Amsterdam: North-Holland (1974).

Diewert, W. E. Functional Forms for Profit and Transformation Functions. *Journal of Economic Theory* 6: 284–316 (1973).

Diewert, W. E. Duality Approaches to Microeconomic Theory. In *Handbook of Mathematical Economics*, eds. K. J. Arrow and M. D. Intriligator, 2: 535–599. Amsterdam: North-Holland (1982).

Diewert, W. Functional Forms for Profit and Transformation Functions. *Journal of Economic Theory* 6: 284–316 (1973).

Dubovitskii, A., A. Milyutin. Extremum Problems in the Presence of Restrictions. *U.S.S.R. Computational Mathematics and Mathematical Physics* 5: 1–80 (1965).

Eilenberg, S., D. Montgomery. Fixed-Point Theorems for Multivalued Transformations. *American Journal of Mathematics* 68: 214–222 (1946).

Enthoven, A. The Simple Mathematics of Maximization. In *The Economics of Defense in the Nuclear Age*, eds. Hitch, C., R. McKean. Cambridge: Harvard University Press (1960).

Fan, K. Extensions of Two Fixed Point Theorems Browder. *Mathematische Zeitschriff* 112: 234–240 (1969).

Färe, R., D. Primont. A Distance Function Approach to Multioutput Technologies. *Southern Economic Journal* 54: 879–891 (1990).

Fenchel, W. *Convex Cones, Sets, and Functions.* Mimeographed notes: Princeton University (1949).

Ferland, J. A. *Quasi-Convex and Pseudo-Convex Functions on Solid Convex Sets.* Technical Report, Stanford University (1971).

Fiacco, A., G. McCormick. *Nonlinear Programming. Sequential Unconstrained Minimization Techniques.* New York: John Wiley and Sons, Inc. (1968).

Fried, H., C. A. Knox Lovel, S. Schmidt (eds.). *The Measurement of Productive Efficiency.* New York: Oxford University Press (1993).

Fuss, M., D. McFadden. *Production Economics: A Dual Approach to Theory and Applications*, 1. New York: North-Holland Publishing Co. (1978).

Gale, D. The Law of Supply and Demand. *Mathematica Scandinavica* 3: 155–169 (1955).

Geoffrion, A. (ed.). *Perspectives on Optimization.* Reading, MA: Addison-Wesley Publishing Co. (1972).

Geoffrion, A. (ed.). Perspectives on Optimization. Addison-Wesley Publishing Company (1972).

Gorman, W.M. Tricks With Utility Functions. In *Essays in Economic Analysis*, M. Artis, A. Nobay, eds. Cambridge: Cambridge University Press (1976).

Graves, R., P. Wolfe (eds.). *Recent Advances in Mathematical Programming.* New York: McGraw-Hill Book Co. (1963).

Graybill, F. *Introduction to Matrices with Applications in Statistics.* Belmont, CA: Wadsworth Publishing Co. (1969).

Hiriart-Urrat, J.-B., C. Lemaréchal. *Fundamentals of Convex Analysis.* New York: Springer (2002).

Hitch, C., R. McKean. *The Economics of Defense in the Nuclear Age.* Cambridge: Harvard University Press (1960).

John, F. Extremum Problems with Inequalities as Subsidiary Conditions. In *Studies and Essays: Courant Anniversary Volume*, K. Friedrichs, O. Neugebauer, J. Stoker, eds. New York: Interscience (1948).

Jorgenson, D. The Theory of Investment Behavior. In Determinants of Investment Behavior, (ed.) Ferber, R. Amsterdam: North-Holland (1967).

Jorgenson, D. Econometric Studies of Investment Behavior: A Survey. Journal of Economic Literature 9:1111–47 (1971).

Jorgenson, D. Investment Behavior and the Production Function. Bell Journal of Economics and Management Science 3: 220–251 (1972).

Jorgensen, D., L. Lau. Duality and Differentiability in Production. *Journal of Economic Theory* 9: 23–42 (1974).

Jorgenson, D., L. Lau. The Duality of Technology and Economic Behavior. *The Review of Economic Studies* 41: 181–200 (1974).

Kakutani, S. A Generalization of Brouwer's Fixed Point Theorem. *Duke Mathematical Journal* 8: 459–459 (1941).

Klee, V. Levay-Schauder Theory Without Local Convexity. *Mathmatische Annalen* 141: 286–296 (1960).

Klein, E. *Mathematical Methods in Theoretical Economics.* New York: Academic Press, Inc. (1973).

Knaster, B., K. Kuratowski, S. Mazurkiewicz. Ein Beweis der Fixpunktsatzes für n-dimensionale simplexe. *Fundamenta Mathematicae* 14: 132–137 (1929).

Koopmans, T. On the Concept of Optimal Economic Growth. In The Economic Approach to Development Planning. Amsterdam: North-Holland (1965).

Kuhn, H. On the Shapley-Sperner Lemma. In *Numerical Solution of Highly Nonlinear Problems*, W. Forster, ed. The Netherlands: North-Holland Publishing Co. (1980).

Kuhn, H. Simplicial Approximation of Fixed Points. *Proceedings of the National Academy of Sciences, U.S.A.*. 61: 1238–1242 (1968).

Kuhn, H., A. Tucker. Nonlinear Programming. In *Proceedings of the Second Berkley Symposium on Mathematical Statistics and Probability*, J. Neyman, ed. New Jersey: Princeton University Press (1951).

Künzi, H., W. Krelle, W. Otelli. *Nonlinear Programming.* Trans. F. Levin. Massachusetts: Blaisdell Publishing Co. (1966).

Lasdon, L. *Optimization Theory for Large Systems.* New York: Macmillan (1970).

Lasserre, J. J. B. Hiriart-Urruty. Mathematical Properties of Optimization Problems Defined by Positively Homogeneous Functions. Journal of Optimization Theory and Applications 112: 31–52 (2002).

Lau, L.J. Applications of Profit Functions. In *Production Economics: A Dual Approach to Theory and Applications*, M. Fuss, D. McFadden, eds., 1, 133–215. New York: North-Holland Publishing Co. (1978).

Lemaréchal, C. Nondifferentiable Optimization. In *Optimization*, G. Nemhauser, A. Rinnooy Kan, M. Todd, eds. New York: Elsevier (1989).

Lindahl, E. Just Taxation – A Positive Solution. In *Classics in the Theory of Public Finance*, R. Musgrave, A. Peacock, eds. London: Macmillan (1958).

Magnus, J., H. Neudecker. *Matrix Differential Calculus with Applications in Statistics and Econometrics*, 2nd ed. New York: John Wiley and Sons (2002).

Mangasarian, O. *Nonlinear Programming*. New York: McGraw-Hill Book Co. (1969).

Mangasarian, O. Sufficient Conditions for the Optimal Control of Nonlinear Systems. *SIAM Journal on Control* 4: 139–152 (1966).

Mangasarian, O., S. Fromovitz. The Fritz John Necessary Optimality Conditions in the Presence of Equality and Inequality Constraints. *Journal of Mathematical Analysis and Applications* 17: 37–47 (1967).

Mangasarian, O. Duality in Nonlinear Programming. Quarterly of Applied Mathematics 20: 300–302 (1962).

Marlow, W. *Mathematics for Operations Research*. New York: John Wiley and Sons (1978).

Mas-Collel, A., M. Whinston, J. Green. *Microeconomic Theory*. New Delhi: Oxford University Press (2006).

McFadden, D. Duality of Production Costs, and Profit Functions. In *Production Economics: A Dual Approach to Theory and Applications*, M. Fuss, D. McFadden, eds. New York: North-Holland Publishing Co. (1978).

McKenzie, L. On the Existence of General Equilibrium for a Competitive Market. *Econometrica* 27: 54–71 (1959).

McKenzie, L. On the Existence of General Equilibrium: Some Corrections. *Econometrica* 29: 247–248 (1961).

McKenzie, L.W. Demand Theory Without a Utility Index. *Review of Economic Studies* 24: 184–189 (1957).

Michal, A. *Matrix and Tensor Calculus*. New York: Dover Publications, Inc. (2008).

Minoux, M. *Mathematical Programming*. New York: John Wiley and Sons (1986).

Nikaido, H. *Convex Structures and Economic Theory*. New York: Academic Press (1968).

Nikaido, H. On the Classical Multilateral Exchange Problem. *Metroeconomica* 8: 135–145 (1956).

Nikaido, H. Introduction to Sets and Mappings in Modern Economics, tr. K. Sato. Amsterdam: North-Holland (1965).

Panik, M.J. *Fundamentals of Convex Analysis*. Boston: Kluwer Academic Publishers (1993).

Pontryagin L.S., V. Boltyanskii, R. Gamkrelidze, E.F. Mishchenko. *The Mathematical Theory of Optimal Processes*. Trans. K.N. Trirogoff. New York: Interscience (1962).

Roberts, D.J. The Lindahl Solution for Economics With Public Goods. *Journal of Public Economics* 3: 23–43 (1974).

Rockafellar, R.T. Augmented Lagrange Multiplier Functions and Duality in Nonconvex Programming. *SIAM Journal on Control and Optimization* 12: 268–285 (1974).

Rockafellar, R.T. *Convex Analysis*. Princeton, N.J.: Princeton University Press (1970).

Ruszczyński, A. *Nonlinear Optimization*. Princeton: Princeton University Press (2006).

Salanié, B. The Lindahl Equilibrium. In *Microeconomics of Market Failure*. Cambridge, MA: MIT Press (2000).

Samuelson, P.A. An Extension of the Le Châtelier Principle. *Econometrica* 28: 368–379 (1960).

Samuelson, P.A. *Foundations of Economic Analysis*, enlarged ed. Cambridge, MA: Harvard University Press (1983).

Samuelson, P.A. Diagramatic Exposition of a Theory of Public Expenditure. *The Review of Economics and Statistics* 37: 350–356 (1955).

Samuelson, P.A. The Pure Theory of Public Expenditure. *The Review of Economics and Statistics* 36: 387–389 (1954).

Scarf, H. *The Computation of Economic Equilibria*. New Haven: Yale University Press (1973).

Schauder, J. Der Fixpunktsetz in Funktional-raumen. *Studia Mathematica* 2: 171–180 (1930).

Shapley, L.S. On Balanced Games Without Side Payments. In *Mathematical Programming*, T.C. Hu, S.M. Robinson, eds. New York: Academic Press (1973).

Shephard, R.W. *Cost and Production Functions*. Princeton, N.J.: Princeton University Press (1953).

Shephard, R.W. *Theory of Cost and Production Functions*. Princeton, N.J.: Princeton University Press (1970).

Silberberg, E. The Le Châtelier Principle as a Corollary to a Generalized Envelope Theorem. *Journal of Economic Theory* 3: 146–155 (1971).

Silberberg, E., W. Suen. *The Structure of Economics*, 3rd ed. New York: Irwin/McGraw-Hill (2001).

Slater, M. Lagrange Multipliers Revisited: A Contribution to Nonlinear Programming. *Cowles Commission Discussion Paper*, Math. 403, Nov. (1950).

Solow, R. A Contribution to the Theory of Economic Growth. Quarterly Journal of Economics 70: 65–94 (1956).

Sperner, E. Neuer Beweis für die Invarianzder Dimensionszahl und des Gebietes. *Abhandlungen aus dem Mathematischen Seminar der Universität Hambur* 6: 265–272 (1928).

Steinitz, E. Bedingt Konvergente Reihen und Konvexe Systeme. *Journal für die reine und angewandte Mathematik* 143: 128–175 (1913).

Stoer, J., C. Witzgall. *Convexity and Optimization in Finite Dimensions*, 1. New York: Springer (1970).

Swan, T. Economic Growth and Capital Accumulation. Economic Record 32: 334–361 (1956).

Takayama, A. *Mathematical Economics*. New York: Cambridge University Press (1985).

Takayama, A. Sensitivity Analysis in Economic Theory. *Metroeconomica* 29: 9–37 (1977). University Press (1970).

Takayama, A. Investment Demand. *Lecture Notes*. Southern Illinois University. October (1987).

Takayama, A. Analytical Methods in Economics. Ann Arbor: The University of Michigan Press (1996).

Taylor, A.E., W.R. Mann. *Advanced Calculus*, 2nd ed. Massachusetts: Xerox College Publishing (1972).

Tucker, A. Dual Systems of Homogeneous Linear Relations. In *Annals of Mathematics Studies*, H.W. Kuhn, A. Tucker, eds, No. 38. Princeton: Princeton University Press (1956).

Turkington, D. *Generalized Vectorization, Cross-Products, and Matrix Calculus*. Cambridge: Cambridge University Press (2013).

Tychonoff, A. Ein Fixpunktsatz. *Mathematische Annalen* 111: 767–776 (1935).

Uzawa, H. A Theorem on Convex Polyhedral Cones. In *Studies in Linear and Non-Linear Programming*, K. Arrow, L. Hurwicz, eds. Stanford, CA: Stanford University Press (1958).

Vial, J.P. Strong and Weak Convexity of Sets and Functions. *Mathematics of Operations Research* 8: 231–259 (1983).

Vial, J.P. Strong Convexity of Sets and Functions. *Journal of Mathematical Economics* 9: 187–205 (1982).

Wolfe, P. A Duality Theorem for Non-Linear Programming. *Quarterly of Applied Mathematics* 19: 239–244 (1961).

Zhao, Y.B., D. Li. A Note on KKT Points of Homogeneous Programs. http://www.optimization-online.org/DB_FILE/2005/06/1161.pdf.

Index